X-ray Lasers 2004

Other titles in the series

The Institute of Physics Conference Series regularly features papers presented at important conferences and symposia highlighting new developments in physics and related fields. Previous publications include:

185 Group Theoretical Methods in Physics
Papers presented at the 25th International Colloquium on Group Theoretical Methods in Physics, Cocoyoc, México
Edited by G S Pogosyan, L E Vicent and K B Wolf

184 Compound Semiconductors 2004
Papers presented at the 31st International Symposium, Seoul, Korea
Edited by J C Woo, H Hasegawa, Y S Kwon, T Yao and K H Yoo

183 Electron and Photon Impact Ionization and Related Topics 2004
Papers presented at the International Conference, Louvain-la-Neuve, Belgium
Edited by B Piraux

182 Light Sources 2004
Papers presented at the 10th International Symposium on the Science and Technology of Light Sources, Toulouse, France
Edited by G Zissis

181 Applied Superconductivity 2003
Papers presented at the Sixth European Conference on Applied Superconductivity, Sorrento, Italy
Edited by A Andreone, G P Pepe, R Cristiano and G Masullo

180 Microscopy of Semiconducting Materials 2003
Papers presented at the Institute of Physics Conference, Cambridge, UK
Edited by A G Cullis and P A Midgley

179 Electron Microscopy and Analysis 2003
Papers presented at the Institute of Physics Electron Microscopy and Analysis Group Conference, Oxford, UK
Edited by S McVitie and D McComb

178 Electrostatics 2003
Papers presented at the Electrostatics Conference of the Institute of Physics, Edinburgh, UK
Edited by H Morgan

177 Optical and Laser Diagnostics 2002
Papers presented at the First International Conference, London, UK
Edited by C Arcoumanis and K T V Grattan

175 Computational Accelerator Physics 2002
Papers presented at the Seventh International Conference, Michigan, USA
Edited by M Berz and K Makino

X-ray Lasers 2004

Proceedings of the 9th International Conference on X-ray Lasers held in Beijing, China, 24-28 May 2004

Edited by
Jie Zhang

Institute of Physics Conference Series Number 186

IoP

Institute of Physics Publishing
Bristol and Philadelphia

British Library Cataloguing in Publication Data

A catalogue record for this book is available from the British Library.

ISBN 0-7503-1010-3

Library of Congress Cataloging-in-Publication Data are available

Published by Institute of Physics Publishing, wholly owned by the Institute of Physics, London
Institute of Physics Publishing, Dirac House, Temple Back, Bristol BS1 6BE, UK
US Office: Institute of Physics Publishing, The Public Ledger Building, Suite 929, 150 South Independence Mall West, Philadelphia, PA 19106, USA

Printed in the UK by Short Run Press Ltd, Exeter

Preface

These proceedings comprise invited and contributed papers presented at the 9th International Conference on X-Ray Lasers (ICXRL2004) in Beijing, China in the period of 24–28 May 2004. Approximately 120 participants from 13 countries and regions attended this conference. When we review the history of this conference, we have to go back to 1986, when the first conference was held in Aussois, France. The subsequent conferences were held in York, UK (1990); Schliersee, Germany (1992); Williamsburg, USA (1994); Lund, Sweden (1996); Kyoto, Japan (1998); Saint Malo, France (2000); and Aspen, USA (2002).

These proceedings review the development made in recent years in X-ray lasers and related fields. It is always a pleasure to see the great progress made since the first demonstration of soft X-ray amplification in 1984. It is an even more enjoyable experience to witness the rapid pace of the development of X-ray lasers and their applications in recent years. The pump energy required by X-ray lasers has been reduced to the milli-joule level, whereas the repetition rate of the X-ray lasers has jumped over the 10 Hz barrier. The pulse duration of X-ray lasers can be as short as sub-picosecond. X-ray laser wavelengths extend from 3.6 to 87 nm. Both the spatial and temporal coherences of present X-ray lasers can meet the requirements for applications. All of these characteristics make X-ray lasers more versatile for wide range of applications and the dream for "table-top X-ray lasers" has already come true.

The 9th International Conference on X-Ray Lasers was sponsored by the National Science Foundation of China, the National ICF Program of China and the Chinese Academy of Sciences. Additional supports by LABest, Coherence gave strong help to attendance of overseas students participants.

On behalf of the organizing committee, I would like to thank the International Advisory Committee for helping to organize the scientific program. I would also like to thank IoP Publishing for their careful reading of the manuscripts and rapid production. My special thanks go to my students at the Institute of Physics, Chinese Academy of Sciences, Quanli Dong, Zhan Jin, Xin Lu, Xiaohui Yuan and other students for their great help in font correction, format re-setting and proof reading of the manuscripts for publication.

Jie Zhang

Conference Organization

Conference Chair
Jie Zhang, Institute of Physics, CAS, Beijing

Conference Co-chairs
Zhizhan Xu, Shanghai Institute of Optics and Fine Mechanics, CAS, Shanghai
Shiji Wang, Shanghai Institute of Laser Plasma, CAEP, Shanghai

International Advisory Board:

H. Daido	D. L. Mathews	J. Dunn
C. H. Nam	D. C. Eder	P. Nickles
E. Fill	G. J. Pert	P. Hagelstein
J. J. Rocca	P. Jaegle	B. Rus
G. Jamelot	S. Suckewer	Y. Kato
G. J. Tallents	M. H. Key	A. V. Vinogradov
V. V. Korobkin	Z. Z. Xu	J. Zhang

Local Organization Committee:

Chairs:
J. Zhang, Z.Z. Xu, S.J. Wang

Members:

J. L. Cao	X. Lu	S. Z. Fu
H. S. Peng	Y. Gu	Q. Wang
R. X. Li	G. P. Zhang	Y. T. Li
W. Y. Zhang		

Conference Secretariat:

Yifei Chen	Xin Lu	Quanli Dong
Fei Yan	Yuan Gao	Jiayong Zhong
Yutong Li	Xiaohui Yuan	Zhan Jin

Organized by Institute of Physics, Chinese Academy of Sciences

Contact Information
ICXRL2004
Institute of Physics, CAS
P.O. Box 603-33
Beijing, 100080
P.R.China

Tel: +86-10-82649356
Fax: +86-10-82649531
E-mail: ICXRL2004@aphy.iphy.ac.cn
Web: http://highfield.iphy.ac.cn/ICXRL2004

Sponsors

National Natural Science Foundation of China

National Hi-tech ICF Program of China

Chinese Academy of Sciences

Conference Schedule

Sunday 23	Monday 24	Tuesday 25	Wednesday 26	Thursday 27	Friday 28
	8:50 - 9:10 **Opening Remarks: Z.Z.Xu** **Session I Review** ***Chairman: G. Jamelot*** 09:10-09:50: **Y. Kato** 09:50-10:20: **G. Pert** 10:20-10:50: **J. Rocca**	**Session V Alternative XRL I** ***Chairman: D. Attwood*** 08:30-09:00: **A. Klisnick** 09:00-09:30: **P. Zeitoun** 09:30-10:00: **V. Shlyaptsev** 10:00-10:30: **B. Rus**	**Session VIII Laser-driven XRLs** ***Chairman: P. Zeitoun*** 08:30-09:00: **S. Sebban** 09:00-09:30: **J. Zhang** 9:30-9:50: X. K. He	**Session X High Order Harmonics** ***Chairman: J. J. Rocca*** 08:30-09:00: **Z. H. Chang** 09:00-09:30:**K. Midorikawa** 09:30-10:00: **C. H. Nam** - 10:00-10:20: Y. P. Huo (cancelled)	**Session XIII Alternative XRL II** ***Chairman: J. Balmer*** 08:30 – 09:00: **M. Kishimoto** 09:00-09:30: **H. Kuroda** 09:30-09:50: W. Hong 09:50-10:10: E. Ivanova 10:10-10:30: V. Vorontsov
	10:50 - 11:20 Photo-taking & Coffee Break	**10:30 - 10:50 Coffee Break**	**9:50 - 10:10 Coffee Break**	**10:20-10:40 Coffee Break**	**10:30-10:50 Coffee Break**
	Session II Transient/Collisional XRL I ***Chairman: J. Nilsen*** 11:20-11:50: **G. Jamelot** 11:50-12:20: **J. Dunn**	**Session VI Capillary Discharge X-Ray Laser** ***Chairman: G. J. Pert*** 10:50-11:10: V. Burtsev 11:10-11:30: Y. P. Zhao 11:30-11:50: A. Ritucci 11:50-12:10: M.G. Capeluto	**Session IX OFI XRLs** ***Chairman: C. Danson*** 10:10-10:40: **S. Hooker** 10:40-11:10: **J. Y. Lin** 11:10-11:40: **K. Lan** 11:40-12:00: T. Mocek 12:00-12:20: K. Janulewicz	**Session XI X-ray laser drives** ***Chairman: J. Dunn*** 10:40-11:10: **C. Danson** 11:10-11:40: **Z. Y. Wei** 11:40-12:00: Y. Zhu (cancelled)	**Session XIV X-ray optics and instrumentation** ***Chairman: A. Vinogradov*** 10:50-11:20: **R. Z. Tai** 11:20-11:40: I. Artyukov 11:40-12:00: Z. S. Wang 12:00-12:20: H. C. Wang 12:20-12:30: **Closing remarks**
9:00-18:00 Registration 18:00- 20:00 Welcoming Reception	**12:20 – 14:00 Lunch**	**12:10 – 13:20 Lunch**	**12:20 – 13:30 Lunch**	**12:00-13:30 Lunch & Advisory Committee meeting**	**12:50-13:50 Lunch**
	Session III **Transient/Collisional XRL II** ***Chairman: Y. Kato*** 14:00-14:30: **J.Balmer** 14:30-15:00: **R. Smith** 15:00-15:30: **S. J. Wang**	**13:20 - 16:00 Poster Session**	**13:30-18:00 Excursion to the Great Wall**	**13:30 - 16:00 Poster Session**	
	15:30-15:50 Coffee Break	**Session VII Applications** ***Chairman: C. H. Nam*** 16:00-16:20: J. Filevich 16:20-16:50: **J. Nilsen** 16:50-17:20: **D. Ros** 17:20-17:50: **G. Tallents** 17:50-18:20: **D. Attwood** 18:20-18:40: H. L. Gao		**Session XII Free-electron lasers** ***Chairman: T.Ozaki*** 16:00-16:30: **A. Vinogradov** 16:30-17:00: **S. Reiche** 17:00-17:30: **Z. R. Huang,** 17:30-17:50: B. Chichkov	
	Session IV **Transient/Collisional XRL III** ***Chairman: G. Tallents*** 15:50-16:20: **P. Nickles** 16:20-16:50: **J. Dubau** 16:50-17:20: **R. X. Li** 17:20-17:40: T. Ozaki 17:40-18:00: Abou-Ali Yousef 18:00-18:20: D. Benredjem				
	18:20-19:20 Dinner	**18:40-19:40 Dinner**	**18:00 –20:30** **Conference Banquet**	**18:00 –19:00 Dinner**	

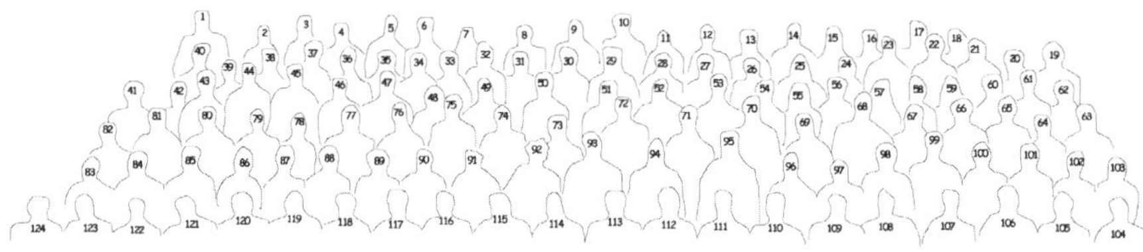

Participants of ICXRL2004 in conference photograph

1 B Chichkov 2 P Möller 3 H C Wang 4 D Benredjem 5 T Mocek 6 J J Park 7 C Menoni 8 V Miroslava 9 V Pavel 10 K Kolacek 11 W Hong 12 E Hotta 13 Y P Zhao 14 W Z Huang 15 L Huang 16 Y Gu 17 Z H Chang 18 W Wang 19 X F Lu 20 D Y Chen 21 Y Zhang 22 H M Li 23 A Jancarek 24 Z Y Zheng 25 C Wang 26 K Namikawa 27 R Z Tai 28 M Kishimoto 29 S Z Fu 30 M Born 31 V Vorontsov 32 X G Song 33 X K He 34 B Shuai 35 C Möller 36 X H Xie 37 Y Zhu 38 W Wang 39 Z H Fang 40 M Chen 41 M Chen 42 W Meng 43 Y F Chen 44 S Hooker 45 I A Artyukov 46 G Faivre 47 K Cassou 48 A Demir 49 Y Abou-Ali 50 Z R Huang 51 N Kalinin 52 V Burtsev 53 N Hasegawa 54 Y Ochi 55 M Tanaka 56 M Nishikino 57 Y P Huo 58 Y Q Liu 59 X J Sun 60 Y Z Cheng 61 W D Zheng 62 Y Q Xia 63 T J Liang 64 X M Qiao 65 C Z Zang 66 Z Q Hao 67 Y T Li 68 W X Liang 69 J W Chen 70 M Suzuki 71 T Ozaki 72 R Ganeev 73 Z Zhang 74 K Mekeown 75 A Ritucci 76 D Ros 77 S Sebban 78 J Zhao 79 F Yan 80 ZH Wang 81 X Y Peng 82 S Q Xiu 83 J R Sun 84 Z Y Wei 85 X Lu 86 Y L Cheng 87 K Lan 88 S Kazamias 89 A Klisnick 90 S Jacquemot 91 K Aggarwal 92 J Filevich 93 R Smith 94 B Motoyoshi 95 H Fiedorowicz 96 Z N Zeng 97 H Y Gao 98 H Xiong 99 J Y Lin 100 Q Z Yu 101 Z Jin 102 H N Han 103 H Xu 104 G P Zhang 105 R X Li 106 S Reiche 107 C Danson 108 G Tomassetti 109 H Kuroda 110 Y Kato 111 J Nilsen 112 G Jamelot 113 Z Z Xu 114 J Dubau 115 G Pert 116 J Balmer 117 J Zhang 118 V Shlyaptsev 119 A Vinogradov 120 D Attwood 121 J Rocca 122 J Dunn 123 G Tallents 124 B Wellegehausen

Contents

Section 7: Theory and simulations of X-ray lasers and plasma media

Section 13: Developments of X-ray laser drivers

Overviews of X-ray lasers research

Inst. Phys. Conf. Ser. No 186
Paper presented at 9th Int. Conf. X-Ray Lasers, Beijing, China, 24–28 May 2004

X-Ray Laser Research: Perspective and Physics Issues

Yoshiaki Kato
Japan Atomic Energy Research Institute,
14-1 Suehiro-cho, Kashiwa, Chiba 277-0842 Japan

Abstract. Historical perspective of the x-ray laser research during these 20 years is briefly reviewed. Emphasis is placed on major achievements, recent progress, and physics issues.

1. Introduction

Since the first demonstration of the laboratory x-ray lasers 20 years ago (Matthews 1985, Suckewer 1985), there has been significant progress in demonstrating x-ray amplification based on various pumping schemes, characterizing and improving their performances, and developing the x-ray laser applications. Especially noteworthy in recent years is the rapid progress in developing compact x-ray lasers. In this report, historical perspective of the x-ray laser research is given, with emphases on major achievements, recent progress and interesting physics issues some of which remain unresolved.

2. Pioneering Works before Demonstration of Laboratory X-Ray Lasers

Lasing in the ultraviolet and shorter wavelengths has been pursued soon after the demonstration of the ruby laser in 1960. Already in 1967, a 37.2 nm laser was proposed based on inner-shell photo-ionization of Na (Duguay 1967). Various pumping schemes have been investigated in the 1970s not only theoretically but also experimentally using the lasers and z-pinch devices as drivers; such as recombination pumping in H-like ions (Pert 1975), collisional pumping in Ne-like ions (Vinogradov 1977), resonant photo-excitation (Bhagavatula 1980), and resonant charge exchange (Dixon 1977). Although the facilities were not quite sufficient at that time to demonstrate lasing, these pioneering works have laid the foundations for developing the x-ray lasers in the 1980s.

The inherent difficulty in x-ray laser development is the requirement for short-pulse and very high-intensity pumping, since the spontaneous emission rate and the pumping power required for significant x-ray amplification scales with the lasing wavelength λ_{ul} as λ_{ul}^{-2} and λ_{ul}^{-4}, respectively (Elton 1990). Fortunately for the x-ray laser development, high performance high power Nd:glass lasers developed for laser fusion research since 1970 became available for x-ray laser experiments in the 1980s. By using the Novette laser at the Lawrence Livermore National Laboratory (LLNL), the first lasing in the soft x-ray region was demonstrated in Ne-like Se by collisional pumping (Matthews 1985).

In parallel with and independently from this approach, lasing with magnetically confined plasmas had been pursued at the Princeton Plasma Physics Laboratory, leading to demonstration of the C H-α laser at 18.2 nm by using a long-pulse CO_2 laser as a pumping source (Suckewer 1985). These results at Livermore and Princeton have driven many scientists interested in the x-ray lasers worldwide to try x-ray lasing in various schemes. It is noted that chirped pulse amplification (CPA), which later played the key role in developing compact x-ray lasers, was conceived at Rochester as a method to generate high power with small Nd:glass lasers that were far more compact than the high energy Nd:glass lasers used for laser fusion research (Strickland 1985).

3. Advances in Collisional Excitation X-Ray Lasers

3.1. Initial Progress with High Energy Lasers

The initial progress in the x-ray laser development was made mainly using the high energy lasers for laser fusion research to pump the x-ray lasers. The collisional excitation x-ray laser was first demonstrated in the $2p^5 3p$-$2p^5 3s$, J=2 to 1 transitions of Ne-like Se at 20.6 and 21.0 nm by illuminating a very thin plastic foil coated with Se with a 450 ps, 0.5 kJ green laser pulse (Matthews 1985). Very quickly the Ne-like Se laser was isoelectronically extrapolated to other ions for lasing and saturated amplification at various wavelengths, achieving the output energy of 7 mJ at 15.5 nm with Ne-like Y (Da Silva 1993). In order to obtain lasing at shorter wavelengths, the Ne-like scheme was extended to the Ni-like ions where population inversion is created in the $3d^9 4d$-$3d^9 4p$ J=0-1 transition. By pumping an exploding foil with two green laser beams of 500 ps-duration and 5.5 kJ total energy, strong amplification was obtained at 4.48 nm in Ni-like Ta and at 4.32 nm in Ni-like W (MacGowan 1990). The wavelength of the Ta x-ray laser, which is slightly longer than the carbon K-edge of 4.37 nm, is suitable for soft x-ray holography of biological samples (London 1989).

The collisional excitation x-ray lasers were pumped initially with long laser pulses with over 1 kJ to several hundred J of energy. Based on the experiment on the curved slab target where the J=0-1 line lased only in the rising part of the 1 ns pumping pulse (Kodama 1995), it was speculated that lasing could be obtained with shorter pulse pumping with which the total pumping energy could be reduced. This has led to the experiment where two laser pulses of 100 ps durations have irradiated a Ge slab target, resulting in strong lasing of the J=0-1 line (Daido 1995). This approach has evolved as the prepulse technique, where a pre-pulse produces a pre-plasma whose scale length is appropriate to reduce the refraction of the x-ray laser beam and to increase the absorption efficiency of the main laser pulse. The pumping efficiency has been improved and the lasing wavelengths have been extended with the prepulse technique, resulting in saturated amplification in Ni-like Dy at 5.9 nm with 60 J, 75 ps pumping (Smith 1999).

3.2. Development of Compact CE X-Ray Lasers

During the x-ray lasers were developed using the high energy lasers as above, the CPA technology matured gradually and became applicable to the development of next generation, compact x-ray lasers. The major breakthrough in this direction was made by the proposal (Afanasiev 1989) and demonstration (Nickles 1997) of the transient collisional excitation (TCE) x-ray laser, where the laser pulse duration is comparable to the collisional relaxation time of the upper laser level, which is 1-10 ps for the soft x-ray lasers. This has lead to significant reduction of the pumping energy and very high gain of the lasing transition as demonstrated by the almost saturated amplification in Ni-like Pd at 14.7 nm with the a table-top ~10 J Nd:glass laser (Dunn 1998). Very recently the pumping energy of the TCE x-ray laser has been further reduced with the grazing incidence pumping due to better coupling of the pumping laser to the gain region, resulting in strong amplification of the 18.9 nm line in Ni-like Mo using a 10 Hz, 300 mJ Ti:sapphire laser (Keenan 2003). By longitudinal pumping of the expanding plasma from a slab target, sub-mrad divergence lasing was obtained in Ni-like Mo by 475 fs, 150 mJ pumping (Ozaki 2002).

There has been very rapid progress in the fast capillary discharge pumped x-ray laser (Rocca 2003a). Although the discharge pumped VUV laser was demonstrated in H_2 at 116 nm in 1970 (Hodgson 1970), there has not been much progress in this approach. The discharge-pumped lasing in EUV was first demonstrated with an Ar-filled capillary in the J=0-1 transition of Ne-like Ar at 46.9 nm by collisional excitation (Rocca 1994). Following this result, saturated amplification with a 20-cm length capillary (Rocca 1996), 4 Hz, 3.5 mW continued operation (Macchietto 1999), full spatial coherence (Liu 2000) and desk-top laser (Rocca 2002) have been reported. When this laser is extended to 13.2 nm as being tested with the Ni-like Cd (Rahman 2002), it will become useful for metrology in the EUV lithography. Lasing at shorter wavelengths could be obtained also by longitudinal laser pumping of the capillary plasma by taking advantage of the fast capillary discharge as a waveguide (Janulevich 2003).

Another approach to developing the compact x-ray laser is the optical-field-ionisation collisional excitation scheme, where highly charged ions generated with high intensity laser in low density gases are excited by hot electrons produced with circularly polarized laser field. Although there have been difficulties in reproducing the first successful demonstration of the 41.8 nm laser in the 5d-5p transition in Pd-like Xe (Lemoff 1995), reproducible results have been obtained recently on saturated amplification in Xe (Sebban 2001) as well as lasing at 32.8 nm in Ni-like Kr (Sebban 2002). Although the output laser energy is limited due to the use of low density gases, this OFI laser can be operated continuously without deterioration of the laser gas and optical components.

3.3 *Physics Issues Related to CE X-Ray Lasers*

There have been many interesting and important issues related to the collisional excitation x-ray lasers. The unexpected performance in the initial Se x-ray laser experiment was the absence of lasing of the J=0-1 line of the 3p-3s transition, contrary to the theoretical expectation where the J=0-1 line should have higher gain since the J=0 level is preferentially populated from the ground state by monopole excitation (Rosen 1985). The clue to resolving this anomaly was obtained at the Naval Research Laboratory where comparable gains were observed for the J=0-1 and J=2-1 lines in Ne-like Ge and Ne-like Cu lasers with long pulse irradiation of slab targets (Lee 1987). Another clue was obtained at Osaka University in irradiation of curved slab targets with a 1 ns pulse where the Ge J=0-1 line lased in the rising part of the pumping laser whereas the 2-1 lines lased mainly after the peak of the pumping pulse (Kodama 1995). These observations have suggested that the 0-1 line lases in the high density plasma during the heating phase in agreement with the theoretical expectation, whereas the 2-1 lines lase in the lower density, equilibrium or recombining phase plasma where collisional relaxation contributes to populating the J=2 levels.

The collisional x-ray lasers generated with slab targets in general have the beam patterns elongated along the slab surface with the beam divergence of typically a few mrad. These properties have been extensively analysed using the hydrodynamic, atomic physics and ray-tracing codes to simulate the beam propagation in two-dimensional gradient density plasmas (Plowes 1996, Nilsen 2003). Although the x-ray laser is basically incoherent at the source since the beam area is too large to select only low order spatial modes, coherence at a far field is established during propagation in free space as observed by the interference patterns (Kato 1991) and the speckles at a far field.

Recently a fully coherent x-ray laser with diffraction limited divergence has been generated with the oscillator-amplifier configuration, where the amplifier located 20 cm away from the oscillator acts also as a mode selector (Nishikino 2003). There is a proper timing for amplification in this case where the density gradient of the gain region is small so that the beam propagation is not affected by refraction in the amplifier. Close to saturated amplification has been achieved recently in this configuration. With relation to spatial coherence, it is now possible to directly measure the wavefront of the x-ray laser beam with a Shack-Hartmann method (Le Pape 2002).

One of the interesting issues is the polarization of the x-ray laser beam. We have found that the 0-1 line of the Ne-like Ge x-ray laser generated with a 100-ps driven slab target was polarized parallel to the target surface with the polarization degree of 53 %. Since the J=0 upper laser level does not polarize, the observed polarization has been ascribed to the polarization of the lower laser level, which is caused by the anisotropy of trapping of the spontaneous emission from the lower laser level to ground state since the the velocity gradient and the optical depth are different along and perpendicular to the target surface (Kawachi 1995). It is expected that stronger polarization will be observed for the 0-1 line in the short pulse driven x-ray lasers which have larger plasma anisotropies. This polarization scheme is not effective in long scale length plasmas as in another experiment where the polarization was not observed for the 2-1 lines of the Ge laser driven by a 650-ps pulse (Rus 1995).

Another important aspect is the line profile of the x-ray laser, which reflects spectral broadening mechanisms in the plasma and gain narrowing/re-broadening due to amplification and saturation. The high resolution spectroscopy experiment on the J=2-1 20.6 nm line of Ne-like Se has shown that the spectral width in the small signal regime is approximately 5 pm, 1.4 times broader than the expected 400 eV Doppler width. The spectral width decreased with the gain length to approximately 1.1 pm, which was close to the instrumental resolution, and did not show saturation re-broadening (Koch 1992). These observations have indicated that homogeneous (collisional and natural) broadening was dominant over inhomogeneous broadening, in disagreement with the expected dominance of the Doppler broadening. The longitudinal coherence length $l_c=c/\Delta\nu$ estimated from the gain narrowed width is approximately 400 μm for this laser line.

Very recently, the longitudinal coherence length was measured directly with a Michelson interferometer for the short-pulse driven Ni-like Pd x-ray laser at 14.7 nm, resulting in l_c of ~400 μm at 1/e half width and a spectral width of ~0.3 pm (Dunn 2003). On the other hand, from the measured 3.7 ps pulse width of the short-pulse driven Ni-like Ag laser, it has been suggested that the x-ray laser pulse is close to transform limited determined from its spectral width (Abou-Ali 2003).

Since a complete set of basic parameters to characterize the x-ray lasers are becoming available, such as amplification and saturation properties, near and far field beam profiles, spatial coherence, polarization, spectral profile, and temporal pulse shape, it is now possible to comprehensively compare the x-ray laser performance with respect to detailed theoretical modelling and computer simulations on the atomic, hydrodynamic and laser physics of the collisional excitation x-ray lasers.

4. Development of Other Schemes of X-Ray Lasers

Although the collisional excitation x-ray lasers have been very successful in generating high power, high quality x-ray lasers, it has an inherent difficulty in scaling to shorter wavelengths since it requires very high intensity pumping of high density plasmas. There is need for continued efforts to develop x-ray lasers with other pumping schemes.

The recombination pumping is conceptually a simple approach for creating the population inversion in highly charged ions. Experimentally, however, there has been much difficulty in realizing a high gain-length product in recombining plasmas. Although indications for population inversions were observed in H-like and Li-like ions in various atomic species, definite evidence for significant amplification was not realized until 1995 where the C H-α laser at 18.2 nm was demonstrated at the Rutherford Appleton Laboratory (RAL) by carefully preparing an optimised condition for adiabatic expansion by irradiating a thin carbon fiber with a 20-J, 2-ps CPA Nd:glass laser (Zhang 1995). The difficulty in observing high gain in the Li-like ions remains unresolved, although it may be ascribed to the kinetics of the Li-like ions (Kawachi 2003).

Amplification in the transition to the ground state is attractive in generating the x-ray lasers at short wavelengths. In the recombination scheme, this requires generating cold electrons sufficient for rapid recombination and de-excitation to the lowest excited state. Amplification in the Ly-α line of Li at 13.5 nm was reported in the optical field ionisation plasma using a slab target (Nagata 1993) or a capillary (Korobkin 1996). There have been difficulties to demonstrate large gain-length products in this approach however, and it has been suggested that the Li micro-particles may have led the apparent gain (Reinhardt 2001).

Another and more direct approach toward the shorter wavelength x-ray laser is the inner-shell ionisation scheme where the population inversion is created between the inner shell vacancy state and the outer shell state (Duguay 1967). There has been continued effort toward realization of the 4.5 nm x-ray laser in the Kα transition of C by pumping with intense and short pulse continuum x-ray radiation (Moon 1998). In order to relax the severe conditions in the self-terminating x-ray lasers described above, a modified scheme was proposed where the lower laser level decays rapidly to other states through the Coster-Kronig processes (Kim 1999). It is expected that progress in these schemes will be made by using the high power, ultrashort pulse Ti:sapphire lasers like the 0.85 PW laser at JAERI shown in Figure 1.

Figure1. The Ti:sapphire laser at the Advanced Photon Research Center, which operates at either 100TW at 10 Hz, or 0.85 PW in every 10 min.

It has been reported that the ~2.9 A emission in the 3d-2p transition of Xe^{+32} ions is enhanced in the forward direction when Xe clusters were irradiated with high intensity, short pulse KrF laser (Borisov 2003). This work remains to be confirmed by reproducing the results with more complete sets of data on basic laser characteristics, such as small beam divergence, spectral enhancement, and large gain-length product.

In the resonant photo-pumping, a particular transition in an ion is pumped with a line radiation whose wavelength matches exactly with the pumped transition. This scheme requires finding the emitter-absorber combination with exact wavelength match, generating the emitter and absorber plasmas with close spatial proximity, and generating a well defined gain region. Although various proposals and experiments have been made on the emitter-absorber combinations, the photo-pumped x-ray lasers realized so far have been only by self-photo-pumping where the lasing ion is pumped with the line emission from the same ion (Nilsen 1996, Fiedorowicz 1996).

5. Applications of the X-Ray Lasers

Since the x-ray lasers have the characteristics of high flux, short pulse and coherent, it is suitable for single shot imaging of transient phenomena. As an example, the fully coherent 13.6nm Ni-like Ag laser generated with the oscillator-amplifier configuration at JAERI has a source size of 50 μm in diameter, output energy of 0.2 μJ, pulse width of 7.0 ps (Ochi 2004), and 0.2 mrad divergence, resulting in the peak brightness of 2 x 10^{26} photons/(s mm^2 $mrad^2$ 0.1%bandwidth) which is significantly higher than the peak brightness of ~10^{20} for the third generation synchrotron radiation sources.

By developing various diagnostic techniques the x-ray lasers have been used to clarify dynamic properties of high density plasmas, such as evolution of imprint due to Rayleigh-Taylor instability by x-ray radiography (Wolfram 1998), laser hole boring into overdense plasmas by x-ray refractometry (Takahashi K 2000), and two-dimensional plasma expansion by x-ray interferometry (Smith 2002).

The application areas of the x-ray lasers are expanding rapidly. The microscopic image of a rat sperm was taken with a Fresnel zone plate as soon as the 4.4 nm Ta laser was developed (Da Silva 1992). Unfortunately the biological applications have not shown much progress since then, mainly because the small size, high brightness water window x-ray lasers have not been developed so far. The material science is the new area of application. Novel results have been obtained using the x-ray lasers while developing new diagnostic techniques: ~10 nm surface change under high static electric field with interferometry (Zeitoun 1998), changes of the micro-domain structures of ferro-electric crystals near the phase transition temperature by x-ray laser speckles (Tai 2002), and core-level and valence band electronic structures of solid surfaces with time-of-flight photoelectron spectroscopy (Nelson 2003). The x-ray lasers, when focused to a small spot, are intense enough to modify the surface structures (Rocca 2003b) and possibly to fabricate nano-structures. Since the x-ray lasers have narrow line widths and close to diffraction limited, it is suitable for application to metrology, such as at-wavelength measurement of the optics for EUV lithography.

6. Summary

There has been remarkable progress in the x-ray laser development and their applications during the 20 years since the first demonstration of the laboratory x-ray lasers. Especially encouraging is the resurgence of the x-ray laser research recently, which has been brought by various innovations on the development and applications of compact and/or new types of x-ray lasers. Since there are now more facilities worldwide available for x-ray laser research, either shared or dedicated such as the one at the Advanced Photon Research Center (Fig. 2), it is hoped that the x-ray lasers will become more user friendly and new types of x-ray lasers will be developed such as the water window or keV x-ray lasers.

Figure 2. The X-Ray Laser Laboratory at the Advanced Photon Research Center, with two 10-J CPA Nd:glass laser beams.

Acknowledgment
The author is indebted to the staffs of the X-Ray Laser Research Group at the Advanced Photon Research Center of JAERI for the help in preparing this manuscript.

References

Abou-Ali Y et al. 2003 Opt. Commun. **215** 397
Afanasiev Yu V and Shlyaptsev V N 1989 Sov. J. Quant. Electronics **19** 1606
Bhagavatula V A 1980 IEEE J Quant. Electr. **QE-16** 603
Borisov A B et al. 2003 J Phys. B: At. Mol. Opt. Phys. **36** 3433
Daido H et al. 1995 Opt. Lett. **20** 61
Da Silva L B et al. 1992 X-Ray Lasers 1992 (IOP London) 387
Da Silva L B et al. 1993 Opt. Lett. **18** 1174
Dixon R J and Elton R C 1977 Phys. Rev. Lett. **38** 1072
Duguay M A and Rentzepis P M 1967 Appl. Phys. Lett. **10** 350
Dunn J et al. 1998 Phys. Rev. Lett. **80** 2825
Dunn J et al. 2003 Proc. SPIE **5197** 43
Elton R 1990 X-Ray Lasers (London: Academic Press)
Fiedorowicz H et al. 1996 Phys. Rev. Lett. **76** 415
Hodgson R T 1970 Phys. Rev. Lett. **25** 494
Janulevich K A et al. 2003 J Opt. Soc. Am. B **20** 215
Kato Y et al. 1991 Proc. SPIE **1551** 56
Kawachi T et al. 1995 Phys. Rev. Lett. **75** 3826

Kawachi T 2003 Phys. Rev. E **67** 016409
Keenan R et al. 2003 Proc. SPIE **5197** 213
Kim D et al. 1999 Phys. Rev. A **59** R4129
Koch J A et al. 1992 Phys. Rev. Lett. **68** 3291
Kodama R et al. 1995 Phys. Rev. Lett. **73** 3215
Korobkin V V et al. 1996 Phys. Rev. Lett. **77** 5206
Lee T N et al. 1987 Phys. Rev. Lett. **59** 1185
Lemoff B E et al. 1995 Phys. Rev. Lett. **74** 1574
Le Pape S et al. 2002 Eur. Phys. J AP **20** 197
Liu Y et al. 2000 X-Ray Lasers 2000 (EDP France) Pr2-123
London R A, Rosen M D and Trebes J E 1989 Appl. Optics **28** 3397
Maccchietto C D et al. Opt. Lett. **24** 1115
MacGowan B J et al. 1990 Phys. Rev. Lett. **65** 420
Matthews D L et al 1985 Phys. Rev. Lett. **54** 110
Moon S J and Eder D 1998 Phys. Rev. A **57** 1391
Nagata Y et al. 1993 Phys. Rev. Lett. **71** 3774
Nelson A J et al. Proc. SPIE **5197** 168
Nickles P V et al. 1997 Phys. Rev. Lett. **78** 2748
Nilsen J 1996 Phys.Rev. A **53** 439
Nilsen J et al. 2003 J. Opt. Soc. Am. B **20** 191
Nishikino M et al. 2003 Phys. Rev. A **68** 061802(R)
Ochi Y et al. 2004, in these Proceedings
Ozaki T et al. 2002 Phys. Rev. Lett. **89** 253902
Pert G J and Ramsden S A 1974 Opt. Commun. **11** 270
Plowes J A et al. 1996 Opt.Quantum Electron. **28** 219
Rahman A et al. 2002 X-Ray Lasers 2002 (AIP NY) 113
Reinhardt C et al. 2001 X-Ray Lasers 2000 (EDP France) Pr2-201
Rocca J J et al. 1994 Phys. Rev. Lett. **73** 2192
Rocca J J et al. 1994 Phys. Rev. Lett. **77** 1476
Rocca J J et al. 2002 X-Ray Lasers 2002 (AIP NY) 125
Rocca J J and Vinogradov A V 2003a Quant. Electr. **33** 7
Rocca J J 2003b Proc. SPIE **5197** 174
Rosen M D et al. 1985 Phys. Rev. Lett. **54** 106
Rus B et al. 1995 Phys. Rev. A **51** 2316
Sebban S et al. 2001 Phys. Rev. Lett. **86** 3004
Sebban S et al. 2002 Phys. Rev. Lett. **89** 253901
Smith R et al. 1999 Phys. Rev. A **59** R47
Smith R F et al. 2002 Phys. Rev. Lett. **89** 065004
Strickland D and Mourou G 1985 Opt Commun. **56** 219
Suckewer S et al 1985 Phys. Rev. Lett. **55** 1753
Tai R Z et al. 2002 Phys. Rev. Lett. **89** 257602
Takahashi K et al. 2000 Phys. Rev. Lett. **84** 2405
Vinogradov A V, Sobel'man I I and Yukov E A 1977 Sov. J. Quantum Electr. **7** 32
Wolfram E et al. 1998 Phys. Plasmas **5** 227
Zeitoun Ph et al. 1998 Nucl. Instr. and Meth. in Phys. Res. A **416** 189
Zhang J et al. 1995 Phys. Rev. Lett. **74** 1335

Inst. Phys. Conf. Ser. No 186
Paper presented at 9th Int. Conf. X-Ray Lasers, Beijing, China, 24–28 May 2004

Refraction and saturation in amplified spontaneous emission X-ray lasers

G.J. Pert
Department of Physics, University of York, Heslington, York YO10 5DD, U.K.

Abstract. Short pulsed solid state lasers have become a common method to drive X-ray lasers. Currently the most popular system involves the irradiation of a plasma formed by a pre-pulse from a solid target by a main pulse generated by a CPA laser. Recent experiments have demonstrated the value of long delays between the pre-pulse and the main pulse as a method of overcoming the limitations of refraction resulting from the density gradient in the plasma plume. With this configuration significant output power may be generated with relatively small input energies. This contrasts with the more conventional experiments in which large input energies are used with relatively short delay. In this note we identify the factors which lead to this improvement, and show that it is key that the ionisation and temperature of the plasma plume remain relatively high at the onset of the main pulse.

1. Thermally pumped collisional lasers

The basic configuration of a thermally pumped collisionally laser is very simple (fig.1.). A slab target of an appropriate element is irradiated along a line focus by a laser pulse with a programmed time sequence of pulses. An initial pulse (or pulses), the pre-pulse, establishes a plasma body expanding away from the target surface with decreasing density and some initial ionisation. The main pulse, usually of significantly higher energy, heats the pre-formed plasma, leading to further ionisation and ultimately to the pumping of the laser transition by electron collision. A number of different pulse configurations have been used, which differ mainly in the form of the main pulse.

One of the most useful systems has proved to be the 'transient pumping scheme' [1, 2], in which the pre-pulse is a relatively long pulse of about 100 ps but the main pulse is only a few ps. The latter being generated by a CPA amplifier/ compressor arrangement. Key to transient system is travelling wave pumping in which the gain pulse moves along the gain axis in phase with the X-ray beam. In this way the limitations of the relatively short gain lifetime are mitigated by maintaining the gain and radiation pulse in synchronism.

2. Basic concepts of refraction

Fundamental to an understanding of these lasers is the process of refraction, which identifies which region of the plasma contributes most strongly to the output.

Consider the plasma configuration sketched in fig.1.. We can imagine that the dominant ray, i.e. the one contributing most strongly to the output is that with the longest gain-length starting

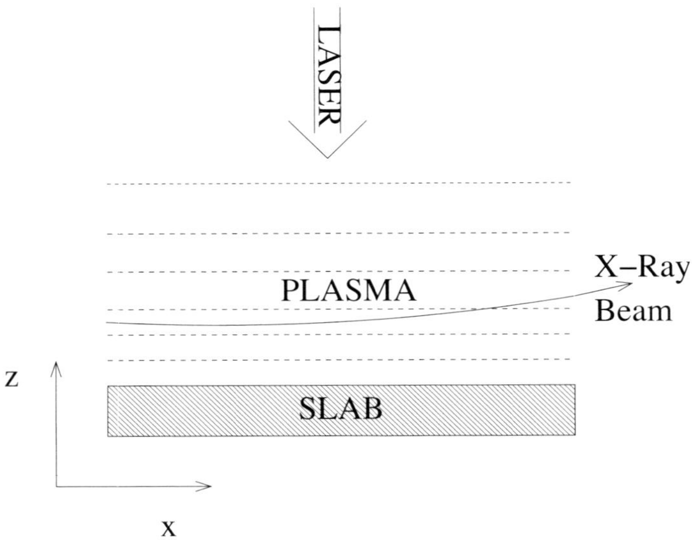

Figure 1: Sketch of the plasma configuration used in the X-ray laser.

in a region of significant spontaneous emission. Assuming the density and gain both decrease into the plume, the ray will have an approximately parabolic profile, as illustrated in fig.2. where the perigee of the ray (turning point) is at the highest density, and therefore highest gain.

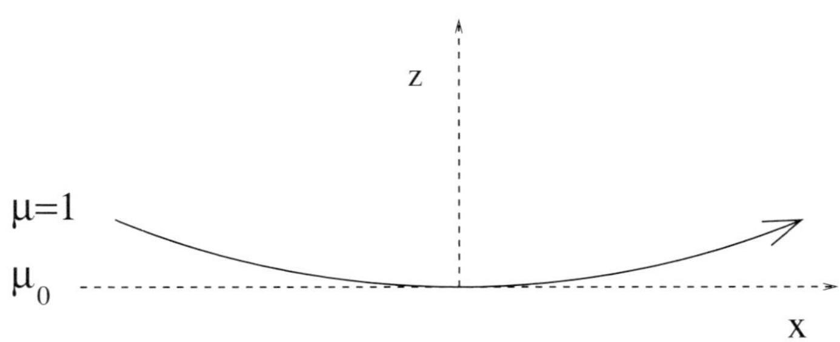

Figure 2: Sketch of the plasma configuration used in the X-ray laser.

If we consider only a two dimensional picture, where the rays lie in the plane containing the normal to the target and the X-ray laser axis, we may identify a simple scaling law for the density at the perigee, i.e. at peak gain, using Snell's law. Since the refractive index surfaces will lie parallel to the target surface, the angle θ between the ray and the plane will be given by:

$$\mu \cos\theta = \mu_0 \tag{1}$$

After completing its transit, the ray will exit into vacuum where $\mu = 1$. Hence if the refractive index is dominated by free electrons $\mu = \sqrt{1 - n_e/n_c}$ where n_c is the critical density at the

X-ray wavelength.

$$n_c = 1.115 \times 10^{25}/\lambda(100\text{Å})^2 \tag{2}$$

Hence

$$n_0 \approx n_c\theta^2 \tag{3}$$

yields the electron density at the perigee. In practice due to the exponential growth nature of amplification, the output is dominated by a narrow pencil of rays centred about the ray with the largest gain length. The experimentally measured deflection therefore gives a good estimate of the density at the peak gain.

If the density gradient is constant, the ray path is given by

$$z = \frac{1}{4}\frac{|\mathrm{d}n/\mathrm{d}z|}{\sqrt{n_c^2 - n_0^2}}\,x^2 \tag{4}$$

from which we may estimate the range of ray in the gain zone as

$$\ell \approx \frac{2\sqrt[4]{n_c^2 - n_0^2}}{\sqrt{|\mathrm{d}n/\mathrm{d}z|}}\sqrt{d} \tag{5}$$

where d is the width of the gain zone.

If the gradient is steep and the gain zone narrow $\ell \ll L$, the illuminated plasma length, then the output becomes the summation of a series of elements of length ℓ each with a gain length

$$X \approx \frac{4}{3}G_0\sqrt{\frac{\sqrt{n_c^2 - n_0^2}\,G_0}{|\mathrm{d}n/\mathrm{d}z|\,\mathrm{d}G/\mathrm{d}Z}} \approx \frac{4}{3}G_0\sqrt{\frac{\sqrt{n_c^2 - n_0^2}\,d}{|\mathrm{d}n/\mathrm{d}z|}} \tag{6}$$

If this is greater than the saturation gain-length $X_s \approx \ln\{I_s/I_0\sqrt{\ln(I_s/I_0)}\}$ where I_s and I_0 are the saturation and spontaneous emission irradiances respectively, non-linear refraction limited saturation occurs [3]

2..1. *Plasma Surface Irregularities*

Thus far we have assumed the plasma isochors (constant density surfaces) to be planar and parallel planes of constant z. Irregularities may take a number of forms:

- Small scale, short length disturbances. In this case the scale length of the disturbance is small compared to the travel length ℓ and the transverse perturbation small compared to the density gradient scale length $n/|\frac{\mathrm{d}n}{\mathrm{d}z}|$, the ray is only weakly perturbed by the refractive index variation, but sees a variation in the gain $G(z) + \frac{1}{2}\frac{\mathrm{d}^2G}{\mathrm{d}z^2}\Delta^2$ where Δ is the R.M.S. height variation.

- Large scale, regular distortions. The isochor may be treated as a locally circularly surface of radius of curvature R inwards. The angle of incidence of the ray path to the isochor is now increased by the deflection of the surface. The effect can be shown to be approximately equivalent to an increase in the local density gradient

$$|\mathrm{d}n/\mathrm{d}z| \Rightarrow |\mathrm{d}n/\mathrm{d}z| + 2[n_c - n_e]/R \tag{7}$$

Since the critical density n_c is much larger than the perigee density n_0 curvature may lead to a significant increase in the deflection even if the radius of curvature is large. We may also note that if the outwards radius of curvature $R = 2n_c/|\mathrm{d}n/\mathrm{d}z$ the curvature compensates for the density gradient - a result obtained many years ago by Lunney [4] and demonstrated experimentally [5].

- Non-parallel isochors. This will lead to a systematic change in the angle of incidence from layer to layer, and may be treated as an effective curvature along the ray path.

In addition to these irregularities in the two dimensional picture, there are of course, corresponding variations in the transverse directions, which will give rise to three dimensional ray effects.

3. Simulations of CPA Pumped Travelled Wave Lasers

Transient X-ray lasers have been developed in two distinct configurations which are designed to generate a shallow density gradient in which laser action can occur without severe loss by refraction over the full length of the pumped medium.

- Short delay between pre-pulse and main pulse. In this case the main pulse starts at or shortly after the peak of the pre-pulse. The pre-pulse and the main pulse have relatively high energy. The pre-plasma has only a short travel but has a fairly flat profile in the neighbourhood of high gain.

- Long delay between main pulse and pre-pulse. In this case the plasma generated by the pre-pulse has a long travel before the main pulse arrives. In consequence there is a reasonably long flat region with relatively high ionisation separated from the critical density region.

Both approaches have been successfully implemented. The latter however offers a lower energy route to lasing in the ~130Å wavelength region. On the other hand the former is probably the more appropriate method at shorter wavelengths.

We will now contrast the two approaches, using simulation with the codes EHYBD and RAYTRACE to examine representative examples of both methods.

3..1. High energy, short pre-pulse delay/main pulse delay

These are exemplified by experiments carried out in many laboratories world wide. Saturated laser action has been observed in both Ne-like and Ni-like ions in a variety of elements and over a range of wavelengths from 326Å to 73Å. As exemplar we will examine the experiments carried out at the Rutherford-Appleton Laboratory and previously reported in [6, 7] with Ne-like Ge ions lasing at 196Å.

The experiments were performed with the pulse configuration

Delay 0 ns	Energy 1.25 J/cm	Half width 280 ps
Delay 2 ns	Energy 12.5 J/cm	Half width 280 ps
Delay 2.21 ns	Energy 36.7 J/cm	Half width 2.9 ps

The energy input used in the simulation has been reduced by a factor of 0.3 to give agreement with experiment [6]. This factor is believed to account for losses by back scattering which are not included explicitly in the calculation.

It can be seen from fig.3 that the gain is established after about 10 ps and lasts for less than 30 ps. Furthermore the spatial extent of the gain is about 10 μm and peaks at about 10 μm from the target surface, i.e. at an electron density of nearly 10^{21} /cc in a region with a shallow density gradient. The measured beam deflection of 13 mrad corresponds to density of 5×10^{20} /cc indicating that due to the limitation set by refraction, the dominant ray accesses plasma of slightly lower gain than the peak. The relatively short gain lifetime leads to a short output pulse of about 10 ps after gain narrowing [7]. An additional important factor found in the simulations

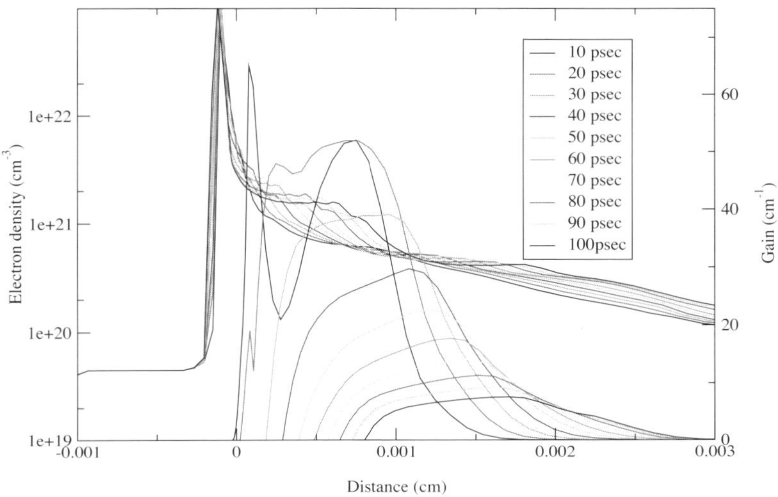

Figure 3: Density and gain profiles calculated for the Rutherford Ne-like germanium laser operating at 196Å. Distances are measured from the target surface, and time from the onset of the main pulse.

[7], but not shown here is that the ionisation at the start of the main pulse is relatively low, and the highest ionisation is located near to the target where high gain will be formed.

3..2. Low energy, long pre-pulse delay/main pulse delay

As an example of these experiments we consider the early set of experiments carried out at Livermore [9] which were performed using Ag and Pd Ni-like ions lasing at 139Å and 147Å rrespectively. In particular we will examine the performance of a silver system using the code EHYBD. The laser characteristics used in the experiments were

Delay 0 ns	Energy 1.5 J/cm	Half width 600 ps	Line width 150μm
Delay 1.0 ns	Energy 5.0 J/cm	Half width 1.0 ps	Line width 80μm

The values used for the simulation correspond to no reduction in the pre-pulse energy and 1/2 in the main pulse compared with that used in the experiment. The initial plasma line width was taken to be 100μm as a compromise between the focal widths. Fig.4 shows the gain profiles obtained in this case. it is immediately apparent that there is a marked difference from the germanium case fig.3. Firstly there are two distinct spatial components of the gain. The inner is found at high density near the critical density of the pump laser at about 20μm. This corresponds to that found in the germanium simulation above. The second component occurs later and progressively moves out through the plume. Although the inner peak has substantially higher gain it is formed in a region of strong refractive index gradient and therefore we anticipate the output will be small. However there is substantial gain late in the expansion formed well out into the plasma plume peaking after about 50 ps at about 40μm. Since the density gradient is small in this region we may expect the effects of refraction to be severely inhibited and that we may obtain significant signal.

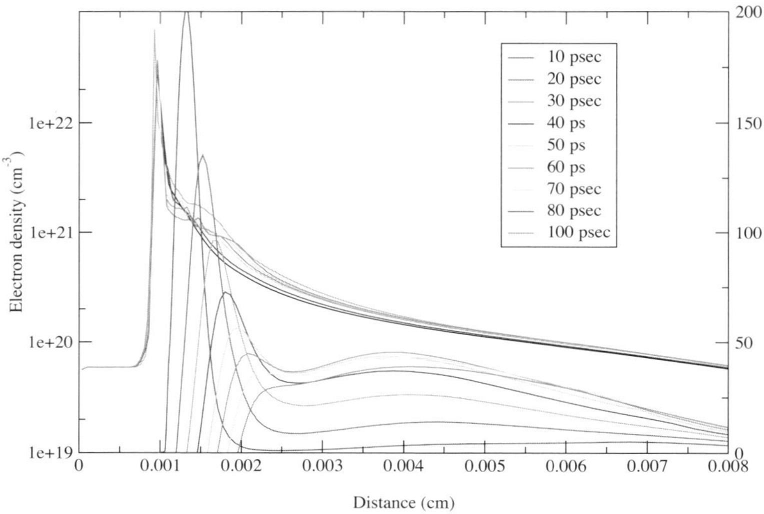

Figure 4: Density and gain profiles calculated for the Livermore Ni-like silver laser operating at 139Å. Distances are measured from the target surface, and time from the onset of the main pulse.

Fig.5 shows that this is indeed the case. The output is calculated by a full three dimensional ray trace taking into account saturation and spectral line profile effects. The output pulse from the inner part of the gain profile can be seen as the small initial spike at 10 ps and that from the outer regions as the long pulse peaking at about 60 ps. The output deflection angle at 50 ps is found to be about 7mrad for a 1.0 cm long plasma and corresponds to an electron density of 2×10^{20} /cc at peak gain, in good agreement with fig.4

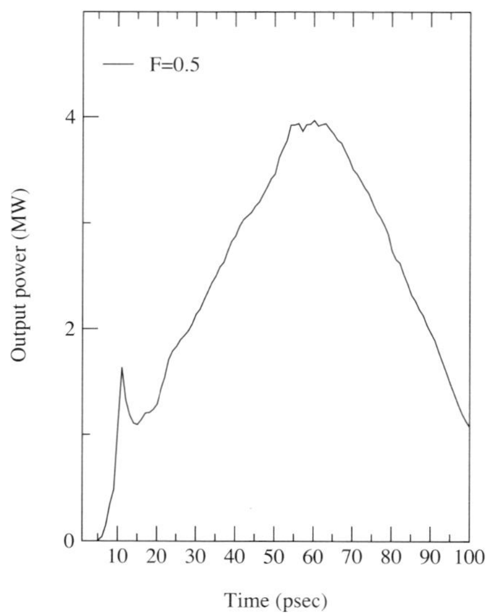

Figure 5: Temporal power profile calculated for the Livermore Ni-like silver laser operating at 139Å. Time is measured from the onset of the main pulse

The question now arises as to the source of this long lived gain in the plume. This is resolved by examining the nature of the plasma at the onset of the main pulse and immediately following. Fig.6 shows the development of the ionisation and the electron temperature at and following the onset of the main laser pulse. It can be seen that at the start of the pulse there is relatively strong ionisation remaining in the plume around 50 μm, in the pre-pulse significantly far downstream of where the critical density will form. As a result substantial collisional absorption of the main pulse takes place in this extended plasma tail at reasonably high density leading to a growth of ionisation and to temperatures sufficiently high for significant pumping. Indeed by 50 ps the ionisation has reached 18 and the temperature 200 eV at 50 μm, ideal values for generating gain. In contrast further in around 20 μm the system over-ionises and the gain is lower despite the higher densities and temperatures. Note the minimum in the electron temperature, which is a consequence of the convection of the original pre-pulse heated plasma downstream and the subsequent stronger heating upstream, and gives rise to a minimum in the gain despite the decreasing densities as we move outwards.

This behaviour is critically dependent on both the level of the pre-pulse and the delay. If either are substantially changed the ionisation in the plume becomes insufficient to allow sufficient absorption to give rise to the simultaneous ionisation and temperature conditions essential to generate reasonable gain. With this arrangement we can identify two modes of operation:

1. A rapid onset of inversion near the critical density of the pump laser. This leads to a narrow gain zone with high gain values in a region of large refractive index gradient. The output is refraction limited, but saturated over this short length due to the high gain. The gain pulse is short, limited by over-ionisation of the plasma, which leads to a short output duration.

2. A slow onset of gain in the plume, which moves progressively outwards. The gain region is therefore wide and the density gain gradient small giving rise to long path lengths. Although the value of the gain is smaller than in the inner region, it is sufficiently large to give rise to substantial gain-lengths and saturation in the output. Since the density is relatively low, beam deflections and divergences are expected to be small.

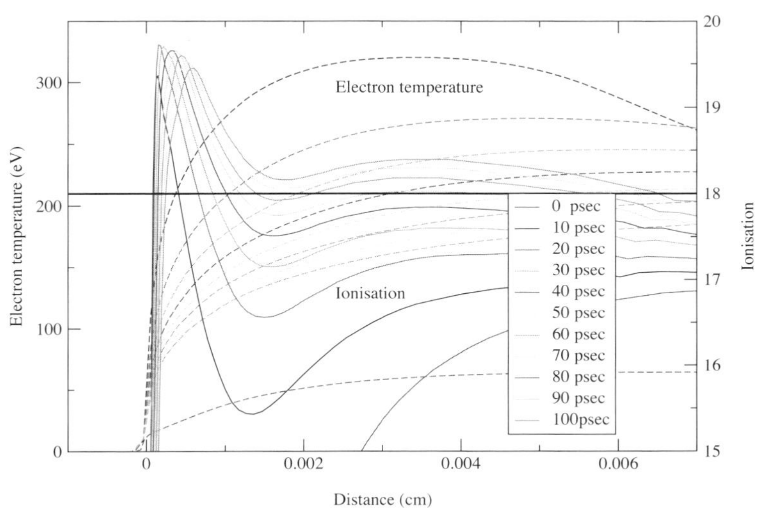

Figure 6: Ionisation and electron profiles calculated for the Livermore Ni-like silver laser operating at 139Å. Distances are measured from the target surface, and time from the onset of the main pulse.

If the system is properly established, the output of type 1 is relatively small and the signal observed comes from the later low density gain region (2) (fig.4). However the profile calculated is of long duration, in contrast to that seen experimentally. This is believed to be due to a failure of EHYBD to correctly model two dimensional motion of the plasma at late times, which lowers both the electron density and temperature of the plasma and sharply reduces the gain.

4. Conclusion

The role of refraction is limiting in the design of collisionally pumped X-ray lasers. In this note we have seen how controlling the density profile of the plasma in which the laser action is generated influences the refractive index gradient and thereby the overall amplification of rays generated by spontaneous emission. It is apparent that the pre-pulse, by determining the plasma structure, thus plays a critical role in regulating the precise way in which the laser will operate. Consequently as we have seen with a short pre-pulse–main pulse delay, the amplification may be generated at high density near to the critical density for the pump laser, whereas with a long delay the laser action occurs at lower density. In the simple form of the simulations presented here the former leads to a shorter gain lifetime, and thus to a relatively short output pulse. The latter on the other hand appears to have a long gain lifetime, which is at variance with experimental measurement of the output pulse duration. This discrepancy is probably due to the onset of two dimensional expansion, and hence greater dilation and cooling, which is poorly treated at present. The main pulse provides the heating and therefore electron temperature necessary to drive the gain. Again the different plasma profile in the two cases gives rise to disparities. In particular the short delay leads to a hot, high density plasma of small spatial extent, whereas the long to an extended hot, lower density plasma. In the former case a significant fraction of the energy is absorbed at high density and effectively lost by thermal conduction into the solid. In latter case about 50% is absorbed by inverse bremsstrahlung into the plasma body.

These results lead to an important conclusion namely that for short wavelengths where pumping at high density is preferred, as refraction is less important and the pumping rate is increased to compensate the faster decays. One the other hand at wavelengths in 100 – 200Å region long delays are preferable allowing greater efficiency and flexibility.

Acknowledgements

I would like to thank Dr R E King and T Ball for useful diiscussions on this topic. This work was supported by EPSRC and EU TMR programme.

References

[1] Afanasiev Y V and Shlyaptsev V N 1989 Sov J Quantum Electron 19 1606

[2] Nickles P V, Shlyaptsev V N , Kalachnikov M P, Schnürer M, Will I and Sandner W 1997 Phys Rev Lett 78 2748

[3] Pert G J 2004 Optics Comm 236, 173

[4] Lunney J G 1986 Appl Phys Lett 73 3215

[5] Kodama R, Neely D, Kato Y, Daido H, Murai K, Yuan G, MacPhee A, Lewis C L S 1994 Phys Rev Lett 73 3215

[6] King R E, 2001 Phys Rev A

[7] King R E and Pert G J 2000 Compte-rendus A Serie IV 1 1093

[8] Dunn J, Osterheld A L , Shepherd R, White W E, Shlyaptsev V N and Stewart R E 1998 Phys Rev Lett 80 2825

[9] Dunn J, Li Y, Osterheld A L, Nilsen J, Hunter J R and Shlyaptsev V N 2000 Phys Rev Lett 84 4834

Inst. Phys. Conf. Ser. No 186
Paper presented at 9th Int. Conf. X-Ray Lasers, Beijing, China, 24–28 May 2004

Demonstration of High Repetition Rate Desk-top and Table-top Soft X-Ray Lasers

J.J. Rocca, B.M. Luther, S. Heinbuch, Y. Wang, M.A. Larotonda, M. Grisham, D. Alessi, M. Berrill, M.C. Marconi, V.N. Shlyaptsev[b], A. Dummer, F. Brizuela, and C.S Menoni.
[a] National Science Foundation ERC for Extreme Ultraviolet Science and Technology, Colorado State University. Fort Collins, CO 80523
[b] University of California, Davis

Abstract. We report the demonstration of the first desk-top size high repetition rate (12 Hz) soft x-ray laser. It consists of a fast capillary discharge emitting ~ 10 μJ laser pulses at 46.9 nm. In a second set of experiments we have demonstrated saturated operation of a table-top 18.9 nm Ni-like Mo laser and its isoelectronic scaling in Ru, Pd and Ag to generate lasers at 16.5, 14.7 and 13.9 nm using grazing incidence pumping with a 5-10 Hz repetition rate Ti:Sapphire table-top laser system. The results impact the development of practical high average power soft x-ray lasers for applications.

1. Introduction

A significant part of the research effort in the field of x-ray lasers in the past decade has been concentrated in reducing the size of soft x-ray lasers to table-top dimensions, and in increasing their repetition rate with the goal of generating substantial laser average powers for applications. Herein we report the first device of a new generation of high repetition rate "desk-top" soft x-ray lasers. This is a capillary discharge pumped laser that fits on a small desk-top and that is capable of emitting ~ 10 μJ pulses in the 46.9 nm line of Ne-like Ar at a repetition rate of 12 Hz. The size of this laser is significantly smaller than that of its table-top capillary discharge laser predecessors [1,2]. We discuss results of the characterization of this laser, including its output pulse energy, average power, and the first lifetime tests at 12 Hz repetition rate operation. In a second set of experiments we have used a 5-10 Hz repetition rate Ti:Sapphire table-top laser system capable of generating compressed laser pulses of up to ~ 1 J to obtain strong lasing in several Ni-like ions at wavelengths between 13.9 and 22.6 nm. The generation of a saturated 18.9 nm Ni-like Mo laser is discussed in detail. The experiments made use of the grazing incidence pumping configuration recently demonstrated to improve the energy deposition efficiency [3-5].

2. Demonstration of a Desk-top 46.9 nm laser operating at 12 Hz repetition rate

Portable soft x-ray laser sources operating at high repetition rate are of significant interest for applications such as the spectroscopy of materials. The particular application that motivated the development of the high repetition rate 46.9 nm capillary discharge laser described here is in the areas of photophysics and photochemistry, and consist in the single-photon ionization spectroscopy studies of metal-oxide nanoclusters. Such a spectroscopy experiment provides a demanding test environment, in which the laser has to operate repetitively at 10 Hz for numerous series of 1500-2000 consecutive shots.

More specifically the targeted goal of this development was the demonstration of a small size 46.9 nm laser with the following characteristics:

- Desktop size
- Portable
- Laser pulse energy > 10 μJ (~ 13 μJ was obtained)
- Repetition rate: 10 Hz (12 Hz was demonstrated)
- External synchronization capability

The approach makes use of capillary discharge excitation. As in other capillary discharge lasers previously demonstrated, laser amplification at 46.9 nm is generated by fast discharge excitation of an Ar-filled capillary. The magnetic force of the current pulse and large thermal pressure gradients near the wall rapidly compresses the plasma to form a dense and hot column with a large density of Ne-like ions, with a very high axial uniformity and a length to diameter ratio of the order of 1000:1. Collisional electron impact excitation of the ground state Ne-like ions produces a population inversion between the 3p $^1S_0$ and 3s $^1P_1^0$ levels resulting in amplification at 46.9 nm [6].

We used aluminum oxide capillary channels 3.2 mm in diameter and 21 cm in length, filled with preionized Ar gas at a pressure of 700 mTorr. The plasma columns are excited by current pulses of ~ 20 kA peak amplitude, with a 10% to 90% rise time of approximately 30 ns. The laser fits on a small desk, and the power supplies and gas system can be accommodated in small racks that fit under a regular optical table. Successful operation of the laser at repetition rates up to 12 Hz with a laser output pulse energy of ~ 13 μJ. Figure 2a shows plots of the variation of the laser output pulse energy as a function of time for 1500 consecutive laser shots. The achieved repetition rate of 12 Hz constitutes an order of magnitude increase respect to the repetition rate of the desk-top soft x-ray laser we reported at the Aspen conference two years ago [7]. The present test also demonstrated for the first time an electrically triggered, relatively low jitter, capillary discharge laser. Lifetime tests were also conducted. A lifetime of about 25,000 shots for laser output energy output decay by 2X was measured (Fig. 3). This is the longest series of laser shot achieved for any EUV laser demonstrated to date. Full output pulse energy can be recovered by changing the used capillary discharge tube by a new one, an operation that demands 30-40 minutes, including the pumping time required to evacuate the system to a pressure of less than 1×10^{-5} Torr.

Figure 1. Photograph of the desk-top 46.9nm laser. This prototype was successfully operated at 12 Hz repetition rate

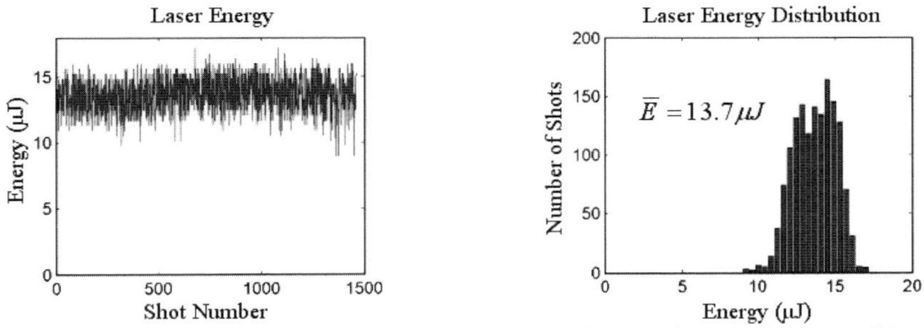

Figure 2. Laser output data corresponding to 1500 contiguous shots at 12 Hz repetition rate operation of the desk-top 46.9 nm discharge-pumped laser, a) laser output energy vs. number of shots, b) statistical distribution of the output pulse energy.

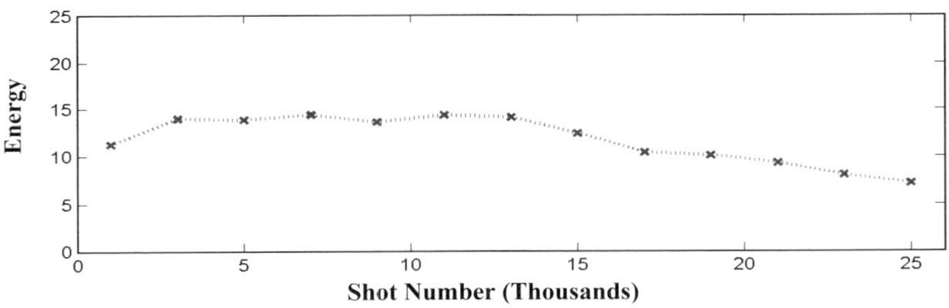

Figure 3. Laser output pulse energy as a function of the number of discharge shots. The laser output energy is observed to decrease by a factor of two in about 25,000 shots.

3. Demonstration of high repetition rate table-top lasers at wavelengths between 13.9 and 22.6 nm

There is also great interest in the development of compact high repetition rate lasers for applications at wavelengths below 20 nm. Several approaches are discussed in papers in these proceedings. Transient collisional electron excitation of targets at normal incidence using 3-10 J of pump energy has produced several saturated lasers in the 12-33 nm spectral range operating at a repetition rate of one shot every several minutes [8,9]. The low repetition rate and the still relatively large laser facilities required to pump these lasers motivates the exploration of more efficient pumping geometries. Longitudinal pumping of a Mo plasma with short pulses from a 10 Hz laser produced non-saturated amplification at 18.9 nm in Ni-like Mo [10]. Transverse excitation of a Mo target with a 150 fs, 300 mJ pulse impinging at 60 degrees from normal incidence resulted in the observation of the 18.9 nm and 22.6 nm laser lines of Ni-like Mo [11]. Recently it was shown that the pumping energy necessary for lasing can be significantly reduced by directing the short pulse onto the target at grazing incidence [3,4]. This inherently traveling wave pumping geometry takes advantage of refraction of the pump beam to increase the path length of the rays in the gain region of the plasma, thereby increasing the fraction of the pump energy absorbed in this region [12]. A gain-length product of ~ 15 in the 18.9 nm line of Ni-like Mo was reported at this conference by J. Dunn et. al. using 150 mJ of total pumping energy from a 10 Hz laser to produce a normal-incidence pre-pulse of 200 ps duration focused into a 15 μm wide line focus, followed by a 1.5 ps short pulse impinging at a grazing incidence angle of 14 degrees [5]. Lasing was observed to occur only over an extremely narrow ~ 50 ps range of prepulse to short pulse time delays. We recently obtained saturated amplification in the 18.9 nm Ni-like Mo and strong lasing in the 22.6 nm line of the same ion using grazing incidence pumping, and a table-top laser [13]. Herein we discuss these results and report lasing in several laser transitions from Ni-like ions (Mo, Ru, Pd and Ag) at wavelengths ranging from 22.6 nm to 13.9 nm using a 5-10 Hz tabletop pump laser that provides picosecond duration laser pulses with energy up to 1 J. Lasing was obtained over a broad range of time delays and pumping conditions. The gain media was created by the combination of two laser pulses, a 120 ps FWHM long pre-pulse followed by a heating pulse with a pulse duration between 1.8 to 8.1 ps FWHM, both generated by the same 800 nm table-top Ti:sapphire laser system with three stages of amplification. A beam-splitter directed 20 percent of the laser energy to the pre-pulse arm. The laser can operate at a repetition rate of 10 Hz. However, for the present experiments the repetition rate was reduced to 5 Hz. An electromechanical shutter was placed on the pump beam of the first amplifier to allow for single shot data acquisition.

A pre-plasma was formed by irradiating the target at normal incidence with the long pre-pulse focused into a line of 30 μm FWHM width and 4.1-5 mm in length using the combination of a f = 67.5 cm focal length spherical lens and a f = 200 cm focal length cylindrical lens. The pre-plasma was rapidly heated with the short pulse impinging at a grazing incidence angle of 14 degrees to generate a transient population inversion. The short pulse was focused into a line of 30 μm FWHM width using a f = 76.2 cm focal length multilayer-coated parabolic mirror placed at 7 degrees from normal incidence (Fig. 4). The off-axis placement of the paraboloid forms an astigmatic focus that results in a line that is further elongated when intercepted at grazing incidence by the target to form a 4.1 mm FWHM line focus. Initial successful experiments were conducted using a f = 2 m spherical mirror to focus the short pulse beam. However, all the results discussed herein were obtained using the parabolic mirror. The overlap of the two line focus on target was monitored and adjusted by imaging the target onto a CCD utilizing digital zoom. The plasma emission was monitored using a flat field spectrograph composed of a 1200 lines per millimeter gold-coated variably space spherical grating placed at 87 degrees grazing incidence and a back-illuminated CCD detector array of placed 48 cm from the target. Taking advantage of the small dimension of the plasma no slit was used. Al filters of 0.3 um and 0.5 μm thickness or their combination with a 0.14 μm thick Zr filter were used. In some of the measurements the beam was further attenuated placing meshes of known transmissivity in front of the diffraction grating.

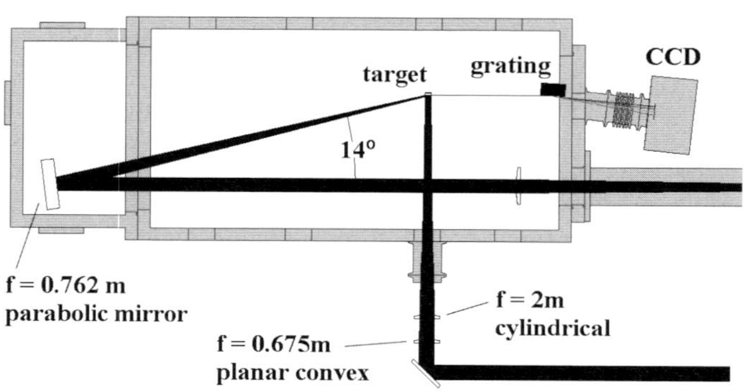

Figure 4. Schematic diagram of one of the two grazing incidence experimental set-ups used to obtain lasing in Ni-ike ions at wavelengths between 13.9 and 22.6 nm.

Figure 5a shows an on-axis spectrum of the Mo plasmas as a function of target length. Strong laser emission is observed in both the 18.9 nm 4d $^1S_0$ – 4p $^1P_1$ and in the 22.6 nm 4f $^1P_1$ – 4d $^1P_1$ lines of Ni-like Mo. Lasing was also observed in plasmas generated using both broader (50 μm FWHM) and narrower (18 μm FWHM) pre-pulse line focus widths, and broader (50 μm FWHM) and narrower (15 μm FWHM) short pulse line widths. However, the best results in terms of output intensity and shot-to-shot reproducibility were obtained using 30μm FWHM width for both the pulses using the optics described above. Lasing was also observed for short pulse durations of 1.8 and 6 ps. However, since with the shorter pulses we did not observe significant improvements of the laser output respect to the results obtained with 8.1 ps pulse duration, the characterization of the laser was conducted with the longer pulses to reduce the possibility of damage to the optics. Lasing at 18.9 nm was observed with pumping energies as low as 140 mJ and 310 mJ for the pre-pulse and short pulse.

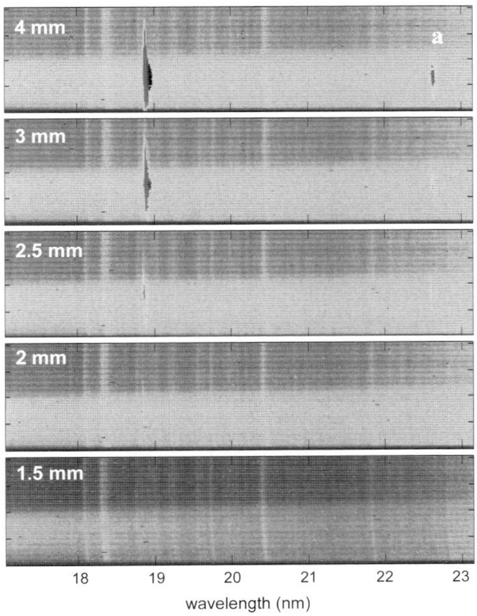

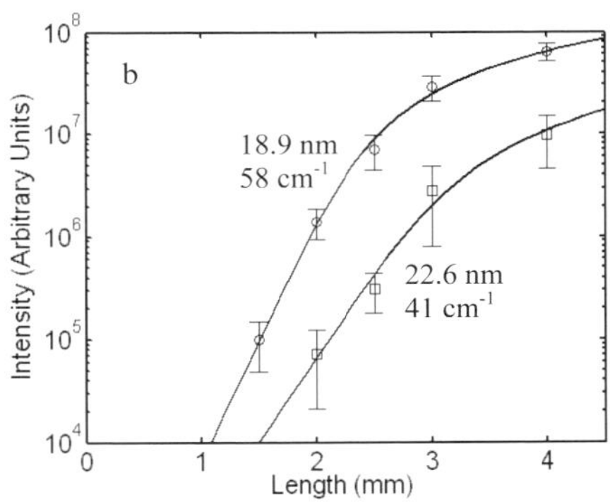

Figure 5. a) Axial spectra of Mo plasma showing amplification in the 18.9 nm and 22.6 nm lines of Ni-like Mo. b) Variation of the output intensity of the 18.9 nm and 22.6 nm laser lines of Ni-like Mo as a function of plasma length. Each of the points is an average of ten or more laser shots.

Figure 5b shows the variation of the intensity of 18.9 nm and 22.6 nm laser lines of Ni-like Mo as a function of target length. The results were obtained exciting the target with a sequence of a 350 mJ long pulse and a ~ 1 J short pulse separated by 700 ps. The intensity of the 18.9 nm Ni-like Mo line is observed to increase by nearly three orders of magnitude when the target length is varied between 1.5 mm, the minimum length for which line emission was observed at this wavelength, and 4 mm. The solid lines in Figure 5 are fits of the data with the expression by Tallents et. al. for the variation of the line intensity with plasma length taking into account gain saturation [14]. The resulting small signal gain coefficient for the 18.9 nm line is 58 cm $^{-1}$ and the gain-length product reaches GL = 15.5, a value that corresponds to saturated operation in collisional laser systems. The laser beam divergence was measured to be approximately 13 milliradians. The energy of the most intense 18.9 nm laser pulses was estimated to exceed 150 nJ from the CCD counts taking into account the losses and the detector quantum efficiency. Assuming a laser pulse width equal to the gain duration divided by

(GL)1/2 (~ 2ps) and a gain region of 35 micrometer diameter, the 18.9 nm laser beam intensity can be calculated to reach the computed gain saturation intensity, 0.5-1 10^{10} W/cm^2, when the pulse energy reaches 100-200 nJ. The 22.6 nm line was measured to amplify with a smaller gain coefficient of 41.1 cm-1, and reach a gain-length product of GL= 13.5.

Figure 6 shows the 18.9 nm laser output intensity as a function of delay between the long pulse and the short pulse. Strong lasing was observed to take place for a broad range of time delays, ranging from about 500 ps to 1200 ps. Model computations show that at significantly shorter times the electron density profile of the pre-pulse plasma is too steep, while at more than 1200-1400 ns the electron density drops substantially below the values for optimum amplification. This time window for lasing is significantly broader than that observed in the experiment by Keenan, Dunn, et. al. (~50 ps) [5]. This suggests the Ni-like Mo lasers demonstrated in these two experiments operate in a different regimes. The larger pump energy and broader long pulse line focus in the present experiment can be expected to create an enlarged gain region that results in the increased laser output pulse energy that we observed.

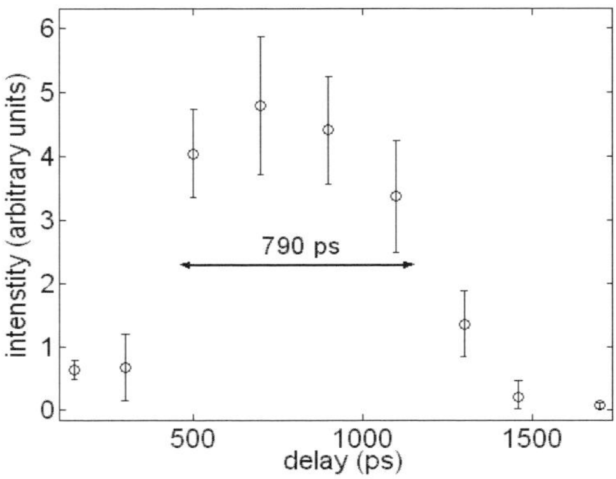

Figure 6. Variation of the intensity of the 18.9 nm line of Ni-like Mo as a function of time delay between the 120 ps pre-pulse and the 8.1 ps grazing incidence short pulse.

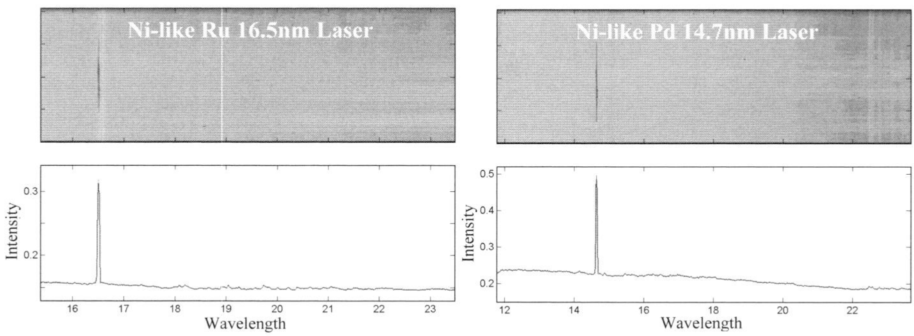

Figure 7. Axial spectra of 4 mm long ruthenium (left) and palladium (right) line focus plasmas with the corresponding integrated spectral intensity distributions showing strong lasing in the 16.5 nm and 14.7 nm lines of the Ni-like ions.

The 18.9 nm Ni-like Mo laser results were scaled to shorter wavelengths using the Ni-like isoelectronic sequence. Figure 7 (left) shows the axial spectrum of a 4 mm long Ru line focus plasma in which the 16.5 nm line of Ni-like Ru completely dominates the spectra. The result was obtained using 340 mJ of pre-pulse and ~ 1 J of short pulse excitation. Figure 7 (right) shows a similar result for Ni-like Pd, obtained using 315 mJ of pre-pulse and ~ 875 mJ of short pulse excitation respectively. Both lasers were measured to have a high gain: 65 cm^{-1} for Ru, and 53 cm^{-1} for Pd (Fig. 8). A gain lasing was observed for a broad range of delays (Fig. 9) between the pre-pulse and the short pump pulse. Finally, lasing was also obtained in the 13.9 nm line of Ni-like Ag, as shown in Figure 10 by the axial spectra of a 3 mm long line-focus Ag silver plasma.

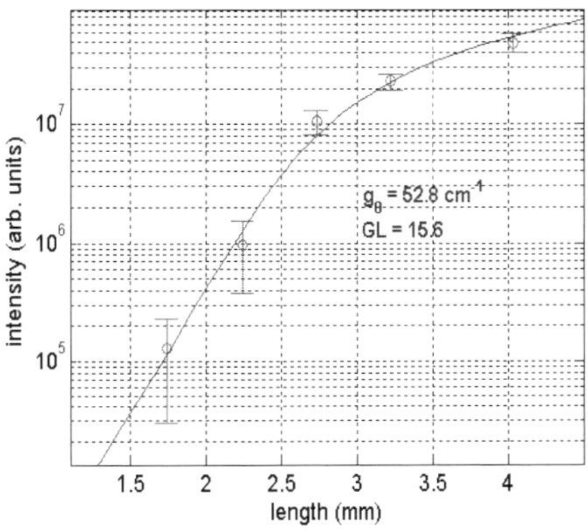

Figure 8. Variation of the output intensity of the 14.7 nm laser line of Ni-like Pd as a function of plasma length.

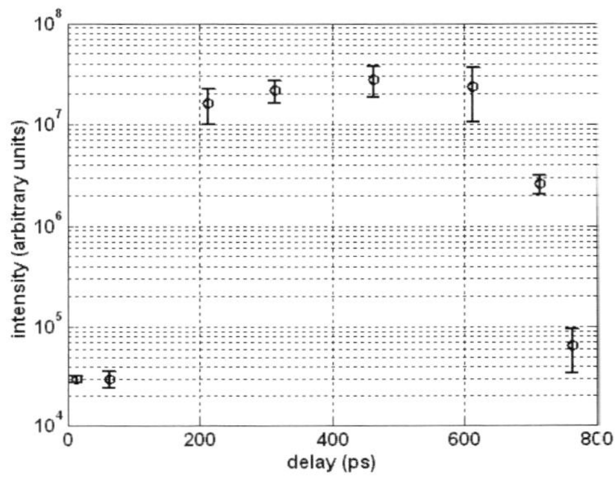

Figure 9. Variation of the intensity of the 14.6 nm line of Ni-like Pd as a function of time delay between the 120 ps pre-pulse and the 8.1 ps grazing incidence short pulse.

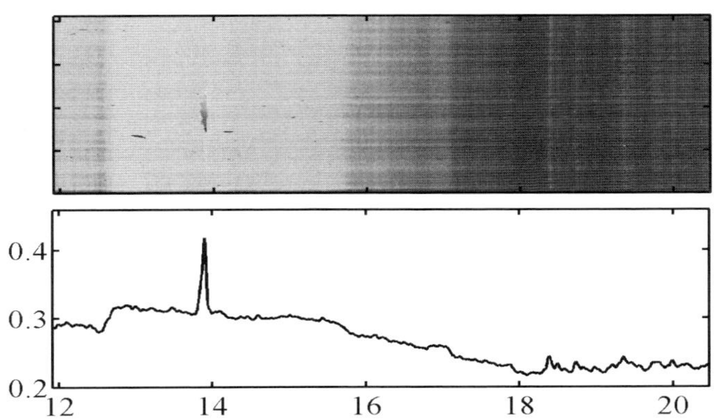

Figure 10. Axial spectra of 3 mm long Ag plasma showing amplification in the 13.9 nm line of Ni-like Ag

In summary, we have demonstrated table-top lasers at wavelengths between 13.9 and 22.6 nm in transitions of Ni-like ions using a grazing incidence pump beam generated by a high repetition rate table-top pump laser. Lasing was achieved over a broad range of time delays and pumping conditions. In a new separate effort to reduce the size of short wavelength lasers a 12 Hz capillary discharge pumped 46.9 nm laser that fits on a small desk-top was demonstrated. The results have significance for the development of practical high average power soft x-ray lasers.

This work was supported by the NSF Center for Extreme Ultraviolet Science and Technology under NSF Award Number EEC-0310717, with equipment developed under NSF grant ECS-9977677. Previous support from DARPA grant DAAD 19-99-1-0279 is acknowledged. We also gratefully acknowledge the support of the W.M. Keck Foundation is also acknowledged. We would like to thank J. Dunn for very useful discussion and Abbey Weith for her assistance with data analysis.

References

1. J.J. Rocca, D.P. Clark, et.al., Phys. Rev. Lett. **77**, 1476, (1996).
2. B.R. Benware, C.D. Macchietto, et.al. Phys. Rev. Lett. **81,** 5804, (1998); C.D. Macchietto, B.R. Benware et.al. Optics Lett. **24**, 1115, (1999).
3. V.N. Shlyaptsev, J. Dunn, et.al., Proc of SPIE **5197** 221 (2003)
4. R. Keenan, J. Dunn, et.al. Proc. of SPIE **5197**, 213 (2003)
4. Dunn, R. Keenan, et.al. (these proceedings); R. Keenan, J. Dunn, et.al. "High Repetition Rate Grazing Incident Pumped X-ray Laser operating at 18.9 nm", Submitted to Phys. Rev. Letters.
6. J.J. Rocca, V.N. Shlyaptsev, et.al., Phys. Rev. Lett., **73**, 2192, (1994).
7. J.J. Rocca et. al. , Lasers 2002, AIP Conference proceedings 641, 125 (2002)
8. P.V. Nickels, V.N. Shlyaptsev,et.al. ,Phys. Rev. Lett. **78**, 2748 (1997)
9. J. Dunn, Y. Li,et.al., Phys. Rev. Lett. **84**, 4834 (2000)
10. T. Ozaki, Ganeev RA,et.al., Phys. Rev. Lett. **89**, 253902 (2002)
11. R. Tommasini, J. Nilson, E.E. Fill, Proc. of SPIE **4505**, 85 (2001)
12. V.N. Shlyaptsev, R. Keenan, et.al (In these proceedings)
13. B.M. Luther, Y. Wang, et.al. "Saturated 18.9 nm laser in a Ni-like Mo plasma excited by a high repetition rate table-top laser", submitted Optics Letters.
14. Tallents, et al, X-Ray Lasers 2002: 8th International Conference on X-Ray Lasers, C641, 291 (2002).

Inst. Phys. Conf. Ser. No 186
Paper presented at 9th Int. Conf. X-Ray Lasers, Beijing, China, 24–28 May 2004

The Progress of Experimental Studies on X-ray Lasers at SILP

S.J. Wang, Ch. Wang, W. Wang, J.R. Sun, Z.H. Fang, W.X. Ma, S.Z. Fu, Y. Gu

Shanghai Institute of Laser Plasma, P.O.Box 800-229, Shanghai, 201800, P.R.China

Abstract. Experimental study on x-ray lasers is a main research subject of Shanghai Institute of Laser Plasma (SILP). Based on SG-II laser system, a series of experimental studies on XRL are conducted over the last several years. Some very good results were obtained.

1. Introduction

Experimental study on x-ray lasers is a main research subject of Shanghai Institute of Laser Plasma (SILP) in China. The study started about 15 year ago, and some classical technique such as cylinderal lens array (CLA) line-focus system[1] and double target coupling[2] were firstly developed. Along with the continual researches, some results were obtained, especially in recent several years.
The most experiments of x-ray lasers were performed on the SG-II laser system. SG-II system can synchronously provide 8 beams of driving laser with 100ps duration and 100J energy at 1.053μm for experimental studies of x-ray lasers.
In this paper, the recent progress on x-ray lasers at SILP was exhibited in brief with several aspects.

2. Generation of Ni-like Ta x-ray laser at 4.48nm

The studies of x-ray lasers in or near *water window* (2.3 to 4.4nm) is an important subject, because of its great application prospects on biologic fields. Especially the Ni-like Ta x-ray laser was the best choice of the application of biologic holography because its wavelength of 4.48nm was just outside the *water window*. Aimed this subject, we started researching the experiments of Ni-like Ta x-ray laser.
However, the drive conditions of Ni-like Ta x-ray laser were stricter than some typical ones such as Ne-like Ge or Ni-like Ag. There were some successful experiments[3], and the maximum GL of about 8, however the required drive energy was too high to afford of SG-II system. It was necessary raising new ideas.
So, we developed a novel scheme of united irradiation of two lasers with different colours (ω: λ_{ω}=1.053μm + 2ω: $\lambda_{2\omega}$=0.53μm) to perform the experiments of Ni-like Ta x-ray laser on SG-II. The experimental setup was shown in Figure 1. There were four laser beams of SG-II were chosen for irradiating double targets, so the total drive energy was about 400J. Among them, two laser beams were red (ω) and the others were green (2ω). Each target was irradiated by a horizontal green laser and an obliquity red laser at one

time. Besides, the classical techniques such as pre-pulse (5% and 1.5ns), uniform CLA line-focus system (6mm×100μm) and double target coupling (5+5mm, 50μm separation in the target normal direction and 3mm at axially one) were applied in the experiments. The total irradiance on the target surface was about 2.5×10^{14}Wcm^{-2}. A 2400 l/mm flat-field grating spectrometer with C_8H_8 filters and a soft x-ray CCD camera were used as the main apparatus to measure x-ray laser spectra.

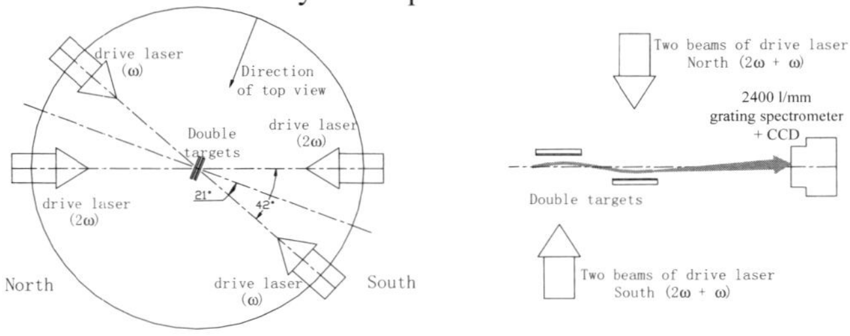

(a) side view (b) top view, inclined 21°

Figure 1. Experimental setup of Ni-like Ta x-ray laser, using the united irradiation of two lasers with different colours.

A typical axially time-integrated spectrum of Ni-like Ta x-ray laser recorded in experiments was shown in Figure 2. The spectrum was clear and its wavelength can be accurately confirmed by the absorption K edge of carbon at 4.37nm. Experiments with different length targets were executed. The data and the fit curve were shown in Figure 3. The maximum GL was about 5.5 for the experiment of double 5mm-long target coupling. On the other hand, the gain behavior seemed as saturation, but it was far from saturation. The reason was probably the strong deflection.

Figure 2. A typical spectrum image of Ni-like Ta x-ray laser record in experiments.

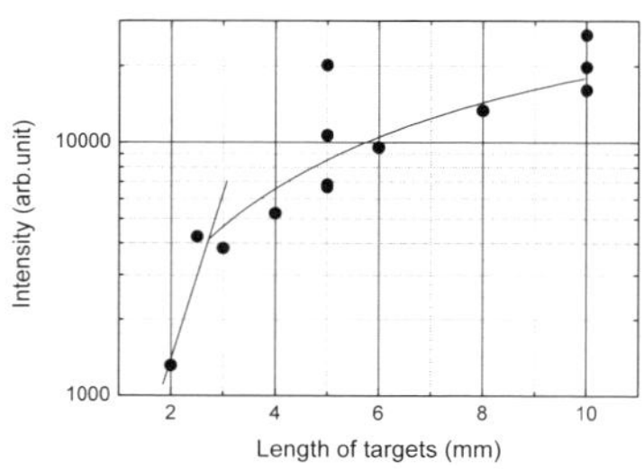

Figure 3. The gain curve of Ni-like Ta x-ray lasers. The maximum GL was about 5.5 and the maximum gain was greater than 20cm^{-1} at experiments with short target lengths.

3. Optimization of Ni-like Ag x-ray laser at 13.9nm

The x-ray laser of Ni-like Ag was a good choice of applications because its wavelength was the most close to the optimal wavelength of Mo:Si multi-layer soft x-ray optical elements. So its mirrors were more effective and easy to manufacture than others. So we

performed and optimized it for applications by using the two methods of measuring spectrum with a grating spectrometer and measuring field patterns respectively.
The Ni-like Ag x-ray laser was produced with two SG-II red laser beams irradiating double 16mm-long target coupling. The techniques of pre-pulse (5% and 3.0ns) and a novel uniform UCLA line-focus system (18mm×80μm, see 6.1 in detail) were used in experiments. The irradiance on the target surface was about $6\times10^{13}Wcm^{-2}$.
Figure 4 showed a typical axially time-integrated spectrum of Ni-like Ag x-ray laser recorded in experiments with a 1200l/mm grating spectrometer. The gain curve was drawn in Figure 5. The maximum GL was about 17 and the output of x-ray laser had been deep saturation.

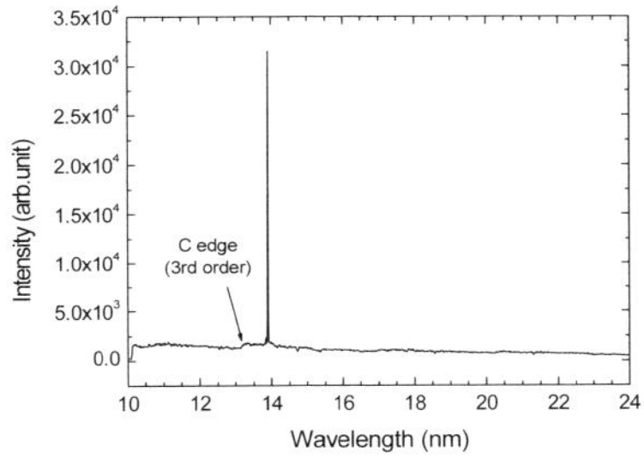

Figure 4. A typical spectrum of Ni-like Ag x-ray laser record in experiments with a 1200l/mm grating spectrometer.

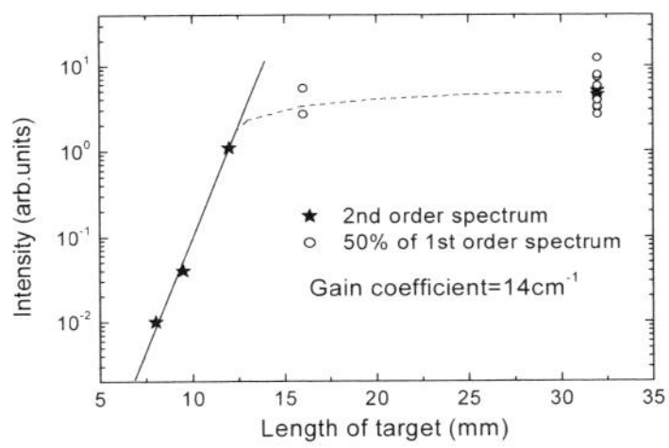

Figure 5. The gain curve of Ni-like Ag x-ray lasers. The saturated GL was about 17 and the small signal gain coefficient was about $14cm^{-1}$.

The method of measuring spectrum with grating spectrometer could give some output characteristics of x-ray laser, however, the information was not sufficient for applications, especially for Ni-like Ag x-ray laser. So we choose the beam cross section intensity distribution (i.e. field pattern) as reference to optimize the Ni-like Ag x-ray laser. The main optimization was the conditions (see Figure 6) of double target coupling. The changed conditions included x (200~300μm) and α (0~4mrad), and the others were almost invariable (L1=L2=16mm, y=3mm).

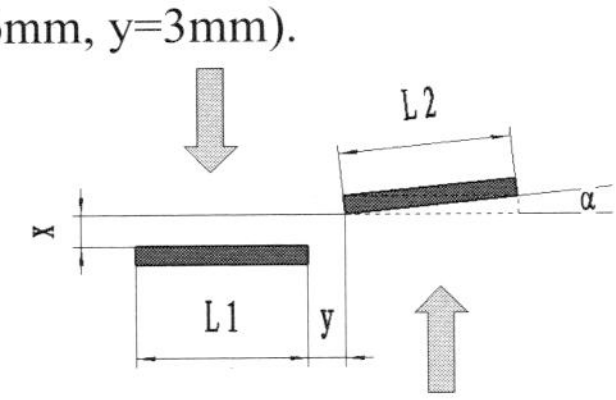

Figure 6. The parameters of double target coupling.

The field pattern with different coupling conditions was shown in Figure 7. It was evident that the pattern was mightily related with the condition. This was the reason why using the method of measuring field pattern. From all the patterns, the last one (i) was the most suitable for application, because of its quasi-homogeneous beam in both horizontal and vertical directions. This was the optimized result.

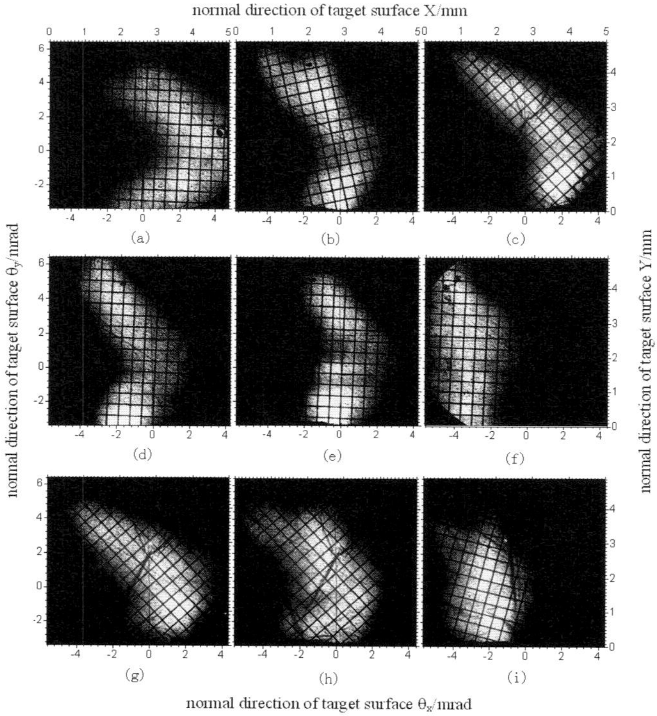

Figure 7. The beam cross section intensity distribution (i.e. field pattern) at 500mm distance from the output end of x-ray lasers with different conditions of double targets coupling. x was: (a)200μm, (b)230μm, (c)240μm, (d)250μm, (e)270μm, (f) 300μm. x and α were: (g)235μm, 2mrad, (h)235μm, 4mrad, (i)265μm, 4mrad.

4. Diagnostics of electron density profiles of a laser-driven plasma

Diagnostics of electron density profiles of laser-driven plasmas were important and significative applications of x-ray laser, since a series of experimental data about laser plasma which were hard to be obtained from theoretical simulations were given. So we performed these applications by using respectively the methods of moiré deflectometry and Mach-Zehnder interferometry.

4.1. Method of moiré deflectometry

Method of moiré deflectometry is a very simple one to diagnose laser plasma electron density profiles[4]. Using a saturated Ni-like Ag x-ray laser as a probe beam, the distribution information of electron density of laser-driven CH plasmas was diagnosed by this deflectometry. The sampling plasma was created by irradiating a polystyrene (C_8H_8) flat target with another SG-II laser beam which was about 80ps duration and 20J energy. The laser spot diameter was about 300μm, so the irradiance on target surface was about 2.5×10^{14}Wcm^{-2}. Probe beam passed through the sampling plasma after 1ns of the end of

drive laser by adjusting the time correlation of drive lasers. The moiré stripe deflectogram was created by a pair of grating with period 28.6μm and rotational offset angle 4.5° to give a fringe spacing of 365μm in the image plane (36.5μm at the object plane). The sensitivity per fringe shift was about 4.6mrad.
The experimental setup and results were shown in Figure 8-10.

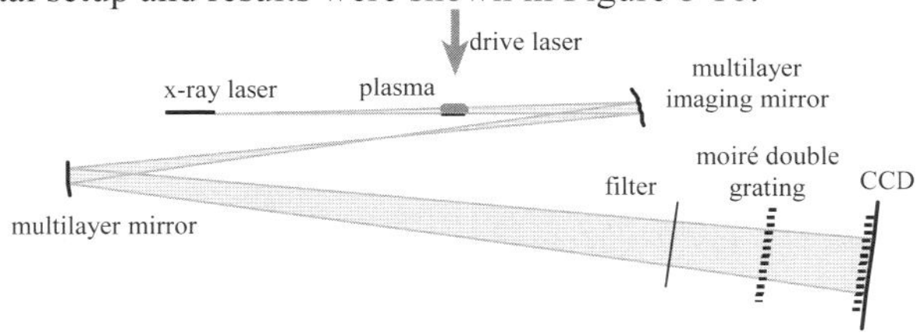

Figure 8. The experimental setup of diagnostics of plasma using moiré deflectometry.

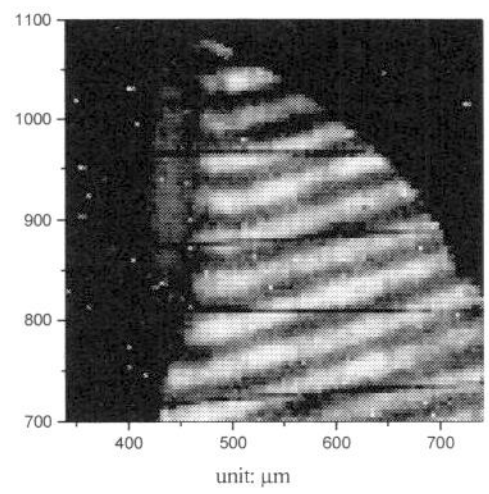

Figure 9. A typical dynamic moiré stripe deflectogram with plasma perturbation recorded in experiments.

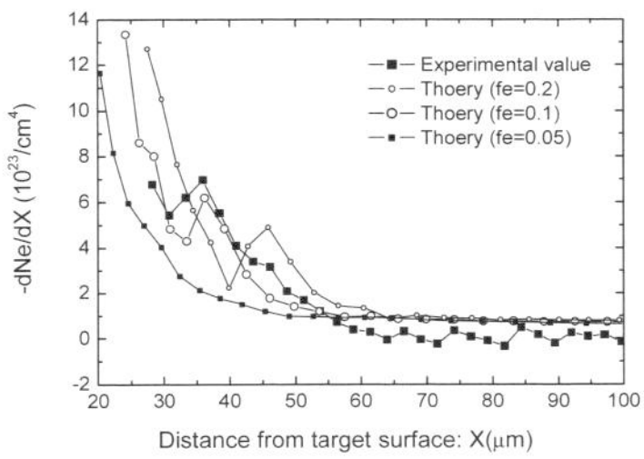

Figure 10. The electron density gradient profiles calculated from the experiments and theoretical simulations. The good fit indicated that 0.1 may be a reasonable choice of flux-limit-factor.

4.2. *Method of Mach-Zehnder interferometry*

Mach-Zehnder (M-Z) interferometry is another method to diagnose laser plasma electron density profiles[5], and its results were more precise and more valuable. More important advantage than the method of moiré deflectometry was this method can measure and give the electron density directly. Using a saturated Ni-like Ag x-ray laser as a probe beam, the distribution information of electron density of a laser-driven CH plasma was diagnosed by this interference method.
The sampling plasma was created by irradiating a polystyrene (C_8H_8) flat target with another SG-II laser beam with line-focus of 450μm×3.5mm and the irradiance of about $6\times10^{13}Wcm^{-2}$. Probe beam passed through the sampling plasma after 1ns of the end of

drive laser. The M-Z interferogram was obtained with an M-Z interferometer constructed of two multilayer beam-splitters and two mirrors. The reflectivity and transmissivity of the beam splitters was respectively about 19% and 24% at the work wavelength of 13.9nm.

The experimental setup and results were shown in Figure 11-13. The magnification times of 3.5× gave a spatial resolution was about 5.7μm. The maximum electron density was about $9.5\times10^{20}cm^{-3}$, and it was very close to the critical density value ($1.0\times10^{21}cm^{-3}$) of irradiated laser.

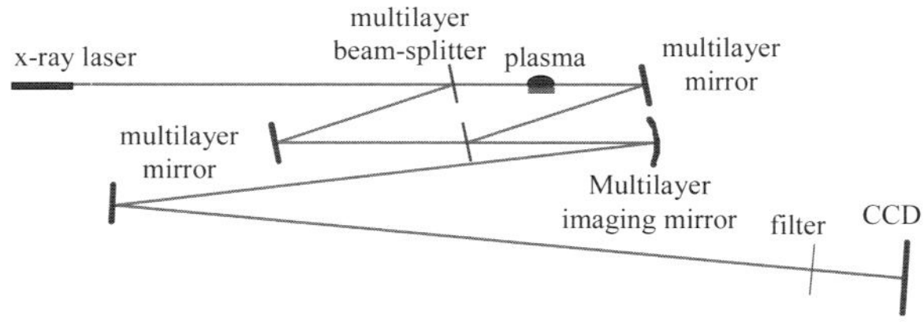

Figure 11. The experimental setup of diagnostics of plasma using M-Z interferometry.

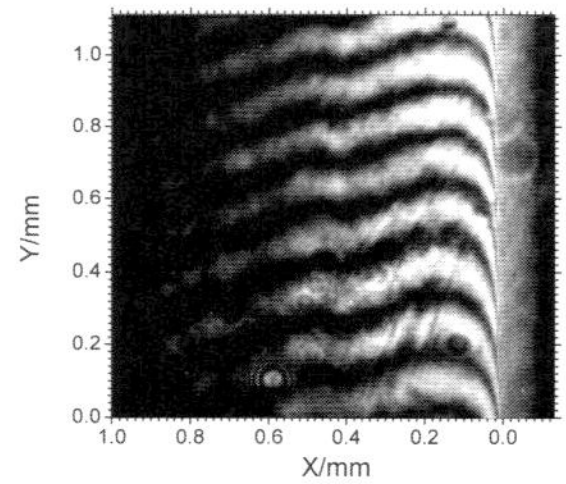

Figure 12. A dynamic M-Z interferogram with plasma perturbation recorded in experiments.

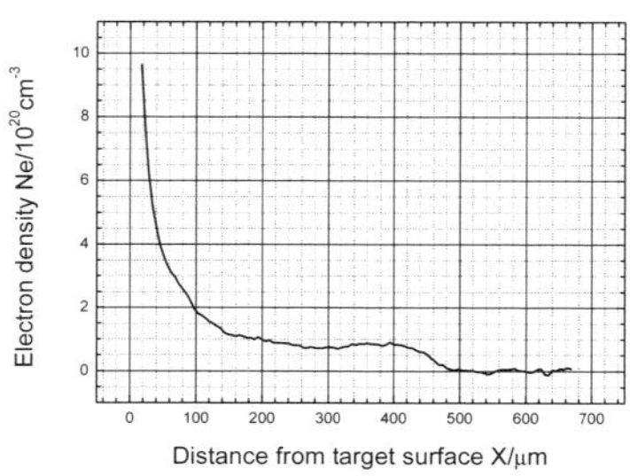

Figure 13. The electron density profiles calculated from the experiments. The data was average results of several similar fringes.

5. Some correlative techniques and preparative works

5.1. Especial line-focus systems made of UCLA

A classical cylindrical lens array (CLA) line-focus system could give a very good line-focus for experiments of x-ray laser. However, it could not work very well under some conditions, especially with special intensity distributions of wave fronts of drive laser, such as SG-II laser system (see Figure 4). So, a novel unequal-width-element cylindrical lens array (UCLA) line-focus system was designed and manufactured based on the special conditions of SG-II laser. The main improvements were as below:

- Choosing the width of middle elements was the same as the diameter of the mid hole, and the other four was the same to avoid the effects of the mid hole.
- Circumvolving the pole of wave fronts of drive laser to horizontal directions to eliminate the effects of the pole.

Figure 15 showed the results of intensity distribution of line-focus with a improved UCLA and a classical CLA system under SG-II laser conditions. The improvement is very obvious. The saturated output of Ni-like Ag x-ray laser mentioned above just based on the good characteristics of this UCLA system.

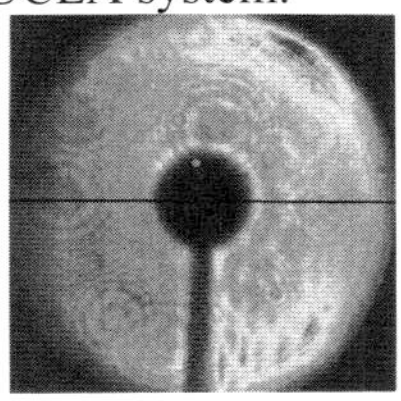

Figure 14. The special wave fronts of SG-II laser.

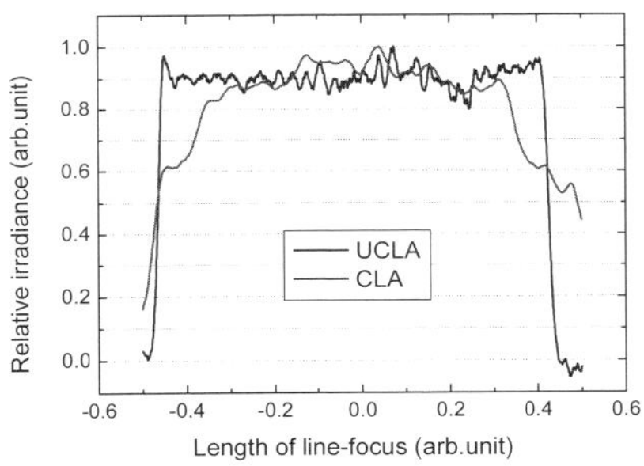

Figure 15. The intensity distribution of line-focus with a improved UCLA and a classical CLA system under SG-II laser conditions.

5.2. Calibration of soft x-ray optical beam splitters in one shot

Beam-splitters were the key elements of M-Z interferometer. High quality multilayer beam-splitters was the requirement of applications of M-Z interferometry. So we manufactured it and also developed a particular method based on laser plasma x-ray sources to measuring both reflectivity and transmissivity of a beam-splitter in one shot. Figure 16 was the experimental setup and Figure 17 was a typical results of a multi-layer beam-splitters.

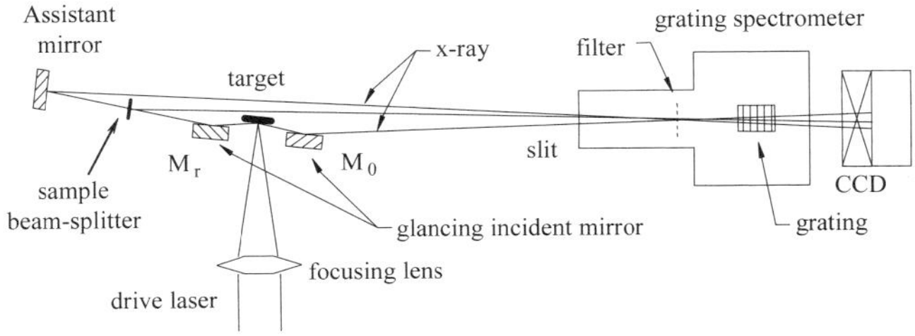

Figure 16. The experimental setup of calibration of a soft x-ray optical beam splitter in one shot.

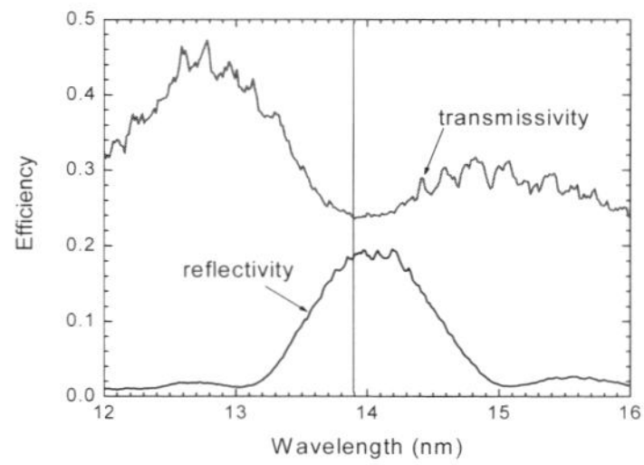

Figure 17. The reflectivity (R) and transmissivity (T) of a beam-splitter measured by using an x-ray source of laser plasma. R~19%, T~24%, R·T~4.5% at 13.9nm.

6. Summary

This paper covered the progress of experimental studies on x-ray laser at Shanghai Institute of Laser Plasma (SILP) over the last several years in brief. The progress include at least the following aspects:

- A certain intensity output (GL~5.5) of Ni-like Ta x-ray lasers were obtained with 400J drive energy by using a novel scheme of united irradiation of two lasers with different colours under SG-II.
- A suitable Ni-like Ag x-ray laser probe beam for applications of x-ray lasers was optimized by using the method of measuring field pattern.
- An application of the diagnostics of laser plasma density profiles was demonstrated by using moiré deflectrometry. The results indicated a appropriate value of flux-limit-factor of about 0.1.
- Another application of the diagnostics of laser plasma density profiles was performed by using Mach-Zehnder interferometry. The maximum density was about $9.5\times10^{20}cm^{-3}$, and it was almost approach the 95% of the critical density value.
- Some correlative techniques and preparative works such as Especial UCLA line-focus system and calibration of soft x-ray optical elements in one shot were researched. These all ensured the success of the experimental studies of x-ray laser.

For more information of these works, a series of papers of Ch. Wang, W. Wang and J.R. Sun in these proceedings were supplemental references.

7. Acknowledgments

Most of the works are supported by National Nature Science Foundation of China (Grant No. 69989801) and National High Technology Development Program of China (Grant No. 863-804-7.3). We acknowledge all the members of SG-II operation group, and other groups for providing the helps.

References

[1] Chen W N et al 1991 Acta Optica Sinica 11: 829 (in Chinese)
[2] Wang S J et al 1992 J.Opt.Soc.Am.B 9 360
[3] MacGowen B J et al 1990 Phys.Rev.Lett. 65 2374
[4] Ress D et al 1994 Science 265 514
[5] DaSilva LB et al 1995 Phys.Rev.Lett. 74 3991

Inst. Phys. Conf. Ser. No 186
Paper presented at 9th Int. Conf. X-Ray Lasers, Beijing, China, 24–28 May 2004

Transient X-ray lasers at LIXAM: temporal coherence and spectral profile of the 13.9 nm laser

A. Klisnick, O. Guilbaud, D. Ros, K. Cassou, S. Kazamias,G. Jamelot, J-C. Lagron
LIXAM, "X-ray Lasers and Applications group", Université Paris-Sud, Orsay, France

D. Joyeux, D. Phalippou, Y. Lechantre
LCF-Institut d'Optique, Orsay, France

M. Edwards, P. Mistry, G. J. Tallents
Dept of Physics, University of York, UK

Abstract. A new Fresnel interferometer is used to investigate the temporal coherence of the 13.9 nm Ni-like Ag X-ray laser. The results obtained, together with a measurement of the pulse duration, are discussed in terms of Fourier-transform limit, spectral linewidth and profile of the lasing line, and far-field pattern of the beam.

1. Introduction

Since the last X-ray laser conference in 2002, the activity of the X-ray laser group at LIXAM has progressed in three complementary directions. Part of our activity has been devoted to the demonstration of applications of both transient and QSS-pumped X-ray lasers. New results have been obtained in the investigation of DNA damage and imaging interferometry of laser-produced plasmas, which are presented by D. Ros et al in these Proceedings [1]. On the other hand we have now started the construction of the pump laser for our future LASERIX user facility. Details on the status of this project and on the new and attractive prospects offered by the Ti-Sa laser technology chosen for the pump laser are presented by Jamelot et al. in these Proceedings [2]. In this paper we will more specifically focus on the third part of our research activity devoted to the development and characterisation of transient X-ray lasers.

Since their first experimental demonstration [3] transient X-ray lasers have attracted much interest world-wide, mostly due to their good prospect for their use as a source for applications. Due to their relatively low pump energy requirement, transient X-ray lasers have been implemented in many laboratories over the last five years. Recently the LIXAM group has collaborated to the first implementation of a Ni-like Zr transient X-ray laser at the new PHELIX laser facility, currently under construction at GSI-Darmstadt [4]. This X-ray laser source will be later used as a spectrocopic diagnostic for heavy-ion beams [5].

Compared to the « original »X-ray lasers pumped in the quasi-steady state (QSS) regime, transient X-ray lasers exhibit specific characteristics and behaviours, related to the ultrashort (~ picosecond) pulse pumping regime. In the recent years the X-ray laser group at LIXAM and collaborators at LCFIO and at University of York have jointly carried out several experiments dedicated to detailed characterisation of transient X-ray lasers [6, 7]. These experiments mostly focussed on the 4d-4p lasing line in Ni-like silver, emitted at 13.9 nm, a wavelength close to optimum for Mo:Si multilayer optics. In particular we have shown that the duration of the 13.9 nm X-ray laser can be very short, a 2 ps pulse was measured for a 1.3 ps CPA glass laser pump [6]. It was then predicted that the duration of the X-ray laser pulse could be close to the Fourier-transform limit [8]. Numerical simulations indicate that the frequency bandwidth of the X-ray laser could be as small as $5.\ 10^{11}$ Hz, but direct experimental measurements are required. However the measurement of such a narrow bandwidth requires a spectral resolution better than 10^5 at the X-ray laser wavelength, which is out of reach of most existing XUV spectrometers.

The spectral bandwidth of the source is inversely related to the temporal coherence, or longitudinal coherence, which can be accurately determined from interferometry. To our knowledge there exist in the literature only two works devoted to the experimental measurement of the temporal coherence of an X-ray laser [9, 10]. Both experiments were based on an *amplitude division interferometer*, in which the optical interfering elements are XUV multilayer beamsplitters. Although significant progress have been made in the fabrication of these optical elements in the recent years, they still have some drawbacks: they are difficult to manufacture, relatively fragile, they have a low reflection and transmission throughput (typically below 20 % at 14 nm). Besides the XUV beamsplitters are chromatic, i.e. they are designed for a single wavelength.

With the aim of being able to investigate the temporal coherence of different X-ray lasers we have used a different approach, based on a *wavefront division interferometer*. In this case the optical elements are grazing incidence flat mirrors, which can be currently manufactured with a very high optical quality, have a high reflectivity (about 70 %) and are achromatic. Hence such interferometers can be easily adapted to any wavelength in the XUV, or even soft X-ray spectral region.

In this paper we will report on the first measurement of temporal coherence of an X-ray laser with a wavefront division interferometer. The inferred bandwidth was compared to the duration of the X-ray laser pulse, which was also measured in the same experiment in order to investigate the transform-limited nature of the pulse. The results are discussed in terms of temporal coherence, and spectral linewidth and profile of the lasing line. Finally we discuss the implication of the temporal coherence of the X-ray laser on small-scale structures which are observed on the footprint images of the beam.

2. Experimental configuration

The experimental campaign was carried out at the 100 TW-CPA LULI laser facility at the beginning of 2004. Apart from investigating the transform-limited nature of the 13.9 nm laser pulse, the general objective of this experiment was to characterise as completely as possible the output parameters of the X-ray laser beam by implementing and operating various diagnostics. In addition to a high-resolution streak camera, and a new Fresnel interferometer, the diagnostics included a footprint monitor (far-field pattern) and a high-resolution imaging mirror (source size, structure). An extensive set of data were obtained a few weeks before the Conference, most of them are presently under analysis. Finally we made a preliminary study of the focalisation of the X-ray laser beam with a spherical multilayer mirror, as described in a companion paper [1].

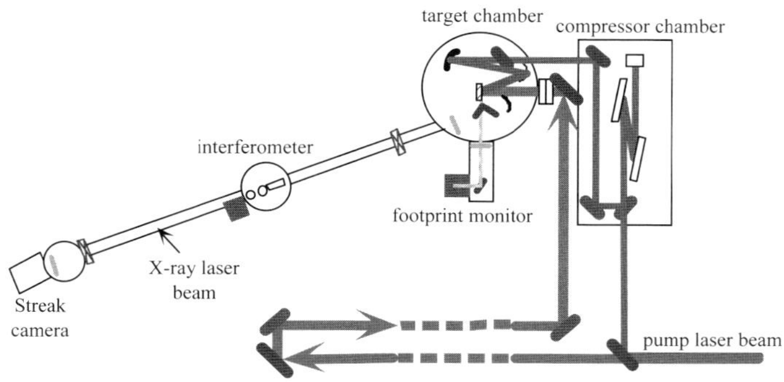

Figure 1. Layout of the experiment, showing the path of the pump laser beams up to the target ,and the X-ray laser beamline to the diagnostics.

Figure 1 is a schematic general view of the experimental set-up. The two pump laser beams originate from one and the same chirped pulse beam and are separated with a beam-splitter. They are then focused independently by two distinct focussing devices, allowing to generate different widths for the corresponding focal lines (200 μm and 80 μm for the long and short pulse respectively). Such a configuration was previously shown to improve the spatial structure of the X-ray laser emitting aperture [7]. The laser irradiation parameters were similar to this previous experiments: a silver slab (maximum length 10 mm) is irradiated by two successive laser pulses separated by 250 ps (peak-to-peak). The intensity of the first 600 ps background pulse was close to 10^{12} W/cm^2 while the short heating pulse was compressed at 400 fs with an irradiance of 10^{15} W/cm^2. Finally a travelling wave (TW) irradiation at v = c was generated for the short pulse beam, by slightly tilting one of the compressor gratings [11].

Due to the presence of the TW the X-ray laser beam at 13.9 nm is preponderantly emitted in one direction [12] along which the diagnostics were aligned. In order to easily shift from one diagnostic to the other from shot to shot, we implemented a motorized system supporting two X-UV multilayer mirrors coated at two distinct incidence angles (55° and 35°). By properly translating and rotating the system it was then possible to direct the X-ray laser beam towards three different paths and diagnostics: (i) when the XUV mirrors were translated off, the X-ray laser beam was sent to a footprint monitor consisting in a 45° Mo:Si mirror and a CCD placed at 1.2 m from the plasma exit; (ii) when the 35° Mo:Si mirror was set in, the X-ray laser beam was sent to a new Fresnel interferometer (described below) placed at 3.5 m from the X-ray laser exit plane for temporal coherence measurements; (iii) when the 55° Mo:Si mirror was set in, the X-ray laser beam was directed to the near-field imaging system. This system, described in [7], includes an aspherical XUV mirror which forms the magnified image (M = 12) of the exit plane of the X-ray laser in a plane located at 6 m from the source, where the photocathode of the Streak camera was placed. The AXIS-Photonique Streak camera was operated in slit mode with a KI photocathode. In order to remain within the linear part of the response of the Streak camera, the level of X-ray incident signal was controlled by slightly translating the near-field imaging mirror so as to vary the size of the X-ray laser beam on the photocathode.

3. A new wave-front division interferometer with variable path difference

The Fresnel-type interferometer shown in figure 2 was specifically designed at LCFIO for the purpose of temporal coherence measurements. The optical interfering elements consist of a pair of roof-shaped mirrors, or dihedrons, which have a common edge and are slightly tilted towards each other. After being spectrally filterered by reflection onto a 55° multilayer mirror (not represented in figure 2), the X-ray laser beam is directed to the dihedron pair, with its incidence axis along the common edge, and a vertical incidence angle of 6°. Because the two dihedrons are tilted, the X-ray laser beam is separated into two half-beams that slightly converge towards each other and overlap. The fringes that are formed in the overlapping region are detected onto an XUV CCD (10^6 pixels, 16 bits, Andor Technology) which is placed at 50 cm from the dihedrons. In order to increase the apparent fringe spacing, the CCD sensor is set at a grazing incidence of 9° relative to the X-ray laser propagation axis. A flat mirror is inserted between the dihedrons and the CCD to redirect and adjust the X-ray laser beam towards the CCD camera.

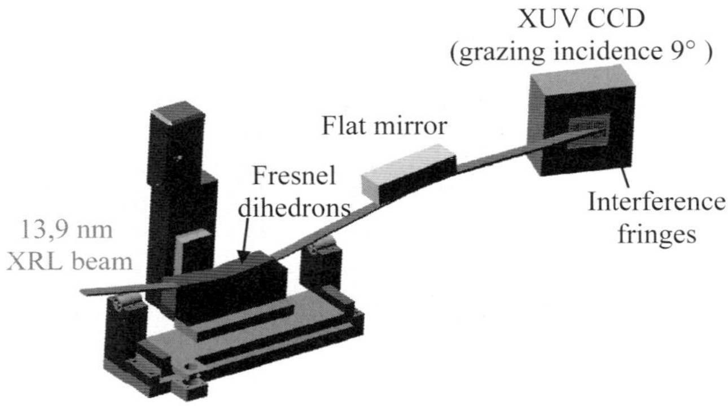

Figure 2. Fresnel interferometer for temporal coherence measurement. One of the dihedron can be translated vertically to introduce controlled path difference between the two X-ray laser half-beams.

The zero path difference of the system (i.e. when the two dihedrons are exactly at the same height) was determined before the experiment, using an optical source, with a very good accuracy (~ 1 μm). The path difference between the two interfering half-beams was then varied from shot to shot by accurately translating one of the two dihedrons vertically. Dihedrons were used rather than simple flat mirrors (as previously in [13]) to ensure that the conditions of spatial coherence are kept constant while changing the path difference. The spatial coherence of the X-ray laser beam is here given by the large distance (3.5 meters) between the source and the interferometer. Figure 3 shows a set of interferograms which were obtained for the zero path difference (top image) and for increasing values of it, up to 1.39 mm (i.e. 10^5 λ). One can see that the visibility of the fringes gradually decreases while increasing the path difference Δ.

In order to account for possible shot-to-shot fluctuations, a minimum of three shots were performed for each value of the path difference. Each interferogram was then numerically processed in order to determine the map of local fringe visibility $V(\Delta)$ on the area of the interferogram.

4. Field autocorrelation and spectral profile of the X-ray laser line

Figure 4 shows the graph of the measured fringe visibility as a function of the path difference. Each data point is an average over several shots. The error bar represents the minimum and maximum values measured for a given path difference. The fringe visibility is equal to the complex degree of coherence of the source, which is defined as the modulus of the autocorrelation function for the electric fields associated to the two interfering phase fronts. The longitudinal coherence length L_C (or the coherence time $t_C = L_C/c$, c is the speed of light) is defined as the path difference for which the fringe visibility is equal to 1/e of its maximum. From figure 4 it can be seen that the measured L_C is close to 1 mm, which corresponds to a coherence time of 3 ps.

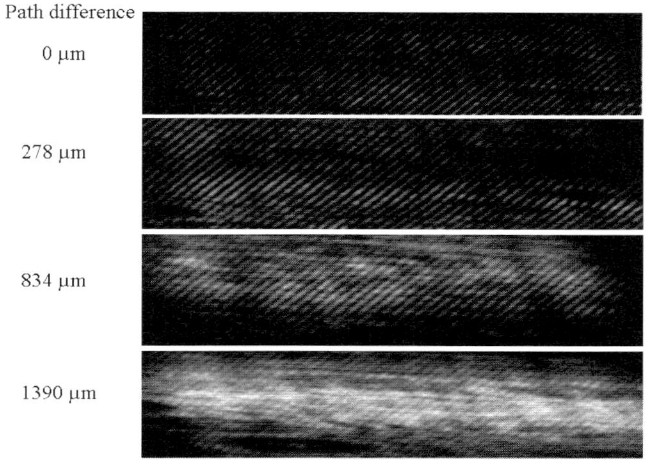

Figure 3. Set of recorded interferograms for increasing values of the path difference (given left). The visibility of the fringes gradually decreases.

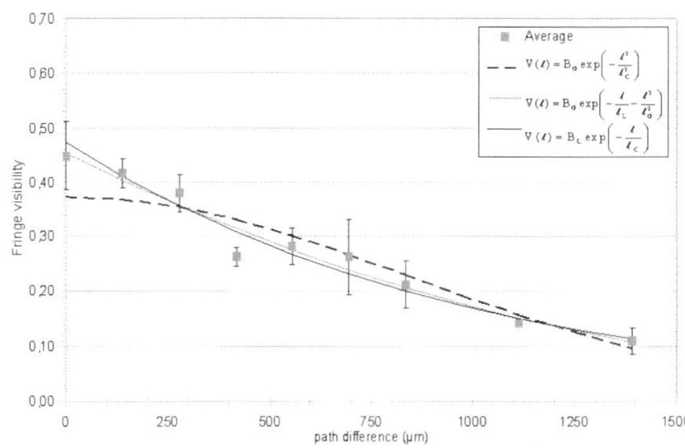

Figure 4. Variation of the measured fringe visibility versus path difference. Three different functions were used to fit the data (see text).

On the other hand the Wiener–Khintchine theorem [14] shows that the complex degree of coherence is related to the power spectral density of the x-ray laser pulse by a Fourier-transform relationship. As a result, the linewidth $\Delta\nu$ and the coherence time t_c are inversely related. The experimental data plotted in figure 4 were fitted with three different analytical functions, corresponding (through a Fourier transform) to usual spectral profiles: (i) a gaussian shape, corresponding to a gaussian spectral profile; (ii) a decreasing exponential, corresponding to a lorentzian spectral profile; and (iii) the product of these two functions, the Fourier transform of which corresponds to the convolution of gaussian and lorentzian functions, i.e. to a Voigt spectral profile. It was found that a gaussian fit is not consistent with our experimental data, which are definitely more adequately described by a decreasing exponential fit. This result is in contrast with the gaussian shape used to fit the fringe visibility data measured in [10]. The longitudinal coherence length given by the exponential fit is $L_c = 976$ μm. The coresponding linewidth associated to the Fourier-transform lorentzian profile is $\Delta\nu_L = 9.8\ 10^{10}$ Hz (or $\Delta\lambda_L = 0{,}63$ mÅ). This -extremely narrow- measured linewidth is a factor almost five smaller than the one measured by Smith et al. in Ni-like Pd [10]. An even better fit of our experimental data was found when using the product of a gaussian and a decreasing exponential functions. The corresponding spectral profile is Voigt shaped, with a gaussian contribution to the linewidth of $\Delta\nu_G = 6.4\ 10^{10}$ Hz, and a lorentzian contribution of $\Delta\nu_L = 6.4\ 10^{10}$ Hz.

It should be outlined that the spectral profile inferred from our measurements relates to the amplified line, i.e. it can not be directly compared to an optically thin line profile, resulting from homogeneous and inhomogeneous broadening mechanisms. In particular, the estimated gaussian contribution $\Delta\nu_G$, if directly compared to a Doppler linewidth, would lead to an unphysically low ionic temperature of less than 1 eV ($T_i \sim 35$ eV is predicted by numerical simulations). Hence a detailed modelling of the spectral profile, including the effect of gain-narrowing and saturation on the overall linewidth, is required to interpret our measurements. Such calculations are currently underway. We note however that a Voigt profile was previously assumed by Koch et al. [15] in order to interpret the absence of saturation rebroadening of the 3p-3s Ne-like Se lasing linewidth observed for the longest target lengths.

5. Measurement of XRL pulse duration and Fourier limit discussion

In order to determine whether the 13.9 nm transient x-ray laser is operated at the Fourier limit we need to compare the measured longitudinal coherence time t_c to the actual duration of the pulse. This duration cannot be shorter than some fundamental limit given by the Fourier-transform of the spectral linewidth. This condition can be expressed by the time-bandwidth product $\Delta\nu.\Delta t$ (both measured at FWHM) being larger than some numerical factor F. The exact value of this numerical factor depends on several conditions among which the shape of the spectral profile. It can be shown that $F = 0.44$ for a gaussian shape and $F = 0.22$ for a lorentzian shape. When the source is completely temporally coherent the temporal and the spectral profile are related by a Fourier-transform relationship, i.e. the temporal profile is gaussian or single-sided exponential for gaussian and lorentzian spectral shapes respectively. As a first estimation for the present analysis we take a lorentzian shape for the spectral profile (hence $F = 0.22$) inferred from the experimental data of figure 4.

As also described in [16], the duration of the 13.9 nm transient laser was measured for the same irradiation conditions than above , using an ultrafast X-ray Streak camera (Axis Photonique). The temporal resolution is limited mainly by the slit of the photocathode width to ~ 1 ps. Figure 5 shows a typical time-resolved image recorded for a 10 mm plasma, as well as a lineout along the time axis. Due to the limited dynamic range of 3-4 previously evaluated for this Streak camera, and to the absence of an image intensifier between the phosphor output and the 16-bit CCD camera, the signal is very low (less than 10 counts) above noise background. As a result the measurement is subject to a ± 1 ps uncertainty and it is difficult to extract information about the shape of the temporal profile. From the temporal lineout we estimate the FWHM pulse duration τ at 6 ± 1 ps.

Using the measured linewidth $\Delta\nu_L$ (FWHM) = 9.8 10^{10} Hz we calculate the time-bandwidth product $\Delta\nu_L.\tau$ = 0.59. This value is 2.7 times the Fourier-transform limit (with F = 0.22). For comparison Smith et al.[10] report a time-bandwidth product $\Delta\nu_G.\tau$ of about 2, i.e. ~ 5 times the transform limit (with F = 0.44) for the transient 14.7 nm laser, with τ = 4.5 ps and a gaussian profile with $\Delta\nu_G$ = 4.7 10^{11} Hz.

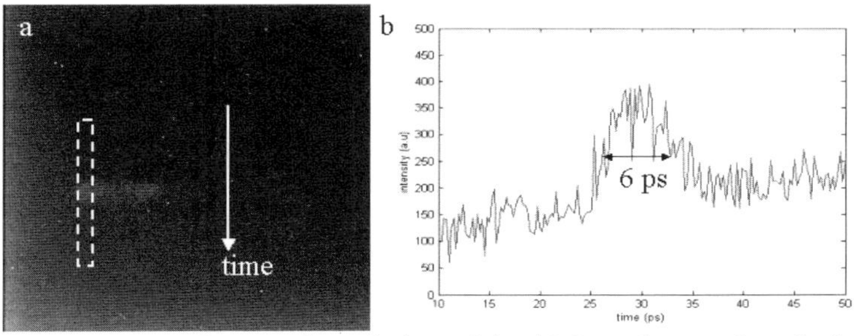

Figure 5. Temporal resolution of the 13.9 nm laser pulse obtained from the Streak camera.

6. Temporal coherence of the XRL beam and observed far-field patterns

The time-bandwidth that we have evaluated thus indicates that the 13.9 nm is operated close to the complete temporal coherence. Such a result has several important implications. First this tends to show that the duration of X-ray lasers pumped in the transient regime cannot be reduced significantly below 1 ps. In our case the transform-limited pulse duration would be of 2 ps.

On the other hand the temporal coherence demonstrated in our experiments helps to interpret the small-scale structures which were reproducibly observed in the footprint images of the X-ray laser beam. A typical example of recorded footprint image is shown in figure 6 (a). We attribute the observed pattern to coherent structures, or speckles. The presence of these speckles can be explained and very well reproduced by numerical simulations, for a source having a high temporal coherence and a poor spatial coherence. Figure 6(b) shows a calculated footprint. The similarity with the experimental speckle pattern is outstanding.

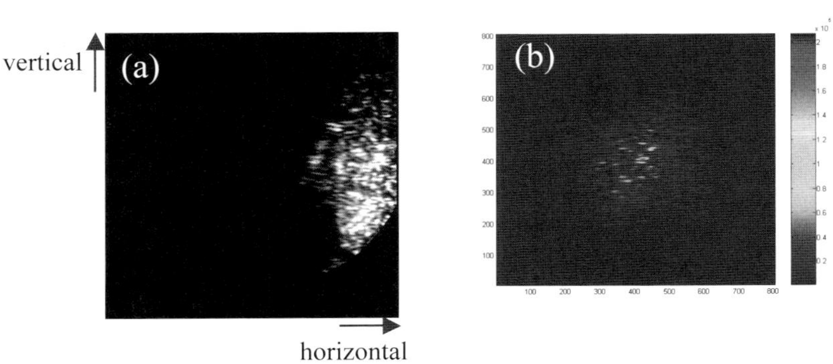

Figure 6. Measured (a) and calculated (b) footprint of the 13.9 nm laser beam at 1.2 m from the source. Speckle patterns are well reproduced by the numerical simulation.

The calculated image results from the diffraction of radiation emitted by a source having characteristics similar to the measured ones: the source size is 15 μm in horizontal and 80 μm in vertical; the pulse duration is 6 ps, the temporal coherence length is 2 ps. The spatial coherence length is assumed to be of 3 μm at the source.

The presence of speckles in the footprint images can thus be considered as a further evidence that our 13.9 nm laser pulse is close to or temporally coherent, while spatially incoherent. More precisely, the phase front of the source at the exit plane of the plasma is composed of many small (a few microns) temporally coherent sources with a random distribution of spatial phases.

7. Conclusion

In conclusion we have reported on the first measurement of the temporal coherence of an X-ray laser with a wavefront division interferometer. The results obtained indicate that the 13.9nm Ni-like Ag laser is highly monochromatic with a linewidth of $\Delta\nu$ (FWHM) = 9.8 10^{10} Hz (or $\Delta\lambda$ = 0,63 mÅ). This is a factor almost five smaller than the one previously measured by Smith et al. in Ni-like Pd [10]. In contrast to that experiment in which a gaussian spectral shape was inferred, we find that the spectral profile of the lasing line is closer to a lorentzian shape, or a Voigt shape with similar contributions of homogeneous and inhomogeneous broadening. Clearly the linewidth is controlled by the amplification processes that tend to narrow the spectral profile. Detailed simulations including Doppler and collision broadening, as well as radiation transfer, will be performed and compared to the experimental profile.

The duration of the X-ray laser was also measured in this experiment and compared to the transform-limited duration inferred from the measured coherence time. We find that the 13.9 nm transient X-ray laser is close to complete temporal coherence, at about 3 times the transform-limited pulse duration.

Acknowledgement: The financial support of the L.E.A. (Laboratoire Européen Associé) is gratefully acknowledged. We would like to thank the technical staff at LULI for their support in conducting the experiment at LULI.

References

[1] Ros D. et al., this Conference
[2] Jamelot G. et al., this Conference
[3] Nickles P.V.et al., 1997 Phys. Rev. Lett. 78 2748
[4] Neumeyer P. et al, 2004 Appl. Phys. B 78
[5] Borneis S. et al., 2000 Hyperfine Interactions 127 537
[6] Klisnick A. et al., 2002 Phys. Rev. A 65 033810
[7] Guilbaud O. et al., 2003 Proc. of SPIE Soft X-ray lasers and applications V, edited. by E. Fill, S. Suckewer Vol. 5197 pp 17-28
[8] Tallents G. et al., ibid [7] pp 9-16
[9] Celliers P. et al., 1995 Opt. Lett. 20 1907
[10] Smith R. et al., 2003 Opt. Lett. 28 1
[11] Chanteloup J-C et al. 2000 J.O.S.A. B 17 151-7
[12] Klisnick A. et al. 2000 J.O.S.A. B 17 1093 - 7
[13] Albert F. et al. 1997, Optics Comm. 142 184
[14] Born M. and Wolf E., 1980 *Principle of Optics*, Pergamon Press p504
[15] Koch J. et al., 1994 Phys. Rev A 50 1877
[16] Tallents G.J. et al., this Conference

Inst. Phys. Conf. Ser. No 186
Paper presented at 9th Int. Conf. X-Ray Lasers, Beijing, China, 24–28 May 2004

X-Ray Laser Research at the Institute of Physics, CAS

J Zhang, X Lu, YJ Li, F Yan, JY Zhong, Z Jin, YQ Liu, J Zheng, YT Li, ZH Wang, ZM Sheng, ZY Wei
Laboratory of Optical Physics, Institute of Physics, Chinese Academy of Sciences, Beijing 100080, People's Republic of China

Abstract. Atomic physics calculations for Ni-like Tungsten ions are performed. The KeV spectrum in laser-produced plasmas of Samarium was studied numerically. The numerical optimizations of picosecond pulse driven Ni-like Mo, Ag, and Y x-ray lasers are carried out. The equivalence relation between some Ni-like x-ray lasers and their partner Ne-like x-ray lasers are found. Possibilities of attosecond x-ray pulses generation by the nonlinear Thomson scattering are studied. A 20 TW Ti-sappher femtosecond laser system and diagnostics are built for x-ray laser experiments. A 5mm long line focus along the laser incidence direction is realized using a zone phase plate.

1. Introduction

The research on x-ray lasers is performed at the Institute of Physics, CAS (Chinese Academy of Sciences). The studies are mainly focused on the investigation of Ni-like x-ray lasers from middle-Z elements. We couple the hydrodynamic code MED103 with sophisticated atomic physics data to simulate Ni-like x-ray lasers and we also develope a ray tracing code to calculate the output intensity of x-ray lasers. The generation of x-ray pulse by the nonlinear Thomson scattering is also considered. We perform atomic physics calculations for Ni-like ions of some middle-Z and high-Z elements. In the experimental side, we built up a 20 TW Ti-sapphher laser system, which can deliver 640 mJ energy in 30 fs pulses. A line focus is formed along the laser incidence direction using a binary zone phase plate, which is very useful for longitudinal pumped x-ray lasers. A flat-field soft x-ray spectrometer is built with high spectral resolution. These form the base of our experimental investigations of x-ray lasers.

2. Numerical study on atomic physics

2.1. Effective collision strengths for Ni-like Tungsten ions

The Ni-like Tungsten x-ray laser is an ideal candidate for biological imaging in the "water-window". In order to understand the complicated physics of the Ni-like Ta x-ray lasers, we performed the atomic data calculation. Here the energy levels, spontaneous radiative decay rates, and electron impact collision strengths are calculated for Ni-like Tungsten ions. These data can be used for simulations of the Ni-like Tungsten x-ray laser. Effective collision strengths, obtained after integrating the collision strengths over the Maxwellian distribution of electron energies, have also been calculated for the electron temperature in the temperature range of 100-2500 eV. The results are compared with other calculations and a good agreement is found. Figure 1 shows the calculated electron excitation rate coefficient as a function of electron temperature for excitation of the ground state to the upper lasing level by our calculation and by Daido et al [1].

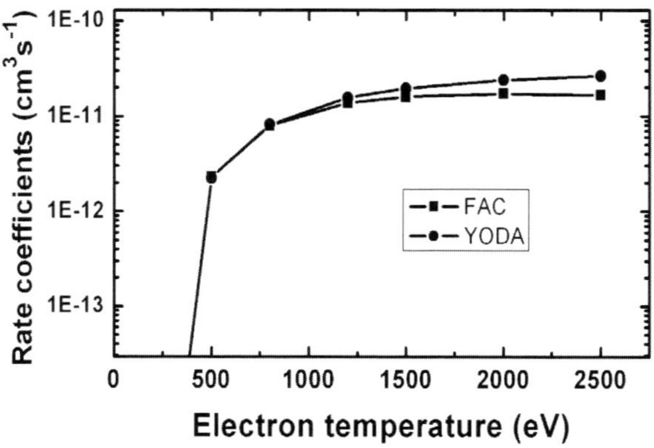

Figure 1. Collisional excitation rates as a function of electron temperature for transitions from the ground level to the upper lasing level. The solid curve with square is calculated using the FAC code and the solid curve with circle from the YODA code [1].

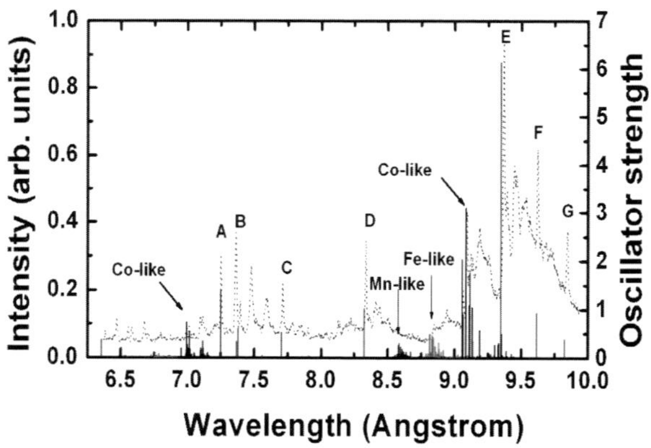

Figure 2. The KeV spectrum from Sm plasma and the calculated spectrum.

2.2. Simulations of keV x-ray spectra from the Samarium plasma

Figure 2 shows a keV spectrum in the 6.5-10 Å range from the plasma produced by the irradiation of slab Sm targets with a single 600 picosecond laser pulse at an intensity near 4×10^{13} Wcm^{-2}. This spectrum are diagnosed by two space-resolving, time-integrating potassium acid phthalate crystal spectrometers and recorded on x-ray charge-coupled device (CCD) detectors. We identify the distinct lines of experimental spectrum based on the oscillator strength in Fig. 2. The atomic calculations are carried out including all coupling of states with a given number of n=3, 4 and n=5 electrons, including 249, 508, 1281 and 2138 levels for Ni-like, Co-like, Fe-like and Mn-like ions, respectively. In Fig.2 we have multiplied the Co-like, Fe-like and Mn-like lines by factors of 0.4, 0.14 and 0.055 respectively, based on the ratio of different ionisation stages.

3. Numerical optimizations of Ni-like Ag, Mo, and Y x-ray lasers.

Ni-like x-ray lasers have a more favorable scaling of laser wavelength with drive energy than Ne-like x-ray lasers. However, the population kinetics of Ni-like ions is much more sensitive to plasma conditions than that of Ne-like ions. This results in difficulties in providing a saturation output for Ni-like x-ray lasers. It is therefore very important to optimize the parameters for experiments.

Here the Ni-like Ag x-ray laser at 13.9 nm (4d→4p, J=0→1) is simulated to calculate the laser-plasma interaction and gain coefficient, using a one-dimensional Lagrangian hydrodynamic code MEDUSA [2], coupled with an atomic physics code and a ray-tracing code. The energy levels, radiative rates and collsional pump rates of Ni-like Ag ions used in the x-ray laser simulations are calculated using the Cowan code [3] in orbital-spin (LS) coupling. High gain (>150cm^{-1}) is predicted using a 1ps, 10^{15}W/cm^2 main drive pulse proceeded by a 2 ns, 6×10^{11}W/cm^2 prepulse with a temporal separation of 1.0 ns between these two pulses, as shown in Fig. 3. The characteristics of the x-ray laser, including the evolution of gain in time and space, the electron temperature and the electron density obtained using the optimized drive configuration is investigated.

A two-dimensional ray tracing code is developed to calculate the x-ray laser output intensity, as a postprocessor to the hydrodynamic code MEDUSA. In the calculation, the saturation and refraction effects are taken into account. A traveling wave excitation is used to keep the spatial profiles of gain and electron density unchanged along the line focus. The total laser output as a function of the plasma length is calculated. Figure 4 shows the profile of spatially integrated laser output intensity as a function of the plasma length. The solid line describes the x-ray laser output intensity when the refraction is taken into account in the calculation. The results show that the refraction on the propagation of x-ray lasers in the plasma plays an important role.

We performed also the numerical simulation of the picosecond pulse dirven Ni-like Mo and Y soft x-ray lasers at 18.9 nm and 23.4 nm respectively, similar high gains (more than 150 cm^{-1}) are obtained using a 1 ps main drive pulse. These results contribute to the further understanding of Ni-like x-ray lasers.

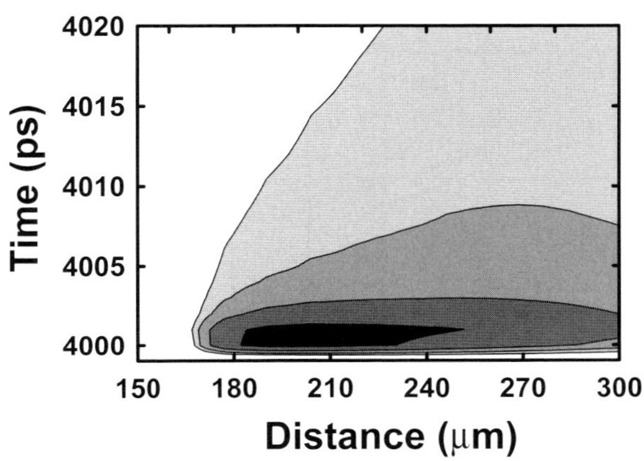

Figure 3 Contours of gain vs space and time. The darkness from light to dark represents the gain over 50, 70, 100, 150 cm^{-1}.

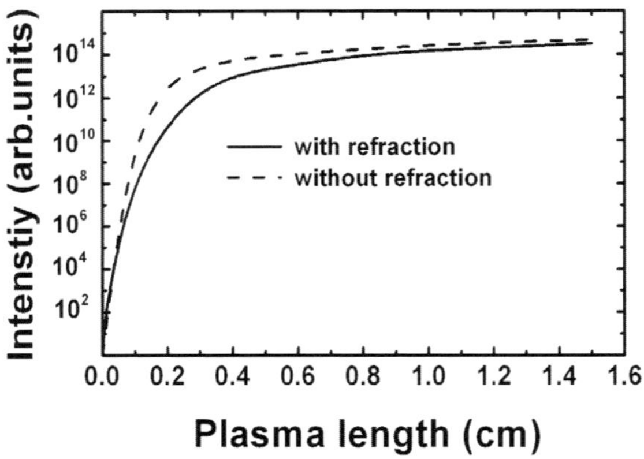

Figure 4 Spatially integrated laser output intensity as a function of the plasma length.

4. Equivalence relation of some Ni-like and Ne-like x-ray lasers.

It is more difficult to calculate atomic data of Ni-like ions because of the complicated level structure. By comparing the hydrodynamic characteristics of plasma conditions for x-ray laser Ni-like and Ne-like plasma conditions, we find a similarity of plasma conditions between some Ni-like x-ray lasers and their partner Ne-like x-ray lasers. Based on this similarity we can predict the performance of some Ni-like x-ray lasers using rather simple hydrodynamic and atomic simulations of their partner Ne-like x-ray lasers.

The plasma conditions of the Ni-like Ag, Cd, In, Sn, Sb x-ray lasers and the Ne-like Fe, Co, Ni, Cu, Zn x-ray lasers are studied. By comparing the hydrodynamic characteristics of ion abundance, ion population density and electron temperature for these Ne-like ions and Ni-like ions, we conclude that there is a similarity of plasma conditions between these Ne-like and their partner Ni-like x-ray lasers. The ionization energy for these Ne-like ions with their partner Ni-like ions is also compared to further understand this similarity. The comparison of the ionization energies supports the conclusion from the hydrodynamic simulation.

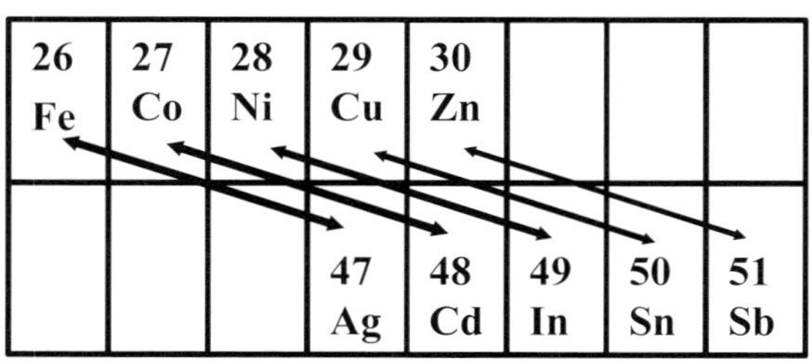

Figure 5. The arrows represent the similarity relation between these Ne-like x-ray lasers and their partner Ni-like x-ray lasers.

However, this similarity is only valid for the elements of medium atomic number and their partner. The degree of similarity is generally weakened from the light elements to the heavy elements, as shown in Fig 5. The main reason is that the Ni-like ions are much easier to be over-ionized to the next ionization stage than the Ne-like ions with the increase of atom number. Although the degree of similarity is weakened generally with the increase of atom number, this still reflects a partnership between some Ne-like plasma conditions and their partner Ni-like plasma conditions under certain drive conditions. Because of the lower value and narrower region of ion abundance, population density and electron temperature for the Ni-like case, we can conclude that the Ni-like x-ray lasers generally have lower gain and smaller high gain region than the Ne-like x-ray lasers under the same drive conditions.

5. Attosecond x-ray pulse generation by nonlinear Thomson scattering

The x-ray generation by the Thomson scattering of relativistic intense laser pulses on free electrons becomes very attractive in recent years [4-8]. We has also investigated the nonlinear Thomson scattering of an electron with certain initial momenta. We calculated the angular distribution of the radiated power detected from the observing point.

Usually, with the increase of the incident laser intensity, the radiation power and its central frequency increase, while the emission duration and angular spread reduce, as shown in Fig. 6. Note that the observer receives two radiated peak during every laser cycle. However, only a single peak is found every laser cycle for a circularly polarized laser pulse (not presented). The radiation produced for a linearly polarized laser pulse has a higher amplitude than for a circularly polarized laser pulse for the same intensity. The initial phase of the laser pulse changes the symmetry of the electron's motion. As a result , it changes the symmetry of the radiation emitted by the electrons.

The polarization state of the radiation is very complicated, which depends on the laser's polarization states and the observation direction as well. Assume that the radiation direction is vector **n,** Fig. 7 shows the relationship between two transverse electric field E_{n2} and E_{n3} for a linearly polarized laser and a circularly polarized field.

The initial velocity of the electron can significantly change the angular distribution of the radiated energy. Whatever the initial energy and moving direction of the electron as well as the laser polarizations are, the radiation generated by a relativistic energetic electron tends to be predominantly along the moving direction of the electron. If the initial moving direction of the electron directs to $-z$ (z is the laser direction), $\pm x$ or $\pm y$ axis, with the increase of initial energy, the radiated pulse duration becomes small.

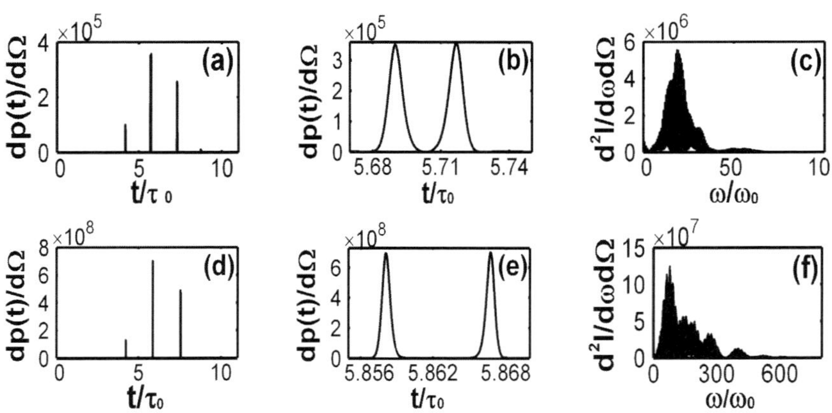

Figure 6. The radiation from the initially stationary electron for various incident laser intensities. The laser is linearly polarized and its duration is 10 laser periods. (a, d) the time history of the radiated power[in unit of $e^2/(4\pi c\tau_0^{\ 2})$] for the laser amplitude parameter a_0=3, 6.8, respectively, at

the direction of maximum radiation($\phi = 0^\circ$, $\theta = \theta_M$). The highest peak in (a) and (d) are magnified in (b) and (e), respectively. (c) and (f) are the spectral intensity [in unit of $e^2/(4\pi c\tau_0^2)$] in the same conditions as in (a) and (d), respectively.

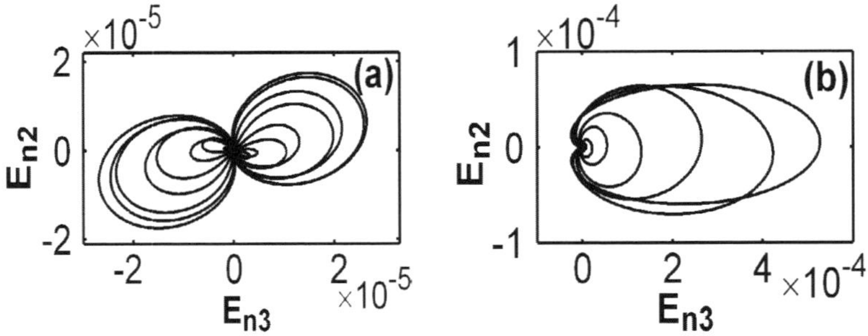

Figure 7. The relationship between two transverse electric field E_{n2} and E_{n3} for (a) a linearly polarized laser field and (b) a circularly polarized laser field.

6. Experimental study on x-ray lasers

A Ti:sapphire chirped pulse amplification laser system is developed (XL-II). The final energy output after the compressor is 600 mJ in 30 fs pulses. This corresponds to a peak power as high as 20TW. This system includes the re-amplifiers, which are adopted for the pre-amplifiers to improve the beam quality. The main amplifiers are pumped with two green pumpers, the amplified chirped pulse before the final compression with a maximum energy up to 900mJ. The compressor is in a vacuum chamber. The compressing grating and stretching grating are taken to be at different densities, which enables to dismiss the higher order dispersion from the materials, to make the compressed pulse more close to the seeding pulse. An adaptive optics system is installed on the laser system. By using this adaptive optics system, the beam quality of the output laser pulse can be improved to 1.5 diffraction limits. Figure 8 shows the laser focus with (a) and without (b) adaptive optic system. Our laser facility XL-II can be used as pump source of ultra-short pulse driven soft x-ray lasers.

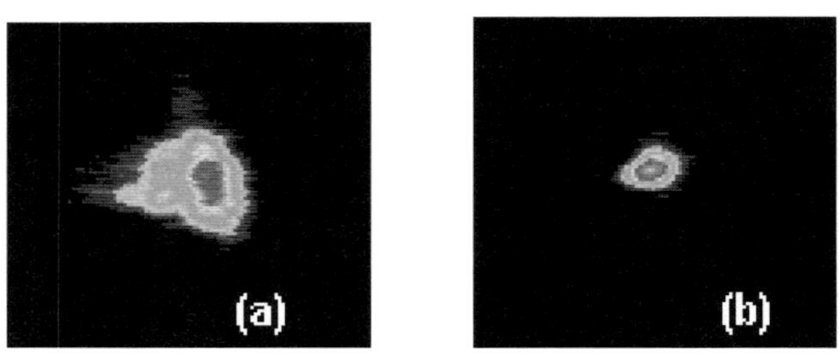

Figure 8. Focal spot of laser beams (a) without the wavefront correction (b) with the wavefront correction

Recently a longitudinal pumping scheme for collisional x-ray laser has been proposed [9] and then realized [10]. For the longitudinal pumping scheme, the line focus in laser axial direction is very beneficial to obtain high gain-length product [11]. Here we used a binary phase Fresnel zone plate to get a uniform line focus, as shown in Fig. 9. The laser beam is focused on a 16 bit, 512×512 CCD camera. The size of the

CCD pixel is 24μm×24μm. By moving the position of the CCD in the laser axis direction (Z directin) one can obtain a sequence of 2D intensity distribution along the line focus. Two conventional lenses with 90mm and 125mm focal length respectively are coupled with the zone plate in experiments.

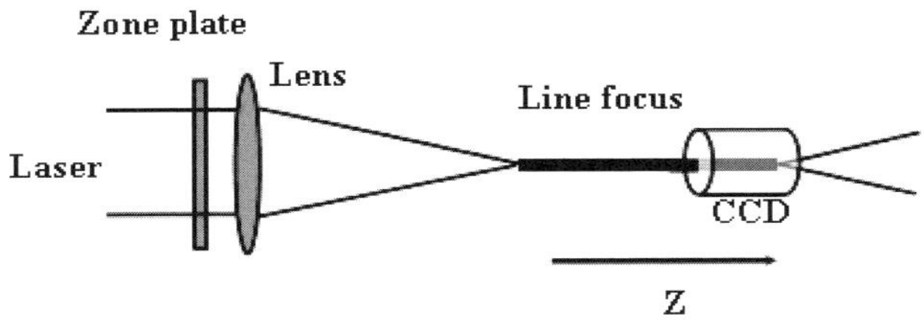

Figure 9. Schematic of experimental setup.

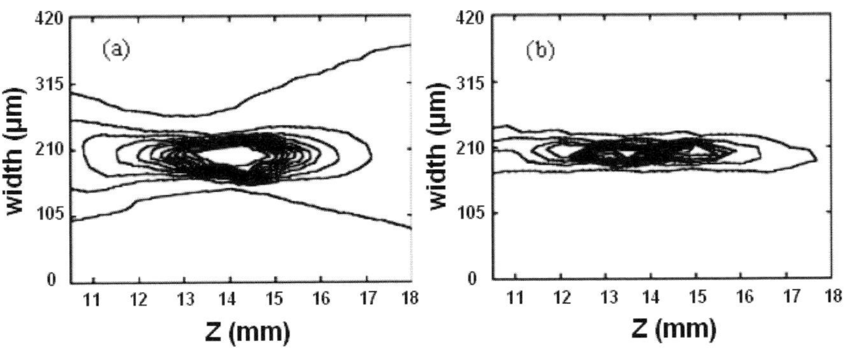

Figure 10. Typical cross-section image of the line focus at the geometrical focus of a *f*=125 mm lens (a) and the line-out along the dash line (b).

Figure 10 shows the 2 dimensional intensity contour without the zone plate (a) and with the plate (b) along the laser axis. The laser is incident from the left. It is obvious that a ~5 mm long line focus is formed when the zone plate is used.

7. Conclusion

In conclusion, we did atomic physics calculations for Ni-like Tungsten ions and simulations of KeV spectrum keV x-ray spectra in the Sm plasma. The numerical optimisation of picosecond pulse driven Ni-like Mo, Ag, and Y x-ray lasers are performed. The plasma conditions of the Ni-like Ag, Cd, In, Sn, Sb x-ray lasers and the Ne-like Fe, Co, Ni, Cu, Zn x-ray lasers are studied. The results suggest that there is a correspondence of plasma conditions between some Ni-like x-ray lasers and their partner Ne-like x-ray lasers by comparing the hydrodynamic characteristics of electron temperature, ion abundance and ion population density. A 20 TW Ti-sappher femtosecond laser system is built as the pump source of x-ray laser. A 5mm long line focus along the laser incidence direction is realized using a zone phase plate. Based on these preparation works, the experiments of x-ray laser has started to produce results at the institute of physics, CAS.

Acknowledgments

This work is jointly supported by the NSFC under Grant Nos. 10374114, 10474134, 60321003, 10425416, and the NKBRSF under Grant No.G1999075206, G1999075203 and the National Hi-tech ICF program.

References

[1] Daido H, Sebba S et al., J. Opt. Soc. Am. B, **16**, 2295 (1999).
[2] Christiansen J P, Ashby D E T et al., 1974 Comput. Phys. Commun. **7** 271-287.
[3] Cowan R D, 1981 The Theory of Atomic Structure and Spectra (University of California Press, Berkeley, CA).
[4] K. Ta Phuoc, A. Rousse, M. Pittman et al., Phys. Rev. L 91, 195001 (2003).
[5] Y. Y.au, Fei He, Donald P. Umstadter, and Richard Kowalczyk, Phys. Plasmas 10, 2155 (2003).
[6] K. Lee and Y. H. Cha, et al., Phys. Rev. E 67, 026502 (2003).
[7] Szu-yuan Chen, Anatoly Maksimchum & Donald Umstadter, Nature 396, 653 (1998).
[8] Eric Esarey, Sally K. Ride, Philip Sprangle, Phys. Rev. E 48, 3003 (1993).
[9] Ruxin Li, Tsuneyuki Ozaki, Teruto Kanai, and Hiroto Kuroda 1998 Phys. Rev. E **57** 7093.
[10] Sebban S, Haroutunian R, Balcou Ph, et al 2001 Phys. Rev. Lett. **86** 3004.
[11] McLeod J H 1960 Journal of the Optical Society of America **50** 166.

Inst. Phys. Conf. Ser. No 186
Paper presented at 9th Int. Conf. X-Ray Lasers, Beijing, China, 24–28 May 2004

Progress in x-ray laser research at IAP-BE

J.E. Balmer, F. Staub, M. Grünig, J. Nilsen[1] and S. Bajt[1]
Institute of Applied Physics, University of Berne
CH-3012 Berne, Switzerland
[1]Lawrence Livermore National Laboratory, Livermore, CA 94551

Abstract. The absolute time of emission of the x-ray laser output with respect to the arrival of the 100-ps pump pulse has been measured with the aid of a calibrated timing fiducial. The results show the x-ray laser to appear up to 60 ps (80 ps) before the peak of the pump pulse in the case of the Sn (Pd) x-ray laser. First results from a x-ray laser microscopy experiment using a set of 11.9-nm multilayer mirrors are reported having a spatial resolution of ~1.2 μm. A proposal is presented for improving the resolution well into the sub-μm region by using a Schwarzschild configuration.

1. Introduction

Rapid progress in the development of collisionally-pumped soft-x-ray lasers in recent years has lead to a number of robust, saturated lasing lines using the Ne-like (e.g. Zn, Ge, Se, etc.) and Ni-like (e.g. Pd, Ag, Sn, etc.) schemes. The output properties of these lasers have been characterized in terms of the most relevant laser parameters including output energy, pulse duration, coherence, near-field and far-field energy distributions, and divergence [1-4]. Of great interest in this context is the precise knowledge of the absolute time of emission of the x-ray laser pulse with respect to the incident drive pulse, as this would permit direct comparison with simulation results and thus help to improve our understanding of the atomic level population dynamics. Measurements for Ne-like schemes using long (~500 ps) drive pulses and the prepulse technique have revealed a marked difference in emission time for the J = 0 - 1 line as compared to the J = 2 - 1 lines which were found to peak 200-300 ps later in time [5,6]. In a double-100-ps pulse experiment using Ne-like Ge, Daido et al. reported the J = 0 - 1 line to be emitted 40 ps earlier than the J = 2 – 1 lines which coincided with the peak of the second pulse [7]. In the case of Ni-like lasers and multiple-150-ps-pulse pumping, J = 0 - 1 lasing was reported to occur during the rising edge of the second pulse, 140 ps before the peak of the continuum emission [8,9]. The absolute timing of the x-ray laser pulse with respect to the drive pulse, however, was not measured.

In section 2, we describe our recent experiments aimed at measuring the absolute timing of the x-ray laser output with respect to the arrival of the drive pulse. This was achieved by implementing a calibrated timing fiducial using a frequency-quadrupled portion of the main pulse. In section 3, we report on a first application of the Sn x-ray laser in a microscopy experiment utilizing a set of 11.9-nm Mo/Y multilayer mirrors with 1.2-μm spatial resolution and ~50-ps temporal resolution.

2. Absolute timing experiment

The experiments were performed using the 1054-nm Nd:glass laser system at the Institute for Applied Physics of the University of Berne (IAP-BE). The laser, having a final amplifier of 90-mm diameter, is capable of delivering up to 30 J at a pulse duration of 100 ps (FWHM). The output beam is focused to give a line focus of 2.5-cm length and approximately 100-μm width using a combination of a 500-mm focal length aplanatic doublet and a –1700-mm focal length cylindrical lens. Targets were 25-mm wide slabs having a diamond-machined surface finish. A 0.5% prepulse irradiated the target 5 ns before the arrival of the main pulse.

The time-resolved diagnostics setup is shown in Fig. 1. The x-ray laser emission is directed onto the 140-μm wide slit of a Kentech ×2-magnification x-ray streak camera via a 1200 lp/mm, aberration-corrected, flat-field Hitachi grating spectrograph. The streak camera cathode is made up of a 200-Δ thick layer of aluminum deposited on a 1200-Δ thick Formvar foil. The streaked signal is read out by an image intensifier coupled to a cooled CCD camera having a pixel size of 23 μm. To avoid saturation of the streak camera, a 10-μm wide slit and several Al-on-Formvar filters were inserted at the entrance of the spectrograph [10]. The (instrumental) time resolution of the streak camera was ~15 ps.

To generate a timing fiducial pulse, a fraction of the main pulse was extracted after the last amplifier turning mirror, frequency-quadrupled in a KTP/KDP crystal pair, and relayed to the streak camera cathode. A two-mirror etalon in the reference beam having a temporal separation of the echo pulses of 187 ps was inserted as a calibration device for the sweep speed of the streak camera on each shot.

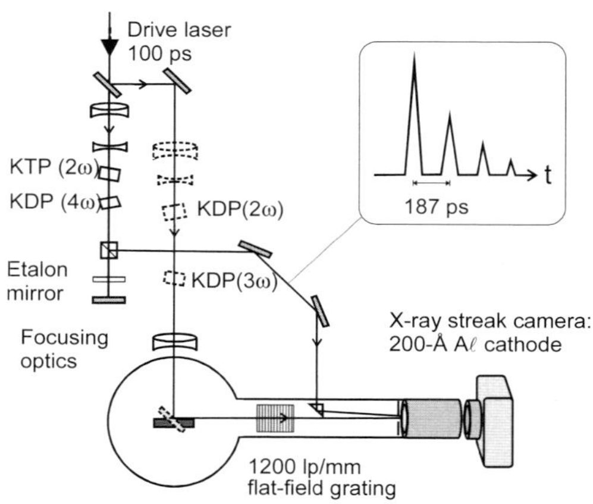

Figure 1. Schematic of the time-resolved diagnostics with 4ω timing fiducial. Dashed components are inserted for the calibration measurements.

The 4ω timing fiducial was calibrated with respect to the main pulse in an auxiliary experiment. For this purpose, a KDP/KDP crystal pair was used in the main beam line to generate a frequency-tripled pulse along the x-ray laser beam path. A 45° turning mirror positioned at the target center served to direct this 3ω "main" pulse to the entrance of the flat-field spectrograph. Using the grating reflection in 0th order and shifting the streak camera to the corresponding position, the 3ω "main" and 4ω fiducial pulses could be recorded on one and the same shot. In this way, the two pulses could be timed with respect to each other with an accuracy of 15 ps. It is important to note, here, that the calibration refers to the *center* of the x-ray laser target. For x-rays reaching the streak camera from the far end of the gain region, the additional time of propagation (~42 ps for a total target length of 25 mm) will have to be accounted for in the analysis of the results.

For the actual x-ray laser timing experiments, the auxiliary components were removed, the focusing optics put back in place, the target inserted, and the streak camera shifted back in position to record the x-ray spectrum in 1st order. Figure 2 shows the streak record (raw data) for a ~10-TW/cm^2 shot on a Sn target. Clearly visible are the 4ω timing fiducial (top) with the echos from the etalon mirror, the Sn x-ray laser (bottom) and the continuum emission.

The temporal profiles obtained from the data of Fig. 2 are shown in Figure 3a) for the drive pulse, derived from the 4ω timing fiducial (dashed line), the Sn x-ray laser (solid line), and the continuum emission (dotted line). It is seen that the 11.9-nm lasing occurs very early during the rising edge of the pump pulse and has a FWHM duration of 37 ps. The time by which the peak of the x-ray laser pulse precedes the peak of the drive pulse is 57 ps for this shot.

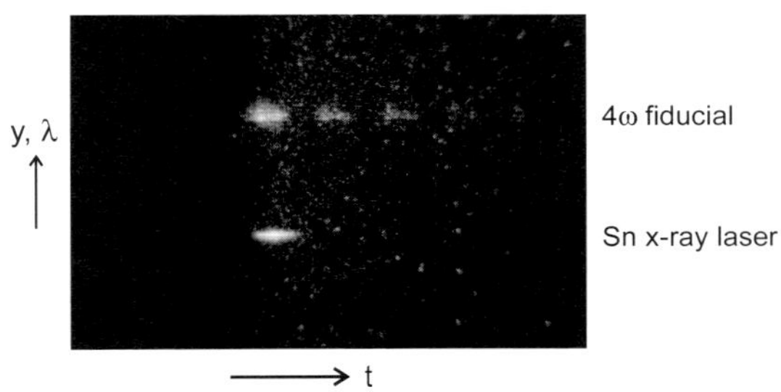

Figure 2. Streak record from a 10 TW/cm^2 shot on a Sn target. Clearly visible are the 4ω timing fiducial (top) with the echos from the etalon mirror, the Sn x-ray laser (bottom) and the continuum emission.

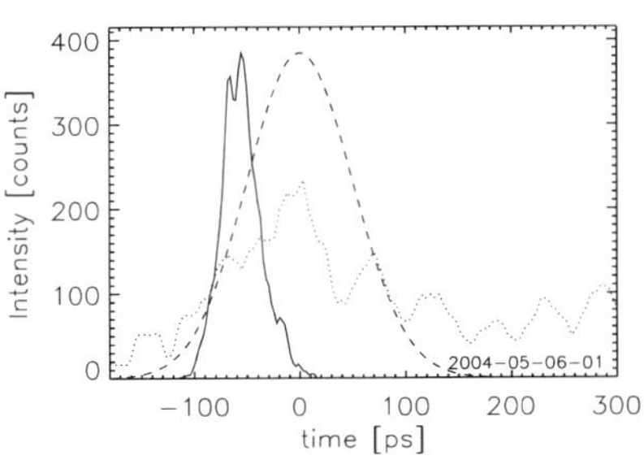

a)

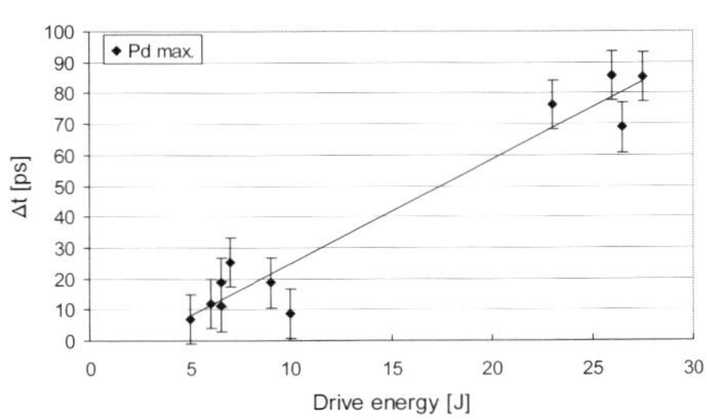

b)

Figure 3. a) Temporal profiles of the drive pulse derived from the 4ω timing fiducial (dashed line), the Sn x-ray laser (solid line), and the continuum emission (dotted line); b) dependence of Δt on the drive energy.

Similar results were obtained for the Pd x-ray laser at 14.7 nm. In order to see how the irradiance of the drive pulse affects the time of emission of the x-ray laser pulse, a series of shots were taken with Pd at reduced drive irradiance. Figure 3b) shows a plot of the $\Delta t = t_{pump} - t_{x\text{-}ray}$, with positive values of Δt meaning that the peak of the x-ray laser pulse is preceding the peak of the drive pulse. It is clearly seen in Fig. 3b) that the higher the drive irradiance, the earlier the x-ray lasing occurs. For the whole range of the measurements, x-ray lasing is seen to occur during the rising edge of the drive pulse, and only at very low irradiance it appears near the peak of the drive pulse. This is found to be in good agreement with preliminary simulation results obtained from the LASNEX and CRETIN codes [11].

3. Microscopy experiments

As a first application of our Sn x-ray laser at 11.9 nm, we have assembled a soft-x-ray microscopy setup consisting of a condensor stage, a test object, and an imaging stage. For the condensor and imaging stages we utilized the concave Mo/Y multilayer mirrors recently developed at LLNL [12] and used for near-field imaging diagnostics of the Sn x-ray laser [13]. The setup used is depicted in Figure 4. The output plane of the x-ray laser is focused by a 50-cm radius-of-curvature (ROC) multilayer mirror to illuminate a test object (12.5-μm mesh) positioned in the object plane. This is then imaged with a 24-cm ROC multilayer mirror via a 45° turning mirror onto our standard P43 phosphor/CCD camera diagnostics, the spatial resolution of which has been previously measured to be 50 lp/mm. Taking into account the imaging magnification of 15×, the spatial resolution of the recording system at the object plane amounts to ~1.3 μm.

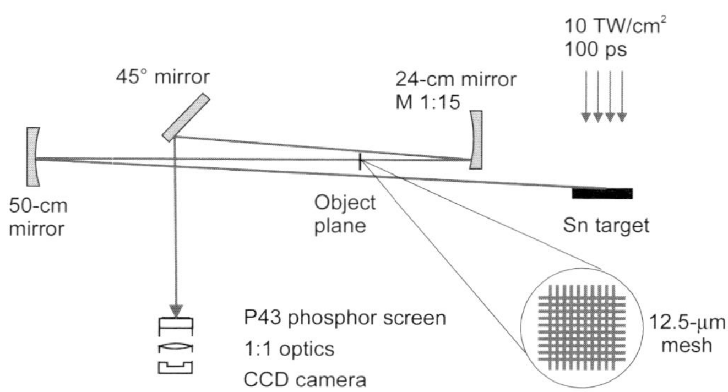

Figure 4. Schematic layout of the x-ray microscopy experiment utilizing the 11.9-nm Sn x-ray laser as the source for illumination.

Figure 5a) shows an image of the 12.5-μm mesh taken in a single exposure with the above setup. Due to the demagnification by the condensor mirror, only a few gridlines of the mesh are illuminated. The spatial resolution deduced from fitting Gaussian profiles to the rising edges of the scanned image is $\Delta \approx 1.2$ μm in the vertical direction, essentially equal to the spatial resolution of the recording system.

If the resolution were limited solely by the recording system, one might attempt to improve it by increasing the magnification. However, spatial resolution of any optical system ultimately depends on diffraction, and thus on the numerical aperture (NA) of the optics. This is shown in Figure 5b), where the limiting spatial resolution, given by $\Delta_d = 0.61\lambda f/R = 0.61\lambda/NA$, where f denotes the focal length of the imaging mirror and R the radius of the x-ray laser beam at the mirror, is plotted as a function of the NA = R/f (solid line). Due to the small divergence of the x-ray laser beam (~3 mrad in the vertical direction), the NA in our experiment is only 0.063, so that the measured spatial resolution is already close to the diffraction limit. One might attempt to improve it by increasing the NA, however this would lead to an increase in longitudinal spherical aberration. This is shown by the dashed line in fig. 5b, which plots the radius of the circle of least confusion (CLC) as a function of the beam NA. Due to the counteracting effect of the two phenomena, the limiting resolution of the single-spherical-mirror imaging amounts to ~0.5 μm.

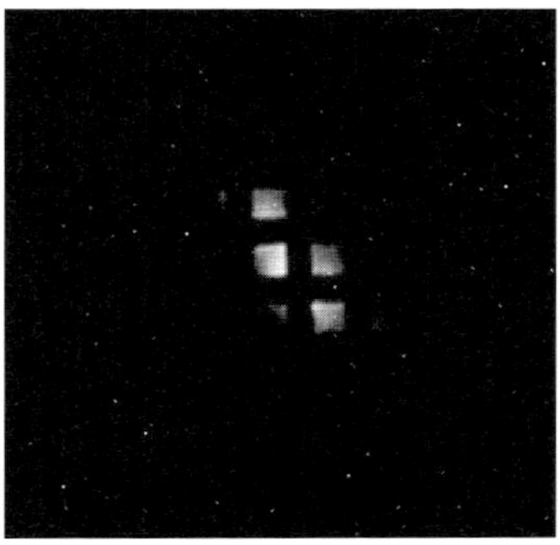

a)

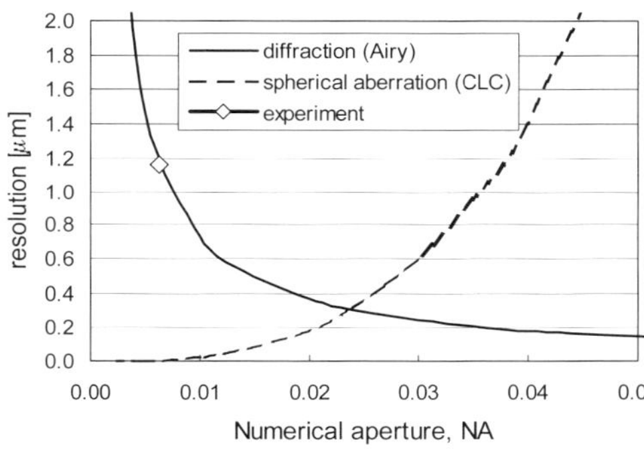

b)

Figure 5. a) Image of the 12.5-μm mesh taken with the setup shown in fig. 4; b) spatial resolution of spherical-mirror optics due to both diffraction and spherical aberration, with (◇) denoting the experimental value from fig. 5a.

Of course, these findings are not new and have been elaborated on in great detail in the context of soft-x-ray projection lithography. In order to resolve features smaller than ~0.1 μm at a wavelength of ~12 nm, the use of a Schwarzschild-type objective having a NA of ~0.1 is required [14]. The standard Schwarzschild configuration is shown by the dashed lines in Figure 6. It consists of two spherical mirrors that are arranged to be nearly concentric, as a result of which the spherical aberration, astigmatism, and coma are eliminated, if the ratio of the radii of curvature is ~0.35 (slightly depending on the magnification) [15]. A Schwarzschild soft-x-ray microscope of this design, operating at a wavelength of ~20 nm, has been reported to yield ~0.2-μm resolution when used with a laser-produced plasma source for illumination [16].

In the case of illumination with a nearly spatially coherent source such as synchrotron radiation or, as in our case, x-ray laser radiation, off-axis illumination of the Schwarzschild objective has proven to give best results in terms of contrast and also to reduce background light in the image plane [14,17]. In this case, the large mirror having a central hole may be replaced by a simple spherical mirror of half the diameter. The design of a Schwarzschild objective consisting of a 25-mm diameter concave mirror (100-mm ROC) and a 10-mm diameter convex mirror (35-mm ROC) and having a NA of ~0.9 is currently under way.

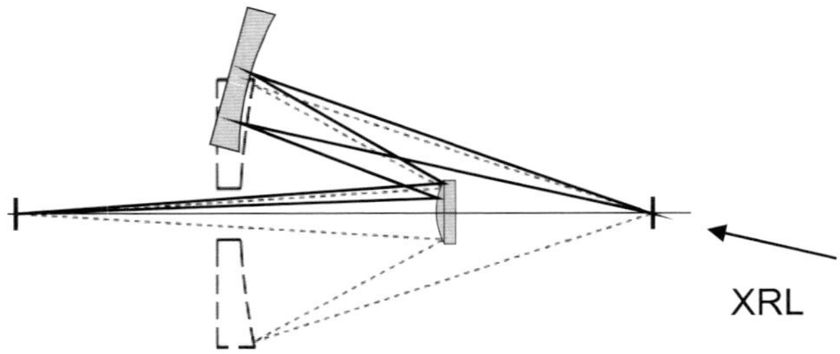

Figure 6. Schematic layout of a Schwarzschild-type, off-axis, soft-x-ray microscope utilizing the 11.9-nm Sn x-ray laser as the source for illumination.

4. Acknowledgements

The authors would like to acknowledge the technical assistance of B. Locher and W. Lüscher for target preparation. This work was supported in part by the Swiss National Science Foundation. The work of the LLNL authors was performed under the auspices of the U.S. Department of Energy by the University of California Lawrence Livermore National Laboratory under Contract No. W-7405-Eng-48.

References

[1] Zhang J et al 1997 Phys. Rev. Lett. 78 3856-3859
[2] Nilsen J et al 1998 Phys. Rev. A 56 3161-3165
[3] Lin J Y et al 1998 Opt. Commun. 158 55-60
[4] Siegel C et al 2002 Opt. Commun. 210 305-312
[5] Nilsen J et al 1993 Appl. Phys. B 57 309-311
[6] Lu P et al 1996 Phys. Rev. A 54 5193-5200
[7] Daido H et al 1995 Opt. Lett. 20 61-63
[8] Nilsen J and Moreno J C 1995 Opt. Lett. 20 1386-1388
[9] Daido H et al 1995 Phys. Rev. Lett. 75 1074-1077
[10] Braud M et al 2002 Proc. 8th Int. Conf. on X-Ray Lasers (Melville/New York: AIP Conference Proceedings, Vol 641)
[11] J. Nilsen (private communication)
[12] Bajt S et al 2003 Opt. Lett. 28 2249-2251
[13] Staub F et al Phys. Rev. A (accepted for publication)
[14] Berreman D W et al 1990 Opt. Lett. 15 529-531
[15] Kingslake R 1978 Lens Design Fundamentals (New York: Academic) p 331
[16] Artioukov I A et al 1995 Opt. Lett. 20 2451-2453
[17] DiCicco D S 1992 Opt. Lett. 17 157-159

Inst. Phys. Conf. Ser. No 186
Paper presented at 9th Int. Conf. X-Ray Lasers, Beijing, China, 24–28 May 2004

Progress on soft-x-ray lasers at LOA

S. Sebban[1], T. Mocek[1], I. Bettaibi[1], Ph. Zeitoun[1], G. Faivre[1], B. Cros[2], G. Maynard[2], J. Dubau[4], A. Butler[3], A.J. Gonzalves[3], C. M. McKenna[3], D.J. Spence[3], S.M. Hooker[3], C. Valentin[1], Ph. Balcou[1], S. le Pape[1], D. Ros[4], L.M. Upcraft[1], S. Kazamias[4], A. Klisnick[4], G. Jamelot[4] and B. Rus[6]

[1-]Laboratoire d'Optique Appliquée (LOA), ENSTA - Ecole Polytechnique, 91761 Palaiseau, France

[2-]LPGP, UMR 8578, CNRS, Université Paris XI, Bât. 210, 91405 Orsay, France

[3-]Department of Physics, University of Oxford , Clarendon Laboratory, Parks Road, OXFORD OX1 3PU, UK.

[4-]LIXAM, Université Paris Sud, 91405 Orsay, France

[5-]Department of X-ray Lasers, Institute of Physics, Na Slovance 2, Prague 8-18221, Czech Republic

Abstract. We report a survey of the Laboratoire d'Optique Appliquée activities in the field of x-ray lasers. The main interest is focussed on the collisional Optical field Ionization (OFI) soft x-ray lasers. We will present recent characterization of the sources as well as dramatic improvement of their performances using the waveguiding technique. We will also show recent results consisting in amplifying a High order Harmonic Generation (HHG) beam into an OFI plasma amplifier ; we produced a highly saturated, energetic, sub-ps, fully coherent and fully polarised tabletop x-ray laser operating at 10 Hz.

1. Introduction

Since the last decade there has been much interest in developing Optical-Field Ionization (OFI) soft-x ray lasers (XRL) in order to reduce the size of XRL systems and to dramatically increase repetition rates. A breakthrough in the collisionally pumped scheme was the demonstration of a gain-length product of 11 of the 5d-5p emission line at 41.8 nm of Xe^{8+}, by focusing a 70 mJ, 40 fs circularly polarized laser pulse into a gas cell filled with xenon [1]. These results followed previous modeling predicting lasing action for three specific laser schemes [2] in the 30-50 nm range that can be driven by an ultra-short pulse focussed to an intensity of 10^{16} to 10^{17} W·cm^{-2}. These systems are based on electron collisional excitation of eight-time-ionized gases created by OFI: Ne-like Ar lasing at 48 nm, Ni-like Kr lasing at 32 nm, and Pd-like Xe at 41 nm. This promising work offered the prospect for high-repetition rate, table-top soft X-ray lasers which would be suitable for widespread applications.

In 1999, our group succeeded in demonstrating lasing on the $4d^95d\,^1S_0$-$4d^95p\,^1P_1$ transition in Xe^{8+} at 41.8 nm by focusing a 330 mJ, 35 fs circularly polarized laser pulse into a gas cell filled with 15 Torr of xenon [3]; saturated amplification was demonstrated for the first time using the collisional OFI x-ray laser technique. More recently, we have extended the OFI x-ray laser technique to the use of Ni-like ions. Strong amplification at 32.8 nm on the $3d^94d\,^1S_0$-$3d^94p\,^1P_1$ transition in Kr^{8+} has been observed [4].

In this article we report recent progress on collisional OFI soft-X ray lasers using Pd-like and Ni-like ions. We show and discuss experimental results concerning the optimization and the characterization of the source. We present investigations of short-wavelength lasing in a capillary waveguide [5, 6]. And finally, we show first results in an original geometry consisting in seeding and OFI amplifier using high order harmonic generation.

2. Experimental set-up

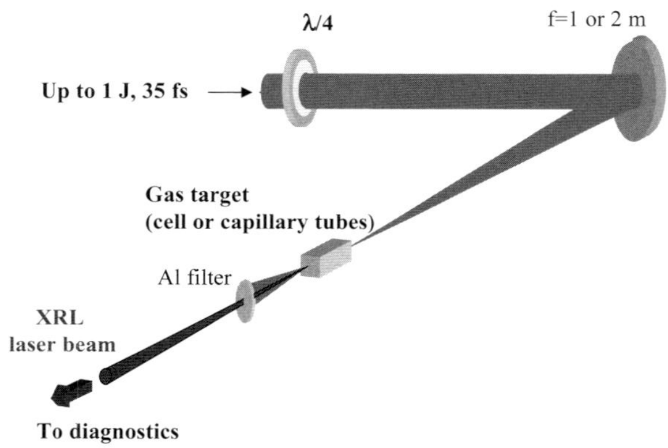

Figure 1. *Schématic description of the experimental set-up*

The experiments were performed at LOA using a 10 Hz multi-terawatt Ti:Sapphire laser system [7] delivering 34 fs pulses (full width at half maximum) with an energy on target of up to 1 J. The circularly polarized laser beam was focused onto the capillary or gas cell entrance by a spherical mirror of focal length 1 m and 2 m. The resulting maximum intensity on target was up to $(1 \pm 0.2) \times 10^{18}$ W cm^{-2} in vacuum.

The spectrometer comprised a grazing incidence gold coated spherical mirror, transmission grating with 2000 lines/mm, and a back-illuminated XUV CCD camera. Stray light from the driving laser was blocked by two 0.3 μm-thick aluminum filters.

The capillaries used were 15 mm long, with an internal diameter of 300 ± 10 μm. Since the channels were laser machined, the walls of the capillaries were not optically smooth, but contained non-uniformities with typical dimensions of 5 μm. Xe gas was flowed into the capillary through channels approximately 500 μm wide and 250 μm deep located 1.5 mm from each end. By injecting the gas in this way, the pressure in the main body of the capillary between the gas slots was uniform.

3. OFI x-ray laser using a gas cell target

3.1. Wavefront measurement

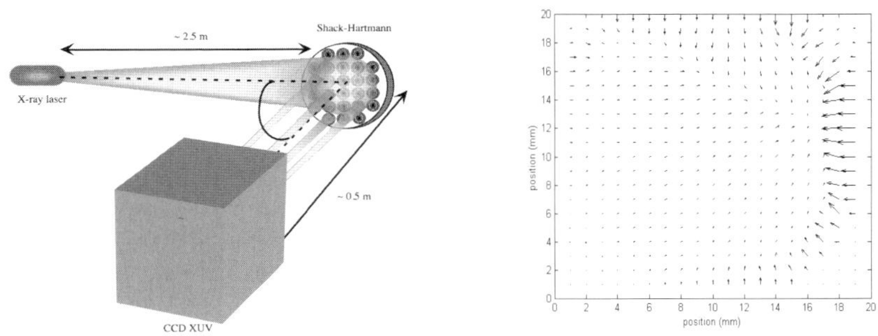

Figure 2. Wavefront measurement using Shack Hartman sensor

The 41.8 nm Xe^{8+} laser beam has been characterized by means of the Shack-Hartmann (SH) soft x-ray wavefront sensor[8]. This technique is based on splitting the incoming aberrated beam by using a lens array into many beamlets, assuming that each beamlet has a plane wavefront. Then after propagation, the positions of these beamlets are recorded on a soft X-ray CCD placed at the focal plane of the lenses. By comparing the position of each spot with the one obtained with a reference wavefront, one may access the wavefront of the aberrated beam. A reference wavefront (diffraction by a pinhole) is necessary for calibrating the sensor in order to suppress the aberrations generated by the wavefront sensor itself. The SH sensor was made using diffractive optic working in reflection (Bragg-Fresnel lenses).

Figure 2 shows the experimental setup used for the wavefront measurement with the Xe^{8+} x-ray laser, and a map of the difference between the x-ray laser wavefront and a perfect spherical wavefront. The reason for this operation is that the spherical aberration is known to have a dominant effect on the wavefront. It is seen that the wave vectors` norms is globally equal to zero and their orientation is homogeneous which indicates that the measured wavefront is almost perfectly spherical. Defects that appear on the beam margins are likely induced by the sensor substrate itself. We found that variation of pressure in the gas cell has almost no effect on the wavefront. This can be easily explained by the very low density of the amplifying medium (5×10^{17} cm^{-3}). In this case the gradient of electron density does not affect the propagation of soft x-ray laser beam. Considering the rather conservative accuracy of the wavefront sensor of about $\lambda/10$ at 42 nm and the nearly perfect spherical wavefront measured, one could speculate that the Xe^{8+} x-ray laser can be focused close to the diffraction limited spot. However, since the footprint measurement indicates highly inhomogeneous energy front, realistic prediction of focusing properties is not possible at the moment.

In order to increase the length of the amplifier and boost the XRL output, the plasma has been generated using longer focal length (2m instead of 1 m previously). As shown by the simulations presented in figure 3, a significant enhancement of the amplifier dimensions is expected in this geometry. Figure 4 presents the measured 41.8 nm signal as a function of the gas cell length using a 2 m focal length focusing optics. As expected from Figure 3, a dramatic enhancement of the XRL output is clearly observed for appropriate conditions (L_{cell}=8 mm). The lasing signal is up to 13 times higher that the one previously measured using f=1m focusing optics. While this geometry is a good alternative for boosting the output of the xenon laser, it was not appropriate to improve the krypton laser at 32.8 nm, the available driving laser intensity being insufficient to generate an elongated plasma column in this geometry.

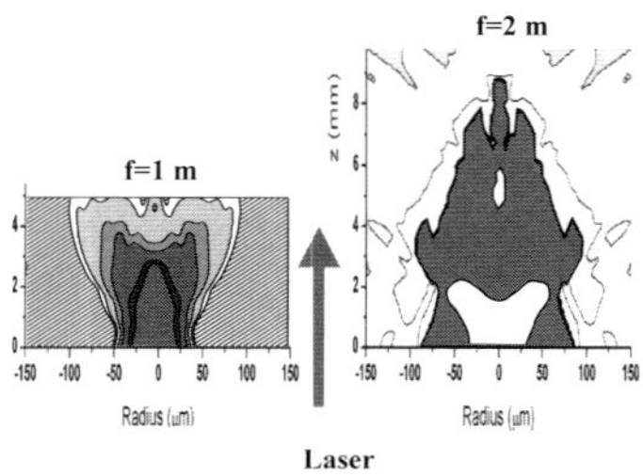

Figure 3. Calculated distribution of Xe charge states following the passage of the pump laser pulse for a 4 mm long (a) cell using 1 m focussing optics, and (b) 15 mm long cell using f=2 m focussing optics.

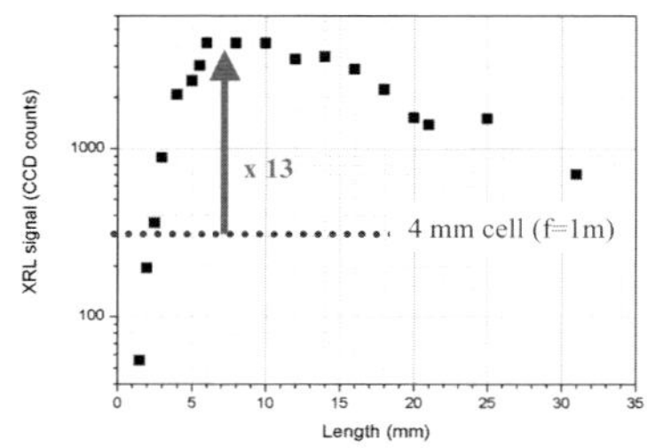

Figure 4. Measured Xe^{8+} laser signal as a function of the gas cell length using f=2 m focussing optic. To allow a comparison with earlier works, we also display typical signal obtained using f=1 m focussing optics and 4 mm gas cell

4. Lasing using waveguides

In order to extend the gain length that can be achieved by longitudinal pumping, it is necessary to guide the pump pulse over many Rayleigh lengths. Several techniques for guiding laser pulses with peak intensities relevant to OFI laser schemes, 10^{16} – 10^{18} W cm^{-2}, have been investigated. These include guiding in hollow capillaries [9], relativistic channelling [10], and several types of plasma waveguides [11-13] . Recently, we have demonstrated lasing at 41.8 nm in Xe^{8+} using a gas-filled capillary discharge waveguide [6].

4.1 Capillary discharge waveguides

The present work utilizes a gas-filled capillary discharge waveguide described in more details in these proceedings (See S. Hooker paper). For H_2-filled capillaries this approach has been shown to form an approximately fully-ionized, parabolic plasma channel. A plasma channel of this form is able to guide a Gaussian beam with a constant spot radius, provided that the spot radius is correctly matched to the curvature of the plasma profile and that the guided pulse does not further ionize the plasma.

Figure 5 shows the peak fluence transmission T_F of the pump laser pulses through the capillary and the normalized Xe^{8+} laser signal as a function of the delay t between the onset of the discharge pulse and the arrival of the pump laser. As shown, the soft x-ray laser signal is correlated to good guiding conditions and was very sensitive to the delay between the onset of the discharge current and the injection of the pump laser pulse.

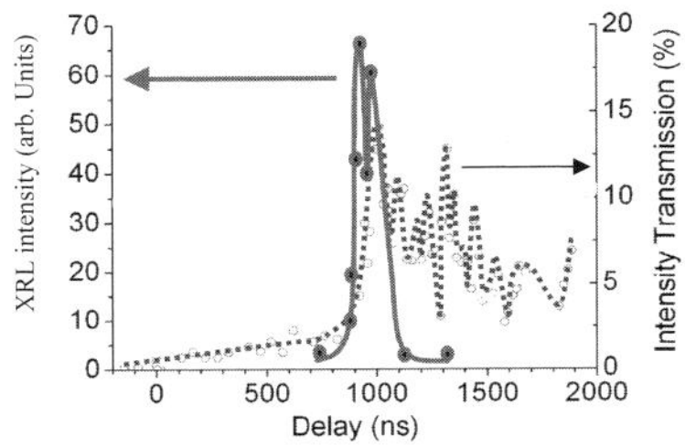

Figure 5. The measured fluence transmission, T_F (circles) as a function of delay t for a 30 mm long capillary filled with 50 mbar of a 1:3 Xe:H gas mixture. The Xe^{8+} laser signal is shown for the same conditions (squares), together with the measured temporal profile of the discharge current (plain line).

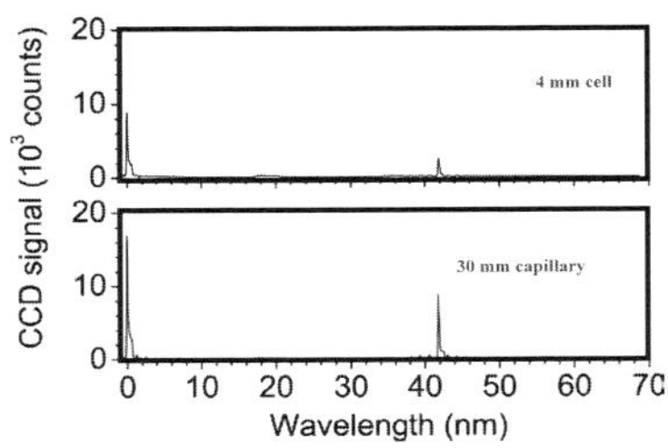

Figure 6 The measured XUV spectrum integrated over 10 pump laser shots for: (a) a 4 mm long gas cell containing 20 mbar of pure Xe; (b) a 30 mm long Xe/H-filled capillary discharge waveguide.

To allow a comparison with earlier works, we also investigated XUV lasing in a 4 mm Xe gas cell. Figure 6 shows the spectrum recorded with the gas cell containing pure Xe at the optimum pressure of 20 mbar. Figure 6 shows the spectrum recorded with a 30 mm long gas-filled capillary discharge waveguide for the optimum conditions: 1:3 Xe:H atom gas mixture, an initial capillary pressure of 120 mbar, and a delay t of 1000 ns. It is seen that strong lasing occurred at 41.8 nm in Xe^{8+} on the $4d^95d$ - $4d^95p$ transition for both the gas cell and waveguide, the signal for the waveguide being approximately 4 times larger. This promising result clearly shows that the performance of the OFI XRL could be significantly improved by guiding the high intensity driving pulse over a few centimetres [14].

4.2 Lasing in capillary tubes

We have investigated an OFI laser driven in a multi-mode gas-filled capillary, as described in detail by Mocek *et al.* in these proceedings. For this experiment 2m focal length off-axis spherical mirror has been used to produce the amplifying plasma. Figure 7 compares optimised output signals from the cell at 17 Torr (squares), and the capillary at 17 (solid triangles) and 30 Torr (open triangles) respectively, as a function of target length. The position of the vacuum focus was 5 (6) mm inside the cell (capillary). The data in Figure 7 were obtained with different filter sets which were cross calibrated using data obtained under nominally identical conditions. The solid curves present results of simulations which will be discussed later. The data clearly shows that the capillary greatly increases the lasing signal compared to that achieved with a gas cell. The strongest signal was observed with the 15 mm long capillary at 30 Torr, being approximately 7 times higher than the largest signal achieved with the cell of the same length, and as much as 43 times higher than that with the 4 mm cell.

To quantitatively evaluate the performance of the studied XUV laser, we undertook simplified modelling of the radiation transfer in the active medium. The model treats amplification of axial emission in an active medium which accounts for the refraction of the XUV laser beam. Results of simulations for the capillary and the cell are shown (solid lines), the normalization constant being the same for both simulations (that which gives the measured XUV laser signal from the capillary at 17 Torr). It is seen that the curve calculated for the cell agrees well with the experimental points for the 4 and 10 mm cells, and the two curves coincide for short lengths, as they should. The main discrepancy with the experimental results is that the simulated signal for the cell does not change after $z = 8$ mm since it is calculated that Xe^{8+} is not generated for values of z greater than this, whilst the measured signal for the 15 mm cell is about 2 times larger than that for the 10 mm cell. This effect can not be explained by the present modelling and might be attributed to differences in the gas densities outside the entrance and exit pinholes as the pinhole sizes are different in the two cases.

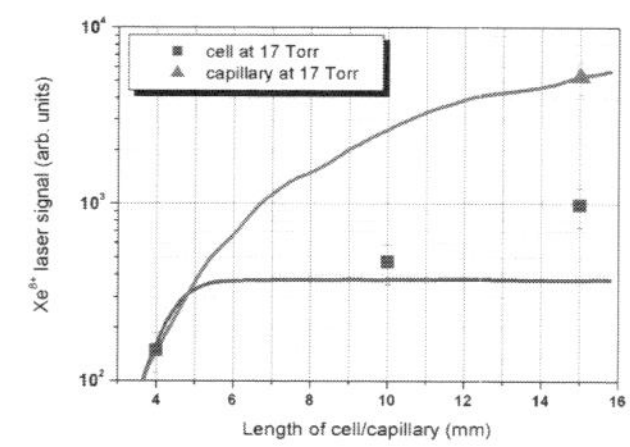

Figure 7. Measured and simulated (solid lines) Xe^{8+} laser signal from the capillary at 17 (solid triangle) and 30 Torr (open triangle), respectively, and the cell at 17 Torr (squares) as a function of the cell or capillary length.

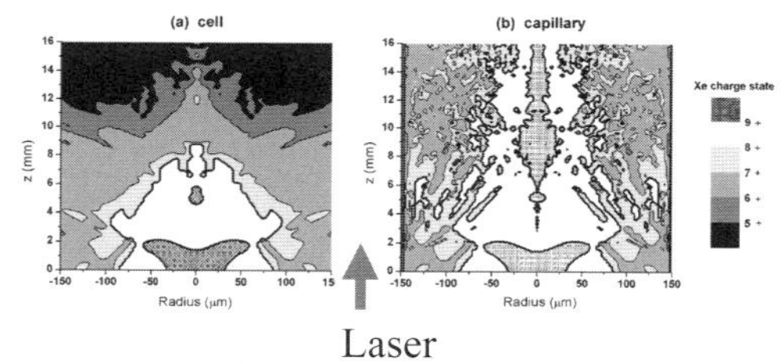

Figure 8. Calculated distribution of Xe charge states following the passage of the pump laser pulse for a 15 mm long (a) cell, and (b) capillary. For both calculations the position of the vacuum focus was 6 mm inside the target, and the pressure was 17 Torr.

Figure 8 shows the calculated distribution of Xe ion stages following the passage of the pump laser pulse through a 15 mm long cell (a) and capillary tube (b) under the conditions found experimentally to produce the largest Xe^{8+} laser signal, i.e. a pressure of 17 Torr and a focusing position of 6 mm inside the cell. It is seen that Xe^{8+} is produced along the whole length of the capillary in an annular region with higher ion stages being generated in an approximately cylindrical region of radius ~20 μm along most of the capillary axis. By comparing Figs. 8(a) and (b) it is clear that the volume of the Xe^{8+} region is increased substantially for the capillary. The size of the Xe^{8+} region was found to be much more sensitive to the focusing than in the case of the cell. For the focusing position of 6 mm inside the capillary the region of Xe^{8+} created was the largest, with much less Xe^{7+} than when focusing at the entrance.

5. Amplification of high order harmonics in an OFI plasma

Finally we have investigated the possibility to make a 2 stages XRL system. By combining the high optical quality of high harmonics as a seed, with an OFI x-ray laser plasma amplifier, we produced a highly saturated, energetic, femtosecond, fully coherent and fully polarised tabletop x-ray laser operating at 10 Hz. For more detailed information are given in this proceedings (see G. Faivre paper). This technique, easily applicable on all existing laser-driven XUV facilities, opens the way to ultra-high intensities studies worldwide.

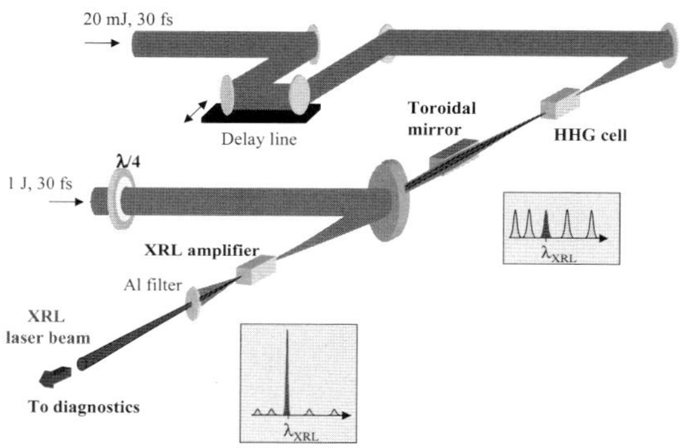

Figure 9. Schematic description of the experiment set-up of the two stages OFI x-ray laser. The seed beam was obtained by focusing a 20 mJ, 30 fs, infrared laser in a gas cell filled with argon[1]. This seed was image relayed onto the entrance of the XRL amplifier (gas longitudinally pumped by a 1 J, 30 fs laser) by means of toroidal XUV mirror.

A traditional laser chain consists in several stages: a seed oscillator, delivering a pulse of perfect optical properties, followed by amplifiers, set to increase beam energy. Extension of this scheme to the XUV range has been studied. The main difficulties found to be the production of optically perfect XUV seed of high energetic output, and coupling of this source with the amplifier, a strongly refractive laser-produced plasma. In our approach, we overcome these problems by, first, using state-of-the-art High Harmonic Generation (HHG) seed, second, focusing it down to increase the level of injection, reaching the energy extraction regime, and finally, using for the first time an optical field ionized plasma as X-ray laser (XRL) amplifying medium, of low density and thus little refraction. The seed beam was obtained by focusing a 20 mJ, 30 fs, infrared laser in a gas cell filled with argon[1]. This seed was image relayed onto the entrance of the XRL amplifier (Kr gas longitudinally pumped by a 1 J, 30 fs laser) by means of toroidal XUV mirror. This XRL operates on the $3d^94d_{J=0}\rightarrow3d^94p_{J=1}$ transition of the Kr^{8+} ion at 32.8 nm, which has a close match with the 25th harmonic of the IR laser. After amplification, the output beam was either analyzed by an XUV spectrometer or a monochromatic cross-section was recorded (see methods section).

Recorded emission spectra are displayed in figure 10 corresponding to (a) high harmonics alone, (b) XRL amplified spontaneous emission (ASE), (c) XRL with HHG seeding long after gain extinction, and (d) HHG synchronised with maximum XRL gain. The amplification period starts 5 ps after the interaction of the IR laser with the gas and lasts about 8 ps. Figure 10 (d) clearly shows that strong amplification of the seed has been achieved. In this case, HHG were seeded at about twice the intensity of the ASE of the XRL, leading to an enhancement of the output signal by a factor of 13. As presented in these proceedings by G. Faivre *et al.* the amplified beam as been characterized in details. This new type of x-ray laser deliver up to 1 μJ for a pulse duration which should be less than 1 ps. The beam conserves the coherence and the polarization of the seed harmonic beam.

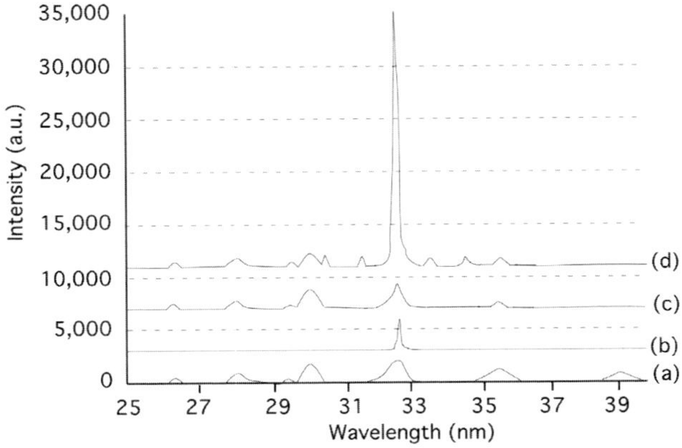

Figure 10. Experimental spectra under conditions : (a) HHG only, (b) ASE XRL only, (c) both HHG and XRL in a non-amplifying timing, and (d) the amplified seeded XRL. Each spectrum corresponds to signal accumulated over 10 laser shots, in order to reduce the shot-to-shot error.

Future work will be to seed an XRL plasma created from a solid target such as to increase the output energy by at least three order of magnitudes (~ 1mJ) thanks to the much higher density of XRL amplifying medium. As this technique matures, future compact XUV laser chains should achieve beams performances close to those of VUV Free-electron lasers.

References

[1] B.E. Lemoff et al. C.L. Gordon III, *Phys. Rev. Lett.*, vol. 74, pp. 1574, 1995.
[2] B.E. Lemoff, et al. *Opt. Lett.*, vol. **19**, pp. 569, 1994.
[3] S. Sebban et al. *Phys. Rev. Lett.*, vol. 86, pp. 3004, 2001.
[4] S. Sebban et al. *Phys. Rev. Lett.*, 2002.
[5] D J Spence et al. *J. Phys. B*, vol. 34, pp. 4103-4112, 2001.
[6] A. Butler et al. *Phys. Rev. Lett.*, vol. 91, pp. 205001, 2003.
[7] M. Pitmann et al.
[8] T. Mocek et al. *Appl. Phys. B*, vol. 00, pp. 1-6, 2004.
[9] F. Dorchies et al. Phys. Rev. Lett.. 82, pp. 4655, 1999.
[10] P. Monot et al. *Phys. Rev. Lett.* **74**, pp. 2953–2956 (1995).
[11] T. Hosokai et al *Opt. Lett.*, vol. 25, pp. 10, 2000.
[12] Y. Ehrlich, et al. *Phys. Rev. Lett.*, vol. 77, pp. 4186–4189, 1996.
[13] G. Durfee and a. H. M. Milchberg, *Phys. Rev. Lett.*, vol. 71, pp. 2409–2412, 1993.
[14] A. Butler, D. J. Spence, and a. S. M. Hooker, *Phys. Rev Lett*, vol. 89, pp. 185003, 2002.

Inst. Phys. Conf. Ser. No 186
Paper presented at 9th Int. Conf. X-Ray Lasers, Beijing, China, 24–28 May 2004

Recent progress in the study of high-order harmonic generation and x-ray laser drivers at SIOM

Zhizhan Xu and Ruxin Li

Laboratory for High Intensity Optics, Shanghai Institute of Optics and Fine Mechanics (SIOM), Chinese Academy of Sciences, P.O. Box 800-211, Shanghai 201800, China.

Abstract. Recent progress in the study of high order harmonic generation and x-ray laser drivers at SIOM is reviewed. We first summarize the recent experimental and theoretical results of high-order harmonic generation at SIOM, which include the phase-matched high-order harmonic generation in atoms, the role of carrier envelope phase in the generation and measurement of atto-second high-order harmonic pulses, and the proposal of high energy atto-second pulse generation by specially designed driving laser pulses. The development of high-intensity ultra-short-pulse laser facilities as x-ray laser drivers at SIOM will be reported. A 120TW/36fs Ti:sapphire laser facility based on the scheme of chirped pulse amplification and a 16.7TW/120fs laser system based on the scheme of optical parametric chirped pulse amplification have been developed. We will also outline the femtosecond petawatt laser project in Shanghai.

1. Introduction

There are several possible approaches to obtain coherent radiation in the spectral region of soft-x-ray. Lasing based on the population inversion in soft-x-ray region [1] and the frequency up-conversion based high-order nonlinear process are two mechanisms which have been investigated intensively. As one example of the latter case, high-order harmonics (HOH) [2] emitted via atoms subjected to a strong laser field have good coherence and tuning ability, work at high repetition rate with a pulse duration down to sub-femtosecond, can be pumped by less than 1mJ laser energy in a femtosecond pulse and is therefore a very compact light source. These properties made HOH ideal for many applications such as the light source for the pump-probe measurement of ultrafast dynamics of materials and the metrology for the optics used in EUV lithographic system. With the development of the extremely short pulse intense laser, HOH have been pushed down to the water-window in spectral region [3] and the subfemtosecond level in time domain [4]. However, because of the phase mismatching in the nonlinear frequency conversion process, the conversion efficiency of HOH from the driving laser is still very low. Phase matched HOH generation is therefore strongly desirable for developing intense soft-x-ray coherent sources [5-7].

In the past years, HOH generation in different gas species, gas density, laser focusing geometry, laser intensity, laser profile, laser chirp conditions and in different gas cells or hollow fibers have been studied systematically, and the optimal conditions to realize phase-matching were found (For a review, see Ref. [5] for example). The highest HOH conversion efficiency is 6.4×10^{-5} for the 15th harmonic generated in xenon gas-filled hollow fiber which was driven by a 35fs, 16mJ, 810nm pulse at 10Hz repetition rate [6]. Besides the phased matched HOH generation in a hollow fiber, those using static gas cells are also attractive due to the simplicity in the experimental setup [5].
The attosecond pulses is another hot topic of HOH generation pursued worldwide. A number of methods have been proposed for attosecond pulse generation [4,5]. Pulse duration down to 650as was achieved at 90eV photon energy [4]. However, the attosecond soft-x-ray pulses are currently too weak to induce measurable non-linearity of atomic media. To produce an attosecond pulse with higher energy is strongly desired for expanding its application.
To realize the lasing at shorter than 2nm by the x-ray laser mechanisms is still a big challenge. Less developed schemes such as inner-shell photoionization pumped mechanism need to be explored, which require new x-ray laser driver at unprecedented parameter region of high than 1PW in power and less than 20 fs in pulse width [8].
In this paper we report the recent progress in the study of high order harmonic generation and x-ray laser drivers at SIOM. We have investigated the phase-matched high-order harmonic generation in argon atoms in a cell, the role of carrier envelope phase in the generation and measurement of atto-second high-order harmonic pulses, and the production of high energy atto-second pulses by specially designed driving laser pulses. We developed a 16.7TW/120fs laser system based on the scheme of optical parametric chirped pulse amplification. We are pursuing the development of a femtosecond petawatt laser and a 120TW/36fs Ti:sapphire laser facility was recently developed as a first step.

2. Phase-matched high-order harmonic generation from argon atoms in a cell

A commercially available all-solid-state Ti: sapphire femtosecond laser system (Spitfire 50, Spectra Physics Co. Ltd.) which delivers 0.55mJ laser energy at 800nm in a 50fs (FWHM) pulse with 1 KHz repetition rate, was used as the HOH driver. The laser beam was focused with a fused silica lens (f/34) and the corresponding laser intensity at the focus is about 2×10^{14} W/cm^2. The laser beam ablated a 0.1-mm-thick stainless steel sealing the glass cell filled with noble gases. The co-propagating HOH and the driving laser beams hit two 400nm thick aluminium foils before entering a stigmatic grazing-incidence flat-field grating spectrograph equipped with a back-illuminated thinned soft-x-ray CCD. One dimensional spatially resolved HOH spectra were recorded. The spectrometer was calibrated with the absorption edge of aluminium and the HOH lines of neon atoms. A typical spectra for the 23rd to 29th HOH generated in 44Torr argon gas are shown in Fig.1. The highest conservation efficiency for the 27th HOH is about 1×10^{-7}. The laser energy used in this work, i.e. 0.55mJ per pulse is the lowest laser energy used to obtain phase-matched HOH, to the best of our knowledge [9].

We found that, for the first time, under the phase matching conditions, the HOH intensity distribution in space behaves different from the Gaussian-like distribution. A flat-top distribution was observed, which is quite different from the experimental results of phased matched HOH in hollow fibers and that in gas cells driven by loosely focused (f/71) laser beam. The reason for this phenomena is mainly the absorption effect and the self-guiding of the driving laser beam in gas cell.

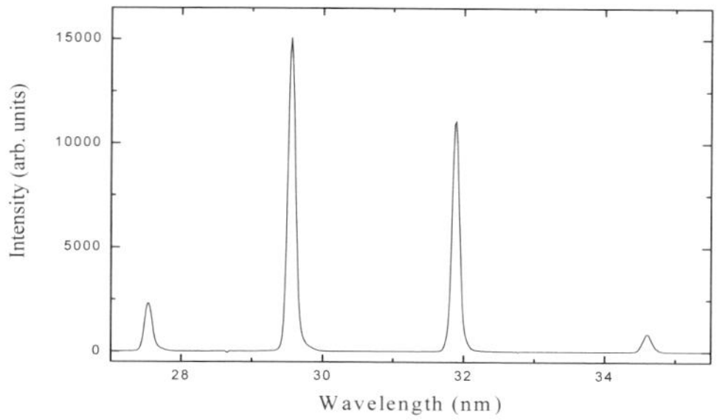

Figure 1. Typical HOH spectra for 44Torr Ar gas, the 23rd to 29th order HOH are shown from right to left.

3. The role of carrier envelope phase in the generation and measurement of atto-second high-order harmonic pulses

We reinvestigated the role of carrier envelope phase (CEP) in the generation and the cross correlation measurement of the atto-second HOH driven by few cycle laser pulses[10]. We calculated the dipole acceleration by solving numerically the time-dependent Schrödinger equation (TDSE). The electric field in the calculation is described by $E(t) = E_0(t)\cos(\omega_L t + \varphi_L)$. ω_L is the frequency of the laser field and φ_L is the CEP. $E_0(t) = E_0 e^{-t^2/\tau^2}$, E_0 is the peak amplitude of the laser field and the duration of the laser pulse is $\sqrt{2\ln 2}\tau$. In the calculation, the wavelength of the laser pulse is 800nm. The dipole acceleration is "Fourier transformed" to get the HOH spectra and a time-frequency analysis (Gabor analysis) generates the time profile of the harmonics. The calculation was performed for helium as model atom whose ionisation potential is about 0.9031a.u. (about 24.6eV). The driving laser pulse is 7fs in width and 3.51×10^{14}W/cm^2 in peak intensity. The HOH emission between 94eV and 100eV which is above the cutoff defined by $I_P + 3.17U_p$ was selected for the analysis of the time profile.

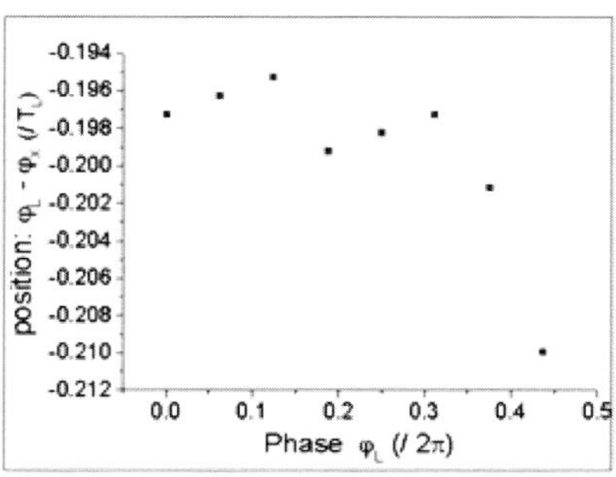

Figure 2. The time position difference between the attosecond HOH pulse peak and the laser field peak as a function of CEP.

Calculations showed that there is a timing jitter between the attosecond HOH pulse peak and the laser field peak when the CEP changes. In Fig.2 we show the time position difference between the attosecond HOH pulse peak and the laser field peak as a function of CEP and the difference is found to be less than $0.018T_L$ (T_L is laser period), i.e. about

48 attoseconds for 800nm laser. This timing jitter set a lower limit of the resolution of the measurement of the attosecond pulse width by the cross correlation between the attosecond pulse and its driving laser field.

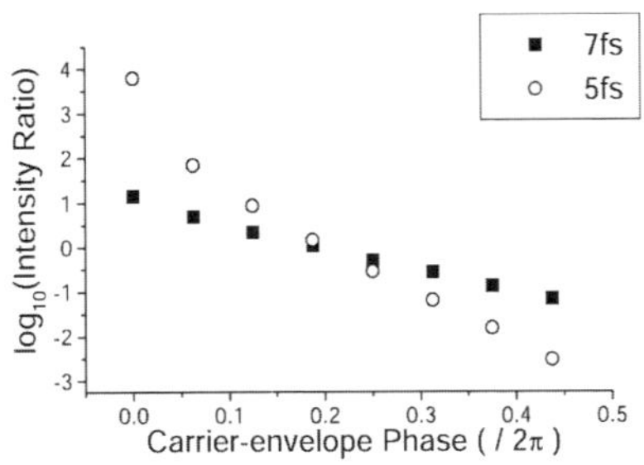

Figure 3. The intensity ratio of the two adjacent attosecond pulses as an exponential function of the CEP.

Calculations also show that, with the 7fs driving laser pulse, two attosecond pulses with a separation of half cycle of the driving pulse are generated with different intensity ratio for different CEP. The intensity ratio is an exponential function of the CEP as shown in Fig.3 for both 7fs and 5fs driving laser pulses. This phenomena suggested that one can in-situ measure the CEP of the laser pulse by monitoring the spatial distribution of the photoelectron generated by the superposition of the attosecond pulse and the laser pulse.

4. High energy atto-second pulses generation by synthesized driving laser pulse

In current experiments of attosecond pulse generation, the few cycle driving laser pulse (typically <1mJ/5fs) was obtained by the post-amplification spectral broadening technique in which the hollow-fiber chirped-mirror high-power compressor was used [5]. However, the energy of the few-cycle laser pulse produced by this method is presently limited in the level of 1mJ due to the ionization of the gas in the hollow fiber, which results in the limitation in the energy and the intensity of attosecond HOH x-ray pulses.

We propose a scheme of generating high-intensity attosecond HOH by using a synthesized laser field. The synthesized laser field is obtained by an appropriate superposition of a few-cycle laser pulse and a relatively long pulse of several tens of femtoseconds whose energy can be increased to the level of several tens of mJ. The energy limitation in the driving laser pulse is therefore overcome. Calculated results show that, the intensity of attosecond HOH pulse in helium driven by the synthesized laser field with a 0.05 atomic unit (a.u.)/5fs laser pulse and a 0.10a.u./50fs laser pulse is several orders higher than that driven by a single 0.05a.u./5fs laser pulse, and it is even stronger than that driven by a single 0.15a.u./5fs laser pulse, although the peak intensities of the driving laser pulses are the same [9].

The numerical calculation approach is similar to those in last section. He atom was selected. The attosecond x-ray pulses are produced in the cutoff regime of HOH generation and the photon energy is 174eV. The CEP and the time delay between two pulses are zero for the synthesized pulse case. Fig.4 shows the typical result of the attosecond pulse generated by synthesized pulse in comparison with the single pulse case. The laser parameters used in the calculation are 0.15a.u./5fs (7.9×10^{14}W/cm^2/5fs) for the single laser pulse case and 0.05a.u./5fs + 0.10a.u./50fs (8.8×10^{13}W/cm^2 /5fs + 3.51×10^{14}W/cm^2 /50fs) for the synthesized pulse case.

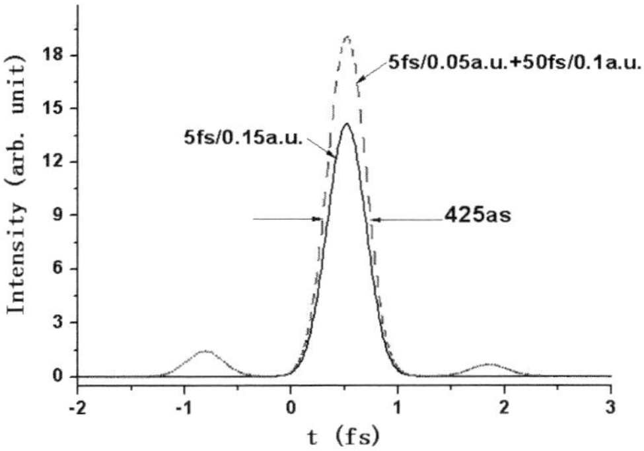

Figure 4. The temporal profiles of the attosecond pulses produced by a single 5fs laser pulse and a synthesized laser pulse of 50fs and 5fs pulse pair.

5. Development of x-ray laser drivers

To explore the undeveloped mechanism for short wavelength x-ray lasers and stimulated by the high-field laser physics research, we are pursuing the development of ultra-high-intensity ultra-short-pulse laser facility. We have launched a femtosecond petawatt laser project recently. The schematic layout of the designed laser system is shown in Fig.5.

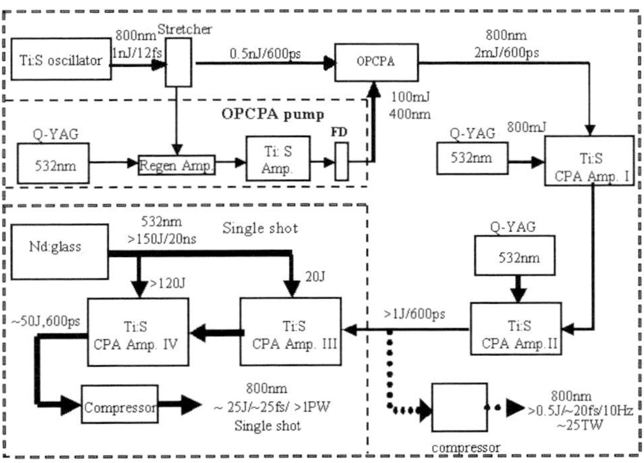

Figure 5. The schematic layout of the 25J/25fs laser system under development.

The output of the petawatt laser system will be 25J/25fs. The facility consists of one optical parametric chirped pulse amplifier (OPCPA) using BBO or LBO crystal as the pre-amplifier and four stages of chirped pulse amplifier (CPA) using Ti:sapphire as the gain media. The use of OPCPA as the pre-amplifier will offer high gain without gain narrowing and promise high contrast ratio.

A unique feature of the design is the OPCPA is pumped by a frequency-doubled spectra-shaped Ti:sapphire laser seeded by the same oscillator as that of the main system, differs from other systems where the OPCPA is pumped by a separate laser system. The new design will reduce the fluctuation of the output. This OPCPA scheme was tested recently and positive results were obtained.

The power amplifiers except the final large aperture Ti:sapphire amplifier were also constructed and tested, an output of 120TW/36fs of the system using a 24fs Ti:sapphire self-mode-locking oscillator as seed was obtained, the last amplifier of which delivers laser energy up to 5.98J with a saturated gain of 20. This is among the highest gain power amplifiers using Ti: sapphire demonstrated up to now.
In addition, OPCPA offers the possibility of high average power, which is attractive for driving X-ray lasers. We have developed a pure OPCPA laser system [11] by using three stages of LBO crystal based OPCPA, an output of 16.7TW/120fs was obtained and this is the highest output power for an OPCPA laser system.

6. Conclusion

Phased matched 27th order harmonic (30nm) driven by 1 KHz 0.55mJ/50fs laser was demonstrated. This is the lowest energy used to obtain phase matched high-order harmonic radiation. Intensity-profile flattening in the space domain for the HOH under phase matching condition was found for the first time. The role of CEP in HOH generation was reinvestigated. We found that the attosecond soft-x-ray pulse is locked to the field of its driving laser with a small timing jitter, which limit the precision of cross-correlation measurement. We proposed a novel method to detect the CEP of the driving laser field by measuring the spatial distribution of the photoelectrons induced by the attosecond soft-x-ray pulse and the driving laser field. We proposed a method to generate high-energy attosecond pulse by a synthesized driving laser field. This new scheme for attosecond pulse generation is possible to pave the way for the realization of high intensity attosecond pulse which can be used in attosecond nonlinear spectroscopy. A 16.7TW/120fs laser system based on the scheme of optical parametric chirped pulse amplification was developed. We have started to develop a femtosecond PW laser system. A 120TW/36fs Ti: sapphire laser facility was developed to test components as a first step. New synchronized pump scheme was used to develop OPCPA pre-amplifier for the PW laser facility.

7. Acknowledgements

We thank the invaluable contribution of the research team with Laboratory for High-Intensity Optics, SIOM, CAS, especially to Drs. Zhinan Zeng, Yuxing Leng and Xiaoyan Liang, Profs. Lihuang Lin and Zhengquan Zhang, Mrs. Xinhua Xie and Haihe Lu. This work is partially supported by the Chinese Academy of Sciences (Grant No.KGCX2-SW-114), Chinese Ministry of Science and Technology (Grant No.G1999075200), Chinese National Natural Science Foundation (Grant Nos.199974058 and 69925513).

References

[1] See, for example, Elton R.C. 1990 X-ray lasers (Academic Press Inc., San Diego)
[2] Mcpherson A. et al. 1987 J. Opt. Soc. Am. B4, 595
[3] Spielmann Ch. et al. 1997 Science 278, 661
[4] Drescher M. et al. 2002 Nature 419, 803
[5] Brabec T, Krausz F 2000 Rev. Mod. Phys. 72, 545
[6] Takahashi E, Nabekawa Y, Midorikawa K 2002 Opt. Lett. 27, 1920
[7] Nam C. et al. 2004 in this proceedings, Midorikawa K. et al. 2004 in this proceedings
[8] See, for example, Key M. 1996 IOP Conference Series Num. 151
[9] Li R. et al. 2004 J. Korean Phys. Society 44, 684
[10] Zeng Z. et al. 2003 Physical Review A 67, 013815
[11] Yang X. et al. 2002 Optics Letters 27, 1135

Inst. Phys. Conf. Ser. No 186
Paper presented at 9th Int. Conf. X-Ray Lasers, Beijing, China, 24–28 May 2004

High quality x-ray laser development at APRC

M. Kishimoto, M. Tanaka, M. Nishikino, T. Kawachi, M. Kado, K. Sukegawa, N. Hasegawa, Y. Ochi, R. Z. Tai, H. Tang, K. Nagashima, A. Nagashima, Y. Kato

X-ray Laser Research Group, Advanced Photon Research Center,
Japan Atomic Energy Research Institute

Abstract. We have demonstrated the 13.9nm soft x-ray laser with full spatial coherence using oscillator-amplifier configuration with two gain media. We have obtained the highly directional x-ray laser beam with a beam divergence of 0.2mrad and confirmed the full spatial coherence by means of Young's double slit interferometry. The gain medium of the second plasma is generated in a low density area for avoiding the influence of refraction, where the electron density is about 1% of the critical density.

1. Introduction

Since the demonstration of x-ray laser in a soft x-ray region in 1984 [1], significant advantages in the soft x-ray laser development have been made [2-4]. Especially, the x-ray laser using transient collisional excitation (TCE) scheme [5] has advantages such as high gain coefficient of a few tens per centimeter, short pulse duration less than 10 ps, and high brilliance, and has been studied as an advanced light source for application research at various laboratories [6-8]. On the other hand, applications such as x-ray interferometry, x-ray holography, and x-ray speckle motivate the development of high quality x-ray laser. However, the properties of the x-ray beam obtained by TCE scheme like beam divergence and spatial coherence have remained quite low, which results from the refraction due to the density gradient in the gain medium plasma and the large Fresnel number due to the high gain [7,8]. Because the small divergence beam has high spatial coherence, it is needed to achieve the beam divergence below the milliradian level to use the x-ray laser for application research.

We have started the x-ray laser research using TCE scheme at Advanced Photon Research Center, Japan Atomic Energy Research Institute since 1998 and demonstrated of lasing in the range from 8.8 to 46.9 nm, in particular succeeded saturated amplification at 12.0 and 13.9 nm [11-14]. Now we are developing compact and high repetitive x-ray lasers with full spatial coherence for applications. In this article, we report the demonstration of full spatial coherence 13.9nm soft x-ray laser using oscillator-amplifier configuration with two gain media, which is called the double target x-ray laser [13, 14].

2. Double target experiment

A schematic diagram of the experimental setup for the double target x-ray laser is shown in Fig.1. The 1st target is illuminated with a pump beam with the traveling wave and a seed x-ray beam is generated. Because the x-ray beam is bent by refraction in the gain medium plasma, the 1st target is rotated so that the seed x-ray can propagate exactly parallel to the second target surface. A coherent part of the seed x-ray is injected into the line-shape plasma generated on the 2nd target by illuminating with a pump beam without the traveling wave. The injection timing of the seed x-ray can be manipulated by the optical delay line of the pump laser. The second target plasma works as both an active spatial filter and an amplifier, and a highly directional x-ray beam is obtained. The generated x-ray beam is reflected with a Mo/Si multilayer mirror and the far-field pattern is taken with a soft x-ray CCD. The target material used in the double target experiment is silver and the wavelength of the x-ray laser is 13.9 nm.

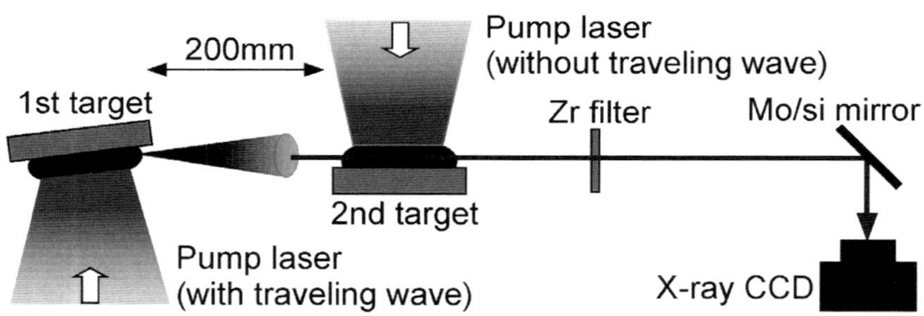

Figure 1 Experimental setup for double target x-ray laser

3. Generation of highly directional x-ray beam

Figure 2 shows the far-field patterns for various seeding timings. Here we make the time of 667ps as the nominal timing, which is the required time for x-ray propagated for the distance between the 1st and 2nd target. We have measured how far-field pattern changes before and after the nominal timing. Figure 2 shows far-field patterns for the seeding timings of 652 (=667-15), 667 and 682 (667+15) ps. In the figure, the dash line and the cross-hatched pattern are the second target surface and a shadow of the Zr filter mesh, respectively. At -15ps, only the shadow of the second target plasma exits and the amplification of the seed x-ray is not observed. At the nominal timing, an intense x-ray beam which is spread in the perpendicular direction to the second target surface appears outside the plasma shadow. This means that the seed x-ray is amplified in the high density region and the x-ray beam is bent by refraction. When the seeding timing is adjusted to be +15ps, an intense, narrow x-ray beam appears near the top of the plasma shadow. The beam pattern of the intense, narrow x-ray beam obtained is close to circular with a nearly Gaussian distribution and the beam divergence of it is 0.2mrad.

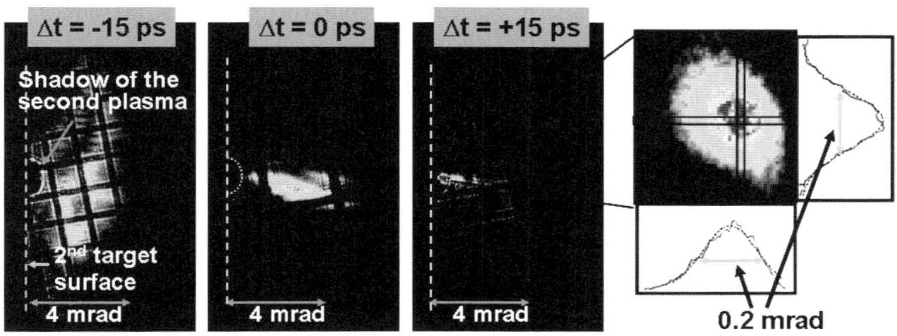

Figure 2 Far-field patterns of the double target x-ray laser for various seeding timings and the blow up and the intensity profiles of the intense, narrow x-ray beam.

4. Amplification property of the double target x-ray laser

Figure 3 shows the amplification property of the double target x-ray laser as a function of gain medium length at the second target. One can see that the output energy of the double target x-ray laser increases with the gain medium length exponentially. The gain coefficient obtained is about 8 cm^{-1} and the maximum gain is about 170 times at the gain medium length of 6.5mm.

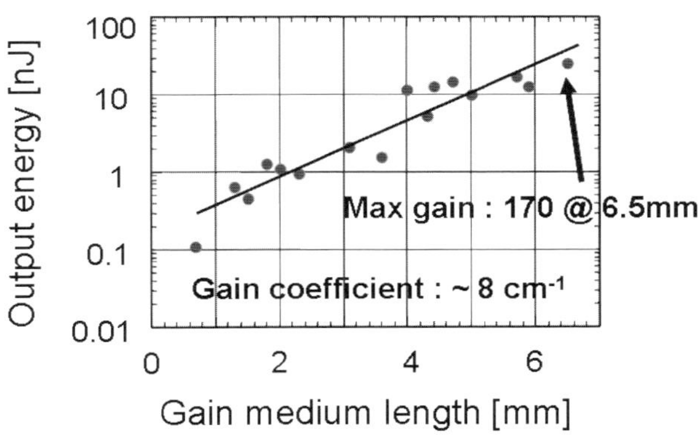

Figure 3 Amplification property of the double target x-ray laser

To summarize the x-ray amplification properties of the double target x-ray laser, one can see that the gain region of the second target plasma is generated in a low density area for avoiding the influence of refraction. Here we have evaluated the electron density at the gain region of the second target plasma. Assuming that the electron density profile is represented by the following equation:

$$n_e = n_{cr} \exp\left(-\frac{x}{L_n}\right), \tag{1}$$

where N_{cr} and L_n are the critical density and the scale length, respectively, we can write the refraction angle as:

$$\theta = \frac{\nabla n}{n} L = \left(\frac{13.9}{1053}\right)^2 \frac{L}{2L_n} \exp\left(-\frac{x}{L_n}\right). \tag{2}$$

Figure 4 shows the relations between the scale length and the refraction angle for various gain positions calculated from the previous equation. Since the position of the gain region roughly evaluated from the far-field patterns is about 100 μm and the refraction angle of the x-ray beam is less than the beam divergence of 0.2mrad, the scale length is estimated to be about 20mm from Figure 4. Therefore, the electron density at the gain region of the second target plasma is estimated to be about 1% of the critical density ($\sim 10^{-19}$).

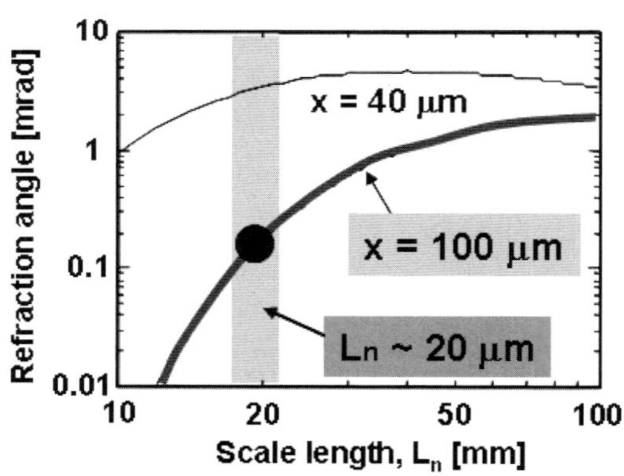

Figure 4 Relations between the scale length and the refraction angle for various gain positions

5. Spatial coherence measurement of the double target x-ray laser

We have conducted the spatial coherence measurement by means of Young's double-slit interferometry. Figure 5 shows the experimental setup. A double-slit is located 2.3m away from the second target and an interference pattern is taken with an x-ray CCD located 3.9m downstream from the double slit. The x-ray beam size at the double slit is about 460 μm. Visibility is calculated from the interference pattern and spatial coherence length is evaluated from the visibility data. The double slits consisted of a pair of 16 μm width slits separated by either 150, 200, 30 or350 μm are prepared. The double slit is placed in either the vertical and horizontal direction to estimate the spatial coherence in both directions.

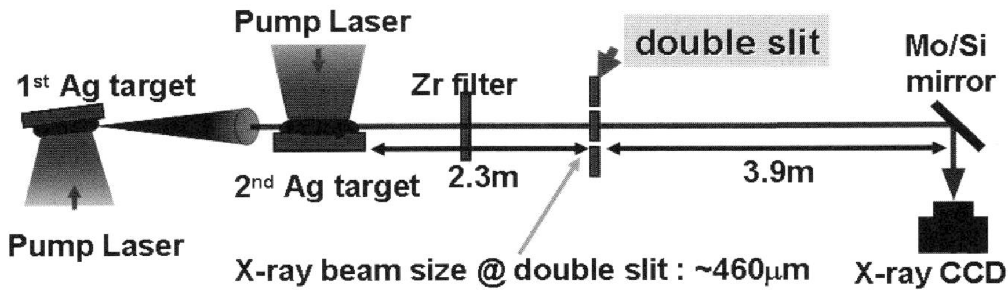

Figure 5 Experimental set up for the spatial coherence measurement using Young's double slit interferometry

Figure 6 shows the intensity profiles of the interference patterns experimentally obtained and the visibilities for the vertical and horizontal direction for various slit separations. From the figure, one can see that visibilities for the vertical and horizontal directions are almost same. This means that the double target x-ray laser beam is

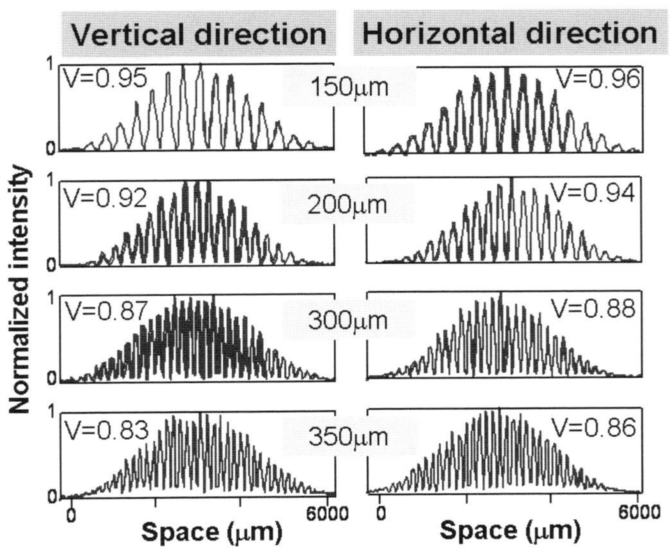

Figure 6 Dependence of visibility on slit the separation in vertical and horizontal direction

isotropic from the view point of spatial coherence.

To evaluate the spatial coherence length from the visibility data, we have employed the Gaussian shell model [15]. According to this model, the relation between the spatial coherence length and the visibility is represented as follows:

$$V(\Delta x) = \exp\left(-\frac{\Delta x^2}{2L_c^2}\right), \qquad (3)$$

where V(Δx), Δx and L_c are visibility, slit separation and spatial coherence length, respectively.

Figure 7 shows the comparisons between the experimental data and the visibility curves calculated from the previous equation. The solid and the dash lines are the visibility curves for a spatial coherence length of 640mm and 580mm, respectively. One can see from the figure that the experimental data are fit well with the visibility curves. The spatial coherence length of the double target x-ray laser evaluated from the calculated results shown in the figure is about 600mm, which is larger than the x-ray beam size at the double slit. This means that the double target x-ray laser is full spatial coherence.

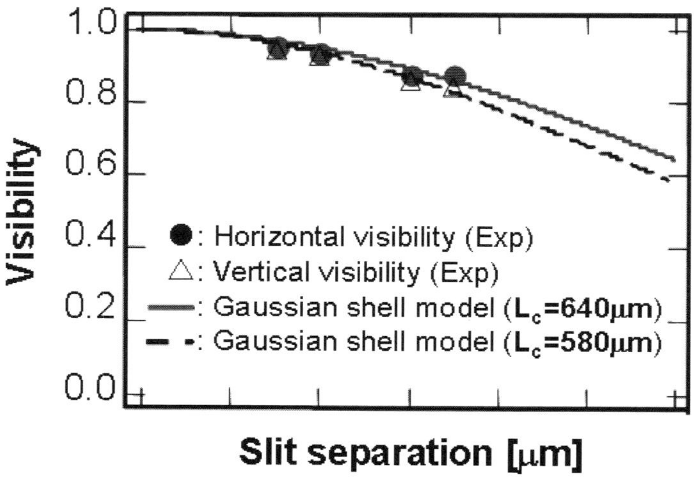

Figure 7 Comparisons between the visibility data experimentally obtained and the visibility curves calculated from the Gaussian shell model..

6. Conclusions

We have demonstrated the 13.9nm soft x-ray laser with full spatial coherence using two gain media. The highly directional x-ray laser beam with a beam divergence of 0.2mrad, which is close to the diffraction limited value within a factor of two [13], has been obtained. The gain medium of the amplifier plasma on the second target is generated in a low density area, where the electron density is about 1% of the critical density, for avoiding the influence of refraction. It is possible to generate an x-ray laser beam with saturated intensity by making the gain medium at the second target longer and using the traveling wave pumping.

References

[1] D. L. Matthews et al.., Phys. Rev. Lett. 54, 100 (1985)
[2] A. Carillon et al., Phys. Rev. Lett. 68, 2917 (1992)
[3] S. Basu et al., Appl. Phys. B 57, 303 (1993)
[4] B. E. Lemoff et al., Phys. Rev. Lett. 74, 1574 (1995)
[5] V. N. Shlyaptsev et al., SPIE 2012, 212 (1993)
[6] P. V. Nickles et al., Phys. Rev. Lett. 78, 2748 (1997)
[7] J. Dunn et al., Phys. Rev. Lett. 80, 2825 (1998)
[8] M. P. Kalachnikov et al., Phys. Rev. A 57, 4778 (1998)
[9] H. Daido et al., Opt. Lett. 20, 61 (1995)
[10] J. Zhang et al., Phys. Rev. Lett. 78, 3856 (1997)
[11] T. Kawachi et al., Appl. Opt. 42, 2198 (2003)
[12] T. Kawachi et al., Phys. Rev. A 66 ,033815 (2002)
[13] M. Tanaka et al., Opt. Lett. 28, 1680 (2003)
[14] M. Nishikino et al., Phys. Rev. A 68, 061802(R) (2003)
[15] M. Born and E. Wolf, Principles of Optics (Wiley, New York) 1984

Collisionally pumped X-ray lasers

Inst. Phys. Conf. Ser. No 186
Paper presented at 9th Int. Conf. X-Ray Lasers, Beijing, China, 24–28 May 2004

Low-energy-single-laser-pulse driven soft X-ray lasers

K A Janulewicz, G Priebe, J Tümmler, R E King[a], G J Pert[a], P V Nickles

Max Born Institute, Max-Born-Str. 2A, D-12489 Berlin, Germany

[a] Department of Physics, University of York, Heslington, York YO10 5DD, UK

Abstract. The pulse shape suitable to produce and pump efficiently a plasma active medium from a metal slab target is discussed. Starting from the experimentally determined pulse shape some of the mechanisms responsible for the energy deposition are discussed. The experimental results are compared with the numerical simulations. Some new effects as plasma re-heating observed in the modelling are well supported by the experimental data.

1. Introduction

Recently an enormous progress has been observed in the development of compact, so called table-top soft X-ray lasers (XRLs) [1-3]. The demonstrated pump schemes of these lasers need the pump energy down to 1 J to emit spectra dominated by the lasing lines. In the laterally irradiated slab target architecture strong reduction in the pump energy was observed in both double-pulse and single pulse arrangements [1,2]. Very recently the arrangement with a skewed incidence (**GR**azing **I**ncidence **P**umping) of the main picosecond pulse on the preformed plasma has been successfully demonstrated [4,5]. This scheme is out the scope of this paper and will not be discussed further.

It was shown that within the pumping scheme applying only one picosecond laser pulse saturation of the XRL emission is possible with the energy of about 3 Joules [2]. This gave a very high figure of merit (defined as a ratio of the gain-length product to the whole applied pump energy) equal to 5 J^{-1} that has been improved only within the GRIP pump arrangement. This result was ascribed to a specific profile of the pump laser pulse combining low-level background and the main picosecond pulse. The low-level background fulfils the role of the pre-pulse in the traditional double-pulse arrangement – it creates the plasma. This plasma is, however, not fully prepared to be heated, as its average ionisation stage is relatively low. Actually the ablation process will be determined by the main pulse. For this reason it is very important from practical point of view that the main pulse is elongated with duration time between 4 and 12 ps. This should give more time for the collision processes during the main part of the pump pulse and increase the ionisation state before heating becomes the dominating process.

The energy deposition and redistribution in such a scheme is one of the most important questions, which should be answered to understand how the system works [6].
In this paper we explain the influence of the low-level structure of the pump laser pulse on the energy deposition and the same on plasma dynamics and kinetics. Parameters of well-characterised X-ray laser in Ni-like silver have been used as an experimental reference for the simulations performed with the numerical hydrodynamics/atomic physics code EHYBRID. The results can easily be transferred to other moderate Z elements as tin, palladium or molybdenum. Some additional measurements of the XRL beam parameters gave look at the application prospects for such sources.

2. Laser output parameters

The Ni-like silver soft X-ray laser pumped by a single "profiled" laser pulse has been demonstrated at the MBI and showed one of the highest figures of merit. The energy of the XRL output pulse was estimated with a synchrotron calibrated CCD and found to be equal to 3 μJ. This corresponds to 10^{11} photons in a single shot and highly directed beam. The output XRL pulse length was unexpectedly long with a length of 32±3 ps [2]. This value is in contradiction to a relatively high gain factor of 33 cm^{-1}, which indicates the transient character of the population inversion, and the latter is usually connected with a short-lived gain. Long-lived and lowered gain is more characteristic for the quasi-stationary working regime. The output beam has shown low divergence of ~3 mrad for targets with a length of 5-6 mm and a deflection angle from the plasma column axis of 8-10 mrad [7].
The beam characteristics as focus, coherence or intensity distribution are deciding for many applications. The emitted radiation possessed a reasonable spatial coherence measured in a double-slit experiment. The equivalent incoherent source with a diameter of 14 μm has been estimated for the X-ray laser with a target length of 6 mm. Such a target length is sufficient to saturate the output of a silver X-ray laser at 13.9 nm. The coherence length estimated at a distance of 16 cm from the target was equal to 60 μm, which is a relatively good result for such a very short target. These are very conservative estimates caused by the limited resolution of the detection systems [8]. This limitation originates from fixed distances required for the correct work of the flat-field spectrometer.
The output beam shape recorded in a near-field is presented in Fig. 1 and shows a strongly curved profile probably determined by the target ablation conditions and refraction on the steep density gradients.

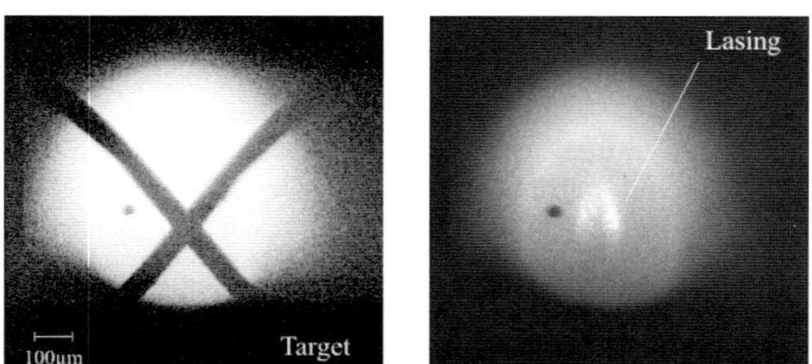

Figure 1. Near-field image of the Ni-like silver soft X-ray laser. Clearly visible is the strongly cured x-ray laser beam in the plasma plume on the target surface. The image on the left shows a 60μm wire cross as a reference.

The focus achieved with this XRL beam is shown in Fig. 2. The spot size of less than 10μm was expected from the geometry of the set up. Focusing down to less than 1μm should be possible.

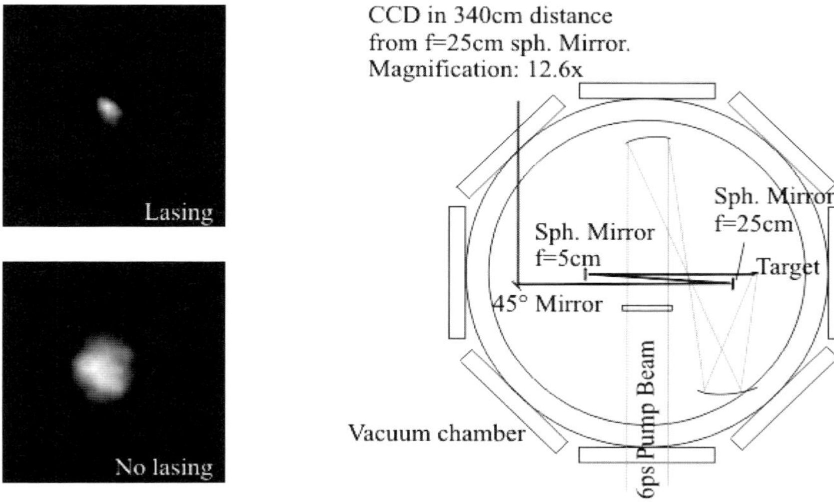

Figure 2. Focus of the XRL beam. The focused XRL spot is less than 10μm and much brighter than the surrounding plasma.

The most important experimental data was delivered by a 3rd-order correlator measurement on the pump laser pulse shape with a dynamic range of 5-6 orders of magnitude. We have built an OPA (optical parametric amplification) correlator, which inherently improves the signal-to-noise ratio. This measurement revealed the structure, which could not be seen with streak cameras, photodiodes or 2nd order correlators. It was clearly seen by comparison between the correlation signals measured at the exit of the regenerative amplifier and at the compressor exit how the amplification process affects the final temporal profile of the pump laser pulse. Additional strongly pronounced pedestal with an amplitude equal to ~5% of the peak intensity and the length between 20 and 40 ps has appeared in the amplified pulse shape [9].

3. Numerical modelling

Modelling was performed using the hydrodynamics/atomic physics code EHYBRID and a ray tracing post-processor both developed at the University of York [6]. Overlapping of three separate pulses of different lengths and peak intensity simulated the experimentally observed temporal pulse structure. Such a pulse is shown in Fig. 3. The length of the background with the intensity of 4 orders of magnitude lower than the peak intensity was determined experimentally as equal to 3.5 to 4 nanoseconds. The second part of the pulse had a form of pedestal with a width of a few tens of picoseconds and variable amplitude within one order of magnitude. The main short pulse with a width of 6 ps and the leading edge of the same length was positioned at the end of the leading edge of the pedestal (Fig. 3). The peak power of the pulse was equal to 2×10^{11} W.

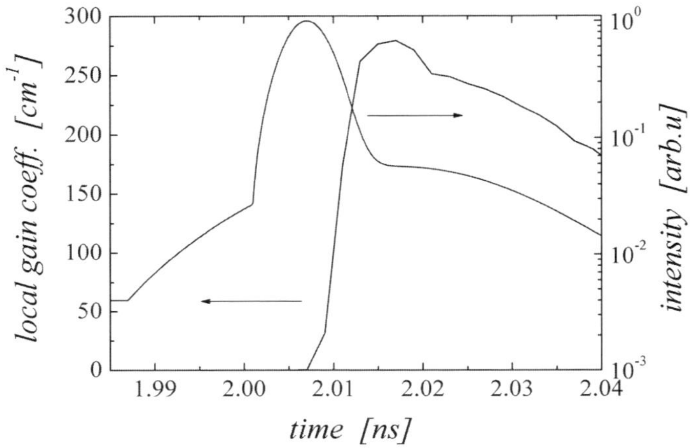

Figure 3. Pump pulse shape (logarithmic scale) and the corresponding temporal dependence of the local gain.

It is seen that the gain depletion rate is strongly reduced and the gain coefficient is after 40 picoseconds still above 50% of its maximum value. This could explain the apparent contradiction between the value of the local gain and the length of the output pulse reported in [2]. The required ionisation stage of the medium with an abundance of the AgXX ions has appeared close to the target and was spread over a few micrometers (Fig. 4).

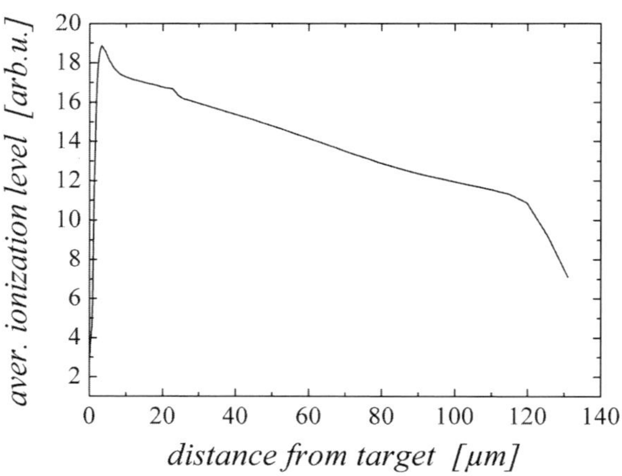

Figure 4. Spatial distribution of the average ionisation stage of the medium under the conditions described in the text.

As a consequence the spatial distribution of the local gain was relatively narrow and concentrated about the maximum ionisation stage (Fig. 5). The maximum value of the gain was also reduced which is connected with a lower electron temperature achieved in the medium (~290 eV).

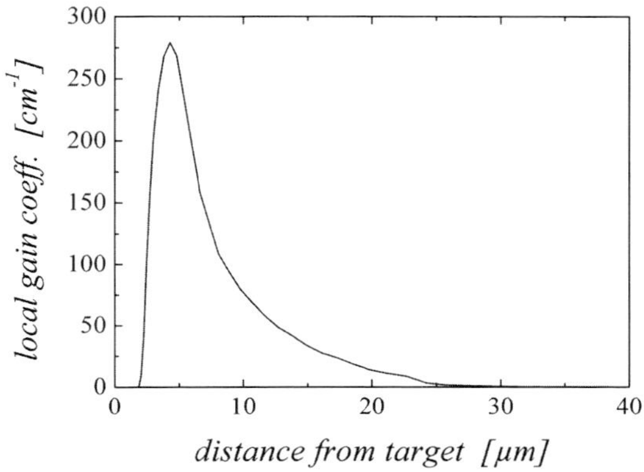

Figure 5. Spatial distribution of the local gain coefficient in the medium pumped by the laser pulse shown in Fig. 2.

We have found that the pedestal on a relatively high intensity level, corresponding well to the pre-pulse technique, is responsible for the gain lifetime extension. The small amount of energy (max. a few hundreds of millijoules) causes the plasma plume to be re-heated. However, it was also found that the pedestal has to be centred behind the main heating picosecond laser pulse to work in the described way.

4. Summary and conclusions

We have demonstrated a full set of the output parameters achieved with the single pulse pump method. The additional measurements on the detailed form of the pump pulse have shown that pulses of more complex structure are able to change in a deciding way the kinetics of the plasma active medium. The problems of increased gain lifetime and the energy redistribution have been indicated and will be described in more detail elsewhere. It is clear that the pulse shape at the contrast level between 10^{-3} and 10^{-2} is decisive for the working regime of such an XRL.

Measurements of the beam focus gave a spot diameter of less than 10μm, which is far from the theoretical diffraction limit. However, the beam was measured at low pump level and showed significant inherent divergence.

References

[1] Dunn J et al., 2000, Phys. Rev. Lett., 84, 4834-4837

[2] Janulewicz KA et al, 2003, Phys. Rev. A, 68, 051802

[3] Ozaki T et al, 2002, Phys. Rev. Lett., 89, 253902

[4] Dunn J et al, these Proceedings

[5] Rocca JJ et al, these Proceedings

[6] Janulewicz KA et al, 2004, Phys. Rev. A, (submitted)

[7] anulewicz KA et al, 2003, Proc. SPIE: Soft X-ray Lasers and Applications, Fill EE, Suckewer S eds, 5197, 90-98

[8] Lucianetti A et al, 2004, Opt. Lett., 29(8), 881-883

[9] Janulewicz KA et al, 2004, J. Sel. Top. Quantum Electron., (submitted)

Inst. Phys. Conf. Ser. No 186
Paper presented at 9th Int. Conf. X-Ray Lasers, Beijing, China, 24–28 May 2004

Improved Energy Coupling into the Gain Region of the Ni-like Pd Transient Collisional X-ray Laser

R.F. Smith, J. Dunn, J. Nilsen, S. Moon, R. Keenan, R. Shepherd, R. Booth, J. R. Hunter
Lawrence Livermore National Laboratory, Livermore, CA 94551

J. Filevich, J.J. Rocca, M.C. Marconi
NSF ERC for Extreme Ultraviolet Science and Technology and Department of Electrical and Computer Engineering, Colorado State University, Fort Collins, Colorado 80523

V.N. Shlyaptsev
University of California Davis-Livermore, Livermore, CA 94551

Abstract. We combine near-field imaging with the recently developed technique of picosecond x-ray laser interferometry to probe the plasma conditions in which the Ni-like Pd transient collisional x-ray laser is generated and propagates. We observe a strong dependence of the laser pump parameters on the effective laser-gain medium coupling. The most effective coupling of laser pump energy occurs when the duration of the main heating pulse is comparable to the gain lifetime (~10ps for mid-Z Ni-like schemes). This can increase the output intensity by more than an order of magnitude relative to the case were the same pumping energy is delivered within a shorter heating pulse duration (< 3ps). The higher intensity heating pulses are observed to be absorbed at higher electron densities and in regions were steep density gradients limit the effective length of the gain medium.

1. Introduction

Lasing at EUV wavelengths is typically generated through single pass amplification along an extended laser-produced plasma column. One of the goals driving x-ray laser research is the prospect of obtaining saturated lasing within the wavelength range between the K absorption edges of carbon (4.37 nm) and oxygen (2.33 nm), the "water window" where carbon-containing biological objects absorb radiation efficiently but water is relatively transparent. This would facilitate powerful new diagnostic techniques of high contrast sub-micron imaging of live biological samples [1]. In recent years saturated lasing has been demonstrated on a number of wavelengths in the 5.7 to 60.8nm range [2]. Currently the most efficient lasing at short wavelengths ($\lambda < 15$nm) has been observed on the $4d^1S_0 \rightarrow 4p^1P_1$ atomic transition in the closed shell Ni-like ion species.

The wavelength output for this transition scales as ~Z^3, where Z is the atomic number of the material. For higher atomic numbers greater ionization, and hence greater input energy is required to achieve the Ni-like ion state. As a result, saturated operation at wavelengths below ~10nm have only been demonstrated on large laser facilities.

In all laser pumped schemes a variation of the prepulse pumping technique is employed whereby a low intensity long pulse preforms a plasma, which is allowed to expand and cool before being heated by a high intensity main pulse. In the transient collisional excitation (TCE) scheme [3] this main heating pulse is temporally compressed by chirped pulse amplification (CPA) techniques down to picosecond durations. The CPA pulse rapidly heats the plasma generating high gain coefficients and saturated x-ray laser output over a few picoseconds duration [4]. In experiments reported on high power laser drivers the pulse duration of the CPA pulse is in the range of 0.5 → 3ps [5, 6, 7]. It has been felt that by maximizing the intensity of the main heating pulse the local gain coefficient will also be maximized. Under these conditions the lowest saturated wavelength currently demonstrated is 7.3 nm for Ni-like Sm [5]. Within this paper we show that high intensity heating of the x-ray laser gain medium can be detrimental to the effective gain-length product. It is observed that by matching the duration of the main pumping pulse to the duration of the gain optimizes the coupling efficiency and increases the x-ray laser output by an order of magnitude. This more effective use of the laser pump energy offers a more efficient route to lower wavelength operation. High intensity heating pulses are shown to deposit their energy at high electron densities were increased levels of refraction limit the gain-length product.

2. Experiment

The Ni-like Pd x-ray laser at 14.68 nm is generated using two laser beams at 1054 nm wavelength from the COMET table-top laser facility at LLNL [8]. Saturated x-ray lasing output of ~10 μJ is achieved with an optical pumping combination of a 600 ps long pulse (1.5 J, ~1×10^{11} W cm^{-2}) and a temporally compressed (0.5 → 27ps) main heating pulse (4 J, 5 → 0.1×10^{14} W cm^{-2}). A 1.6 cm x 140μm line focus employed in a traveling wave geometry is used to irradiate a 1.25 cm long polished Pd slab target [8]. A series of shots were taken in which the output plane of the x-ray laser plasma column was imaged by a 11.75cm focal length spherical multilayer mirror and relayed onto a charged coupled device (CCD) camera with a magnification of ~22 giving a two-dimensional intensity map of the x-ray laser 'footprint' as a function of distance away from the target surface. The peak-to-peak separation between the long and short pulse (Δt) was varied in intervals from 0 → 1ns, the CPA pulse duration was varied from 500fs → 27ps and the resulting x-ray laser output intensity map was recorded.

3. Results

By integrating over the area of the footprint and taking into account the measured pulse durations [4] we can determine the relative output intensity of the x-ray laser as a function of laser pump parameters (Fig. 1). The highest output is observed for the main heating pulse durations of 6.7 and 13.4ps and for a Δt of ~700ps. Under these pumping conditions previous measurements have determined the x-ray lasing transition to be saturated [8]. The outputs for the shorter duration CPA pulses (1.7ps and 0.5ps), however, are more than an order of magnitude less than for the 13ps case and are therefore considered to be unsaturated. As the CPA pulse duration is increased to ~27ps the x-ray laser output intensity falls off from the peak value. For a given heating pulse duration there is a strong dependence on the x-ray laser output as a function of Δt.

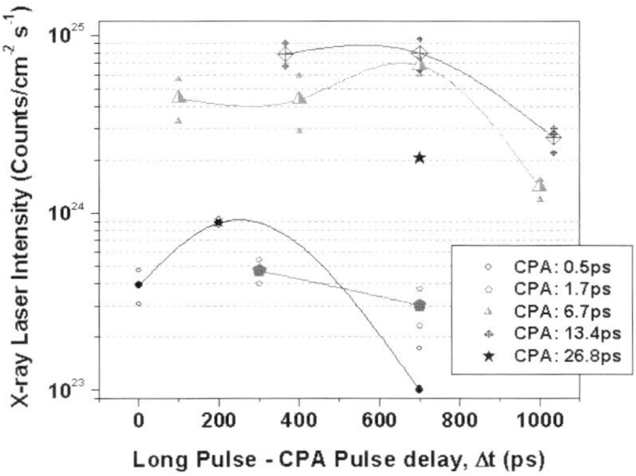

Fig 1. X-ray laser output intensity as a function as laser pumping parameters. The standard operating conditions for the x-ray laser on the COMET laser facility is for a CPA pulse duration of 13.4ps with a Long pulse – Short pulse delay, Δt, of 700ps. The duration of the x-ray laser pulse as reported in Ref [4] is shown in brackets when measured for a Δt of 700ps.

Simulations with the 1D Lasnex code [9] shown in figure 2 show the calculated gain profile for the $4d^1S_0 \rightarrow 4p^1P_1$ gain transition as a function of distance away from the target surface (x-axis) and time relative to the peak of the CPA heating pulse. The duration of the gain is ~10 ps irrespective of the duration of the heating pulse. This is supported by recent measurements in which the output pulse durations which only change by a factor of two (~ 4 → 8ps) when the duration of the main heating pulse is changed by a factor of 55 (0.5 → 27.8ps) [4]. For short CPA pulse durations (< 3ps) the plasma is heated very rapidly, the energy is deposited at higher densities and in regions of larger density gradients than for longer CPA heating pulses (see below). We can see from the contour plot in Fig. 2 (a) the peak of the gain in the plasma heated by the 0.5ps heating pulse occurs after the laser irradiation is switched off. The higher short pulse intensity is not coupled efficiently and as a result the effective gain coefficient is less. It is possible to drive the shorter heating pulsed driven x-ray laser into saturation but it requires higher intensities to do so [5, 6, 7]. Pumping pulse durations < 3ps are typically used on large laser facilities. The shorter duration heating pulses exhibit slightly enhanced output for reduced values of Δt. This is consistent with ref [3] which showed optimized Δt of ~150ps for Ni-like Sm when pumped with a 1ps CPA pulse. This is not a result of more effective beam propagation but is expected to be representative of the higher local gain coefficient for early Δt due to the presence of higher temperatures and densities. Refraction is expected to limit the extent to which this high gain region is sampled. For the lower intensity longer CPA pulses shown in Fig 2(b) (6.7ps and 13.4ps) the heating occurs over the same time scale and during the same time period as the gain duration and the energy is coupled more effectively into the gain region, resulting in enhanced output. For the 27ps (Fig. 2(c)) case the gain has dissipated before most of the energy has even been delivered on target. Therefore the coupling effectivity is poor and the x-ray laser output intensity drops. For all the CPA pulse durations there are variations in the output with Δt. It is expected that at later times (Δt: 700ps) the density gradients have relaxed sufficiently to optimise the beam guiding of the x-ray laser photons. As a result the energy output of the x-ray laser is maximised.

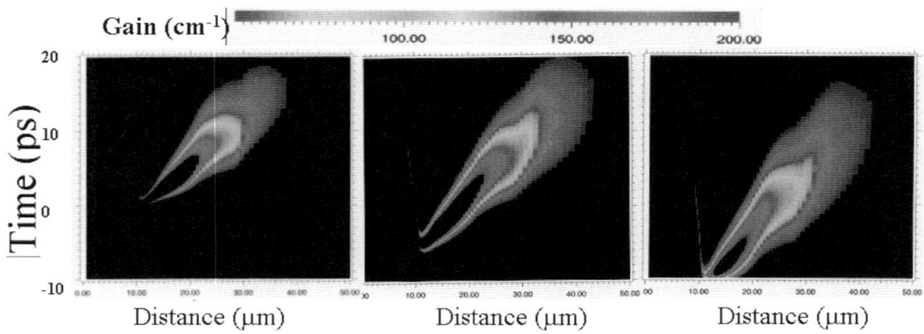

Fig 2. 1D Lasnex simulations show that the gain profile within the x-ray laser medium is centered in time around the optical pumping of a 10ps+/-3ps heating pulse. For CPA pulse durations < 3ps the peak of the gain occurs after the optical heating pulse is turned off. For the 26.8ps heating pulse the gain occurs at the leading edge of the pulse and is extinguished before the peak of the energy is delivered. The color table for gain saturates in the centre of the contours at values of 200cm^{-1}.

Previous work has shown that the combination of the Ni-like Pd transient collisional x-ray laser with a diffraction grating interferometer offers unique diagnostic capabilities in accurately measuring two-dimensional electron density maps of laser-produced plasmas [8]. The picosecond timescale and short wavelength output mitigate effects due to refraction, absorption and motion blurring making this technique appropriate for accurately diagnosing the long millimeter scale plasmas, which are characteristic of the x-ray laser gain medium. Simulations of the x-ray laser medium have shown [7] that the electron density profile setup by the long pulse remains relatively unperturbed (± 5%) during the interaction of the main CPA heating pulse and the time period over which gain and x-ray laser propagation takes place. Therefore, by interferometrically measuring the 2D electron density profile at a chosen delay after the peak of the 600ps long pulse, we are observing the density conditions in which the x-ray laser gain is generated and in which the x-ray laser pulse subsequently propagates.

Within one arm of the interferometer a plasma was formed from a Pd target using the same irradiation conditions as those implemented for the 600ps long pulse used to generate the x-ray laser. A 3.2mm x 140μm line focus was generated on the plane of a 2mm long polished Pd target via the combination of a cylindrical lens and off-axis parabola. The energy delivered in the line focus was 300mJ which gives an on-target intensity of $\sim 1\times10^{11}$ W/cm^2. The x-ray laser probe beam travels longitudinally through the resultant Pd plasma. The output plane of the plasma is imaged via a 25cm focal length Mo/Si multilayer optic and relayed to a 13μm pixel CCD camera ~5.5m away giving an average magnification of ~22. The number of fringe shifts, N_{fringe}, is related to the electron density n_e by [10], $N_{fringe} \cong (n_e L)/(2 n_{crit} \lambda)$, where n_{crit} is the critical density for the probe wavelength λ and L is the plasma length. For the 2mm target used, one fringe shift is equivalent to an electron density of $\sim 7.6\times10^{19}$ cm^{-3}. Using this technique the 2D electron density profile was measured at different delays relative to the peak of the 600ps plasma forming beam. Within the x-ray laser context these times can be understood as different long pulse → CPA pulse delays (Δt). The average ionisation for the Pd plasma from Δt = 0 → 1ns was calculated to be between 14 to 16. Under these conditions the contribution for the bound electrons to the refractive index within the Pd plasma is expected to be small.

By overlaying the near field images with the density maps obtained through interferometry we can determine the density at which the x-ray laser footprint exits the plasma column. As the x-ray laser is generated in a traveling wave geometry the electron density through which the x-ray propagates is effectively frozen over the ~ 4 picosecond pulse duration. The extracted density profiles, therefore, represent a good picture of the conditions within the frame of the propagating x-ray laser. Simulations predict that the region which contributes most effectively to the x-ray laser amplification lies at an electron density of $\sim 1\times10^{20}$ cm^{-3} [10]. Shown in figures 3(a) and (b) are the combined density and intensity maps at the output of the x-ray laser gain medium for a Δt of 0.7ns and a CPA pulse duration of 0.5 and 6.7ns, respectively. The 500fs heating pulse produces an x-ray laser output centered at $\sim 5\times10^{19}$ cm^{-3} while for the same density conditions the low intensity 6.7ps heating pulse generates an x-ray laser output centered at 1×10^{20} $cm^{-3.}$ The angle of refraction may be approximated to be inversely proportional to the density scale length of the plasma [11]. The extent of this deflection gives an indication of how long the x-ray laser photons remain within the gain region, which ultimately determines the extent of amplification. This suggests that the region contributing most the gain for the short CPA case is at a higher density were larger density gradients exist than for the 6.7ps CPA case. This is consistent with Lasnex simulations, which predict that the shorter CPA pulse durations deposit their energy preferentially within higher density regions. For the 6.7 ps heating pulse the energy is deposited in a region with more relaxed density gradients yielding reduced refraction and as a consequence the x-ray laser photons spend a long time within the gain region and experience greater amplification. At this image plane the x-ray laser has traversed

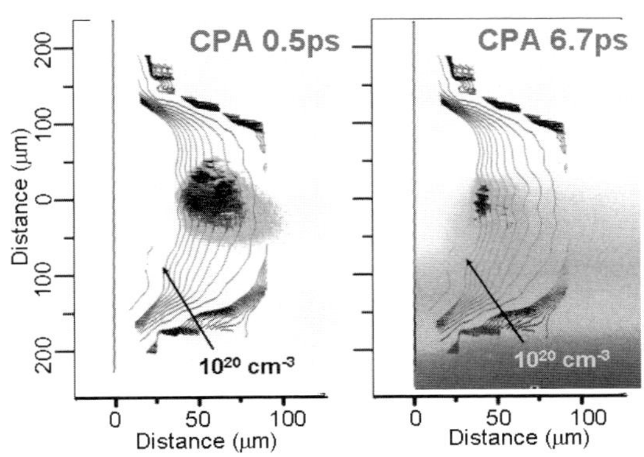

Fig 3. The footprint of x-ray laser is overlaid on interferometrically determined electron density maps. The CPA pulse duration is 0.5ps and 6.7ps for figure (a) and (b), respectively. In both cases the value for Δt was 0.7ns. The color scales have been optimized for each particular image in order to enhance detail. The relative outputs are shown in figure 1. The extracted electron density contour map is shown in a logarithmic scale with each line connecting regions of equal density. The 10^{20} cm^{-3} contour is highlighted, as this is the density at which the most effective gain contribution is predicted to originate from.

through the 12.5 mm long amplifying medium with very small deflection. It is apparent from these images that the dimensions of the x-ray laser output are influenced greatly by the laser parameters. We note that as the x-ray laser becomes more saturated the footprint becomes more concentrated in space, which is a result of the localized region of high gain dominating the amplification process.

We have also observed from the near field images that the beam profile is more structured for the shorter duration CPA pulses. These observations are consistent with recent measurements that show that the spatial coherence of the x-ray laser also increases for longer duration CPA drivers [12]. As the near field monitor is time integrated there will be some smoothing effect from the longer pulse durations. However as the x-ray laser pulse duration is largely insensitive to the main heating pulse width this does not explain the observations. The reduced structure at longer heating pulse durations may be due to smoothing effects by the optical beam. In the 500fs case the drive beam is turned off many picosecond before the x-ray laser is generated. Spatial inhomogeneities in the intensity profile are imprinted instantaneously [13] and without any further energy being deposited into the plasma, non-uniformities in the gain profile may result. For longer CPA pulse drivers imprinting also occurs at the leading edge of the pulse but at lower intensities. In addition, as energy is being deposited for many picoseconds, smoothing from continued coupling of the drive laser is likely to aid dissipation of the initial imprinted non-uniformities.

We have conducted a series of experiments studying the effect of varying the laser pump parameters on the x-ray laser performance which has showed that by matching the duration of the CPA pulse with the gain duration (~10 ps) an enhancement in the x-ray laser output intensity by more than an order of magnitude was achieved over the short CPA pulse case (< 2 ps). By keeping the density conditions constant, we infer from the increased levels of refraction that short CPA pulses deposit their energy at densities higher than for the longer CPA pulses (as predicted by simulations).

The support of Al Osterheld and Andy Hazi is greatly appreciated. The authors are pleased to acknowledge the technical contributions from Carl Bruns and Al Ellis. This work was performed under the auspices of the U.S. Dept. of Energy by the University of California Lawrence Livermore National Laboratory, through the Institute for Laser Science and Applications, under Contract No. W-7405-Eng-48 and by US Department of Energy Grant No. DE-FG03-98DP00208. This research was sponsored by the National Nuclear Security Administration under the Stewardship Science Academic Alliances program through DOE Research Grant # DE-FG03-02NA00062. The CSU researchers also gratefully acknowledge the partial support of the NSF ERC Center for Extreme Ultraviolet Science and Technology, award number EEC-0310717 and the W.M. Keck Foundation.

References

[1] Solem J C and Baldwin G C, Science **218**, 229 (1982).
[2] Daido H, Reports on Progress in Physics, vol.65, no.10, pp.1513-76, (2002).
[3] Shlyaptsev V N, Nickles P V, et.al., Proceedings of the SPIE - The International Society for Optical Engineering, (Ultrashort Wavelength Lasers II, San Diego, CA, USA, 12-13 July 1993.) **2012**, pp.111-18 (1993).
[4] Dunn J, Smith R.F., et.al., these proceedings.
[5] King R E, Pert G J, et.al., Phys. Rev. A **64**, 053810 (2001).
[6] Klisnick A, Zeitoun P, et.al., J. Opt. Soc. Am. B **17**, *pp.1093-7* (2000).

[7] Tang H, Daido H,et.al., Japanese Journal of Applied Physics Part 1-Regular Papers Short Notes & Review Papers **42**, no.2A, pp.443-8. (2003).

[8] Dunn J, Li Y, et.al., Phys. Rev. Lett. **84**, 4834 (2000).

[9] Zimmerman G B and Kruer W L, Comments on Plasma Physics for Controlled Thermonuclear Fusion **2**, 51-61 (1975).

[10] Smith R F, Benredjem D, et.al., SPIE-Int. Soc. Opt. Eng. Proceedings of Spie - the International Society for Optical Engineering **4505**, pp.107-15 (2001).

[11] Smith R F, Dunn J, et.al., JOSA B **20**, Issue 1, p. 254 (2003).

[12] We measured the spatial coherence of the x-ray laser output with a Young's double slit experiment. The distance between the slits was systematically varied and backlit by the x-ray laser generated with different CPA pulse durations. It was observed that the spatial coherence improved as a function of increasing CPA pulse duration. This data is unpublished.

[13] Glendinning S G, Weber S V, et.al., Phys. Rev. Lett. **69**, 8, pp.1201-4, (1992).

Inst. Phys. Conf. Ser. No 186
Paper presented at 9th Int. Conf. X-Ray Lasers, Beijing, China, 24–28 May 2004

Pumping laser energy absorption in X-ray laser experiments

Y Abou-Ali[1], Q L Dong[1], A Demir[2], G J Pert[1] and G J Tallents[1]
[1]Department of Physics, University of York, York YO10 5DD, U.K.
[2]Department of Physics, University of Kocaeli, Kocaeli 41000, Turkey

Abstract. The effective pumping laser energy absorbed when producing soft X-ray lasing is investigated. The EHYBRID fluid and atomic physics code is used to simulate X-ray laser output. Comparisons are made with published experimentally measured Ne-like nickel X-ray laser output at 23.1 nm. We show that with ~ ps pulses, the pumping laser energy employed in the EHYBRID code needs to be reduced to ≈ 0.3 of the experimental laser energy for the simulated results to agree with experiment. A particle-in-cell (PIC) code is used to simulate the laser energy absorption and resulting electron energy distribution. It is shown that some of the absorbed energy is distributed to fast electrons that do not produce collisional excitation of the population inversion.

1. Introduction

Plasmas created by high power optical lasers focussed to a line onto solid targets have produced saturated amplification of soft X-rays in the 5.9–30 nm wavelength range [1-3]. Saturation of a laser occurs when stimulated emission starts to deplete the population inversion and is important in laser development as output is maximised and pumping efficiency is optimised. Since the initial observations of unsaturated X-ray lasing created using single pulses of ~ nanosecond duration and high (~ kilojoule) energy [4, 5], experiments are now increasingly aimed at reducing the size of the driver laser needed to achieve saturated lasing [6, 7]. Using short pulse (~ picosecond) lasers has enabled the necessary laser energy to pump lasing in ions such as nickel-like silver to be reduced to a few joules [8, 9].

With picosecond pulse pumping of X-ray lasers, the irradiances on target are much greater (up to 10^{16} Wcm^{-2}) than needed with experiments using higher energy, longer pulse pumping lasers (typically up to 2×10^{14} Wcm^{-2}). In addition, the high irradiance pulses are incident into long scalelength plasmas due to the use of a pre-pulse which means that although the laser absorption is high, parametric processes [10] may cause some of this absorbed energy to produce fast electrons that will not efficiently collisionally excite the population inversion. We examine this indirect effect of parametric processes on the production of population inversions in X-ray laser experiments in this paper.

Simulations of X-ray laser experiments sometimes require a reduction of the assumed pump laser energy absorbed to obtain agreement with experiment [11, 12]. A correction factor, $f \leq 1$, is utilised to multiply the actual incident laser energy to give a laser energy for use in the code simulations. In this paper, we determine the necessary value of f needed to model X-ray laser experiments involving picosecond duration pumping by comparing the fluid and atomic physics code EHYBRID [13] output with published experimental results. To understand the cause of the reduced "effective energy" absorbed in the experiment, we model the laser energy absorption and resulting electron energy distribution using a particle-in-cell (PIC) code [14]. The electron energy distribution calculated by the PIC code is used to calculate collisional excitation rates for comparison to the rates used in EHYBRID where Maxwellian distributions are used.

2. Experiment

X-ray lasing was produced in Ne-like nickel at 23.1 nm ($2p^5 3p\ ^1S_0 \rightarrow 2p^5 3s\ ^1P_1$) by irradiating solid nickel slabs with two beams of wavelength 1.06 μm from the VULCAN glass laser at the Rutherford Appleton Laboratory. A pre-plasma was formed with a controlled pre-pulse of duration 280 ps and peak irradiance 2×10^{13} Wcm^{-2}. The main pulse irradiance had a duration of 1.2 ps and peak irradiance 7×10^{15} Wcm^{-2}. The peak-to-peak delay between the pre-pulse and main pulse was 300 ps. The main pulse was incident at travelling wave speed equal to the speed of light c along the horizontal target length and was polarised vertically. The experimental set-up and measured X-ray laser output has been presented by Abou-Ali *et al* [15]. In this paper, these experimental results are compared to simulatons in order to determine the absorbed laser energy.

3. PIC code simulations

The PIC code simulated the 1 ps pulse interaction with a pre-formed plasma of profile and temperature as predicted by the EHYBRID code for the experimental conditions. An initial exponential density profile with an e^{-1} scale-length of 30 μm and temperature of 150 eV is used. For pumping laser irradiances greater than 10^{15} Wcm^{-2}, the PIC code shows the production of a two-temperature electron energy distribution with laser energy absorbed dropping with increasing laser irradiance. The electron phase-space distribution calculated by the PIC code indicates that the laser absorption at irradiances above 10^{15} Wcm^{-2} is due to inverse Bremsstralung and parametric instabilities around the region of the quarter critical density. These two processes produce a lower and higher temperature electron group respectively.

For a specific electron energy distribution, the collisional excitation rate coefficient from the ground state to the upper lasing level of the Ne-like ion can be expressed as:

$$K = <\sigma(\upsilon)\upsilon> = \int_{\Delta E}^{\infty} \sigma(\upsilon) f(\upsilon) \upsilon d\upsilon, \qquad (1)$$

where $\sigma(\upsilon) = \Omega/E$ is the collisional excitation cross-section, Ω is the collision strength of the transition, while E and υ are the energy and the velocity of electrons. ΔE is the excitation energy and $f(\upsilon)$ is the electron energy distribution function. With the Maxwellian distribution used in the EHYBRID code, the calculated collisional excitation rates are higher than the values calculated with the two-temperature electron distribution calculated in the PIC code. An estimate of the f-factor can be obtained from figure 1 by comparing $<\sigma\upsilon>/\Omega$ from the PIC and EHYBRID codes. The ratio of irradiances required to produce comparable $<\sigma\upsilon>/\Omega$ gives the value of f. It can be seen that an irradiance of 7×10^{15} Wcm^{-2} implies an f-factor of 0.22 and 0.16 for Ne-like Ge and Ne-like Ni respectively (see the arrows in figure 1).

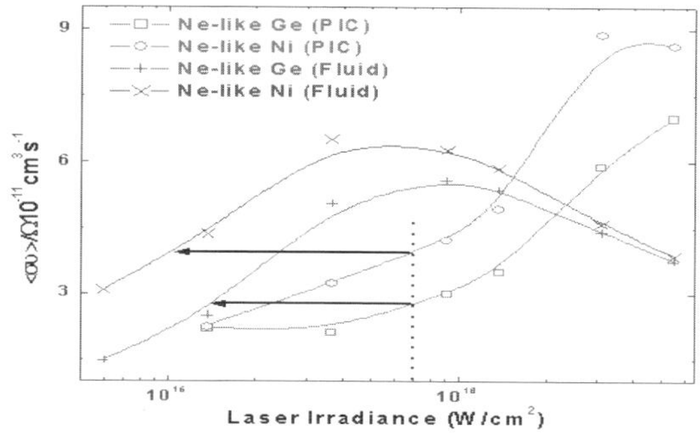

Figure 1. The reduced collisional excitation rate coefficient of the upper lasing energy level for Ne-like Ni and Ge calculated assuming a Maxwellian electron distribution from the EHYBRID code and the electron distribution from PIC simulations at the end of the 1 ps pulse.

4. Fluid code simulations

Figures 2 shows the simulated temporal variation of the Ne-like nickel X-ray laser output at 23.1 nm with different values of the correction factor *f* equal to 0.2, 0.3 and 0.5 superimposed on the experimental temporal variation of the Ne-like nickel output recorded with the streaked spectrometer as presented by Abou-Al *et al* [15].

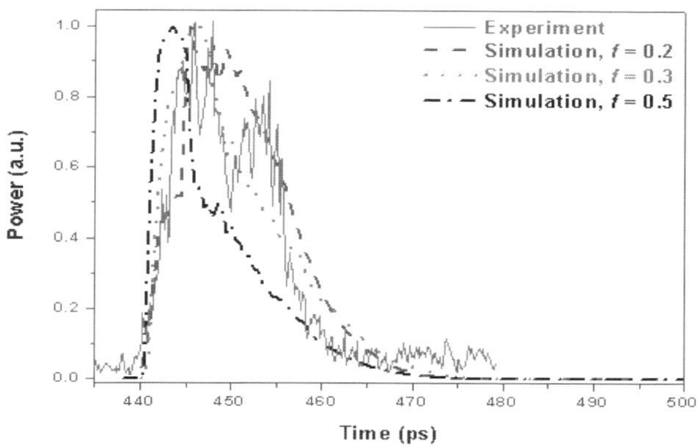

Figure 2. The simulated temporal variation of the Ne-like nickel X-ray laser output at 23.1 nm after ray tracing for a target length of 6 mm superimposed on the experimental temporal variation of the Ne-like nickel output recorded using spectrometer and a fast streak camera.

Figure 3a shows the Ne-like Ni X-ray laser energy output dependence on the arrival time of the peak of the main pulse relative to the peak of the pre-pulse from experiment [15] and a numerical simulation with RAYTRACE post-processing of the space-time output from the EHYBRID code. Different values of the correction factor *f* equal to 0.2, 0.3 and 0.5 have been used in the EHYBRID code. Figure 3b shows the simulated output of the Ne-like Ni X-ray laser with different values of *f* equal to 0.2, 0.3 and 0.5 superimposed on some single shot experimental Ne-like Ni X-ray laser outputs. Good agreement between the simulations and the experimental points within the estimated experimental errors is obtained with *f*= 0.3 (figure 3a and b).

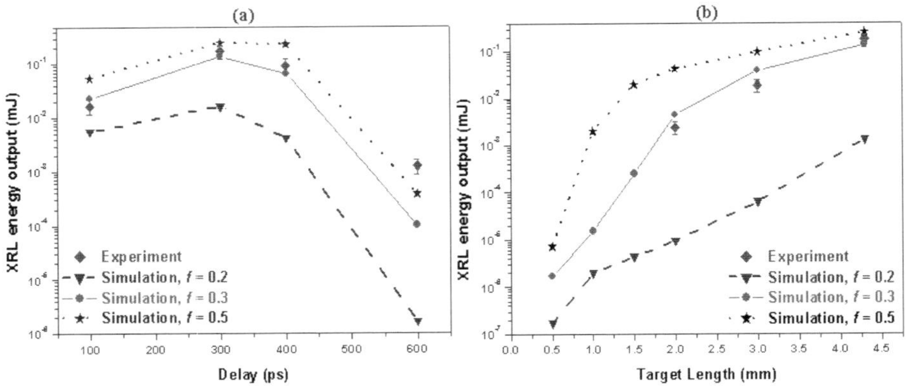

Figure 3. The simulated energy output of the Ne-like Ni X-ray laser at 23.1 nm after ray tracing (a) as a function of the separation between the pulses (b) as a function of target length with f=0.2, 0.3 and 0.5 superimposed on the experimental Ne-like Ni X-ray laser.

5. Conclusion

The EHYBRID fluid and atomic physics code has been used to compare simulations to published experimentally measured Ne-like nickel X-ray laser output at 23.1 nm. A particle-in-cell (PIC) code has been used to simulate the laser energy absorption and resulting electron energy distribution. Some of the absorbed energy is distributed to fast electrons that do not produce collisional excitation of the population inversion. We have shown that the pumping laser energy employed in the EHYBRID code needs to be reduced to $\approx$ 0.3 of the experimental laser energy for the simulated results to agree with experiment. The fast electron production predicted by the PIC code explains the need to reduce the laser energy absorbed in the fluid code. It can be concluded that X-ray lasers created from pre-formed plasmas can be more efficient if the driving laser irradiance is kept below 10^{15} Wcm^{-2} so that the effect of fast electrons on gain is reduced.

References

[1] Tallents G J 2003 J. Phys. D: Appl. Phys. 36 R 259
[2] Carillon A et al 1992 Phys. Rev. Lett. 68 2917
[3] Lin J Y et al 1999 J. App. Phys. 85 672
[4] Matthews D L et al 1985 Phys. Rev. Lett. 54 110
[5] Lee T N, McLean E A and Elton R C 1987 Phys. Rev. Lett. 59 1185
[6] Nickles P V et al 1997 Phys. Rev. Lett. 78 2748
[7] Kuba J et al 2000 Phys. Rev. A 62 43808
[8] Dunn J et al 2000 Phys. Rev. Lett. 84 4834
[9] Janulewicz K A et al 2002 AIP Conference Proceedings. 641 26-30
[10] Baldis H A et al 1991 Handbook of Plasma Physics (Amsterdam: Elsevier) 361
[11] Cairns G F et al 1996 J. Phys. B 29 4839
[12] King R E et al 2001 Phys. Rev. A 64 053810
[13] Pert G J 1983 J. Fluid Mech. 131 401
[14] Sheng Z M et al 2000 Phys. Rev. Lett. 85 5340
[15] Abou-Ali Y, Tallents G J et al 2003 Optics Commun. 215 397

Inst. Phys. Conf. Ser. No 186
Paper presented at 9th Int. Conf. X-Ray Lasers, Beijing, China, 24–28 May 2004

Pulse duration measurements of a picosecond laser-pumped 14.7 nm x-ray laser

J. Dunn, R.F. Smith, R. Shepherd, R. Booth, J. Nilsen, J.R. Hunter
Lawrence Livermore National Laboratory, Livermore, CA 94551, USA

V.N. Shlyaptsev
University of California, Davis, Livermore, CA 94551, USA

Abstract. The temporal dependence of the 14.7 nm Ni-like Pd ion x-ray laser is measured as a function of the laser drive conditions with a fast sub-picosecond x-ray streak camera. The chirped pulse amplification laser beam that pumps the inversion process is varied from 0.5 – 27 ps (FWHM) to determine the effect on the x-ray laser pulse duration. The average x-ray laser pulse duration varies by a relatively small factor of 2.5 times from 3.6 ps to 8.1 ps with travelling wave (TW) irradiation conditions. Slightly shorter pulse durations approaching 2 ps are observed with the x-ray laser operating below saturation. The x-ray laser is found to be 4 – 5 times transform-limited for 6 – 13 ps laser pumping conditions.

1. Introduction

We report results of the Ni-like Pd 4d – 4p x-ray laser pulse duration as a function of the short pumping pulse duration at a fixed delay of 700 ps peak-to-peak between the two laser pulses. These experiments in common with many transient x-ray lasers use a combination of two laser pulses where a long pulse (0.2 – 1 ns) forms a pre-plasma that is further heated by a second, shorter picosecond, high intensity pulse [1]. The precise laser pumping conditions can dramatically affect the x-ray laser characteristics. For example, the x-ray laser intensity can be optimized as a function of the short pulse duration with a maximum achieved for 6 – 13 ps pumping [2]. Recent temporal measurements of Ni-like Ag 4d – 4p 13.9 nm x-ray laser pumped by 1.3 ps laser pulses have demonstrated that the x-ray laser pulse duration can be as short as 2 – 3 ps (FWHM) [3, 4]. The objective of this work is to determine the sensitivity of the x-ray laser duration to the short pulse pumping duration when the laser pulse is varied from 0.5 ps to 27 ps.

2. Experimental Description and Results

The Ni-like Pd 14.7 nm x-ray laser was generated using two 1054 nm wavelength laser beams from the Compact Multipulse Terawatt (COMET) facility at LLNL This laser can be fired at a repetition rate of 1 shot every 4 minutes [5]. A 600 ps (FWHM) pulse forms the pre-plasma and is followed by a short pulse after a delay of 700 ps peak-to-peak. The long pulse was focused to 1.6 cm × 150 μm (FWHM) (L × W) with 1.25 ± 0.25 J energy on target (EOT). The short pulse beam was focused to 1.6 cm × 100 μm (FWHM) and contained 3.75 ± 0.25 J EOT. Travelling wave (TW) line focus geometry was implemented before the focusing optics by using a high-reflectivity, 0° dielectric-coated reflection echelon consisting of seven flat vertical mirror segments. The TW line focus irradiation geometry is stepped and set for *c* along the target. The picosecond beam was set by adjusting the compressor gratings to give a duration of 0.5, 1.75, 3.4, 6.7, 13.4, and 26.8 ps (FWHM). Polished Pd slab targets up to a maximum of 1.25 cm were irradiated.

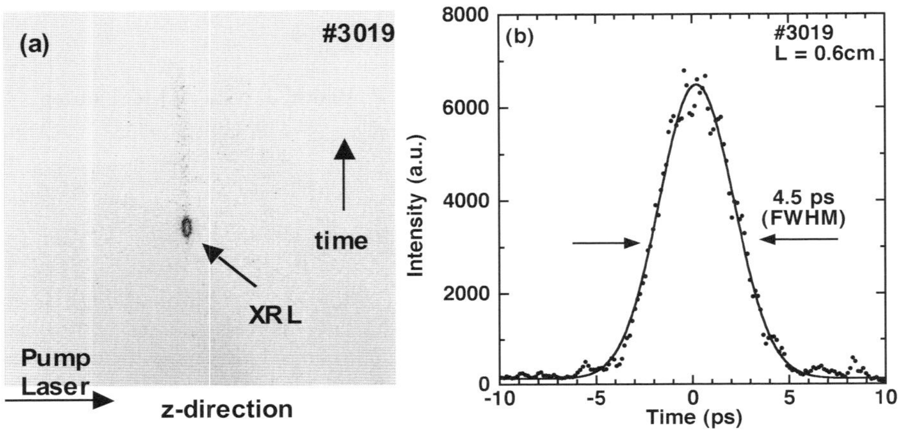

Figure 1. (a) X-ray laser streak image for 6.7 ps (FWHM) short pulse pumping incident on a 0.6 cm target. Laser pump is incident from left and time is increasing up. (b) Intensity lineout through X-ray laser image. Experimental data (solid circles) fit by Gaussian shape (solid line) with 4.5 ps (FWHM).

The experimental geometry was chosen to spatially and temporally resolve the x-ray laser emission relative to the target. The x-ray laser beam leaving the plasma column was imaged by a Mo:Si multilayer-coated, high reflectivity spherical mirror with 11.75 cm focal length. This was relayed by a 45° Mo:Si multilayer-coated flat mirror onto either a streak camera or back-thinned CCD camera with 22× magnification. Both detectors were at normal incidence to the x-ray laser beam and equidistant to the Pd target to within ±0.05 cm. A second Mo:Si flat motorized mirror could intercept the x-ray laser beam and direct it to either detector. A time-integrated 2-D near-field image or a 1-D time-resolved image of the x-ray laser beam was recorded. The streak camera entrance slit 100 μm × 2.5 cm (W × L) was orientated in the horizontal direction, with the long axis aligned along the z-axis of the near-field image perpendicular to the Pd target. This is swept in time as shown in Fig. 1(a). The streak camera has been used previously with a measured temporal response of 500 fs [6]. In this experiment the photocathode consisted of 120 nm CsI/75 nm Al coated on a thin, etched 40 nm Si_3N_4 window in a 250 μm Si wafer substrate. A combination of three filters was used to attenuate the x-ray laser fluence up to a maximum of 10^4 to keep the x-ray signal within the linear range of the detectors. This was particularly important for the x-ray streak camera where instrumental space charge saturation effects can produce pulse broadening and loss of temporal resolution. The streak camera characteristics were measured during the experiment with the instrumental temporal resolution determined to be 1.1 ± 0.3 ps (FWHM) [7]. More experimental details can be found elsewhere [7]. Figure 1(a) shows the 14.7 nm x-ray laser from a 0.6 cm Pd target pumped with a 6.7 ps short pulse where the output is operating in the gain saturation regime. The x-ray laser emission is short-lived, less than 5 ps, with a low intensity continuum lasting for many tens of picoseconds and beyond the streak camera window. An intensity lineout, Fig. 1(b), shows the x-ray laser pulse shape is symmetrical and is well-fitted by a Gaussian. Some variation in the pulse shape is observed from shot-to-shot.

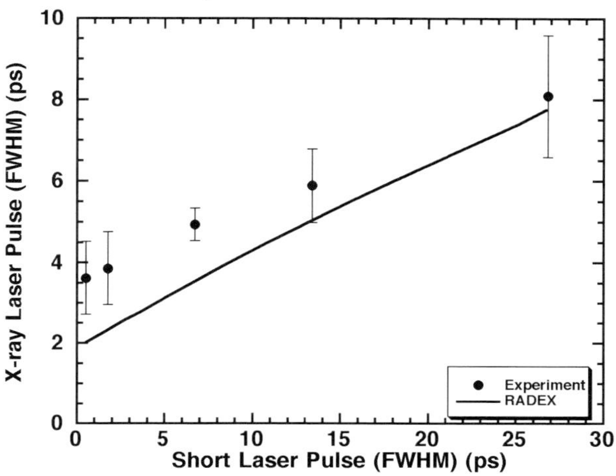

Figure 2. X-ray laser duration plotted as a function of short pulse laser drive from 0.5 to 27 ps. The mean value, of 4 – 6 experimental data points, (closed circles) is shown with error bars of 1σ. RADEX simulations are shown for the experimental pumping conditions.

A study covering the Pd pre-formed plasma density profile, x-ray laser near-field imaging, temporal duration and output intensity was reported for a large range of pumping conditions [8]. Here, the measurements were taken for a delay of 700 ps between the two laser pulses with all other laser pumping conditions, including the laser energy, kept constant. The measured x-ray laser pulse duration is plotted versus the short pulse drive of 0.5 to 27 ps, Fig. 2. The short pulse irradiance varies by 50 times. The x-ray laser output intensity is also changing for the optical pump drives: The 6.7 ps and 13 ps pumping pulses typically operated in the gain saturation range with highest output while the shorter pumping pulses had lower output. The plotted experimental measurements are the mean value for several data points corrected for the instrumental response of 1.1 ± 0.3 ps by subtracting in quadrature. From Fig. 2 the trend is clear that the x-ray laser pulse duration decreases with the laser drive, but only by a factor of 2.2× compared with the 50× decrease in the laser drive duration. The longer laser drives generate x-ray pulses of 8.1 ps while the shortest drive gives 3.6 ps. The shortest x-ray laser pulse durations observed (not plotted) are 2.4 ± 0.1 ps for drives below 1.75 ps. It is also noted that in some cases when the laser pump energy is lowered close to the threshold for lasing, sub-3 ps x-ray laser durations are measured for a 6.7 ps laser drive. The x-ray laser is operating below the gain saturation regime in both cases. These shorter x-ray laser pulse durations are within ~0.5 ps of the results reported recently for the Ni-like Ag laser for different pumping conditions [3, 4].

The experimental values shown in Fig. 2 are slightly higher than RADEX simulations modelled with a perfect travelling wave [9]. These simulations do not include ray-tracing but assume from earlier beam deflection measurements [5] that the optimum propagation for the x-ray laser beam is at a maximum electron density of 1.3 – 1.9 × 10^{20} cm^{-3}. There is good agreement at the longer pulse drive with some deviation at the shorter durations. Part of this difference is from the decoupling of the different gain regions along the line focus from the stepped mirror TW irradiation that results in a lengthening of the pulse.

The longitudinal coherence of the Pd x-ray laser was measured for the 6 ps and 13 ps drives using a Michelson interferometer. It was found to have a 1/e half width (HW) of 342 μm ± 24 μm and 400 μm ± 35 μm corresponding to a spectral line width of 0.34 pm and 0.29 pm (FWHM) respectively [10]. The fringe visibility measurements around the spectral line center had a Gaussian shape primarily explained by gain-narrowing of a low temperature thermally-broadened line. As previously discussed [10] this did not give a complete explanation and other contributions to the line shape would have to be taken into account. Nonetheless, the time-bandwidth product $\Delta\nu\ \Delta t$ = 0.441 for Gaussian shapes, e.g. Fig. 1(b), suggests the transform limit is 0.93 – 1.09 ps for the above spectral line widths. Therefore, the Pd x-ray laser is 4 – 5× transform limited for these pumping conditions. Recent measurements of the longitudinal coherence for the Ag x-ray laser have better dynamic range that indicate more Lorentzian profile in the wings [11]. It should be possible to make a more complete study that includes full analysis of all of the spectral line-broadening (and narrowing) contributions. When combined with the temporal pulse shape data a better understanding of the conditions determining the transform limit may be achieved.

3. Conclusions

We have reported measurements of the Pd 14.7 nm x-ray laser duration as a function of the short pulse laser drive. The 14.7 nm pulse duration is largely determined by the gain lifetime influenced by collisional and ionization times within the plasma. The trend shows that ps-driven x-ray lasers are limited to ~1 ps operation by the transform limit.

Acknowledgments

The continued support and encouragement of Al Osterheld is greatly appreciated. This work was performed under the auspices of the U.S. Dept. of Energy by the University of California Lawrence Livermore National Laboratory under Contract No. W-7405-Eng-48.

References

[1] P.V. Nickles *et al.*, *Phys. Rev. Lett.* **78**(14), 2748-2751 (1997).
[2] J. Dunn, *et al.*, *X-ray Lasers 2000*, ed. G. Jamelot, C. Möller, and A. Klisnick, J. Phys. IV **11**, Pr2-19 (2001).
[3] A. Klisnick *et al.*, *Phys. Rev. A*. **65**, 033810 (2002).
[4] Y. Abou-Ali *et al.*, *Opt. Comm.* **215**, 397 (2003).
[5] J. Dunn *et al.*, *Phys. Rev. Lett.* **84**, 4834 - 4837 (2000).
[6] P. Audebert, R. Shepherd, *et al.*, *Phys. Rev. Lett.* **89**, 265001-1 - 4 (2002); R. Shepherd, R. Booth, private communication (2002).
[7] J. Dunn *et al.*, *Soft x-ray lasers and Applications V*, SPIE Int. Soc. Opt. Eng. Proc, vol. **5197**, ed. E.E. Fill and S. Suckewer, 51-59 (2003).
[8] R.F. Smith *et al.*, *Soft x-ray lasers and Applications V*, *ibid.*, 155-167 (2003).
[9] V.N. Shlyaptsev *et al.*, *Soft x-ray lasers and Applications V*, *ibid.*, 221-228 (2003).
[10] R. F. Smith *et al.*, *Opt. Lett.* **28**(22), 2261 (2003); J. Dunn *et al.*, *Soft x-ray lasers and Applications V*, *ibid.*, 43-50 (2003).
[11] A. Klisnick *et al.*, these proceedings (2004).

Inst. Phys. Conf. Ser. No 186
Paper presented at 9th Int. Conf. X-Ray Lasers, Beijing, China, 24–28 May 2004

Simulation study of a Ni-like Mo x-ray laser at 18.9 nm driven by ultra-short laser pulses

X Lu[1], F Yan[1], Y J Li[1,2], H M Li[1,2], J Y Zhong[1,3] and J Zhang[1]

[1]Laboratory of Optical Physics, Institute of Physics, Chinese Academy of Science, Beijing 100080, People's Republic of China

[2]Department of Physics, China University of Mining & Technology, Beijing 100083, People's Republic of China

[3]National Astronomical Observatories, Chinese Academy of Sciences, Beijing 100012, People's Republic of China

Abstract. The Ni-like Mo x-ray laser at 18.9 nm was investigated numerically using a one-dimensional hydrodynamic code MED103 coupled with an atomic data package. Simulations are performed for a Mo planar target irradiated by a 1 ns prepulse followed by a 1 ps, 10^{15}W/cm^2 drive pulse from a Ti:sapphire laser. The energy levels, radiative transition rates, and collisional excitation rates are calculated using the Cowan code. The optimization calculations are performed in terms of the intensity of the prepulse and the delay time between the prepulse. The relative output intensity of the x-ray laser is predicted using the ray-tracing method. We predicted that the Ni-like Mo soft x-ray laser at 18.9 nm could possibly be generated using a table-top Ti:sapphire laser system.

1. Introduction

The use of ultra-short pulse pumping has shown to be able to save significantly the drive energy of collisional x-ray lasers [1-12]. In present days, the high power compact Ti:sappher CPA lasers at 800 nm are available for many university level laboratories. Such laser systems have great potential to be used as the pump source of Ne-like or Ni-like table-top x-ray lasers with a high repetition rate. In this work we simulate a Ni-like Mo x-ray laser of 4p $\rightarrow$ 4d, J=0 $\rightarrow$ 1 transition at 18.9 nm driven by a Ti:sappher CPA laser. The pump geometry is set to be the standard line focus on a slab target. The Ni-like Mo ions are modeled for the ground state 3d^{10}, and four most important exited levels, i.e., $3d^9 4p\ ^3D_1$, $3d^9 4p\ ^1P_1$, $3d^9 4d\ ^1S_0$, $3d^9 4f\ ^1P_1$, and the ground state of next ionization stage (Co-like state). The modified version of 1D Lagrangian hydrodynamic code MED103 [13] coupled with atomic physics code is used to predict the time evolution of laser-plasma interactions and gain coefficient.

2. Atomic physics calculation

The energy levels of Mo XV ions used in the x-ray laser simulations are calculated using the Cowan code [14] in orbital-spin (LS) coupling and presented in Table I. Table II lists the wavelengths and the radiative rates of dipole transitions between levels. The calculated wavelength of the strongest lasing transition of $3d^9 4d\ ^1S_0 \rightarrow 3d^9 4p\ ^1P_1$ is in good agreement with the experimental measurement of 18.9 nm. The collisional excitation rates between levels as a function of electron temperature are shown in Figs. 1(a) and 1(b). Figure 1(a) presents the excitation rates from the ground state to exited states, and the collisional mixing rates between exited states are shown in Fig. 1(b). The collisional mixing rates of transition $3d^9 4p\ ^1P_1 \rightarrow 3d^9 4p\ ^3D_1$ and $3d^9 4p\ ^3D_1 \rightarrow 3d^9 4f\ ^1P_1$ are negligibly small and are not plotted in the figure.

TABLE I. Energy levels of Ni-like Mo XV

Index	State Configuration	RCN/RCG (eV)
1	$3d^{10}\ ^1S_0$	0
2	$3d^9 4p\ ^1P_1$	245.2
3	$3d^9 4p\ ^3D_1$	247.9
4	$3d^9 4d\ ^1S_0$	311.0
5	$3d^9 4f\ ^1P_1$	349.8

TABLE II. Wavelength and radiative transition rates in Ni-like Mo XV

Transition ($j \rightarrow i$)	Wavelength (nm)	A_{ji} (10^{10} s^{-1})
$2 \rightarrow 1$	5.06	45.0
$3 \rightarrow 1$	5.00	8.7
$5 \rightarrow 1$	3.55	760.0
$4 \rightarrow 2$	18.86	10.1
$4 \rightarrow 3$	19.65	1.6
$5 \rightarrow 4$	31.96	1.2

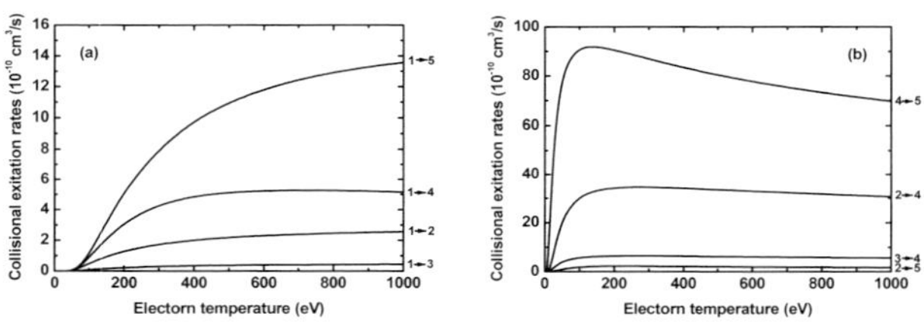

Figure 1. Collisional excitation rates as a function of electron temperature for transitions from ground state to excited levels (a), and between excited levels (b).

3. Preparation of an optimum preplasma

Simulations were performed for 100 μm thick niobium planar targets irradiated by a 1 ns prepulse and a picosecond or sub-picosecond main pulse at 800 nm wavelength from a Ti:sappher CPA laser. It is well known that for an ultrashort pulse puming x-ray laser, prepulse should produce a properly ionized preplasma with large scale length, the main pulse is used to rapidly heat the preplasma and make a sudden jump of electron temperature. In order to obtain an optimum intensity of prepulse, simulations have been performed for a target irradiated by 1 ns FWHM Gaussian prepulse with various peak intensities. Figure 2 shows the spatio-temporal profile of the ground state population of Ni-like Mo ions fraction in plasmas produced by 1 ns prepulses with peak intensities of 2.4×10^{11} W/cm^2(a), 2.6×10^{11} W/cm^2(b), 2.8×10^{11} W/cm^2 (c), 3.0×10^{11} W/cm^2 (d). The prepulses reach their peak at 1.2 ns in the time scale used in Fig. 2. The prepulse with a peak intensity of 2.8×10^{11} W/cm^2 and 3.0×10^{11} W/cm^2 can produce a high fraction of Ni-like ions in preplasmas. But in Fig. 2(d) we can see an over-ionized region near 50 μm distance at 1400–1600 ps. The prepulse with 2.8×10^{11} W/cm^2 peak intensity can produce high fraction of Ni-like ions without over ionization.

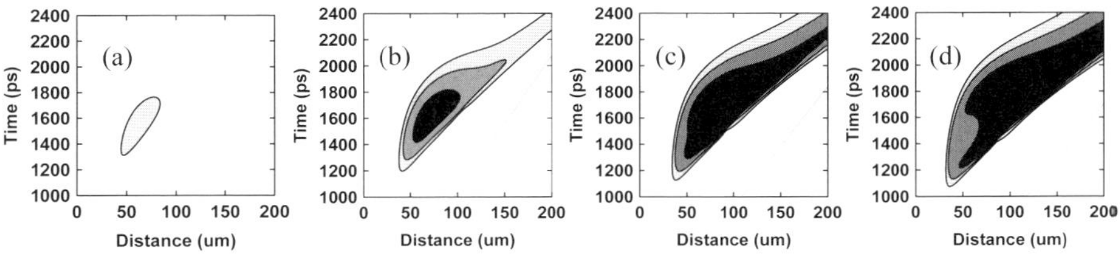

Figure 2. The Ni-like Mo ion fraction versus space and time in the plasma generated by 1 ns prepulse with peak intensities of 2.4×10^{11} W/cm^2 (a), 2.610^{11} W/cm^2 (b), 2.8×10^{11} W/cm^2 (c) and 3.0×10^{11} W/cm^2 (d) . The fraction in plot is from >60% (blackest) to <20% (white) by a step of 20%.

4. X-ray laser simulation

In our simulations, the x-ray laser gain is generated by a 1 ps, 1×10^{15} W/cm^2 peak intensity main pulse after a 1 ns, 2.8×10^{11} W/cm^2 intensity prepulse. Before the main pulse, the preplasma should have a rich population of Ni-like ions and large scale length, so the suitable delay time can be determined from Fig. 2(c). The time region at 1.6–1.8 ns seems to be a good starting time for the main pulse. To investigate the effect of the delay time, we modeled the x-ray laser for different delay times of 0.2, 0.4, 0.6, and 0.8 ns. Figures 3(a), 3(b), 3(c) and 3(d) show the x-ray laser gain at 18.9 nm versus space and time for 0.2, 0.4, 0.6, and 0.8 ns delays, respectively.

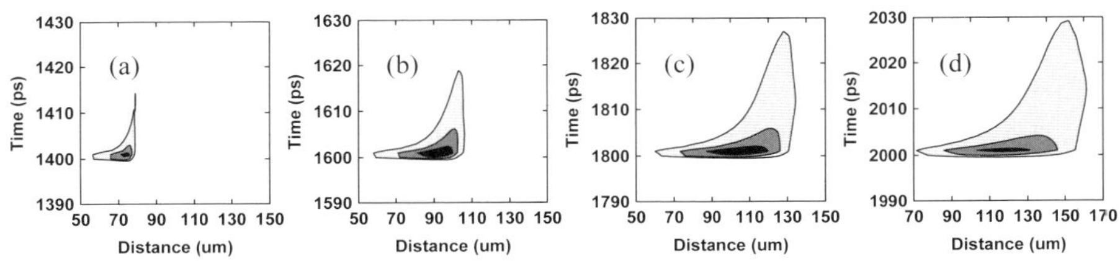

Figure 3. Contours of Ni-like Mo x-ray laser gain at 18.9 nm generated by a 1 ps, 1×10^{15} W/cm^2 drive pulse with delay time of 0.2 ns(a), 0.4 ns(b), 0.6 ns(c) and 0.8nm(d). The gain in the plot is from >150 cm^{-1} (blackest) to <50 cm^{-1} (white) by a step of 50 cm^{-1}.

5. Ray-tracing calculations

The refraction and saturation effect of x-ray laser in the plasma can strongly affect the laser output. In order to model the propagation and amplification of the x-ray laser in the plasma medium, we developed a two-dimensional ray tracing postprocessor to calculate the relative output intensity of x-ray laser for delay times of 0.2, 0.4, 0.6, and 0.8 ns. We assumed that the traveling wave excitation is used, so the spatial profiles of gain and electron density can keep their form unchanged along the line focus. For simplification, the time evolution of the x-ray laser output was not included. In our ray-tracing calculation the spatial profiles of gain and electron density are in the moment of peak temporal and spatial gain, taken from Figs. 3(a) – 3(d). Fig. 4 shows the dependence of the spatially integrated laser output intensity on the plasma length for illustration of the role of refraction and saturation effect. In ideal conditions, the effective gain length production reaches 14 at 1.2 mm target length for 0.6 ns delay. That means if the line focus is 50 mm wide, then the total pump energy for saturated x-ray laser output can be less than 1 J.

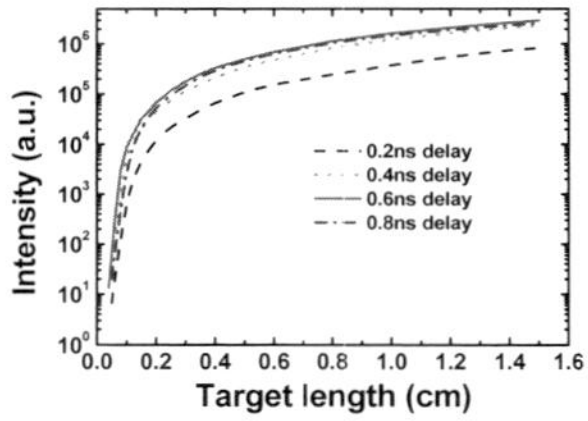

Figure 4. Spatially integrated laser output intensity versus the target length at the peak gain time for the time delays between prepulse and main pulse of 0.2, 0.4, 0.6 and 0.8ns. The optimum time delay should be 0.6ns.

6. Summary

In Summary, we have investigated the Ni-like Mo x-ray laser at 18.9 nm numerically using a one-dimensional hydrodynamic code MED103 coupled with an atomic data package for Mo planar target irradiated by a 1 ns prepulse followed by a 1 ps, 10^{15} W/cm^2 drive pulse from a Ti:sapphher laser. The optimization calculations are performed for intensity of prepulse and delay time between the prepulse and main pulse. The relative output intensity of x-ray laser was predicted using ray-tracing method. We calculated the possibility to generate a saturated Ni-like Mo soft x-ray laser at 18.9 nm using less than 1 J pump energy on a table-top Ti:sapphher laser system.

Acknowledgements

This work was supported by the NSFC under Grant Nos. 10374114, 60321003, 10474134 and 60128505, and the NKBRSF under Grant No. G1999075206, and the National Hi-tech ICF Program.

References

[1] Yu. V. Afanasiev *et al.* 1989 Sov. J. Quant. Electron **19** 1606.
[2] J. Nilsen *et al.* 1993 Phys. Rev. A. 48 4682.
[3] P. V. Nickles *et al.* 1997 Phys. Rev. Lett. **78** 2748.
[4] J. Dunn *et al.* 1998 Phys. Rev. Lett. 80 2825.
[5] P. J. Warwick *et al.* 1998 J. Opt. Soc. Am. B **15** 1808.
[6] M. P. Kalachnikov *et al.* 1998 Phys. Rev. A **57** 4778.
[7] J. Y. Lin *et al.* 1999 Opt. Commun. **166** 211.
[8] Y. L. Li *et al.* 2000 J. Opt. Soc. Am. B **17** 1098.
[9] J. Dunn *et al.* 2000 Phys. Rev. Lett. **84** 4834.
[10] T. Ozaki, *et al.* 1998 Phys. Rev. Lett. 89 253902.
[11] Y. J. Li *et al.* 2001 Phys. Rev. E. **63** 036410.
[12] Y. J. Li *et al.* 2002 Phys. Rev. E. **66** 046501.
[13] J. P. Christiansen *et al.* 1974 Computer Phys. Commun. **7** 271.
[14] R. D. Cowan, The Theory of Atomic Structure and Spectra, (U. of Calif. Press, Berkeley, Calif., 1981).

Inst. Phys. Conf. Ser. No 186
Paper presented at 9th Int. Conf. X-Ray Lasers, Beijing, China, 24–28 May 2004

Demonstration of transient collisional excitation x-ray lasers for Ne-like argon and Ni-like Xenon

Peixiang Lu

State Key Laboratory of Laser Technology, Huazhong University of Science and Technology, Wuhan 430070, China

T. Kawachi, M. Kishimoto, K. Sukegawa, M. Tanaka, N. Hasegawa, M. Suzuki, R. Tai, S. Namba, M. Kado, Y.Ochi, M. Nishikino, K.Nagashima, H. Daido, and Y. Kato

Advanced Photon Research Center, Japan Atomic Energy Research Institute, 8-1 Umemidai Kizu-cho, Souraku-gun, Kyoto 619-0215 Japan

H. Fiedorowicz, and A. Bartnik

Institute of Optoelectronics , Military University of Technology Poland Kaliskiego 2, 00-908 Warsaw, Poland

Abstract: We demonstrated high gain TCE x-ray lasers for Ne-like Ar and Ni-like Xe, using picosecond-laser-irradiated gas-puff targets. The elongated X-ray laser plasma column was produced by irradiating the gas puff target with line-focused double picosecond laser pulses in a travelling-wave excitation scheme. Strong soft x-ray lasing on 3p $^1S_0$ - 3s $^1P_1$ transition for Ne-like Ar at 46.9 nm was observed with ~9 J total energy. A high gain of 18.7 cm^{-1} and a narrow beam divergence of less than 3.7 mrad were measured on the laser transition for targets up to 0.45 cm long. By irradiation of a xenon gas puff target with double 3 ps driving laser pulses, strong Ni-like Xe ion lasing line at a shorter wavelength of 9.98nm was clearly observed with a high gain coefficient of 17.4 cm^{-1}, and a weak lasing line at 9.64 nm was also observed with a gain coefficient of 5.9 cm^{-1}.

1. Introduction

One of the most promising ways of developing table-top x-ray lasers for applications is a scheme based on transient collisional excitation (TCE). This scheme has the advantage that ~10 J of laser energy from a table-top laser generating picosecond high-power laser pulses with chirped-pulse amplification (CPA) is sufficient for lasing. The first transient gain x-ray laser was demonstrated at 32.6 nm in Ne-like Ti[1]. Saturated operation of transient gain x-ray lasers was demonstrated for both Ne-like[2] and Ni-like[3,4] ions with low pump energies. This opened the way for a practical table-top soft x-ray laser suitable for various applications. In this paper, we present results on transient collisional excitation x-ray lasers in which the line-shaped gain medium is produced by laser irradiation of a gas puff target. The use of a gas puff instead of a solid target offers several advantages. From the viewpoint of practical applications, the gas puff target provides a high repetition rate x-ray laser without production of target debris. In principle, the gas puff also allows better control over the density and minimizes gradients in the plasma if we can precisely characterize the interaction processes. X-ray lasers using a gas puff target were first demonstrated with a 0.5 ns duration, 500 J energy laser pulse by a German-Polish collaboration[5]. In our experiment, we used two 1.5

picosecond laser pulses with a total energy of 9 J to create soft x-ray lasing in Ne-like Ar gas puff targets, and a transient high gain of 18.7 cm^{-1} and a narrow beam divergence of less than 3.7 mrad for the Ne-like Ar 3p-3s lasing line were successfully achieved[6]. To extend this gain to the shorter-wavelength region, Ni-like Xe x-ray lasing experiment was carried out by irradiation of a xenon gas puff target with double 3ps driving laser pulses. Strong Ni-like Xe ion lasing line at a shorter wavelength of 9.98nm was clearly observed with a high gain coefficient of 17.4 cm^{-1}, and a weak lasing line at 9.64 nm was also observed with a gain coefficient of 5.9 cm^{-1}.[7]

2. Experimental setup

The experiment was performed on a CPA hybrid laser at APRC of JAERI[4]. This laser provides up to 20 J of light at 1054 nm in a 1.5-3.0 ps pulse. For this work, a prepulse had to be applied before the main pulse, therefore we used the double picosecond pulses(1.5ps for the Ne-like Ar laser and 3.0ps for Ni-like Xe lasers) as the driving laser pulse for soft x-ray lasing experiments. The separation time between the double picosecond pulses was set to be about 1.2 ns, and the energy ratio of the first pulse to the second pulse was about 1 : 6.5(1:8 in the case of Ni-like Xe lasers). The line focus with a length of 0.55 cm and a width of 20 μm was achieved by using a combination of an off-axis parabolic mirror and a spherical mirror[4]. For a typical total output energy of 9 J in the Ne-like Ar experiment(18J for Ni-like Xe lasers), the irradiances in the line focus are of about 7.0×10^{14} Wcm^{-2} (6.0×10^{14} Wcm^{-2} for Ni-like Xe lasers) for the first pulse and about 4.7×10^{15} Wcm^{-2} (4.8×10^{15} Wcm^{-2} for Ni-like Xe lasers)for the main laser pulse. Travelling wave geometry was used to irradiate the target by using a stepped mirror technique[4].

The gas puff target was formed using a solenoid valve developed at the Institute of Optoelectronics. It was equipped with a 0.6-cm-long nozzle with a 500-μm-wide slit[8]. Argon and xenon gases were used to form gas puff targets. The laser beam illuminated the gas puff target in the transverse direction with respect to the flow of gas. To achieve a high gain, a high gas density around 10^{20} cm^{-3} is required. Therefore, the laser line focus position was placed about 150 μm (80 μm for Ni-like Xe lasers) above the slit nozzle output. The gas backing pressure in the valve was set to be 10 bar and the valve time delay between the opening of the valve and the laser pulse was 400 μs. The maximum gas density in the interaction region was roughly estimated to be $\sim10^{20}$ cm^{-3} using the x-ray backlighting method[8]. The maximum plasma column length was set to be 4.5 mm so that the line focus was overfilling the gas puff target in order to avoid residual absorption of the x-ray laser in the cold gas in the direction of the main detector. For the gain measurement, the length of a plasma column was changed by blocking part of the gas puff from being irradiated by means of a thin silver plate. The main diagnostic, aligned to the axis of the line focus, was a 1200-line/mm flat-field grating spectrometer with a back-illuminated charge-coupled device (CCD). The spectrometer used a gold-coated spherical mirror collection optic to observe the on-axis x-ray emissions. An additional 2400-line/mm flat-field grating spectrometer was aligned 45^0 off the axis to monitor the ionization balance of the plasma column.

3. Results and Discussions

Figure 1 presents typical on-axis emission spectra for the laser-irradiated argon gas puff targets with two different plasma column lengths of 0.25 and 0.45 cm. The driving laser pulse with a combination of a ~1.2 J, 1.5 ps prepulse and a ~7.8 J, 1.5 ps main pulse was used to irradiate a target, and the total energy was 9.0 ± 1.0 J. One can clearly see from Fig. 1b that the Ne-like argon *3p* - *3s* laser line at 46.9 nm dominates the spectrum for a gas puff target length of 0.45 cm. A very large increase in the x-ray lasing intensity with increasing gas puff target lengths from 0.25 to 0.45 cm strongly indicates a very high gain for the Ne-like Ar 3p-3s x-ray laser. In our experiment, we did not observe the self-photopumped 3d-3p laser line at 45.1 nm, which was observed to be very strong in the experiment using a combination of a long 600 ps prepulse followed by a 6 ps driving pulse[9]. A shorter pumping pulse of 1.5 ps used in our experiment might be the reason, and our result is advantageous for possible applications.

In Fig.2, we show the intensity versus plasma column length for the Ne-like Ar $3p$ - $3s$ laser line. These data points were obtained under the condition of a total driving laser energy of 9.0 ±1.0J. Using the Linford formula[10], we obtained a high gain coefficient of 18.7 ± 1.0 cm^{-1} for the Ne-like Ar 3p - 3s, J = 0 - 1 transition lasing. This corresponds to a gain-length product of 8.4 for the 0.45-cm-long plasma column.

From the registered spectral distribution, we could also obtain the angular profile of the x-ray laser line and determine the beam divergence angle. Figure 3 shows the angular distribution of the Ne-like argon $3p$ - $3s$ x-ray laser for a gas puff target of 0.45 cm. From Fig.3, we can estimate the divergence of the x-ray laser beam to be less than 3.7 mrad. However, the angular profile is not so regular and the divergence for the central part is even less than 1 mrad. Tentatively we suppose that this is a result of modes in the amplifier. Also, this value is significantly smaller than that obtained with a combination of a long 600 ps prepulse followed by a 6 ps driving pulse, indicating a larger beam divergence of about 9 - 12 mrad[9]. The small beam divergence achieved in our experiment is mainly due to the 1.5 ps shorter pumping pulse resulting in a more transient gain in the narrower high-density gain region.

To extend the scheme to shorter wavelength x-ray lasers, we performed an experiment for the Ni-like Xe x-ray laser at ~10 nm with two 3-ps driving laser pulses with a total energy of 18 J. Fig. 4 presents on-axis x-ray emission spectra for the laser-irradiated xenon gas puff targets with two different plasma column lengths of 0.26 and 0.40 cm. The driving laser pulse with a combination of a ~2J, 3 ps prepulse and a ~16 J, 3 ps main pulse was used to irradiate xenon gas puff targets with the total energy of 18± 2 J, and the backing pressure in the valve of xenon gas puff target was set to be 4 bar. From Fig. 4b, one can clearly see a strong lasing line as well as another weak lasing line at a shorter wavelength. The Ni-like Xe 4d – 4p, J=0-1 laser line at 9.98 nm dominates the spectrum for a longer gas puff target length of 0.40 cm. A very large increase in the 9.98-nm x-ray lasing intensity with increasing gas puff target lengths from 0.25 to 0.40 cm strongly indicates a very high gain for the Ni-like Xe 4d-4p, J=0-1 x-ray laser. The weaker lasing line at the wavelength of 9.64nm may be identified as belonging to the lasing transition of $(3d_{3/2}, 4d_{3/2})_{J=0} - (3d_{3/2}, 4p_{1/2})_1$.

The x-ray lasing was fairly reproducible except for one strongest shot with a target length of 4 mm. Therefore, we performed a gain measurement by moving the thin titanium plate to vary the plasma lengths from 0.26 to 0.45 cm. Fig.5 shows the intensity versus plasma column length for the Ni-like 4d – 4p, J=0-1 laser lines at 9.64 nm and 9.98 nm. These data points were obtained under the condition

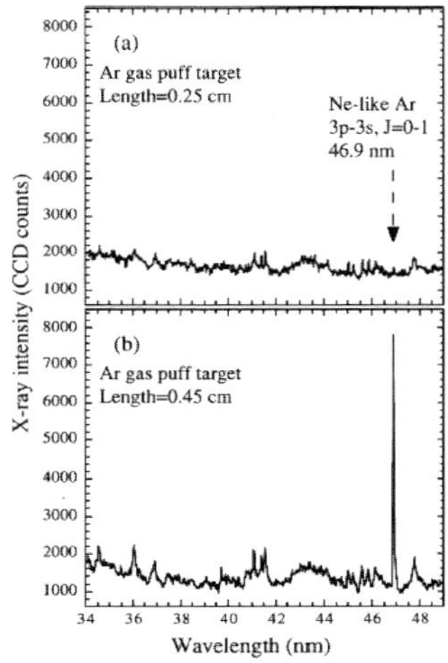

Fig. 1. Typical on-axis emission spectra for (a) 0.25- and (b) 0.45-cm-long argon gas puff targets.

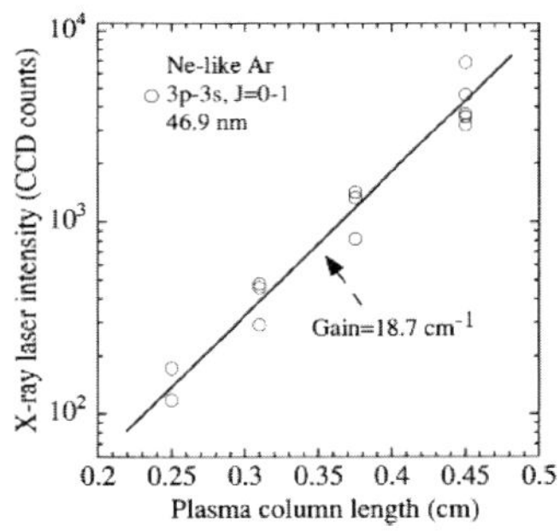

Fig. 2. Ne-like Ar $3p\ ^1S_0$ - $3s\ ^1P_1$ x-ray laser line intensity (open circles) versus length of the plasma column.

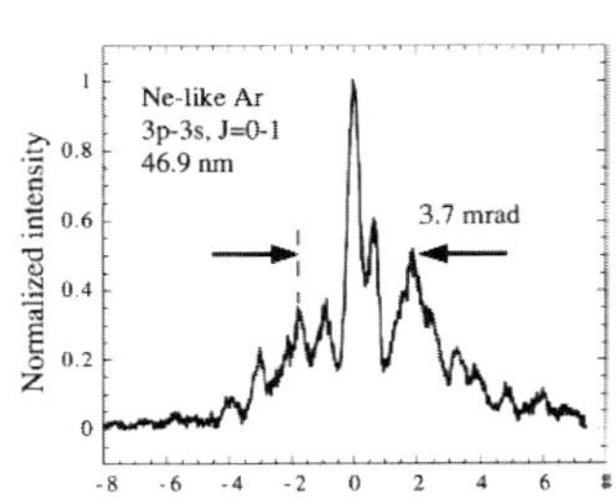

Fig. 3. Angular distribution of the Ne-like Ar $3p\ ^1S_0$ - $3s\ ^1P_1$ X-ray laser for a 0.45-cm-long gas puff target.

of a total driving laser energy of 18.0 ± 2.0 J. Using the Linford formula[10], we obtained a high gain coefficient of 17.4 ± 2.7 cm^{-1} for the strong Ni-like Xe laser at 9.98 nm, and a gain coefficient of 5.9 ± 1.8 cm^{-1} for the weak lasing line at 9.64 nm. These correspond to gain-length products of 7.8 and 2.6 for the 0.45-cm-long plasma column, respectively. In Fig.5b, we can see one exceptional data point is far away from the gain curve, and its value is 5~6 times larger than those of other data points for the same target length of 4 mm, which implies that the current x-ray laser did not work mainly under the optimum condition. From the intensity difference between this data point and others for a target length of 4 mm, we can estimate that the gain coefficient for the 9.98-nm lasing line could reach about 22 cm^{-1} if the laser could work under the same condition as the best shot.

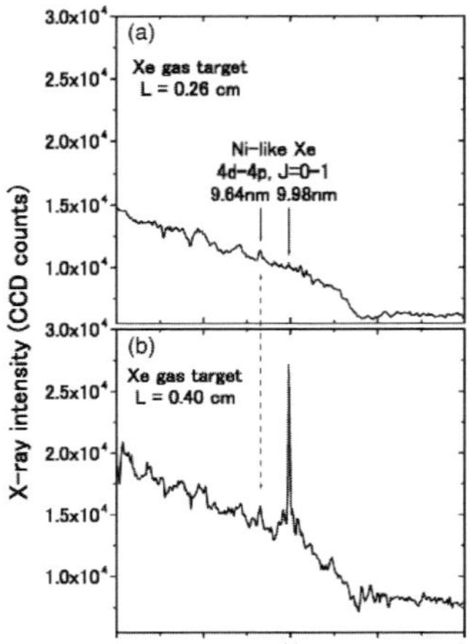

Fig. 4. On-axis x-ray emission spectra for (a) 0.26- and (b) 0.40-cm-long xenon gas-puff targets.

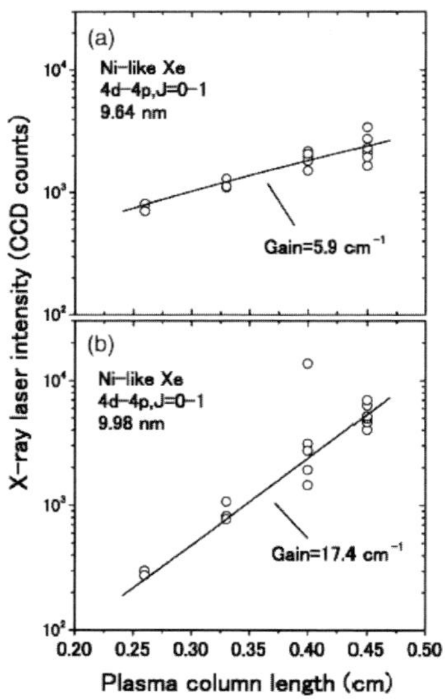

Fig. 5. X-ray laser intensity versus length of the plasma column for Ni-like Xe $4d$ $4p$, J 0 1 x-ray lasers at (a) 9.64 nm and (b) 9.98 nm. Note that an exceptional data point for the target length of 4 mm was excluded from the gain fitting.

4. Summary

In summary, we have observed large soft x-ray amplification in Ne-like Ar and Ni-like Xe ions by irradiation of gas puff targets with two picosecond driving laser pulses in a travelling-wave geometry. Strong soft x-ray lasing on $3p\ ^1S_0$ - $3s\ ^1P_1$ transition for Ne-like Ar at 46.9 nm and $(3d_{3/2}, 4d_{3/2})_0$ - $(3d_{5/2}, 4p_{3/2})_1$ transition for Ni-like Xe at 9.98 nm was observed. The result yields a positive prospect for the demonstration of a high-repetition-rate saturated x-ray laser with no target debris production at a shorter wavelength of ~10 nm for x-ray laser applications.

Acknowledgments

This work was supported by the National Key Basic Research Foundation under grant No. TG1999075206-2 and the National Natural Science Foundation of China under grant No.10375083.

References

1. P. Nickles et al., Phys. Rev. Lett. **78**, 2748-2751 (1997).
2. M.P.Kalachnikov et al, Phys. Rev. A **57**, 4778(1998).
3. J. Dunn et al., Phys. Rev. Lett. **84**, 4834(2000).
4. T. Kawachi et al Phys. Rev. A **66**, 033815(2002).
5. H. Fiedorowicz et al., Phys. Rev. Lett. **76**, 415-418(1996) .
6. Peixiang Lu et al., Jpn. J. Appl. Phys. 41, L133(2002).
7. Peixiang Lu et al., Opt. Lett. 27, 1911(2002).
8. H. Fiedorowicz et al., Opt. Commun. **163**, 103(1999).
9. H. Fiedorowicz et al., Opt. Lett. **26**, 1403 (2001).
10. G. J. Linford et al., Appl. Opt. **13**, 379(1974).

Inst. Phys. Conf. Ser. No 186
Paper presented at 9th Int. Conf. X-Ray Lasers, Beijing, China, 24–28 May 2004

Analysis of gain generation of the TCE x-ray laser through temporal duration

Yoshihiro Ochi, Tetsuya Kawachi, Akira Sasaki, Keisuke Nagashima, Noboru Hasegawa, Momoko Tanaka, Maki Kishimoto, Masaharu Nishikino, and Yoshiaki Kato

Advanced Photon Research Center, Kansai Research Establishment, Japan Atomic Energy Research Institute, Kyoto, Japan

Abstract. Pulse duration of the TCE x-ray lasers were measured using an x-ray streak camera coupled with a flat field grating spectrometer. The XRL was driven by 4.8-ps CPA Nd:glass laser pulse with 400-ps prepulse at 1053 nm. Three target materials, Ag (Z=47), Pd (Z=46), and Mo (Z=42), were used.. Obtained duration was 7.0 ps, 5.6 ps, and <3.3 ps for Ni-like Ag-, Pd-, and Mo-XRLs respectively, which showed XRL pulse duration became longer with the atomic number increasing.

1. Introduction

Since the first experimental demonstration of the transient collisional excitation (TCE) [1], several x-ray lasers (XRLs) pumping by use of table-top class lasers have been demonstrated. At the Advanced Photon Research Center, we have demonstrated the saturated gain with Ni-like 13.9 nm Ag- and 12.0 nm Sn-XRLs [2] and substantial gain with 8.8 nm La-XRL [3] using CPA Nd:glass laser. Recently, we demonstrated a spatially full-coherent XRL with Ni-like Ag by means of the double targets scheme [4]. This high quality XRL beam will be provided to application researches such as x-ray interferometric diagnostics, non-linear x-ray processes and nano-processing. Hence it is important to determine a pulse duration and temporal coherence of the XRL. In this paper, we obtained pulse duration of the Ni-like Ag-, Pd-, and Mo-XRLs by means of temporally and spectrally resolving measurements. The gain duration derived from the pulse duration suggested that a collisional ionization is a dominant process to degenerate the inversion population density in the plasma medium.

2. Experimental description

The pulse duration measurements of the Ni-like XRLs were performed with compact CPA Nd:glass laser system at the Advanced Photon Research Center [5]. The central wavelength is 1053 nm and the spectral bandwidth after the amplifier chain is 3.5 nm. The laser light after a regenerative amplifier is divided by a beam splitter into a prepulse and main pulse chain. The prepulse is stretched by an additional stretcher and the main pulse goes through a variable delay line, then these pulses are combined coaxially.

In the present experiment, 400-ps prepulse and 4.8-ps main pulse with separation of 500 ps was chosen for pumping the XRL. The energy ratio of the prepulse to the main pulse was manipulated to be 1:8 by means of the first beam splitter. A 4-mm length and 20-μm width line focus was made by an off-axis parabolic mirror and a spherical mirror. A quasi-travelling wave using 6-step mirror was employed in order to compensate the short gain duration in the plasma medium [2]. The travelling wave velocity was 1.03c measured using an optical streak camera, where c represents the light speed [6]. It is noted that the minimum pulse duration to be measured is determined by means of reflection from each block in the line focus, i.e. 4.0 ps (1.2 mm)..

At first, we measured temporally integrated spectrum of the Ni-like Ag XRL (Z = 47, $\lambda_{4d\text{-}4p}$ = 13.9 nm) using a back-illuminated (BI) CCD camera coupled with a flat field spectrometer equipping with 1200 lines/mm grating in order to investigate the XRL output intensity. As a result, the XRL output intensity was estimated to be 7.5 x 10^8 W/cm^2, which was one or two orders of magnitude less than the saturated intensity. Therefore we can say that we made XRL duration measurements in not-saturated condition. Then temporal variations of Ni-like Ag-, Pd- (Z = 46; $\lambda_{4d\text{-}4p}$ = 14.7 nm) and Mo- (Z = 42; $\lambda_{4d\text{-}4p}$ = 18.9 nm) XRLs were measured using an x-ray streak camera instead of the BI-CCD camera. Instrumental width was decided by temporal response of the streak camera and temporal smearing due to the grating in the spectrometer resulting 3.1 ps (Ag), 3.2 ps (Pd), and 3.9 ps (Mo). Shown in fig. 1 are obtained streak images (bottom) and temporal profiles integrated over the indicated region (top). In the figure, Gaussian curve fitted for the temporal profile is also presented by a solid line.

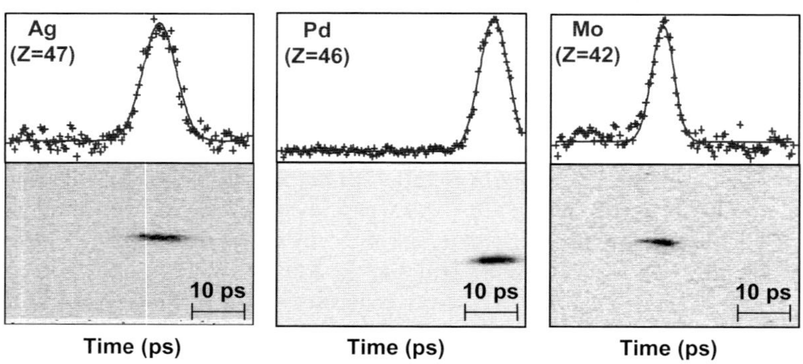

Figure 1. Streak spectrograph of Ni-like Ag-, Pd-, and Mo-XRLs (bottom) and their spectrally integrated profiles (top).

3. Discussion

Pulse duration (FWHM) is obtained by deconvoluting the instrumental width from the data shown in Fig. 2. The duration is 7.0 ps, 5.6 ps, and <3.3 ps for Ag-, Pd-, and Mo- XRLs respectively. Since the Mo-XRL duration is shorter than the minimum measurable value due to the block width of the step mirror, we can not decide it quantitatively. However it is clearly seen that the XRL pulse duration becomes longer as the atomic number increasing. If the plasma heating was occurred at the density region given by the plasma scaling low [7], i.e. $Ne \propto Z_{eff}^{7}$ and $C \propto Z_{eff}^{-3}$, where C means the rate coefficient for the collisional process and Z_{eff} the effective atomic charge, the duration should be scaled by $\propto Z_{eff}^{-4}$. However the experimental results show the contrary trend. This suggests that the pumping laser heated the same density region in any case not a suitable gain region.

It is noted that the gain duration is longer than the XRL duration. The output XRL profile *I(t)* is corresponded to the gain profile *g(t)* by a relationship as $I(t) = \exp(g(t) \cdot l)$, where l is the gain length. Assuming the Gaussian gain profile $g(t) = g_0 \cdot \exp(-4\ln 2(\frac{t}{\tau_{gain}})^2)$, where τ_{gain} is the FWHM duration, *I(t)* is expressed as $I(t) = \exp\left[g_0 l \cdot \exp(-4\ln 2(\frac{t}{\tau_{gain}})^2)\right]$. In the present experiment, the gain-length product $g_0 l$ is estimated to be ≈ 12 from the XRL output energy obtained from the temporally integrated spectrum. By comparing the gain profile and the XRL output profile, it is resulted that the gain duration becomes 3.3 times longer than the XRL duration. Therefore the gain duration in the present experiment are longer than 10 ps. This suggests that the gain duration is mainly determined by the collisional ionization, since the collisional *l*-mixing time calculated by use of the HULLAC code [8] is sub-ps. We consider two collisional ionization processes; one is the ionization directly from the ground state and another one is through excited levels. The schematic diagram is shown in Fig.2, in which *S(1)* means the collisional ionization from the ground state, *C(1,q)* and *F(q,1)* are the collisional excitation and de-excitation, and *A(q,1)* is radiative de-excitation. Hence the collisional-radiative ionization rate coefficient S_{CR} is given as,

$$S_{CR} = S(1) + \sum_{q \geq 2} \left[C(1,q) - \{F(q,1) \cdot Ne + A(q,1)\} \cdot n(q) \cdot \frac{1}{Ne} \right], \tag{1}$$

where *n(q)* is the population density of the level *q*. We calculate the *S(1)* using the empirical formula given by Landshoff and Perez [9],

$$S(1) = 1.24 \times 10^{-6} \xi \cdot (kTe)^{-3/2} \cdot \frac{\exp(-E/kTe)}{(-E/kTe)^2} \cdot \left[\frac{0.915}{(1 + 0.064\frac{kTe}{E})^2} + \frac{0.42}{(1 + 0.5\frac{kTe}{E})^2} \right], \tag{2}$$

where ξ is the number of bound electrons, kTe and E is electron temperature and the ionization energy in eV respectively. The contribution of ionizations through the excited levels, i.e. the second term in the eq.(1), to *S(1)* is evaluated using the HULLAC code, resulting in 0.15. The calculated S_{CR} is shown in Fig. 3. It is shown that the rate coefficient is lager for lower atomic number material because of the smaller ionization energy, which is consistent with the experimental data.

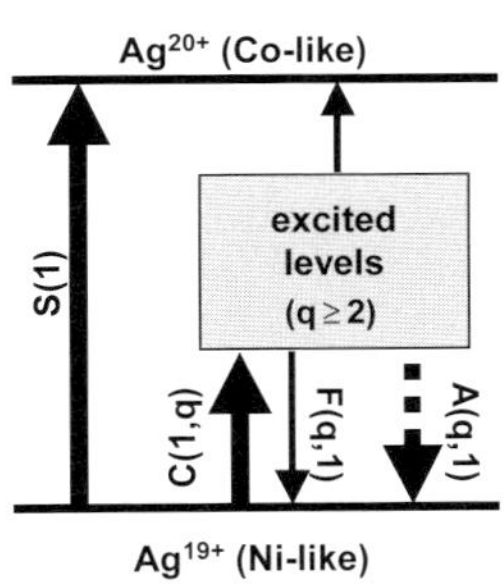

Figure 2. Schematic flow of the collisional ionization.

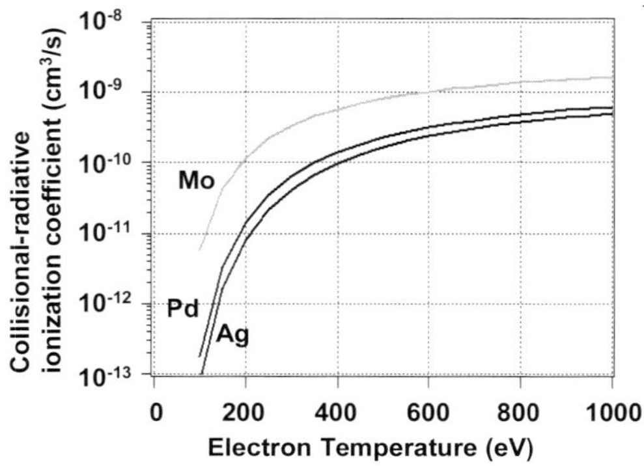

Figure 3. Collisional-radiative ionization rate coefficient for Ag, Pd, and Mo. Collisional ionization from the ground state but also through excited levels are considered.

4. Summary

Pulse duration of the Ni-like Ag-, Pd-, and Mo- XRLs in the transient collisional excitation scheme were measured using the x-ray streak camera coupled with the flat field spectrometer. The obtained duration, 7.0 ps for Ag-XRL, 5.6 ps for Pd-XRL, and <3.3 ps for Mo-XRL showed that the XRL pulse duration increased with increasing atomic number. This suggested that the pumping laser energy was deposited at same density region just under the critical density in any case and the XRL pulse duration was mainly determined by the collisional ionization.

References

[1] Nickles P V, et al, 1997 Phys. Rev. Lett.78 2748-2751
[2] Kawachi T, et al, 2002 Phys. Rev. A 66 033815-1-7
[3] Kawachi T, et al, 2004 Phys. Rev. A 69 033805-1-4
[4] Nishikino M, et al, 2003 Phys. Rev. A 68 0261802-1-4
[5] Kawachi T, et al, 2003 Appl. Opt. 42 2198-2205
[6] Ochi Y, et al, 2004 Appl. Phys. B 78 961-963
[7] Fujimoto T, 1985 J. Phys. Soc. Jpn 54 2905-2914
[8] Klapisch M and Bar-shalom A, 1997 J. Quant. Spectrosc. Radiat. Transf. 58 687
[9] Landshoff R K and Perez J D, 1976 Phys. Rev. A 13 1619-1632

Inst. Phys. Conf. Ser. No 186
Paper presented at 9th Int. Conf. X-Ray Lasers, Beijing, China, 24–28 May 2004

Modeling picosecond-laser-driven Ne-like Ge 19.6nm X-ray laser experiments

Xiumei Qiao[1], Guoping Zhang[2],Tianxin, Zhang[2]
[1]Graduate School of China Academy of Engineering Physics, Beijing, China
[2]Institute of Applied Physics and Computational Mathematics, Beijing, China

Abstract: A simulation model of the TCE (Transient Collisional Excitation) x-ray laser has been developed. With this model, we simulate the lasing of TCE Ne-like Ge 19.6nm soft x-ray laser experiment, which shows qualitative agreement. We also model the near future experiment which will be done on a two-rays laser later this year in Shanghai, by irradiation of 1cm long Ge slab target with two 250ps, 1.053-μm Nd-glass laser which is followed 80ps later by a 1ps, 10J high intensity drive pulse. Calculations with different delays between the long and the short pulses, and with different durations for the short pulse are presented to provide a better insight into optimization of the laser output.

1. Introduction

TCE(transient collisional excitation) enables low-Z Ne-like and Ni-like ions to lase, driven by small lasers ,with only 10 J of energy ,which greatly improves the efficiency. This approach uses a low intensity long prepulse to produce a preplasma with rich population of Ni-like or Ne-like ions in the ground state, after the prepulse, a high intensity short main pulse is used to make a jump of electron temperature (up to several Kev). We have developed a simulation model of the TCE x-ray lasers. The one dimensional hydrodynamic code Jb19 is used to calculate the hydrodynamic evolution of the plasma and to provide the temperatures and densities to the recently developed N61 code, which performs the kinetics calculations to determine the gain. The 37 fine structure levels in the Ne-like Ge ,113 fine structure levels in F-like Ge with main quantum number n up to 3 and 12 fine structure levels in Na-like Ge with main quantum number n up to 4 are considered, for which the level energies and electron collisional excitation rates and radiation rates are provided by the Atomic Group of Institute of Applied Physics and Computational Mathematics(IAPCM), for higher excited states as well as other ions ,the energies are calculated using the screened hydrogenic model, while the collisional and radiative rates are estimated using empirical formulas.

In section two, we model the Ne-like Ge 19.6nm soft x-ray laser experiment by irradiation of 0.9cm long Ge slab target with two 280ps long prepulses and after some delay time, with another 2.7ps short pulse, the short pulse irradiance was 1.5×10^{15} w/cm^2, the main long pulse irradiance was 5×10^{12} w/cm^2, and the first prepulse arrived 2 nsec early and the pulse energy ratio between the two pulses was 0.1 and simulations show good agreement. In section three, we model the future experiment on a two-ray laser in Shanghai and present results with different conditions to provide optimization of the Ne-like Ge soft x-ray laser.

2 Modeling

2.1 Modeling the RAL experiment[1]

We calculated Ne-like Ge 19.6 nm soft x-ray laser experiment[1], but with different delay times between the second prepulse and the third prepulse,figure.1 shows the variation of the output energy of Ne-like Ge 19.6nm x-ray laser versus the delay time, the horizontal axis stands for the CPA peak time relative to the peak time of the second prepulse in ps ,and the vertical axis stands for the output x-ray laser energy in μJ, and the target length in this figure is 5mm, it indicates that the output energy decreases as the delay time increases and that the output energy gets the largest value at 0ps delay time, comparing this figure with figure 2 in [1], we can see that except for 350ps delay time, our results is lower than experimental values, which can be attributed to the uncertainty in measuring the output energy and the best delay time is 0ps but not 50ps in reference [1],pursuing the reason for that we give figure 2.

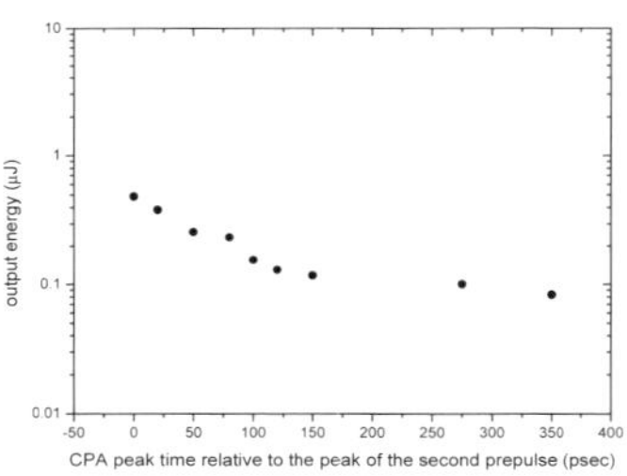

Figure 1. X-ray laser output dependence on the delay time between the short pulse and the second prepulse

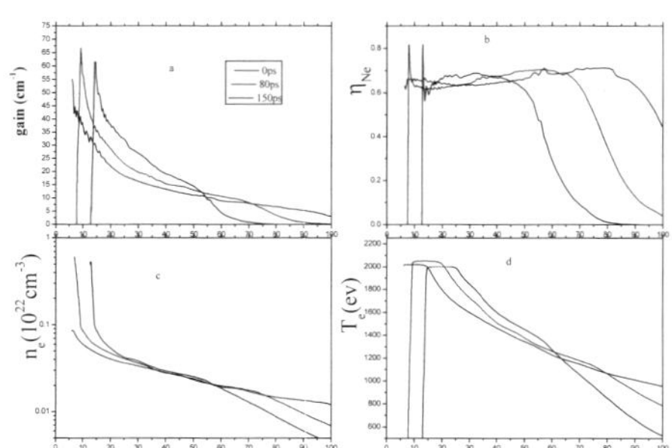

Figure2. Gain, fraction of Ne-like ions, electron density and temperature of the 19.6nm line at the time of peak intensity as functions of distance from the target surface for 0ps (solid line), 80ps (dot line) and 175ps (dash-dot line) delay time. The zero at the horizontal axis is the position of the target surface.

Figure 2 shows the variation of gain, fraction of Ne-like ions and electron density and temperature versus distance from target surface at the time of peak intensity of the 19.6nm x-ray laser line. We can see that the gain and the fraction of Ne-like ions jump to peak value because of the sharp increase of the electron temperature near the critical surface and although for 80ps delay time ,it has the largest peak gain near the critical surface because of highest electron temperature ,it contributes little to the effective gain-length product because of refraction , for the shortest delay time 0ps, the gain has the largest value during the main gain region, which explains why the 0ps delay time works best, and outside the main gain region, it decreases more quickly to 0/cm because of the quick decrease of the electron density. And the electron temperature only gets peak value in a very narrow spatial region and then falls slowly with the distance from target surface because in TCE, the short pulse directly heats up the preplasma, whereas the electron heat conduction process contributes little, and as the electron density decreases with the distance from target surface in the under-dense region, the absorb efficiency of the short pulse drops correspondingly. We also note that the electron temperature gets higher value in a very thin layer over the critical surface which can be attributed to slight difference between the electron densities from the hydrodynamic code Jb19 and from the atomic kinetics code N61.

To give more detailed comparison, we present analytic growth curves based on the Linford formula [2] for a range of small signal gain coefficients with delay time 80ps as shown in figure 3,in this figure and the rest figures in this section, the target length takes 9mm and the delay time takes 80ps, the horizontal axis stands for target length in μm and the vertical axis stands for the total output energy in mJ, comparing it with figure 3 in [1], the peak gain coefficient achieved is >30/cm .The total output energy from the 9mm target is ~ 40.5μJ which is a bit less than the value given by experiment[1] ,within the uncertainty in measuring the x-ray laser energy, the results are believable.

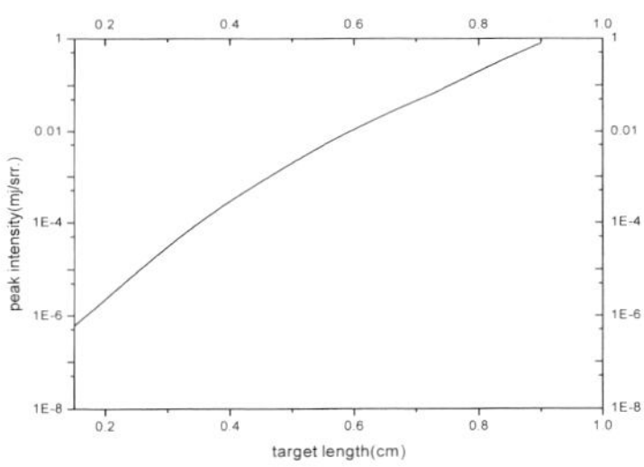

Figure 3 Peak intensity (mJ/srr.) output versus target length (cm) for 19.6 nm line

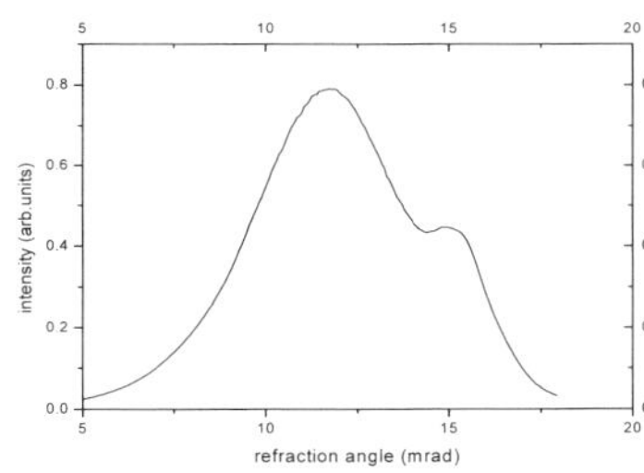

Figure 4 .Refraction angle for the intensity from 9mm target of Ge

Figure 4 shows angular distribution for the output intensity from 9mm Ge target, The beam refraction angle is ~ 11.5 mrad and the divergence angle is ~6.2 mrad, comparing them with the experimental value 10 mrad and 9mrad,we can see that the divergence angle in our results is smaller, which in part explains the lower output energy calculated in our simulation and simulations indicate that the x-rays which contribute most to the output start at 48.3μm from the target surface where the electron density is $2.4 \times 10^{20} cm^{-3}$ with the electron temperature is 1.4keV and the fraction of Ne-like ions is 0.7.

2.2 Simulation of Future Experiment

According to the capacity of a two-ray laser in Shanghai, we designed a series of experiments, and did some calculations with different parameters, and in this section we will show our simulations to provide optimum conditions for experiments.

Simulations were carried out for Ne-like Ge 19.6nm x-ray laser by irradiation of Ge slab target with two 250ps long pulses which is followed after 80ps,by a 1ps short pulse producing a ~100μm wide and ~1cm long focus line, all the pulses are Gaussian at 1.053μm in wavelength, and the target length is ~0.9cm, the total energy of the two long pulses is 10 J, and the energy of the short pulse is also 10 J, the peak to peak delay time between the first prepulse and the second prepulse is 450ps, and the energy ratio between the first pulse and the second one is 0.3, and the quasi traveling wave pump for the short pulse is achieved by a step-mirror.

We did calculations with different delay times between the second prepulse and the short pulse, simulation predicts that for the 40 ps delay time, the output energy achieved is 9.52μJ, which is the largest and after 40ps delay time, the output energy drops to 2.74μJ as the delay time increases to 120ps.Figure 5 shows the dependence of gain, fraction of Ne-like ions, electron density and temperature on the distance from target surface at the time of peak intensity for 20ps, 40ps, 80ps and 120ps delay times. As

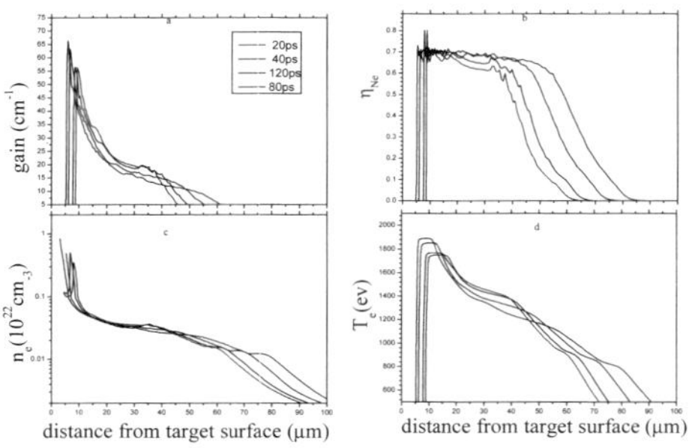

Figure5. Gain, fraction of Ne-like ions, electron density and temperature for the 19.6nm line at the time of peak intensity as functions of distance from the target surface for 20ps(dot line), 40ps (solid line), 80ps (dash-dot-dot line) and 120ps (dash-dot line) delay time The zero at the horizontal axis is the position of the target surface.

figure 2 in section 2, for the 40ps delay time, it has proper fraction of Ne-like ions, electron density, and higher electron temperature and gain in the main gain region, which makes it works best.

3 Conclusions

In this paper ,we have modeled the experiment[1] to show good agreement and have modeled the future experiment performed with a two-ray laser in Shanghai to irradiate ~0.9cm-long slab target of Ge with two 250ps prepulses that is followed 80ps later by a ~1ps drive pulse. Different delay times between the short pulse and the second prepulse are studied and simulations indicate that for the 40 ps delay time, the output energy can be 9.52μJ which is the largest.

The authors thank the Atomic Group of IAPCM for their atomic data .And thank the workmates in High Computation Center of IAPCM for the convenience they provided .

Reference:

[1] R. E. King , G. J. Pert ,etc. , Phys. Rev. A, Vol. 64 ,053810 (2000)
[2] G. J. Linford, E. R. Peressini, et.al. ,Appl. Opt. 13,379 (1974)

Inst. Phys. Conf. Ser. No 186
Paper presented at 9th Int. Conf. X-Ray Lasers, Beijing, China, 24–28 May 2004

Generation of the coherent x-ray laser using two gain media

M.. Tanaka, M. Nishikino, T. Kawachi, N. Hasegawa, M.. Kado, M.. Kishimoto, K. Sukegawa, Y. Ochi, K. Nagashima, Y. Kato
Advanced Photon Research Centre, Japan Atomic Energy Research Institute,
8-1-1-2, Umemidai, Kizu-cho, Souraku-gun, Kyoto, 619-0215, Japan

Abstract. We have succeeded in generating nickel-like silver x-ray laser with full spatial coherence. The highly directed x-ray beam with a divergence of 0.2 mrad is generated by use of two gain media, where the seed x-ray laser from the first medium is amplified the second medium. The observed beam divergence is close to the diffraction-limited value within a factor of two.

1. Introduction

Recently, x-ray lasers are extensively used as soft x-ray sources for several measurements, such as soft x-ray scattering [1] and interferometry [2, 3]. For these measurements, pulse duration, intensity and coherence of the x-ray laser are important. The transient collisional excitation (TCE) x-ray lasers have been intensively studied, because they can achieve a high gain of a few tens per centimeter [4, 5] and picosecond pulse duration [6, 7] using a tabletop laser system. However, the beam divergence of the TCE x-ray laser is large and spatial coherence is quite low, due to refraction from the density gradient of the plasma and the large Fresnel number of the gain medium.

We have used the double target configuration to improve the coherence of the TCE x-ray laser. A part of the x-ray laser generated in the first medium is used as the seed x-ray laser, and it is injected into the successive second medium, which is used as an amplifier with calm density gradient. Here we describe experimental results of the generation of nickel-like silver x-ray laser (λ=13.9nm) using this technique [8, 9].

2. Experiment

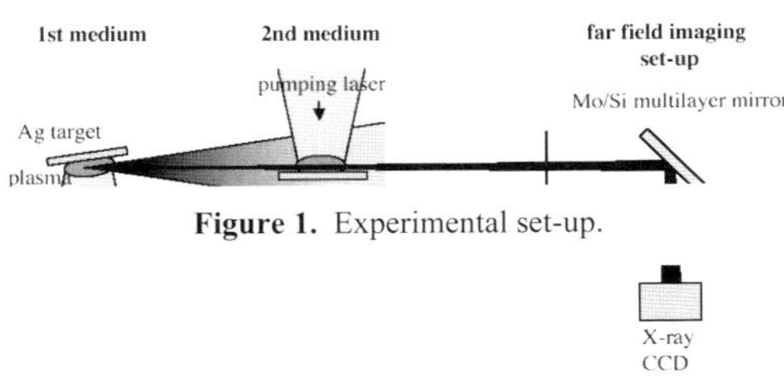

Figure 1. Experimental set-up.

Figure 1 shows a schematic diagram of the experiment. The gain medium plasmas are generated with a CPA Nd:glass laser system with two beam lines [10]. Each beam consisted of two pulses; a pre-pulse with a pulse duration of 300 ps and a picosecond heating pulse with a temporal delay of 600 ps from the pre-pulse. The pulse duration of the heating pulse for the first target is 4 ps with the quasi-travelling wave arrangement [5] and that for the second target is 12 ps without the travelling wave. The total pumping energy is set to be 14 J for the first target and 11 J for the second target. The energy ratio of the pre-pulse and the heating pulse is 1:8. The laser pulses are focused with a line shape of 6.5 mm × 20 μm on flat silver targets with an irradiance of the heating pulse of $\sim 10^{15}$ W/cm^2. The distance between the two targets is 20 cm. Since the x-ray laser from the first target is bent by refraction, the first target is rotated by 7 mrad in order to make the direction of the seed x-ray laser exactly parallel to the surface of the second target. The injection timing of the seed x-ray laser with respect to the amplifier plasma is varied by changing the optical delay of the pumping laser of the first target. The time integrated energy and the far field pattern of the x-ray laser are measured using back-illuminated soft x-ray CCD.

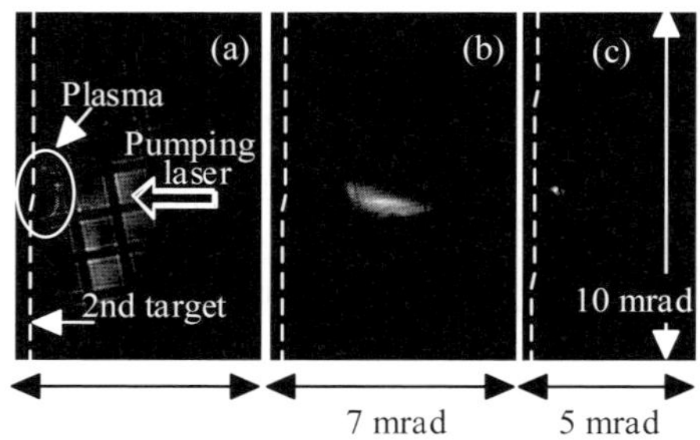

Figure 2. Far field patterns obtained 74 cm from the second target with the target length of 4.0 mm. The injection timing of the seed x-ray for each figure is (a) ; -15 ps, (b) ; 0 ps, (c) ; +15 ps. The pumping laser is incident from the right hand side. The area encircled with a solid line in (a) is the shadow of the second plasma.

3. Results and discussion

Shown in Fig. 2 (a) is the far field pattern where the pumping time of the first medium is 687 ps before than that of the second medium. We express this timing as -15 ps in this paper, since the required time for the x-ray propagated for 20 cm is 667 ps. The longitudinal shadow indicated by a broken line on the left hand side of the pattern is the surface of the second target. Cross-striped shadow patterns are due to the backing mesh of a zirconium filter. The size of the mesh shadow depends on the position of the light source; small and large cross stripes are due to x-rays from the first medium and the second medium, respectively. At -15 ps, only the seed x-ray laser from the first target is observed. The x-ray laser generated at the second target independent on the first target is too weak to be observed. A shadow of the plasma backlit by the seed x-ray can be observed on the target surface. The beam divergence of the

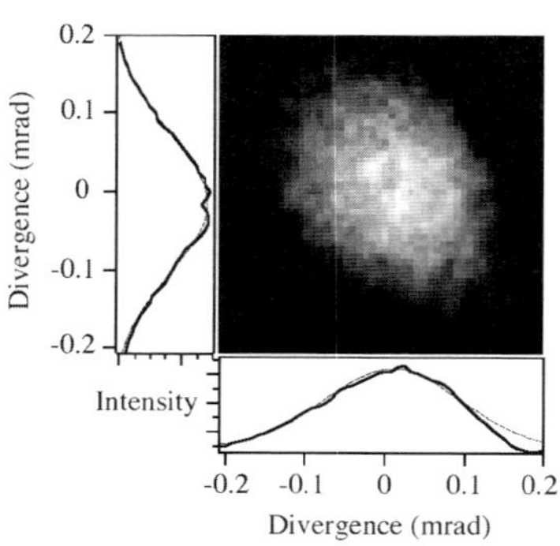

Figure 3. A far field pattern obtained 281 cm from the second target with the target length of 6.5 mm at the timing of +15 ps. The line graphs are the spatial intensity distributions in the horizontal and vertical directions. The dashed line shows experimental results and the solid line shows the curve fit by a Gaussian.

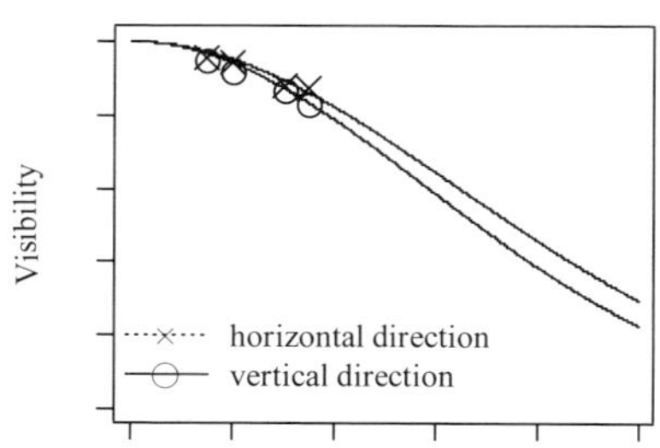

Figure 4. The interference pattern of the horizontal and the vertical direction for the slit separation of 150 μm at 3.9 m away from the double slit.

seed x-ray laser is approximately 6 mrad in full width at half maximum (FWHM). The total energy of the seed x-ray laser is 270 nJ.

Figures 2 (b) and (c) show far field patterns for 0 ps and +15 ps, respectively. The intense beams observed in Fig. 2 (b) and (c) are x-ray lasers amplified in the second medium. When the timing is adjusted to be 0 ps, an intense x-ray laser beam with little divergence of 0.5 mrad in direction parallel to the target surface is observed outside the shadow of the second plasma. The beam is spread in the direction perpendicular (2.5 mrad). This means that the seed x-ray is amplified in the high gain region, where the refraction is dominant. Delaying the pumping time of the seed x-ray from the timing of 0 ps, the spreading x-ray laser becomes weaker, and an intense and narrow spot appears inside the shadow of the plasma. At the timing of +15 ps, only the narrow spot is observed. A magnification of the far field pattern of this spot is shown in Fig. 3. It is fit well with a Gaussian profile. The beam divergence is obtained to be a much reduced value of 0.20 mrad (FWHM) for the directions perpendicular and parallel to the target surface. The diffraction limited beam divergence for a coherent Gaussian beam with a source size of 50 μm [5] is analytically calculated to be 0.11 mrad. The observed divergence is only 1.8 times greater than this value. This implies that the obtained x-ray beam is quite close to the diffraction limited condition.

In order to characterize the spatial coherence of the x-ray laser beam, we carried out a Young's double slit experiment [9]. A pair of slits with width of 16 μm and separation of 150-350 μm is placed 2.3 m away from the second target where the beam diameter is 460 μm. Figure 4 shows the interference pattern for the slit separation of 150 μm recorded by a CCD at 3.9 m away from the double slit. The spatial coherence of the quasi-monochromatic light is characterized by visibility of the fringe pattern defined as $V = (I_{max}-I_{min})/(I_{max}+I_{min})$, where I_{max} and I_{min} are maximum and minimum values of the fringes, respectively. Assuming the Gaussian shell model for the x-ray laser source, the spatial coherence length L_c can be determined by $V(\Delta x) = \exp(-\Delta x^2/2L_c^2)$, where Δx is the separation of slits [11]. Figure 5 shows the dependence of the visibility on the slits separation. The spatial coherent length is estimated to be about 600 μm, which is longer than the beam diameter at the position of the double slit. It means that this narrow beam is spatially full coherent.

Shown in Fig.6 is the total energy of the narrow beam as a function of the length of the second medium. The maximum energy of 25 nJ is measured with the longest medium by integrating the far field pattern over the region with intensity higher than 1/e of the maximum intensity. The total energy will increase with increasing the energy of the seed x-ray and the second target length, because amplification in

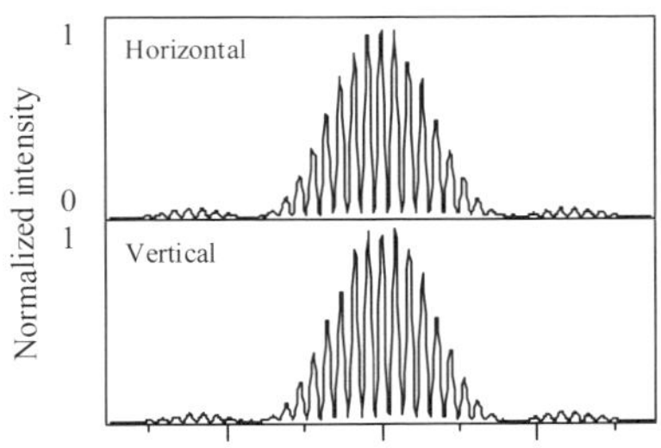

Figure 5. Visibility of the interference pattern plotted with the slit separation. Lines show fitting with a Gauss function.

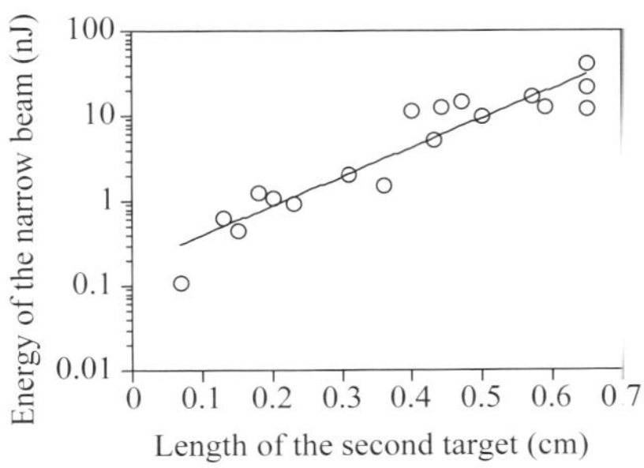

Figure 6. Total energy of the narrow beam as a function of the second target length. The solid line shows fitting with an exponential function.

the second medium is not saturated as can be seen in the figure. The gain coefficient of the amplifier plasma at the timing of +15 ps is estimated to be 7.9 cm^{-1} with exponential fitting to the data. This value is smaller than the gain coefficient of the single target, which is observed to be 35 cm^{-1} in a previous study [5]. This small gain coefficient and small beam divergence suggest that the gain region of this highly directed beam is located in a low density area, where the influence of refraction is negligible.

4. Conclusion

We have succeeded in generating an x-ray laser beam at a wavelength of 13.9 nm with full spatial coherence and nearly diffraction limited divergence of 0.2 mrad. This beam is generated by amplifying the seed x-ray laser from the single gain medium in an additional medium, which is separated 20 cm from the first medium.

References

[1] Tai R Z et al 2002 Phys. Rev. Lett. 89 257602
[2] Filevich J et al 2003 Rhys. Rev. E 67 56409
[3] Tang H et al 2004 Appl. Phys. B 78 975-977
[4] Li Y et al 2000 J. Opt. Soc. Am. B 17 1098-1101
[5] Kawachi T et al 2002 Phys. Rev. A 66 33815
[6] Klisnick A et al 2002 Phys. Rev. A 65 033810
[7] Ochi Y et al 2004 Appl. Phys. B 78 961-963
[8] Tanaka M et al 2003 Opt. Lett. 28 1680-1682
[9] Nishikino M et al 2003 Phys. Rev. A 68 61802R
[10] Kawachi T et al 2003 Appl. Opt. 42 2198-2205
[11] Mandel L and Wolf E 1995 Optical coherence and Quantum Optics (New York: Cambridge University Press)

Inst. Phys. Conf. Ser. No 186
Paper presented at 9th Int. Conf. X-Ray Lasers, Beijing, China, 24–28 May 2004

Equivalence relation between some Ni-like x-ray lasers and their partner Ne-like x-ray lasers

F. Yan, J. Zhang, X. Lu
Laboratory of Optical Physics, Institute of Physics, Chinese Academy of Science, Beijing 100080, People's Republic of China

J.Y. Zhong
National Astronomical Observatories, Chinese Academy of Sciences, Beijing 100012, People's Republic of China

Abstract. The plasma conditions of the Ni-like Ag, Cd, In, Sn, Sb x-ray lasers and the Ne-like Fe, Co, Ni, Cu, Zn x-ray lasers are studied. By comparing the hydrodynamic characteristics of ion abundance, ion population density and electron temperature for these Ne-like ions and Ni-like ions, we conclude that there is a similarity of plasma conditions between these Ne-like and their partner Ni-like x-ray lasers. The ionization energy for these Ne-like ions with their partner Ni-like ions is also compared to further understand this similarity. The comparison of the ionization energies supports the conclusion from the hydrodynamic simulation.

1. Introduction

Due to high quantum efficiency, Ni-like x-ray lasers, in principle, have a more favorable scaling of laser wavelength with drive energy than Ne-like x-ray lasers [1]. However, the population kinetics of Ni-like ions in general is much more sensitive to plasma conditions than that of Ne-like ions. Moreover, it is more difficult to calculate atomic data of Ni-like ions because of the complicated level structure. By comparison, some codes have been developed to successfully simulate Ne-like x –ray lasers, for example, LASNEX [2], JB-19 [3] MEDUSA [4]. In this paper by comparing the hydrodynamic characteristics of plasma conditions for x-ray laser Ni-like and Ne-like plasma conditions, we find a similarity of plasma conditions between some Ni-like x-ray lasers and their partner Ne-like x-ray lasers. Based on this similarity we can predict the performance of some Ni-like x-ray lasers using rather simple hydrodynamic and atomic simulations of their partner Ne-like x-ray lasers

2. Simulation of plasma conditions of the Ne-like Fe and Ni-like Ag x-ray lasers

Simulations are carried out in this paper using the code MEDUSA, which is a one-dimensional Lagrangian hydrodynamic code. In the simulations we use Gaussian pulses at 1.053 μm and 100-μm-thick slab targets. The drive pulse duration is set at 2ns and the peak irradiance is optimized to produce optimum plasma condition.

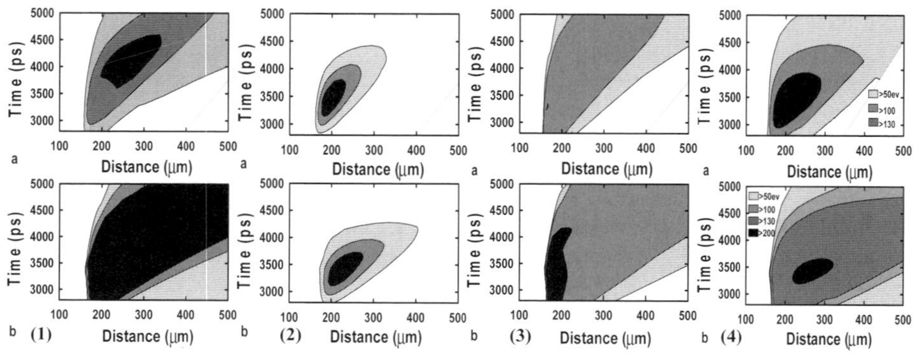

Figure 1 Ni-like ion abundance (1), Co-like ion abundance (2), ion population density (3) and electron temperature (4) for the Ni-like Ag (a) and the Ne-like Fe (b) x-ray lasers are given.

An optimized plasma condition was achieved with a large Ne-like Fe ion abundance and a high gain of the x-ray laser [5]. The same drive condition is used in this paper to calculate the plasma condition of the Ni-like Ag x-ray laser. Fig. 1(1) gives the contours of ion abundance vs space and time for Ni-like Ag (a) and Ne-like Fe (b) ions, respectively. The target surface is located at 100 μm and the drive pulse reaches its peak at 3000 ps. Both the Ne-like Fe and the Ni-like Ag x-ray lasers can reach high ion abundance (>70%) under the same drive conditions. Their distribution contours in space and time are similar, although for the Ne-like Fe case, the area for high ion abundance is lager and appears earlier in time. The next higher ionization stage of the Co-like ions and F-like ions are also calculated under the same drive condition. As shown in Fig.1 (2), the ion abundance of the Co-like Ag ions reaches 10% while the F-like Fe ions only reaches 5%. In order to avoid over-ionization, we carry out another calculation at reduced drive irradiance. When the maximum ion abundance of the Co-like Ag ions equals to that of the F-like Fe (5%) ions, the Ni-like Ag ion abundance is too low to achieve an optimum plasma condition for high gain operation. This implies that it is necessary to tolerate somewhat over-ionization for an optimum operation of the Ni-like Ag x-ray laser. Fig.1 (3) gives the contours of ion population density vs space and time for the Ni-like Ag and the Ne-like Fe ions. These two contours in space and time are similar. The ion density of the Ni-like Ag ions at 1×10^{19}cm^{-3} has less space and time extent than that of the Ne-like Fe ions, due to the larger Co-like Ag ion population than the F-like Fe ion population. Fig.1 (4) shows the contours of the electron temperature (Te) for the Ne-like Fe and the Ni-like Ag x-ray lasers vs space and time. A similarity also exists in distribution shape though the electron temperature for the Ni-like Ag case is lower than that for the Ne-like Fe x-ray laser.

3. Simulation of other partners of Ne-like and Ni-like x-ray lasers

Similarly, we simulate the preplasma conditions for the Ne-like Co, Ni, Cu, Zn x-ray lasers. The laser pulse duration is fixed at 2 ns and the peak irradiance is optimized in order to generate optimum conditions for high gain operation. The optimized intensities are 0.8, 1.2, 1.7, 2.1×10^{12}W/cm^2, respectively. The plasma conditions under the corresponding drive irradiance for the above elements have high Ne-like ion abundance (70%) and low F-like ion abundance. Comparing with the plasma conditions for the Ne-like Fe x-ray laser we conclude that high gain operation can occur under these plasma

conditions. Using the optimized drive intensities for Ne-like ions, plasma conditions for their partner Ni-like x-ray lasers are simulated. By analyzing ion abundance, electron temperature and ion population density, our simulations show that the plasma conditions for the Ni-like Cd, In, Sn, Sb x-ray lasers under the corresponding drive intensities (0.8, 1.2, 1.7, 2.1 ($\times 10^{12}$W/cm^2)) are also suitable to produce high gain.
Comparison of the plasma conditions of these Ne-like x-ray lasers with their partner Ni-like x-ray lasers show that there is a similarity of plasma conditions between some Ne-like x-ray lasers and their partner Ni-like x-ray lasers. The similarity also applies to plasma conditions of Ne-like Co x-ray laser and Ni-like Cd plasma conditions, etc, as shown in Figure 2. The degree of similarity is generally weakened from the light elements to the heavy elements. The main reason is that the Ni-like ions are much easier to be over-ionized to the next ionization stage than the Ne-like ions with the increase of atom number. Although the degree of similarity is weakened generally with the increase of atom number, this still reflects a partnership between some Ne-like plasma conditions and their partner Ni-like plasma conditions under certain drive conditions. Because of the lower value and narrower region of ion abundance, population density and electron temperature for the Ni-like case we can predict that the Ni-like x-ray lasers generally have lower gain and smaller high gain region than the Ne-like x-ray lasers under the same drive conditions.

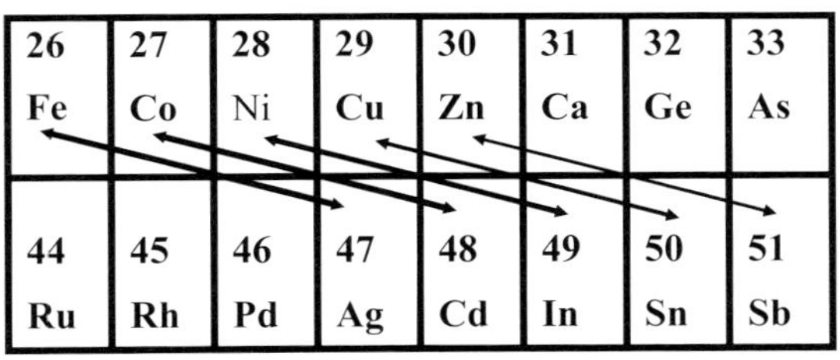

Figure 2 The arrows represent the similarity relation between these Ne-like x-ray lasers and their partner Ni-like x-ray lasers.

4. Comparision of the electron configuration of some Ne-like ions and their partner Ni-like ions

Ne-like ions have ten bound electrons, which configure a ground state with $1s^2 2s^2 2p^6$ closed shells. Population of the 3p levels is provided through collisional excitation from the ground state, cascading from higher-n states, and a combination of three-body radiative and dielectron recombination from the F-like ions. Electrons in the n=3 shell will usually rapidly decay back to the ground state by fast 3s→2p or 3d→2p dipole transitions. The 3p electrons are radiatively metastable to the ground state. Inversion is hence created on various 3p→3s transitions in the Ne-like ions. Similar to Ne-like x-ray ions, Ni-like ions also have a closed outer shell. In Ni-like x-ray lasers, the upper laser level $3d^9 4d\ ^1S_0$ is populated through the monopole collisional excitation from the ground state $3d^{10}\ ^1S_0$, the strong lasing lines are dominated by the $3d^9 4d\ ^1S_0 \rightarrow 3d^9 4d\ ^1P_1$ and $3d^9 4d\ ^1S_0 \rightarrow 3d^9 4d\ ^3D_1$ transitions. The energy level interval between 4d and 4p levels for the Ni-like ions is larger than that between 3p and 3s levels for the Ne-like ions while the bound energy of the ground state for Ni-like and Ne-like ions is similar. This leads to higher quantum efficiency for Ni-like x-ray lasers than Ne-like x-ray lasers.
To further understand this similarity, we also compare the ionization energy for these Ne-like ions with their partner Ni-like ions used in the above sections. The ionization energy

in Table I is calculated by Cowan and MCDF atomic-physics packages. The ionization energies for Fe^{15+} ions and Ag^{18+} ions are 489.26eV and 499eV, respectively. And the ionization energy of Fe^{14+} ions is close to that of Ag^{17+} ions. So we conclude that the pumping condition for Ne-like and Ni-like x-ray laser is similar. According to the values in table 2, there is a similar correspondence between the Ne-like x-ray laser ions and Ni-like ions, in which the ionization energy for Ne-like Fe, Co, Ni ions are more similar to that of Ag, Cd, In ions than the paris of Ne-like Cu, Zn and that of Ni-like Sn, Sb ions. Therefore the conclusion drawn from the comparison of the ionization energy supports the conclusion from the hydrodynamic simulation.

Table I. Ionization energy for some ions (eV)

Fe	Fe^{15+}: 489.3	Fe^{14+}: 457	Ag	Ag^{18+}: 499	Ag^{17+}: 469
Co	Co^{16+}: 546.8	Co^{15+}: 512	Cd	Cd^{19+}: 544	Cd^{18+}: 513
Ni	Ni^{17+}: 607.2	Ni^{16+}: 571	In	In^{20+}: 592	In^{19+}: 559
Cu	Cu^{18+}: 671	Cu^{17+}: 633	Sn	Sn^{21+}: 641	Sn^{20+}: 607
Zn	Zn^{19+}: 737	Zn^{18+}: 689	Sb	Sb^{22+}: 691	Sb^{21+}: 657

5. Conclusions

The plasma conditions of the Ni-like Ag, Cd, In, Sn, Sb x-ray lasers and the Ne-like Fe, Co, Ni, Cu, Zn x-ray lasers are studied. By comparing the hydrodynamic characteristics of ion abundance, ion population density and electron temperature for these Ne-like ions and Ni-like ions, we conclude that there is a similarity of plasma conditions between these Ne-like and their partner Ni-like x-ray lasers. The ionization energy for these Ne-like ions with their partner Ni-like ions is also compared to further understand this similarity. The comparison of the ionization energies supports the conclusion from the hydrodynamic simulation.

However, this similarity is only valid for the elements of medium atomic number and their partner. The degree of similarity between the Ne-like and Ni-like x-ray lasers is generally weakened for elements with higher atomic number. The main reason is the Ni-like ions are much easier to be over-ionized to the next ionization stage than the Ne-like ions with the increase of atom number. This similarity enables us to predict the performance of some Ni-like x-ray lasers using rather simple simulations of their partner Ne-like x-ray lasers.

Acknowledgements

This work was supported by the NSFC under Grant Nos. 10374114, 60321003, 10474134 and 60128505, and the NKBRSF under Grant No. G1999075206, and the National Hi-tech ICF Program.

References

[1] Zhang J et al., 1996 Phys. Rev. A **54** R4653

[2] Christiansen J P, Ashby D E T et al., 1974 Comput. Phys. Commun. **7** 271

[3] Peng H M, Zhang G P et al., 1990 Chinese Journal of Computing Physic*s* **7** 91

[4] Zimmerman G B and Kruer W L, 1975 Comments Plasma Phys. Control. Fusion **2** 85

[5] Yan F and Zhang J 2002 Acta. Phys. Sin. **11** 2524

Inst. Phys. Conf. Ser. No 186
Paper presented at 9th Int. Conf. X-Ray Lasers, Beijing, China, 24–28 May 2004

Improvement of output fluence of the fully coherent x-ray laser at 13.9 nm

Masaharu Nishikino, Momoko Tanaka, Noboru Hasegawa, Yoshihiro Ochi, Masahiko Ishino, Maki Kishimoto, Masataka Kado, Keisuke Nagashima, and Yoshiaki Kato
Advanced Photon Research Center, Kansai Research Establishment, Japan Atomic Energy Research Institute
8-1 Umemidai, Kizu, Kyoto, 619-0215 Japan

Naofumi Ohnishi
Department of Aeronautics and Space Engineering, Tohoku University
01 Aramaki-Aza-Aoba, Aoba-ku, Sendai 980-8579, JAPAN

Abstract. We have recently reported the successful development of a fully coherent x-ray laser (XRL) at 13.9 nm by the double-target configuration. The observed XRL beam has the full spatial coherence and 0.2 mrad of the nearly diffraction limited divergence. In order to improve the output fluence, the amplification properties of the XRL beam have been investigated with various plasma length of the second amplifier target. The output fluence have been improved by an order of ten increasing the length of the gain region to 10 mm, resulting about 0.2 μJ output energy.

1. Introduction

X-ray lasers (XRLs) with saturated intensity and a higher gain coefficient of a few tens per centimeter have been demonstrated by using the transient collisional excitation (TCE) scheme[1, 2]. The saturated XRL pulse has considerable amount of photons compared to the other present x-ray source and a pulse length of a few picosecond[3, 4]. The properties of small beam divergence and high spatial coherence are important issues for application research, such as interferometry[5] and speckle measurement[6] have also progressed. However, the properties of the laser-driven XRL have remained quite low, due to refraction from the density gradient in the medium plasma with high gain coefficient. As a result, the laser-driven XRL has large beam divergence of over 1 mrad.[7-9] Recently, a full coherent XRL at 13.9 nm is generated by the double-target configuration.[10] In this experiment, the output energy was 25 nJ, and the gain coefficient g was 7.9 cm^{-1}. The output energy and the output intensity are lower than the saturation intensity by two-orders[4], but the gain region is close to circular and the refraction influence is very small. An increase in the output energy is expected by increasing the injection energy and the length of the gain medium. In order to improve the output energy, the amplification properties of the XRL beam have been investigated with longer plasma length of the second amplifier target. Since, the coupling efficiency to the second medium was about 0.1%, which was evaluated from the beam divergence, in the double-target configuration, about 1000 times amplification is required in the plasma on the second target. Assuming the gain coefficient is about 7 cm^{-1}, the required length of the gain medium is estimated at about 10 mm.

2.Experimental Conditions

The double-target configuration in this experiment is similar to the coherent XRL generation experiment (see, for example, this proceeding [11]). We have carried out experiments of the double-target configuration by using a silver (Ag) XRL at the wavelength of 13.9 nm. Two slab Ag targets were irradiated with two pumping laser beams from a pumping laser system operating at 1.053 μm.[12] The pulse width, energy and relative timing between the two beams are independently controlled. The laser pulse irradiated for the first target consists of a 300-ps prepulse and a 4-ps main pulse with the quasitraveling-wave pumping.[3] The irradiation intensity of the prepulse and the main pulse on the first target were 6×10^{12} and 8×10^{14} W/cm^2, respectively. The laser pulse irradiated for the second target consists of a 300-ps duration prepulse and a 14-ps main pulse without a quasitraveling-wave pumping. The irradiation intensity of the prepulse and the main pulse on the first target were 5×10^{12} and 6×10^{14} W/cm^2, respectively. Each pumping laser beam was focused in a line shape with a parabolic mirror and a spherical mirror. The focusing line width is about 20-μm and the length is about 6.5-mm on the first target, and 11-mm on the second target. The irradiation of the pumping laser for the second target was delayed from that for the first target, and the delay time was changed by using an optical delay line. When the delay time Δt between the two lasers is 667 ps, which is value of the target distance of 200 mm divided by light speed, the amplification takes place in the second amplifier. The optimum condition for the generation of a coherent XRL beam has been found by carefully varying the delay time between two pumping lasers.

3. Experimental Results and Discussion

The XRL beam with the nearly diffraction limited divergence of 0.2 mrad and full spatial coherence was generated in optimum condition of the gain-length of under 6.5 mm at Δt = 667-15 ps. In the length of the second gain medium over 6 mm, a narrow divergence XRL beam is generated at Δt = 667-10 ps. Figure 1(a) shows the far-field pattern (FFP) of the XRL beam after amplification with the second medium length of 11 mm. In Figure 1(a), the beam divergences along the vertical and horizontal directions are 0.5 and 0.3 mrad, respectively, at FWHM. The FFP of the amplified XRL is affected by the refraction in the longer amplifier medium. The average beam divergence of the XRL beam, which is affected by the refraction in the gain medium length of 11 mm, is 0.4 and 0.5 mrad (FWHM) for the vertical and horizontal directions, respectively. The double slit was placed in either the horizontal or vertical direction to characterize the coherence along both directions. Figure 1(b) shows the lineout of the fringe pattern along the vertical direction for the slit separation of 300 μm. The visibilities in the horizontal and vertical directions were observed the almost same value. The visibilities averaged over the horizontal and vertical directions are 0.98, 0.94, and 0.91 for the slit separations of 100, 300, and 400 μm, respectively. Fringe visibility is defined as $V=(I_{max}-I_{min})/(I_{max}+I_{min})$, where I_{max} and I_{min} are the maximum and minimum intensities of the fringe pattern. The spatial coherence of a quasimonochromatic light source is characterized by the normalized degree of coherence, which becomes equal to the fringe visibility in the case of a uniformly illuminated double slit.[14,15] Assuming a Gaussian Shell-model for the source region, both the intensity distribution and the degree of coherence are given by Gaussian distributions.[15] The coherence length L_c can be determined from $V(\Delta x) = |\mu(\Delta x)|$, where $|\mu(\Delta x)| = \exp[-(\Delta x^2)/2L_c^2]$ is the degree of coherence, and Δx is the slit separation. From the comparison between the calculated degree of coherence and the observed visibilities, the spatial coherence length L_c is estimated to be 900 μm, and it is longer than or comparable with the beam diameter. Therefore, the XRL beam affected by the refraction satisfies the condition for full spatial coherence.[16] Due to this high spatial coherence, we conclude that the divergence of this XRL beam is very close to the diffraction limit. The output energy of the XRL beam after amplification is plotted with the second medium length in Fig. 1(c). The output energy increases exponentially with the gain medium length until about 9 mm. The output energy seemed the saturation behavior for gain medium length longer than 9-mm of the gain length on the second target, because the gain duration is shorter than the duration of the x-ray propagation in the amplifier medium over 9-mm region. In Fig. 1(c), the gain coefficient is estimated to be 7.6 cm^{-1} in the exponentially region.

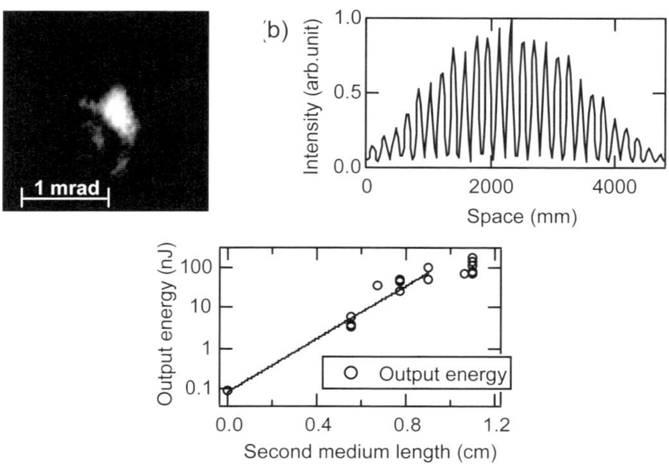

Fig. 1.(a)Far Field pattern (FFP) of the XRL beam after amplification. (b) The interference pattern and the fringe pattern. (c) Output energy of the XRL beam after amplification plotted with the second medium length. The solid line shows a fit with an exponential fitting.

This gain coefficient was 7.6 cm^{-1} and the gain medium length was $L = 9$ mm, resulting in a gain length product $gL = 6.8$ and total gain of $G = \exp(gL) \sim 900$. In this experiment, the maximum output energy of the XRL beam was about 170 nJ at the gain medium length of 11mm, corresponding to the flux of 1×10^{10} photons/pulse, when a small fraction of the seed XRL beam (about 150 nJ) is amplified. Together with the expected bandwidth of $\Delta\lambda/\lambda \leq 1\times10^{-4}$ and the pulse length of 8 ps[4], the spectral peak brilliance is estimated to be $B = 2\times10^{26}$ photons/ (s mm^2 $mrad^2$ 0.1 % bandwidth). The near-field pattern (NFP) was measured using a near field imaging system with magnification 8.5.[17] The gain region was about 40-50 μm distance from the target surface, and the size of the gain region was about 20 μm at end of the gain medium. Based on the gain area of 20-30 μm diameter, the saturation intensity of approximately 5×10^{10} W/cm^2 for the Ag x-ray laser, and the gain duration of 8 ps[4], we can expect the output energy of a few μJ at the saturated amplification.

4. Conclusion

We have demonstrated about 1000 times amplification in the second amplifier and the maximum output energy of about 0.2 μJ. The divergence of the XRL is slightly affected by the refraction influence in the longer amplifier medium. We have started hydrodynamic simulations to simulate the generation of the gain medium plasma by using the 2-D Eulerian code. One order of magnitude further increase of the brilliance is expected by the injection XRL beam with saturated intensity.

References

[1] Nickles P V *et al* 1997 Phys. Rev. Lett. 78 2748-2751

[2] Dunn J *et al* 1998 Phys. Rev. Lett. 80 2825-2828

[3] Kawachi T *et al.* 2002 Phys. Rev. A66 033815

[4] Klisnick A *et al.* 2002 Phys. Rev. A65 033810

[5] Tang H *et al.*2004 to be published in Applied Phyics. B

[6] Tai R Z *et al.* 2002 Phys. Rev. Lett.89 257602.

[7] Zhang J *et al.* 1997 Phys. Rev. Lett. 78 3856-3859

[8] Daido H, *et al.*1995 Opt. Lett. 20 61-63

[9] R. Kodama *et al.*, 1994 Phys. Rev. Lett. 73 3215-3218

[10] Tanaka M *et al.*2003 Opt. Lett. 28 1680-1682

[11] Tanaka M *et al.* 2004 in this proceedings

[12] Kawachi T *et al.* 2003 Appl. Opt. 42 2198-2205

[13] Burge R E *et al.*1997 J. Opt. SOC. Am. B 14 2742-2751

[14] Goodman 1985 J Statistical optics (Wiley, New York)

[15] Born M and Wolf E 1986 Principles of Optics (Pergamon, Oxford)

[16] Mandel L and Wolf E 1995 Optical coherence and quantum optics (Cambridge University Press, New York)

[17] Tanaka M *et al.* 2002 Surface Review and Letters 9641-644

[18] Ohnishi N *et al.* 2001 J. Quant. Spectros. Radiat. Transfer. 71 551

Inst. Phys. Conf. Ser. No 186
Paper presented at 9th Int. Conf. X-Ray Lasers, Beijing, China, 24–28 May 2004

Optimization of Ni-like Ag 13.9 nm x-ray laser by measuring field pattern for applications

W. Wang, Ch. Wang, J. R. Sun, Z. H. Fang, J. Wu, J. Q. Dong, R. R. Wang, J.J. Ye, W. X. Ma, S. Z. Fu, Y. Gu, S. J. Wang
National Laboratory on High Power Laser and Physics, Shanghai Institute of Laser Plasma, Shanghai, 201800, China

G. P. Zhang, W. D. Zheng, T. X. Zhang, H. M. Peng
Beijing Institute of Applied Physics and Computational Mathematics, Beijing, 100088, China

P. Shao, Z. Q. Lin
National Laboratory on High Power Laser and Physics, Shanghai Institute of Optics and Fine Mechanics, Shanghai, 201800, China

Abstract. Due to its special field pattern, Ni-like Ag 13.9 nm x-ray laser (XRL) from double slab targets driven by 100 ps pre main laser pulses was optimized by measuring its field patterns instead of measuring its spectral line with spectrometer. The double targets coupling conditions were truly optimised to produce a quasi-homogeneous beam in both the horizontal and vertical directions with high pulse energy. This XRL is successfully applied as a probe to the diagnosis of laser produced plasma.

1. Introduction

Considerable progress has been made in recent years on the generation of x-ray laser (XRL) under different pumping regimes. These different regimes lead to a range of saturated operating XRL's, which have specific requirements and advantages. The fast capillary discharge scheme produced intense Ne-like Ar 46.9 nm XRL at high repetition rate, high average power operation [1]. The fast heating of a preformed plasma with a ~ 1 ps, several J laser pulse was used to generate XRL's under transient collisional excitation (TCE) scheme [2, 3]. The repetition rate of these XRL's is currently limited to 1 shot every ~10 min, and the output energy is typically 10 μJ in a short, picosecond pulse. Optical-field ionization (OFI) scheme has led to saturated amplification Pd-like Xe soft x-ray lasing with a repetition rate of 10 Hz [4], and clear lasing at 32 nm in the Ni-like Kr has also been demonstrated [5]. These XRL's deliver a low energy (typically 10 nJ) pulse, but they operate at a relatively high repetition rate. The quasi-steady-state (QSS) XRL's are pumped by longer (~ 100 ps), higher energy (~ 100 J) pulses [6]. They thus rely on large laser facilities operating at low repetition rate. However, QSS XRL's have much higher pulse energy, up to a few mJ in a several tens of ps pulse. All these types of XRL have complementary properties and are able to cover a large variety of applications.

We have focused our studies on the Ni-like Ag 13.9 nm XRL based on QSS scheme. This wavelength is particularly attractive for applications not only because it has high output pulse energy, but also because it is steady and the wavelength is close to the maximum reflectivity for soft x-ray optics based on Mo/Si multilayer mirrors. In this paper, we present the experimental result of optimization of double targets coupling conditions by measuring the field patterns of the XRL. Beam profile at the position of object to be diagnosed which is needed for applications has been investigated.

At SILP, much experimental effort on the SG-II laser facility has developed this Ni-like Ag 13.9 nm XRL. Demonstrations of various applications, including x-ray laser deflectometry and interferometry of dense plasmas, have been carried out, and are discussed in detail by Ch. Wang [7, 8].

2. Experimental Setup

in our earlier experiments of XRL's, output energy of XRL's has been estimated by the spectral intensity, which has been measured using a flat-field spectrometer (FFS) with a 1200 lines mm^{-1} aperiodic ruled grating, coupled to a back-illuminated 1024×1024 CCD detector with 24×24 μm^2 pixels, and double targets coupling conditions have been optimized according to the output energy. This method works well for some Ne-like XRL's. As for Ni-like Ag XRL, the field pattern is inhomogeneous under most double target coupling conditions (see fig.2). It is impossible to determine the output pulse energy of XRL based on the spectral intensity. Therefore, the optimization of the double targets coupling conditions can not be achieved in this way. In this paper, the pulse energy of XRL is determined by measuring of beam field pattern. XRL Beam profiles are also necessary for applications.

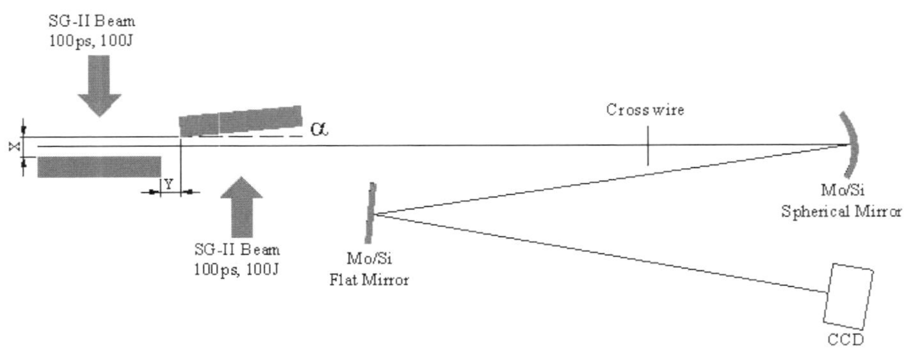

Figure 1. Experimental setup (not to scale). The XRL is produced by double Ag targets driven by two SG-II beams, each beam delivers 100 J in 100 ps pulse with a prepulse (5% intensity ratio, 3ns separation), generates a 18 mm long line focus with a 80 μm focal width. The second beam is delayed for quasi-traveling wave pumping. Field pattern at the position of cross wire is imaged by a Mo/Si multilayer spherical mirror and relayed by a Mo/Si flat mirror to a CCD detector at a 5× magnification. See that x and α denote double targets surface distance and angle, respectively.

Experimental setup is illustrated in figure 1. The XRL is produced by double Ag targets driven by two SG-II beams, each beam delivers 100 J in 100 ps pulse with a prepulse (5% intensity ratio, 3ns separation), generates a 18 mm long line focus with a 80 μm focal width on the target surface. A novel cylindrical lens-array technique is used to obtain uniform line focus, this work is described in detail by Ch. Wang in these Proceedings [9]. The driving conditions have been optimized in our earlier XRL experiments. The second beam is delayed for quasi-traveling wave pumping. Field pattern at the position of cross wire is imaged by a Mo/Si multilayer spherical mirror and relayed by a Mo/Si flat mirror to a CCD detector at a 5× magnification. See that x and α denote double targets surface distance and angle, respectively. Mo/Si multilayer mirrors and filters used in the experiment are all calibrated using laser plasma x-ray source. This work is described in detail by J. R. Sun in these Proceedings [10].

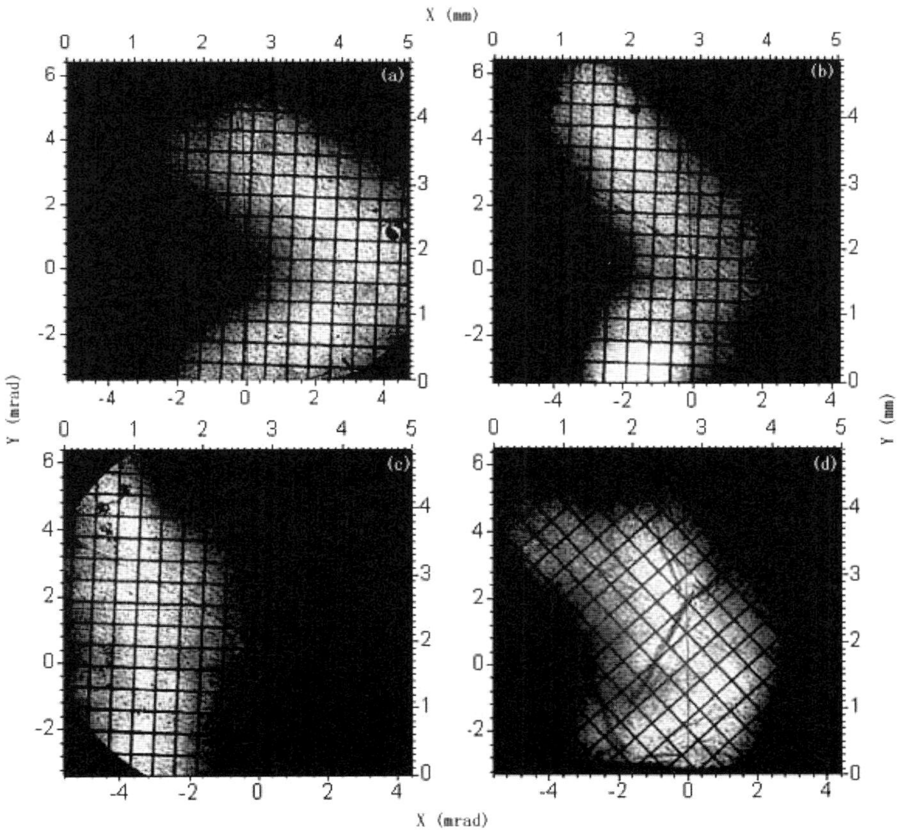

Figure 2. Footprint imaging at a distance of 50 cm from end of target under different double targets coupling conditions. Target length L1=L2=16mm, y=3mm, (a) x=200 μm, α=0, (b) x=250 μm, α=0, (c) x=300 μm, α=0, (d) x=235 μm, α=4 mrad. A high power pulse with a quasi-homogeneous beam in both the horizontal and vertical directions has been obtained (d).

3. Results

XRL far field patterns at a distance of 50 cm from the amplifying plasma exit plane are monitored using an imaging system illustrated in Figure 1. Driving conditions are as mentioned above, which are the optimized conditions under our circumstance, and remain approximately the same from shot to shot. And then, by varying the conditions of double targets coupling, the pulse energy and beam profile can be optimized. Results under different double targets coupling conditions are showed in Figure 2. Target length L1=L2=16mm, y=3mm, (a) x=200 μm, α=0, (b) x=250 μm, α=0, (c) x=300 μm, α=0, (d) x=235 μm, α=4 mrad. A high power pulse with a quasi-homogeneous beam in both the horizontal and vertical directions has been obtained (d).

Using this XRL as a probe, variety of applications has been carried out on SG-II laser facility. Laser produced plasma has been diagnosed using soft x-ray laser moiré deflectometry and Mach-Zehnder interferometry. Results in detail are given by Ch. Wang [7, 8].

4. Conclusion

Through imaging XRL beam profile the double targets coupling conditions were truly optimised to produce a quasi-homogeneous beam in both the horizontal and vertical directions with high pulse energy. This XRL is successfully applied to the diagnosing of laser produced plasma as a probe.

Acknowledgments

The authors would like to thank Lingyan Chen, Zhanshan Wang, Bin Zhou and Hongchang Wang at Tongji Univ., Xinglong Xie and Kui Yi at SIOM, Chunshui Jin at CIOM, and Y. K. Ho at Fudan Univ. for their continued support. The expert assistance from laser and target area staff at the SG-II and SPS Laser Facility is, as always, essential to progress in these complex experiments and we acknowledge all their help. The work has been supported by the NNSF of China under contract 69989801 and the National High Technology (863) Foundation of China under contract 863-804-7-3.

References

[1] J. J. Rocca et al. Phys. Rev. Lett. **77** 1476 (1996)
[2] Yu.V.Afanas'ev and V.N.Shlyaptsev Sov. J. Quantum Electron **19** 1606 (1989)
[3] P.V. Nickles et al. Phys. Rev. Lett. **78** 2748 (1997)
[4] S. Sebban et al. Phys. Rev. Lett. **86** 3004 (2001)
[5] S. Sebban et al. Phys. Rev. Lett. **89** 253901 (2002)
[6] J. Zhang et al. Science **276** 1097 (1997)
[7] Ch. Wang et al. Acta Phys.Sin. **51** 847 (2002) (in Chinese)
[8] Ch. Wang et al. "Electron Density Measurements of Laser Plasma Using Soft X-ray Laser Interferometry" these proceedings
[9] Ch. Wang et al. "The Uniform Line-Focus System Used for Special Laser Beam Intensity Distribution of SG-II laser facility" these proceedings
[10] J. R. Sun et al. these proceedings

Inst. Phys. Conf. Ser. No 186
Paper presented at 9th Int. Conf. X-Ray Lasers, Beijing, China, 24–28 May 2004

Experimental Studies of Ni-like Ta x-ray laser at 4.48nm near water window

Ch. Wang[1], W. Wang[1], J.R. Sun[1], Z.H. Fang[1], W.X. Ma[1], J. Wu[1], J.Q. Dong[1], R.R. Wang[1], S.Z. Fu[1], Y. Gu[1], G.P. Zhang[2], W.D. Zheng[2], T.X. Zhang[2], P. Shao[3], Z.Q. Lin[3], S.J. Wang[1]

[1] Shanghai Institute of Laser Plasma, P.O.Box 800-229, Shanghai, 201800, P.R. China

[2] Beijing Institute of Applied Physics and Computational Mathematics, Beijing, 100088, P.R. China

[3] Shanghai Institute of Optics and Fine Mechanics, P.O.Box 800-211, Shanghai, 201800, P.R. China

Abstract. The experimental studies of Ni-like Ta x-ray laser at 4.48nm near the *water window* on SG-II laser facility were reported. By using a new scheme of united irradiation of two lasers with different colours, the distinct spectrum signal of Ni-like Ta x-ray laser with the efficient gain-length product of about 5.5 was obtained with the total drive energy of 400J.

1. Introduction

The studies of x-ray lasers in or near *water window* (2.3 to 4.4nm) is an important subject, because of its great application prospects on biologic fields. The attempts of extending the wavelength of x-ray laser to shorter were not given up, since the first experimental demonstration of x-ray laser. There were many saturated outputs of x-ray laser had been obtained in many experiments, however the outputs of x-ray lasers in or near *water window* were far from gain saturation.

Firstly, a series of experiments of Ni-like ions x-ray lasers at and near water window were conducted with a several thousand joules energy green laser irradiating an exploding foil target at Lawrence Livermore National Laboratory (LLNL). The typical results of gain-length products for Ni-like Yb (5.0nm), Ta (4.5nm) and W (4.3nm) were about 2, 6 and 7 respectively[1,2]. The required drive energy of those experiments was too high to study widely. Then, a new technique of pre-pulse application was raised by Rutherford Appleton Laboratory (RAL), and the required drive energy was largely reduced. They obtained the saturation output of Ni-like Sm x-ray laser at 7.3nm with only several hundred joules drive energy[3]. Afterwards, using the same pre-pulse and double-target-coupling technique, Institute of Laser Engineering (ILE) and National Laboratory on High Power Laser and Physics (NLHPLP) cooperatively demonstrated x-ray lasers output of Ni-like Yb, Hf (4.6nm) and Ta with several hundred joules drive energy[4,5]. The gain-length products for Yb and Hf were about 11 and 6, but output of Ni-like Ta x-ray laser was only weakly observed.

Along with the development of water window x-ray lasers, some applications in biologic fields were demonstrated, such as holography[6] and microscopy[7]. The Ni-like Ta x-ray laser at 4.48nm was the best choice for the application of biologic holography[8], so we paid attention to it especially. In this paper, we reported an experimental demonstration of the electron collisional excitation scheme of Ni-like Ta x-ray laser using united irradiation of two lasers with different colours. A distinct x-ray laser signal at 4.48nm with gain-length product of 5.5 was observed with the total drive energy of 400J.

2. Scheme of united irradiation by two lasers with different colours

The drive conditions of Ni-like Ta x-ray laser were stricter than some typical ones such as Ne-like Ge or Ni-like Ag because of its short wavelength and high ionization rate. Some theoretical calculation indicated that the optimal electron density of amplified plasma for Ni-like Ta x-ray laser was about $2.0\times10^{21}cm^{-3}$, and this value was just between the critical density of red laser and the one of green laser. How to transfer energy from the critical density regions to the gain region by electron thermal transmission was a master problem. The effects of gain plasma heating might be not good, whether only using a red laser or a green one. So a new scheme of united irradiation by two lasers with different colours (ω+2ω) was raised to improve this, shown in Figure 1.

Generally, the laser energy was deposited around the critical electron density region in laser plasmas. A red laser heated the area of ⑤ in Figure 1 and a green laser heated ⑥. There were two processes (③ and ④) of transferring energy to the gain region at one time, hence, the heating of plasma might be more effective in the gain region and the higher gain would be obtained.

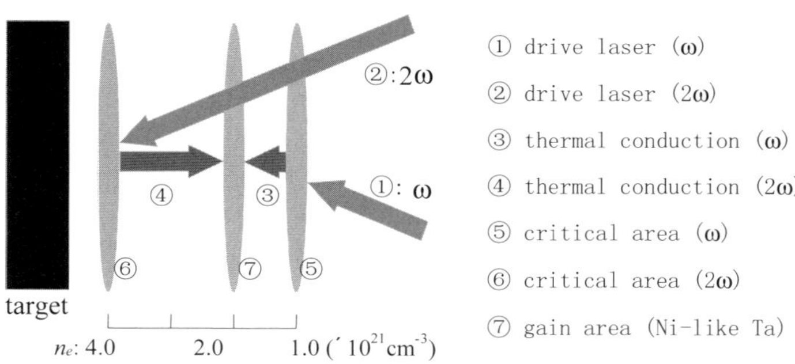

Figure 1. The scheme of united irradiation of two lasers with different colours.

3. Experimental setup

The experiments of Ni-like Ta x-ray laser using the scheme of united irradiation by two lasers with different colours (ω+2ω) were performed at the high power laser facility, SG-II. The laser facility delivers eight laser beams with 100 ps pulse width, 100 Joule energy at 1.053μm wavelength. By with appropriate crystalloids, each laser beams can be transformed to green laser (2ω: 0.53μm) with efficiency of about 60%.

In this experiment, there were four laser beams were chosen for irradiating double targets, so the total drive energy was about 400J. Among them, two laser beams were red (ω) and the others were green (2ω). Each target was irradiated by a horizontal green laser and a obliquity red laser at one time, the angle was about 42°. The target was inclined by 21° so that the two beams of drive laser could symmetrically irradiate the target (see Figure 2 a). Besides, the classical techniques such as pre-pulse application, uniform line-focus system and double target coupling were also applied in experiments. The intensity ratio of pre-pulse and main-pulse red laser was about 5% and the temporal separation was about 1.5ns. The united line-focus with size of about 6mm×100μm was obtained by using cylindrical lens array (CLA) line-focus systems. The total irradiance on the target surface was about $2.5\times10^{14}Wcm^{-2}$. The target was made of about 1μm thickness Ta coated on a flat glass, and the maximum length was 5mm to avoid the production of cool plasma at the ends of targets. The separations between two targets were about 50μm in the target normal direction and 3mm at axially one.

A 2400 l/mm flat-field grating spectrometer and a soft x-ray CCD camera were used as the main apparatus to measure x-ray laser spectra. The spectrometer and CCD were also slantways placed so as to ensure the slit being parallel to the target surface normal direction. Several 50μm diameter fiducial wires were setup in front of the CCD detector to determine the angle distribution of x-ray lasers. Some C_8H_8 filters were placed in the light route.

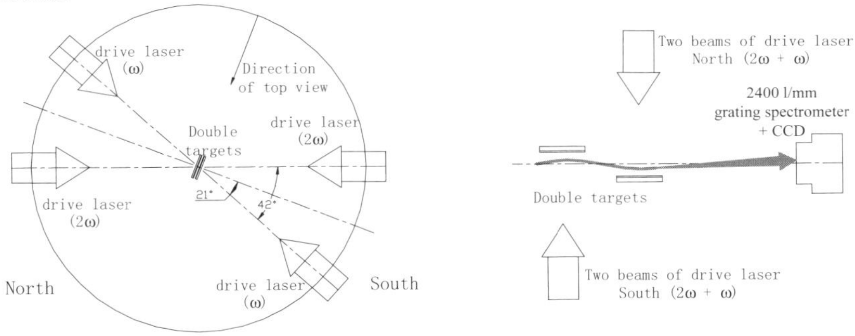

(a) side view (b) top view, inclined 21°

Figure 2. Experimental setup of Ni-like Ta x-ray laser, using the united irradiation of two lasers with different colours.

4. Experimental results and discussion

A typical axially time-integrated spectrum of Ni-like Ta x-ray laser recorded in experiments was shown in Figure 3 and 4. The spectrum was very clear and its wavelength can be accurately confirmed by the absorption K edge of carbon at 4.37nm

In order to study the gain property, some experiments with different length targets were executed. The data and the fit curve were shown in Figure 5. The behavior seemed as saturation, but it was far from saturation. The gain length product was only about 5.5. The reason of this analogue saturation behavior was probably that the x-ray laser beam departure from gain area too early because of the strong deflection.

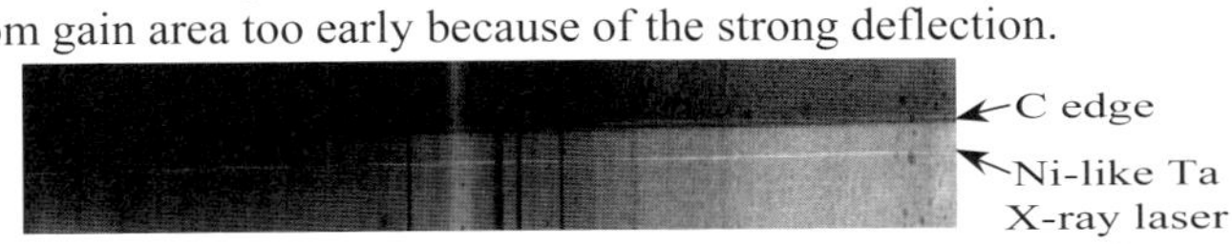

Figure 3. A typical axially time-integrated spectrum of Ta x-ray laser.

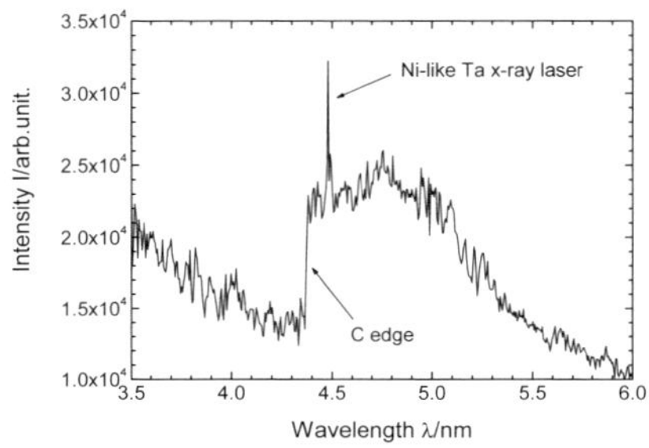

Figure 4. A typical spectrum of a Ni-like Ta x-ray laser at 4.48nm.

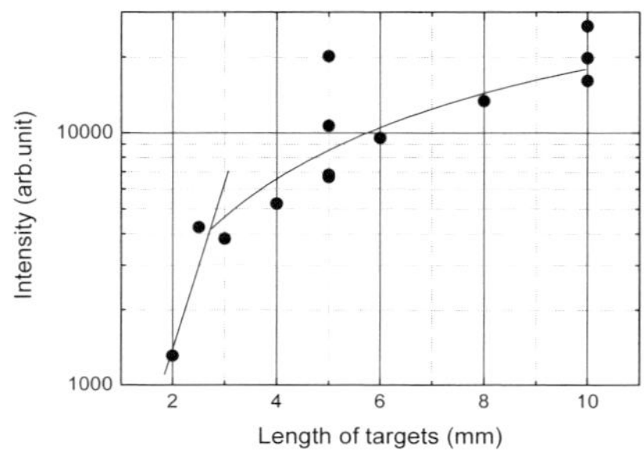

Figure 5. The gain curve of output intensity as a function of plasma length. The maximum GL was about 5.5.

5. Conclusion

A new scheme of united irradiation by two lasers with different colors was used in experimental studies of Ni-like Ta x-ray laser. Certain intensity output of Ni-like Ta x-ray laser with the efficient gain-length product of about 5.5 was obtained with the total drive energy of 400J. The strong deflection restrained the more magnification of Ni-like Ta x-ray laser, and how to reduce the effects of deflection might be an important point of next researches.

6. Acknowledgments

This work is supported by National High Technology Development Program of China (Grant No. 863-804-7.3). We acknowledge all the members of SG-II operation group, whose excellent works made the experiment going on wheels. We also thank L.Y. Chen's group for providing the targets and filters.

References

[1] MacGowan B J et al 1988 J.Opt.Soc.Am.B 5 1858
[2] MacGowan B J et al 1990 Phys.Rev.Lett. 65 420
[3] Zhang J et al 1997 Science 276 1097
[4] Daido H et al 1995 Phys.Rev.Lett. 75 1074
[5] Daido H et al 1999 Soft X-ray Laser and Applications III, SPIE 3776 54
[6] Trebes J E et al 1987 Science 238 517
[7] DaSilva L B et al 1992 Science 258 269
[8] London R A et al 1989 Appl.Opt. 28 3397

Inst. Phys. Conf. Ser. No 186
Paper presented at 9th Int. Conf. X-Ray Lasers, Beijing, China, 24–28 May 2004

Experimental studies on X-ray laser driven by double pre-pulse

W.Z.Huang [1], Y.J .Li [2,3], J.J.Hao [2,3] , Y.L .You [1], Y.Q .Gu[1], Y.L.He[1], Y.B.Chen [1], B.Xu[1]
[1] Institute of nuclear physics and chemistry, P.O.Box 919-986 Mianyang, 621900,China
[2] Institute of physics, Chinese academy of sciences, Beijing 100080, China
[3] Department of Physics, China University of Mining & Technology, Beijing 100083,China

Abstract. Using ~40J of drive energy in a 170-ps main pulse and double pre-pulse technique, we demonstrated lasing on the $J = 0 \rightarrow 1$, $3p \rightarrow 3s$ transition in Ne-like Cr at 28.5nm. Theoretical simulation is presented using MED103 code at the same time. The results show that the intensity of XRL driven by double pre-pulse is not remarkbly higher than that driven by single pre-pulse.

1. Introduction

Since the first demonstration of X-ray laser in 1984[1], great progresses had been made in studying soft X-ray laser. In resent years, by the use of pre-pulse technique and multi-target coupling technique, saturated X-ray laser output has been achieved from collisionally pumped Ne-like Ge [2], Ne-like Se [3], Ne-like Fe[4] and Ni-like Ag[5], and the wavelength of Ni-like Ta is approaching "water window". The range of X-ray laser's applications is very wide as a result of it's main properties of X-ray laser: short wavelength, high power, coherence and high brilliance. X-ray laser probe technique is a vital means of diagnosing the state of the high-temperature plasma, so it plays a great role in Inertial Confinement Fusion (ICF).

Electron collisional-excitation scheme is a main principle on producing X-ray laser. Resent studies had been testified that pre-pulse technique was an efficient way to improve output intensity of the irradiating slab target XRL experiment. In 1999, Balmer et al. had observed the saturated XRL output from Ne-like Fe at 25.5nm in an experiment which carried out at Institute for Applied physics of Berne university using double-prepulse technique and 18mm-long curve target[4].

With the aim to grope for a more effective means to enhance the XRL output, the experiment on Ne-like Cr using double pre-pulse technique was performed. We achieved soft X-ray lasing at 28.5nm, as well as some other valuable results.

2. Experimental Setup

The experiments were performed on a relatively small Nd-glass laser facility, "Xingguang II", which delivers up to 40 Joules of energy at 1053 nm in 170 ps on targets. A schematic diagram of the experimental setup is shown in Fig1. The two spectrometers are fixed on the two sides of the target chamber. A slit camera set above the target's front to inspect the uniformity of the line focus. A line focus of 2.2cm3 120μm is produced by focusing system consists of cylindrical lens array and non-sphere lens. This allows a maximum irradiance on flat target of $1.8TW/cm^2$. The intensity inhomogeneity of line focus is less than 15% in the effective length of 15mm. The target is a polished chromium flat slab with 2cm long and 2mm thick. The surface roughness of it is less than 0.6μm.

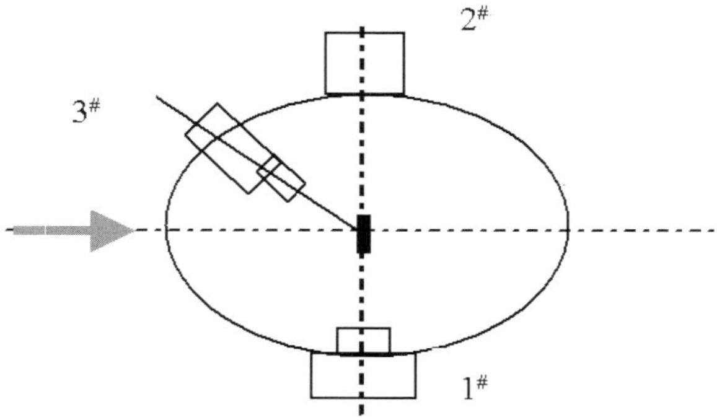

Fig 1. Experiment setup, 1# and 2# spectrometer ,3# Slit Camera

3. Results and Analysis

From Fig 2 we could find the intensity of XRL is obviously higher than that of spontaneous emission. Y-axis means the intensity of emissions at different wavelength, X-axis means length of scanning direction. Fig 3 and Fig 4 show the angular intensity distribution measurement for double pre-pulse scheme and single pre-pulse scheme. In spite of the peak intensities of the two graphs is similar, the difference of them is obvious. There are two reasons to explain the difference: one is the gradient of gain area produced by double pre-pulse is smoother than that done by single, the other is the gain area of the double pre-pulse scheme is wider than that of single pre-pulse. Table 1 is the measurement on different energies of pre-pulses.

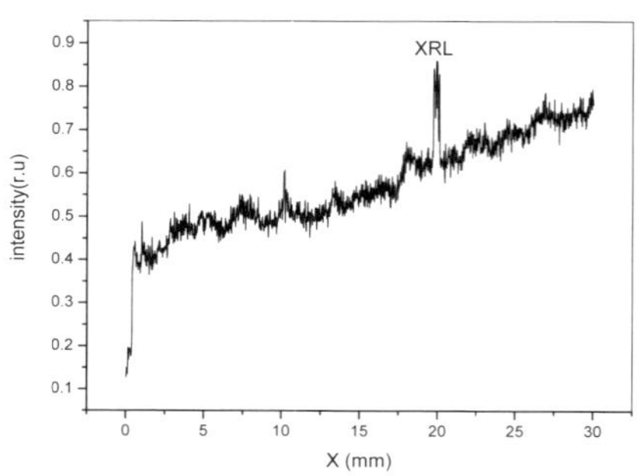

Fig 2. Lineout of the Ne-like Cr lasing

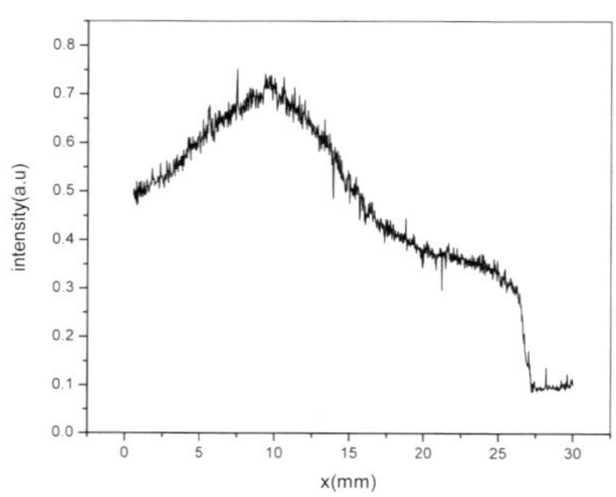

Fig 3. Divergence angles in single pre-pulse.

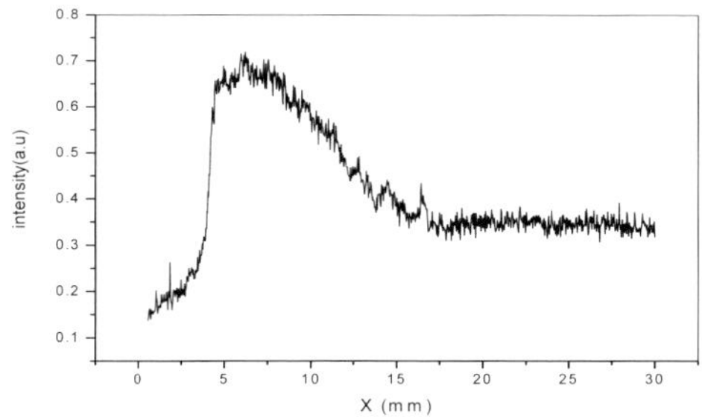

Fig 4. Divergence angles in double prepulse.

Shot number	Drive energy (J)	Pre-pulse 1 level	Pre-pulse 2 level	Intensity *I*(a.u)
51402	37.9	0.58%	0.58%	3.86
51501	35.2	0.17%	0.99%	4.54
51502	31.5	0.17%	0.99%	4.40
51701	47.4	0.17%	0.69%	4.49
52005	33.9	0.17%	0.69%	4.42

Table 1. Output intensity on different energy of pre-pulses

The analysis shows that the double pre-pulse scheme shows no clear advantage over the single pre-pulse scheme.

4. Simulation

In the simulation, the peak intensity of the first pre-pulse is set to be $1.0*10^{12}W/cm^2$, and that of the main pulse is $1.8*10^{13}W/cm^2$. The first pre-pulse precedes the main pulse by 5ns. We presume that the delay of the second pre-pulse is variable, so as to find a special pulse delay which can greatly increase the output. The influence of the time and the intensity of pre-pulse on the gain coefficient of XRL is through carefully numerical simulation, as well as the influence of the fluid mechanics character of the laser plasma. The calculated result shows that there are two pulse structures, which can optimize the spatial and temporal distribution of the gain coefficient.

The simulation is under the best condition of the "XingGuang II" facility, 40 Joules of energy, 170 ps of full-with at halt-maximum (FWHM), the target in simulation is presume to be Fe slab. The calculated contour of the gain coefficient versus the spatial and temporal is shown in Fig 5(b). The gradient of (b) is smoother that that of (a), which can prevent the rays from escaping out of the gain area as result of refraction effect. On the other hand, the gain coefficient of (b) change unevidently, so we think the output intensity is similar.

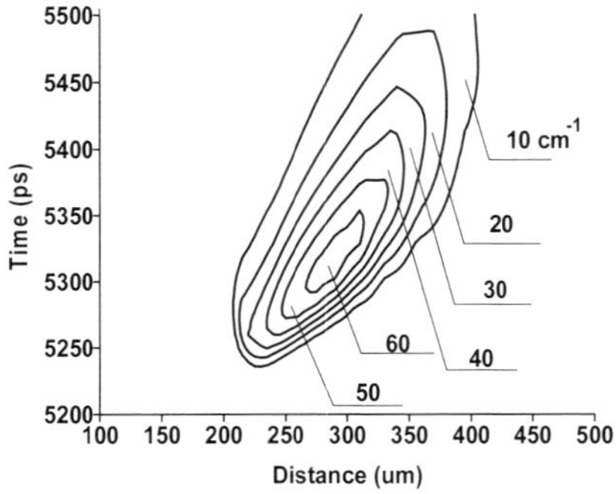

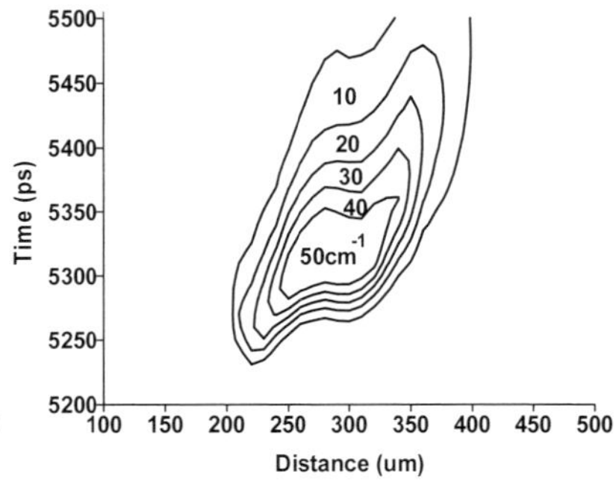

(a) (b)

Fig 5. Contours of gain versus space and time Ne-like Fe. (a) Simulation result of the single pre-pulse scheme; (b) The result of the double pre-pulse scheme.

5. Conclusion

Using double pre-pulse technique and flat target, we achieved the Ne-like Cr XRL at 28.5nm($J=0\rightarrow1$), and compared the intensity to the output of single pre-pulse scheme. From the results we could conclude that the double-prepulse scheme could not greatly enhance the XRL intensity. This result is coincidence with the simulation, but it is different from Balmer's, so to seek the physical reason of it will stimulate us to make a further research.

6. Acknowledgement

The authors would like to thank the technical support of all the members of the "XingGuang II" operating group.

References

[1] D.L. Matthews, P.L. Hagelstein, M.D. Rosen, et al. Demonstration of a soft X-ray amplifier [**J**]. Phys. Rev. Lett. **54**,110(1985)

[2] A. Carillon, H.Z. Chen, P. Dhez, et al., Phys. Rev. Lett.**68,**2917(1992)

[3] J.A. Koch, B.J. MacGowen, L.B. Dasilva, et al., Phys.Rev.Lett.**68**,3291(1992)

[4] J.E. Balmer, R. Tommmasini, and F. Lowenthal, IJSQEN, 1999,5(6), p1435

[5] Wang Chen， Gu Yuan， Wang Shi-ji， High Power Laser and Particle Beams[J]， **14**， 3（2002）

[6] J. Zhang, S.T. Chunyu, Y.L. You, et al., Phys.Rev.A **53**, 3640(1996)

[7] Huang W.Z, You Y.L, Chun Yu S.T, Zhang J, et al., Chinese Science， A26， 565（1996）

[8] Huang W.Z， Gu Y.Q, Li Y.J, et al., High Power Laser and Particle Beams[J]， 68(2), 155(2003)

Capillary discharge-pumped X-ray lasers

Inst. Phys. Conf. Ser. No 186
Paper presented at 9th Int. Conf. X-Ray Lasers, Beijing, China, 24–28 May 2004

Electrodischarge radiation source of capillary type

V.A. Burtsev, E.P. Bolshakov, A.S. Ivanov, N.V. Kalinin, V.A. Kubasov, R.F. Kurunov, V.G. Smirnov, V.I .Chernobrovin
Efremov Scientific Research Institute of Electrophysical Apparatus
196641 Санкт-Petersburg, Metallostroy, NIIEFA, Russia

Abstract. Results of experimental, analytical and numerical researches of a model source of electromagnetic radiation, based on the capillary discharges, supplied from a generator of high-voltage pulses by a long transmission line are informed. It is shown, that application of the running wave of the sliding discharge allows to breakdown long gaps in a wide range of initial gas pressure and to create near-wall plasma layer, sufficient for ignition of main high-current discharges at moderate charging voltage.

1. Introduction

Application of fast capillary discharges for last years becomes one of the basic ways for creation of compact coherent EUV sources on plasma of multiply ionized ions. In connection with that in single-pass amplification of spontaneous radiation, aspect ratio of an active plasma column should be high enough ($l/d \sim 1000$ [1]), there is rather serious problem of preliminary ionization of working gas. Gravity of this problem is aggravated with the requirement of high azimuthal and axial uniformity of the preliminary plasma, following of necessity of homogeneous compression of plasma.

Necessity of the application of small rise times of discharge current ($\tau \leq 10$ ns), caused by inadmissibility of injection in the discharge erosion products and development of MGD instabilities, imposes the rigid requirement to the minimal inductance of capillary load. At close placing of the coaxial return conductor relatively the dielectric tube, the limited penetration of an electric field into the capillary excludes the application of additional gas discharges because of impossibility of electric breakdown of long gas gaps at moderate voltage. Just in these conditions the ignition of sliding surface discharge, developing as a running wave becomes rather effective [2].

In the given work results of researches of the model source of electromagnetic radiation on a base of low-inductive fast capillary discharges, supplying through the long transmission line are informed. The basic attention is given to the questions connected to breakdown and preliminary ionization of gas in extensive low-inductive capillaries with the help of the running wave of the sliding discharge, and subsequent ignition of the basic high-current discharges.

2. The approach and the experimental equipment

The approach developed in the given work is based on the following positions:

- Optimization of pumping processes by matched temporal programming of entered electromagnetic power in dependence from the concrete mechanism of inversion.

- Application of DFL and transporting lines with the double travel time of voltage wave exceeding duration of pumping pulse $2l(\varepsilon\mu)^{-1} \geq \tau_{incU}$ and excluding the influence of reflected waves.
- Minimization of capillary inductance up to value at which front time of current pulse is less than front time of incident wave, and the steady-state value of discharge current equals double size of current in incident wave (a mode of closed line), i.e. $L_{cap} \leq \rho\tau_{incU}$, $I_{cap} \approx 2U_{inc} / \rho$.
- Application and optimization of the running wave of the sliding discharge at voltage on the capillary, equal to double voltage of incident wave $U_{cap} \approx 2U_{inc}$ (a mode of open-circuit line).

At creation of experimental installation 8-channel generator of high-voltage pulses on base of DFL with oil-paper isolation [3] has been used. The line has output wave resistance ~ 9 Ohm, total capacity C=8C_0 = 20,6 nF and is designed on charging voltage up to 100 kV. Output of DFL is connected through sharpening gas switch to transmitting 8-channel cable line with the length of 10 m. The generator provides on matched load pulses with amplitude up to 100 кV, FWHM ~ 100 ns, rice time ~ 8 ns.

The cable transmitting line with the help of coaxial inputs is connected to a current-collecting disk of a power unit of radiation source. The high-voltage electrode of the special form is connected to the central part of current-collecting disk. The output tubular electrode is connected to earthed return coaxial conductor. As the dielectric discharge chamber, an alumina tube serves with internal diameter 5.5 mm. Internal diameter of return coaxial conductor – 10.5 mm. Thus, linear inductance of the capillary equals 1.4 nH/cm.

For modelling and research of the formation of the sliding discharges and the ignition of the main discharge, one of cables of transporting line was connected to described above capillary load (Fig. 1). The basic experimental results, submitted below, have been received on this model source with the tube length of 50 mm and at using of argon.

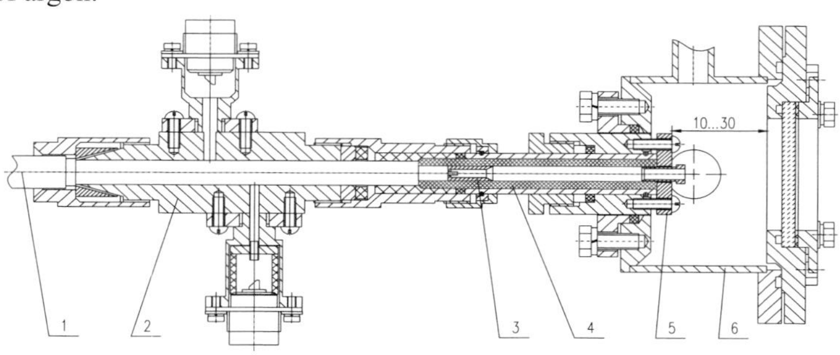

Figure 1. One-cable model source.1-coaxial cable, 2- block of detectors, 3- high voltage electrode, 4- capillary, 5- earthed electrode, 6- gas chamber.

3. Research of running wave of sliding discharges

The model source has been equipped with a magnetic detector, the signal with which is proportional to discharge current and a detector of the capacitor type, giving the information about pulse voltage on the entrance of the capillary (Fig.1). As it is possible to see on Fig. 2, on which signals from these detectors are presented, all process can be divided into two phases. The first phase is caused by developing of the sliding discharge with the high impedance. After achievement by sliding sparks of the output electrode, the main discharge develops with small resistance that results in essential falling of voltage, despite of increase in discharge current.

As carried out researches have shown, the dependence of travel time of running wave (delay time of main discharge) τ versus initial argon pressure showed the presence of wide minimum in the range of 1-100 Torr (Fig.2). The rate of the spread of sliding discharge front reaches the maximal value ~ 10^9 cm/s at voltage of incident wave of U_{inc} = 60 кV. At increasing gas pressure, the delay of the main discharge grows, but near 350 Torr it falls again. The left branch of the curve after growth passes in saturation at pressure below 10^{-1} Torr.

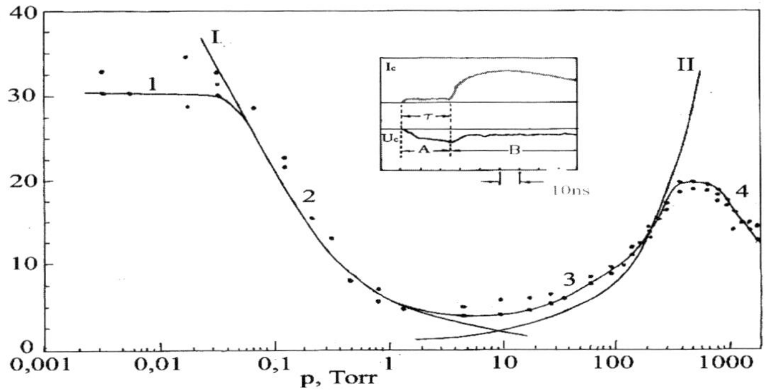

Figure 2. Dependence of travel time of sliding discharge front versus argon pressure. 1-4 – branches of experimental curve. I, II – dependences, calculated on formulas (3), (2) correspondingly. U_{inc}= 60 кV

On the base of analysis of elementary processes, occurring in discharge tube, it can try to explain observed character of this dependence.

As calculations of an electric field in the capillary have shown, the tangential component of field intensity E_t falls in e time on the distance approximately equal to internal diameter of return conductor D. We replace this exponential distribution by uniform distribution with value of E_0. In this case electrons, emitted from the sharp edge of the ring high-voltage electrode, will drift along the tube with the velocity of u_e. Time of the spread of an avalanche (Townsend) discharge up to the front of field distribution will be D/u_e. We shall consider, that in this moment charging of the ceramic coaxial condenser with length of D begins with a constant of $\rho C_l D$, where C_l – linear capacity. As the result of this charging, the peak of E_t moves on the depth D, and the field will fill the following interval. Then electron avalanches will continue to drift along the tube and similar process will repeat, while the avalanche front will reach the output electrode. In the given model the full time of the sliding discharge spread will be defined by expression $\tau = l/u_e + \rho C_{cap}$.

The first term of this expression having the drift nature, depends on pressure and kind of working gas, and also from field. Drift electron velocity can be expressed through the average thermal electron velocity $\overline{\upsilon}_e$ and the cross-section of the interaction of average electron with atoms σ_{ea}. In result we receive the expression for this drift term.

$$\tau_{dr} \approx 20 l \overline{\upsilon}_e \sigma_{ea} \frac{p}{E}, \qquad (1)$$

Now it is necessary to concretize σ_{ea}. First we shall consider a branch 3 of the dependence $\tau_{dr} = f(p)$. The basic process of electron interaction with atoms in this range of pressure at the chosen field intensity should be elastic collisions. Using experimental data on $\sigma_{tr} = f(\overline{\varepsilon_e}), \overline{\varepsilon_e} = f(E/p)$, resulted in [4, 5], and receiving from them approximation formulas, it is possible to find following simple expression:

$$\tau_{dr} = 3.10^{-6} l \frac{p}{E} \tag{2}$$

As it is possible to see on Fig.2, rather good coincidence of calculated and experimental results (curves II and 3 correspondingly) is observed.

Applying similar procedure, it is possible to receive expression for the description of branch 2of experimental curve $\tau_{dr} = f(E_0/p)$.

$$\tau_{dr} = 0{,}54.10^{-9} l \left[10^{-2} \left(\frac{E_{02}}{p}\right)^{0.5} - 13\left(\frac{p}{E_{02}}\right)^{0.5} \right] \tag{3}$$

At receiving of this expression it has been taken in account, that the basic kind of electron interaction with atoms are ionization impacts. Slow electrons, moving in the zone of the weak field, are responsible for the observed character of the spread of the avalanche discharge at high *E/p*. Their average thermal energy corresponds to the left branch of $\sigma_i = f(\overline{\varepsilon_e})$. Fast electrons in the zone of high field move in the mode of running away and, hitting against the wall, give secondary electrons, initiating avalanches.

4. Research of main capillary discharges

Survey research of the main discharge also was carried out in the wide range of changing gas pressure. Radiation of plasma in longitudinal direction in visible range of spectrum (220-650 nm) was registered with the help of a high-speed coaxial photo element. Results of measurements are submitted on Fig.3 as dependences of amplitudes of light signals J_m versus gas pressure. In the range of pressure $p \approx 60$-300 Torr, a short light pulse was observed before main pulse.

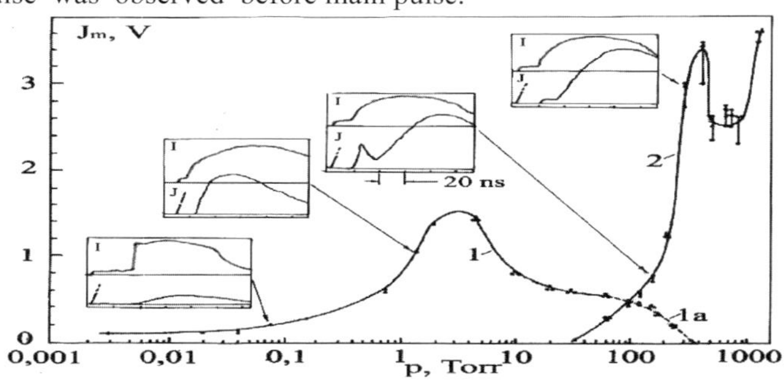

Figure 3. Dependences of amplitudes of first (1) and second (2) light pulses versus argon pressure and also oscillograms of current (upper rack) and signals from fast photo element (lower tracks). $U_{inc} = 60$ kV

In the range of pressure $p \approx 0.1$-10 Torr, the most interesting at creation of EUV-source, it is possible to see the maximum of light intensity in area of ~ 3 Torr when minimal time of the spread of the sliding discharge is observed. The temporal form of the light signal in this range is distinctive that the maximum of intensity takes place in the first part of current pulse. As against it in area $p \geq 10$ Torr the maximum of light intensity takes place at the second part of current pulse.

It is necessary to note changing structure of current front with changing of gas pressure. This front has a fast part with duration ~ 4 ns that is much less than duration of front of incident wave (takes place the effect of current sharpening). However the height of this part falls with increasing in pressure from ~ 100 % of current amplitude at $p \sim 0.01$ practically up to zero at $p \sim 300$ Torr. Application of optical filters has allowed to find out dominating UV component of light at smaller pressure and red - orange component at the bigger pressure.

Integrated photographing of discharges has shown, that at small pressure the main discharge comes off walls of the tube while at the big pressure it remains in contact to walls all time (Fig.4b,c). These facts enable to speak about week-compressional discharge in the first case and about radiative low-temperature discharge in dense gas in the second. X-ray radiation in these regimes did not observe. In opposite, at $p = 0.01$ Torr the radiation in visible range was very week, but x-ray radiation was been registered (Fig.4a). The last was generated by sliding discharge as was shown by temporal measurements.

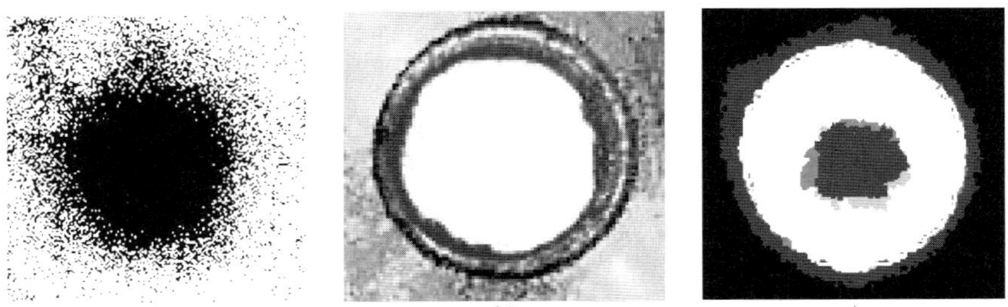

a) b) c)

Figure 4. Time integrated images of discharges
a- x-ray picture of sliding discharge at $p = 0.01$ Torr,
b- photo of main discharge at $p = 3$ Torr,
c- photo of main discharge at $p = 760$ Torr

5. Numerical modeling of capillary discharges

The program has been developed and applied for research of dynamics and radiation of capillary discharges on a base of 1D 2T magnetic radiative hydrodynamics with the detailed account of ionization kinetics. The basic results of the carried out numerical research are resulted in the report submitted on XRLC-04 [6]. Let's note only some results concerning above mentioned experimental materials.

The carried out calculations of main discharges in the capillary, preliminary filled by argon at homogeneous and ring near-wall initial distribution of temperature, have shown, that in the second case higher parameters of plasma are reached. At that more high efficiency of energy input occurs due to several factors: localization of all current in the ring near-wall area and tearing off plasma from walls, more optimum combination of magnetic compression and Joule heating caused, in particular, by turbulence of plasma.

The main discharge at initial pressure of gas ~ 1 Torr carries low-compressional character. At the initial pressure which is coming nearer to atmospheric, takes place low-temperature discharge keeping the ring form during all process. It is in the good consent with experimental results, received in the model source.

6. Conclusion

For the first time it has begun possible to realize separately the regimes of preliminary ionization of gas in low-inductive extended capillary (~ 1,5 nH/cm, 150 mm) as a running wave of the sliding discharge and ignition of main discharge arising after reaching output electrode by running wave at rather low working voltage (charging voltage of DFL up to ~ 70 кV).

The transmitting line works on the stage of the sliding discharge as on opened end one with doubling of voltage and on the stage of main high current discharge as on shorted end one with doubling of discharge current.

Opportunities of combined use of various kinds of the discharges, which are supplied from one pulse power system, are being opened for optimization of plasma preparation and the active medium pumping in compact EUV-sources of capillary type.

References

[1] Vinogradov AV and Rocca J J 2003 Kvantovaya Electronika v.33 1 7-17 (In Rus)
[2] Dashuk P N and Kulakov S L 1979 Pis'ma ZhTF 5 2 69-72 (In Rus)
[3] Bol'shakov E P Burtsev V A et al 2003 Int. Conf. on High power modulators June 30 - July 3, 2002, Hollywood, CA pp. 394-395.
[4] Raizer Yu 1987 Physics of gas discharge(Moscow: Nauka). In Rus.
[5] Korolev Yu D and Mesyaz G A 1991 Physics of pulse breakdown of gases (Moscow: Nauka). In Rus.
[6] Burtsev V A and Kalinin N V 2004 Int. Conf. on XRL, 23-28 May Beijing (in publishing).

Inst. Phys. Conf. Ser. No 186
Paper presented at 9th Int. Conf. X-Ray Lasers, Beijing, China, 24–28 May 2004

Recent results on a 46.9nm capillary discharge laser and applications

G. Tomassetti[1], A. Ritucci[1], L. Reale[1], F. Flora[2], L. Arrizza[1]
[1]Dept. of Physics, University of L'Aquila, gc LNGS of INFN, INFM - 67010 Coppito, L' Aquila, Italy
[2]ENEA Dipart.. Innov., Div. Fisica Applicata, CRE Frascati - C.P. 65, 00044, Italy

Abstract. Recent results regarding the development of a fully saturated 46.9 nm capillary discharge laser are reported. The output energy of 300-400 μJ/pulse, the almost full spatial coherence and the pulse repetition rate of 0.2 Hz make this laser an attractive source for the development of the applications. The use of this laser for the processing of optical dielectric materials, which are of interest in the field of photonics and the production of miniaturized optical devices, is here considered. As an example, we report the holographic encoding of photo-luminescent patterns of color-centers on the surface of LiF crystals.

1. Introduction

A 46.9 nm soft X-ray laser in Ar Ne-like pumped by a capillary discharge was first developed by the group of Rocca since 1994. The compactness of the experimental device, the high energy per pulse (0.1-1 mJ/pulse), the high spatial coherence and the pulse repetition rate as high as 0.1-10 Hz make the 46.9 nm capillary discharge laser actually suitable for the development of applications [1]. In this paper we briefly describe the present state of development of our 46.9 nm capillary discharge laser. A highly saturated output in 45-55 cm long capillaries with pulse energy of 300-400 μJ and a full spatial coherence is reported. A particular regime, in which the laser beam has an extremely low divergence ≤1 mrad and a quasi-gaussian distribution of the intensity, has also been found.

The use of this kind of 46.9 nm laser for the diagnostics and the metrology, such as for the interferometry of high-density plasmas and the testing of soft X-ray optics and detectors has been already demonstrated experimentally in refs. [2-4]. In this contribution we consider the possibility to employ such a laser as a powerful tool for the processing of materials. In this regard a large class of optically transparent dielectrics having a large optical band gap (E ≈10-14 eV), such as SiO_2, sapphire or lithium fluoride (LiF), can be considered. These materials are of great interest in the field of photonics for the development of the integrated optical circuits, but they are difficult to process by the conventional optical laser light. The conventional techniques for the processing of these materials are based on the use of high-energy e-beams, ion implantation, X- and gamma ray irradiation. Generally, these methods do not allow an easy encoding of patterns with the high spatial-resolution and the low-dimensionality required for the future integrated optical technology. A soft X-ray laser, on the other hand, has enough large photon energy, enough high intensity and the low penetration depth to allow the processing of large crystal areas with sub-micrometer spatial resolution in very short exposure times. Moreover, due to the high spatial coherence of the X-ray laser beam, holographic techniques can be used for the direct encoding of fine grating structures. Here, we demonstrate, by use of a Lloyd mirror interferometer, the encoding of photo-luminescent lines of stable color-centers (CCs) on the surface of LiF crystals.

2. The laser

2.1. The laser pulse characteristics

A detailed description of our capillary discharge laser can be found in recently published papers [5]. The amplification of the 46.9 nm laser line is produced inside a high-temperature and a high density Ar plasma, on the 3p-3s transition of the Ne-like Ar ions. The plasma column is formed by the Z-pinch compression of a pre-ionized Ar gas, induced by a 50-ns rise-time current with amplitude of 20 kA. The discharge is produced inside a 3-mm diameter alumina capillary tube, which initially contains the Argon at the pressure of 350-550 mTorr. The amplification is produced by the single pass propagation of the lasing radiation through the plasma medium. At the capillary length of 45 cm the laser operates in a fully saturated regime (20 x I_{SAT}) with the energy per pulse of about 300 μJ (see ref. [5]). Further improvement is obtained in 55 cm long capillaries, where the energy increases to an average value of ≈ 400 μJ (see Fig. 1). The laser pulse duration is of 1.7 ns, which provides a peak power of 0.2 MW and output intensity of $8\cdot10^8$ W/cm^2 (table 1). The pulses are produced at the repetition rate of 0.2 Hz, which, at present, allows achieving an average power of 0.06 mW. The pulse-to-pulse fluctuation of the output intensity remains in the range of ±20%.

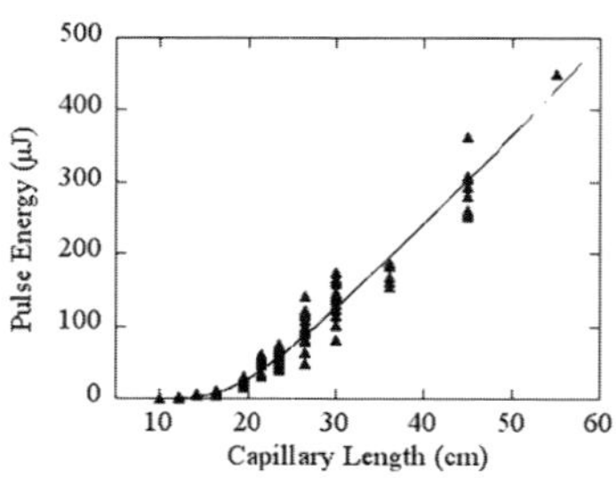

Wavelength	46.9 nm
Pulse Energy	400 μJ
Pulse duration	1.7 ns
Peak power	0.2 MW
Beam divergence	5-0.6 mrad
Pulse rep. rate	0.2 Hz
Average power	0.06 mW
Total efficiency (E_{laser}/Electr. Stored Energy)	5-2 10^{-6}
Peak spectral brightness	10^{24} ph/(s mm^2 mrad^2 in 0.01 % bw)

Figure 1. Plot of the laser pulse energy versus the capillary length in the saturation regime. **Table** 1 (on the right) summarizes the characteristics of our laser device.

2.2. The Laser coherence

Regarding the intensity distribution of the beam and its spatial coherence, we describe here the regime of operation, typical of our laser [6], which is free from the effect of the refraction defocusing. This regime is found, in our discharge conditions, using the

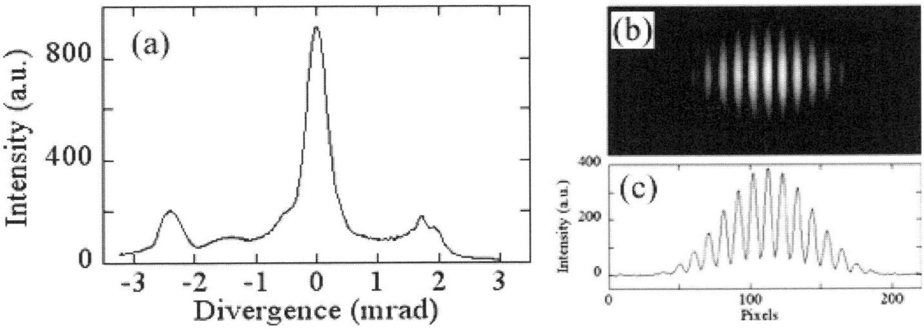

Figure 2. (a) Profile of the intensity distribution in far-field condition. (b) Interference fringes and (c) the corresponding profile obtained by the double slit experiment with the slit separation of 130 μm placed at 42 cm from the capillary output.

initial gas pressure of 0.43-0.48 Torr. In this range of pressure the beam reaches a highly collimated and axially peaked structure (Fig. 2 (a)) with the divergence of only 0.6 mrad. This value is close to the limit of diffraction. We can attribute the formation of the low-divergent beam to the fact that the laser amplification is produced before the plasma collapses on the capillary axis, when the electron density distribution has still an almost flat plateau close the capillary axis [7]. By our ray-tracing simulations it results that the full transverse dimensions of this plateau, where the collimated beam can be formed without the effects of the refraction, should be of the order of 90 μm. The small diameter of this region and the large capillary length (45 cm) provide for the gain medium a Fresnel number ($F = ka^2/L$) of ≈1.5, ensuring the amplification of the lowest order modes and the large spatial coherence.

The almost full transverse coherence of the sub-milliradiant emission has been demonstrated experimentally by a classical Young's double slit experiment. In this kind of experiments, provided the slits are illuminated with the same intensity, the modulus of the coherence degree $|\mu_{12}|$ is equal to the fringe visibility $V=(I_{max}- I_{min})/(I_{max} + I_{min})$. Figure 2 (b) shows a typical single shot interferogram obtained using the slits with separation of 130 μm at 42 cm from the capillary output and produced by the only contribution of the central peak. In these conditions we found an excellent fringe visibility (0.78) as it is demonstrated in the corresponding lineout (Fig. 2 (c)). The coherence degree experimentally decreases from 0.9 to 0.78 with the increase of the slit separation from 90 μm to 130 μm. Considering these data and assuming a planar Gaussian-Schell model source to determine the coherence length (L_c = 195 μm at 42 cm from the capillary output) and the generalized Van Cittert-Zernike theorem (see more detail in [6]), we have determined the equivalent source dimension $\sigma_S \approx 28$ μm, and the intrinsic coherence length $\sigma_G \approx 30$ μm. The beam quality factor of the beam, estimated as $M \approx \sqrt{1+\sigma_S^2/\sigma_G^2}$ according to [7], is of about 2. This value provides evidence of the quasi-Gaussian properties of this laser beam.

3. Photo-luminescent pattern encoding in LiF crystals

In recent years, a great interest has been devoted to the materials exhibiting active optical properties for their potential applications to the device miniaturization and the development of integrated optical circuits [9-11]. Among these materials, alkali-halide crystals containing color-centers (CCs) have been extensively studied for the realization

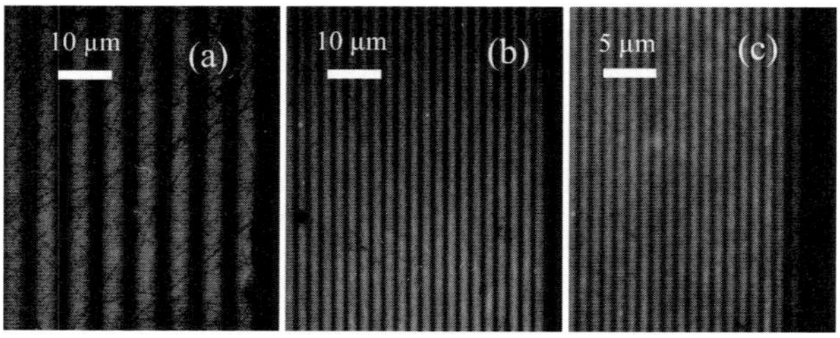

Figure 3. Periodic fluorescent patterns encoded on LiF crystals by the 46.9 nm soft X-ray laser. The patterns have line spacing of (a) 6.8, (b) 2.4 and (b) 1.2 μm, respectively.

of devices such as solid-state lasers operating in the visible and in the near-infrared spectral ranges. A particular interest has been focused on lithium fluoride (LiF), as it can host different types of CCs (for instance the F_2 and F_3^+), which can be laser active also at room temperature, and for its high photo-thermal stability and good physical and chemical properties. Moreover the observed increase of the refraction index of the crystal in the presence of the F_2 and F_3^+ centers distributions in LiF is suitable for miniaturized optical waveguides, diffraction gratings or Distributed Feedback (DFB) lasers.
To demonstrate the potentialities of our soft X-ray laser in this field, we have encoded fine grating structures of CCs with line spacing down to the sub-micrometer region using a Lloyd mirror interferometer. The LiF samples consisted of commercially available 3 mm thick crystals with 75 mm^2 area and optically polished surface. To observe the fluorescence of CCs after irradiation with the soft X-ray laser, we have used an optical microscope (ZEISS Axioplan 2) operating in fluorescence mode, with the fluorescence induced by the 445-495-nm emission band of a mercury lamp. Using the maximum optical magnification of 100^X, the microscope allowed to read out the printed fluorescent patterns with the maximal spatial resolution of ~500 nm. In our experimental configuration, 10 laser shots (energy flux on LiF of 25 mJ/cm^2) where sufficient to produce permanent and well-observable photo-luminescent patterns on area up to 10 mm^2. The fluorescence spectrum of the irradiated samples, with the broad-band emission between 500 and 700 nm and the two peaks at 640 and 540 nm, shows the formation of the F_2 and F_3^+ centers, respectively. Due to the high absorption coefficient of LiF at 46.9 nm, about ~71 μm^{-1}, the thickness of the colored patterns is estimated to be of the order of 15-20 nm. In Figs. 3 (a), (b) and (c) we show small regions of three irradiated LiF crystals where periodic lines with interline distances of 6.8, 2.4 and 1.2 μm, respectively, can be clearly observed. Due to the high coherence of the X-ray laser a large number of lines, from several tens up to 100, depending on the line spacing, are encoded with a very good visibility (~0.5). Fringes with the estimated line spacing of 0.5 μm, limit of resolution of the used optical microscope, have been observed by the further increase of the grazing angle of the mirror but, in principle, no limitation is given to the printing of fringes with period down $\lambda/2$.

4. Conclusions

In conclusion we have reported on the performances of a 46.9 nm soft X-ray laser. The laser, produced in 55 cm long capillary channels, shows energy up to 400 μJ per pulse, an almost full spatial coherence and can operate in a regime free from the effect of the refraction defocusing with divergence as low as 0.6 mrad. We have applied this laser, operating at the pulse repetition rate of 0.2 Hz, to the processing of optically dielectric materials. As an example, we have shown that the laser beam can efficiently produce, by linear absorption, permanent, photo-luminescent patterns of F_2 and F_3^+ color centers in lithium fluoride crystals. Fine grating structures with interline distances down to the sub-micrometer resolution have been efficiently produced by holographic encoding.

References

[1] B. R. Benware, C. D. Macchietto, C. H. Moreno, J. J. Rocca, Phys. Rev. Lett. **81**, (1998) 5804.

[2] J. J. Rocca, J. Filevich, E. C. Hammarsten, E. Jankowska, B. Benware, M. C. Marconi, B. Luther, A. Vinogradov, I.. Artiukov, S. Moon, V.N. Shlyaptev, Nucl. Instr. and Meth. in Phys. Res. A, **507** (2003) 515.

[3] B. R. Benware, A. Ozols, J. J. Rocca, I. A. Artiukov, V. V. Kondratenko, A. V. Vinogradov, Opt. Lett., **24**, (1999) 1714.

[4] G. Tomassetti, A. Ritucci, A. Reale, L. Palladino, L. Reale, L. Arrizza, G. Baldacchini, F. Bonfigli, F. Flora, L. Mezi, R. M. Montereali, S.V. Kukhlevsky, A. Faenov, T. Pikuz, J. Kaiser, Europhys. Lett. **63**, (2003) 681.

[5] G. Tomassetti, A. Ritucci, A. Reale, L. Palladino, L. Reale, S.V. Kukhlevsky, F. Flora, L. Mezi, A. Faenov, T. Pikuz, A. Gaudieri, Opt. Comm., **231**, (2004) 403.

[6] Ritucci, G. Tomassetti, A. Reale, F. Flora, L. Mezi, (in print in Phys. Rev. A).

[7] A. Siegman, New developments in Laser Resonators, in D. Holmes (Ed.), Optical Resonators, Proc. of SPIE, Vol. 1224, USA, 1990, pp. 2-14.

[8] A. Ritucci, G. Tomassetti, A. Reale, L. Palladino, L. Reale, F. Flora, L. Mezi, S.V. Kukhlevsky, A. Faenov, T. Pikuz, Appl. Phys. B, **78**, (2004) 965.

[9] T. Kurobori, K. Inabe, N. Takenuchi, J. Phys. D, **16**, (1983) L121.

[10] R. M. Montereali, M. Piccinini, E. Burattini, Appl. Phys. Lett., **78**, (2001) 4082.

[11] E. J. Caine S. D. Miller J. Vac. Sci. Technol. B, **16**, (1998) 3232.

Inst. Phys. Conf. Ser. No 186
Paper presented at 9th Int. Conf. X-Ray Lasers, Beijing, China, 24–28 May 2004

Effects of pressure on the lasing time and intensity of soft x-ray laser pumped by capillary discharge

Yongpeng Zhao*, Yuanli Cheng, Qi Wang
National Key Laboratory of Tunable Laser Technology, Harbin Institute of Technology, Harbin, China, 150001

Yasushi Hayashi, Eiki Hotta
Department of Energy Sciences, Tokyo Institute of Technology, Nagatsuta, Midori-ku, Yokohama 226-8502, Japan

Abstract. The laser spike of Ne-like argon 46.9nm, which is pumped by capillary discharge with predischarge current of 20A and main current of ~30kA, is measured with XRD (x-ray diode). Changing the pressure of Ar gas filled in the ceramic capillary, we observe the effects of pressure on amplitude and time of laser spike. The experimental results show that the lasing time, which is the time of laser spike after the onset of main current, increases monotonously between 35ns and 40ns when the pressure increases from 220mTorr to 420mTorr. Nevertheless, there is maximum amplitude of the laser spike about 320mTorr. The main current at the lasing time is 70~75% peak current.

1. Introduction

Since the first demonstration of the soft x-ray laser, one of the major goals in the laser research is the development of practical laser sources that could be widely utilized in several important applications such as surface science, photophysics, photochemistry and so on[1]. High power lasers are commonly used as pumping sources for ablating the target to produce soft x-ray lasers in suitable plasmas[2]. These schemes have a low efficiency and the widespread use has been hindered by their large scale, complexity and high cost. Other pumping sources must be applied to develop compact and practical soft x-ray laser. For this reason, direct discharge excitation has been studied for many years. In 1994, J.J.Rocca et.al realized the first demonstration of large soft x-ray amplification using a fast discharge excitation of a capillary channel filled with uniformly preionized Ar[3]. The gain coefficient of 0.6cm^{-1} and gain-length product of 7.2 were obtained in the *J=0-1* line of Ne-like Ar at 46.9nm. In the experiments, the predischarge current is about 10A and several microseconds duration and main current is about 40kA with 10%-90% rise time of 20ns. The soft x-ray laser has the advantages of a relatively large gain volume, long gain duration, high efficiency and compact size. In 1996, using an iridium mirror, they realized the double-pass amplification at 46.9nm

* e-mail zhaoyp3@yahoo.com.cn; Phone 86-451-6412643; Fax: 86-451-6416060

and achieved laser pulse energies up to 30μJ and a gain-length product greater than 25[4]. In 1999, they demonstrated the generation of a millijoule-level laser pulse with repetition frequency of 4Hz, the highest average output pulse energy of 0.88mJ and average power of 3.5mW in 34.5mm aluminum oxide capillary channel[5]. From then on, they have finished many application experiments with the soft x-ray laser, specially the diagnostics of dense plasma[6]. They predicted that fast capillary discharge excitation would lead to amplification in Ne-like and Ni-like ions[7]. Presently, they have obtained the amplification of the *J=0-1* line of Ne-like sulfur and Ne-like chlorine at 60.8nm and 52.9nm respectively[8, 9]. Other research groups also realized the Ne-like Ar 46.9nm soft x-ray laser and many useful results have been obtained in experiment and theory[10-12].

In order to enhance the efficiency and energy of 46.9nm soft x-ray laser, the optimum parameters, which include the material, inner diameter and length of the capillary, the amplitude of the predischarge current, main current and so on, are selected in experiment[7,10,11] and the laser beam characteristics are also studied experimentally[13, 14]. The experimental results show that the optimum time for lasing is around the peak of the main current, when the first compression shock wave reaches the axis[4]. However, in our experiments, the current at the lasing time is only 70%~75% peak current. In order to delay the lasing time to the time of peak current, we study the effects of pressure in the capillary on the lasing time. These results should be useful to achieve the optimum parameters of the laser and improve its energy conversion efficiency.

2. Experimental Setup

The experimental device used in the experiments is shown in figure 1. The device has been described elsewhere in detail[10, 15-16]. The pulsed power supply consists of three-stage Marx generator, a 2:30 step-up transformer and a pulse-forming line (PFL) that is made of a water capacitor with capacitance of 3nF. The water capacitor is discharged through a pressurized SF_6 spark gap switch into a capillary channel, which is made of Al_2O_3 ceramic and has an inner diameter of 3mm and a length of 15cm. A molybdenum rod with a length of 3mm is inserted into the capillary as the electrode. A hole with a diameter of 3mm drilled in the other electrode enable the penetration of the Ar gas into the capillary and the output of soft x-ray light. The capillary is inserted in a plastic cylinder with a wide hole. The capillary is filled with argon and differentially pumped with a turbo molecular pump through a pinhole of 1mm diameter. The pinhole is located at a distance of 15mm from the end of the capillary.

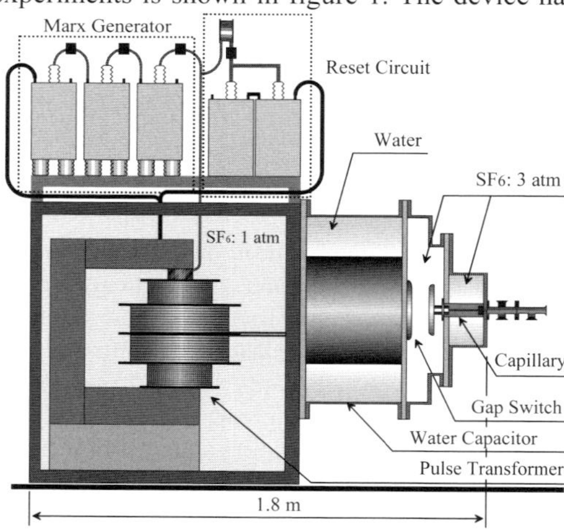

Figure 1. Schematic diagram of the experimental setup

The capillary is predischarged by a current pulse of about 20A and several microseconds duration to produce a uniform preionization. The main current pulses in the experiments have amplitude of ~30kA, 10%~90% rise time of 52ns and duration of

~120ns. The temporal evolution of the laser pulse is measured with a XRD and the laser spike is observed in the experiment. The XRD consists of a photocathode coated with a gold layer and a wire mesh anode. The distance between the photocathode and anode is 2mm and a negative bias voltage of –500V is applied to the photocathode. Because the lasing time scale is of ns order, the calibration on the time delay of the measurement system has been carried out and the error is controlled to less than 0.1ns.

3. Experimental Results and Discussion

The typical discharge current waveform and XRD output waveform are shown in figure 2. The filling pressure of argon gas is 300mTorr and the amplitude of pre-discharge current and main current are 20A and 33kA respectively. The directivity and spectroscopy of the spike has been studied elsewhere and experimental results show that the spike is the Ne-like 46.9nm soft x-ray laser[15, 16]. Other research group also measures the similar laser spike and its duration is about 1ns[4, 5]. The results in figure 2 show that the lasing time, which is the time of spike after the onset of the main discharge current is about 38ns. At this time, the amplitude of the main current is 24kA. In fact, the maximum amplitude of the main current is 33kA after the lasing time 26ns. According to the theory, it would be benefit to improve the energy conversion efficiency, enhance the stability of the plasmas and reduce the density gradients in the plasma columns, if the laser spike could be produced at the time of the peak current. Hence, the lasing time should be postponed near the time when the amplitude of the main current is maximum. The theoretical results indicated that the compression of the plasma become more difficult for higher pressure. Thus, increasing the pressure, we study the laser spike in the experiment. In order to postpone the lasing time, the

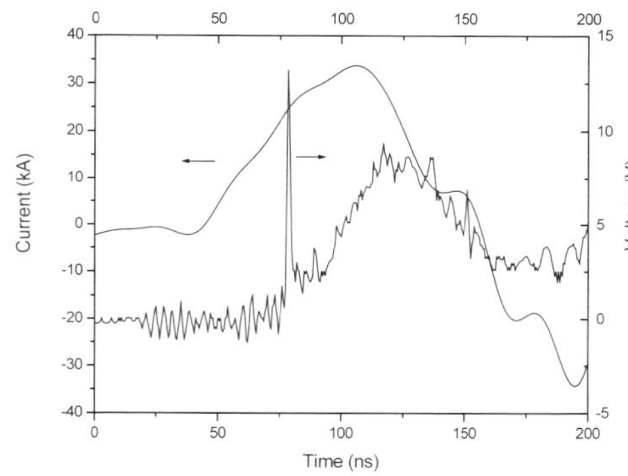

Figure 2. The waveforms of main discharge current and XRD output signal corresponding to predischarge current of 20A and 3mm-diam capillary filled with argon of 300mTorr.

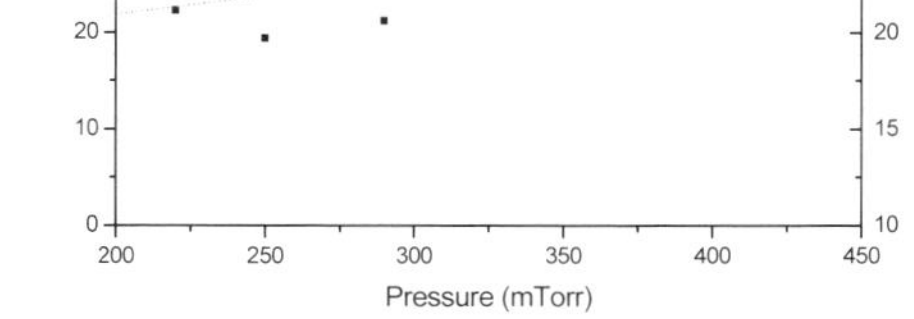

Figure 3. Lasing time and the amplitude of main current at lasing time as a function of pressure. The experimental data of lasing time is indicated by triangle and the current at the lasing time is indicated by square. The predischarge current is 20A, main discharge current is 33kA and inner diameter of the capillary is 3mm.

pressure of argon gas filled in the capillary is changed from 220mTorr to 420mTorr. In figure 3, the plot of the lasing time and the amplitude of main current at lasing time as a function of pressure is shown. The results indicated that the lasing time increases monotonously between 35ns and 40ns with the increase of pressure, and the amplitude of main current at lasing time increases from 21kA to 25kA, which is 65%~75% of peak current. The error in the figure is mainly attributed to the breakdown fluctuation of the spark gap switch.

If the pressure is more than 420mTorr, the laser spike can't be measured in the experiment. The plot of spike amplitude as a function of pressure is shown in figure 4. There is maximum amplitude of the spike about 320mTorr. When the pressure is changed from 320mTorr to 420mTorr, the intensity of the laser decreases monotonously. For delay the lasing time, the pressure should be increased to higher pressure. However, there is optimum pressure for the intensity of laser. Consequently, only by increasing the pressure, it is difficult to postpone the lasing time and improve the laser energy at the same time. If the laser spike could occur near the time when the main current is maximum in the optimum pressure, the shape of main current pulse or the parameters of the capillary should be changed.

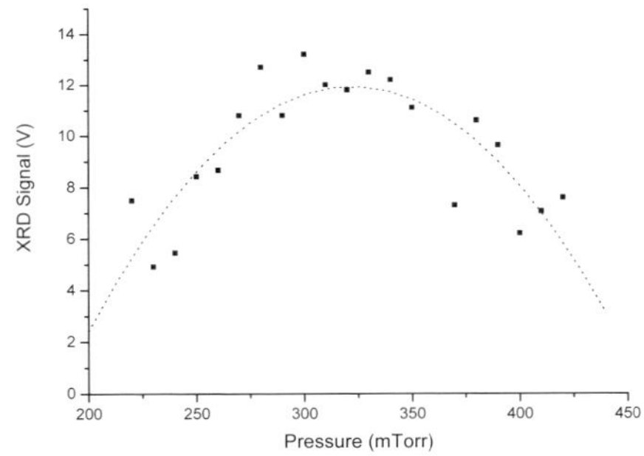

Figure 4. Amplitude of laser spike as a function of pressure. The predischarge current is 20A, main discharge current is 33kA and inner diameter of the capillary is 3mm.

4. Conclusions

We study Ne-like Ar 46.9nm laser pumped by capillary discharge. A XRD is used to measure the laser spike in the experiments. Changing the pressure in the capillary, we observe the effects of pressure on amplitude and time of laser spike. According to the experimental results, we can draw the following conclusions:

(1) We can postpone the lasing time for several nanoseconds through increase the pressure of argon filled in the capillary.

(2) When the pressure is increased, the laser intensity don't increase monotonously. There is optimum pressure for the laser intensity.

(3) Only by increasing the pressure, it is difficult to postpone the lasing time and improve the laser energy at the same time.

(4) According to the experimental results, we could also obtain the laser spike as figure 2, if the 10%~90% rise time of the main current pulse is ~30ns and the amplitude of the current is ~25kA. Therefore the pulse shape, specially the rise time, of the main current and the parameters of the capillary should be further studied in the near future experiments.

References

[1] Benware B R, Moreno C H et al 1997 Opt.Lett. 22 796-798

[2] Ben-Kish A, Shuker M et al 1999 Inst. Phys. Conf. Ser. No.159 191-194

[3] Rocca J J, Shlyaptsev V et al l 1994 Phys.Rev.Lett. 73 2192-2195
[4] Rocca J J, Clark D P et al 1996 Phys.Rev.Lett. 77 1476-1479
[5] Macchietto C D, Benware B R et al Opt.Lett. 1999 24 1115-1117
[6] Rocca J J, Filevicha J, et al 2003 Nuclear Instruments and Methods in Physics Research A 507 515-522
[7] Rocca J J, Tomasel F G. et al 1995 Phys.Plasma. 2 2547-2554
[8] Tomasel F G, Rocca J J et al 1997 Phys.Rev.A 55 1437-1440
[9] Frati M, Seminario M et al 2000 Optics Letters 25 1022-1024
[10] Niimi G, Hayashi Y et al 2002 IEEE Transactions on Plasma Science 30 616-621
[11] Ben-Kish A, Nemirovsky R A et al 1999 Proc.SPIE 3776 166-174
[12] Ritucci A, Tomassetti G et al 2003 Contrib. Plasma Phys. 43 88-93
[13] Tomasel F G, Shlyaptsev V N et al 1996 Phys.Rev.A 54 2474-2478
[14] Moreno C H, Marconi M C, et al 1998 Phys. Rev. A 58 1509-1514
[15] Niimi G, Hayashi Y et al 2001 J.Phys.D:Appl.Phys. 34 2123-2126
[16] Hayashi Y, Xiao Y et al 2003 Jpn. J. Appl. Phys. 42 5285-5289

Inst. Phys. Conf. Ser. No 186
Paper presented at 9th Int. Conf. X-Ray Lasers, Beijing, China, 24–28 May 2004

Divergence Measurement of Ne-like Ar Soft X-ray Laser Beam Generated by Capillary Z-pinch Discharge

Yasushi Hayashi, Hamid Ghomi, Masato Watanabe, Akitoshi Okino, Kazuhiko Horioka and Eiki Hotta
Department of Energy Sciences, Tokyo Institute of Technology, 4259 Nagatsuta, Midori-ku, Yokohama 226-8502, Japan

Abstract. We present results and a method of measuring the divergence of a Ne-like Ar soft X-ray laser beam generated by a capillary z-pinch discharge. The simple method allows us to measure the laser beam at different section, which is required to obtain the divergence of the beam, without opening the device. The diameter of the laser beam measured by this method is in a range of 4.3 to 5.6 mm, which is corresponding to a range of divergence from 2.99 to 3.89 mrad.

1. Introduction

In the recent decades, the research on soft X-ray (SXR) has been sharing the spotlight, due to its short wave length that enables the fabrication of nano-scale structure. Additional advantages can be found for SXR laser, because it exhibits extraordinary characteristics both as a laser and as a soft X-ray source [1]. However, one of the problems that exist in this kind of laser is the extreme large size of the device. A Z-pinch using a capillary discharge is considered as the solution of this problem, because its lasing equipment and power source can be reduced to a "table-top" size. In 1994, J J Rocca et al. has firstly observed this kind of lasing, which appeared as a spike on the output of the X-ray diode (XRD)[2,3]. The same spike output has also been reported in our past works [4,5], which has been proven as Ne-like Ar lasing (J=0-1, λ=46.9 nm) by spectroscopic measurement [6]. In this paper, a method of measuring the divergence of the laser beam has been proposed, which allows us to get the information of the beam spot without opening the device. Using this method, the beam divergence of the SXR laser has been measured.

2. Experimental Apparatus

The experimental device used in the present study is shown schematically in Fig. 1, and the detailed description of the device can be found elsewhere [4-6]. Briefly, the discharge system is driven by a pulsed power supply, which consists of a 3-stage Marx

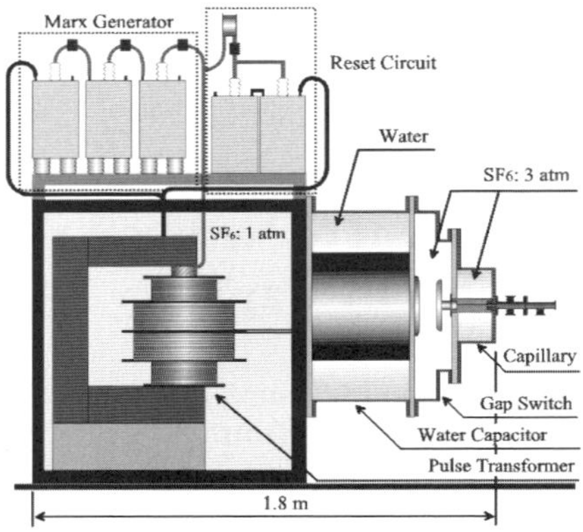

Figure 1. The experimental apparatus used in the SXL experiment.

generator, a 2:28 step-up transformer and a pulse forming line (PFL) made of a water-capacitor. The capillary is pre-discharged by a current of up to 20 A, which is provided by a separate circuit. The output voltage of the Marx generator is 40 kV, which subsequently yields a PFL output voltage of 560 kV. With this power supply system, it is able to obtain a discharge current pulse of up to 32 kA. The capillary is made of Al_2O_3 ceramic with an inner diameter of 3 mm and a length of 200 mm. The SXR laser and background emission were measured by an X-ray diode (XRD) installed at a position 720 mm downstream from the end of the capillary.

Figure 2 shows schematically the measurement system of the SXR laser. The photocathode of the XRD is made of gold, which enables a sensitive detection of X-ray ranging from 10 to 100 eV. An aperture of diameter 4mm has been adopted in front of the XRD to eliminate the SXR beam. A bellows is adopted at the upstream end of the drift tube, which enables adjustment of the location of the XRD without breaking vacuum.

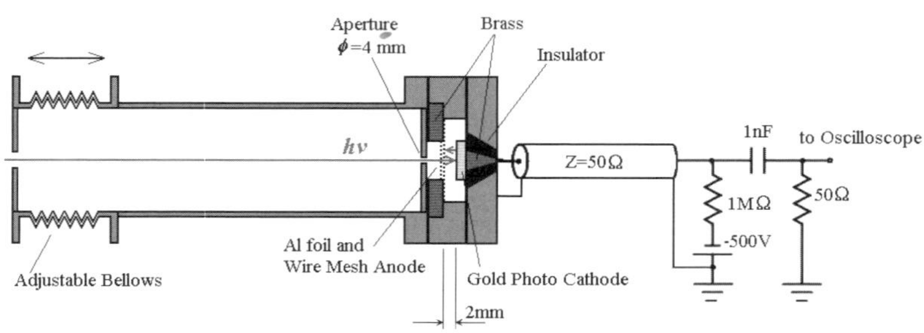

Figure 2. XRD measurement system.

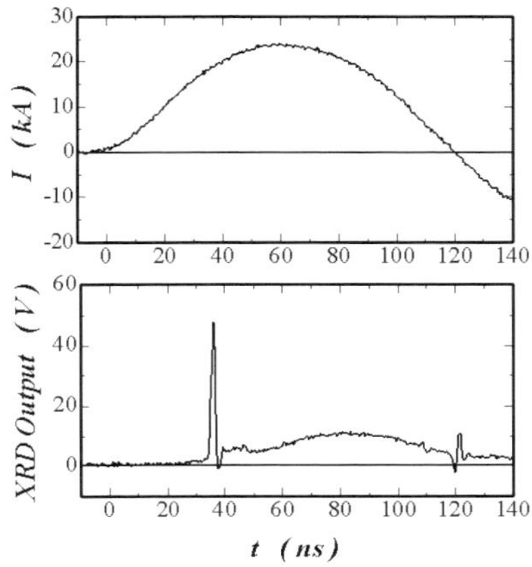

Figure 3. Discharge current and the XRD output.

3. Experimental Results

Figure 3 shows the SXR laser outputs measured when the XRD is set on the axis of the SXR laser, which corresponds to the maximum output, without using the aperture. Here, the filling Ar gas pressure is 300 mTorr. The discharge current has a peak value of 23 kA and a pulse width of 100 ns. The amplitude of the XRD output that corresponds to the laser spike and the background radiation are 45.8 and 10.6 V, respectively. The energy of the laser spike is found to be 2.1 μJ. It has been reported in our previous experiment that the laser output is due to the 3p-3s Ne-like Ar lasing at wavelength of 46.9 nm.

Two series of the experiments have been conducted to measure the beam size of the SXR laser by changing the location of the XRD horizontally and vertically, where the center of the beam is set initially on the vicinity of origin of the coordinate. Figure 4(a) shows the SXR laser measured by the XRD by changing the location of the XRD horizontally. The laser signal is observed in a range from –4 to 4.3 mm, which corresponds to a range of 8.3 mm and is relatively symmetrical to the center of the beam axis. This indicates that the beam spot has a diameter of 4.3 mm, since the radius of 2 mm of the aperture has to be subtracted from both sides. Figure 4(b) plots the results measured by changing the position of the XRD vertically. It is seen that the axis symmetry becomes worse than in the case of changing the location of the XRD vertically. The laser signal is observed in a range of 9.6 mm from –3.8 to 5.8 mm, which corresponds to a beam size of 5.6 mm. This indicates that the beam diameter varies from 4.3 to 5.6 mm, and is not of a very circular shape. The range of the above beam diameter corresponds to a range of beam divergence from 2.99 to 3.89 mrad.

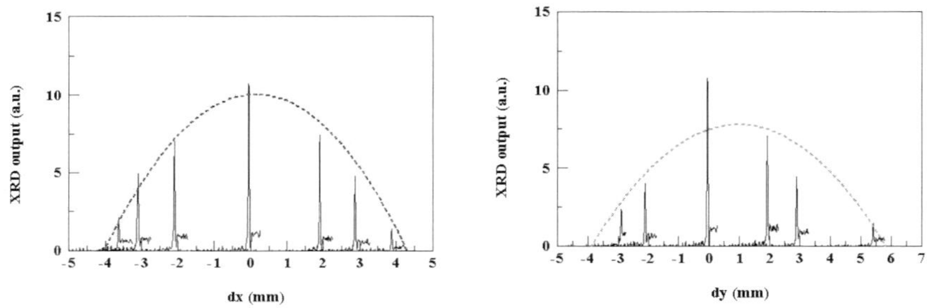

(a) **(b)**

Figure 4. XRD output measured by changing the lateral location of XRD (a) horizontally and (b) vertically.

In the present study, the resolution of the measurement is not very high, since a relatively large beam aperture of 4mm in diameter has been used. The reason for this is to assure a relatively high output of the SXR laser signal, especially at the edge of the beam spot. We cannot obtain the beam diameter at half height of the beam intensity, because it is not possible to obtain the radial distribution. Efforts will be made in the future to reduce the diameter of the aperture or using a slit.

4. Conclusion

We have presented a method of measuring the beam size and divergence of SXR laser. The simple method allow us to measure the laser beam at different section, which is required to obtain the divergence of the beam, without breaking the vacuum. Using this method, we have measured the divergence of the Ne-like Ar soft X-ray laser beam generated by a capillary z-pinch discharge. The diameter of the laser beam measured by this method is in a range from 4.3 to 5.6 mm, which corresponding to a range of divergence from 2.99 to 3.89 mrad.

References

[1] Attwood D 1999 Soft X-rays and Extreme Ultraviolet Radiation (Cambridge University Press UK).

[2] Rocca J J, et al. 1994 Phys.Rev.Lett. **73** 2192.

[3] Rocca J J, et al. 1995 IEEE Journal of Selected Topics in Quantum Electronics, **1** 945.

[4] Niimi G, et al. 2001 J. Plasma and Fusion Research **77** 1239 (in Japanese).

[5] Niimi G, et al. 2002 IEEE Trans. Plasma Sci. **30** 616.

[6] Hayashi Y, et al. 2003 Jpn. J. Appl. Phys. **42** 5285.

Inst. Phys. Conf. Ser. No 186
Paper presented at 9th Int. Conf. X-Ray Lasers, Beijing, China, 24–28 May 2004

Numerical research of dynamic and radiation of plasma in capillary discharges

V.A. Burtsev, N.V. Kalinin
D.V. Efremov Scientific Research Institute of Electrophysical Apparatus
196641 Metallostroy, St. Petersburg, Russia

Abstract. The report presents the results of numerical analysis of fast Z-discharges in a low-inductive capillary channel (1.4 nH/cm) with the operating conditions varying in wide ranges. The methods are analyzed to correlate the plasma dynamics and radiation with the processes in the electric chain with a high-voltage pulse generator.

1. Introduction

To obtain radiation with the required high characteristics in a powerful pulsed electrical discharge it is necessary to correlate the plasma dynamics with the physical processes in the electric power supply circuit of the discharge, including the primary energy storage. This multifactor problem can be solved by a comprehensive numerical simulation of the discharge using a self-consistent models which take into account the mutual effect of the physical processes determining pulse generation of electric power introduced into the discharge, the plasma dynamics and radiation. These models and results of the comprehensive numerical analysis of capillary discharges, as applied to the experiments performed on the facilities developed by NIIEFA, are presented below.

2. Physical models of capillary discharges

To analyze numerically capillary discharges one-dimensional (1D), one- and two-temperature (1T and 2T) MRHD-models were developed which take into account peculiar features of the operating conditions (for example, high initial gas pressure, wall material evaporation, pre-ionization character, etc.). The peculiar features of the operating conditions are taken into account in the initial and boundary conditions.

The base model of the capillary discharge is, for the most, similar to that described in [1]. The plasma dynamics is considered in one-dimensional (1D) single-liquid two-temperature (2T) approximation with regard for the following:

- magnetic field diffusion into the plasma with consideration for the real electrical circuit of the discharge;
- Joule plasma heating, including heating caused by anomalous plasma electric conduction;
- magnetic compression and shock waves;
- dissipation processes in the plasma, i.e. heat conduction and viscosity;
- radiation transfer caused by lines emission and absorption, photo-ionization and recombination, bremstrahlung processes resulting in emission and absorption of photons.

The following equations and models are applied:

- wide-range equations of state and interpolation models for calculation of transport properties of the working gas plasma and wall materials [2-8];
- thermodynamic and radiation-collision models for calculation of the plasma charge composition [4,9-10] to organize most efficiently the calculation algorithm for different operating conditions;
- different approximations to calculate radiation transfer in the plasma - approximation of optically «semi-transparent» plasma, as well as diffusion approximation.

The pulse power supply system was considered on the basis of the equivalent circuit for a transporting line connected to a high-voltage generator generating voltage pulses with a double amplitude of an incident voltage wave and the resistance equal to the line wave resistance.

The effect of evaporated material of capillary walls on the plasma dynamics was taken into account in the same way as in [1].

3. Calculation results and their discussion

Discharges in argon-filled heat-resistant capillary ceramic tubes 5.5 mm in diameter were analyzed. The capillary length, l, varied from 50 mm to 150 mm. A voltage pulse of an incident wave had the following parameters: maximum voltage – 140 kV, pulse length at half-height – 100 ns, pulse leading-edge time – 10 ns. The initial gas pressure in the capillary, P_0, varied from 0.01 to 700 Torr. Different initial distributions of gas temperature and density in the capillary were used.

To test the models the capillary discharges were calculated, as applied to the conditions described in the references. Two versions differing in the method for plasma production in the capillary were considered. In the first version the capillary is initially empty, the plasma is produced by evaporation of the wall material. In the second version the capillary was considered to be filled with pre-ionized gas. These versions correspond to the well-known experiments and calculations [1]. According to our calculations the pattern of capillary discharge evolution in the initially empty capillary correlates well with [1]. In a quantitative sense, our calculations correlate with [1] but only satisfactorily. The plasma dynamics, as observed in our calculations, in the capillary pre-filled with the weakly ionized plasma correlates well with [1] in both quantitative and qualitative senses.

To estimate the optical thickness of the capillary discharge plasma we calculated the photon run lengths averaged over Rosseland $l_R(T,\rho)$ (for optically thick plasma) and Planck $l_P(T,\rho)$ (for optically thin plasma), with the bremstrahlung and photo-ionization absorption taken into account. The calculations make it possible to define the plasma states which are realized in the capillary with $d_0 = 5.5$ mm and $l_0 = 50$ mm, at which plasma radiation is of volume character $l_P(T,\rho) >> a$, or the plasma radiates as a black body $l_R(T,\rho) << a$, here $a = d_0, l_0$ - the characteristic scale of the discharge length. The calculations showed that in the analyzed range of temperatures $1 \le T \le 200$ eV and densities $10^{15} \le N \le 10^{19}$ cm^{-3} the optical thickness of the plasma can vary essentially in both radial and axial directions. A sizable intermediate area of the «semi-transparent» plasma lies between the optically dense and optically transparent states. The electric discharge in such a «semi-transparent» plasma is of interest, as it has positive properties both as an opaque discharge with maximum integral radiation and as an optically transparent discharge characterized by radiation selectivity and high efficiency. Let us define the boundary radiation wavelength $\lambda_0(T,\rho)$ for the «semi-transparent» plasma on condition that the optical discharge thickness $\tau = \chi_\lambda(\lambda_0)\cdot d = 1$. The plasma is considered completely opaque for photons with the wavelength $\lambda > \lambda_0$ and completely transparent for short-wave photons with $\lambda < \lambda_0$. Rosseland and Plank photon track length as a function of the plasma state parameters is shown in Fig.1.

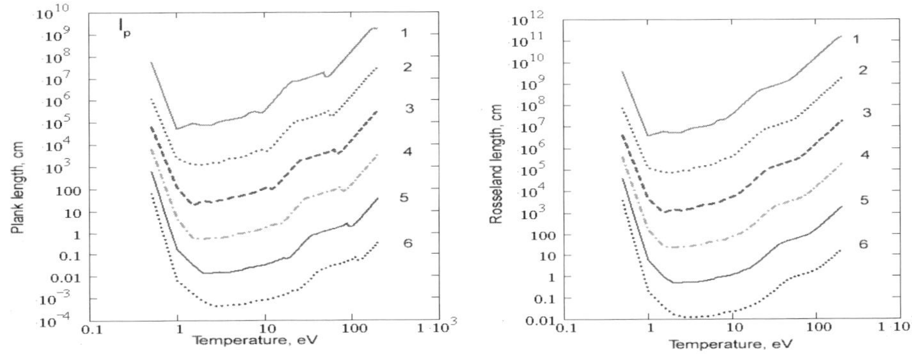

Fig.1. Dependences of photon run lengths versus electron temperature and different densities. N, 10^{17} cm^{-3}: 1- 0.01, 2- 0.1, 3- 1, 4- 10, 5- 100, 6- 1000

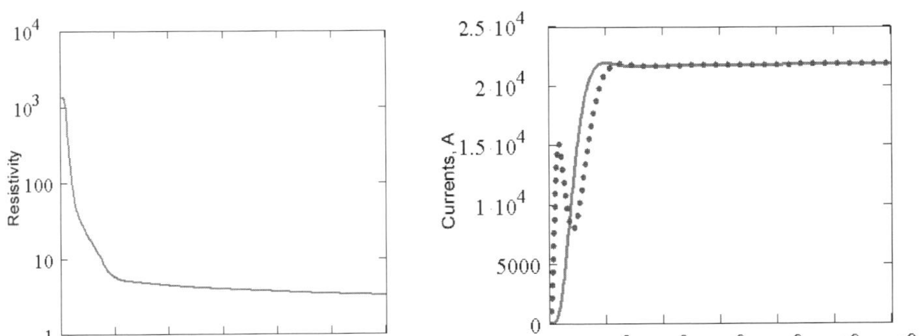

Fig.2. Temporal dependences of discharge resistivity and current, and also condenser current (dot curve)

Fig. 2-4 shows the results of discharge calculation in the capillary pre-filled with gas (Ar), with the temperature initially distributed nonuniformly. Temperature in the ring-shaped plasma layer produced by a surface discharge on the inner wall of the capillary was taken equal to 1 eV. The value and distribution of plasma density in the layer and its thickness were considered as the task parameters. The plasma parameters were the highest at such an initial pre-ionization.

The calculations showed that the plasma dynamics is mainly the same in the case of both the above-considered nonuniform pre-ionization and uniform initial distribution of pre-ionization, as in [1]. At sufficiently high rates of discharge current rise, plasma heating can be more effective due to localization of entire current in the discharge layer and plasma detachment from the walls, optimal combination of magnetic compression and Joule heating caused, in particular, by plasma turbulence.

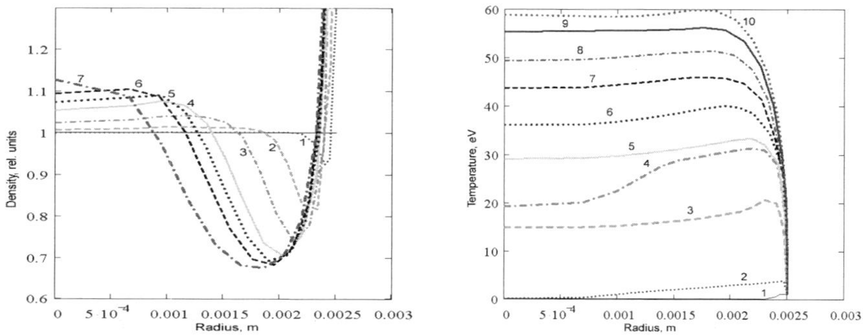

Fig.3. Radial distributions of density and temperature in different moments. Density, t, ns: 1- 0, 2- 20, 3- 30, 4- 40, 5- 45, 6- 50, 7- 60
Temperature, t, ns: 1- 0, 2- 2.5, 3- 7.5, 4- 12, 5- 20, 6- 30, 7- 40, 8- 45, 9- 50, 10- 60

4. Conclusion

The calculations performed show that changing in the operating conditions, i.e. initial gas pressure, pre-ionization character, changing in the current pulse shape, makes it possible to obtain radiation pulses with the intensity and spectral composition varying in sufficiently wide ranges. To improve the plasma parameters it is appropriate to strive for producing the uniform plasma shell near the capillary walls and improving the correlation between the magnetic compression dynamics and Joule heating with account to plasma turbulization.

References

[1] Bobrov N.A., Budanov S.V., Raznikova T.L., Sasorov P.V. 1996 Fizika plazmy 22 5 387 – 402 (in Russian).

[2] Ermakov V.V., Kalitkin N.N. 1979 Fizika plasmy 5 3 650 – 658 (in Russian)

[3] V.Ya. Karpov, A.P. Fadeev, G.V. Shpatakovskaya, 1982 V.V. Keldysh Institute of Applied Mathematics Preprint № 147 (in Russian)

[4] V.Ya. Karpov, A.P. Fadeev, G.V. Shpatakovskaya 1982 V.V. Keldysh Institute of Applied Mathematics Preprint № 110 (in Russian).

[5] Bespalov I.M., Polischuk A.Ya. 1989 Pisma v ZhTF 15 2 4 - 8 (in Russian)

[6] LeeY.T., More R.M. 1984 Phys. Fluids 27 5 1273 - 1286

[7] Basko M.M. 1985 Teplofizika vysokikh temperatur 23 3 483 - 491 (in Russian)

[8] Polischuk A.Ya. 1990 Teplofizika vysokikh temperatur 28 5 877 - 885 (inRussian)

[9] Zeldovich Ya.B., Raizer Yu.P 1966 Fizika udarnykh voln i vysokotemperaturnykh gidrodinamicheskikh yavlenii (M.: Nauka in Russian)

[10] Derzhiev V.I., Zhidkov A.G., Yakovlenko S.I. 1986 Ion radiation in the dense non-equilibrium plasma (M.: Energoatomizdat, in Russian).

Inst. Phys. Conf. Ser. No 186
Paper presented at 9th Int. Conf. X-Ray Lasers, Beijing, China, 24–28 May 2004

Technical issues of the gas-filled-capillary driver for Ne-like Ar soft X-ray laser

K Koláček, J Schmidt, V Prukner, J Štraus, V Boháček, O Frolov and P Vrba

Institute of Plasma Physics, Academy of Sciences of the Czech Republic, Za Slovankou 3, P.O. Box 17, 182 21 Prague 8, Czech Republic

Abstract. There are presented various historical and contemporary approaches to proper adjustment of both the boundary conditions (capillary radius, wall material) and the initial conditions (gas filling, internal and external pre-pulse generation) favourable for amplification on Ne-like Ar line at 46.88 nm.

1. Introduction

The soft X-ray lasers based on gas-filled-capillary discharge – as a rule – work according to electron collisional excitation pumping scheme. This scheme requires an efficient heating and ionisation of adequate number of particles in the gap between electrodes. To control the number of heated particles the main discharge-creating electrical pulse with fast rise rate is applied to a plasma-filled capillary (i.e. to a gas-filled capillary with some pre-ionisation) [1]. This pre-plasma ensures that the main current pulse starts (due to skin effect) in a thin cylindrical layer near the capillary wall [2], rapidly detaches from it (due to Z-pinch effect), and wall-ablation remains limited. Then the plasma column is compressed into a thin, dense, hot, straight plasma fibre, in which there are favourable conditions for amplification of spontaneous emission in soft X-ray region and for lasing.

2. Boundary and initial condition of the discharge

2.1. Capillary wall material

In the beginning plastic (polyacetal, teflon, …) capillaries were used [3]. In this case the main current pulse had to have high rise rate (approaching $5x10^{12}$ A/s) to ensure a rapid plasma detachment from the capillary wall and to limit plasma ablation of the wall (at least during the first discharge current half-period). Thus several hundred shots did not significantly deteriorate the capillary structure [3]. However, numerical modelling of plasma dynamics takes ablation of plastic capillaries into account [4-6]. The whole process is described as follows: the current sheet is generated near the capillary wall (due to skin effect); the fast current rise in the beginning generates a radially propagating shock wave that compresses and heats the plasma; however, the main part of the current remains at the capillary wall (due to dense cold plasma ablated from the capillary wall); it was supposed that namely this current-less compression was responsible for plasma column stability; in the moment of maximum compression the shock wave is reflected and expands. Despite the plastic capillaries are cheap, they are a rich source of soot (extremely harmful to all diagnostic equipment); therefore, later on (since 1998 [7]) these plastic capillaries were substituted gradually in all laboratories by alumina (ceramic) ones.

The alumina (Al_2O_3) capillaries dramatically reduce the mass of ablated material (in comparison to polyacetal capillaries); an indication is much smaller pressure increase at the capillary exit after the discharge ($<10^{-2}$ Torr instead of several Torr [7]). This fact enables significantly to reduce the discharge current. Recently created numerical model [8] describes plasma dynamics in non-ablating capillaries according to the Z-pinch scenario – the main current starts (due to skin-effect) near the capillary wall; when the current increases, increasing magnetic field compresses and heats the plasma column maintaining Bennett equilibrium between magnetic a kinetic pressures.

2.2. *Filling gas pressure*

The filling gas pressure p is together with the capillary radius r and the main current amplitude $I_{\max}$ one of the main factors that influence pinching time t_{pinch} (time interval between the main current onset and the maximum plasma column compression). Earlier it was analytically deduced and experimentally confirmed a similarity relation [5, 6]

$$t_{pinch} \sim r.I_{\max}{}^{-1/2}.p^{1/4}. \qquad (1)$$

Originally the capillary was first evacuated down to the pressure $2\text{x}10^{-5}$ Torr and then Ar gas was injected by fast valve shortly before the initiation of the current pulse [3]. Later on this pulse filling was replaced by continuous Ar gas flow and a differential pumping [7]. The most accurate system [9] evacuates the closed capillary, fills it with argon, and fast pulse valve opens a small on-axis hole to the detection system just for the time of shot.

When continuous filling and pumping assembly is designed, it should be considered not only that there is a certain pressure range favourable for building of population inversion (typically in the range of 10^{-1}-1 Torr), but also that the absorption of argon gas of the pressure 0.5 Torr and of the length 1 cm is nearly 42 per cent at the laser-wavelength 46.88 nm. Therefore, the design has to minimise the path-length of the output radiation through the not- or weakly-excited filling gas/plasma. An efficient separation of the capillary from the evacuated diagnostic can be achieved either by a "long thin pipe" or by a "hole in thin wall". Fluid conductivities of these two types of connections give a theory of viscous and molecular flow. We used as a reasonable compromise either "long thin pipe" of ϕ1.5x20 mm or "hole in thin wall" ϕ0.8 mm.

2.3. *Pre-pulse generation*

The pre-pulse of 20-40 A and duration of approximately 2 μs was mentioned already in the first papers dealing with argon filled capillary for soft X-ray amplification (e.g. [3]). However, no reason and no details were provided. Only recently it was shown [10] (with the help of on-axis X-ray imaging by quadro-pinhole camera and with soft X-ray spectroscopy) that the suitable pre-pulse current I_p lies in limits $10\ \text{A} < I_p < 50\ \text{A}$. A more detail study of the pre-pulse and its influence on the main discharge was based on side-on photographing of the pre-discharge and of the main discharge in Pyrex glass capillary by high-speed framing camera (exposition 10 ns) [11]. We interpret the limits for pre-ionisation current in the following way.

In case of small pre-ionisation current (creating weakly ionised plasma) the plasma conductivity (Gauss system of units)

$$\sigma_w \sim n{\cdot}e^2 / (m_e{\cdot}\nu) \qquad (2)$$

(where n is plasma density, e is electronic charge, ν is collision frequency and m_e is electron mass) is proportional to plasma density [12] and, because the shape of radial density distribution in cylindrical capillary is the first order Bessel function J_0 with maximum on the discharge axis, the main current starts to flow along the axis, heats plasma there, but does not cause any compression.

On the other hand sufficiently high pre-ionisation current generates fully ionised plasma with conductivity

$$\sigma_f \sim \{3\cdot m_e / (16\cdot\pi^{1/2}\cdot Z\cdot e^2\cdot \ln\Lambda)\}(2\cdot\kappa\cdot T / m_e)^{3/2} \tag{3}$$

(where $Z\cdot e$ is mean ion charge, $\ln\Lambda$ is Coulomb logarithm, κ is Boltzmann constant, and T is plasma temperature), which is not proportional to plasma density [12]. For temperatures 1-2 eV it has value $\sigma_f \sim 0.4\text{-}1.0\cdot10^4\ \Omega^{-1}\cdot\text{m}^{-1}$ (which is about 200-500 times higher than σ_w) and the skin depth d

$$d = c / \{2\cdot\pi\cdot\mu\cdot\sigma\cdot\omega\}^{1/2} \tag{4}$$

(where c is speed of light in vacuum, μ is permeability, σ is conductivity, and ω is characteristic frequency of the process) is of the order of 1 mm. Therefore, in case of fully ionised pre-plasma the main current starts in the vicinity of capillary wall and compresses and heats the plasma (by Z-pinch effect). Because for plasma modelling it is simpler to work with only one conductivity, an approximate expression σ_a

$$\sigma_a \sim [(1/\sigma_f) + (1/\sigma_w)]^{-1} = (\sigma_f.\sigma_w)/(\sigma_f + \sigma_w) \tag{5}$$

for arbitrary degree of ionisation can be used.

Hence, the condition to create a fully ionised pre-plasma imposes the lower limit on the pre-ionisation current. The higher is the pre-ionisation current the higher is pre-plasma temperature and the smaller is the skin depth, but also the worse this pre-plasma can be compressed. Therefore, the upper limit for pre-ionisation current is a result of optimisation.

The pre-plasma can be generated either internally or externally. At internal pre-pulse the high voltage electrode floats on the potential $U_{\text{line}}.(C_{\text{cap}}/C_{\text{sg}})$, where U_{line} stands for voltage on the pulse forming line, C_{cap} for capacitance of high voltage electrode to ground, and C_{sg} for capacitance of the spark gap; at some potential of the high voltage electrode the capillary breaks down and if the capacitance (and resistance) of the spark

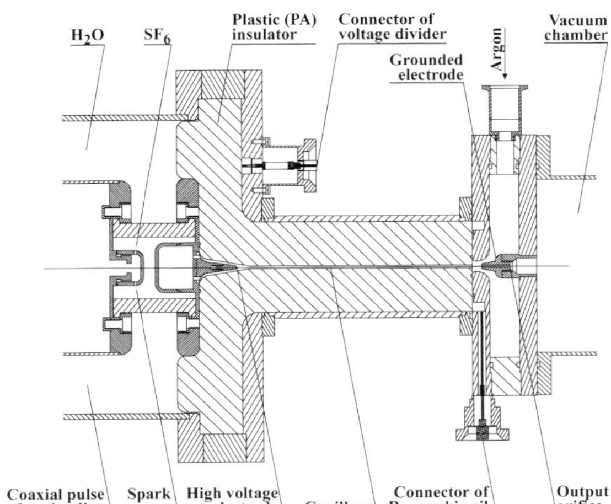

Figure 1. End of coaxial pulse forming line, spark gap, capillary, working gas (argon) input and radiation output (to vacuum chamber)

gap is sufficient to supply at least a discharge self-sustaining current then the pre-pulse current flows through the capillary. Such spark gap has to have large capacitance, electrodes of which may extend beyond insulator casing (filled with pressurised SF_6 gas) to water region (see Fig. 1 and in more detail [13]). At external pre-pulse the high voltage electrode is connected through an inductance (coil) and/or resistance (of a total impedance ~1 kΩ) to an external circuit. Namely this connection has to have an adequate isolation and power dimensioning.

3. Conclusion

This paper not only reviews historical and contemporary approaches to adjustment of favourable conditions in gas-filled capillaries for amplification in soft X-ray region (at Ne-like Ar line – 46.88 nm), but also offers some physical reasons for this choice.

- Plastic capillaries are cheap, but they are a rich source of soot (extremely harmful to all diagnostic equipment); that is why now all laboratories use alumina capillary with substantially smaller wall ablation. Ablating and non-ablating capillaries have totally different compression mechanism (current-less compression by radial shock wave in the former case, classical Z-pinch compression in the later case).
- Gas filling geometry have to ensure not only a suitable gas pressure for proper compression (according to actual capillary diameter, discharge current quarter-period, etc.), the maximum of which should take place just before the discharge current maximum, but also as good as possible evacuation of the diagnostic part.
- Pre-pulse should provide a fully ionised pre-plasma in which the main current pulse is skinned since the very beginning at the outer surface of the plasma column. The externally generated pre-pulse enables a full control of pre-plasma parameters.

Acknowledgement

The experimental part of this work was supported by the Grant Agency of the Czech Republic under Grant 202/03/0711 and the theoretical part by the Ministry of Education, Youth, and Sports of the Czech Republic under Grant 1P2004LA235.

References

[1] Rocca J J, *et al.* 1994 Phys. Rev. Letters **73** 2192-2195

[2] Kolacek K, *et al.* IEEE ICOPS 2003, The 30th Int. Conf. on Plasma Science, Jeju, Korea, June 2-5, 2003, IEEE Conf. Record – Abstracts, p 456

[3] Rocca J J, *et al.* 1993 Phys. Rev. E **47** 1299-1304

[4] Bobrova N A, *et al.* 1998 Plasma Phys. Reports **24** 1-5

[5] Vrba P, *et al.* 2001 Int. Symp. on Plasma Research and Application, PLASMA 2001, Warsaw, Poland, Sept. 19-21, 2001, Paper 3.23

[6] Koláček K 2003 Proc. SPIE, **5228** ECLIM 2002: 27th Europ. Conf. on Laser Interaction with Matter 557-573

[7] Benware B R 1998 Phys. Rev. Letters **81** 5804-5807

[8] Vrba P, *et al.* 2004, Proc. 21st Symp. on Plasma Physics and Technology, June 14-17, 2004, *to be published*

[9] Tomassetti G. 2002 Eur. Phys. J. D **19** 73-77

[10] Ben-Kish, *et al.* 2001, Phys. Rev. Lett. **87** 015002

[11] Niimi G, *et al.* 2002 IEEE Trans. Plasma Science **30** 616-621

[12] Krall N A, and A W Trivelpiece A W, 1973 Principles of Plasma Physics (New York: McGraw-Hill Book Company)

[13] Kolacek K, *et al.* 2004, Proc. 21st Symp. on Plasma Physics and Technology, June 14-17, 2004, *to be published*

Inst. Phys. Conf. Ser. No 186
Paper presented at 9th Int. Conf. X-Ray Lasers, Beijing, China, 24–28 May 2004

Simulation Study of Nitrogen Soft X-ray Capillary Discharge Laser

P Vrba[+], M Vrbova[*], N A Bobrova[&], P V Sasorov[&]

[+]Institute of Plasma Physics AS CR, Prague 8, Czech Republic
[*]Czech Technical University in Prague, Prague 1, Czech Republic
[&]Institute of Theoretical and Experimental Physics, Moscow, Russia

Abstract. We performed computer modelling of a pinching discharge in a nitrogen-filled alumina capillary in order to discover discharge system parameters that may lead to efficient recombination pumping of H-like nitrogen ions during the pinch decay. The inversion population was found at the Balmer α- transition of N^{6+} ions. A gain factor of 1 cm^{-1} spanning the time interval of 1 ns was found for an optimised arrangement with capillary radius 1.5 mm, peak current 50 kA, quarter period 40 ns and filling gas pressure 0.5 kPa. It is pointed out that even higher values of the gain factor may be achieved with thinner capillaries and shorter current pulses.

1. Introduction

The computer modelling is performed in the following three stages:

(1) The spatial and time dependences of plasma density, plasma electron and ion temperatures and electron density are evaluated by means of the code NPINCH [1].

(2) The time dependences of ionisation fractions and energy level populations for lithium-, helium- and hydrogen-like nitrogen ions are evaluated by means of kinetic code FLY used as a post-processor [2].

(3) Gain factor for Balmer α transition is assessed according to [3].

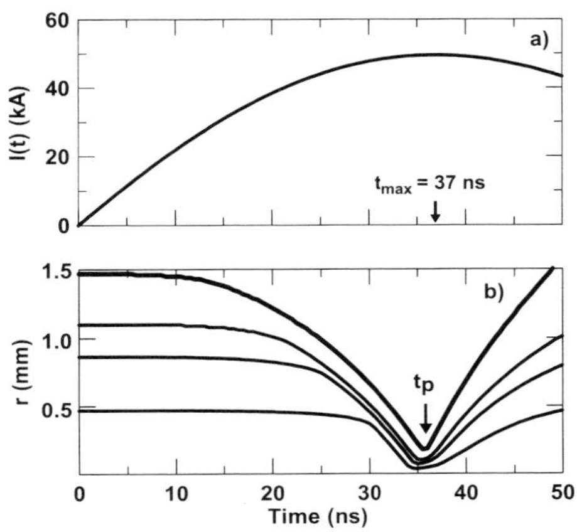

Figure 1. a) Time dependence of the capillary current, b) r-t diagram of plasma trajectories.

2. Plasma dynamics

One-dimensional two-temperature and one-fluid MHD approximation is used. The wall ablation is neglected in plasma dynamics simulations. The current pulse passing through the capillary is presumed in the form:

$$I(t) = I_1 \sin(\pi t / 2t_1) \exp(-t / t_2) ,$$

I_1, t_1, t_2 are input parameters. Figure 1b) presents calculated temporal

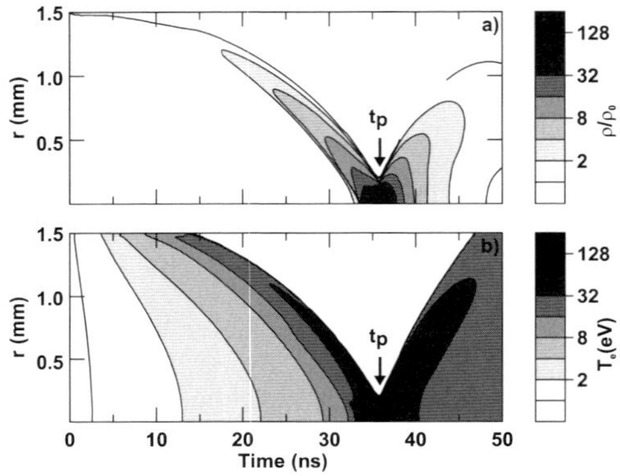

Figure 2. Radial and time evolution of a) mass compression ratio, b) electron temperature.

evolution of the trajectories of plasma elements inside the channel if $I_1 = 60$kA, $t_1 = 40$ ns, $t_2 = 200$ ns (Figure 1a), alumina capillary with inner r_0 = 1.5 mm pre-filled with pure molecular nitrogen at pressure p_0 = 0.5 kPa. (atom density N_0 = 2.4×10^{17} cm^{-3}) and the initial plasma electron temperature T_e = 0.5 eV are taken into account. During the first 14 ns the plasma is not detached from the wall. Then a vacuum region appears near the wall and the pinch is detached. The converging shock wave propagates from the wall to the channel axis. The time of the first contraction or pinching of the plasma is $t_p = 36$ ns. During the compression stage Joule heating plays a minor role. The converging shock wave finally produces high density and temperature plasma on the axis, as shown in Figure 2 a, b). A hot plasma core with the radial dimension r_{core} = 0.125 mm and central mass density $\rho_{core} = \rho(0,t_p) = 5.7\text{x}10^{-4}$ g/cm^3 is formed. The core electron temperature is 150 eV. The electron density is of the order of $1.5\text{x}10^{20}$ cm^{-3}.

After the pinch, the plasma column begins to expand due to the high pressure inside. In expanding plasma the electron temperature decreases from 150 eV to 35 eV during a very short time interval Δt =3.5 ns. It is well known that the first compression of the pinch is stable and results in a compact dense and hot plasma column on the capillary axis. At the stage of expansion the MHD instabilities may break the uniformity of the plasma column, leading to turbulent mixing, enhanced heat transport and resistance. However, as we know from the theory of the classical pinch effect, MHD instabilities begin to evolve when the expansion velocity begins to decrease. If the lasing occurs during the few first nanoseconds after the reflection of the shock wave, MHD instabilities have no time to evolve. Our description of the plasma behaviour in the capillary discharge ends at the moment when the diverging shock wave reaches the capillary wall.

3. Nitrogen ionisation and energy level populations

Quick changes of electron temperature during the plasma pinch and decay lead to ionisation and excitation non-equilibrium, which may result in active medium creation [4,5]. The FLY code [2] is used as a post processor to determine the time dependences of the ionisation state and energy level populations of selected lithium-, helium- and hydrogen-like nitrogen ions. The files containing the temporal history of axial plasma electron temperature $T_e(0,t)$ and mass density $\rho(0,t)$, resulting from MHD simulations, serve as input history files for the code.

The results are shown in Figure 3a). Substantial changes of ionisation and population densities are found in the time interval from 33 ns to 37 ns, i.e. during the pinch collapse. The inversion population $N^{6+}{}_3 > N^{6+}{}_2$ for the quantum transition n = 3 → n = 2 arises at 36.8 ns, when the temperature T_e is decreasing very fast and the fully stripped ions N^{7+} recombine efficiently.

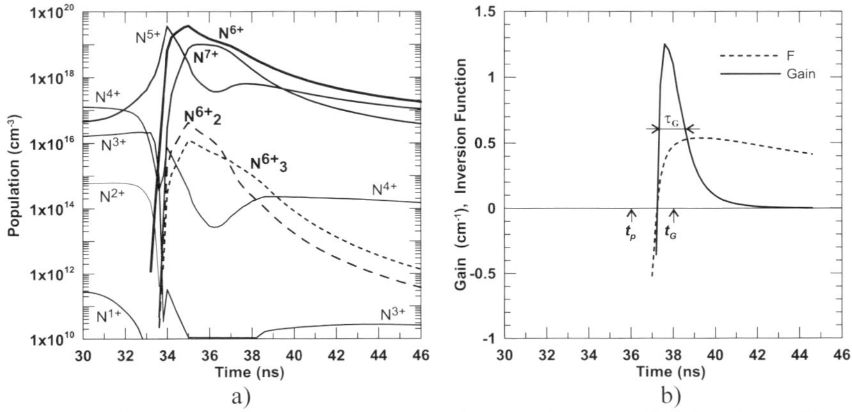

Figure 3. Time dependences of a) nitrogen ion densities (full lines) and densities of excited states of N^{6+} (dashed lines), b) gain factor G and inversion function F.

4. Gain factor time dependence

The instantaneous gain factor G for Balmer alpha transitions is evaluated as

$G = \sigma_{stim} N_3^{6+} F$, where σ_{stim} = 2.38 .10^{-15} cm^{-3} , $F = 1 - \frac{g_3}{g_2}\frac{N_2^{6+}}{N_3^{6+}}$ and $g_2 = 8$, $g_3 = 18$

are taken into account [3]. The maximum value $G_{max} = 1.25$ cm^{-1} is obtained at the time $t_G = 37.6$ ns, see Figure 3b), being delayed almost 2 ns after the pinch time t_p. The active laser medium with a reasonable gain is created for a relatively short period $\tau_G = 1.3$ ns.

5. Laser system optimisation

In the frame of our model the optional experimental parameters are capillary radius r_0, initial filling pressure p_0 (or atom concentration N_0), electric current peak value I_{max} (or current amplitude I_1) and quasi-quarter period $T_{1/4} \cong t_1$. Hence, the optimisation is performed in four-dimensional space (r_0, N_0, I_{max}, $T_{1/4}$) taking the peak value of the gain factor G_{max} as the basic criterion.

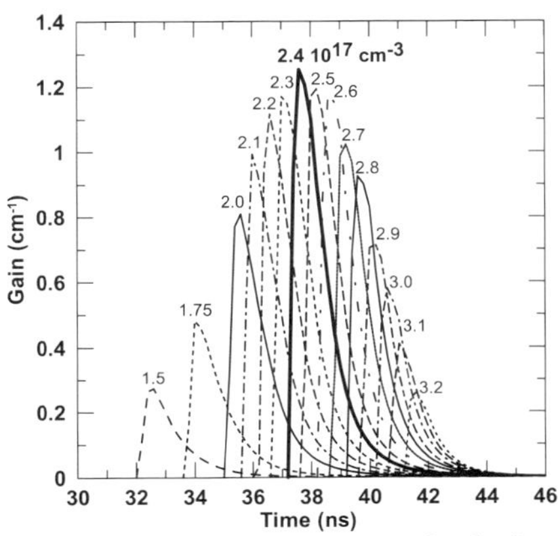

Figure 4. Time dependences of gain factor for $r_0 = 1.5$ mm, $T_{1/4} = 40$ ns, $I_{max} = 50$ kA and various initial atom densities N_0.

5.1. Gain factor gas pressure dependence

Gain factor time dependences have been calculated for various initial atom densities N_0 by repetitive simulations (Figure 4). The pinch time t_p is tuned by the choice of initial gas filling density N_0. For lower initial values of N_0 the pinch time t_p is shorter. The highest peak value of the gain factor is achieved for $N_0 = 2.4\text{x}10^{17}$ cm^{-3} and corresponds to the case, when the pinch time t_p is approximately equal to the current peak time t_{max}. The peak value of electron temperature $T_{e,max}(t_p)$ = 141.1 eV, which assigns the ionisation degree, results in highly ionised plasma with a non-negligible concentration of N^{7+} (Figure 3a). The electron density at gain time $N_e(t_G)$ = 6,39 10^{19} cm^{-3} is lower than the limit value for laser level population collisional mixing.

5.2. Dependence of the pressure optimised gain factor on the current peak value

To every driving current pulse shape and capillary radius an optimum gas density may be found. Results obtained for r_0 = 1.5 mm, $T_{1/4}$ = 40 ns and various current peaks I_{max} are summarized in Table 1. The pressure optimised gain factor increases remarkably with increasing peak current, whereas the electron density $N_e(t_G)$ at the gain time t_G increases slowly. It is interesting to note that in all the pressure optimised cases the peak values of electron temperature are approximately the same and equal to 140 eV.

Table 1: Pressure optimised gain factor G vs. peak current I_{max} : $T_{1/4}$ = 40 ns, r_0 =1,5 mm

I_{max} [kA]	$N_{0,opt}$ [cm^{-3}]	$T_{e,max}$ [eV]	ρ/ρ_0	t_G [ns]	$T_e(t_G)$ [eV]	$N_e(t_G)$ [cm^{-3}]	$G(t_G)$ [cm^{-1}]
49,52	2,50E+17	141,1	138,5	38,0	68,9	6,39E+19	1,070
45,40	2,1E+17	144,3	95,4	38,4	55,6	3,51E+19	0,616
41,27	1,8E+17	140,3	86,8	39,0	57,9	3,05E+19	0,296
33,01	1,2E+17	138,5	73,4	40,0	54,3	1,52E+19	0,029
24,76	7,0E+16	138,1	58,5	41,4	46,2	5,13E+18	6,1E-04

5.3. Pressure optimised gain factor for narrower capillaries and shorter pulses

We observed that the best laser pumping is obtained if the pinch time t_p is approximately equal to the current peak time t_{max}. Consequently, all input parameters have to be carefully matched. We performed supplementary simulations for $T_{1/4}$ = 15 ns, r_0 = 0.5 mm. In this case the optimised gain factor $G(t_G)$ = 6.1 cm^{-1} is forecast for I_{max} = 55.7 kA. We conclude that conditions for lasing are more favourable with thinner capillaries and shorter current pulses. It is interesting to compare these results with results of experiment and computer simulations recently published in [5, 6]. The authors proved efficient Z-pinch plasma heating induced by current pulses with rise time of ~10 ns and peak amplitude of ~200 kA passing through 3 mm diameter argon filed capillary.

Conclusions

The optimisation was performed in the region of parameters that are easily achieved in laboratories. We hope that the results may be taken as a proposal for experimental implementation, and that they will support the general effort to develop a scheme for a tabletop soft X-ray laser with recombination pumping.

Acknowledgement

This research has been supported by grants from the Ministry of Education, Youth and Sports of the Czech Republic, under projects LN000A100 and KONTAKT ME 609.

References

[1] Bobrova N A et al 1996 Plasma Physics Reports **22** 387-402
[2] Lee R W and Larsen J T 1996 J.Q.S.R.T. **56** 535-556
[3] Elton R C 1990 X-Ray Laser (New York: Academic Press)
[4] Rocca J J 1999 Rev. Sci. Instr. **70** 3799-3822
[5] Lee K, Kim J H and Kim D 2002 Z-Scaling for the H-like Recombination Laser in the Capillary Discharge in X-Ray Lasers 2002 (AIP C641) 97-101
[6] Gonzalez J J et al 2002 Phys. Rev. E **65**

Inst. Phys. Conf. Ser. No 186
Paper presented at 9th Int. Conf. X-Ray Lasers, Beijing, China, 24–28 May 2004

Soft X-ray Emission from Nitrogen Capillary Discharge

M Vrbova, A Jancarek, P Vrba*, A Fojtik, L Scholzova, R Havlikova, S Palinek

CTU, FNSPE Brehova 7, CZ-115 19 Prague 1, Czech Republic
*IPP AS CR, CZ-182 21 Prague 8, Czech Republic

Abstract. We report measurements of soft X-ray spectra emitted by pinching capillary plasma created by a fast electrical discharge in an alumina capillary filled with nitrogen. The observed spectral lines are identified according to the synthetic spectra generated by means of the SAPPHIRE code. Peak plasma electron temperature of about 30 eV at the time of weak pinch collapse is estimated for the peak of current pulse 10 kA and the quarter of quasi-period 50 ns.

1. Introduction

Our aim is to realize soft X-ray laser recombination pumping during pinch decay and to get lasing on the Balmer alpha transition of N^{6+} ions. For this purpose, the plasma should be heated sufficiently during the pinch collapse and than quickly cooled during the pinch decay. We looked for the optimum experimental arrangement that results in proper capillary plasma pinch. Spectral diagnostics is used as a principal diagnostic tool.

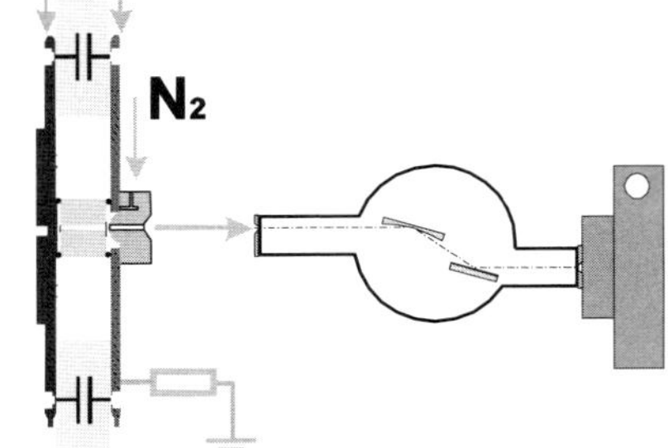

Figure 1. Experimental setup.

2. Experimental setup

Our system is manually controlled by a dual-channel pulse generator which triggers a two-stage Marx generator and diagnostic system. The Marx generator stage capacity is 15 nF and is charged up to 40 kV. The capacitor bank of 7.5 nF is charged at a 0.5 kV/ns rate until capillary self-breakdown takes place. Capillaries with 1 or 3 mm diameter made from RF alumina were used. Nitrogen pressure

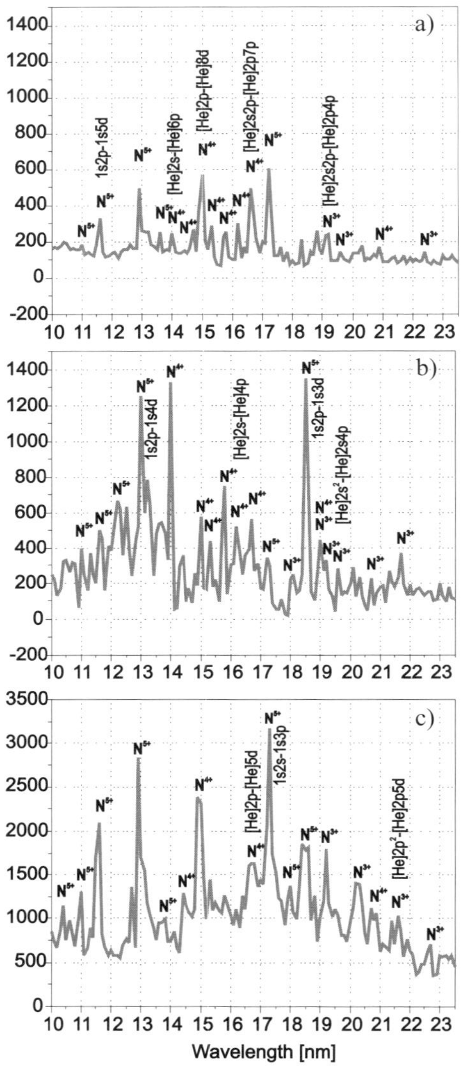

Figure 2. Time integrated X-ray spectra for a) capillary diameter $2R_0 = 3$ mm and initial pressure $p_0 = 1.3$ mb, b) capillary diameter $2R_0 = 3$ mm and initial pressure $p_0 = 2.6$ mb, c) capillary diameter $2R_0 = 1$ mm and initial pressure $p_0 = 2.6$ mb.

inside the capillary was varied within 0.5-15 mbar. The capillary voltage profile corresponded to an under-critically dumped RLC circuit with a quarter period of 50 ns and a current peak value of about 10 kA. Time integrated EUV emission spectra in the range 10 – 24 nm were studied by means of the Jobin-Yvon monograph PGM-PGS 200 and a Reflex BICCD camera RXBIC512c in repetitive shots.

3. Time integrated spectra measured

We observed many spectral lines in the studied spectral range. These lines should belong to three or more times ionised nitrogen atoms. Their intensities varied if the initial filling pressure was varied and if the capillary diameter was changed (see Figure 2). To identify the spectral lines observed, we estimated the time dependence of the plasma electron temperature by means of the MHD code [1] and then evaluated ion abundances by means of FLY code [2]. Unfortunately, the last code could not be used for evaluation of N^{3+} spectral lines that might have been observed in the investigated spectral range. To identify all lines, we use the SAPPHIRE code [3].

4. MHD simulations

The measured current pulse shape is approximated as

$$I(t) = I_1 \sin(\pi t / 2t_1) \exp(-t / t_2)$$

The optional experimental parameters for the code were a capillary radius R_0, an initial filling pressure of p_0 (or atom concentration of N_0), and electric current amplitude of I_1 and a quarter period of t_1. It has been shown that pinch dynamics is very changeable, particularly with respect to pressure variations. From the results which correspond to our experimental conditions (Figure 3), we can see that the pinch compression ratio is rather small and the highest plasma electron temperature is about 30 eV. Plasma reflection from the wall is not taken into account, and the results are valid only in the limited time interval until the plasma reaches the capillary wall.

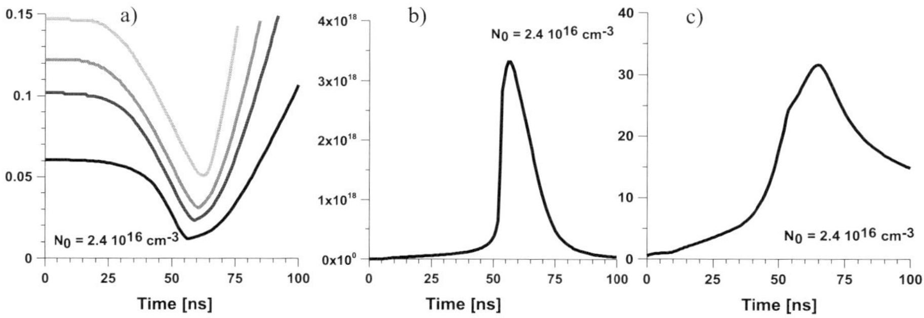

Figure 3. Results of MHD simulations for capillary diameter $2R_0$ = 3 mm, current amplitude I_1 = 11 kA, quarter period t_1 = 50 ns, damping time t_2 = 170 ns and initial pressure 1.0 mb; a) Plasma element trajectories r (in cm), b) Electron density $N_e(0,t)$ (in cm^{-3}) on the capillary axis, c) Electron temperature $T_e(0,t)$ (in eV) on the capillary axis.

5. Ion kinetics

Files containing the temporal history of axial plasma electron temperature $T_e(0,t)$ and electron density $N_e(0,t)$ (Figures 3b) and 3c)), were used as input history files for the FLY code [2]. At the very beginning of the current pulse, the plasma ionisation is small. During the pinch compression, the ionisation stage of plasma was quickly changed (Figure 4). The densities of forth and five times ionized atoms quickly grow and achieve their peak values approximately at the pinch time t_p= 57 ns. During the pinch expansion stage, the helium-like ions (N^{5+}) prevail. Time dependences of population densities of energy levels and line intensities corresponding to selected quantum transitions for N^{6+}, N^{5+} and N^{4+} were evaluated using the subprogram FLYSPEC. All these lines have steep increase of the intensities at the time of the pinch. These time dependencies were observed during our measurements in the UV spectral range (200- 300 nm). The rapid increase of their intensities is one main feature of the pinch.

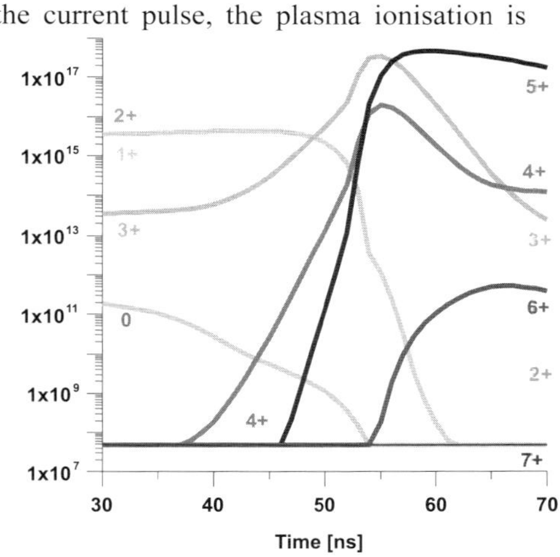

Figure 4. Time dependences of nitrogen ion densities.

6. Synthetic spectra

To identify the quantum transition corresponding to the spectral lines observed in the experiment in the EUV range from 10 to 23 nm (Figure 2), and also in the UV range from 200 to 300 nm, synthetic spectra were generated separately for helium-, lithium- and beryllium-like ions (Figure 5) by means of the SAPPHIRE code [3]. In this case we have taken into account the plasma equilibrium temperature of 30 eV as it follows from the MHD code (see Figure 3c)). As is well known, the lines with wavelength shorter than 13 nm may belong to helium- and hydrogen- like ions only, whereas, in the range 13-16 nm, lithium- like lines may also be observed. For even longer wavelengths, the

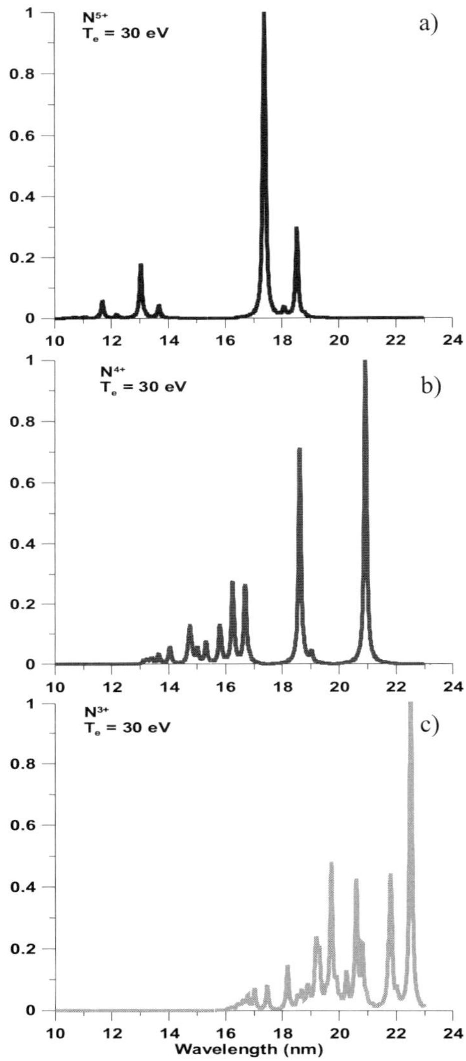

Figure 5. Synthetic nitrogen spectra for plasma electron temperature T_e = 30 eV for a) N^{5+}, b) N^{4+}, c) N^{3+} ions .

lines belonging to the N^{3+} also should be judged. The identified quantum transitions, corresponding to the most intensive spectral lines observed in our time integrated spectra (Figure 2), are summarized in Table 1.

Table 1: EUV quantum transitions

λ [nm]	E_L [eV]	Ion	L. L. Config.	U. L. Config.
11.7	426.3	N^{5+}	1s(2S)2p	1s(2S)5d
13.0	426.33	N^{5+}	1s(2S)2p	1s(2S)4d
14.0	0.00	N^{4+}	[He]2s	[He]6p
15.0	9.976	N^{4+}	[He]2p	[He]8d
17.3	426.42	N^{5+}	1s(2S)2s	1s(2S)3p
17.5	0.00	N^{3+}	[He]2s2s	[He]2s6p
18.5	430.70	N^{5+}	1s(2S)2p	1s(2S)3d

L.L. – lower level, U.L. – upper level

In the UV range (200-300 nm) the strongest lines were identified as follows from Table 2.

Table 2: UV quantum transitions

λ [nm]	E_L [eV]	Ion	L. L. Config.	U. L. Config.
248	63.8	N^{3+}	[He]2s4d	[He]2s5f
265	64.1	N^{3+}	[He]2s4f	[He]2s5g
288	84.1	N^{4+}	[He]5p	[He]6d
297	38.3	N^{2+}	[He]2s2p3p	[He]2s2p3d

7. Conclusion

Practical realization of recombination laser pumping scheme is conditioned by a fine careful tuning of all system parameters. To this purpose detail spectral diagnostics but time resolved should be developed.

Acknowledgement

This work was supported by grants from the Ministry of Education Youth and Sports CR under contracts MSM 210000022 and INGO-1P2004 LA235.

References

[1] Bobrova N A et al. 1996 Plasma Physics Reports 22, 387-402

[2] Lee R W and Larsen J T 1996 J.Q.S.R.T. 56 535-556

[3] Sapphire Professional, version 1.0.3.2 (Cavendish Instruments Ltd. 2003)

Inst. Phys. Conf. Ser. No 186
Paper presented at 9th Int. Conf. X-Ray Lasers, Beijing, China, 24–28 May 2004

Guiding of intense laser beams in highly ionized Ar plasma columns generated by a fast capillary discharge

B. M. Luther, Y. Wang, M.Berrill, D. Alessi, M.C. Marconi , M. A. Larotonda, and J.J. Rocca
National Science Foundation Engineering Research Center for Extreme Ultraviolet Science and Technology, and Electrical and Computer Engineering Department, Colorado State University, Fort Collins, CO 80523

Abstract. We report the characterization of multiply ionized plasma waveguides created by a fast capillary discharge and the demonstration of guiding of laser pulses with peak intensity up to 2.2 x10^{17} W/cm^2 in a 5.5 cm long plasma column containing highly charged Ar ions. These discharge plasmas can reach the Ne-like or Ni-like stage of ionization for several atoms and are of interest for the generation of efficient soft x-ray lasers by longitudinal laser excitation. The guides were characterized using plasma interferometry, near and far-field imaging, and transmission measurements. A rapid discharge-driven hydrodynamic compression is observed to guide progressively lower order modes through the plasma with increasing density and degree of ionization, until the guide collapses on axis. The lowest order mode (FWHM ~50 μm) is guided shortly before the plasma reaches the conditions for lasing in Ne-like Ar. The subsequent rapid plasma expansion forms a significantly more leaky and absorbent guide.

1. Introduction

The guiding of intense laser beams in plasmas has attracted significant attention [1-6]. This is motivated by the need of extending the interaction length between intense laser pulses and plasmas beyond the limitations set by diffractive defocusing and ionization-induced refraction. An application of considerable interest for plasma waveguides is the longitudinal excitation of soft x-ray lasers, that can potentially result in saturated amplifiers with reduced laser pump energy and increased efficiency [7-10]. Lasing at 60.8 nm by collisional electron excitation of S VIII [8], collisional recombination in Li III at 13.5 nm [9], and optical-field ionization driven collisional excitation in Pd-like Xe at 41.8 nm [10] has been reported in laser-driven plasma channels created in wall ablated [8,9] and gas filled [10] micro-capillaries excited by relatively slow discharge current pulses.

Herein we report the characterization of multiply ionized plasma waveguides created by a fast Ar capillary discharge of the type used to develop discharge-pumped collisional soft x-ray lasers, and the demonstration of guiding of laser pulses with peak intensity up to 2.2x10^{17} Wcm^{-2}. In contrast to slow capillary discharges these discharges can reach the

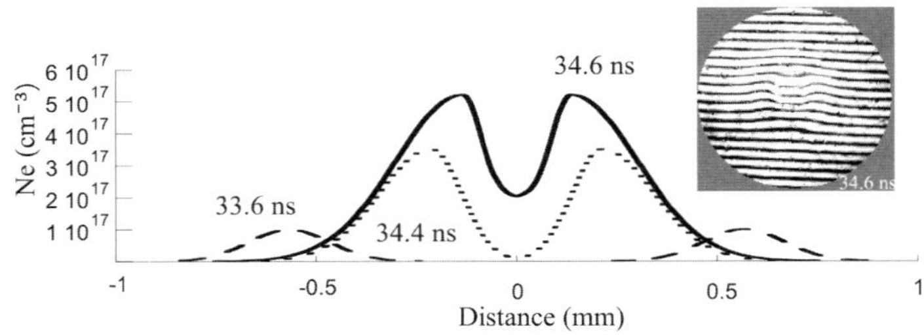

Figure 1. Ne profiles obtained from interferometry of a 1.4 cm long capillary. Times are with respect to the beginning of the current pulse.

Ne-like or Ni-like stage of ionization for several atoms of interest. The experiments made use of a fast desk-top Ar capillary discharge [11] and a Ti:Sa short pulse laser system containing up to three stages of amplification. Most of the experiments used ~ 2.5 ps pulses, except for the demonstration of high intensity beam guiding that was done using ~100 fs pulses. The discharge utilized 3.2 mm diameter, 5.5 to 11 cm long Al_2O_3 capillaries filled with 180 to 250 mTorr of Ar. A pressure on the order of $1x10^{-4}$ Torr outside the discharge channel was maintained using differential pumping. The discharge current pulses had a peak amplitude of 15 to 21 kA and a half-period of ~ 120 ns. Synchronization between the capillary discharge and the laser pulse injection was obtained by triggering a pressurized spark-gap with a focused ~ 1 mJ, 800 nm, laser pulse compressed to ~ 50 fs. For most of these experiments the jitter was on the order of a few ns, sub-ns jitter is possible [12]. The plasma compression is accompanied by an increase in the degree of ionization, reaching Z = 8 shortly before the pinch, as corroborated during these experiments by the observation of lasing at 46.9 nm in Ne-like Ar. Maximum amplification occurred 1.5 ±0.3 ns before the kink in the current pulse. Interferometry shows the evolution of an annular plasma shell with density minimum on the center that constitutes a waveguide of continuously decreasing diameter and increasing density. A high degree of radial symmetry is observed. Figure 1 shows a sequence of electron density (Ne) profiles obtained from the interferograms of a 1.4 cm long capillary discharge at a pressure of 250 mTorr. They correspond to the early part of the plasma evolution, when the diameter of the column is large and the electron density is in the range of $1\text{-}5x10^{17}$ cm^{-3} and refraction of the 267 nm beam is negligible. Latter in time the waveguide diameter continues to shrink and the density increases until the shockwave collides on axis destroying the waveguide.

2. Nearfield and far field images

For near and far field imaging and transmission measurements a capillary length of 5.5cm was chosen after experimentally verifying that it does not obstruct the exit beam even at the time of the pinch when strong refraction occurs. Pulse energies between 1 and 100 mJ were focused at the entrance of the capillary to a measured spot of ω_0 = 30μm. Figure 2 shows the progression of the output mode profiles as a function of time with respect to the initiation of the current pulse. For times less than 30 ns the plasma density is low and the peak of the electron density is far from the axis, resulting in no guiding of the beam. When the plasma column compresses to ~800 μm in diameter the wings of the pulse begin to experience a phase shift that results in output modes with concentric circles as shown in Figure 2a. As the column continues to compress a variety of lower order modes, such as that illustrated in Figure 2b, are seen. Figure 2c shows an

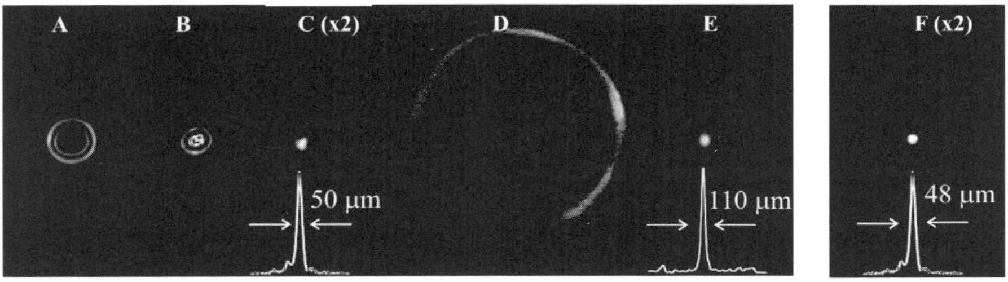

Figure 2. (A-E) Sequence of images showing the evolution of the exit mode profile as a function of time (see times in Fig. 3). F) Exit intensity profile for a laser pulse with a peak input intensity of 2.2 $\times 10^{17}$ Wcm^{-2} injected 1.8 ns before the pinch. C and F are magnified by 2X.

approximately gaussian output mode with a FWHM of ~ 50 μm. Subsequently the guide is lost as the plasma pinches and the electron density profile becomes convex, resulting in the beam being strongly refracted as shown in Figure 2d. The development of a second guiding phase was observed to take place during the first few ns of the expansion that follows the pinch, as seen in Figure 2e. However the post-pinch guides are significantly more leaky.

Channels composed of Ar VIII or Ar IX ions are immune to tunnel ionization for intensities < 3 $\times 10^{16}$ Wcm^{-2} and < 1.5$\times 10^{18}$ Wcm^{-2} respectively [13]. We verified that the pre-pinch matched guides are maintained at high intensities by injecting pulses with a measured average energy of 500 mJ and pulse duration of 112 fs, corresponding to a peak input intensity of 2.2$\times 10^{17}$ Wcm^{-2}. The exit intensity distribution patterns were observed to be similar to those recoded for the lower intensity beams, with slightly larger pedestal, as illustrated in Fig. 2f.

3. Beam transmission measurements

Transmission experiments were preformed for ~5 mJ laser pulses placing a Joule meter at the end of the capillary. In the far field, coupling and leakage losses show up as a diverging ring around the guided mode. This allows for the separation of the guided mode from the leaked and uncoupled light using an iris. The solid curve in Figure 3 shows the results for the total transmission (iris open) of the beam. The dashed curve represents the amount of energy transmitted with the iris closed, corresponding the amount of guided beam energy transmitted. The difference between the solid and dashed curves is indicative of the amount of light lost due to poor coupling and leakage from the guide, or by refraction near the time of the pinch when the waveguide is lost. For early times when the electron density is less than 1 $\times 10^{18}$ cm^{-3} the total transmission is greater than 90%. For times a-c as the column compresses towards the pinch and guides increasingly lower order modes the guided transmission is seen to drop slightly from about 85% to 75%. Since at the same time the total amount of transmitted light is observed to remain relatively unchanged, the drop in guided light transmission can be attributed to an increase in uncoupled light due to the decreasing size of the guide and associated increased sensitivity to shot to shot beam walk. After time c the amount of transmitted guided light rapidly drops due to inverse bremsstrahlung absorption, to decreased coupling as the guide shrinks below the 60 μm diam. of the input beam, and increased leakage as ΔNe decreases. As the time of the pinch is approached the guide collapses and the amount of guided light decreases to zero. The electron density rapidly

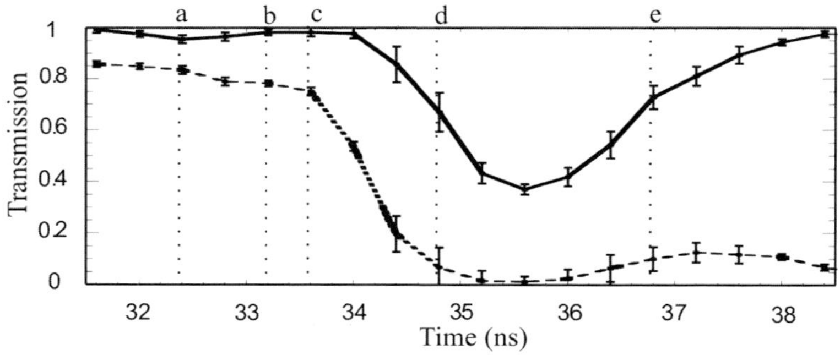

Figure 3: Transmission of guided beam with respect to time from the beginning of the current pulse. The solid curve is total transmission, dashed curve is guided transmission. The vertical lines show the timings of the images shown in Fig. 2.

increases to a maximum and the total transmission (solid curve) decreases to ~ 40% due to increased inverse bremsstrahlung absorption. Shortly after the pinch the guided transmission increases to 10-15% with the formation of the leaky post- pinch guide corresponding to Figure 2e. At the time of this late guide, occurring ~ 1.5 ns after the pinch, the column remains very highly ionized, and the solid curve indicates the absorption is about 25%. This absorption, which is larger than that observed at the time of the pre-pinch waveguides, can be estimated to correspond to an electron density of ~ 2-3x10^{18} cm^{-3} assuming an electron temperature of 50 eV and Z=8 at this time.

In conclusion we have characterized two distinct regimes of optical guiding in highly ionized argon plasma channels generated by a fast capillary discharge, and demonstrated the ability to propagate pulses with peak intensities up to 2.2x10^{17} Wcm^{-2}. The high degree of ionization of these plasma guides and their scalability to longer lengths make them attractive for applications such as the development of compact and efficient collisional transient soft x-ray lasers. Full details are given in reference [11].

We gratefully acknowledge the supported of NSF grant ECS-9977677, the NSF ERC for Extreme Ultraviolet Science and Technology, and the W.M. Keck Foundation.

References

1. C.G. Durfee III and H.M. Milchberg, Phys Rev. Lett. **71**, 2409, (1993)
2. Y. Ehrlich, C. Cohen, A. Zigler, et al., Phys. Rev. Lett, **77**, 4186, (1996)
3. M.C. Marconi, C. H. Moreno, J.J. Rocca, *et al.*, Phys. Rev. E, **62**, 7209, (2006)
4. T. Hosokai, M. Kando, H. Dewa *et al.*, Opt. Lett, **25**, 10, (2000)
5. C. Fauser and H. Langhoff. Appl. Phys. **B 71,** 607, (2000)
6. A. Butler, D.J. Spence and S.M. Hooker, Phys. Rev. Lett, **89,**185003, (2002)
7. H. Milchberg, C.G. Durfee III and J. Lynch. JOSA B, **12,**731, (1995)
8. K. A. Janulewicz, J. J. Rocca, F. Bortolotto, *et al.,* Phys Rev. A **63**, 033803 (2001).
9. D. Korobkin, C.H. Nam, S. Suckewer, *et al.,* Phys. Rev. Lett. **77**, 5206, (1996)
10. A. Butler, A.J. Gonsalves, C.M. Kenna *et al.*, **91**, Phys. Rev. Lett., 205001, (2003)
11. B.M. Luther, Y. Wang, M.C. Marconi, *et al.*, Phys. Rev. Lett., in press
12. B. Luther, L. Furfaro, A. Klix and J.J. Rocca, Appl. Phys. Lett., **79**, 3248 (2001).
13. B. M. Penetrante, J.N. Bardsley, Phys. Rev. A., **43**, 3100 (1991)

Inst. Phys. Conf. Ser. No 186
Paper presented at 9th Int. Conf. X-Ray Lasers, Beijing, China, 24–28 May 2004

Dense Capillary Discharge Plasma Waveguide for a Ni-like Ag Soft X-Ray

Y. Wang [1], B. M. Luther [1], F. Pedacci [1], M. Berrill [1], F. Brizuela [1], M. A. Larotonda [1], M. Marconi [1], J. J. Rocca [1],and V. N. Shlyaptsev [2]

[1]National Science Foundation Engineering Research Center for Extreme Ultraviolet Science and Technology, and Electrical and Computer Engineering Department, Colorado State University, Fort Collins, CO 80523

[2]Department of Applied Science, University of California Davis-Livermore, Livermore, CA 94551

Abstract. The generation of dense plasma waveguides containing a large concentration of silver ions of interest to the development of a longitudinally pumped Ni-like Ag laser is reported. The plasma guide is created by means of a fast (~ 55 ns first half-cycle) micro-capillary discharge. Discharge ablation of Ag_2S capillaries 330 μm or 440 μm in diameter with 3-5.5 kA current pulses was measured to form concave plasma density profiles with axial electron density > 1 x10 ^{19} cm^{-3}. The dynamic of this mid-to high Z plasma waveguide was studied with interferometry, absorption measurements, and hydrodynamic model simulations.

1. Introduction

The development of longitudinally excited laser-pumped soft x-ray lasers requires extending the interaction length between intense laser pulses and plasmas beyond the limits imposed by diffraction and ionization-induced refraction. Several schemes for the generation of plasma waveguides based on either laser or discharge excitation have been utilized in the development of soft x-ray lasers [1-5]. However, all of these soft x-ray laser amplification experiments have made use of plasmas containing either gases or relatively low Z atoms ablated from the capillary walls. Herein we report the use of a fast (~ 55 ns first half-cycle) micro-capillary discharge of larger peak current (3-5.5 kA) for the generation of dense plasma waveguides containing a large concentration of silver ions. The results are relevant to the development of a longitudinally excited Ni-like Ag 13.8 nm laser that takes advantage of transient collisional excitation.

2. Generation and characterization of the capillary plasma

The plasmas were generated by discharge ablation of the walls of Ag_2S capillaries 330 μm or 440 μm in diameter and 2 to 4 mm in length. The discharge current pulses had a peak amplitude of 3 to 5.5 kA and a half-period of ~ 55 ns. To enhance the uniformity of the plasma column and its shot-to-shot reproducibility a pre-plasma was generated by

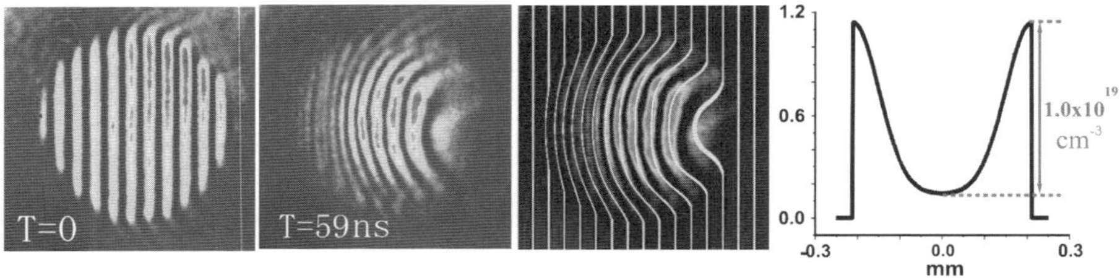

Figure 1. Interferograms of the plasma created in a 440 μm diameter, 2 mm long Ag_2S capillary discharge plasma by a 5.2 kA current pulse of 55 ns first half-cycle duration. a) no plasma present. b) 59 ns after the initiation of the current pulse. c) Overlay of measured and best-fit simulated interferogram. d) Corresponding radial variation of the electron density distribution. Notice that due to the absence of reliable reference fringes the interferograms yields the radial variation of the electron density, but not its absolute value. The on-axis electron density value was instead determined from absorption measurements.

filling the capillary channels with 0.6 Torr of He and pre-ionizing the gas with a low current pulse. The plasma density distribution was measured by sub-picosecond laser interferometry using the third harmonic (267 nm) from a Ti:Sapphire laser system (~ 1 mJ, 800 nm). A reference interferogram acquired prior to the initiation of the current pulse is shown in the left of the figure 1. The second frame shows an on-axis interferogram corresponding to a 440 μm diameter 2 mm long Ag_2S capillary excited by a 5.2 kA current pulse, obtained 59 ns after the initiation of the current pulse. The third frame of the figure shows an overlay of the experimental interferograms and interferograms computed from an electron density profile that was adjusted to fit the experimental data. The resulting electron density profile found by this procedure is shown in the far right of the figure. The electron density difference between the axis of the channel and the vicinity of the walls is measured to be $\Delta Ne \sim 1.0 \times 10^{19}$ cm^{-3}. Electron density profiles for 330 μm diameter capillaries excited by a 3.3 kA current pulses 57 ns after the initiation of the current pulse have density differences between the axis and the periphery of $\Delta Ne = 1.3 \times 10^{19}$ cm^{-3}.

Figure 2 shows a sequence of interferograms that map the temporal evolution of the electron density distribution within the 400 μm diameter Ag_2S capillary. It is observed that at these discharge conditions the time span during which concave electron density profiles are generated ranges from about 50 to 70 ns from the beginning of the current pulse. For times longer than ~ 70 ns, after the end of the first half-cycle of the current, the concave density profile starts to degrade rapidly. It is of interest to notice that late in time, after a few half-periods of the current (e.g. 220 ns frame in Fig. 2) the curvature of the fringes reverse direction, indicating the presence of a convex electron density profile with maximum density on axis.

The absolute value of the electron density on axis was determined measuring the absorption of a 267 nm wavelength (3rd harmonic of a sub-picosecond Ti:Sa laser) probe beam of ~100 μm diameter propagated along the capillary axis. The electron density was computed from the absorption values assuming an electron temperature of

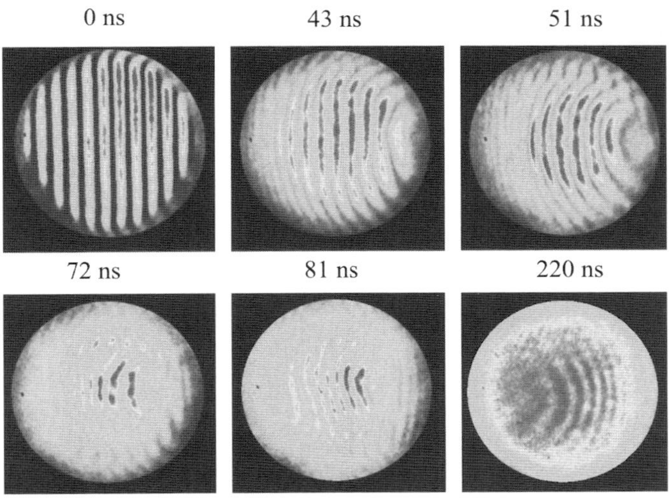

Figure 2. Sequence of interferograms corresponding to a 440 μm diameter Ag_2S capillary. The times are measured respect to the beginning of the current pulse.

20- 30 eV and a mean ionization of Z= 7 - 8.9 estimated from hydrodynamic model computations done using the code RADEX. The measured absorptions for the majority of the shots within the time interval corresponding to the formations of strongly guiding density profiles ~ (50-70 ns) amount to axial electron densities of ~ 2 - 3.5×10^{19} cm^{-3}. The discharges through smaller diameter capillaries were measured to generate plasmas with higher electron density on axis and steeper density walls.

3. Guiding characteristics

Modeling of the guiding characteristics of these plasmas was accomplished by first determining a radially dependent phase delay based on the interferometrically measured electron density profiles and subsequently propagating the beam using a fast-Fourier transform. Figure 3 shows the modeled propagation of a ω = 27 μm beam using the measured electron density profile for a 440 μm diameter capillary shown in Fig. 1.

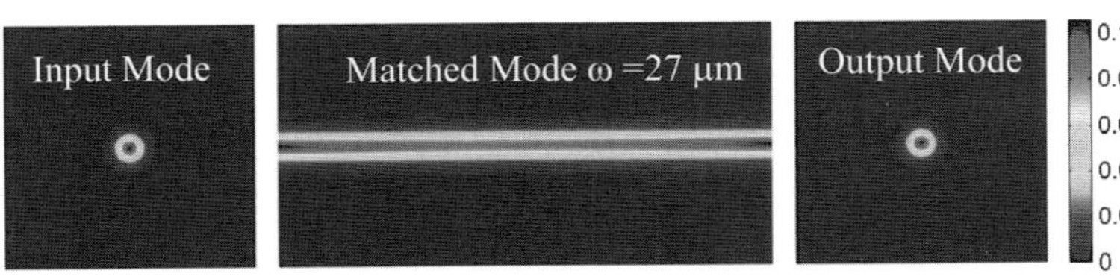

Figure 3. Modeled propagation of a ω = 27 μm input pulse in a 4mm long 440 μm diameter capillary using a radially dependent phase delay based on the interferometrically measured electron density profile from fig. 1.

Beam propagation experiments were conducted focusing 180 ps duration, 100 mJ Ti:Sapphire laser pulses at the entrance of the capillary channel into a $2\omega_0 \sim 60$ μm spot. The output plane of the capillary was imaged onto a CCD camera. Fig.4 shows the exit mode patterns corresponding to a 440 um diameter, 4 mm long capillary, acquired 55 ns after the initiation of the current pulse. Fig. 4a shows an output mode pattern with a well defined peak of ~ 54 μm FWHM and a relatively low pedestal. However, we also observed shot-to shot variations in the exit beam intensity distribution, with some shots showing a significantly more spread exit beam distribution such as that in Fig. 4b.

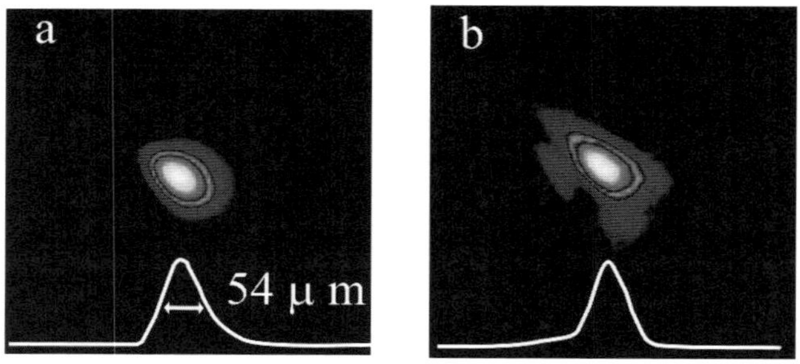

Figure 4. Exit beam intensity distribution patterns corresponding to a Ti:Sapphire laser beam propagated through a 440 um diameter, 4 mm long Ag_2S capillary. The 180 ps laser pulse had an energy of 100 mJ

In summary, we demonstrated for the first time the generation of dense plasma waveguides in a relatively high Z metal vapor plasmas of interest for collisional Ni-like soft x-ray lasers.

4. Acknowledgments.

This work was supported by the NSF ERC for Extreme Ultraviolet Science and Technology using equipment developed under NSF grant ECS-9977677. We also gratefully acknowledge the support of the W.M. Keck Foundation.

References

1. K. A. Janulewicz, J. J. Rocca, et al, Phys Rev. A **63**, 033803 (2001).
2. D. Korobkin, C.H. Nam, et al. Phys. Rev. Lett. **77**, 5206, (1996)
3. A. Butler, A. J. Gonsalves, et al, Phys. Rev. Let. **91** 205001 (2003)
4. S. Hooker et al. in these proceedings.
5. K.A. Jenulewicz et al. (In these proceedings)

OFI and photo-pumped X-ray lasers

Inst. Phys. Conf. Ser. No 186
Paper presented at 9th Int. Conf. X-Ray Lasers, Beijing, China, 24–28 May 2004

Optical Field Ionization Lasers Driven in Waveguides

S M Hooker, D J Spence, C M McKenna, A Butler and A J Gonsalves
Department of Physics, University of Oxford, Clarendon Laboratory, Parks Road, Oxford OX1 3PU, United Kingdom

T Mocek, S Sebban, V. Vorontsov and I Bettaibi
Laboratoire d'Optique Appliquée (LOA), ENSTA-École Polytechnique, Chemin de la Huniére, 91761 Palaiseau, France

B Cros and G Maynard
Laboratoire de Physique des Gaz et des Plasmas, Université Paris Sud, Bâtiment 210, 91405 Orsay, France

Abstract. We demonstrate increased energy output from collisionally-excited OFI lasers driven within two types of waveguide. We also present calculations which show that the cold electrons found in the plasma channel of a gas-filled capillary discharge waveguide could increase the gain coefficient of recombination OFI lasers significantly, and discuss the operation of OFI lasers in a regime of extreme saturation.

1. Introduction

In the early experimental work on lasing at soft x-ray wavelengths, large visible and infra-red driving lasers were employed to generate and heat the plasma gain medium, leading to the formation of population inversions by collisional excitation or electron-ion recombination. This pioneering work led rapidly to a host of new laser transitions being realized through collisional excitation of neon-like or nickel-like ions [1] or formation of hydrogen-like ions by recombination [2].

Following this early work, considerable effort has been devoted to reducing the size, and increasing the pulse repetition rate of the driving laser. Significant progress towards these goals has been achieved by using shorter duration, lower energy pump laser pulses than the nanosecond long, kilojoule energy pulses used originally. For example, several collisionally-excited lasers have been driven by using a relatively low energy, nanosecond pulse to form a uniform plasma containing a large proportion of the lasant ion stage, followed by an intense picosecond laser pulse which rapidly heats the lasant ions to produce very high transient gain [3, 4].

The use of femtosecond pulses allows the high pump powers required to form the population inversion to be achieved with even lower pulse energies, which enables the pulse repetition rate of the driving laser to be increased. Femtosecond pump pulses are too short to heat and ionize the target by inverse bremsstrahlung heating and collisional ionization. Instead the target is ionized by optical field ionization, in which the high electric field of the electromagnetic wave grossly distorts the intra-atomic potential, leading to rapid ionization. The ionization rate exhibits a threshold behaviour with the intensity of the driving laser, which affords considerable control over the ion stage that may be generated. Further, the energy of the ionized electrons following the passage of the driving laser pulse may be controlled by adjusting the polarization: linearly polarized radiation forms relatively cold electrons, suitable for driving recombination lasers; circularly-polarized light produces hot electrons which can drive collisionally-excited laser transitions [5].

The prospect of developing compact OFI lasers operating at high pulse repetition rates provided the impetus for a great deal of research following the first demonstrations of these systems in the mid 1990s. However, this work has identified several problems that must be overcome: it is difficult to generate sufficiently long gain lengths to reach saturation, or even threshold, by longitudinal pumping; for recombination laser schemes inverse bremsstrahlung (IB) and above threshold (ATI) heating of the plasma significantly reduces the population inversion achieved; the saturation intensities of OFI lasers are low, so that the output energy is relatively small at the onset of saturation.

In the present paper we discuss how these problems might be overcome by driving OFI lasers within waveguides. We present the results of recent experiments to demonstrate driving a collisional OFI laser within the channels of two types of waveguide, and discuss how one of these could enhance the gain coefficient as well as the single-pass gain of OFI recombination lasers. We also discuss the prospect of using waveguides to extend the gain length of OFI lasers so that they operate under conditions of extreme saturation, at which point relatively high energy output could be achieved.

2. Waveguides for high-intensity laser pulses

The duration of the population inversion in OFI lasers is usually limited by the lifetimes of the laser levels to a few tens of picoseconds, so that the pumping must be in a travelling-wave configuration. Of these the most convenient is longitudinal pumping, in which case the length over which the intensity remains close to the focal intensity is of the order of the Rayleigh range $Z_R = \pi W_0^2 / \lambda$, where W_0 is the spot radius ($1/e^2$ radius in intensity), and λ the wavelength of the beam. In addition, further ionization of the plasma by the pump radiation generates additional electrons, the density of which, $N_e(r)$, decreases with radial distance r from the beam axis. Since the refractive index of a plasma is given by $\eta(r) = (1 - \lambda_0^2 N_e(r) e^2 / 4\pi^2 \varepsilon_0 m_e c^2)^{1/2}$, the additional electrons cause the refractive index to increase with r and hence the beam is defocused.

A large number of techniques for guiding high-intensity laser pulses has been investigated, including multi- and mono-mode guiding in hollow capillaries [6, 7], relativistic and ponderomotive channelling [8, 9], and several types of plasma waveguide

[10-13]. Here we briefly describe the operation of one of these, the gas-filled capillary discharge waveguide [14], since it was used in the work described in Section 3.1 and plays a central role in the proposed recombination laser discussed in Section 4.1.

2.1 The gas-filled capillary discharge waveguide

Figure 1 shows schematically the gas-filled capillary discharge waveguide. An alumina capillary, with an inner diameter of a few hundred micrometres is filled with hydrogen gas by flowing the gas through narrow holes located approximately 4 mm from each end of the capillary. In the design illustrated in Fig. 1, a pulsed discharge current is flowed from a central electrode to earth electrodes located at each end of the capillary. The discharge current, of order a few hundred Amperes, ionizes the hydrogen gas and forms the plasma channel.

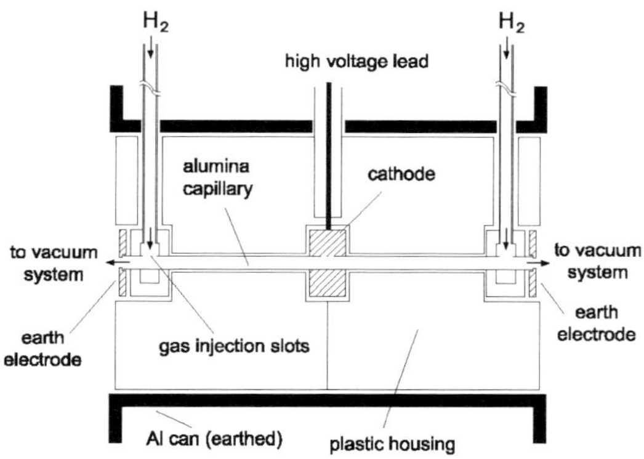

Fig. 1 Schematic diagram of a gas-filled capillary discharge waveguide.

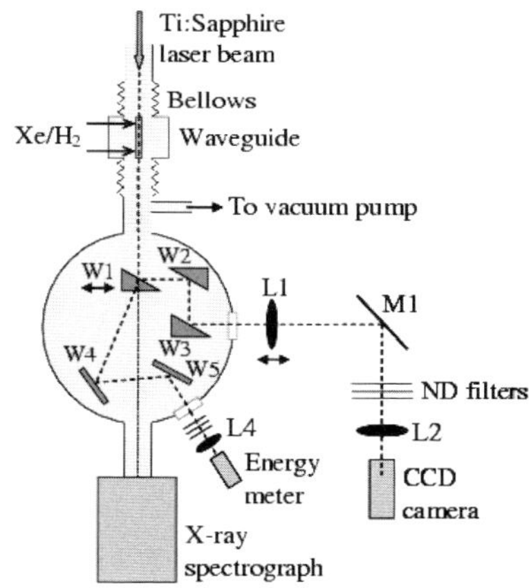

Figure 2 Layout employed in the experiments to demonstrate lasing in Xe^{8+} waveguides.

The plasma channel has been measured to be fully-ionized, and approximately parabolic [14]. Magnetohydrodynamic simulations [15] confirm this and show that the channel is formed by heat conduction to the cold capillary wall.

We have demonstrated guiding of laser pulses with a peak intensity of up to 1×10^{17} W cm^{-2} through hydrogen-filled capillary discharge waveguides up to 50 mm long [16, 17]. For example, for a 30 mm long capillary initially filled with H_2 at a pressure of 330 mbar, the energy and peak intensity transmission were approximately 80% and 40% respectively for laser pulses injected at the optimum delay after the onset of the discharge pulse.

3. Driving the collisionally-excited Xe^{8+} OFI laser in waveguides

Here we briefly describe two recent experiments to drive OFI lasers within waveguides. For these demonstration experiments we investigated lasing on the 5d-5p transition at 41.8 nm in Xe^{8+} because this transition has been well characterized [18, 19] since its first demonstration by Lemoff et al. In that earlier work the gain region was generated by focusing circularly-polarized pump laser pulses into a gas cell, the length of the Xe^{8+} column produced being limited to less than 8.5 mm owing to refraction and diffraction of the pump laser radiation.

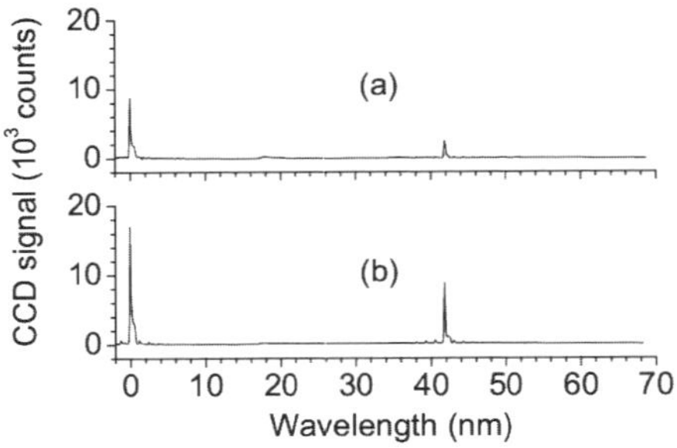

Figure 3 The measured XUV spectrum for (a) a gas cell (b) a Xe/H gas-filled capillary discharge waveguide. The feature at '0 nm' is the zero-order.

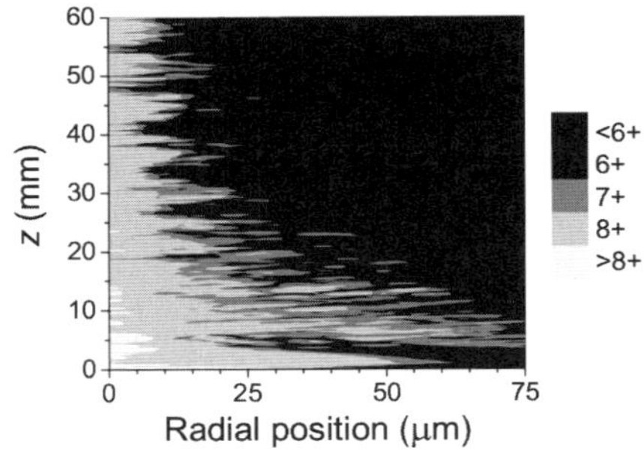

Figure 4 The calculated Xe ion stage as a function of propagation distance z for the experimental conditions of Fig. 3 for a plasma channel preionized to Xe^{5+} by the discharge.

3.1 OFI lasing in a gas-filled capillary discharge waveguide

Figure 2 shows schematically the layout employed in experiments performed at LOA [20]. Circularly-polarized laser pulses of 240 mJ energy and 37 fs duration were focused with an off-axis paraboloid to a focal spot radius of 34 μm in the entrance plane of a 30 mm long gas-filled capillary discharge waveguide. The spectrum of the XUV radiation leaving the capillary could be recorded by driving wedge W1 out of the beam, so as to allow the radiation to enter an XUV spectrograph. In order to introduce lasant gas into the plasma channel formed by the capillary discharge, a mixture of Xe and H_2 gas was flowed into the capillary.

To enable comparison with earlier work, the waveguide could be replaced with a 4 mm long gas cell, similar to that described previously [19]. The cell was positioned such that, in vacuum, the pump beam was focused 2.5 mm after the front pinhole.

Figure 3 compares the spectrum recorded with the gas cell containing pure Xe at the optimum pressure of 20 mbar with the spectrum recorded using the waveguide for the optimum conditions: 1:3 Xe:H atom gas mixture at an initial pressure of 120 mbar, and a delay $t = 1000$ ns. It is seen that for both the gas cell and waveguide strong lasing occurred at 41.8 nm, the signal recorded with the waveguide being approximately 4 times larger than achieved with the gas cell.

The Xe^{8+} laser signal recorded with the capillary discharge waveguide was found to depend critically on the delay at which the pump laser pulses were injected into the capillary, and to be closely correlated to the conditions required to achieve good guiding of the pump laser pulse. As such it was concluded that the Xe^{8+} lasing depended not only on the plasma conditions established by the discharge, but also on good guiding of the pump laser pulses over the 30 mm length of the waveguide.

Figure 4 shows the results of a simulation of the propagation of the pump laser pulses through a plasma channel formed in a 1:3 Xe:H atom gas mixture at an initial pressure of 120 mbar, assuming that the discharge initially ionized the Xe to Xe^{5+}. It is seen that with the plasma channel Xe^{8+} is formed over almost the entire 30 mm length of the channel, compared to the 4 mm achieved in the gas cell.

3.2 OFI lasing in a multi-mode gas-filled capillary

We have recently investigated an OFI laser driven in a second type of waveguide: a multi-mode gas-filled capillary, as described in detail by Sebban *et al.* in these proceedings. This experiment was also performed at LOA, using an experiment layout very similar to that illustrated in Fig. 4. For these experiments a spherical mirror used at f/33 focused 1 J laser pulses with a duration (FWHM) of 34 fs to a focus containing about 40% of the energy in a spot of radius 28 μm. The capillaries were 15 mm long and had an internal diameter of 300 μm.

It was found that much greater signals were recorded with the waveguide, the optimum signal being almost an order of magnitude greater than the strongest signal recorded with the gas cell. In addition, the waveguide allowed lasing to be achieved at higher Xe pressures, and over a wider range of pressures, than in the gas cells.

With this waveguide, channelling of the laser pulses occurs as a result of reflection at the capillary wall. It is important to note that for these experiments the spot radius of the pump pulse in the entrance plane of the capillary was approximately 10 times smaller than the capillary diameter, so that the guiding is expected to be highly multi-mode. Calculations show that reflections from the walls refocus diffracted and refracted pump laser radiation, thereby extending the region of Xe^{8+} formed along the whole length of the capillary.

3.3 Discussion

The plasma conditions within the multi-mode gas-filled capillary waveguide described in Section 3.2 will be identical to those generated in a Xe gas cell. However, the conditions within a gas-filled capillary discharge waveguide will be quite different in that the pump pulse interacts with a partially ionized plasma that also contains a large density of electrons from discharge ionization of hydrogen. This could reduce the small-signal gain coefficient of the Xe^{8+} laser transition in a number of ways, as discussed previously [21]. For similar partial pressures of Xe, the ratio of the Xe^{8+} laser signal obtained with a waveguide to that achieved with a 4 mm long gas cell was greater for the 15 mm long multimode capillary than the 30 mm long gas-filled capillary discharge waveguide. This suggests that the gain coefficient for Xe^{8+} laser transition is indeed smaller in the plasma waveguide than in a gas cell or gas-filled capillary. However, further work is necessary to clarify this point, and to determine whether this is also the case for other collisionally-excited OFI lasers.

4. Future directions

The experiments described demonstrate clearly that waveguides can be used to extend the gain region of OFI lasers. The use of techniques of this type is expected to increase the output energy, decrease the divergence, and improve the transverse coherence of OFI and other short-wavelength lasers substantially.

In this section we briefly examine how the properties of gas-filled capillary discharge waveguides could be used to increase both the *gain coefficient* and *length* of the gain region of recombination OFI lasers, and we discuss the operation of ASE lasers under conditions of extreme saturation.

4.1 Recombination lasers driven in gas-filled capillary discharge waveguides

It has been recognized for some time that, since the rate of recombination scales with the electron temperature as $T_e^{-9/2}$, the gain coefficient achieved in recombination lasers driven by OFI is very strongly reduced by IB and ATI heating of the OFI plasma [22].

Grout et al. [23] have shown that the electron temperature can be strongly reduced by diluting the lasant gas with a high density of hydrogen, leading to a large increase in the small-signal gain coefficient. For example, for KrF laser pump pulses at 248 nm, those authors calculate that the small-signal gain of the 3p – 4s transition in Ar^{7+} at 23.2 nm does not exceed 10 cm^{-1} for OFI of pure Ar. However, for 4 mbar of Ar doped into 68 mbar of H_2, the resulting decrease in electron temperature was calculated to increase the small-signal gain coefficient to 350 cm^{-1}. The need for a high-power KrF is a strong disadvantage of this scheme; pumping with an 800 nm laser pulse, however, provides a gain of only 6 cm^{-1}, owing to greatly increased ATI heating at the longer wavelength.

The gas-filled capillary discharge waveguide offers an extension of this idea in two ways: (i) the waveguide is able to extend the length of the gain region produced; (ii) the plasma channel itself contains a dense pool of relatively cold electrons which can enhance the recombination rate and so increase the gain coefficient.

We have calculated the gain coefficient produced within the plasma channel of Ar/H-filled capillary discharge waveguides. We have shown that pre-ionisation of these gases by the discharge leads to a greatly reduced electron temperature by removing the ATI heating of the hydrogen atoms. The small signal gain is predicted to increase ten-fold to over 60 cm^{-1} for 4 mbar of Ar doped into 400 mbar of H_2. This increase makes it feasible to pump the laser with 800 nm pulses from a Ti:Sapphire laser; pumping with the second harmonic would generate even larger gain.

In addition to the increase in gain coefficient produced by the favourable plasma conditions in the capillary waveguide, we also predict an increase in the length of the gain region owing to channelling of the pump laser pulses. Optical field ionization of Ar ions in the discharge plasma will cause ionization-induced refraction, and so perturb the plasma waveguide. However, since the fraction of Ar in the plasma is low (much lower than the fraction of Xe used in the experiments described in Section 3.1), the waveguide can channel the pump pulse almost as effectively as a pure hydrogen waveguide: calculations show that the Ar^{8+} ion could be produced over 20 cm using only a 1.5 TW Ti:Sapphire laser pulse. This extremely long gain length would result in a small-signal gain of over exp(1000): in this regime the XUV laser would operate under conditions of extreme saturation.

4.2 *ASE lasers operating under conditions of extreme saturation*

The experimental and theoretical results presented above suggest that it should be possible to drive OFI lasers over long lengths, with very high values of the single-pass, small-signal gain. Under such conditions the laser transition will be very strongly saturated such that the population inversion is almost zero for most of the duration of the ASE pulse. It may be shown [24] that in this limit the energy extracted by the ASE pulse per unit volume of gain medium, $U_s^{\text{ex.sat}}$, is given by,

$$\left(\frac{\hbar\omega_L}{\beta}\right)\int\left[R^*(\tau)-\gamma_2^*\frac{\exp(-\overline{\gamma}_{21}\tau)}{1+g_1/g_2}\int_{-\infty}^{\tau}\left[R_2(\tau')+R_1(\tau')\right]\exp(\overline{\gamma}_{21}\tau')d\tau'\right]d\tau \quad (1)$$

where the outer integral is evaluated over the interval for which the integrand is positive. Here we have considered the gain region to lie along the $z-$axis, and the laser transition to have an angular frequency ω_L between the upper (2) and lower (1) levels with fluorescence lifetimes τ_i and degeneracies g_i. The rate of pumping per unit volume of level i is R_i, $\tau = t - z/c$, $\beta = 1 + g_2/g_1$, $\overline{\gamma}_{21} = [g_2(1/\tau_2 - A_{21}) + g_1(1/\tau_1 - A_{12})]/(g_1+g_2)$, $\gamma_2^* = 1/\tau_2 + (g_2/g_1)A_{21} - (1/\tau_1 + (g_1/g_2)A_{12})$, where A_{ij} is the rate per unit volume of non-stimulated transitions from $i \to j$, and $R^* = R_2 - (g_2/g_1)R_1$.

This result emphasizes that the maximum energy extracted by the ASE pulse is independent of the optical gain cross-section of the laser transition, and hence independent of the saturation intensity. Instead the energy extraction is determined by the pump and decay rates of the laser levels.

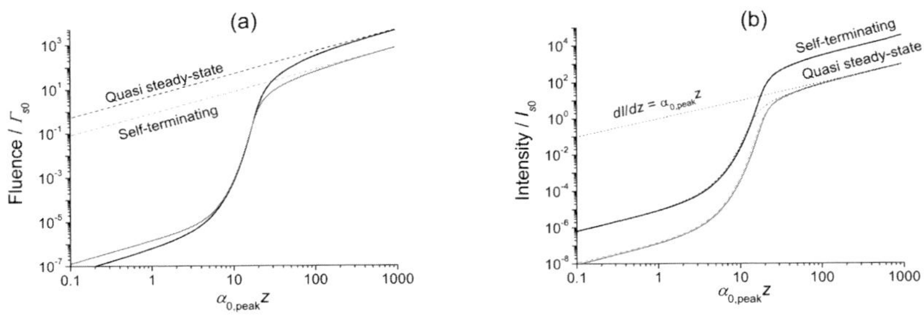

Figure 5 Calculated growth of (a) the fluence, (b) the peak intensity of the ASE pulse in a quasi steady-state and a self-terminating system. In (a) the limit set by eqn. (1) is given by the dashed line. In (b) the results of a steady-state analysis are given by dashed lines.

In order to demonstrate the approach to extreme saturation, we have solved numerically the equations describing the laser level populations and the growth of the ASE pulse for pump pulses of the form $R_i^{\text{peak}}\exp[(-\tau/\Delta\tau_i^{\text{pump}})^2]$. Figure 5(a) shows the calculated growth in the fluence of the ASE pulse in terms of the peak small-signal gain coefficient $\alpha_{0,\text{peak}}$ produced by the pumping for two cases of interest: a system operating in the quasi steady-state with $\tau_2 : \Delta\tau_2^{\text{pump}} : \Delta\tau_1^{\text{pump}} : A_{21}^{-1} : \tau_1 = 1{:}3{:}3{:}10{:}0.1$, and a self-terminating transition with $\tau_2 : \Delta\tau_2^{\text{pump}} : \Delta\tau_1^{\text{pump}} : A_{21}^{-1} : \tau_1 = 1{:}1000{:}1000{:}10{:}50$. It is seen that in both cases the growth in the fluence saturates for $\alpha_{0,\text{peak}} z$ greater than approximately 20, and tends to the limit set by eqn. (1) for $\alpha_{0,\text{peak}} z$ larger than around 100. Note that the fluence grows to be nearly an order of magnitude larger than the saturation fluence Γ_{s0}. In Fig. 5(b) we compare the peak intensity of the ASE pulse with that calculated by a steady-state analysis [25]. It is seen that the growth in intensity for the quasi-steady state system agrees well with the steady-state analysis, but that there is substantial disagreement for the self-terminating transition. The numerical solution shows that the peak intensity of the ASE pulse grows to be more than four orders of magnitude greater than the saturation intensity of the laser transition.

5. Conclusions

We have demonstrated that waveguides may be employed to increase the output energy of OFI lasers, and it is expected that these techniques should also improve their transverse coherence. We have described calculations that show that the cold electrons found in the plasma channel of a gas-filled capillary discharge waveguide could increase the gain coefficient of OFI recombination lasers significantly, offering the prospect of very large single-pass small-signal gains. If realized, OFI lasers would operate in a regime of extreme saturation with output intensities many orders of magnitude higher than the saturation intensity of the transition.

6. References

[1] Macgowan, B J, et al. 1992 Phys. Rev. Lett. 4. 2326-2337.
[2] Kim, D, et al. 1989 Opt. Lett. 14. 665-667.
[3] Nilsen, J, et al. 1993 Phys. Rev. A 48. 4682-4685.
[4] Daido, H, et al. 1995 Opt. Lett. 20. 61-63.
[5] Corkum, P B and Burnett, N H 1988 OSA Proceedings on Short Wavelength Coherent Radiation 225-232.
[6] Jackel, S, et al. 1995 Opt. Lett. 20. 1086-1088.
[7] Dorchies, F, et al. 1999 Phys. Rev. Lett. 82. 4655-4658.
[8] Monot, P, et al. 1995 Phys. Rev. Lett. 74. 2953-2956.
[9] Borisov, A B, et al. 1992 Phys. Rev. Lett. 68. 2309-2312.
[10] Nikitin, S P, et al. 1999 Phys. Rev. E 59. R3839-R3842.
[11] Volfbeyn, P, Esarey, E, and Leemans, W P 1999 Phys. Plasmas 6. 2269-2277.
[12] Gaul, E W, et al. 2000 Appl. Phys. Lett. 77. 4112-4114.
[13] Ehrlich, Y, et al. 1996 Phys. Rev. Lett. 77. 4186-4189.
[14] Spence, D J and Hooker, S M 2001 Phys. Rev. E 63. 015401(R).
[15] Bobrova, N A, et al. 2002 Phys. Rev. E 6501. art. no.-016407.
[16] Spence, D J, Butler, A, and Hooker, S M 2001 J. Phys. B-At. Mol. Opt. Phys. 34. 4103-4112.
[17] Butler, A, Spence, D J, and Hooker, S M 2002 Phys. Rev. Lett. 89. 185003.
[18] Hooker, S M, Epp, P T, and Yin, G Y 1997 J. Opt. Soc. Am. B-Opt. Phys. 14. 2735-2741.
[19] Sebban, S, et al. 2001 Phys. Rev. Lett. 86. 3004-3007.
[20] Butler, A, et al. 2003 Phys. Rev. Lett. 91. art. no.-205001.
[21] Spence, D J, Butler, A, and Hooker, S M 2003 J. Opt. Soc. Am. B-Opt. Phys. 20. 138-151.
[22] Burnett, N H and Corkum, P B 1989 J. Opt. Soc. Am. B-Opt. Phys. 6. 1195-1199.
[23] Grout, M J, et al. 1997 Opt. Commun. 141. 213-220.
[24] Hooker, S M and Spence, D J Manuscript in preparation.
[25] Pert, G J 1994 J. Opt. Soc. Am. B-Opt. Phys. 11. 1425-1435.

Inst. Phys. Conf. Ser. No 186
Paper presented at 9th Int. Conf. X-Ray Lasers, Beijing, China, 24–28 May 2004

Investigation of optical field driven x-ray lasers in cluster gases

H.-H. Chu
Department of Physics, National Taiwan University, Taipei 106, Taiwan
H.-E. Tsai
Graduate Institute of Electro-Optical Engineering, National Taiwan University, Taipei 106, Taiwan
S.-Y. Chen, J. Wang
Institute of Atomic and Molecular Sciences, Academia Sinica, Taipei 106, Taiwan
C.-H. Lee
Research Center of Applied Science, Academia Sinica, Taipei 115, Taiwan
J.-Y. Lin
Department of Physics, National Chung Cheng University, Chia-Yi 621, Taiwan
E-mail: jylin@phy.ccu.edu.tw

Abstract. By using a prepulse with an adjustable delay to control the spatial and temporal density distribution of nanoplasmas, we show that Pd-like xenon collisional excitation optical field ionization x-ray lasing signal at 41.8 nm can be efficiently produced in Xe cluster gases. The influence of the pumping laser polarization, prepulse energy, prepulse delay, and backing pressure of the gas jet on the x-ray lasing output is investigated systematically.

1. Introduction

Optical-field-ionization (OFI) plasmas have been suggested as the efficient gain media to produce tabletop x-ray lasers through collisional excitation or recombination processes [1-3], because the electron energy distribution and the ionization state in OFI plasma are mainly controlled by the laser intensity and polarization. A linear polarized laser pulse can minimize the mean electron temperature and promote strong recombination. The hot electrons required for the collisional excitation pumping can be produced by a circular polarized laser pulse during the ionization. Moreover, the appearance intensity for various ion species in the optical-field-produced plasma is discrete, and the generation of a specific ion state can be very selective. Those properties enable us to make the best use of ions and therefore reduce the pumping energy required for generating efficient OFI soft x-ray lasers.

Recently, the saturation operation of collisional excitation OFI x-ray laser has been demonstrated by focusing a circular-polarized laser pulse into a gas cell filled with pure Xe[4]. Although a high gain is able to be produced in such optical-field-driven plasmas, it still requires a method to suppress the ionization-induced refraction in plasma and keep the ionization pulse above threshold intensity for full ionization of the correct ionization stage, over a reasonable length for lasing. Otherwise, the OFI x-ray laser can only be operated in low density gases and the x-ray lasing output is therefore limited. In addition, the alignment of the gas cell is also critical in order to avoid the damage of the gas cell or pinholes.

To suppress the ionization defocusing effect, the production of a preformed plasma waveguide is an efficient way to guide a high power laser pulse over several confocal lengths in plasma without changing its intensity significantly. A high gain-length product in Pd-like Xe OFI x-ray laser is demonstrated by Butler *et al.* in a capillary discharge waveguide[5]. However, the damage of the capillary tube may limit its practical operation rate. Recently, a 1.2 cm-long high efficient plasma waveguide is demonstrated in Ar gas with only 100 mJ of the pumping energy by using an axicon and ignitor-heater pumping scheme[6]. This method may be applied to increase the gain length of OFI x-ray lasers.

To further increase the OFI x-ray laser output, the increase of the gas density is advantageous if the propagation of the pumping laser beam can be well controlled. An alternative experimental arrangement for providing high-density gas and high-repetition-rate operation would involve a gas jet operated at a high backing pressure. The jet can be produced by a nozzle designed to give a few millimeter long, constant pressure channel of gas/cluster with sharp boundary. It has been shown that clusters absorb laser energy efficiently to form highly charged hot plasmas and emit strongly in the spectral range from extreme ultraviolet to hard X-ray due to its high local density. Moreover, the transient lens effect can be achieved in cluster gases if the temporal and spatial evolution of nanoplasmas can be well controlled[7,8]. The laser propagation in a nanoplasma gas is therefore determined by the balance of diffraction, ionization defocusing and transient lens effects.

In this paper, we show that OFI soft x-ray lasers can be produced by focusing a prepulse and a high-intensity main pulse onto a Xe clustered gas jet. A strong lasing signal at 41.8 nm from Pd-like Xe ions is detected, and about 5×10^{8} photons are

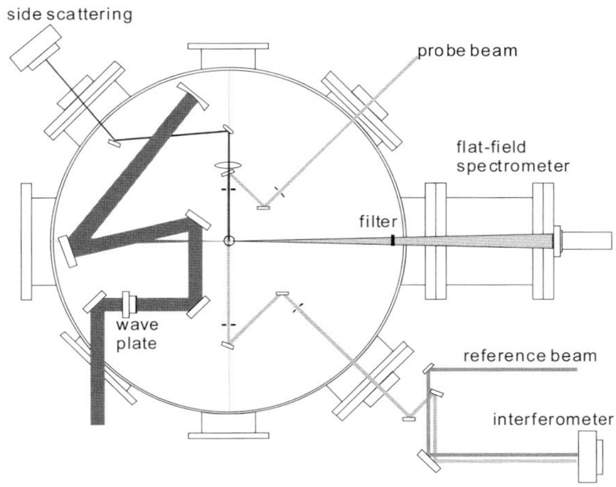

Figure 1. Schematic diagram of the experimental setup.

produced on the $4d^95d$-$4d^95p$ transition. The influence of the pumping laser polarization, prepulse energy, prepulse delay, and backing pressure of the gas jet on the x-ray lasing output is investigated systematically.

2. Experimental Setup

A 10 TW, 55 fs, 810 nm, 10 Hz Ti:sapphire laser system based on chirped-pulse amplification was used for our experiment[9]. The laser contrast is $>10^8$ at -10 ns, $>10^6$ at -50 ps, and $>10^4$ at -1 ps. After the final-stage amplifier, the laser beam is expanded to 40 mm in clear aperture. The beam is then split into two, each part going through an energy tuner and a pulse compressor. With this double compressor configuration, the pulse duration, chirp, energy, beam size, and delay of these two laser beams can be controlled independently with a combined maximal energy of 550 mJ. After being recombined by using a thin-film polarizer and a half-wave plate, the two beams copropagate and are focused with a spherical mirror of 1 m focal length onto a xenon gas jet. The focal spot of the main pulse fits to a Gaussian profile of 25 μm in full width at half maximum (FWHM), with 50% energy enclosed, and the focal spot of the prepulse is enlarged to 50 μm FWHM by inserting a 2 cm diameter hard aperture in the path of the prepulse before recombination. The gas jet is produced with a pulsed valve and a supersonic nozzle, producing a flat-top region of ~5 mm in length at 1 mm above the output of the nozzle. Clusters are formed inside this jet as a result of rapid adiabatic cooling. Side imaging of Rayleigh scattering from nanoplasmas, bandpass filtered at 810 nm, is used to observe the propagation of laser beam in the bulk nanoplasma gas, and to measure the spatially resolved temporal evolution of the magnitude of nanoplasma polarizability.

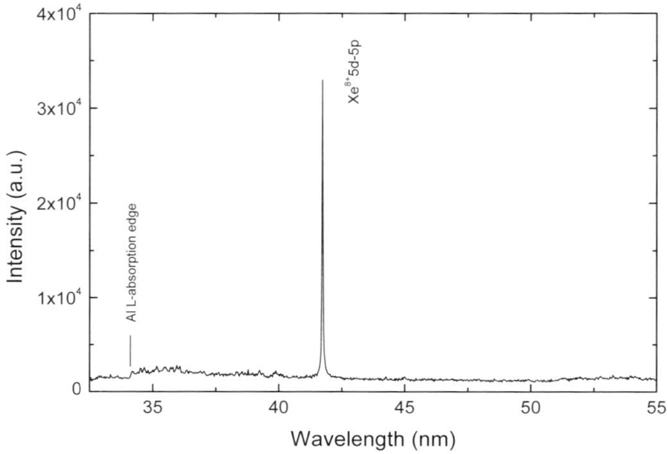

Figure 2. Axial spectrum of Pd-like Xe soft x-ray laser pumped by two circularly polarized laser pulse with a 3.8 ns delay.

Mach-Zehnder interferometry with a 10% probe beam split from the main beam passing transversely through the gas jet is used to observe the spatially and temporally resolved propagation of the prepulse and main pulse. The on-axis diagnostic is a flat-field spectrometer (FFS) made of an aperiodic grazing-incident grating with an average groove density of 1200/mm and a back-illuminated 16-bit soft X-ray CCD camera. The spectral range of this FFS is 30–60 nm and the resolution is 0.05 nm in this experiment. A 1.05 μm-thick Al filter is used to block the pump light and to attenuate X-ray emission. The schematic diagram of the experimental setup is shown in figure 1.

3. Experimental Results

Figure 2 shows a time-integrated spectrum obtained with two circularly polarized laser pulses. The energies of the prepulse and the main pulse are 17 mJ and 350 mJ, respectively, and the temporal delay between two pulses is 3.8 ns. The line emission of the 5d-5p transition in Xe^{8+} ions at 41.8 nm completely dominates the spectrum. The second and third order of Al L-absorption edge and several incoherent transition lines from Xe^{7+}ions can also be found in the spectrum, but are much weaker than the lasing line. The lasing signal totally disappears if the polarization of the pump beams becomes linear.

To obtain the optimal experimental conditions for soft x-ray amplifications, the energy of the prepulse is varied systematically at various backing pressures for a fixed delay of 3.8 ns. The dependence of the x-ray laser output on prepulse energy for various backing

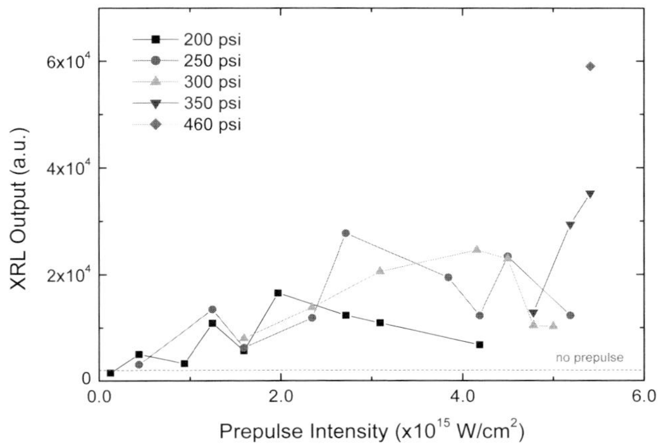

Figure 3. The x-ray laser output versus the prepulse intensity for various backing pressures of the gas jet. The delay between the prepulse and main pulse is 3.8 ns.

pressures is shown in figure 3. The output signal for the case using a prepulse is about 30 times more efficient than that without a prepulse. This result is opposite to that obtained using pure gas target. Hooker *et al.* suggests that the existence of prepulses will decrease the OFI laser output dramatically due to higher ion temperature and decrease of the ion density [10]. In contrast, the prepulse in our experiment is used to break up Xe clusters to produce a uniform plasma before the arrival of the main pulse. Although the on-axis ion density of the preformed plasma decreases as a result of hydrodynamic expansion of the plasma channel, it is still high enough for producing a high gain since the initial gas density ($> 10^{18}$ cm^{-3}) used are higher than the optimal density used in the gas cell. For example, the average ion density upon the incidence of the main pulse is about $5–7\times10^{17}$ cm^{-3} for a backing pressure of 460 psi, measured by the interferometer. An optimal energy of the prepulse is presented as a result of the tradeoff between a higher on-axis density and a longer gain length. The general trend shows that stronger x-ray lasing can be obtained by using a higher backing pressure and a corresponding higher prepulse energy. This results from the higher on-axis density and longer preformed uniform plasma channel. This trend remains the same even when the pulse delay is changed to 2 ns or 6 ns, as expected. The absolute photon number produced for a backing pressure of 460 psi in figure 3 is estimated to be about 5×10^8 by taking account of the filter transmission, the grating efficiency and the CCD response. This number is close to the margin of saturation for OFI x-ray lasers using pure gas target [4].

In order to investigate the feasibility of generating the recombination OFI x-ray laser, we

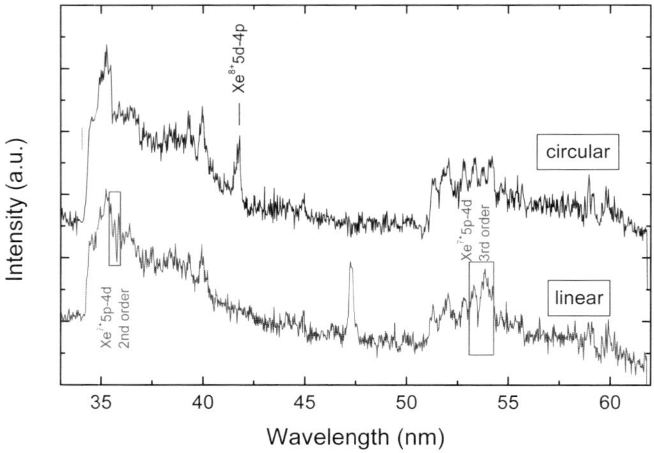

Figure 4. The output spectra produced by the linear and circular polarized laser pulses. The intensity of the prepulse and the main pulse are 4×10^{14} W/cm^2 and 3×10^{17} W/cm^2 respectively. The pulse delay is 3.9 ns and the backing pressure is 180 psi.

rotate the laser polarization from circular to linear. In this study, a smaller nozzle producing a ~ 3-mm-long flat-top density region was used. Figure 4 shows the comparison of the output spectra produced by the laser pulses with linear and circular polarization. When the polarization of the pump laser is switched from circular to linear, we find the lasing signal at 41.8 nm pumped by the collisional exciation process vanishes. An unknown emission line with a small divergence angle of ~ 2 mrad at 47.2 nm is observed instead. Besides, the intensity of the recombination lines on the transition of 5p-4d in Xe^{7+} ions becomes relatively stronger. This indicates that the plasma produced by the prepulse has cooled down to a level that does not destroy the condition of strong recombination upon the arrival of the main pulse. The recombination OFI x-ray laser is possibly produced from the 6s-5p transition in Xe^{7+} ions. It is also important to note that, although the gas density produced by the small nozzle is higher for the interaction region, the laser output at 41.8 nm decreases due to a shorter amplification length.

4. Summary

By using a prepulse with an adequate delay to control the spatial and temporal density distribution of nanoplasmas produced from cluster gases, we are able to produce high efficient xenon collisional excitation OFI x-ray lasers at 41.8 nm. About 5×10^{8} soft x-ray photons are produced with a long pulse delay. The laser output is enhanced significantly for backing pressure up to 460 psi and limited by the pressure of the gas reservoir. Further increase of the x-ray laser output is possible if a higher backing pressure and a more intense prepulse can be used. Although a possible signature of the recombination OFI x-ray laser at 47.2 nm is observed, it still requires a systematic study to find the optimal condition for high gains. The influence of the transient lens effect on the x-ray laser output must be investigate further in order to improve the laser absorption efficiency and produce high efficient OFI x-ray lasers,

References:

[1] Lemoff B E, Yin G Y et al. 1995, Phys. Rev. Lett. 74, 1574.
[2] D. V. Korobkin, C. H. Nam et al., 1996, Phys. Rev. Lett. 77, 5206.
[3] Grount M J, Janulewicz K A et al., 1997, Opt. Commun. 141, 213.
[4] Sebban S, Haroutunian R et al., 2001, Phys. Rev. Lett. 86, 3004.
[5] Butler A, Gonsalves A J et al., 2003, Phys. Rev. Lett. 91, 205001.
[6] Xiao Y-F, Chu H-H et al., 2004, Phys. Plasmas 11, L21.
[7] Alexeev I, Antonsen T M et al., 2003, Phys. Rev. Lett. 90, 103402.
[8] Chu H-H, Tsai H-E et al., 2004, Phys. Rev. E. **69**, 035403(R).
[9] Chu H-H, Huang S-Y et al., 2004, Appl. Phys. B in press.
[10] Hooker S M, Epp P T et al., 1997, J. Opt. Soc. Am. B, 14, 2735.

Inst. Phys. Conf. Ser. No 186
Paper presented at 9th Int. Conf. X-Ray Lasers, Beijing, China, 24–28 May 2004

Strong line enhancement at 24.7 nm in a boron-nitride capillary discharge irradiated by a femtosecond laser pulse

B Greenberg[a], K A Janulewicz, G Priebe, J Tümmler, M Schnürer, M Levin[a], A. Pukhov[a], P Mandelbaum[a], P V Nickles, A Zigler[a]

Max Born Institute, Max-Born-Str. 2A, D-12489 Berlin, Germany
[a] Racah Institute of Physics, Hebrew University, Jerusalem 91904, Israel

Abstract. Investigations of plasma produced by a boron-nitride capillary discharge irradiated with a guided 20 TW Ti-Sapphire laser pulse at peak intensity of $4x10^{18}$ W/cm^2 are presented. The guided laser radiation in the plasma channel generates He-like ions which subsequently, subject to suitable plasma temperature, recombine to Li-like nitrogen ions by the three-body recombination process. Intense short-wavelength radiation at a wavelength of 24.77 nm was observed, indicating a possible lasing at the $3d_{5/2} - 2p_{3/2}$ transition in Li-like nitrogen.

1. Introduction.

Short-wavelength lasers pump schemes combining electric discharge and laser irradiation in one, frequently termed as hybrid, have already been demonstrated [1-3]. The important advantage of this inherently longitudinal arrangement is an elongated medium in the form of a plasma channel efficiently pumped during the guiding phase [4,5]. The possibility to compress the pumping beam into a tight focus facilitates reaching of high peak intensity at low energy level. However, the elongated medium increases absorption and ionisation losses during the propagation. Some aspects of this problem were discussed in [6]. It was concluded, that some excess of the intensity in comparison to the ionisation threshold, depending on the elements included in the excitation process and the medium length, is necessary to ensure efficient and relatively uniform pumping along the plasma channel. It is well known that both recombination and collisional pumping schemes can be realized in such a plasma pipe provided that the optical field ionisation (OFI) is the main mechanism in creation of highly ionised medium. The polarisation of the ionising radiation enables control over the plasma residual temperature and decides about the choice of the working scheme. Detailed analysis of the control over the plasma parameters from the perspective of the OFI-X-ray lasers was given in [7]. The lithium-like nitrogen (NV) is especially interesting as an active medium due to its expected resistance to the overionisation process. The intensity gap between the ionisation thresholds of He-like and H-like nitrogen covers more than two decades. The high ionisation level of the pre-plasma reduced the beam distortion caused by very intense ionisation.

Significant progress was observed in development of the collisionally pumped OFI lasers after its demonstration reported in [8]. The recombination scheme was tested in H-like lithium but mostly with a limited output from a short capillary due to the loss mechanisms mentioned above. Here we concentrate our interest on the recombination scheme working in a Li-like nitrogen (NV). The capillary discharge formed the plasma and a short linearly polarised laser pulse ionised it by tunneling. It will be shown that the plasma irradiation with 20 TW laser pulses is accompanied by a strong increase in intensity of a line at 24.77 nm, which corresponds to the transition $3d_{5/2} - 2p_{3/2}$ in Li-like nitrogen. It will also be shown that this increase in the line intensity could be explained by lasing. The set up of the experiment is shown in Fig. 1.

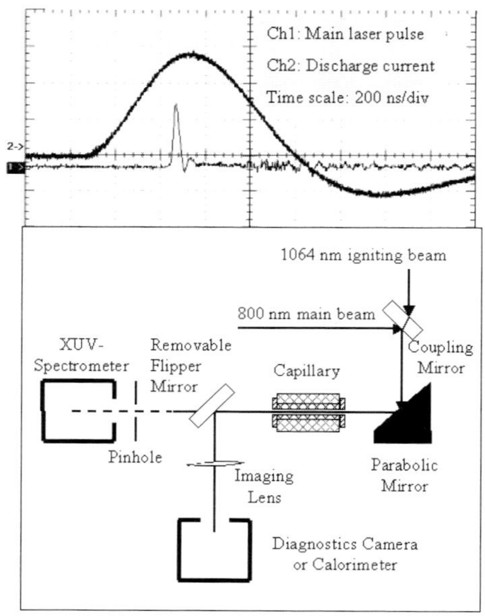

Figure 1. Experimental set up. The inset shows temporal history of the electrical parameters [10]

A boron-nitride capillary with a length of 15 mm and a diameter of 350-400 μm was fired in vacuum. The battery consisted of a 0.05 μF capacitor charged in this experiment up to 4-5 kV. An auxiliary Nd-YAG laser delivered pulses with a length of about 10 ns and an energy of 20-30 mJ, which after co-linear alignment with the main laser pulse were used for discharge triggering [9]. This reduced the jitter between the discharge and the main laser pulse to less than 5 ns. The discharge current rise time was equal to 400 ns with the full width of the first half-cycle period of 800 ns and the peak current between 500 and 650 A. The temperature of the discharge plasma equal to 4 to 6 eV was estimated by spectroscopic measurements. The spectrum of the plasma in a short-wavelength range (XUV) was recorded with a flat-field spectrometer consisting of a Harada-type diffraction grating (1200 l/mm), a multi-channel plate (MCP), a phosphor screen and a cooled 16-bit CCD camera. Gating of the MCP with an electric pulse with a width of 40 ns allowed improving significantly the signal-to-noise ratio. The main pulse was focused with an off-axis f/5 parabola (OAP) onto the capillary entrance.

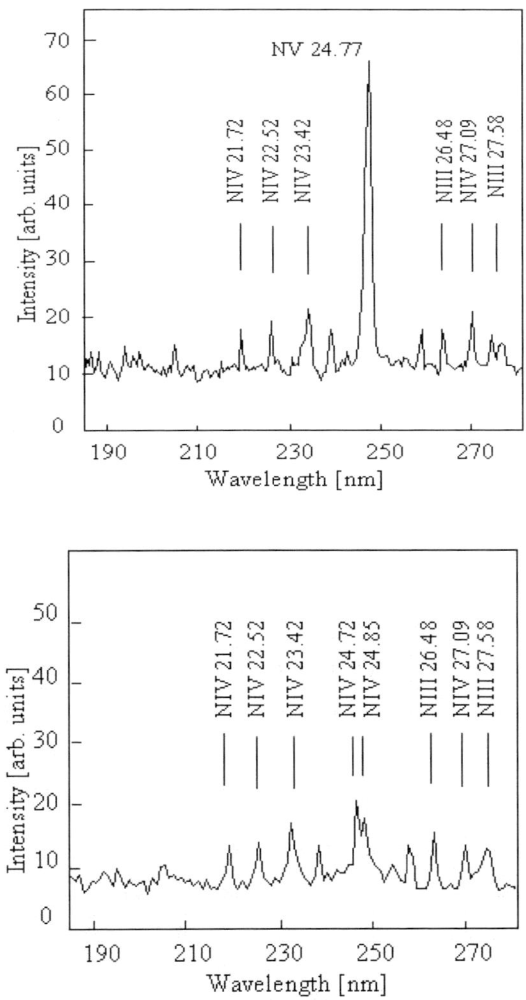

Figure 2. Spectrum of the capillary discharge irradiated by a femtosecond laser pulse (a); Plasma emission spectrum for pure capillary discharge (U = 4 kV, 50 nF) (b). [10]

The laser pulse applied in the described experiment had a width of about 40 fs and was sufficiently short to avoid significant collisional heating even if the electron density was about 2×10^{19} cm^{-3}. The pulse profile at the contrast level of about 10^2 can, as was shown in [10], dramatically change the residual temperature. The extremely high intensity applied in the experiment ensured the required ionisation level along the whole capillary length as the possible but quantitatively not exactly determined losses were easily compensated. As a consequence the energetic transmission of the ionising beam measured in the experiment showed a maximum throughput of ~70%. The beam guiding effect within a time gap of 20 ns was confirmed in a separate measurement and the beam diameter (FWHM) was estimated to be equal to 20 μm. These measurements together with a monitored laser energy gave a reliable estimate of the peak intensity of the ionising beam at the capillary entrance I_L as equal to ~(3-4)x10^{18} W/cm^2. Such high intensity being well above the relativistic limit has been used for the first time in experiments on recombination OFI-X-ray lasers.

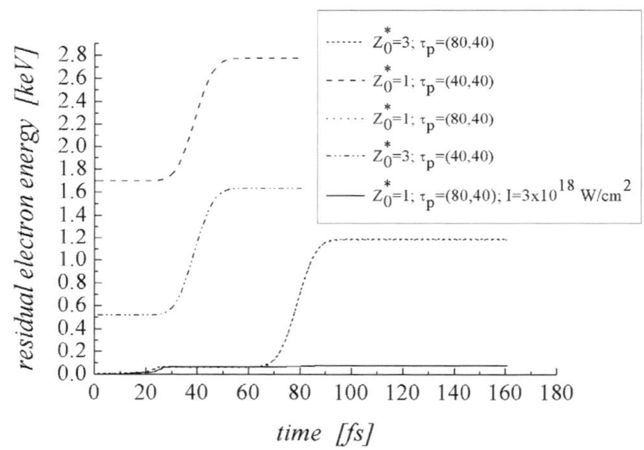

Figure 3. Dependence of the electron residual energy on the pulse shape and length combined with its intensity and plasma performing. If not declared the pulse peak intensity is equal to 5×10^{18} W/cm^2.

A high-intensity line at 24.77 nm corresponding to the transition $3d_{5/2} - 2p_{3/2}$ in Li-like nitrogen (NV) dominated the observed emission spectrum (Fig. 2a) if the plasma channel was irradiated with an ultra-short pulse. This line was absent in the discharge spectrum (Fig. 2b). This observation and the fact that the line at 26.63 nm corresponding to the transition 3p-2s in lithium-like nitrogen does not appear as a strong line suggest that the observed emission at 24.77 nm could be caused by lasing. If the origin of this intensity enhancement was of thermal nature the lines at 20.93 and 26.63 nm having the upper level below the upper level of the 3d-2p transition had to be at least comparably strong. This is not the case and both lines are absent in the spectrum.

The experimental results were also modelled numerically. For simplicity sake only nitrogen atoms being dominantly represented in the spectrum were taken into account in the simulations of the plasma residual temperature. Significant intensity surplus above the ionisation threshold increased enormously the residual temperature (Fig. 3). This

temperature level deteriorates the recombination process. Reduction in the intensity about one order of magnitude would decrease the temperature to the level below 100 eV and such a temperature makes the radiative-collisional cascade reasonably efficient if the decoupling effect is taken into account [11, 12].
The experimentally estimated pre-plasma energy between 6 eV and 9 eV and the initial ionisation stage of $Z_0^*=2.3$ were used in the simulations. However, the initial ionisation stage is important only for the pulses with a very high contrast and a steep leading edge. The influence of the pulse shape, its length, intensity and the initial ionisation stage on the electron residual temperature are shown in Fig.3. It is seen that a small elongation of the pulse by increase of the pulse leading edge works at lower Z^* similarly to the intensity reduction. As a result the residual energy is about 70 eV. OFI plasma shows an electron energy distribution function (EEDF) of bi-Maxwellian character and the de-coupling effect could reduce the plasma temperature to 20 - 30 eV, which is already acceptable for the recombination pump scheme.
In summary we have demonstrated strong enhancement of the 24.77 nm line in lithium-like nitrogen. Some spectroscopic features indicated clearly that the enhancement is not of thermal origin and suggested we have observed lasing.

References

[1] Goltsov A, Morozov A, et al., 1999, J. IEEE Sel Top Quantum Electron, 5, 1453-1459
[2] Janulewicz K A, Rocca J J, et al., 2001, Phys Rev A, 63 033803
[3] Butler A, Gonsalves AJ, et al., 2003, Phys Rev Lett, 91, 205001
[4] Ehrlich Y, Cohen C, et al., 1998, J Opt Soc Am B, 15, 2416-2423
[5] Hosokai T, Kando M, et al., 2000, Opt. Lett., 23, 10-12
[6] Janulewicz K A, Bortolotto F, et al., 2003, J Opt Soc Am B, 20, 215-220
[7] Pert G J 2001 J. Phys. B, 34, 881- 908
[8] Lemoff B E, Yin G Y, et al., 1995, Phys Rev. Lett., 74, 1574
[9] Greenberg B, Levin M, et al, 2003, Appl. Phys. Lett., 83, 2961.
[10] Janulewicz K A, Greenberg B, et al., 2004, Opt. Lett. (submitted)
[11] Donnelly T D, Lee R W, et al., 1995, Phys. Rev. A, 51, R2691-R2694
[12] Janulewicz K A, Healy S B, et al., 1997 Opt. Commun., 140, 165-178

Inst. Phys. Conf. Ser. No 186
Paper presented at 9th Int. Conf. X-Ray Lasers, Beijing, China, 24–28 May 2004

Significant improvement of the 41.8 nm Xe^{8+} laser using gas-filled capillary tubes

T Mocek, S Sebban, I Bettaibi and V Vorontsov
Laboratoire d'Optique Appliquée (LOA), ENSTA-École Polytechnique, Chemin de la Huniére, 91761 Palaiseau, France

B Cros and G Maynard
Laboratoire de Physique des Gaz et des Plasmas, Université Paris Sud, Bâtiment 210, 91405 Orsay, France

C M McKenna, D J Spence, A J Gonsalves and S M Hooker
Department of Physics, University of Oxford, Clarendon Laboratory, Parks Road, Oxford OX1 3PU, United Kingdom

Abstract. We report on significant improvement of the 41.8 nm Xe^{8+} collisionally excited OFI XUV laser output achieved by means of multi-mode guiding of high-intensity, femtosecond laser pulses in a gas-filled dielectric capillary tube. Capillaries of various designs and lengths have been investigated and compared to gas cells of the same length. Under optimum conditions the lasing signal from the capillary was about an order of magnitude higher than that from a comparable gas cell. Numerical simulations of the propagation of the pump laser pulse in the capillary revealed that this enhancement is due to reflections from the capillary wall which made it possible to increase the length of the Xe^{8+} plasma column over the whole length of the waveguide. The far-field pattern of the capillary-driven 41.8 nm laser has been measured.

1. Introduction

In the past few years, several inversion schemes for recombination and collisionally excited X-ray lasers (XRL) based on optical-field-ionization (OFI) have been proposed [1], experimentally demonstrated [2-5], characterized [6], and improved [7]. For both classes of XRL, the lifetime of the population inversion is very short and so the pumping must be in the form of a travelling wave. This can be achieved most easily by longitudinal pumping. However, the most serious problem with this pumping geometry is the formation of a sufficient length of plasma for amplification of the X-ray photons to occur, since the Rayleigh length naturally limits the length of the amplifying medium to at best a few millimeters. In order to extend the gain length that can be achieved by longitudinal pumping, it is necessary to guide the pump pulse over many Rayleigh lengths. Several techniques for guiding laser pulses with peak intensities relevant to OFI laser schemes, 10^{16} – 10^{18} W cm^{-2}, have been investigated [8-13]. Recently, we have demonstrated lasing at 41.8 nm in Xe^{8+} using a gas-filled capillary discharge waveguide [7]. The 41.8 nm laser signal observed with a 30 mm long capillary discharge waveguide filled with 120 mbar of a 1:3 Xe:H atom gas mixture was ~4 times larger than that achieved with a 4 mm long gas cell containing 20 mbar of pure Xe [7]. The presence of hydrogen is expected [14] to decrease the small-signal gain coefficient of the Xe^{8+} laser transition through partial pre-ionization of the Xe by the discharge, reduction of the gain cross-section resulting from the increased transition linewidth, and an increased rate of de-excitation of the laser levels. As a consequence, whilst the gain length of collisionally-excited OFI lasers can be greatly increased by driving them within a gas-filled capillary discharge waveguide, the single-pass gain may not be increased by the same factor.

In this contribution we present a significant improvement of the output of a 41.8 nm Xe^{8+} laser achieved by means of multi-mode guiding of high-intensity, femtosecond laser pulses in a gas-filled dielectric capillary tube. The capillary is filled with pure lasant gas, and no discharge is employed. Under optimum conditions the output of the Xe^{8+} laser from a gas-filled capillary is nearly an order of magnitude stronger than that achieved with a gas cell of the same length. The experimental results are explained with the help of numerical simulations.

2. Experimental setup

The experiment was performed at LOA using a 10 Hz multi-terawatt Ti:Sapphire laser system [15] delivering 34 fs pulses (full width at half maximum) with an energy on target of up to 1 J. Figure 1 shows a schematic view of the experimental setup. The circularly polarized laser beam was focused onto the capillary or gas cell entrance by a spherical mirror of focal length 2 m. The measured transverse intensity distribution was close to a Gaussian distribution in the focal plane, with a central spot corresponding to a waist (the radius at which the intensity falls to $1/e^2$ of the axial intensity) of 28 ± 2 μm containing ~40 % of the energy in the plane. The laser beam was measured to be 1.6 times diffraction limited, the measured Rayleigh length was 3 mm, and the resulting maximum intensity on target was (1 ± 0.2) x 10^{18} W cm^{-2} in vacuum. The infrared radiation leaving the capillary or cell was reduced in intensity by reflections from two optically flat wedges, rendered parallel by a lens, and imaged onto a 12-bit charge-coupled device (CCD) camera by another lens and a microscope objective. The on-axis XUV radiation was recorded by moving the first wedge out of the beam path, thereby allowing the radiation to enter the transmission grating spectrometer. The spectrometer comprised a grazing incidence gold coated spherical mirror, transmission grating with 2000 lines/mm, and a back-illuminated XUV CCD camera. Stray light from the driving laser was blocked by two 0.3 μm-thick aluminum filters.

Two types of capillary tubes have been used: sapphire and glass. The sapphire capillaries were 15 mm long, with an internal diameter of 300 ± 10 μm. Since the channels were laser machined, the walls of the capillaries were not optically smooth, but contained non-uniformities with typical dimensions of 5 μm. Xe gas was flowed into the capillary through channels approximately 500 μm wide and 250 μm deep located 1.5 mm from each end. By injecting the gas in this way, the pressure in the main body of the capillary between the gas slots was uniform. The glass tubes were 15 – 56 mm long, with an internal diameter of 300 μm, the walls were of good optical quality, and the gas slots were located 2-3 mm from each end.

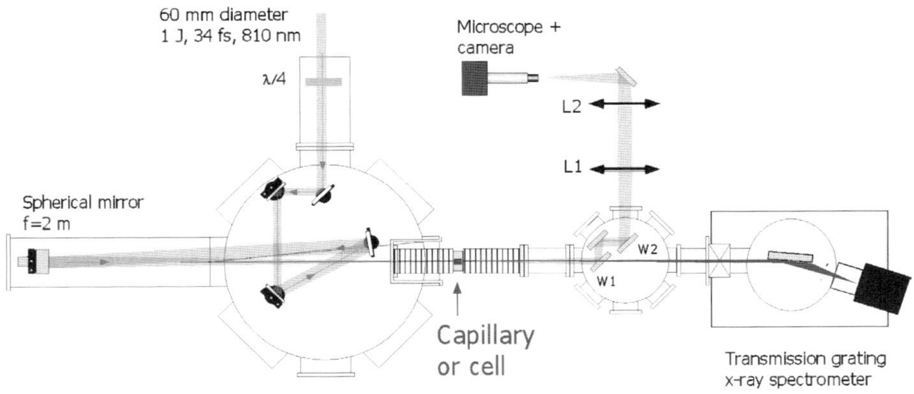

Figure 1. Schematic of the experimental setup.

3. Experimental results with sapphire capillary

To compare the output lasing signal from the capillary with that obtained in previous work [16], we also investigated lasing in Xe cells of 4, 10 and 15 mm lengths. The entrance and exit windows of the cells were made of thin glass plates through which pinholes (~0.5 mm in diameter) were drilled by the pump laser at the beginning of each experimental session. The XRL signals obtained from the waveguide and gas cells were optimized by varying the gas pressure and the position of the vacuum focus of the pump laser. Figure 2 compares optimized output signals from the cell (squares) and the capillary (solid triangles) at 17 Torr as a function of target length. The position of the vacuum focus was 5 (6) mm inside the cell (capillary). The solid curves present results of simulations which will be discussed later. The data clearly shows that the capillary greatly increases the lasing signal compared to that achieved with a gas cell. The strongest signal was observed with the 15 mm long capillary at 30 Torr, being approximately 7 times higher than the largest signal achieved with the cell of the same length, and as much as 43 times higher than that with the 4 mm cell.

Figure 3 shows the measured pressure variation of the Xe^{8+} laser signal for a 15 mm long capillary (triangles) and a 15 mm long cell (squares). The position of the vacuum focus was at the centre (entrance) of the cell (capillary). It is seen that in both cases the XUV laser signal exhibits a distinct optimum pressure: 30 ± 5 Torr for the capillary compared to 17 ± 2 Torr for the cell. Apart from a significantly higher optimum pressure, the other main feature of the capillary lasing is that it occurred over a much broader range of pressures than for the gas cell: 10 – 70 Torr for the capillary compared to 10 – 30 Torr for the cells. There was no significant difference in the pressure dependence of the Xe^{8+} laser signal from the 4, 10, and 15 mm cells, the present results being consistent with earlier results for 4 mm long cells [16].

The laser signal was also dependent on the position of the vacuum focus of the pump laser relative to the entrance of the target. For the 15 mm long cell the optimum position of the vacuum focus was found to be 5 ± 1 mm inside the cell, though the dependency was weak. By contrast, in the case of the 15 mm long capillary the signal was found to be quite sensitive to the focal position with a distinct optimum at 6 mm. Further, when the laser focus was set at 8 mm the capillary was destroyed in a relatively low number of pump laser shots. In the case of the 4 mm long cell we did not find any clear optimum and the output signal was rather insensitive to the focusing position, in contrast to earlier work with a 1 m focal length optic [16]. This difference arises from the fact that use of a longer focal length optic extends the Rayleigh range, and hence the spot size and divergence of the beam at the entrance to the target is less sensitive to the position of the vacuum focus.

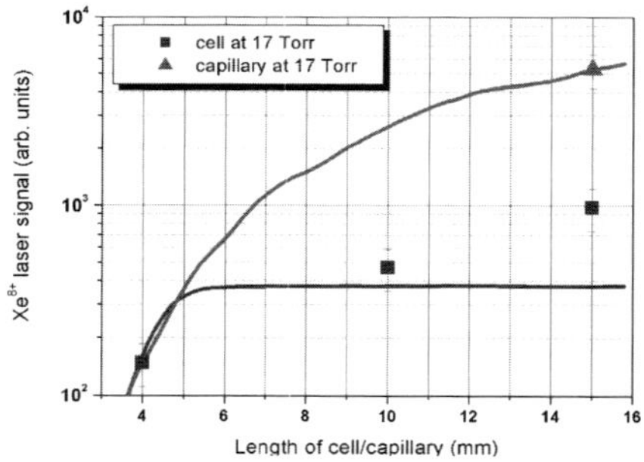

Figure 2 Measured and simulated (solid lines) Xe^{8+} laser signal from the sapphire capillary (solid triangle) and the cell at 17 Torr (squares) as a function of the cell or capillary length.

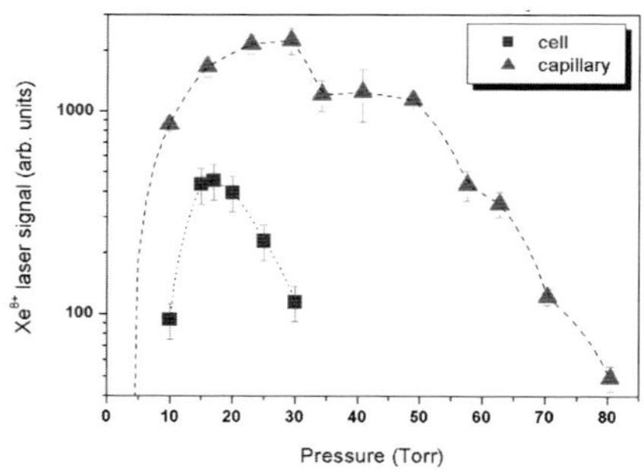

Figure 3 Pressure dependence of the 41.8 nm lasing line for the sapphire capillary (triangles) and the 15 mm long cell (squares); the dashed and dotted lines are a guide to the eye.

4. Simulations

To model the propagation of high intensity laser pulses through Xe gas undergoing OFI we used a numerical simulation code. The code simulates the time dependent propagation of the driving pulse in a 3-D cylindrically symmetric geometry and has been developed from that in [17]. The laser pulse is assumed initially to have Gaussian spatial and temporal profiles. The paraxial wave equation is solved in cylindrical geometry to describe its propagation through the initially neutral gas, and the gas ionization is taken into account self-consistently. The pulse may be arbitrarily polarized and the effects of ionization-induced refraction as well as relativistic self-focusing are included. Hydrodynamic effects within the plasma are assumed to be insignificant on the femtosecond timescales of propagation, and as such are not taken into account.

The pump laser pulse was assumed to be circularly polarized with a Gaussian temporal profile, a FWHM duration of 34 fs and a peak input intensity of 1×10^{18} Wcm^{-2}. The incoming laser field was taken as a sum of several (typically 9) Gaussian propagating modes as the best fit of the experimentally measured fluence profile in several planes before and after the best focal plane in vacuum. In the case of propagation inside the capillary, a boundary layer with a dielectric constant larger than 1 was introduced to simulate the capillary wall. Its value was adjusted so as to reproduce the measured value of the transmitted pump energy which is about 40% at a pressure of 25 Torr, and ~80% at 17 Torr.

The simulations showed [18] that in the case of the 15 mm long cell at a pressure of 17 Torr and a focusing position of 6 mm inside, Xe^{9+} and higher ion stages are created in a small region around the axis for distances of up to ~2 mm from the entrance. The lasant ion stage, Xe^{8+}, is generated in an annular region up to z = 8 mm, and outside this Xe^{8+}

region the degree of ionization decreases. Simulations performed for focusing positions less than 6 mm exhibited similar distributions of Xe ion stages but the length and volume of Xe^{8+} was progressively smaller, in agreement with the experimentally observed slow decrease of the output laser signal. No further enlargement of the Xe^{8+} region was found for focusing positions beyond 6 mm.

In the case of the capillary under the same pump laser conditions as above, we found that Xe^{8+} is produced along the whole length of the capillary in an annular region with higher ion stages being generated in an approximately cylindrical region of radius ~20 μm along most of the capillary axis. Clearly the volume of the Xe^{8+} region increased substantially for the capillary. The size of the Xe^{8+} region was found to be much more sensitive to the longitudinal position of the focus than in the case of the cell, in agreement with experimental data. For the focusing position of 6 mm inside the capillary the region of Xe^{8+} created was the largest, with much less Xe^{7+} than when focusing at the entrance.

To quantitatively evaluate the performance of the studied XUV laser, we undertook simplified modeling of the radiation transfer in the active medium. The model treats amplification of axial emission in an active medium which accounts for the refraction of the XUV laser beam. It is assumed that both the small-signal gain and the emissivity are constant along the amplifier axis and for the calculations we used values of gain and saturation intensity measured previously [16].

In Fig. 2 results of simulations for the capillary and the cell are shown (solid lines), the normalization constant being the same for both simulations. It is seen that the curve calculated for the cell agrees well with the experimental points for the 4 and 10 mm cells, and the two curves coincide for short lengths, as they should. The main discrepancy with the experimental results is that the simulated signal for the cell does not change after $z = 8$ mm since it is calculated that Xe^{8+} is not generated for values of z greater than this, whilst the measured signal for the 15 mm cell is about 2 times larger than that for the 10 mm cell. This effect can not be explained by the present modelling and might be attributed to differences in the gas densities outside the entrance and exit pinholes as the pinhole sizes are different in the two cases.

The guiding of the pump radiation observed in this work is quite different from the monomode guiding observed previously for evacuated and gas-filled capillaries [9, 19]. In order to achieve monomode guiding the spot size of the pump laser pulse at the capillary entrance must be carefully matched to the diameter of the capillary. Abrams [20] has shown that optimum coupling into the lowest-order mode of the capillary occurs for a Gaussian beam when the input waist size is 0.6435 times the capillary radius. In the present work the laser spot radius at the capillary entrance was ~0.2 times the capillary radius, so highly multi-mode guiding is expected to occur. Our experimental results suggest that capillary targets of this type, which are robust and of a simple design, might be very useful in the search for lasing on other OFI XRL transitions, and thus complement other guiding techniques.

5. Results with glass tubes

In the experiment with glass capillaries we succeeded in (multi-mode) guiding over 56 mm and performed detailed characterization [21] of the generated XRL. The far-field pattern of the 41.8 nm laser was obtained by placing a removable 45°, XUV mirror in the beam path, redirecting the beam towards a cooled, thin, back-illuminated CCD. The multilayer mirror was made from a stack of 30 $Mo/B_4C/Si$ tri-layers. Figure 4 shows a typical imprint of the 41.8 nm laser from a 8 mm gas cell at 15 Torr, and a 25 mm glass capillary at 5 Torr. For these optimal lasing conditions the measured divergence is ~8 mrad for the cell whilst only ~4 mrad for the capillary.

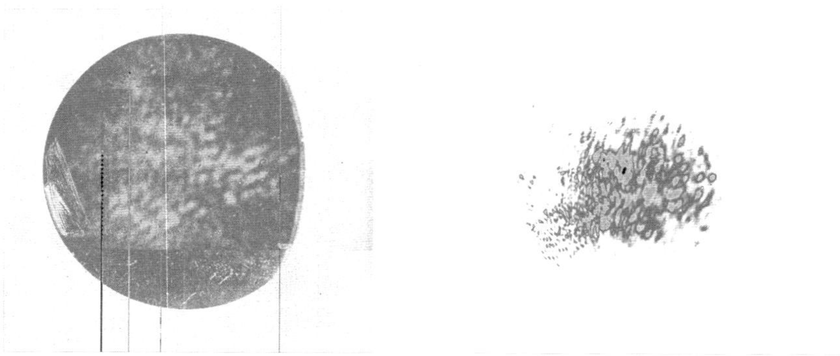

Figure 4. Measured far-field patterns of the Xe^{8+} laser from a 8 mm long cell at 15 Torr (left), and a 25 mm long glass capillary at 5 Torr (right).

6. Conclusions

We have achieved a significant improvement of the OFI Xe^{8+} laser performance by employing multi-mode guiding of femtosecond laser pulses at an intensity of $\sim 10^{18}$ W cm^{-2} through a gas-filled capillary. Simulations confirmed that this enhancement is due to reflections from the capillary wall, which resulted in a substantial increase of the length and volume of the Xe^{8+} plasma column. The capillary driven XRL at 41.8 nm currently provides up to 1.5 x 10^{11} photons/pulse (0.8 μJ) at 10 Hz, in 4 mrad of divergence.

References

[1] Lemoff B E *et al.* 1994 Opt. Lett. 19 569-571
[2] Nagata Y. *et al.* 1993 Phys. Rev. Lett. 71 3774-3777
[3] Chichkov B N *et al.* 1995 Phys. Rev. A 52 1629-1639
[4] Lemoff B E *et al.* 1995 Phys. Rev. Lett. 74 1574-1577
[5] Sebban S *et al.* 2002 Phys. Rev. Lett. 89 art. no. 253901
[6] Mocek T *et al.* 2004 Appl. Phys. B 78 939-944
[7] Butler A *et al.* 2003 Phys. Rev. Lett. 91 art. no. 205001
[8] Jackel S *et al.* 1995 Opt. Lett. 20 1086-1088
[9] Dorchies F *et al.* 1999 Phys. Rev. Lett. 82 4655-4658
[10] Monot P *et al.* 1995 Phys. Rev. Lett. 74 2953-2956
[11] Durfee C G and Milchberg H M 1993 Phys. Rev. Lett. 71 2409-2412
[12] Ehrlich Y *et al.* 1996 Phys. Rev. Lett. 77 4186-4189
[13] Hosokai T *et al.* 2000 Opt. Lett. 25 10-12
[14] Spence D J *et al.* 2003 J. Opt. Soc. Am. B 20 138-151
[15] Pittman M *et al.* 2002 Appl. Phys. B 74 529-535
[16] Sebban S *et al.* 2003 J. Opt. Soc. Am. B 20 195-202
[17] Rae S C 1993 Opt. Commun. 97 25-28
[18] Sebban S *et al.* this Proceedings
[19] Cros B *et al.* 2000 IEEE Trans. Plasma Sci. 28 1071-1077
[20] Abrams R L 1972 IEEE J. Quantum Electron. 8 838
[21] Cros B *et al.* in preparation

Inst. Phys. Conf. Ser. No 186
Paper presented at 9th Int. Conf. X-Ray Lasers, Beijing, China, 24–28 May 2004

Investigations on short-pulse laser pumped xuv laser schemes

M. Born, S. Passinger, V. A. Vorontsov, I. F. Shaikhislamov*, B. N. Chichkov** and B. Wellegehausen

Institut für Quantenoptik, Universität Hannover, Welfengarten 1, 30167 Hannover, Germany
*Institute of Laser Physics, pr. Lavrentyeva 13/3, Novosibirsk, 630090 Russia
**Laser Zentrum Hannover e.V., Hollerithallee 8, 30419 Hannover, Germany

Abstract. Investigations on reported optical field ionization (OFI) - type xuv recombination laser schemes are performed to improve the gain or gain-length product. To enhance the recombination rate, specifically plasma cooling by metal surfaces in gases is studied. Results on He^+ 2p–1s at 30.4 nm, N^{2+} 3s(^{2}S)-2p(^{2}P) at 45.2 nm and O^{2+} 3s(^{3}P)-2p(^{3}P) transition at 37.4 nm are reported.

1. Introduction

The realization of short pulse laser pumped xuv lasers seems to be attractive, as realiable compact powerful pump laser systems are available. However, success with the optical field ionization (OFI)-type lasers so far is restricted to the collisionally excited schemes such as Xe^{8+} or Kr^{8+}, where saturated amplification has been achieved [1][2], while little or no progress has been possible with the recombination schemes, where the achieved gain-length-products are still below 5 [3]. Obviously, the required plasma conditions for the recombination schemes, a high electron density at a low electron temperature, are difficult to realize in the OFI-process. The electron temperature may be reduced by applying specific cooling mechanisms, as heat conduction cooling or cooling by mixing the target gas with a gas of low ionization energy. In this contribution we will present and discuss continued investigations on such cooling mechanisms to improve specific systems where a weak gain has already been demonstrated as for He^+ at 164 nm [4], N^{2+} at 45.2 nm [5] and O^{2+} at 37.4 nm [6].

2. Experiments

As pump laser system we use a 10 Hz Ti:sapphire laser with 100 fs pulses and up to 200 mJ pulse energy. The laser radiation is focussed into a pulsed gas jet, emerging from a slit nozzle. The short pulse laser induced plasma radiation is spectrally selected with a focussing grating monochromator (Jobin Yvon LHT30). With a MCP detector and a fast oscilloscope time resolved measurements are performed. A typical time resolved signal of a plasma line can be seen in fig 1, with a short electron collisional peak and a longer recombination signal. The collisional peak corresponds to the 100 fs laser pulse generating some energetic electrons by OFI which collisionally excite ions. The recombination signal occurs in the expansion phase of the plasma, when the electrons are slow enough to recombine with ions. The delay between both peaks can be considered as a measure of the electron temperature of the plasma. A shorter delay thereby corresponds to a stronger decrease of the temperature.

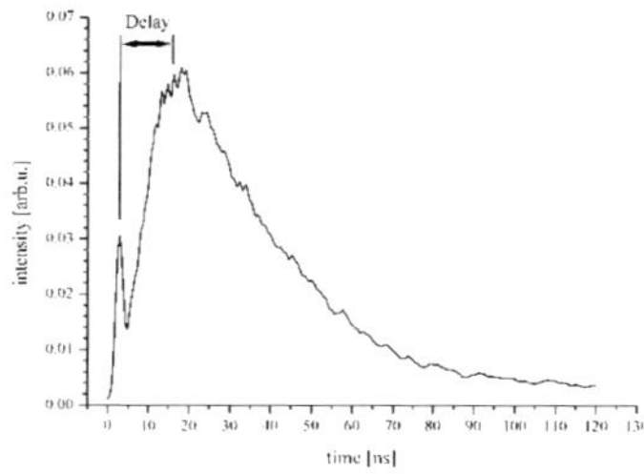

Figure 1. A typical time resolved signal of the OFI plasma emission line (He^+ 2p-1s at 30.4 nm).

To slow down the electrons we use heat conduction cooling by a 0.1x1x10 mm piece of aluminum that was translated towards the focus spot of the laser pulse. Fig 2 shows the signal intensities of the electron collisional and the recombination peak and the delay between the two in dependance of the distance of the cooling metal surface. The signals correspond to the a) He^+ 2p–1s transition at 30.4 nm, b) N^{2+} 3s(^{2}S)-2p(^{2}P) transition at 45.2 nm c) and d) O^{2+} 3s(^{3}P)-2p(^{3}P) transition at 37.4 nm.

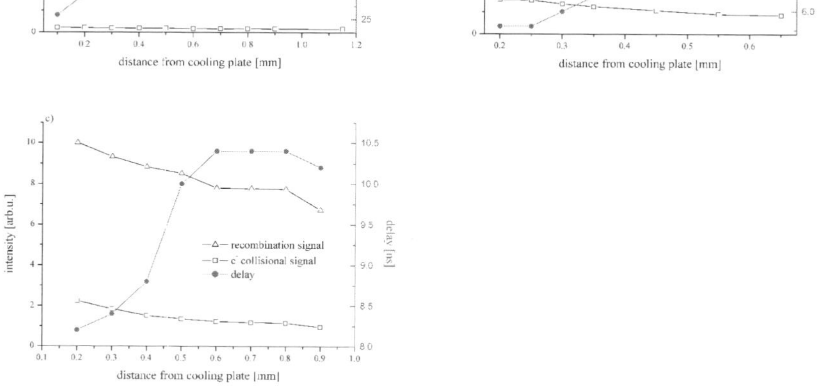

Figure 2. Signal intensity of the electron collisional, the recombination signal and the delay between both in dependance of the distance to the cooling surface for a) He^+ at 30.4 nm, b) N^{2+} at 45.2 nm and c) O^{2+} at 37.4 nm.

As can be seen from fig 2, the approach of the metal surface strongly reduces the delay for all considered transitions which indicates better conditions for the recombination process. However, the expected increase of the recombination component is much less pronounced and there is even an increase of the collisional signal. This is explained by the disturbance of the gas flow due to the metal piece, which affects the particle density and with it the signal strength. In fact, with a thicker metal plate this enhancement is even stronger.

In future experiments we plan to eliminate such effects by specific constructions and further enhance cooling by metal plates on both sides of the plasma or by using metal capillaries, which will allow larger interaction lengths. Further possibilities to improve the recombination conditions by adding cooling gases and by using shorter pump wavelengths will also be considered. All these improvements of the plasma properties should lead to larger gain-length-products.

References

[1] S. Sebban et al.; Phys. Rev. Lett. **86**, (2001)

[2] S. Sebban et al.; Phys. Rev. Lett. **89**, (2002)

[3] Proc. 8th Int. Conf. on X-Ray Lasers, Aspen, Colorado, AIP Conf. Proc. **641**, (2002)

[4] A. Egbert, D. M. Simanovskii, B. N. Chichkov and B. Wellegehausen; Phys. Rev. E **57**, 7138(1998)

[5] P. Lu, H. Nakano, T. Nishikawa and N. Uesugi; Proc. of SPIE **3886**, 294 (2000)

[6] A. Egbert, D. M. Simanovskii, B. N. Chichkov and B. Wellegehausen; Phys. Rev. E **59**, 2305 (1999)

Inst. Phys. Conf. Ser. No 186
Paper presented at 9th Int. Conf. X-Ray Lasers, Beijing, China, 24–28 May 2004

Study of X-Ray Lasers in OFI plasmas

Yuanqin Xia, Jianxin Chen, Deying Chen, Xingfa Lu, Qi Wang
Research Institute of Opto-Electronics, Harbin Institute of Technology, Harbin China 150001

Abstract This paper theoretically and experimentally investigates the X-ray laser in B-like-N and Ni-like-Kr by OFI scheme. After one coefficient error in ADK model was corrected, simple model to describe the electron energy distribution in plasma produced by OFI was constructed and ionization parameters including ionization rate, threshold intensity and original electron energy distribution are calculated. After Doppler broadening and natural broadening are considered, the theoretical peak gain coefficients of Ni-like Kr and B-like N are calculated. In experiment, the soft X-ray spectra in N_2 and Kr driven by 105fs laser pulses with linearly-polarization and near circularly-polarization in a static gas target are obtained.

1 . Introduction

In 1994, Lemoff et al.[1] published for the first time the theoretical simulation of Ne-like Ar (Ar IX)、Ni-like Kr (Kr IX) and Pd-like Xe (Xe IX) of OFI-based electron collision excitation. The results show that each system may have a very high gain, which are 17cm^{-1} 、94 cm^{-1}and 107 cm^{-1} respectively in theory. In 1995 they successfully achieved Pd-like Xe 41.8nm $4d^9 5d^1 S_0$-$4d^9 5d^1 P_1$ transition of the amplification of X-ray[2].The value of gain length gl reached 11. Then Hooker and Harris[3] put forward a new scheme using Be-like ion as laser medium in OFI-based electron collision mechanism and gave the result of theoretical calculation of Be-like N (NIV) and Be-like Ne (NeVII), and the results of theoretical calculation of schemes mentioned are 5.5cm^{-1} and 3 cm^{-1} respectively. In 2001 Pd-like Xe 41.8nm amplification with gl=15 and Ni-like Kr 32.8nm amplification were successfully achieved with 330mJ/35fs and 700mJ/35fs Ti:sapphire laser in gas cell [4-5].

In this paper, we report some theoretical results and primary experimental results based on OFI Ni-like Kr and B-like N X ray lasers. In section II we study the theoretical calculation of ionization parameters of Ni-like Kr and B-like N plasmas. In section III we describe the laser parameters of Ni-like Kr and B-like N and present some primary experimental results in N_2 and Kr with linearly-polarized and near circularly-polarized femtosecond laser pulses.

2. Calculation of ionization parameters of OFI Ni-like Kr and B-like N plasmas

Among the many theoretical models to describe ionization in the intense field, the quasistatic tunneling ionization model (ADK model) by Ammosov, Delone, Krainov et al in 1986[6] is not only in good agreement with experiments, but is also simply compared with some theories such as nonperturbative theory. The quasistatic tunneling ionization model indicates that when the ionization potential of ionized atoms or ions is greatly less than the quiver energy in the laser field, namely when $E_{ion} << U_p$, the ionization rate is independent of optical frequency, and the ionization process can be described with tunneling ionization. Furthermore, the ionization rate can be calculated in the quasiclassical approximation, namely the ionization rate in an alternating field can be obtained by averaging the ionization rate in a static field within an optical cycle [6,7]. We can get the ionization rate formula in an alternating linearly-polarized field by substituting E_s in the formula in a static field for $E_0 \cos \omega t$ and averaging it in an optical cycle. The ionization rate of complex atoms or ions in an alternating field can be described as follows [7]:

$$W_{Lin} = C_{n^*l^*}^2 \left(\frac{3}{\pi}\right)^{\frac{1}{2}} \left(\left(\frac{E_i}{E_h}\right)^{\frac{3}{2}} \frac{E_b}{E_0}\right)^{-\frac{1}{2}} \frac{1}{2} \frac{E_i}{E_h} \frac{(2l+1)(l+|m|)!}{2^{|m|}(|m|)!(l-|m|)!} \times \left(2\left(\frac{E_i}{E_h}\right)^{\frac{3}{2}} \frac{E_b}{E_0}\right)^{2n^*-|m|-1} \exp\left[-\frac{2}{3}\left(\frac{E_i}{E_h}\right)^{\frac{3}{2}} \frac{E_b}{E_0}\right]\omega_0 \tag{1}$$

in the condition of quasiclassical approximation $n^* >> 1$, and $l << n$, we can get [6]:

$$C_{n^*l^*} = \left(\frac{2e}{n^*}\right)^{n^*} \frac{1}{(2\pi n^*)^{1/2}} \tag{2}$$

when l and m are 1 and 0 respectively, putting(2) in(1) ,we can rewrite

$$W_{Lin} = \left(\frac{3e}{\pi}\right)^{\frac{3}{2}} \frac{Z^2}{n^{*4.5}} \left(\frac{4eZ^3 E_b}{n^{*4} E_0}\right)^{2n^*-1.5} \times \exp\left[-\frac{2}{3}\left(\frac{E_i}{E_h}\right)^{\frac{3}{2}} \frac{E_b}{E_0}\right]\omega_0 \tag{3}$$

Substituting $\left(\frac{3e}{\pi}\right)^{\frac{3}{2}} = 4.182$ and $4e = 10.873$ into(3) , we can get

$$W_{Lin} = 4.182 \frac{Z^2}{n^{*4.5}} \left(10.873 \frac{Z^3 E_b}{n^{*4} E_0}\right)^{2n^*-1.5} \times \exp\left[-\frac{2}{3}\left(\frac{E_i}{E_h}\right)^{\frac{3}{2}} \frac{E_b}{E_0}\right]\omega_0 \tag{4.a}$$

As the magnitude of vector composed by two circularly-polarized hefts holds the line in an optical cycle, the ionization rate in the static field can be substituted for that in alternating when laser field is circularly-polarized [1,6]:

$$W_{Cir} = W_{st} = \frac{3e}{2\pi} \frac{Z^2}{n^{*4}} \left(\frac{4eZ^3 E_b}{n^{*4} E_0}\right)^{2n^*-1} \times \exp\left[-\frac{2}{3}\left(\frac{E_i}{E_h}\right)^{\frac{3}{2}} \frac{E_b}{E_0}\right]\omega_0 \tag{4.b}$$

We can discover based on the induction and analysis that the formula of tunneling ionization rate of atoms and ions in a linearly polarized laser field which has been widely discussed and applied in reference [6] is described as follows by mistake.[8,9]

The ratio of formula (4.a) and formula (5) is about 2.60 .

$$W_{Lin} = 1.611 \frac{Z^2}{n^{*4.5}}\left(10.873 \frac{Z^3 E_b}{n^{*4} E_0}\right)^{2n^*-1.5} \times \exp\left[-\frac{2}{3}\left(\frac{E_i}{E_h}\right)^{\frac{3}{2}} \frac{E_b}{E_0}\right]\omega_0 \quad (5)$$

Now we discuss the effect on the calculation of threshold laser intensity on account of the error in formula (5). The threshold laser intensity can be estimated with 1D quasiclassical coulomb Barrier-Suppression Ionization (BSI) model by Augst [10]

$$I_{th} = 4.0\times10^9 E_i^4 / Z^2 [W/cm^2] \quad (6)$$

Table 1 Calculated threshold laser intensities for Xe I to Xe VIII using the formula (4) and (5) when the ionization rate is $10^{12}s^{-1}$

Ions of Xe	Xe^{1+}	Xe^{2+}	Xe^{3+}	Xe^{4+}	Xe^{5+}	Xe^{6+}	Xe^{7+}	Xe^{8+}
IE(eV)	12.127	21.2	32.1	45.5	57.0	68.0	96.0	110.0
I_1 (W/cm^2)	4.79E13	1.41E14	3.55E14	8.30E14	1.35E15	1.96E15	5.53E15	7.44E15
I_2 (W/cm^2)	5.49E13	1.57E14	3.89E14	9.01E14	1.46E15	2.10E15	5.93E15	7.95E15
$(I_2 - I_1) / I_1$	14.54 %	11.06 %	9.49 %	8.62 %	7.87 %	7.29 %	7.28 %	6.90 %

Table 1 shows that the laser threshold intensities I_1、I_2 are needed to produce Xe atom's eight charge ions when the ionization rate calculated by formula (4) and (5) is $10^{12}s^{-1}$. The error $[(I_2 - I_1) / I_1]$is greater than 10% for Xe I and Xe II due to the mistake in formula (5) when the ionization rate is about $10^{12}s^{-1}$ in table 1, and that the error$[(I_2 - I_1) / I_1]$ is less than 10% for more than two charge ions. When the ionization rate is $10^{14}s^{-1}$, the errors are about 23.06 %, 10.08 % and 8.54 % respectively for each、five and eight charge ions due to the mistake in formula(5). Thus the calculated results have been affected for low charge ions because of the error when the ionization rate and threshold laser intensities are calculated in some references which cite the formula (5), and the impact on high charge ions is small due to the small errors.

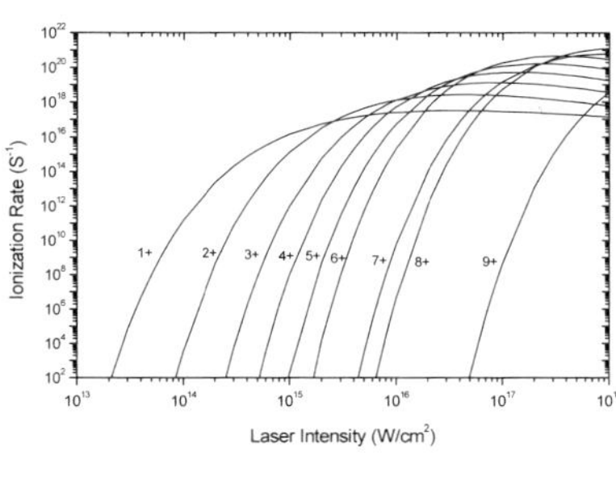

Fig.1 Calculated ionization rate of Kr I to Kr IX as the function of laser intensity

The depending relationship between the ionization rate of Kr I to Kr IX in a circularly polarized laser field and laser intensity is calculated according to formula4.b.The ADK-model-calculated ionization rate of Kr I to Kr IX as the function of laser intensity is presented for laser pulse width τ_p=105fs in Fig.1.

The establishment of electron energy distribution model is simplified as follows: 1) In the low electron density($<10^{19}cm^{-3}$), the time of laser pulse duration(usual tens of femtoseconds ~ 500 femtoseconds) is shorter than that of collision between electrons (>1ps)[11], the plasma heating caused by collision between electrons is neglected in the calculation of electron energy distribution; 2) The inverse bremsstrahlung absorption originated from the collision between electrons and ions is neglected; 3) The ionization rate can be calculated with ADK model because that depends little on the laser frequency in an altering field; 4) The atom in a field is usually ionized order by order.[12] First, suppose the highest order charge state is Z_{max} in the ionization, the relative ion numbers of charge state *j* (ion density) is $N_j(t)$ and the ionization rate from charge state *j-1* to charge state *j* is $W_j(E(t))$ the values of ionization rate can be got with the formula of quasistatic tunneling ionization. The change of every charge state relative population with time can be given with first order coupling differential equations in the case of atom ionization order by order.

$$\begin{cases} \dot{N}_0(t) = -W_1(E(t))N_0(t) \\ \cdots\cdots \\ \dot{N}_j(t) = W_j(E(t))N_{j-1}(t) - W_{j+1}(E(t))N_j(t) \\ \cdots\cdots \\ \dot{N}_{Z_{max}}(t) = W_{Z_{max}}(E(t))N_{Z_{max}-1}(t) \\ \sum_{j=0}^{Z_{max}} N_j(t) = 1 \end{cases} \tag{7}$$

Supposing that in the time of ionization *t* the relative electron number is n(t), according to formula(7), we can get the change expression of n(t) per unit time from the change rate of every charge state relative population with time.

$$\frac{dn(t)}{dt} = \sum_{j=1}^{Z_{max}} W_j(E(t))N_{j-1}(t) \tag{8}$$

Second, the expression of electron energy change per unit time. For circularly polarized laser, namely for polarization parameter $\alpha=1$, the electron quiver is

$$U_p(t) = \frac{e^2E_0^2(t)}{2m_e\omega^2} = \frac{e^2E_0^2 \sec h^2\left[\frac{1.76(t-t_{max})}{\tau_p}\right]}{2m_e\omega^2} \tag{9}$$

and the electron residual energy in a circularly polarized laser field equals to the quiver energy in the time of ionization, so the electron residual energy can also be described as fellows:

$$\varepsilon(t) = U_p(t) = \frac{e^2E_0^2(t)}{2m_e\omega^2} \tag{10}$$

From formula（10）we know that the electron residual energy generated at a given instant is determined only by the laser intensity or electric intensity of laser field in the time of ionization. The expression of electron change in per unit time from formula (10) is as follows：

$$\frac{d\varepsilon(t)}{dt} = \frac{3.52}{\tau_p}\varepsilon(t)\cdot th\left[\frac{1.76(t_{max}-t)}{\tau_p}\right] \tag{11}$$

Where th( ) is the hyperbolic secant function. From formula (8) and formula (11) we know that the change numbers and change energy of electrons per unit time can be determined by the time of ionization *t,* so the numbers of electrons per unit time can also be determined by the time of ionization *t*. From formula (10) we also know that the time of ionization *t* uniquely determines the electron residual energy here. So taking the time of ionization *t* as variable, we can calculate the curve of $dn(t)/d\varepsilon(t) \sim \varepsilon(t)$, namely the initial ionized electron energy distribution of function $f_e(\varepsilon)$. From formula (8) and formula (11) we can get the function of the electron numbers per unit energy or the ionized electron energy distribution as follows:

$$\frac{dn(t)}{d\varepsilon(t)} = \frac{dn(t)}{dt} \Big/ \frac{d\varepsilon(t)}{dt} = 0.284 \frac{\sum_{j=1}^{Z_{max}} W_j(E(t)) \cdot \tau_p \cdot N_{j-1}(t)}{\varepsilon(t) \cdot th\left[1.76(t_{max} - t)/\tau_p\right]} \quad (12)$$

The ionized electron energy distribution function can be obtained by use of formula (12) after calculation of the ionization rate and each order of charge state relative population

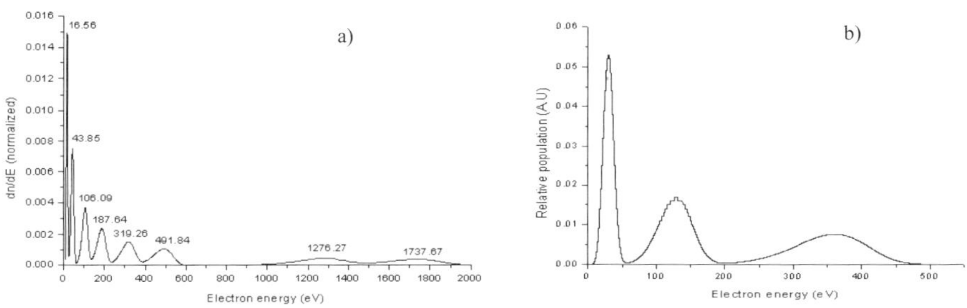

Fig.2 Calculated electron energy distribution in a) Kr IX, and b) NIII plasma with peak optical field intensity of 3.5×10^{16} W/cm^2 and 5×10^{15} W/cm^2 respectively

evolution vs time. Fig.2 gives the calculated electron energy distribution in Kr IX and NIII plasma with peak optical field intensity of 3.5×10^{16} W/cm^2 and 5×10^{15} W/cm^2 respectively (for τ_p=105fs、λ = 800 nm). As shown in Fig.2 the calculated electron energy distribution in Kr IX plasma, eight peaks exist and they correspond to the eight ionized electrons. The electron energy that corresponds to eight peaks are 16.56eV, 43.85eV, 106.09eV, 187.64eV, 319.26eV, 491.84eV, 1276.27eV and 1737.67eV respectively. In NIII plasma, The electron energy that corresponds to three peaks are 29eV、129eV、357eV respectively.

3. Calculation of laser parameters and experimental results

For the computation of peak gain efficiency of Ni-like Kr and B-like N based on OFI, the Doppler and natural broadening are considered. We used the former values of Kr IX electron energy that corresponds to eight peaks to get the spectra half width of Doppler broadening half width of $\Delta\lambda_d$. Substituting the values of $\Delta\lambda = 1.412\times10^{-4}$ nm into the formula of peak gain efficiency in references[1], the peak gain coefficients are 14cm^{-1} and 141cm^{-1} respectively when the ion densities are 1×10^{17}cm^{-3} and 3.2×10^{17}cm^{-3} respectively. With the same methods, we can get the peak gain coefficients of B-like N are 7.5cm^{-1}, 30 cm^{-1} when the ion densities are 3.24×10^{16}cm^{-3}, 6.48×10^{16}cm^{-3} respectively.

Ti:sapphire fs laser made in American Spectra-Physic laser Inc. is employed in this experiment. The output from this 10Hz laser system is 50 mJ /105fs, linearly polarized. An f_0=75mm focus lens is applied and the diameter of the focal spot is about 80 μm .

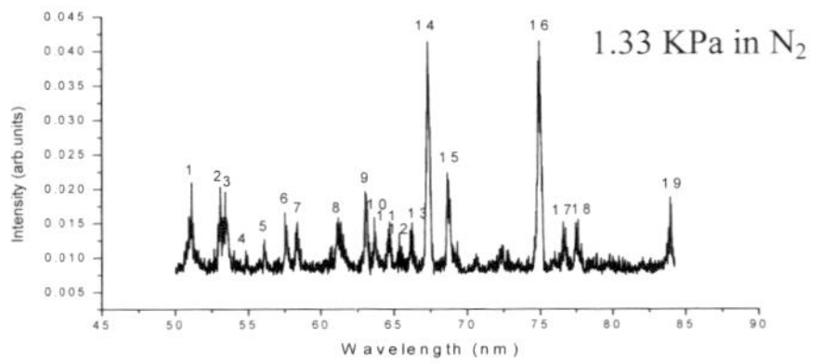

Fig.3 The soft X-ray spectra in N_2 driven by 105fs laser pulses with linearly-polarized laser energy of 34mJ

The focused laser beam bores an incidence pinhole and an out pinhole on the two lattens, which are 3mm away from each other in the target. Circularly-polarized laser beam can be got by using a 1/4 waveplate. The X ray laser from the target room is directed by an XUV spectrometer produced by American Acton Inc. and received by a detector. An R595 Electron Multiplier Tube whose central wavelength is about 70nm is manufactured in Japan. Finally the time integral spectrum of X-ray radiation is measured by the 4400 signal analysis and process system. We do the primary experiment with nitrogen by using linearly polarized laser. The incident pulse energy is about 34mJ and the pressure in the target chamber is about 8 torr. The experiment result is shown in Fig.3. By adding a $\lambda/4$ plate, we changed the polarization of the output laser from TSA, and got near circularly polarized laser beam. After focusing this laser beam and substitute the gas sample with krypton, 7th -13th high order harmonics are observed although the driving laser polarization was changed. No transition spectrum near 32nm was observed in this experiment.

4. Conclusion

We discuss some questions of the theoretical calculation of Ni-like Kr and B-like N in OFI. We think there is a small coefficient error in the formula of the ADK model, and calculate with the corrected formula of ADK model. We give a simplified model of the electron energy distribution and calculate the peak gain coefficients of Ni-like Kr and B-like N. In experiment we observe the fluorescence spectrum of OFI N_2 and soft x-ray emission spectrum that contains HHG in krypton driven by 0.6TW fs laser pulses.

Acknowledge:

This work is supported by the National “863” High Technology of China

Reference

[1]. Lemoff B E，Barty C P J et al 1994 Opt. Lett.19 569~571
[2]. Lemoff B E, Yin G Y et al 1995 Phys. Rev. Lett. 74 1574~1577
[3]. Hooker S M, and Harris S. E. 1995 Opt. Lett. 20: 1994~1996
[4]. Sebban S, Haroutunian R et al. 2001 Phys. Rev. Lett. 86: 3004~3007
[5]. Sebban. S, Mocek T et al 2001 Proc. SPIE 4505 195~2036.
[6]. Ammosov M V, Delone N B et al 1986 Sov. Phys. JETP. 64 1191~1194
[7]. Burnett N H and. Corkum P B 1989 J. Opt. Soc. Am. B 6 1195~1199
[8]. Penetrante B M and Bardsley J N 1991 Phys. Rev. A 43(6): 3100~3113
[9]. Djacui A and Offenberger A A 1994 Phys. Rev. E 50: 4961~4968
[10]. Augst S, Strickland D et al 1989 Phys. Rev. Lett. 63: 2212~2215
[11]. Murphy M, Glasheen C et al 1991 Phys. Rev. A 55 R2543~R2546
[12]. Corkum P B 1993 Phys. Rev. Lett. 71 1994~1997

Inst. Phys. Conf. Ser. No 186
Paper presented at 9th Int. Conf. X-Ray Lasers, Beijing, China, 24–28 May 2004

High-density gas puff target for laser-driven x-ray lasers

H Fiedorowicz, A Bartnik, R Jarocki, J Kostecki, R Rakowski, M Szczurek

Institute of Optoelectronics, Military University of Technology
Warsaw, Poland

Abstract. A new high-density gas puff target for experiments on laser-driven x-ray lasers is presented. The target is based on a double-stream gas puff target approach, which makes possible to form elongated gas sheet with steep density gradient and high density of gas in the interaction region. In the paper the valve system to produce the gas puff targets and the results of the target characterization measurements performed using x-ray backlighting technique are presented.

1. Introduction

The gas puff targets for experiments on laser-driven x-ray lasers are created by pulsed injection of gas through a slit nozzle using a high-pressure electromagnetic valve. In this way an elongated sheet of gas is formed which is suitable for the creation of an x-ray laser active medium (Fiedorowicz et al 1994). Usefulness of a gas puff target for investigations on laser-driven x-ray lasers has been confirmed in many experiments, both for perpendicular or longitudinal pumping (Fiedorowicz et al 1996, Morozov et al 1997, Ros et al 1998, Mocek et al 1998, Fiedorowicz et al 2001, Lu et al 2002). In the experiments various laser pulses, with time duration from nanoseconds to femtoseconds, were used for irradiation of the targets. The elongated gas puff target has been also successfully used for generation of bright low-divergence high-order harmonics (Lee et al 2002). In these experiments the gas puff targets were produced with the use of the gas puff valve systems developed and characterized at the Institute of Optoelectronics, Military University of Technology.

The gas puff targets offer a number of substantial differences and advantages when compared to solid targets. A fundamental difference between gas puff and solid target is that the gas puff provides the gain medium at a density closer to the lasing conditions. The laser-plasma coupling is also much different than with a solid, since the plasma starts below critical density. Moreover, use of a gas puff target instead of a solid target offers the advantage of developing a high-repetition rate x-ray laser with no target debris production. The gas puff target provides better control over the density and minimizes gradients in the plasma.

In the paper, a new gas puff target for experiments on laser-driven x-ray lasers is presented. The target is based on a double-stream gas puff target approach (Fiedorowicz et al 2000). It is formed by pulsed injection of working gas through the central nozzle in a form of a slit, which is surrounded with an additional outer nozzle. The outer nozzle is supplied with light gas. The stream of working gas injected through the central nozzle is confined by the outer stream of light gas, thus the central stream poses high density of gas at relatively large distance from the nozzle output. It allows efficient absorption of laser radiation in the plasma. Moreover, the central stream has a steep density gradient, which is advantegous for x-ray laser experiments.

2. Solenoid valve system for the gas puff target formation

The valve system to produce elongated double-stream gas puff target for x-ray laser experiments has been developed. Schematic of the valve system is presented in Fig. 1.

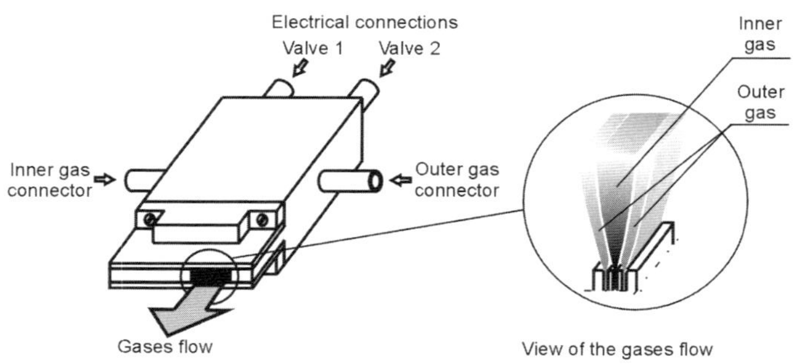

Figure 1. Schematic of the valve system to form an elongated double-stream gas puff targets for x-ray laser experiments.

The valve system is composed of two separate electromagnetic valves combined in a common body. The system is equipped with the nozzle setup that is composed of the central slit nozzle and the surrounding nozzle. The central nozzle is 0.25 mm wide and up to 9 mm long. The width of the outer nozzle is 0.25 mm. Each nozzle is supplied with different gas separately from the valves. The valves are driven by the electric pulses synchronously with the x-ray laser-driving pulse. The gas density spatial profiles of the target can be controlled by changing the gas backing pressure and the time delays between the laser pulse and the opening of the valve.

3. Characterization measurements with x-ray backlighting technique

The produced elongated double-stream gas puff targets have been characterized with the use of x-ray backlighting technique. A laser-plasma x-ray source based on a krypton gas puff target irradiated with a Nd:YAG laser generating 4 ns and 0.8 J laser pulses was used as a backlighter. X-ray radiation at the wavelength of about 0.7 nm was selected by 20 μm Si filter combined with 10 μm Be filter. The x-ray shadowgrams of the gas puff targets were registered with the use of a CCD camera (Reflex s.r.o.).

The typical x-ray shadowgram of the elongated xenon-helium double-stream gas puff target is shown in Fig. 2 b. For comparison the shadowgram of the ordinary xenon gas puff target (without injection of helium) is presented in Fig. 2 a. The confinement of working gas (xenon) injected from the central nozzle by the outer gas (helium) is clearly seen.

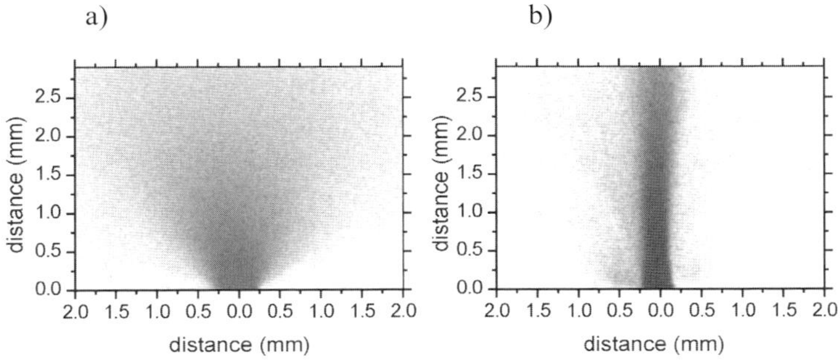

Figure 2. Typical x-ray shadowgram of the ordinary xenon gas puff target (a) and the elongated xenon/helium double-stream gas puff target (b) for comparison.

The corresponding target characteristics (gas density spatial profiles) for the xenon and the xenon/helium targets are presented in Fig. 3.

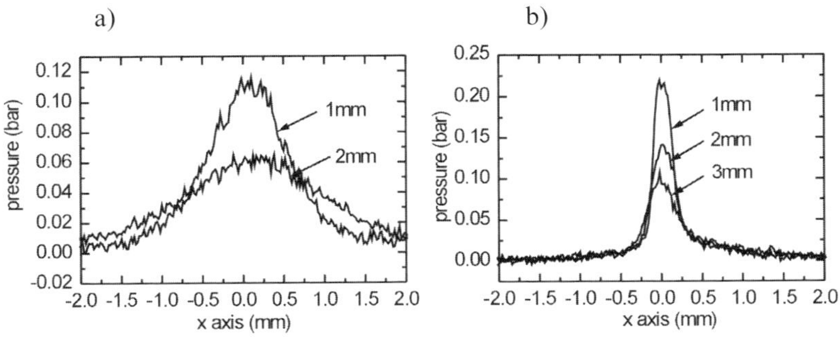

Figure 3. The target characteristics (gas density spatial profiles) for the ordinary xenon gas puff target (a) and the double-stream xenon/helium gas puff target (b).

4. Conclusions

The proposed target can be useful for various laser-driven x-ray laser experiments, including the collisionally excited and recombination, both side- and longitudinally-pumped x-ray lasers. It can be also used in the experiment on high harmonics generation. The collaborative experiments on these subjects are planned.

5. Acknowledgements

This work was supported by the Ministry od Science and Information Technologies of Poland under grant No 3 T11B 031 26.

We would like to thank J. Firak for manufacturing the 20 μm Si filters used in the experiments for selection of x-ray radiation.

References

Fiedorowicz H, Bartnik A, Gac K, Parys P, Szczurek M, Tyl J 1994 Proceedings of the 4th International Colloquium on X-ray Lasers, Williamsburg, VA, 1994 (AIP Press, Washington, 1995) p 538

Fiedorowicz H, Bartnik A, Li Y, Lu P, Fill E 1996 Phys. Rev. Lett. 76 415

Fiedorowicz H, Bartnik A, Jarocki R, Rakowski R, Szczurek M 2000 Appl. Phys. B 70 305

Fiedorowicz H, Bartnik A, Dunn J, Smith R, Hunter J, Nilsen J, Osterheld A, Shlyaptsev V N 2001 Opt. Lett. 24 101

Lee D G, Kim H T, Hong K H, Nam C H, Choi I W, Bartnik A, Fiedorowicz H 2002 Appl. Phys. Lett. 81 3726

Lu P, Kawachi T, Kishimoto M, Sukegawa K, Tanaka M, Hasegawa N, Suzuki M, Tai R, Kado M, Nagashima K, Daido H, Kato Y, Bartnik A, Fiedorowicz H 2002 Opt. Lett. 27 1911

Mocek T, Shin H J, Cha Y H, Lee D G, Hong K H, Nam C H, Fiedorowicz H, Bartnik A, Szczurek M 1998 Proceedings of the 6th International X-ray Lasers Conference, Kyoto, Japan, 1998 (IOP Publishing, Bristol, 1999) p 329

Morozov A, Ping Y, Suckewer S, Bartnik A, Fiedorowicz H 1997 Proceedings of SPIE vol. 3156 Soft X-ray Lasers and Applications II (SPIE Press, Bellingham, 1997)

Ros D, Fiedorowicz H, Rus B, Bartnik A, Szczurek M, Jamelot G, Albert F, Carillon A, Jaegle P, Klisnick A, Sebban S, Zeitoun P 1998 Optics Communications 153 368

High-order harmonics XUV radiation

Inst. Phys. Conf. Ser. No 186
Paper presented at 9th Int. Conf. X-Ray Lasers, Beijing, China, 24–28 May 2004

Spatio-temporal optimization of high harmonic generation in a long gas jet

C H Nam[1], H T Kim[1], I J Kim[1], C M Kim[1], V Tosa[1,2], and H Fiedorowicz[3]
[1]Dept. of Phys., CXRC, KAIST, Korea
[2]NIRDIMT, Rumania
[3]Inst. of Optoelect., Mil. Univ. of Technol., Poland

Abstract. The optimization of high harmonic generation in space and time domains was performed. The characteristics of intense femtosecond laser pulse propagating though a long neon gas jet were analyzed both in theory and in experiments. Bright harmonics were generated around 13-nm wavelength with low beam divergence and narrow bandwidth by applying self-guided and chirped pulses to the long neon gas jet.

1. Introduction

High-order harmonic generation is a noble method to obtain an extreme ultraviolet (EUV)/ soft x-ray source with unique features such as excellent coherence [1], ultrashort pulse duration [2], continuous wavelength tuning capability [3] and tabletop scale. The generation of bright high-order harmonics is essential to practical applications of such harmonic sources to interferometry and ultrafast spectroscopy. When an intense femtosecond laser pulse impinges a high-density long gas medium to generate bright harmonics, the ionization of gas medium becomes so severe to disturb harmonic generation. The ionizing gas medium causes plasma defocusing induced by higher electron density in the center than in the outer region, which results in the reduction of effective harmonic generation volume [4]. In addition, the temporal characteristics of the laser pulse are modified by the self-phase modulation (SPM) in a rapidly ionizing medium, which can cause spectral broadening of generated harmonics [5]. Hence, the effects of ionization should be controlled both in space and time domains in order to maximize the harmonic generation volume and to generate spectrally sharp harmonics.

In this proceeding, we report the spatio-temporal optimization of high harmonic generation achieved by controlling the ionization effects of the laser pulse [6]. A long neon gas jet was selected for bright harmonic generation due to its simple target alignment and the capability of coupling a large laser energy into a large interaction volume [7]. To utilize this long and dense medium for bright harmonic generation, the ionization effects on the laser propagation should be understood and controlled in both the space and time domains. We have analyzed, both experimentally and theoretically, the propagation characteristics of laser pulses in an ionizing medium and obtained the conditions for self-guided propagation and laser chirp control. With self-guided and chirped laser pulses, the generation of harmonics was optimized to have high brightness, low beam divergence, and narrow spectral bandwidth. At the optimized condition, the 61st harmonic at 13.4 nm from a 9-mm long neon gas jet had the brightness over $1x10^{15}$ $W/cm^2/srad$ with divergence of 0.5 mrad and bandwidth of 0.07 nm.

2. Laser propagation in an ionizing long gas medium

Intense femtosecond laser pulses propagating through an ionizing gas medium is affected by the spatio-temporal modification of the refractive index. In the space domain, the plasma defocusing, caused by the high electron concentration in the central part of the medium, makes the laser beam quickly diverge and shortens the effective harmonic generation length. Thereby, the plasma defocusing significantly reduces the harmonic conversion efficiency. In the time domain, SPM can be induced by the rapid change of refractive index during ionization. The harmonic spectrum becomes broad and complex due to the SPM-induced chirp of the laser pulse. Therefore, these two effects of ionization in space and time must be properly understood and controlled for bright harmonic generation with narrow bandwidth.

To investigate the laser pulse propagation in an ionizing medium, a three-dimensional (3D) simulation of laser pulse propagation was performed. The refractive index and absorption of neutral atoms, optical Kerr effect, and the change of refractive index with electron density were considered [8]. The spatial and temporal changes of the laser profile during propagation were calculated for the case of a negatively chirped 42-fs laser pulse with 5-mJ energy. The medium was a 9-mm Ne gas jet with a density of $1.4x10^{18}$ cm^{-3}. The reason for using a negatively chirped pulse will be discussed later in detail. Truncated Gaussian pulses, which were focused with a power of f/85, were used in the calculation to closely follow actual experimental conditions. The gas jet position is defined as z=0 when the center of the gas jet is positioned at the laser focus.

The spatial changes of laser beam in ionizing medium were investigated in the 3D simulation. Figure 1 shows the evolution of laser beam profiles at the time of pulse peak along the laser propagation at the two gas jet positions of z=0 and z=-18 mm. The negative sign means that the gas jet is located before the laser focus. At z=0 (Fig. 1(a)), the laser beam spreads quickly due to plasma defocusing. After the propagation, the laser intensity decreases to $3x10^{14}$ W/cm^2, though the initial intensity is more than $1x10^{15}$ W/cm^2. However, when the gas medium is placed 18 mm before the laser focus, a flattened laser profile is formed along the propagation and the laser beam is guided in the flattened region as shown in Fig. 1(b).

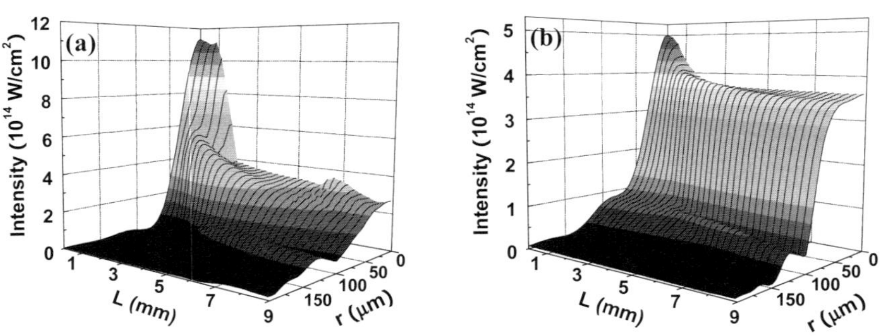

Figure 1. Evolution of the laser beam profile in an ionizing medium calculated by 3D simulation. L is the propagation distance measured from the entrance plane of a 9-mm gas jet. The gas jet positions are (a) z= 0 and (b) z= -18mm.

The profile flattening and guiding of the laser pulse, when the gas jet is placed before the laser focus, arises due to the balance between plasma defocusing and the convergence of the focused laser beam. When a gas medium is positioned before the laser focus, the central part of laser beam is refracted outwards due to higher electron concentration at the center, while the less-affected outer beam still converges. This behaviour of laser pulses in ionizing medium redistributes the laser beam so as to form a flattened profile. After formation of the flattened laser profile, the electron density profile also has flattened profile with the same radius but rapidly drops at the boundary. This makes a flattened refractive index profile with a sharp boundary, which act as a wave-guide where the laser beam is self-guided.

The self-guiding and profile flattening was experimentally observed by capturing visible plasma images along the propagation direction and radial profiles of the laser beam at the exit of the medium using charge-coupled device (CCD) detectors. The experiments were performed using chirped femtosecond laser pulses from a chirped pulse amplification (CPA) Ti:sapphire laser operating at 10-Hz repetition rate [9]. The visible plasma images were captured from a transverse direction and the laser beam profile at the end of medium was obtained by imaging it on a CCD. The experiments performed with same conditions as those for the 3D calculations. The visible plasma image changed drastically with gas jet positions. At the gas jet position of z=0 (Fig. 2(a)), the plasma image was bright only for the first part of the medium. As the gas jet position moved away from the laser focus, the bright part of plasma image was extended and a nearly uniform plasma image over the entire gas jet was formed at z=-18 mm. These are well matched with the 3D calculation results in Fig. 1, which shows that laser intensity rapidly decreased for z=0 and self-guiding occurs for z=-18 mm. The laser beam profile at the end of the medium provides a direct evidence of the profile flattening as shown in Fig. 1(b) and (c). At z=0 (Fig. 2(b)), the laser profile was severely distorted and weakened after the propagation. On the other hand, at z=-18 mm (Fig. 2 (c)), the laser beam formed a flattened profile with a radius of 60 μm, which closely matched the 3D calculation. These results clearly shows that a proper selection of the gas jet position is critical condition to form the profile flattening and self-guiding of laser pulses.

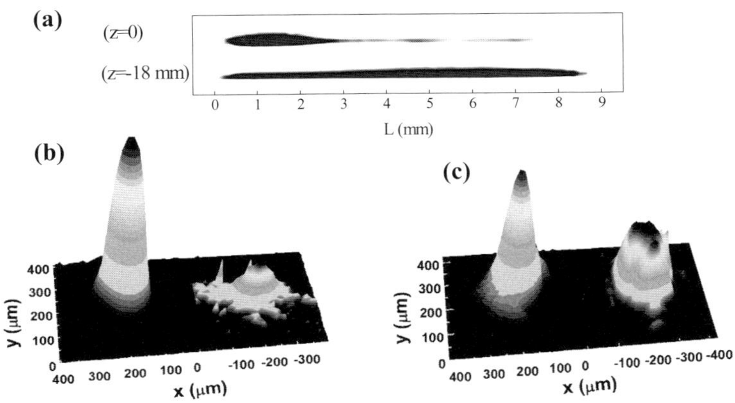

Figure 2. (a) Visible images of Ne plasma obtained at two gas jet positions z=0 and -18 mm. Laser beam profiles at the entrance (left) and exit (right) of the gas jet for the cases of (b) z=0 and (c) z=-18 mm.

We then analyzed the chirp structures of laser pulses in ionizing medium by a time-frequency analysis using the spectrogram methods [10]. A time-frequency distribution function can be a powerful tool to analyze chirp structures, which cannot be revealed by a Fourier analysis [11]. We examined the spectrograms of the laser pulse during the propagation though an ionizing medium. The time evolution of laser field was obtained by 3D simulation of laser pulse propagation with same parameters as used in Fig. 1(b). Figure 3 shows the change of laser chirp structure during the pulse propagation. The dotted lines indicate the mean frequency at a given time. It clearly shows the temporal modification of laser chirp structure caused by SPM. After 1-mm propagation, the initial negative chirp was modified by the SPM and laser pulses became nearly chirp-free in the leading edge. This positive chirp in the leading edge of the laser pulse increased with the propagation length and it overcame the initial negative chirp after 4-mm propagation.

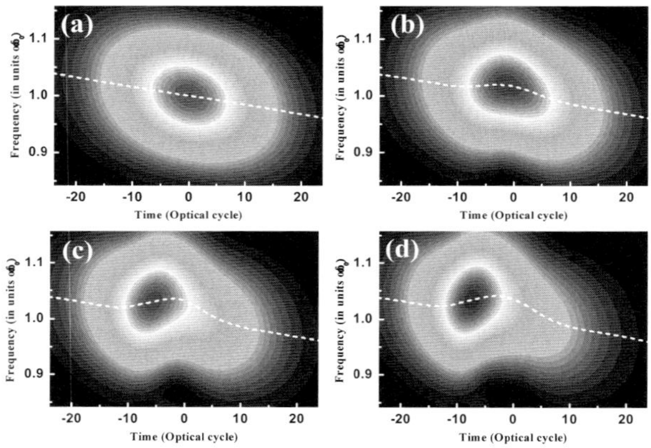

Figure 3. Spectrograms of a propagating laser pulse through a 9-mm Ne medium with propagation length of (a) L = 0, (b) L = 1 mm, (c) L = 4 mm, and (d) L = 9 mm

We used the negatively chirped pulses for reducing SPM effect on high harmonic generation. In the case of chirp-free or positively chirped laser pulse, SPM-induced positive chirp in the leading edge of the laser pulse can be severe enough to broaden the generated harmonic spectrum as the phase-matched harmonics are mainly generated in the leading edge. When we use negatively chirped pulses, the induced positive chirp in the leading edge can be reduced and matched with dynamically induced negative chirp of harmonics, which results in generating sharp and bright harmonics. This chirp control can be achieved while maintaining, at the same time, the self-guiding conditions. Thus we can control the ionization effects on high-order harmonic generation both in space and in time.

3. Spatio-temporal optimization of high harmonic generation

To optimize the harmonic generation in space and time domains, we investigated harmonic spectra from a long neon gas jet at various gas jet positions and different laser chirps in the wavelength region of 10.0 – 18.0 nm. The generated harmonics were observed with a flat-field extreme ultraviolet (EUV) spectrometer equipped with a back-illumination X-ray CCD (Princeton Instruments) [12]. Two Zr filters with a total thickness of 400 nm were installed in front of the CCD to block scattered laser light.

Laser energy and gas medium parameter were the same as in Fig. 1. The harmonic spectra were calibrated by accounting for the transmittance of spectrometer and Zr filter.

Figure 4(a) shows harmonic spectra from a long neon gas jet driven by negatively chirped 42-fs laser pulses at various gas jet positions. The harmonic signal was drastically increased when the gas jet was positioned before the laser focus. The optimized harmonics were obtained at the gas jet position z=-18 mm, at which profile-flattening and self-guiding occurred as shown in Fig. 2. The 61st harmonic, at 13.4 nm, was the strongest harmonic at this position due to harmonic cutoff in higher orders and gas absorption in lower orders. At the optimum condition, the 61st harmonic obtained with 5-mJ laser pulse was strong enough to saturate the x-ray CCD in a single shot.

The self-guided and profile-flattened laser pulse can provide a good phase matching between generated harmonics and a driving laser pulse. Since the profile flattening and self-guiding of the laser pulse makes a nearly uniform intensity and phase distribution in the radial and axial directions, the phase matching is mainly determined by the dispersion due to neutral atoms and free electrons. The 61st harmonic satisfies the phase matching condition at the ionization level of about 1 %. In the 3D calculation, 1 % ionization level occurs at the leading edge of the pulse. The profile flattened and self-guided laser pulses provide uniform phase matching conditions over a large cross section by overcoming the problems associated with the variation of phase-matching conditions along the radial direction. Therefore, the self-guiding and profile flattening provided an ideal condition for well phase-matched strong harmonic generation in a large volume.

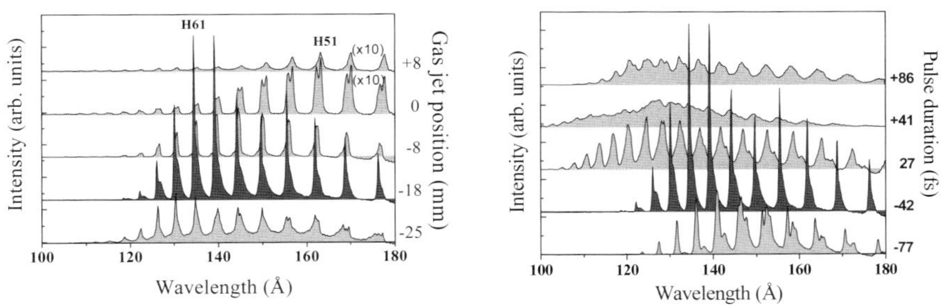

Figure 4. (a) Harmonic spectra at different gas jet positions. First two spectra are multiplied by 10. *'Hn'* denotes *n*th order harmonic. (b) Harmonic spectra with different laser chirp at the gas jet position of z=-18 mm. The sign of the pulse duration refers to the sign of applied laser chirp.

The beam divergence of the 61st harmonic was measured by replacing the toroidal mirror in the EUV spectrometer with a cylindrical mirror. The measured beam divergence of 61st harmonic was only 0.5 mrad at the gas jet position of z=-18 mm. We also measured the energy of harmonics by using an EUV silicon photodiode (International Radiation Detectors AXUV-100) in place of the x-ray CCD. The energy of the 61st harmonic at optimum conditions was estimated to be 0.4 nJ by assuming nominal values for the filter transmission, the diffraction grating efficiency, reflectivity of the toroidal mirror, and quantum efficiency of the EUV photodiode. This is a very conservative value when we consider the oxidation and contamination of optical components and detector with time. The harmonic brightness, defined as the power of a harmonic per unit area per unit solid angle subtended at the source, was estimated to be about 1×10^{15} W/cm^2/srad (also a conservative value). Thus, we have accomplished bright high-order harmonic generation by establishing the profile-flattened and self-guided laser pulses in the long gas jet medium.

We then performed a further analysis on the spectral structure of high-order harmonics by controlling the initial laser chirp [13]. It is well known that high-order harmonics can have dynamically induced negative chirp [14], which originates from the temporal variation of the harmonic phase that is dependent on laser intensity. As discussed in Sec. 2, the SPM from ionized medium induces a positive laser chirp at the leading edge of the driving pulse [5], which is also the time duration when high harmonics are mainly generated. This positive chirp can be reduced by using a negatively chirped pulse and matched with dynamically induced harmonic chirp. This is substantiated by the strong dependence of harmonic spectrum on the laser chirp as shown in Fig. 4 (b). In the case of a positively chirped 41-fs pulse a quasi-continuous harmonic spectrum was obtained while the negatively chirped 42-fs pulse generated the sharpest and brightest harmonics. With negatively chirp 42-fs laser pulses, the bandwidth of the 61st harmonic was measured to be 0.07 nm. The spectral brightness was enhanced by a factor of four over that obtained from a chirp-free 27-fs pulse. Since the laser propagation does not depend sensitively on the laser chirp near the chirp-free region, the generation efficiency was not significantly changed around the chirp-free, while the spectral brightness of the harmonics was very sensitive to the laser chirp. Therefore, the chirp control technique can be applied to generate sharp and bright harmonics without much disturbing the self-guided propagation.

Since the high-order harmonic generation can be simultaneously controlled by adjusting the gas jet position and laser chirp, an optimized condition of harmonic generation can be found by varying the gas jet position and laser chirp [15]. Figure 5 shows the spectral brightness of the 61st harmonic with respect to gas jet position and laser chirp. It clearly shows that the maximum spectral brightness of the 61st harmonic occurs at the gas jet position of about z=-18 mm and for a negatively chirped 42-fs laser pulse. Thus, the spatio-temporal optimization of harmonic generation was achieved by appropriately selecting the gas jet position and laser chirp.

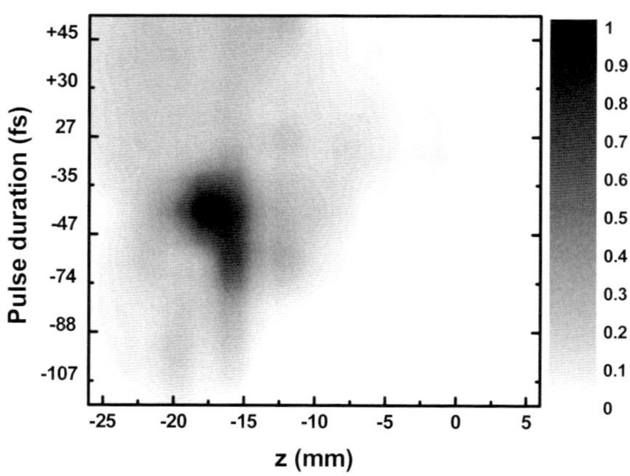

Figure 5. Spectral brightness of 61st harmonic with respect to gas jet position and laser chirp.

4. Conclusion

We have presented methods to optimize harmonics from a long gas jet both in the space and time domains by using self-guided and chirped laser pulses. We investigated and controlled the characteristics of intense femtosecond laser pulses propagating in the ionizing medium. In the space domain, the plasma defocusing shortens the effective harmonic generation volume, which causes significant reduction of harmonic efficiency. We found a method to overcome the effect of plasma defocusing on harmonic generation by using profile-flattening and self-guiding of laser pulses, which can be achieved by adjusting the gas jet position. In the time domain, SPM-induced positive chirp of laser pulse at the leading edge can broaden generated harmonic spectrum. We obtained sharp and bright harmonics by using negatively chirped laser pulses for reducing SPM effects. By the combination of self-guiding and laser chirp control, the harmonic generation was optimized in space and time. At the optimized condition, bright 61st harmonic at 13.4 nm was produced from a 9-mm Ne gas jet with low beam divergence of 0.5 mrad and narrow spectral bandwidth of 0.07 nm. Since high-reflectivity Mo:Si mirrors are available at 13.4 nm, this method can be applied for the harmonic source development for EUV metrology.

Acknowledgment

This work was supported by the Ministry of Science and Technology of Korea through the Creative Research Initiative Program.

References

[1] Lee D G, Park J J, Sung J H, and Nam C H 2003 Opt. Lett. **28** 480-482.
[2] Kienberger R, Goulielmakis E, et.al. 2004 Nature **427** 817-821.
[3] Kim H T, Lee D G, et.al. 2003 Phys. Rev. A **67**, 051801(R);
Shin H J, Lee D G, et.al. 1999 Phys. Rev. Lett **83** 2544-2547.
[4] Bellini M, Corsi C, and Gambino M C 2001 Phys. Rev. A **64** 023411.
[5] Kim J –H and Nam C H 2002 Phys. Rev. A **65** 033801.
[6] Kim H T, Kim I J, et.al. 2004 Phys. Rev. A **69** 031805(R).
[7] Lee D G, Kim H T, et.al. Appl. Phys. Lett. **81** 3726-3728.
[8] Tosa V, Takahashi E, et.al. 2003 Phys. Rev. A **67** 063817.
[9] Cha Y H, Kang Y I, and Nam C H 1999 J. Opt. Soc. Am. B **16** 1220-1223.
[10] Cohen L 1989 Proc. IEEE **77** 941-981.
[11] Kim J –H, Lee D G, et.al. 2001 Phys. Rev. A **63** 063403.
[12] Choi I W, Lee J U, and Nam C H 1997 Appl. Optics **36** 14570-1466.
[13] Lee D G, Kim J –H, et.al. 2001 Phys. Rev. Lett. **87** 243902;
Kim H T, Kim I J, et.al. 2004 J. Phys. B: At. Mol. Opt. Phys. **37** 1141-1152.
[14] Chang Z, Rundquist A, et.al. 1998 Phys. Rev. A **58** 30-33(R).
[15] Kim H T, Kim I J, et.al. 2004 Appl. Phys. B **78** 863-867.

Inst. Phys. Conf. Ser. No 186
Paper presented at 9th Int. Conf. X-Ray Lasers, Beijing, China, 24–28 May 2004

High-intensity high-order harmonics and multiphoton process in the soft X-ray region

Katsumi Midorikawa, Yasuo Nabekawa, Eiji J. Takahashi, Hirokazu Hasgawa, Hiroki Mashiko, and Akira Suda
Laser Technology Laboratory, RIKEN, Hirosawa 2-1, Wako, Saitama 351-0198, Japan

Abstract. We report our efforts on the observation of multiphoton process in the XUV region, including high intensity high harmonic generation and its application technology such as beam splitters and multilayer mirrors.

1. Introduction

High-order harmonic generation using femtosecond high-power lasers is very attractive as an intense coherent short wavelength source in the XUV region [1]. Its high peak power is promising to induce nonlinear optical phenomena. Multiphoton ionization of rare-gas atoms in the VUV region has been reported already using high harmonics [2, 3]. However, observation of mutliphoton and/or nonlinear processes in the soft x-ray region (< 30 nm) is still a challenging issue.

We have already reported the investigation of the energy scaling of high-order harmonics under the phase-matched condition using a long interaction length and a loosely focused pumping geometry [4]. In those experiments, we have generated peak powers of 130 MW at 62.3 nm in Xe [5] and 10 MW at 29.6 nm in Ar [4], assuming the same pulse width as the pump pulse of 35 fs. Recently, we have applied this harmonic energy scaling procedure for Ne and successfully generated a peak power of 1 MW at 13 nm [6].

Although we have generated intense harmonics by the energy scaling procedure, it caused severe problem for the application of the harmonics. Because the high power fundamental beam coaxially propagating with the harmonics degrades or destroys XUV optics or a sample. In order to avoid this problem, we have developed high-throughput, high-damage-threshold beam splitters for separation of the harmonics from the fundamental pulse [7]. Combination of this beam splitter and a multilyer mirror allows us to produce high intensity pulses in the XUV region. When we focused the 27th harmonic pulses with appropriate multilayer mirrors, the focused intensity attained to 10^{13}-10^{14} W/cm^2 [8]. This high intensity is expected to give rise to nonlinear optical phenomena in the XUV region [9].

In this paper, we report our efforts on the observation of multiphoton process in the soft x-ray region including high intensity high harmonic generation and its application technology such as beam splitters and multilayer mirrors. Following to high intensity soft x-ray pulse generation, we will discuss its application to observation of mulitphoton ionization processes of helium in the soft x-ray region

2. High-throughput, high-damage-threshold broadband beam splitter for high-order harmonics

The spatial separation of the coaxial propagating high energy pumping beam from the harmonics is a serious problem for applications of high harmonics (HH). We have developed a high-throughput, high damage-threshold beam splitter (BS) for HH using pump pulse's Brewster angle Si, SiC and Mo plates.

To generate the high-order harmonic (HH), we used Ar and Ne for target gas. Details of our HHG experimental setup have been presented in a previous publication [4, 5]. The generated harmonics illuminated a slit of the spectrometer, which was placed at 3.5 m from the exit pinhole of the interaction cell. We used two BSs arranged in parallel because of easy alignment. To perform the correct measurement of the BS efficiency, we first measured the HH spectrum without BSs. Secondly, we inserted the two BSs in front of the spectrometer, and the spectrometer was moved perpendicularly to the optical axis. Thus, we can compare the HH spectra with and without BSs. Figure 1 shows the typical HH spectra in Ar obtained by 500-shot accumulation with and without Si BS. The incident angle θ_B was set to 75 degrees for Si and 69 degrees for Sic, which correspond to Brewster angle θ_B of the 800 nm pump pulse. At these θ_Bs, the attenuation (reflection) rate of the pump pulse was measured to be 10^{-4} for Si and 6×10^{-5} for SiC per plate. As is shown in Fig.1, the HH spectrum was not modulated on passing through the two Si BSs. The spatial profile of the 27th harmonic after the BS showed a perfect beam quality (see insets of Fig.1). The power density of the 30 fs pump pulse was estimated to be 0.8 TW/cm^2 (average power density 0.25 W/cm^2) at the BS surface. No damage was observed on optical surfaces of Si and SiC after several hours of irradiation at this power density level.

By comparing the HH spectra obtained with and without BSs, we can obtain the reflection efficiency of the BS. Figure 2 shows the measured BS reflection efficiency for Si, SiC and Mo. For the measurement of the Mo BS, we used Ne gas for generation of

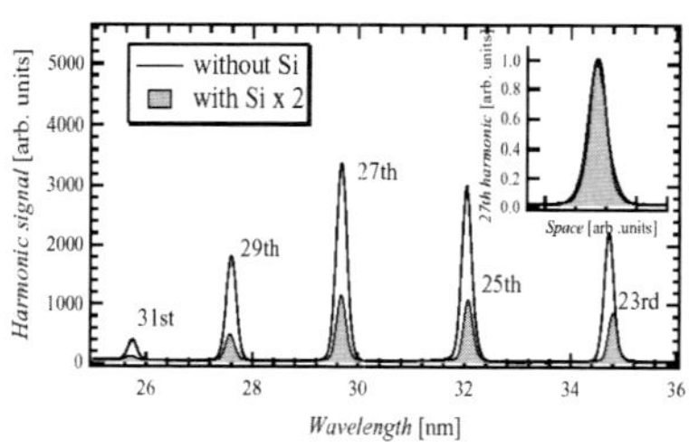

Figure 1. The spectra with and without Si beam splitters arranged in parallel for θ_B =75 degrees. The insets show normalized far-field one-dimensional spatial profiles of the 27th harmonic (29.6 nm) obtained with and without BSs.

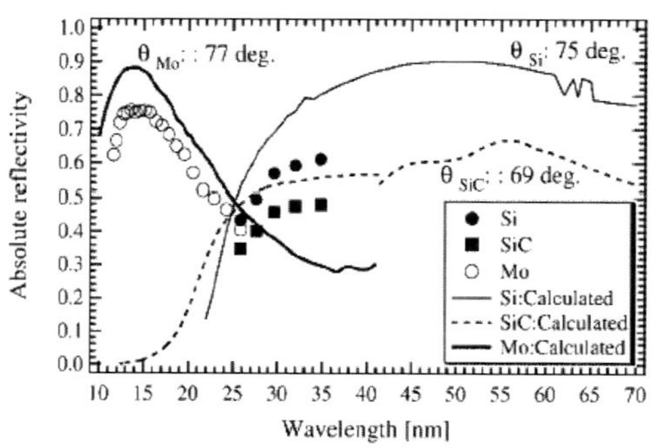

Figure 2. Si, SiC and Mo reflection efficiencies at the pump pulse's Brewster angle. Squares and circles and triangles correspond to the experimentally obtained reflection efficiencies.

HH around 10-25 nm. The wavelength dependence of the measured efficiency was in good agreement with the calculated reflection efficiency.

By introducing these BS, other optics such as gratings, thin metal foils, and multilayer x-ray mirrors can be freely used without being damaged by the high-energy pumping pulse. This device is very simple and highly useful for not only HHG, but also longitudinally pumped x-ray lasers.

3. Focusing of high-harmonics

Although the advent of HH havng excellent beam quality and spatial coherence, together with high output energy, opens up the possibility for focusing XUV beams to micron-order spot size with intensities of greater than 10^{14} W/cm^2, measured intensities are currently lower than 10^{12} W/cm^2 [10]. Here, we report a technique for focusing an intense coherent soft x-ray beam to a 1 μm spot size with a peak intensity as high as 10^{14} W/cm^2.

The experiment was carried out using a Ti:sapphire laser chirped-pulse amplification system. This system produces an output of 14 mJ, with a full-width at half maximum (FWHM) pulsewidth of 22 fs at a repetition rate of 10 Hz and a center wavelength of 785 nm. The output pulse was loosely focused into an interaction cell filled with 12-mbar Ar to stimulate harmonic generation by a plano-convex lens with a focal length of 4000 mm. The $1/e^2$ diameter at the focus was 260 μm, leading to an intensity of 1.2×10^{15} W/cm^2. The interaction cell has 1-mm-diameter pinholes in both ends (which are separated by 12 cm) in order to isolate the vacuum and gas-filled sections. The generated harmonics were sent to a Si beam splitter for the separation of the harmonic wave from the fundamental wave. A 200-nm-thick Al filter, which could be moved into the beam path, was used to block the residual fundamental beam and to reduce the intensity of the high-order harmonics. The transmittance was 14% at 29.6 nm (the 27th harmonic wave). The harmonic spectrum was monitored using a grazing-incidence spectrometer with a micro channel plate (MCP). A CCD camera detected the two-dimensional images of the spectrally-resolved far-field profiles of the harmonics on a phosphor screen placed behind the MCP. The 27th harmonic beam has a nearly Gaussian profile. The absolute energy of the high-order harmonics was measured by positioning a calibrated silicon XUV photodiode (IRD Inc. XUV-100) in front of the spectrometer. Typical measured energies for the 27th harmonic wave with and without an Al filter were 23 nJ and 160 nJ respectively. At the exit of the interaction cell, the generated energy was estimated to be 320 nJ with a conversion efficiency of 2.3×10^{-5}.

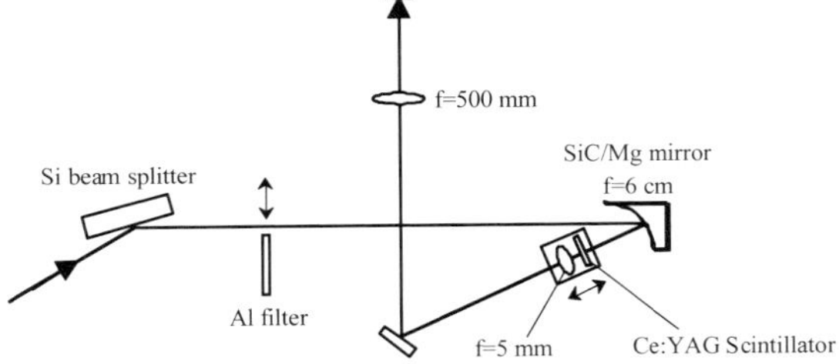

Figure 3. Experimental setup for imaging the focused profiles of high-order harmonics.

We used an imaging system as shown in Fig. 3 to measure the focal spot of the soft x-ray beams. The 27th harmonic wave was selected and focused by an off-axis parabolic mirror (f=6 cm, θ=24°) with a SiC/Mg multilayer coating. The peak-to-valley surface accuracy of the mirror substrate over an area within a 1-inch diameter was $\lambda/8$ at 633 nm corresponding to 2.7λ at 29.6 nm. The RMS surface roughness was 0.7 nm, which is good enough to support the quality of the multilayer coating. The reflectivity of the multilayer coated mirror was 40% at 29.6 nm with a FWHM bandwidth of 2 nm. We estimated that the diameter of the beam on the mirror surface was 2 mm from the MCP spectrograph. In order to visualize the focal spot profile, the 27th harmonic wave was converted to visible light at 550 nm using a 100-μm-thick Ce:YAG scintillator [10, 11] placed at the focal point. A set of image relay optics were constructed with an objective lens (NA=0.65, f=5 mm) and a convex lens (f=500 mm), which transferred the visible light image onto an image-intensified CCD camera (42 dB dynamic range) installed outside the vacuum chamber. The scintillator and the objective lens were fixed on a single stage that could be translated along the beam axis to observe the variation of the beam profiles. The resolution of the imaging system was 0.5 μm, which was limited by the NA value of the objective lens. The magnification factor was 80.

Figure 4 shows the $1/e^2$ beam spot size of the 27th harmonic beam as a function of distance from the off-axis parabolic mirror. The open circles and the filled squares correspond to the horizontal and vertical spot sizes, respectively. There is little difference between the horizontal and vertical diameters, showing that circularity is maintained during focusing. The solid line is the best-fit curve based on the equation, $\omega(z) = \omega_0[(1+z/R)^2+(M^2\lambda z/\pi\omega_0^2)^2]^{1/2}$, where ω_0 and $\omega(z)$ are the beam radii on the parabolic mirror and at a distance z from the mirror, respectively, R is the wavefront curvature of the soft x-ray beam just after reflection, λ is the wavelength and M^2 is the beam quality parameter. Consequently, the best-fit curve provided a figure of M^2=1.4.

Figure 5 shows a typical focused image taken in a single shot. The dimensions of the spot in the vertical and horizontal directions are 1.0 μm and 0.9 μm respectively, both of which are less than 1.7 times the diffraction-limited value. Because the high-order harmonics have excellent spatial coherence and beam quality, the focusability might be dominated by the performance of the mirror.

We have also measured the spot size from an ablation pattern produced on a gold-coated mirror placed at the focal position. The patter was prouced by single shot irradiation of the focused 27th harmonic pulse. The diameter of the circular hole is

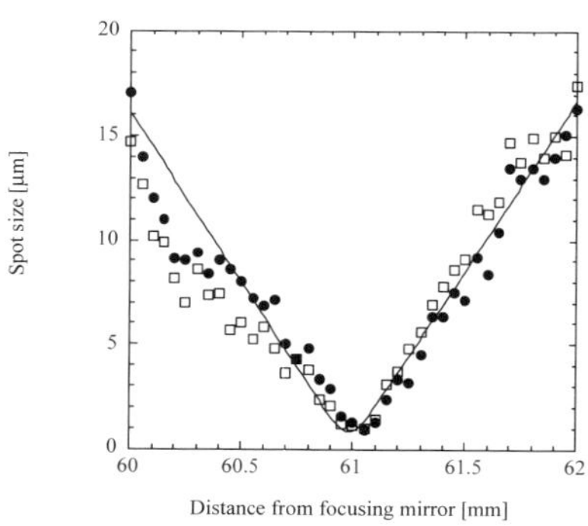

Figure 4. Focused beam spot size of the 27th harmonic wave as a function of distance from the focusing mirror. The open circles and filled squares correspond to the horizontal and vertical directions respectively.

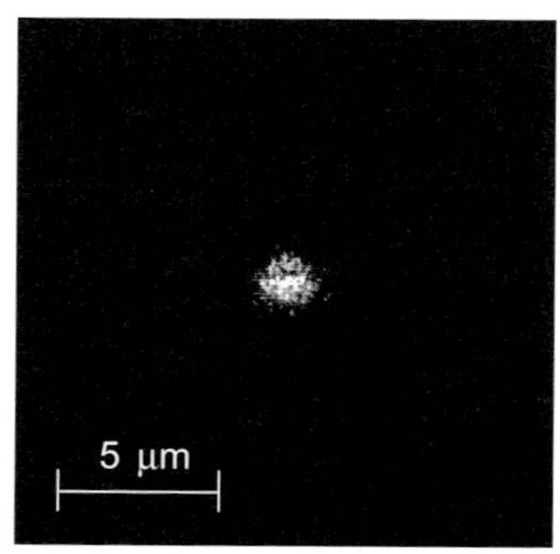

Figure 5. Typical focal spot image for the 27th harmonic beam

approximately 2 μm, which agrees well with the scintillator experiment.

Finally, we estimated the focused intensity of the 27th harmonic wave from the energy, the spot size, and the pulse width. Assuming that the pulse width of the 27th harmonic wave was the same as the fundamental pulse, the maximum focused intensities with and without the Al filter were 1.4×10^{13} W/cm^2 and 1.0×10^{14} W/cm^2, respectively.

4. Multiphoton ionization of He by soft x-ray photons

In this session, we report on the observation of two-photon ionization processes using 41.8 eV soft x-ray photons for the first time and also discuss the cross sections of two-photon double ionization (TPDI) [12, 13] and above-threshold ionization (ATI) [14]. High-intensity soft x-ray generated by phase-matched HH enables the investigation of these nonlinear optical processes, which were beyond the reach of conventional light sources.

The relevant energy diagram of He, He^+, and He^{2+} and ionization pathways using 41.8 eV photons is shown in Fig. 6 when two-photon absorption occurs. Nonsequential two-photon absorption is required for the production of doubly ionized helium ions. Therefore, the observation of the doubly charged ion of He provides clear evidence for the nonlinear interaction in soft x-ray spectral region. The ATI process also occurs by two-photon absorption. However, since the final charge state of this process is single ionized one, it is impossible to detect it separately from the single ionized ions produced by one-photon absorption. Therefore, we used time of flight (TOF) electron energy spectrometer for detection of the ATI process instead of ion TOF mass spectrometer. Although the TPDI process also produces characteristic electrons, the TPDI signal of produced electron is difficult to be detected due to low-energy electrons produced with residual weak harmonics.

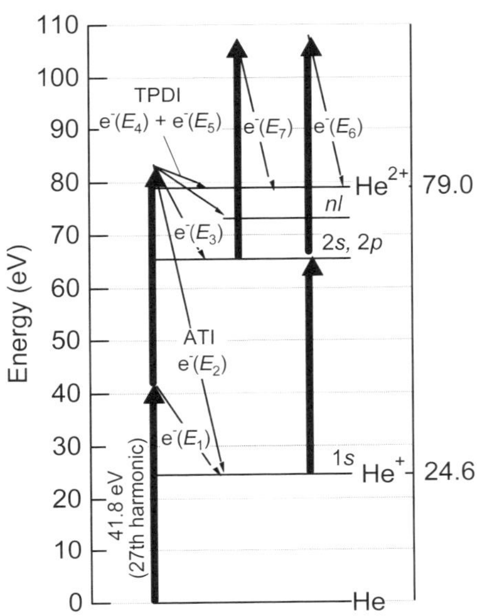

Figure 6. Relevant energy diagram of He, He^+, and He^{2+} and ionization pathways of 41.8 eVphoton.

In the experiment, we measured the mass spectra of ions with a conventional TOF mass spectrometer. Several harmonics around the 27th harmonic generated with a femtosecond pulse of a Ti:sapphire laser (a pulse width of 23 fs, a central wavelength of 800nm, a pulse energy of 20 mJ/pulse and a repetition rate of 10 Hz) were separated from the intense fundamental pulse with beam splitter(s) of Si or SiC, then sent to a spherical mirror multilayered coat of SiC/Mg with a radius of curvature of 100 mm in the interaction chamber. The energy of the 27th harmonic was estimated to be 24 nJ/pulse in the interaction region. The spot size was estimated to be $\omega_0 = 3.1$ μm from a separate experiment. Thus, the intensity of the 27th harmonic was estimated to be $I = 7 \times 10^{12}$ W/cm^2. Here, we regard the pulse width of the 27th harmonic to be the same as that of the fundamental. The confocal parameter was also determined to be 250 μm from the same experiment.

The TOF spectra were recorded on a oscilloscope and/or a computer, then averaged to assign the strong peaks (of the spectra) which originated from the single-photon process and also processed for counting the signals which exceed the threshold voltage to find the weak signals in the spectra. The isotope ^3He was used as the sample gas in the ion experiment to avoid the overlap of signals between $^4He^{2+}$ and H_2^+, the latter of which originated from residual water in the vacuum chamber (the estimated density of the sample gas in the interaction region, $d = 3.2 \times 10^{12}$ c m^{-3}). The number of shots accumulated is 10000.

Figure 7(a) shows the averaged spectrum of the ions produced by the interaction between the focused harmonic pulses and ^3He. Because the TOF of each mass can be determined from this spectrum, we can assign the masses for the small peaks in the accumulated spectrum of the counted signals in Fig. 7(b). The strongest peak of 2.15 μs can be assigned to $^3He^+$. The production of $^3He^+$ results dominantly from the one-photon absorption of ^3He. The effects of the consumption of neutral He is negligible under our experimental conditions. The peak at 1.53 μs of the TOF spectrum, shown in Fig. 7(b), can be assigned as doubly charged $^3He^{2+}$. The peaks at the TOF of 1.24 μs and 1.75 μs correspond to the signals of H^+ and H^+ ions originating from residual H_2O molecules in the vacuum chamber, respectively. The signal of $^3He^{2+}$ clearly appears between the H^+ signal and H_2^+ signal. The generation of doubly charged He^{2+} confirms the observation of a nonlinear optical process (two-photon absorption) in the soft X-ray region.

We carried out a subsequent experiment to measure the photoelectron spectra so that we could estimate the contribution of TPDI to the ionization process responsible for the production of doubly charged ions. Unfortunately, a few harmonics other than the 27th were partially reflected by the SiC/Mg mirror, thus disturbing the photoelectron spectra in the energy range of 0 eV to 4.6 eV, in which TPDI electrons are expected to appear, due to the large signals of the single-photon ionization processes. Accordingly, while we

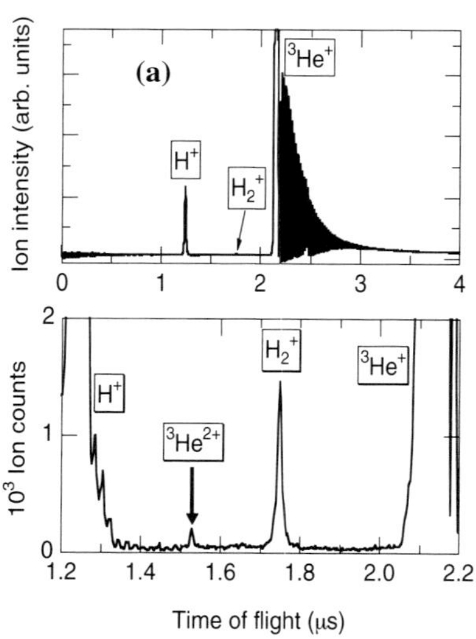

Figure 7. TOF mass spectra of ^3He.

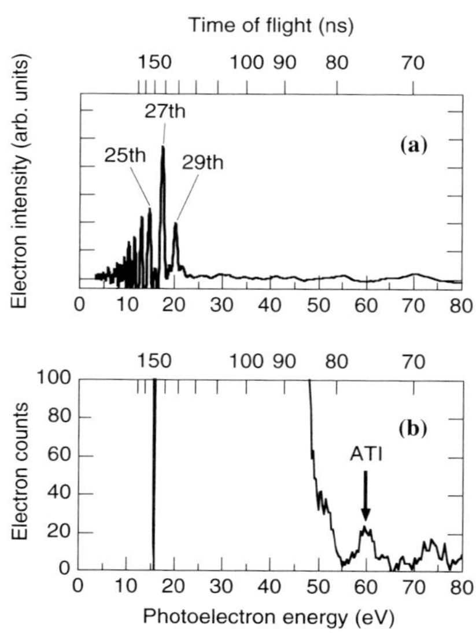

Figure 8. Photoelectron spectra of He.

could not observe electrons from the TPDI process, we succeeded in detecting the ATI peak. The averaged spectrum of the TOF experiment for electrons is shown in Fig. 8(a). The setup and signal processing used in this experiment are similar to those used in the ion TOF experiment except for the use of the TOF chamber for the electrons. The strongest peak at 142 ns corresponds to the photoelectron which accompanies the single-photon ionization from He to $He^{+}(1s)$ of the 27th harmonic. The other photoelectron signals, which are originated from single-photon ionization, appear at 158 ns for the 25th harmonic and 131 ns for the 29th harmonic. With the counting measurement of the spectrum, we found the remarkable feature of photoelectron at the TOF of 76 ns as shown in Fig. 8(b) indicated with an arrow. By converting the TOF to kinetic energy using the assigned peaks, this fast (high-energy) peak was ascribed to photoelectrons with an energy of 59 eV produced via the ATI of the 27th harmonic. This result is more evidence of the nonlinear optical process.

5. Conclusion

We have developed high intensity soft x-ray source based on high-order harmonics, including XUV optics for application. The focused intensity of the 27th harmonic attained to 1×10^{14} W/cm^2. Using this high intensity pulses, the multiphoton ionization of He was investigated and then nonlinear optical processes were observed for the first time in the soft x-ray region.

References

[1] Salières P, L'Hullier A, et.al. 1999 Adv. At. Mol. Opt. Phys. **41** 83-142
[2] Kobayashi Y, Sekikawa T, et.al. 1998 Opt. Lett. **23** 64-66
[3] Papadogiannis N A, Nikolopoulos L A A, et.al. 2003 Phys. Rev. Lett. **90** 133902
[4] Takahashi E, Nabekawa Y, et.al. 2002 Phys. Rev. A **66** 021802
[5] Takahashi E, Nabekawa Y, and Midorikawa K 2002 Opt. Lett. **27** 1920-1922
[6] Takahashi E, Nabekawa Y, and Midorikawa K 2004 Appl. Phys. Lett. **84** 4-6
[7] Takahashi E, Hasegawa H, et.al. 2004 Opt. Lett. **29** 507-509
[8] Mashiko H, Suda A, and Midorikwa K 2004 (to be published)
[9] Ishikawa K and Midorikawa K 2002 Phys. Rev. A **65** 043405
[10] Valentin C, Douillet D, et.al. 2003 Opt. Lett. **28** 1049-1051
[11] Yoshitomi D, Shimizu T, et.al. 2002 Opt. Lett. **27** 2170-2172
[12] Colgan J and Pindzola M S 2002 Phys. Rev. Lett. **88** 173002
[13] Nakajima T and Nikolopoulos 2002 Phys. Rev. A **66** 041402
[14] Agostini P, Fabre F, et.al. 1979 Phys. Rev. Lett. **42** 1127-1130

Inst. Phys. Conf. Ser. No 186
Paper presented at 9th Int. Conf. X-Ray Lasers, Beijing, China, 24–28 May 2004

Two-color field assisted atto-second XUV photo-ionization and its application

Yiping Huo, Zhinan Zeng, Ruxin Li and Zhizhan XU
Shanghai Institute of Optics and Fine Mechanics, Chinese Academy of Sciences

Abstract. The high order harmonic emission in atoms driven by intense femto-second laser pulses can lead to a single atto-second XUV pulse or an atto-second XUV pulse train in which the pulse period is half of the fundamental laser period, depended on the driving pulse duration and the pulse's carrier-envelop phase (CEP). The driving laser field assisted atto-second XUV photo-ionization was used successfully to measure the duration of the atto-second pulse based on the cross-correlation method. However, this method in principle can not tell single atto-second pulse from atto-second pulse train. In this paper, we propose a technique for directly telling single atto-second pulse from atto-second pulse train based on the photo-ionization of atoms by the atto-second XUV pulse in the presence of two-color strong laser pulse.

1. Introduction

Atto-second physics has become a new reality which promises unprecedented application in different areas of science and technology [1-6]. Isolated atto-second bursts of radiation are required to reliably trace inner-shell electronic movement which take place within a fraction of the oscillation period of visible radiation. But in high order harmonic generation, the production not only can be single atto-second pulse but also can be atto-second pulse train consisting of two or more comparably strong pulses separated by half of the fundamental laser period[7]. Present methods for characterizing atto-second pulse are mostly based on the laser assisted atto-second XUV photo-ionization [1,2,8-11]. These methods failed when it comes to distinguish between single atto-second pulse and atto-second pulse train. In this paper, we propose a variant of driving laser assisted XUV photo-ionization for directly telling single atto-second pulse from atto-second pulse train.

2. Two-color field assisted atto-second XUV photo-ionization

The effect of the laser field on the x-ray photoelectron energy spectrum is now well understood in a quantum-mechanical framework [8]. In the strong field limit, the most conspicuous features can be explained by a quasi-classical model as two-step process [2]: first, the photoelectron is ejected by an atto-second XUV pulse; and second, the photoelectron is subsequently accelerated as a classical particle by the electric field. In accordance with the quantum treatment, the model predict that, the momentum along the electric field vector Vp_x transferred from the laser field to the photoelectron depends on the phase of the laser field at the moment the photoelectron is born. The momentum transfer results in a shift of the photoelectron redistribution in momentum space. When we detected within a small solid angle aligned orthogonally to the propagating direction and electric field vector of the linearly polarized laser pulse, the momentum transfer gives rise to a spread of the final kinetic energy of photoelectrons. In this paper, we always detect photoelectron spectrum from orthogonal detection.

In the previous work, assistant field is an intense linearly polarized laser pulse, which can be given by

$$\mathbf{E}_L(t) = \mathbf{E}_0(t)\sin(\omega_L t + \varphi), \tag{1}$$

where ω_L is the frequency of the laser pulse, φ is the carrier-envelope phase, and $\mathbf{E}_0(t) = \mathbf{E}_0 \exp(-t^2/\tau^2)$ denotes the envelope of the laser pulse, where $\mathbf{E}_0$ is the peak amplitude of the laser pulse and $\sqrt{2\ln 2}\tau$ is its duration. In high-order harmonic generation, XUV pulse is produced at the peak of the laser pulse and the period between two XUV pulses is $T_0/2$. Consider an atto-second XUV pulse train contained two pulses named the first pulse and the second pulse separately, the effect of the laser field on the photoelectron freed by the first pulse is the same as the effect on the photoelectron produced by the second pulse. Scanning the birth instant of the photoelectron through the laser field oscillations by changing the relative delay t_d between the laser pulse and the XUV pulse, both single atto-second pulse and atto-second pulse train produced photoelectron energy spectrum are modulated by the delay t_d at half fundamental laser period $T_0/2$. So it can not be justified convincingly by conventional laser assisted atom ionization that what we measured is single atto-second pulse or not.

To solve this problem, we adopt two-color field composed of the fundamental field and its double frequency radiation as assistant laser, which can be expressed as

$$\mathbf{E}_c(t) = \mathbf{E}_0(t)\left(\sin(\omega_L t) + \varepsilon \sin(2\omega_L t)\right), \tag{2}$$

where $\varepsilon (0 < \varepsilon \le 1)$ is the ratio of peak amplitude of the double frequency field against the fundamental one. For simplicity, we assumed the carrier-envelop phases of two fields as zero. For the introduction of two-color field, the electric field evolvement is different from the fundamental one. At some instant, the photoelectron energy spectrum produced by the first pulse will separate from the photoelectron energy spectrum produced by the second pulse with sufficiently intense field.

3. Simulation results and discussion

Following [8,12], the amplitude of populating a state $|\mathbf{v}\rangle$ with kinetic momentum $\mathbf{v}$ at a moment T after the end of both the laser and the x-ray pulse is

$$a_{\mathbf{v}}(T) = -i \int_{-\infty}^{T} dt \mathbf{d}_{\mathbf{p}(t)} \mathbf{E}_X(t) e^{-i\int_t^T dt' \left(1/2[\mathbf{p}(t')]^2 + I_p\right)}, \tag{3}$$

where $\mathbf{p}(t) = \mathbf{v} + \mathbf{A}(t)$ is the instantaneous kinetic momentum, and $\mathbf{d}_{\mathbf{p}(t)}$ is the dipole transition matrix element [12] from the ground state to the continuum with kinetic momentum $\mathbf{p}(t)$. For a sufficiently short XUV pulse, Eq. (3) can be approximated using the saddle point method [8], so we have

$$|a_{\mathbf{v}}|^2 \propto \frac{\left|\mathbf{d}_{\mathbf{p}(t_0)} \mathbf{E}_X(t_0)\right|^2}{\mu} \exp\left[-\left(\frac{\mathbf{p}^2(t_0)}{2} - W_0\right)^2 \frac{\tau^2}{\mu^2}\right], \tag{4}$$

$$\mu = \sqrt{1+\eta^2}\,, \qquad\qquad \eta = \mathbf{E}_L(t_0)\mathbf{p}(t_0)\tau^2\,,$$

Equations (3) and (4) apply for all directions of observation and laser polarizations.

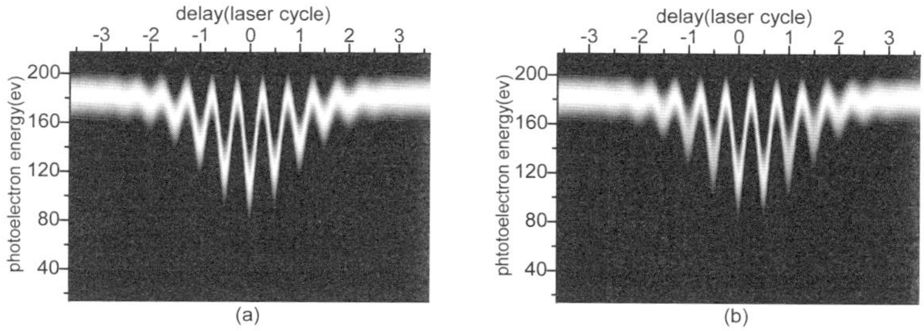

Figure 1. Photoelectron spectrum(vertical axis) produced by (a) single atto-second XUV pulse, (b) atto-second pulse train contained two pulses as a function of delay t_d between fundamental laser pulse and XUV pulse (horizontal axis).

Our simulations are based on Eq. (4). For comparison, firstly we simulated fundamental field assistant atto-second XUV photo-ionization. Fig. 1a depicts the photoelectron energy spectrum produced by single atto-second pulse as a function of delay t_d, obtained with a laser pulse whose duration is 220.3857a.u. (about 5fs) and peak intensity is 0.14a.u. ($6.86\times10^{14}\,\mathrm{W/cm^2}$). The atom considered is helium, whose ionization potential is 0.8996 a.u. (about 24.5 eV). The energy of the XUV photoelectron is 7.1324 a.u. (194 eV). Both the full width of half maximum (FWHM) of spectrum and downshift of the centre gravity of the spectrum are subject to a modulation versus t_d at a period of $T_0/2$, in agreement with the result of Ref.[1]. In Fig. 1b, we show the total photoelectron energy spectrum, which is produced by atto-second pulses train, modulated by the delay t_d. From Fig. 1, we can not distinguish single atto-second pulse from atto-second pulse train.

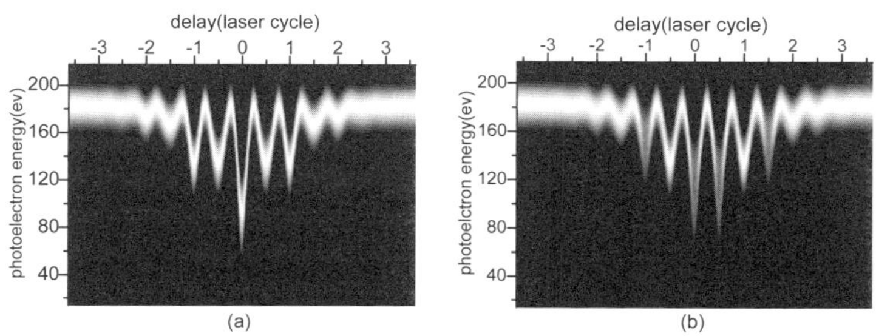

Figure 2. Photoelectron spectrum (vertical axis) produced by (a) single atto-second XUV pulse, (b) atto-second pulse train contained two pulses as a function of delay t_d between two-color laser pulse and XUV pulse (horizontal axis).

In Fig. 2, we show two-color field assisted atto-second XUV photo-ionization produced respectively by single atto-second pulse and atto-second pulse train. The laser pulse in the calculation is described by Eq. (2) of $\varepsilon = 0.2$. Other conditions are the same as in Fig. 2. Comparing Fig. 2b with Fig. 2a, we can see for some periods of delay, the photoelectron spectra are distinct in intensity and FWHM. For the adoption of two-color field as assistant field, the field the photoelectron freed by the first pulse felt is different from the field the photoelectron freed by the second pulse felt, especially at some delays two spectra separate from each other. From Fig. 2 we can directly tell single atto-second pulse from atto-second pulse train based on the appearance of spectrum separation at some positions.

Fig. 3a shows the total photoelectron spectrum of driving field assisted ionization at delay $t_d = 0$. The triangles correspond to the total spectrum, the circles correspond to the photoelectron spectrum produced by the first pulse and the crosses correspond to the photoelectron spectrum produced by the second pulse. Two spectra are almost superposed. Because of the ultra-short duration of 5-fs laser pulse, the difference of peak intensity between two nearby peaks is large, which results in the small displacement of two photoelectron spectra. Fig. 3b shows photoelectron spectrum of two-color field assisted ionization at $t_d = 0$. Two spectra are detached from each other obviously.

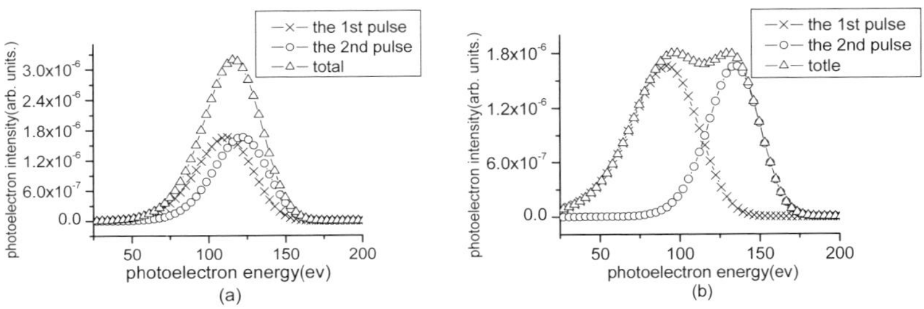

Figure 3. (a)Driving laser, (b) two-color laser assisted atto-second XUV photo-ionization. Triangles: total spectrum; circles: photoelectron energy produced by the first pulse; crosses: photoelectron spectrum produced by the second pulse.

4. Summary

In conclusion, we introduced a method named two-color field assisted atto-second XUV photo-ionization for directly distinguishing single atto-second pulse from atto-second pulse train. We have shown that the total photoelectron spectrum separated into two parts at some delays between the atto-second XUV pulse and the two-color assisted field, which can be exploited for the telling single atto-second pulse from atto-second pulse train.

References

[1] Hentschel M et al 2001 Nature **414** 509-513
[2] Drescher M et al 2001 Science **291** 1923-1928
[3] Kienberger R et al 2002 Science **297** 1144-1148
[4] Niikura H et al 2002 Nature **417** 917-922
[5] Bandrauk A D et al 2002 Phys. Rev. A **66** 031401 (R)
[6] Bandrauk A D et al 2002 Phys. Rev. Lett. **89** 283903
[7] Zeng Z et al 2003 Phys. Rev. A **67** 013815
[8] Itatani J et al 2002 Phys. Rev. Lett. **88** 173903
[9] Queere F et al 2003 Phys. Rev. Lett. **90** 073902
[10] Andre D B et al2003 Phys. Rev. A **68** 041802
[11] Scrinzi A et al Phys. Rev. Lett. **86** 412-415
[12] Lewenstein M et al 1994 Phys. Rev. A **49** 2117-2132

Inst. Phys. Conf. Ser. No 186
Paper presented at 9th Int. Conf. X-Ray Lasers, Beijing, China, 24–28 May 2004

Second generation X-ray lasers

Ph. Zeitoun, G. Faivre, S. Sebban, T. Mocek, S. Hallou, Ph. Balcou, F. Burgy, D. Douillet, T. Lefrou, P. Mercère¶, A. S. Morlens, J. P. Rousseau, and C. Valentin
Laboratoire d'Optique Appliquée, chemin de la hunière, 91128 Palaiseau, France
Imagine Optic, 18, rue Charles de Gaulle 91400, Orsay, France

M. Fajardo
GoLP, Instituto Superior Technico, Lisbon, Portugal

S. Kazamias
Laboratoire d'Interaction du Rayonnement X avec la Matière, bât. 350, Université Paris-Sud, 91405 Orsay, France

D. Aubert, G. de Lachèze-Murel
CEA/ DIF, BP 12, 91680 BRUYERES-LE-CHATEL, France

S. le Pape
Laboratoire pour l'Utilisation des Lasers Intenses, Ecole Polytechnique, 91128 Palaiseau, FRANCE

Abstract. We studied seeding of two x-ray lasers at 32.8 nm (Kr^{8+}) and 41.6 nm (Xe^{8+}). We succeeded to increase from a factor 10 to 200 the harmonics energy, without decreasing their optical qualities. The resulting beam was polarized, coherent and we estimate its duration to 500 fs. This structure seems to be the beginning of a new generation of X-ray laser (XRL), with considerable improvements compare to the actual XRL. Furthermore this technique can help for the measurement of many plasma parameters.

Since some years x-ray laser community try to promote this source for applications. In this aim, it is necessary to offer an optically good beam to the users. Nevertheless most of x-ray laser (XRL) are using amplified spontaneous emission, i. e. laser has no cavity to select a spatial mode. Consequently the beam can not be polarized or spatially coherent and the wave front is not regular. Thanks to the most recent works on x-ray laser and on high order harmonics generation (HHG), it is now possible to produce an energetic beam (from 0.1 to 1mJ) having at the same time the required optical

properties. The solution consists in seeding the XRL amplifier medium with another beam (HHG). This experiment was successfully realized in LOA. We studied seeding of two x-ray laser transitions, 32.8 nm (Kr^{8+}) and 41.6 nm (Xe^{8+}).
The XUV amplifying laser chain was made of three parts: a high harmonics generation (HHG) seed, a focusing system, and an optical field ionised x-ray laser (XRL) amplifier medium. The seed beam was obtained by focusing a 20 mJ, 30 fs, infrared laser in a gas cell filled with argon [1]. This seed was image relayed onto the entrance of the x-ray laser (XRL) amplifier (Kr gas longitudinally pumped by a 1 J, 30 fs laser) by means of toroidal XUV mirror. This X-ray laser[2] operates on the $3d^94d_{J=0} \rightarrow 3d^94p_{J=1}$ transition of the Kr^{8+} ion at 32.8 nm, which has a close match with the 25th harmonic of the IR laser (815 nm). After amplification, the output beam was either analysed by an XUV spectrometer or a monochromatic cross-section was recorded.

The experiment has been performed in LOA with the multiterawatt, 30 fs, 10 Hz, Ti:sapphire laser emitting at 815 nm. The laser delivers two chirped beams having separate pulse compressor enabling to optimise independently the XRL or the HHG drivers. One beam, lately called "pump", contains most of the energy (about 1 J on target) and is used for creating the amplifying XRL medium. The second beam, called "probe", has energy around 20 mJ on target and generates the HHG (Figure 1)

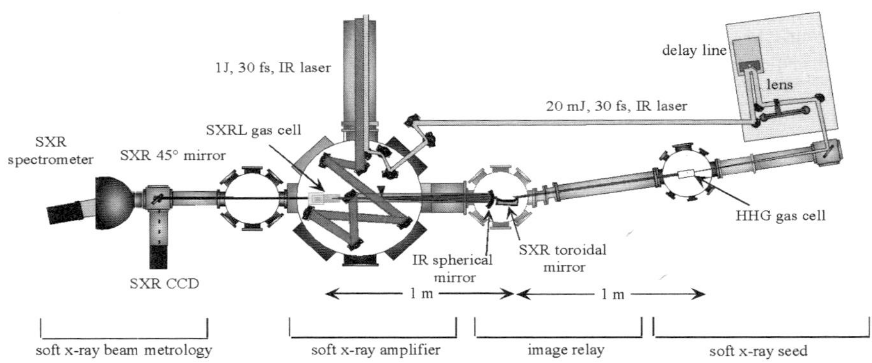

Figure 1 : setup of the experiment

Harmonics were produced by focusing the probe beam into a gas cell of variable pressure and length, filled with argon. The linearly polarised beam was focused with a f=1.5 m, aspherical lens. The focal spot has been measured to be about 150 μm leading to an intensity of about 10^{15} Wcm^{-2}. The harmonic flux was optimised by changing the conditions of laser interaction with the active medium, such as moving the focal spot position, changing the gas pressure, and the cell length. The highest HHG flux was obtained for a pressure of 25 mbar, 10 mm long cell, and focal plane situated 4 cm behind the gas cell. The main optimisation was the fine adjustment of the wavelength of one harmonic (25th) so as to coincide with the wavelength of Kr^{8+} XRL (32.8 nm). This was done on a day-to-day basis by adjusting the chirp of the laser generating the HHG [2,3].

The amplifying plasma was created by focusing the pump beam into another gas cell of variable length (from 0 to 6 mm) and pressure. The cell was filled with krypton held at uniform pressure. Fundamental radiation from the Ti:Sapphire laser was circularly polarised by a quarter-wave plate and focused with an f=0.85 m, spherical on-axis mirror into the gas cell. Circular polarisation ensures the creation of energetic free electrons that collisionally pump the Kr^{8+} lasing ions [4]. The measured focal spot was around 50 μm in diameter leading to a net intensity of about 5×10^{17} Wcm^{-2}.

The output of the harmonic source was re-imaged on the entrance of the amplifier plasma, using a 1:1 grazing incidence (4°) gold coated toroidal mirror. The theoretical reflectivity is around 80%. The focal spot of HHG has been measured to be around 150 μm leading to a spatial coupling of about 1/9 between HHG and the XRL gain region. Special attention has been paid to the temporal and spatial overlapping between the harmonic and the X-ray laser beams. As the OFI XRL is longitudinally pumped, the position of the gain region coincides with the position of the IR driving laser. Since also the HHG are collinear with the IR beam, the superposition of the HHG and the XRL gain region has been done in IR. Temporal synchronisation of the HHG with the period of XRL amplification could not be done in such a simple way since the plasma starts to amplify after the interaction of the driving laser with the gas. Based on previous experiment with another OFI x-ray laser (Xe^{8+} emitting at 41.8 nm), we may assume that the gain will appear around 5 ps after laser interaction and will last about 8 ps [5]. Since HHG are emitted only during the interaction of the IR laser with the gas, the synchronisation of the seed and the XRL period of amplification was done by adjusting the delay between the IR pump and probe beams. A delay line with an accuracy of 30 fs was used for the synchronisation of the seed with the period of amplification.

The XUV beams generated in this experiment were analysed by an on-axis XUV spectrometer [6] and by making monochromatic images of the beam cross-section. These images were obtained by placing a removable 45°, XUV mirror in the beam path, redirecting the beam towards a cooled, thin, back-illuminated charge-coupled device (CCD). A 300 nm thick aluminium filter has been used to remove the IR beams. The interferential multilayer mirror has been designed and manufactured at the Institut d'Optique Théorique et Appliquée (France) so as to achieve a high reflectivity (estimated at 50%) and a strong polarisation at 45°. The XUV mirror is made of a stack of 30 $Mo/B_4C/Si$ tri-layers.

The delay between XRL plasma creation and HHG injection was variable to reach optimum amplification conditions, synchronizing HHG seeding with maximum gain. Four emission spectra are displayed in figure 2, corresponding to (a) high harmonics alone, (b) x-ray laser alone, (c) XRL and HHG seeded long after gain extinction, and (d) HHG synchronised with maximum XRL gain. The amplification period starts 5 ps after the interaction of the IR laser with the gas and lasts about 8 ps. Figure 2 (d) clearly shows that a strong amplification of the seed has been achieved. The HHG were seeded at about twice the intensity of the ASE x-ray laser, leading to an enhancement of the output signal by a factor of 13. Since the level of seeding may influence many laser parameters (intensity, pulse duration, ratio of the energies in the seeded beam to this in the ASE beam), it is crucial to understand how is the seeding level related to the saturation intensity.

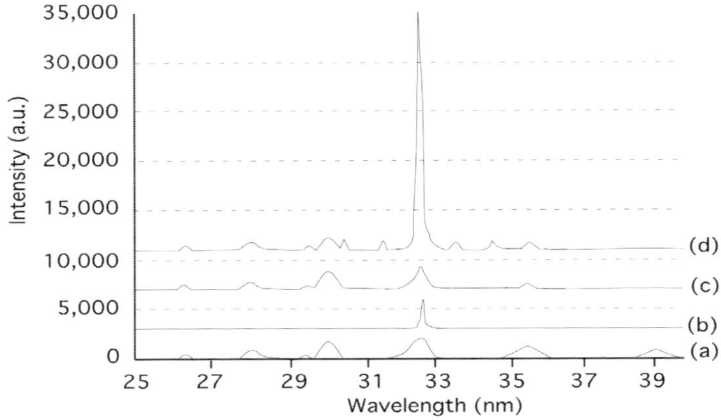

Figure 2 : Experimental spectra under conditions : (a) HHG only, (b) ASE XRL only, (c) both HHG and XRL in a non-amplifying timing, and (d) the amplified seeded XRL. Each spectrum corresponds to signal accumulated over 10 laser shots, in order to reduce the shot-to-shot error. The spectra have been artificially shifted from the bottom to the top in order to help for comparison. On spectrum (d), small spikes are visible mainly around the seeded XRL peak as coming from its diffraction on the grating holder.

The amplification law (output intensity versus amplifier length) of the amplified spontaneous emission (ASE) XRL has been measured to clarify the level of seeding. The gas cell length ranged from 0 to 4 mm and the intensities have been fitted taking into account saturation effects [6]. A small-signal gain coefficient of 80 cm^{-1} has been inferred, reaching saturation above 1.7 mm plasma length. Clearly, the spectrum in Fig. 2(d) was obtained for an HHG seeding above saturation intensity (about 4.5 times higher). In the second part of the experiment, the seeding intensity was reduced down to $1/50^{th}$ of the saturation intensity. The output signal was 38 times higher than the saturation intensity; the amplification factor became as large as 200 i.e. 15 times higher than for the amplification achieved with above-saturation-intensity seeding. Surprisingly, for both high and low seed intensities we measured an intensity output 10 to 20 times higher than the best ASE signal, while from classical theory of laser amplification the seeded XRL intensity should be about twice the ASE intensity [7]. This result can be explained only by considering the influence of amplification on the seeded x-ray laser spectrum, while estimating the pulse duration.

Since HHG has much shorter pulse duration (20 fs) than the X-ray laser gain duration (~8 ps), the final output pulse duration is solely limited by the amplified spectral line shape within the Fourier limit. This is not the case for ASE x-ray lasers, where the duration of the spontaneous emission plays a major role in pulse duration [8]. Two phenomena may modify the line width: gain narrowing and saturation re-broadening [9]. Using experimental parameters (seeding and saturation intensities as well as XRL [10] and HHG line widths) we modelled the seeded XRL line shape during the amplification process up to the end of the plasma. Insert in Figure 3 shows the intrinsic XRL and HHG line shapes, respectively, (Lorentzian and Gaussian), and the seeded XRL line shapes after amplification over 4 mm of plasma. The main part of Fig. 3 is a zoom of the insert to highlight the differences between the two XRL line shapes corresponding to the two regimes of seeding (low or high intensity). In both cases it appears that due to very strong saturation, the line widths (at half-maximum) reach $\Delta\lambda \sim 2$ pm implying that the Fourier-limited seeded XRL duration is about 500 fs. To our knowledge the seeded X-ray laser reported here is the only femtosecond-range x-ray laser demonstrated to date. Figure 3 also shows that for the lowest seeding intensity, the amplified spectrum keeps the Lorentzian shape because saturation is reached only at the very end of the plasma. On the contrary, for the above-saturation-intensity seeding the centre of the line is weakly amplified while the line wings staying unsaturated are strongly amplified. This is only possible because the HHG seed has a very wide line width, about 200 times larger than the XRL width. Such a phenomenon could not appear for any ASE X-ray laser in the role of a seed. Considering the energy contained in the wings, the output signal while seeded above saturation intensity is estimated to be higher by a factor of about 15 compared to the ASE X-ray laser. This is in good agreement with the experimental data showing an enhancement by a factor of 23. As demonstrated both theoretically and experimentally, seeding an x-ray laser with a source having a much wider bandwidth is very beneficial for energy extraction. Ions with stimulated transitions that are normally too much spectrally shifted for amplifying incident radiation may now participate to the amplification.

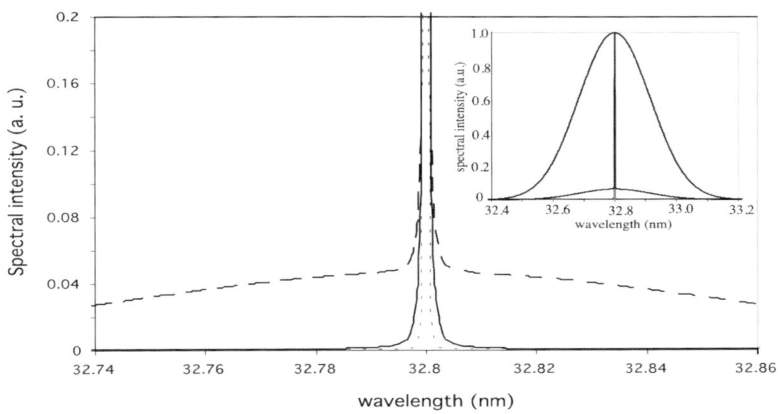

Figure 3 : Calculated line shapes for the HHG (seed), the ASE XRL, and the seeded XRL for two levels of seeding. The insert shows the line shapes at the real scale pointing out the intense line wings of the seeded XRL (red) as compared to the ASE XRL (black) and HHG (blue). The main part of the figure is a zoom of the insert comparing the seeded XRL line wings for the highest (dashed) and the lowest (dotted) seeding intensities to the intrinsic XRL (plain).

We also measured the gain duration for several pressures. These measurements were done for both gases, krypton end xenon. For the Xe^{8+} , at 41.6nm (fig4a), the maximum gain was obtained for a pressure of 15 Torr and the duration of the gain was about 8ps. For Kr^{8+} at 32.8 nm (fig4b) the best pressure was 15 Torr and for this pressure the gain duration was 4ps. At our knowledge it was the first time that the duration of the gain was measured.

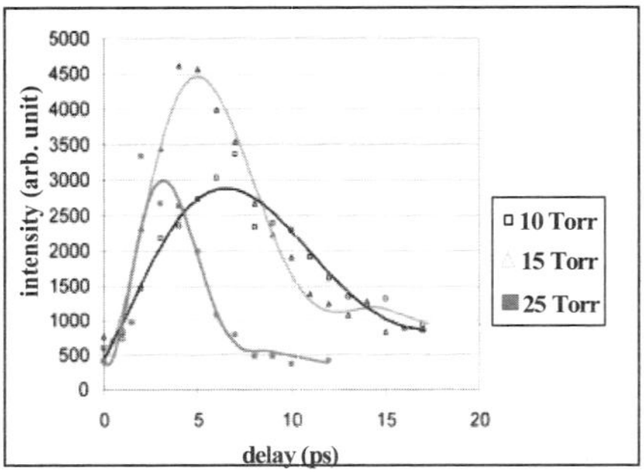

Figure 4a : intensity in fonction of the delay for Xe for 3 pressure

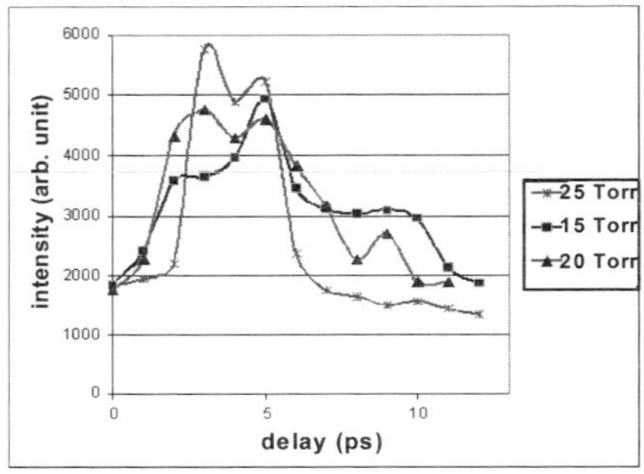

Figure 4b : intensity in fonction of the delay for Kr, for 3 pressure.

The seeded XRL beam was also characterised in terms of optical parameters such as divergence, spatial coherence and polarisation. These parameters were directly inferred from images of the beam cross-section (fig. 5 and 6). Theoretically, the ratio of the mirror reflectivity for the 25th harmonic (32.8 nm) to the 23rd (35.65 nm) or to the 27th (30.37 nm) is about 10. Therefore only the 25th harmonic and/or the XRL might be reflected towards the detector. The seed divergence is imposed by both the geometry of the toroidal mirror and the HHG intrinsic divergence. The divergence of the seeded XRL was slightly lower than those of the seed. The brightest part of the image in Fig. 4 corresponds to the seeded XRL while the weak surrounding, circular signal is the ASE x-ray laser. The divergence of the ASE x-ray laser was found to be much larger (>12 mrad) than that of the seeded beam (1 mrad).

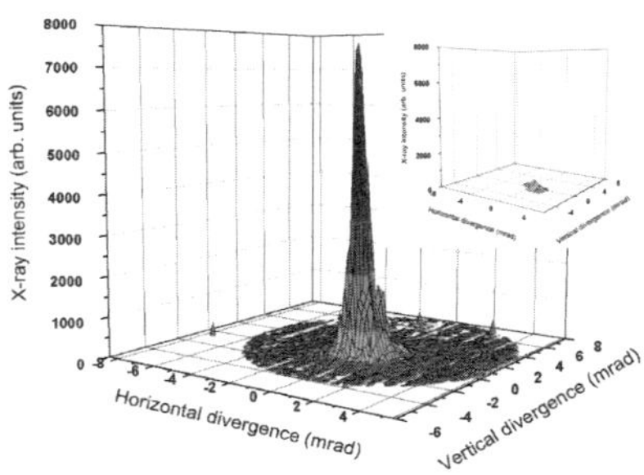

Figure 5 : 3D images (false color) of the seeded XRL cross-section as recorded by the XUV CCD detector after reflection on the monochromatic XUV mirror. The central strong signal (gray) belongs to the seeded XRL while the ASE XRL illuminating essentially the whole filter generated a circular signal (purple). The insert shows the seeding HHG signal recorded with the same set-up and displayed on the same scale as the main image.

The multilayer mirror has been designed to be highly polarising. The extinction ratio (intensity peak/lowest intensity) has been investigated by measuring the reflected signal versus the angle of polarisation of the HHG. Figure 6 displays the variation of both the HHG and the seeded XRL intensity versus the angle of polarisation The extinction ratio of the mirror is as high as 20. From these curves, it appears that the seeded XRL is polarised and follows well the initial polarisation of HHG. No sign of depolarisation has been observed. Inserts in Fig. 6 show the beam cross-section when the HHG were s- polarised (peak reflectivity) and p- polarised (minimum reflectivity). As expected, the ASE XRL alone is unpolarized.

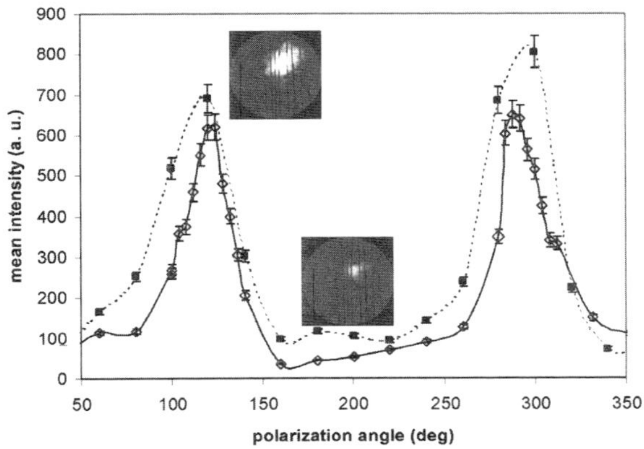

Figure 6 : Variation of the HHG (open diamond) and seeded XRL (black square) intensities versus the angle of polarization. Note that the angle "0" has no physical meaning here. The image of the XRL beam cross-section clearly shows the high degree of polarization achieved with the seeded XRL. Indeed, for the p- polarization (bottom image), only the unpolarized ASE X-ray laser is reflected towards the XUV CCD.

By comparing the images in Fig. 6, one may estimate the ratio between the seeded XRL and the ASE total energies. When the seed was heavily saturated, the energy contained in the seeded XRL was about 17 times higher than the energy of the ASE XRL. Note that the ASE level might be easily reduced to a negligible level by placing an aperture selecting only the seeded beam. Knowing the ASE level, we estimated the seeded XRL photon number to be about 1.5×10^{11} (0.7 μJ) per pulse. On all the images of seeded XRL cross-section, strong diffraction structures appeared close to alignment wires. For HHG and seeded XRL the diffraction was highly contrasted showing that the seeded XRL kept the high degree of coherence of the HHG. ASE x-ray laser alone generated hardly visible diffraction fringes confirming the expected weak coherence. From previous measurements of the ASE XRL wave front [5] we may expect to maintain the seeded XRL as good as the HHG wave front [11]. Then, assuming a pulse duration of 500 fs and a focal spot close to the diffraction limit (0.1 μm), such a seeded x-ray laser may achieve an intensity as high as 1.5×10^{16} Wcm^{-2}.

Future work will be in a near future to seed in an OFI capillary (ASE 30 times higher compare to OFI in a gas cell). With this kind of amplifier, the expected output energy is about 30μJ at 30nm). Later we plan to seed an XRL plasma created from a solid target such as to increase the output energy by at least three order of magnitudes (~ 1mJ) thanks to the much higher density of XRL amplifying medium (see proceeding G. Faivre et al). This rise in density (and temperature) will also lead to a spectral width broadening (10 times) and consequently to even shorter pulse duration (100 fs). It seems to be realistic that within few years, we may expect to produce intensities around

10^{20} Wcm^{-2} with lasers emitting around 10 nm. As this technique matures, future compact XUV laser chains should achieve beams performances close to those of VUV Free-electron lasers [12].

REFERENCES

[1] Kazamias S. et al, Phys. Rev. Lett., **90**, 193901 (2003)
[2] Lee D. G., Kim J. H., et al., Phys. Rev. Lett, **87**, 24, 243902 (2001)
[3] Reitze D. H., Opt. Lett.**, 29** 1, p. 86 (2004)
[4] Sebban S. et al., Lett., **89**, 253901 (2002).
[5] T. Mocek et al, Appl. Phys. B, in press
[6] Sebban, S. et al, Phys. Rev. Lett. **86**, 3004 (2001).
[7] Siegman A. E, Lasers, University Science Books, Mill Valley, California, 1986.
[8] R. Smith et al, Opt. Lett, **28**, 22, 2261 (2003)
[9] Casperson L. W. and Yariv A., IEEE Jour. Quant. Electr., **8**, 2, 80 (1977)
[10] Lemoff B.E., Barty C.P.J. et al., Opt. Lett., **19**, 8, 569 (1994)
[11] Takahashi E. et al, J. Opt. Soc. Am. B, **20**, 1, 158 (2003)
[12] Ayvazyan A. et al, Eur. Phys. J. D, **20**, 149 (2002)

Inst. Phys. Conf. Ser. No 186
Paper presented at 9th Int. Conf. X-Ray Lasers, Beijing, China, 24–28 May 2004

Phase dependence of relativistic electron dynamics and emission spectra in the superposition of a ultraintense laser field and a strong uniform magnetic field

Xinkui He, Ruxin Li, Bin Shuai, Xiaochun Ge, and Z. Z. Xu
Laboratory for High Intensity Optics, Shanghai Institute of Optics and Fine Mechanics, Chinese Academy of Sciences, Shanghai 201800, China

Abstract. The phase dependence of the dynamics and emission spectra of a fully relativistic electron in the superposition of a ultra-intense plane wave laser field and a strong uniform magnetic field has been investigated. For circular polarization only the axis of the helical trajectory is changed with the variation of the initial laser field phase. However, for linear polarization, the effect of changing initial phase is contrary in the two parameter regions divided by the resonance condition. Phase dependence of the electron's energy and velocity components was also studied. Some beat structure is found when the initial laser phase is zero and this structure is absent when the initial laser phase is a quarter of a period.

1. Introduction

The development of ultra-intense lasers has led to much interest in laser-matter interaction because of the new possibility to study high-field and ultra-high-field physics and new method to generate efficient sources of x-ray and high-energy particles [1]. An important model for electron acceleration and x-ray emission is the interaction of free-electrons with ultra-intense laser, such as nonlinear Thomson scattering [2,3].

In this letter, we investigate the interaction of a relativistic free-electron and the ultra-intense laser with an added uniform magnetic field parallel to the laser propagation direction. This problem is important for the understanding of laser plasma interaction and the related problems of high-energy electron emission [4,5], as well as for the study of inertial confined fusion using fast ignition concept [6]. In recent papers [7,8], Salamin presented the exact analytic solution of the problem. He found that, there is a very important parameter r which affects the behavior of the system critically. Here r stands for the frequency ratio

$$r = \frac{\omega_c}{\omega_l}\sqrt{(1+\beta_0)/(1-\beta_0)} \tag{1}$$

where $\beta_0 = v_0 / c$ is the initial electron speed normalized by c, and $\omega_c = eB_s / mc$ is the cyclotron frequency of the electron moving in the magnetic field B_s, ω_l is the frequency of laser field. When $r = 1$, the electron can gain maximum energy from the laser field, so $r = 1$ is called the resonance condition [8]. When $r \neq 1$, for observation along the same direction of laser propagation and the magnetic field, light at frequency of the laser and another frequency Ω_0 is scattered, and Ω_0 dependent on the electron initial velocity, the intensity and frequency of the laser and the strength of the magnetic field, which is called magnetic peak [7].

Because of the ultrahigh intensity of the laser field and the ultrashort duration of the laser pulse, the phase at which the electron experience the laser pulse will be important. All of the previous work of this model considered that the initial laser phase is zero. In this letter, we investigate the influence of the initial laser phase on the dynamics of the relativistic electron and its emission spectra. In the problem of free electrons interacting with intense laser field, different electron will enter the laser field at different time or different position, so in the initial position different electron will see different laser field phase. In this work we suppose that the initial position of the electrons is the same point (0,0,0), but the electrons enter the laser field at different time.

2. Formulation

Considering a relativistic electron with mass m and charge -e, in the simultaneous presence of an intense plane-wave laser field and a strong uniform magnetic field. The vector potential of the fields can be represented by

$$\mathbf{A} = \mathbf{A}_l\left[\hat{\mathbf{i}}\delta\cos\eta + \hat{\mathbf{j}}\sqrt{1-\delta^2}\sin\eta\right] - \frac{\mathbf{B}_s}{2}\left(\hat{\mathbf{i}}y - \hat{\mathbf{j}}x\right) \tag{2}$$

The first term in Eq. (2) represents a linearly polarized plane-wave with field strength A_l, where $\eta = \omega_l t - \mathbf{k}\bullet\mathbf{r} + \eta_0$ is the phase of the laser field with frequency ω_l, η_0 is the initial laser field phase at $t = 0$, and k is the wave vector of the laser field, pointing in the positive z direction. The second term represents the uniform magnetic field in the same direction. The electric field and magnetic field can be derived from equations

$$\mathbf{E} = -\frac{1}{c}\frac{\partial \mathbf{A}}{\partial t}, \qquad \mathbf{B} = \nabla\times\mathbf{A}\,. \tag{3}$$

The behaviors of the election in the superposed fields is governed by equation

$$\frac{d\mathbf{P}}{dt} = -e(\mathbf{E}+\boldsymbol{\beta}\times\mathbf{B}) \qquad \frac{d\varepsilon}{dt} = -ec\boldsymbol{\beta}\bullet\mathbf{E}, \tag{4}$$

where $\varepsilon = \gamma mc^2$, $\mathbf{P} = \gamma mc\boldsymbol{\beta}$, $\boldsymbol{\beta}$ is the electron velocity normalized by c, the speed of light, and $\gamma = \left(1-\beta^2\right)^{-1/2}$.

3. phase dependence of the electron's dynamics

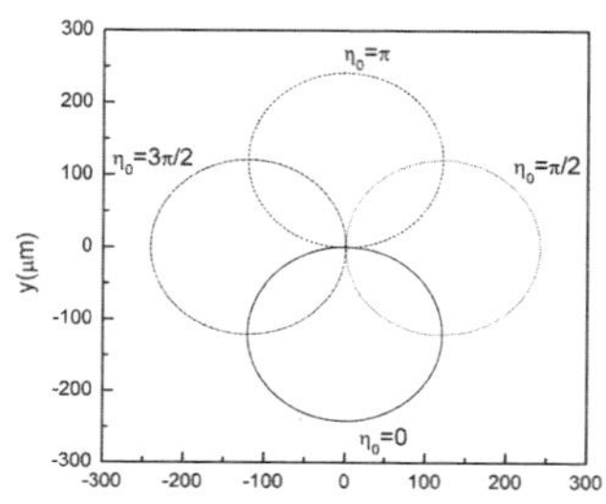

Figure 1. Projection of the electron trajectory of different initial laser field phase η_0 on the xy plate for circular polarization. It is plot over 450 field cycles with $\beta_0 = 0$. $q = 3$ and $\lambda = 0.8\mu m$, the magnetic field $B_s = 30T$.

The trajectory equation and the normalized components of velocity of the relativistic electron with arbitrary initial laser field phase can be derived from Eq. (4). Following, we give numerical results of the electron dynamics graphically.

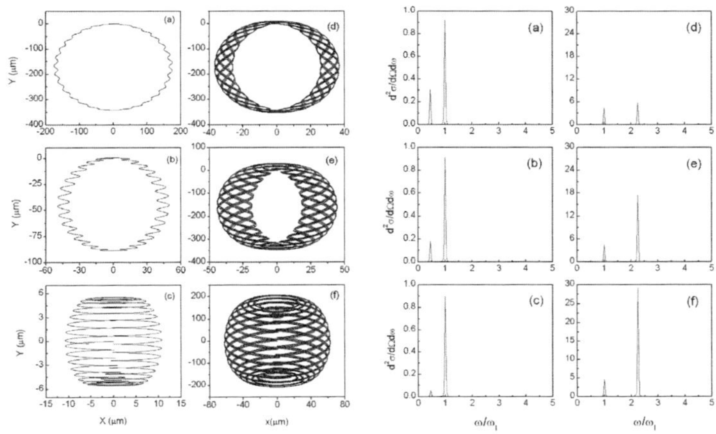

Figure 2. Projection of the electron trajectory on the xy plate (left: The trajectory is plot over 32 field cycles) and the corresponding spectra (right) for linear polarization. Left: (a)-(c) for r=0.0316 with β₀ =0.99, and (d)-(f) for r=1.4173 with β_0=0.999995, (a) (d) η_0=0, (b) η_0=π/2.4, (e) η_0=π/8, (c) (f) η_0=π/2. Right: (a)-(c) r=0.4482, (d)-(f) r=2.2409. (a) (d) η_0=0, (b) (e) η_0=π/4, (c) (f) η_0=π/2.

That can be learned from Salamin's work [7], the electron will follow a wiggly helical path whose axis is parallel to common direction of $\mathbf{B}_s$ and $\mathbf{k}$. So we just present the projection of the trajectory on the plane perpendicular to the laser propagation. In the case of circular polarization, we can see from Fig.1 ($q = eA_l / mc^2$ is the dimensionless intensity parameter), for different initial laser phase the axis of the helical trajectory is different. It also can be found that, other physical characteristics will not change with the initial laser field phase for the case of circular polarization. For linear polarization case, from Figure 2 one can see that, for $r<1$, when the initial laser phase change from 0 to $\pi/2$, the radius of the helical trajectory is decreased, with a minimum value at $\pi/2$. Note that the amplitude of the wiggles does not change, so it seems that the wiggles looks more violent. When $r>1$, the radius of the helical trajectory is increased with the increase of the laser field phase from 0 to $\pi/2$, and the amplitude of the wiggles is increased too. We found that the helical axis also changes with the initial laser phase.

In Figure 3 (left) we show the velocity components $\beta_x, \beta_x, \beta_x$ as a function of the laser phase for the initial laser field phase 0 and $\pi/2$. Figure 3 (right) shows the energy as a function of the laser phase at the same parameters. The figures clearly show that β_x, β_x and γ exhibit beat structures as $\eta_0 = 0$, and β_x also exhibit some periodical structure, but all these structures absent when $\eta_0 = \pi/2$, instead the curves exhibit oscillation almost at the uniform amplitude (except for β_x). We also found that the maximum value of β_x and the average value of γ is obviously smaller at $\eta_0 = \pi/2$ than that at $\eta_0 = 0$. It seems that the electron exchanges less energy with the laser field when the initial laser field phase is $\pi/2$ than it does when $\eta_0 = 0$

4. radiation spectra

The spectrum that would be observed along the laser propagation direction is shown in Figure 2 (right), Consistent with Salamin's work, there are only two peaks in the spectrum, the Thomson peak and the magnetic peak. When $r < 1$, we can find that, the height of the magnetic peak decreases with the increase of the laser field initial phase from 0 to $\pi/2$. The contrary result is found when $r > 1$. This also indicates that the magnetic peak is decided by the helical trajectory of the electron's movement.

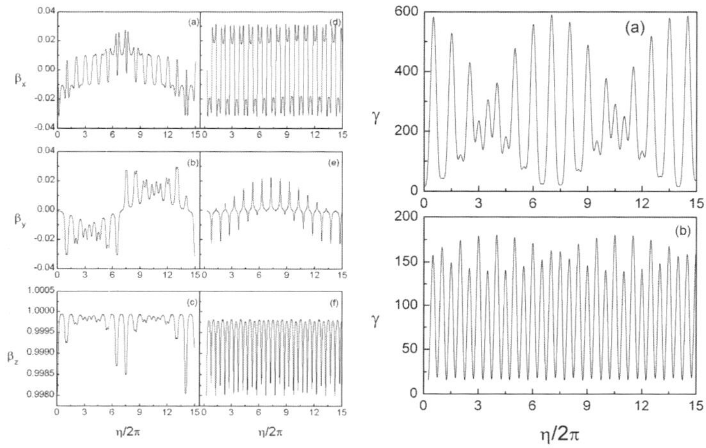

Figure 3. Components of the scaled velocity vector (left) and the scaled energy (right) of the electron is shown over 15 field cycles. $\beta_0 = 0.998$. The parameter of laser field and uniform magnetic field is the same as Figure 1. (a)-(c) $\eta_0 = 0$, (d)-(f) $\eta_0 = \pi/2$.

4. Summary

we have shown that, the dynamic behaviors of the relativistic electron and the emission spectrum vary dramatically with different initial laser field phase. For linear polarization, the effect of changing initial phase is contrary in the two parameter regions divided by the resonance condition $r = 1$ for both the radius of electron's helical trajectory and the height of magnetic peak. This indicates that the magnetic peak is decided by the helical trajectory of the electron's movement. We also studied the phase dependent of the electron's energy and velocity components, and found they show some beat structure when $\eta_0 = 0$ and this structure absent when $\eta_0 = \pi/2$.

References

[1] Umstadter D and Norris T 1997 IEEE J. Quantum Electron 33 1878
[2] Sarachic E S and Schappert G T 1970 Phys. Rev. D 1 2738
[3] Chen S-Y, Maksimchuk A and Umstadter D 1998 Nature (London) 396 653
[4] Boreham B W and Hora H 1979 Phys. Rev. Lett 42 776
[5] Cicchitelli L, Hora H and Postle R 1990 Phys. Rev. A 41 3727
[6] Mima K, Tahabe H and Nakai S 1989 ibid. 7 249
[7] Salamin Y I and Faisal F H M 1998 Phys. Rev. A 58 3221
[8] Salamin Y I and Faisal F H M 2000 Phys. Rev. A 62 53809

Inst. Phys. Conf. Ser. No 186
Paper presented at 9th Int. Conf. X-Ray Lasers, Beijing, China, 24–28 May 2004

Optical harmonics generation in molybdenum plasma

R A Ganeev, M Suzuki, M Baba, J Zhang, I A Kulagin [*], H Kuroda

The Institute for Solid State Physics, The University of Tokyo,
5-1-5 Kashiwanoha, Kashiwa, Chiba 277-8581, Japan
[*] NPO Akadempribor, Tashkent 700125, Uzbekistan

Abstract. We investigated the harmonic generation (from 3rd to 9th) in molybdenum plasma using 5 mJ, 300 ps, 795 nm heating radiation and 3 mJ, 130 fs, 795 nm probe radiation. The conversion efficiency of 3rd harmonic generation was estimated to be 5×10^{-4}. The conversion efficiency was decreased by one order for each next order of harmonics.

1. Introduction

Nonlinear optical effects (harmonic generation, CARS, etc.) in a weakly excited laser-produced plasma have a number of distinctive features from those in a dense laser plasma produced by high-intensity (10^{17}-10^{19} W cm^{-2}) optical fields. In particular, harmonic generation in the visible and VUV ranges has been investigated in a laser plasma formed by optical breakdown in gases and on metal surfaces at the laser intensity of 10^{11}-10^{13} W cm^{-2} [1,2].

However, in most of these studies the low-order harmonics were observed. At the same time some predictions for "optimal" plasma conditions show the opportunity of the efficient generation of high-order harmonics using commercially available lasers operated at moderate power.

Below we present the results of investigation of harmonic generation (from 3rd to 9th) in molybdenum plasma using 5 mJ, 300 ps, 795 nm heating radiation and 3 mJ, 130 fs, 795 nm probe radiation.

2. Experimental set-up

We used the Ti:sapphire laser operating at different lasing conditions. The Ti:sapphire laser (seeding oscillator, Tsunami + CPA TSA-10F regenerative amplifier with double passed linear amplifier, Spectra Physics) delivering femtosecond radiation (t = 130 fs, λ = 795 nm, E = 10 mJ) with 10 Hz pulse repetition rate was used in most studies presented below.

The part of uncompressed radiation separated by beam splitter before the compression stage was used for the excitation of Mo plasma. This radiation (t=300 ps, E=5.2 mJ) was focused by 100 mm focal length lens into the hole of Mo target (Fig. 1). The divergence of this radiation was increased using additional lens pair to achieve the broad beam waist size (400μm). Second beam (t=130 fs, E=3.5 mJ) was used as probe radiation interacted with molybdenum plasma at different time delay with respect to picosecond pulses. The

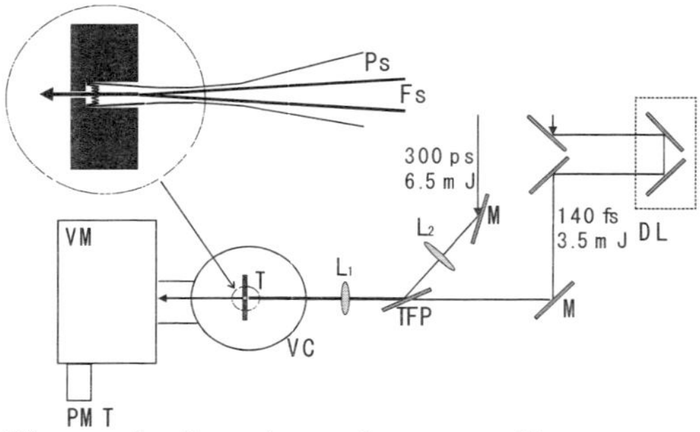

Figure 1. Experimental set-up. T: target; TFP: thin film polarizer; M: mirrors; L_1: focusing lens; L_2: telescope; VC: vacuum chamber; VM: vacuum monochromator; PMT: photomultiplier tube.

beam waist diameter of femtosecond radiation was measured to be 20 μm. This radiation and generated harmonics were propagated through the Mo plasma and thin hole in Mo target. We analysed the characteristics of generated radiation using monochromator (VM-504, Acton Research Corporation) and photomultiplier tube with sodium salycilate-coated glass plate.

The target was a 1.8 mm thick Mo plate with step-shaped 120 μm and 500 μm holes (see inset in Fig.1). Most part of picosecond radiation was interacted with the inner surface of 500 μm hole to ignite the plasma. The femtosecond radiation propagated then through the 120 μm hole without the interaction with the surface of Mo target.

3. Results and discussion

Figure 2 presents the harmonic spectrum registered at 12 ns delay between picosecond and femtosecond pulses. 3rd (265 nm), 5th (159 nm), 7th (114 nm), and 9th (88 nm) harmonics were observed in these studies. The conversion efficiency of 3rd harmonic generation was estimated to be 5×10^{-4}. The conversion efficiency was decreased by nearly one order of magnitude for each next number of harmonics. We investigated the dependences of harmonic intensities on the time delay between pulses. Figure 3 presents these dependences for 3rd and 5th harmonics. We analysed these dependences on the base of simple model of plasma. This model allows estimating some characteristics of plasma and harmonics and comparing them with those revealed from the experiment. From our experimental data we estimated the plasma temperature of 500 eV, the relaxation time of plasma formation of 120 ns, and the ratio between phase mismatching for 3rd and 5th harmonics of 0.81.

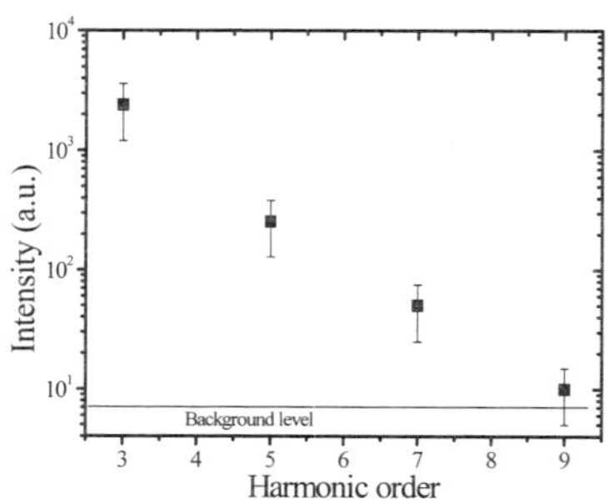

Figure 2. Harmonic spectrum.

In our case the dependences of harmonic intensity on femtosecond radiation intensity were close to predicted ones revealed from the simple non-perturbative theory of harmonic generation [3]. Figure 4 presents these dependences for two harmonics, and the slopes of these curves were close to 2.5 and 4.5. The saturation at femtosecond pulse energies above 1 mJ ($I_0=1.7\times10^{15}$ W cm^{-2}) was observed in the case 3rd harmonic generation, and was, probably, caused by competing multiphoton processes such as absorption and ionisation. The influence of heating energy had less steep dependences (Fig. 5). However, at the energy of picosecond pulses of 4 mJ the growth of the slope of conversion efficiency was observed for 3rd harmonic, whereas no considerable change of the slope of efficiency of 5th harmonic was registered in the whole range of picosecond pulse energies used.

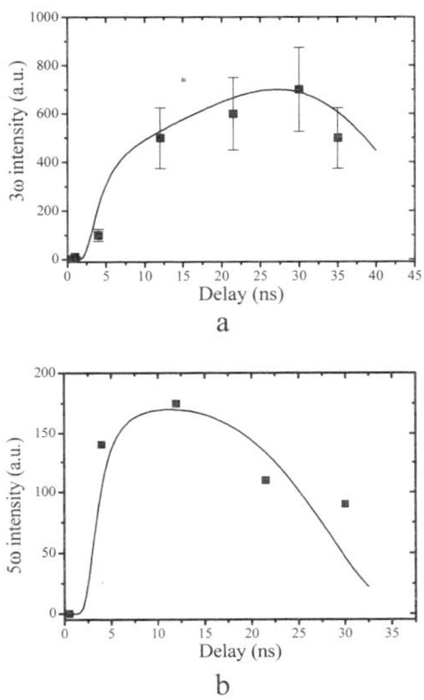

Figure 3. The intensities of (a) 3rd, and (b) 5th harmonics versus delay between heating and probe beams. Solid lines are the theoretical calculations.

Our experiments were carried out in the conditions when the confocal parameter of the beam (1 mm) was longer than plasma sizes even at time delays of 30 ns, so the nonlinear frequency conversion was performed in the conditions of plane wave interaction. It is well known that the focusing geometry is the main factor for the phase-matching conditions and harmonic efficiency. With appropriate alignment and focusing, one can get the optimal vapour densities and control them by varying the time delay between the ultrashort pulse and the preceding ablation pulse.

For intense, ultrashort pulses optical field ionisation rapidly reduces the refractive index within the focal volume, forming electron density and refractive index gradients. The time dependence of the index change generates spectral modulation predominantly toward shorter wavelength. The electron density gradients produced in the ionised region cause refractive divergence, which can rapidly reduce the on-axis field amplitude.

We did not observe the broadening and spectral shift of harmonics at $I_0<5\times10^{14}$ W cm^{-2}. However, at higher intensity the 2 nm red-shift with comparing to low-intensity case was registered for 3rd harmonic generation. We did not observe any variation of transmitted spectra of 795 nm radiation at different delays.

Both the 5th, 7th, and 9th harmonics were considerably depended on vacuum conditions inside the vacuum chamber. We investigated the influence of air pressure on harmonic generation and found that it cannot be explained by a decrease of registered VUV radiation due to simple consideration of absorption by air in the chamber and monochromator. It means that both nitrogen and oxygen can play a significant role in the disruption of optimal conditions for high-order harmonic generation in Mo plasma.

The growth of harmonic generation in the case of low-density, low-excited plasma can be caused by the conditions of closeness of multi-photon transitions with quasi-resonance transitions of molybdenum plasma. The population of these transitions can be increased at observed conditions, thus leading to the growth of high-order nonlinear susceptibilities responsible for frequency conversion. The ionization energy is around of 7 eV for neutral Mo atoms and of 16 eV for single-ionized Mo. So there should be an abundance of population of excited states of Mo even at average plasma temperatures.

The degree of excitation of the surface plasma affected significantly the nonlinear optical processes occurring in the plasma, which was particularly marked in the case of high (7th and 9th) harmonics. The intensity of the 9th harmonic was two to four time higher comparing with the intensity of the plasma radiation in a nearby range.

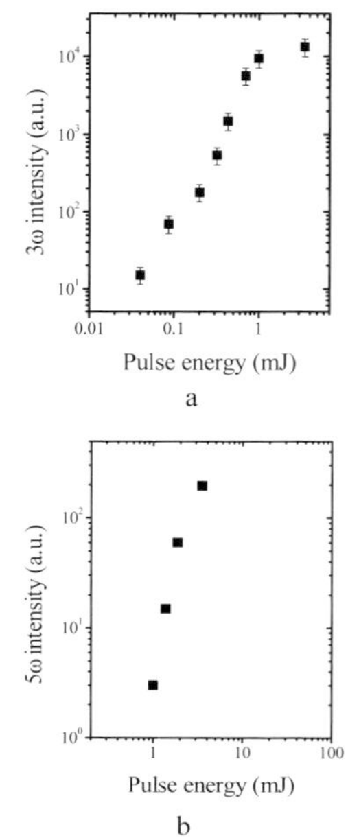

Figure 4. (a) 3rd and (b) 5th harmonic intensities as functions of probe pulse energy.

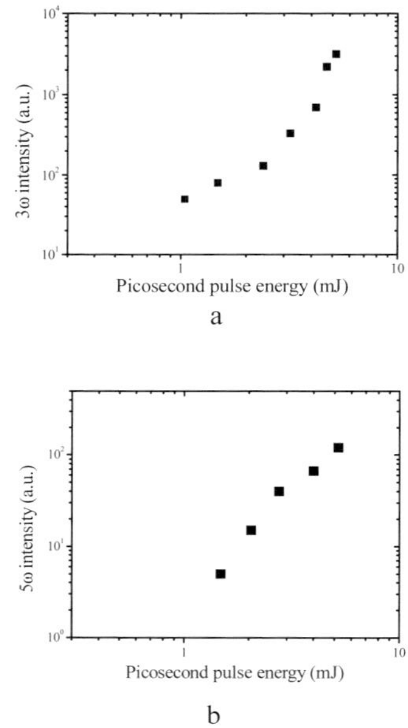

Figure 5. (a) 3rd and (b) 5th harmonic intensities as functions of heating pulse energy.

4. Conclusions

In conclusion, optical harmonic generation in Mo plasma was investigated. We analysed the harmonics generated after the propagation of the probe femtosecond radiation through Mo plasma created by picosecond radiation. We observed 3rd (265 nm), 5th (159 nm), 7th (114 nm), and 9th (88 nm) harmonics generation and analysed their spectra, conversion efficiencies, and other characteristics, as well as plasma parameters, depending on delay time between the heating and probe radiation and their energies. The conversion efficiency of 3rd harmonic generation was estimated to be 5×10^{-4}. The conversion efficiency was decreased by one order for each next number of harmonics.

References

[1] Zheltikov A M, Koroteev N I and Fedotov A B 1994 Laser Phys. 4 569-573

[2] Ganeev R A, Redkorechev V I and Usmanov T 1997 Opt. Commun. 135 251-256

[3] Reintjes J F 1984 Nonlinear Optical Parametric Processes in Liquids and Gases (New York: Academic Press)

Inst. Phys. Conf. Ser. No 186
Paper presented at 9th Int. Conf. X-Ray Lasers, Beijing, China, 24–28 May 2004

Amplified high order harmonic light for high coherent x-ray laser

Noboru Hasegawa, Tetsuya Kawachi, Maki Kishimoto, Kouta Sukegawa, Momoko Tanaka, Masataka Kado, Tai Renzhong, Yoshihiro Ochi, Masaharu Nishikino, Keisuke Nagashima, and Yoshiaki Kato.

Advanced Photon Research Center, Kansai Research Establishment, Japan Atomic Energy Research Institute, 8-1 Umemidai, Kizu, Kyoto, 619-0215 Japan

Abstract. We obtained the amplified high order harmonic light with neon-like manganese x-ray amplifier at the wavelength of 26.9 nm. 29th harmonic light of Ti;Sapphire laser was injected into the x-ray amplifier as a seed light for the purpose of generation of transform limited x-ray laser. X-ray amplifier selected the temporal coherent mode of the seed light and the beam divergence of amplified harmonic light was equal to the non-amplified harmonic light. It seems the generation of transform limited x-ray laser was succeeded and the pulse duration was estimated to 360 fs.

1. INTRODUCTION

The development of x-ray optics makes it possible application experiments such as x-ray interferometer with x-ray laser (XRL). Therefore, the coherent volume of XRL becomes important. The coherent volume is decided by the spatial and temporal coherence of XRL. Now, the spatial coherence has improved by using the double target experiment[1],[2]. But the temporal coherence is not so good. Example, for the nickel-like silver XRL, the temporal coherent ratio in its pulse was estimated to only 6 %[3]. The next step of the development of XRL is improvement the temporal coherent ratio, in other words, generation the transform limited (T.L.) XRL.

In this paper, the high order harmonic light was injected into the x-ray gain medium (we call it x-ray amplifier) as a seed light for the XRL. The x-ray amplifier works as the active mode selector, it selects and amplifies only temporal coherent mode of the seed light. The x-ray amplifier has quite narrow gain spectral width[4] compared with the seed light, so each mode of the seed light in narrow width have nearly equal phase even if the seed light is a chirped pulse such as high order harmonic light.

2. EXPERIMENTAL SETUP

Figure 1 shows the experimental setup for the generation of the T.L. XRL. This setup consists of three parts, seed light generator, x-ray amplifier and x-ray detector. We chose the 29th harmonic light of the Ti;Sapphire laser and neon-like manganese XRL medium (λ= 26.9 nm) such as the seed light and the x-ray amplifier respectively.

The high order harmonic light was generated with Ti;Sapphire laser and Ar gas target. A 20 mm beam diameter fundamental laser was focused into the Ar gas target with a 2000 mm focal length lens to a spot of about 100 μm. The central wavelength of fundamental laser was 790.1 nm, spectral width was 13nm, pulse duration was about

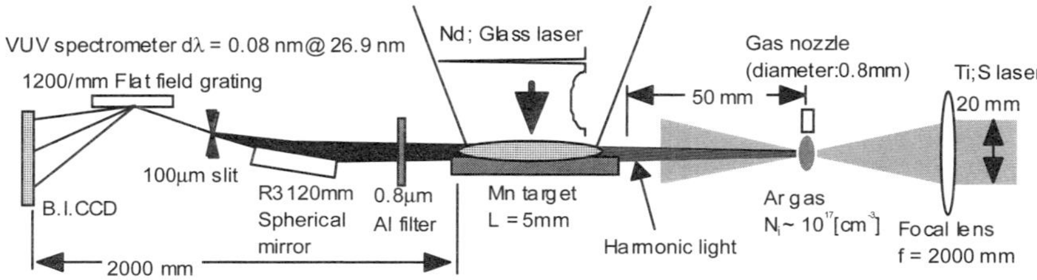

Figure 1:Experimental setup. The 29th harmonic light of the Ti;Sapphire laser was injected into the neon-like manganese x-ray amplifier (λ= 26.9 nm) generated with CPA Nd;glass laser.

80 fs, and peak intensity at the focal point was about $2x10^{14}W/cm^2$.

The x-ray amplifier was generated with single shot CPA Nd:glass laser[5] and Mn slab target. Mn slab target was irradiated by a line-focused Nd:glass laser at a wavelength of 1053 nm. The laser light consists of a prepulse for ionization of target and a mainpulse for excitation of ions. The total irradiation energy was 9 J, the energy ratio of the prepulse to the main pulse was 1:8, each pulse duration were 400ps and 4.8 ps respectively, and the pulse separation was about 500 ps. The width and the length of the line focus were 200 μm and 5 mm respectively, and the intensities of the prepulse and the main pulse are $1.4x10^{13}$ W/cm^2 and $1.1x10^{14}$ W/cm^2, respectively.

The oscillators of each laser systems were synchronized to the 80MHz frequency master signal with the phase synchronizer (Spectra-Physics ‘Lock-to-clock’). Timing jitter at the target chamber between both lasers was measured to less than 5 ps with the streak camera (HAMAMATU ‘FESCA200’).

The x-rays were observed by a grazing incidence spectrometer set on the axis on the target surface along with the harmonic light. A back-illuminated charged coupled device (B.I.CCD, Princeton SXTE-1024) was coupled with this spectrometer as the x-ray detector.

3. EXPERIMENTAL RESULT AND DISCUSSION

3.1. Seed light characterization

The x-ray amplifier amplifies a part of the harmonic light. The effective seed light energy and intensity are written by $E_{seed} = E_{29th}\ \varepsilon_V\ \varepsilon_s$, $I_{seed} = E_{29th}\ \varepsilon_V\ \varepsilon_s\ /\Delta\tau_{seed}\ /S_{seed}$ respectively. ε_v and ε_s are the spectral and the spatial coupling efficiency between the seed light and x-ray amplifier, $\Delta\tau_{seed}$ and S_{seed} are the effective seed light pulse duration and beam size at the edge of x-ray amplifier respectively. Figure 2 shows the 29th harmonic light spectrum and the gain spectrum of Ne-like Mn x-ray amplifier. The open circles were data points of 29th harmonic light spectrum. The 29th harmonic light energy and spectral width were measured to 0.1 nJ/shot and 0.12 nm respectively. Ne-like Mn x-ray amplifier gain spectral width was estimated to 56 mÅ from the result of spectral width of Ne-like selenium XRL[4]. Therefore the spectral coupling efficiency was 4.7 %. The 29th harmonic light beam divergence was measured to 1.8 mrad and the beam diameter at the edge of x-ray amplifier was estimated to 90 μm. The spatial coupling efficiency was estimated to 1.0 because the seed light beam diameter was less than glass laser focal width. From these result, effective seed light energy and intensity were 4.7 pJ and 4.1×10^5 W/cm^2 respectively. The pulse duration of effective seed light was supposed to 180 fs from x-ray amplifier gain spectral width, because the effective seed light was mode selected pulse, in other words, nearly T.L. pulse.

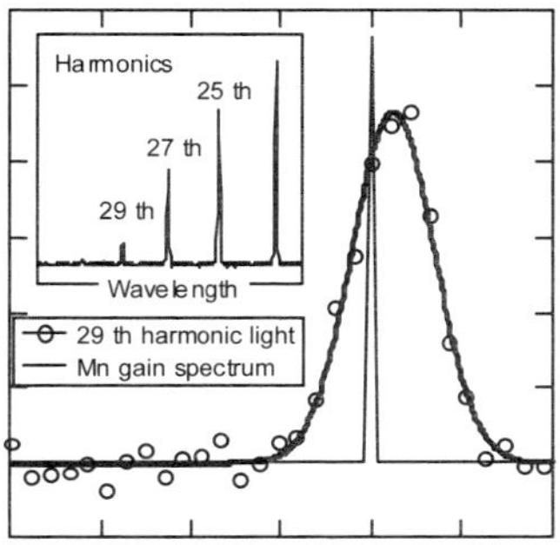

Figure 2:Seed light spectrum. The 29th harmonic light was tuned to the wavelength of Ne-like Mn (λ= 26.9 nm). The energy and spectral width were measured to 0.1 nJ and 0.12 nm respectively. The spectral coupling efficiency was 4.7 %.

In this experiment, a spontaneous emission generated from the edge of x-ray amplifier is amplified at the same time. Therefore the effective seed light intensity has to be larger than a spontaneous emission intensity. The spontaneous emission intensity was calculated from the Ne-like Mn 3s-3p spontaneous transition probability and the geometry of x-ray amplifier[6]. It was 6.7×10^3 W/cm^2 under the condition of the gain coefficient supposed to 13 cm^{-1} and it was negligibly small compared with the effective seed light intensity. Therefore the 29th harmonic light had enough intensity as the seed light for XRL.

3.2. Result of seed light amplification

Figure 3 shows the experimental result of XRL beam divergence under the condition of the seed light was injected into the x-ray amplifier after 40ps from the glass laser irradiation on to the Mn target. In order to compare the beam divergence under the same condition, 27th harmonic light was selected as the non-amplified seed light, instead of 29th harmonic light. These harmonic lights have nearly equal beam divergence. In this case, seeded XRL had nearly equal beam divergence to the non-amplified seed light and these beam divergences were smaller than no seeded XRL. It shows the beam divergence of the seeded XRL was decided from the seed light. In term of the spectrum, the spectral width was measured to 0.04 nm (= instrumental broadening of this spectrometer) and it was not changed between seeded and no seeded XRL. The output energy of seeded XRL was measured to about 3 nJ and it was larger than the seed light (= 0.1 nJ). From these results, the reason of beam divergence narrowing compared with no seeded XRL was the seed light amplification. The magnification of this x-ray amplifier was estimated to 640. The gain length product and gain coefficient were gl = 6.5 and g = 13 cm^{-1} respectively.

3.3. Temporal coherence of seeded XRL

Temporal coherence of the seeded XRL is decided by the effective seed light temporal coherence and gain narrowing effect with x-ray amplifier. The effective seed light temporal coherence was calculated from the spectral coupling efficiency. Harmonic light was supposed a Gaussian spectral and temporal profiles, the temporal coherence of mode selected harmonic light was calculated by Fourier transformation of the waveform function. Figure 4 shows the effect of mode selection with x-ray amplifier. Closed circle point shows the condition of this experiment, 29th harmonic light had about 3 times of the T.L. pulse. It was estimated from the result of pulse compression experiment of harmonic light[7]. The effective seed light was calculated to 1.0002 times of T.L. pulse. After amplification, x-ray spectral width became narrow to 29 mÅ with the gain narrowing effect at gl = 6.5, therefore, the seeded XRL pulse duration was estimated to 360 fs.

In this experiment, T.L. XRL was obtained but the spectral coupling efficiency was only 4.7 %. Figure 4 shows the spectral coupling efficiency can be improved to 20 % under the condition to obtain 1.01 times of T.L. pulse. It seems harmonic light with narrow spectral width is required, in other words, sub-pico second laser is suitable for the seed light generation.

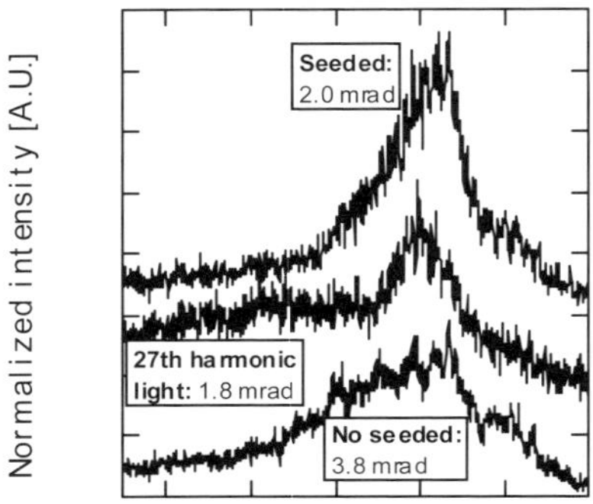

Figure 3: Experimental result of beam divergence. Seeded XRL had nearly equal beam divergence to non-amplified seed light (27th harmonic light) and they were smaller than no seeded beam divergence.

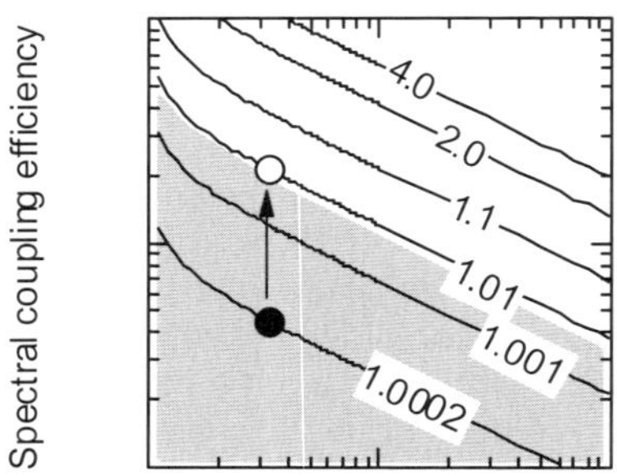

Figure 4: Mode selection effect when the input pulse has **n** times of T.L. pulse. Each line and number shows the temporal coherence of mode selected pulse (**m** times of T.L. pulse). Hatched area shows the condition to obtain T.L. pulse.

4. SUMMARY

We obtained the amplified high order harmonic light with neon-like manganese x-ray amplifier at a wavelength of 26.9 nm. It seems the generation of T.L. XRL was succeeded and now we can generate the full coherent XRL.

References

[1] Momoko Tanaka *et al.*, Opt. Lett. **28**, 1680 (2003).

[2] Masaharu Nishikino *et al.*, Phys. Rev. A **68**, 061802 (2003).

[3] Yoshihiro. OCHI *et al.*, Appl. Phys. B **78**, 961 (2004).

[4] Jeffrey A. Koch *et al.*, Rev. A **50**, 1877 (1994).

[5] Tetsuya. Kawachi *et al.*, Appl. Opt. **42** 2201 (2003).

[6] Noboru. Hasegawa, *et al.*, Proc. SPIE **4505**, 204 (2001).

[7] Taro Sekikawa, *et al.*, Phys. Rev. Lett. **83**, 2564 (1999).

Inst. Phys. Conf. Ser. No 186
Paper presented at 9th Int. Conf. X-Ray Lasers, Beijing, China, 24–28 May 2004

Low energy femtosecond laser driven phase-matched high-order harmonics in argon gas cell*

Xinhua Xie, Zhinan Zeng, Ruxin Li*, Haihe Lu, Dingjun Ying, Zhizhan Xu
Shanghai Institute of Optics and Fine Mechanics, Chinese Academy of Sciences

Abstract. Phase-matched high-order harmonic generation (HHG) in Ar gas-filled cell was investigated experimentally. We obtained phase-matched 27th order harmonic driven by a commercially available solid-state femtosecond laser system at 0.55mJ/pulse energy level and 1 KHz repetition rate. Moreover, we found that the spatial distribution of intensity of HHG was flat-top profile other than a Gaussian one under the condition of optimized conversion efficiency in the static gas cell.

HHG is generated by coherent interaction of an intense laser and atoms or molecules [1]. With the development of the intense ultrashort pulse laser, the research of HHG has reached the water-window in spectral region [2] and subfemtosecond in time domain [3]. Especially, the generation and application of subfemtosecond pulse led the study of HHG into a completely new world [4, 5]. It has made the study of ultrafast science from the femtochemistry to the attophysics. Presently, HHG can be easily generated on a tabletop laser system. Because of their high spatial and temporal coherence and low divergence, HHG can be used in such as atomic and molecular spectroscopy and nonlinear optics as an excellent EUV or soft X-ray source, and can be used in lithography and holography [6].

However, because of the phase mismatching, the conversion efficiency of HHG is still very low. The main reason of phase mismatching is generation of free electrons that ionized from atoms when the high intense laser focused into the gas. In the past years, HHG in different gas species, gas density, focus position, laser intensity, laser profile, laser chirp conditions and in different gas cells or hollow fibers have been studied systematically, and the optimal conditions to reach phase-matching were found in both experiments and theories (For a review, see Ref. [1] for an example). Among these experiments, the report of the highest harmonics conversion efficiency is 6.4×10^{-5} for the 15th harmonic generated in xenon gas-filled hollow fiber using a 35fs, 16mJ, 810nm pulse at 10Hz repetition rate [7].

The experimental setup is illustrated in Fig. 1. In the experiment, we used a 1 KHz repetition rate, 800nm, 50fs (FWHM) pulse duration, and 0.55mJ/pulse commercially available all-solid-state femtosecond laser system. The spatial profile of output laser beam is nearly a Gauss distribution. Focused by a fused silica lens (f/34), the Gauss laser beam fought its way thought the 0.1-mm-thick stainless steel seals of the glass cell that filled with noble gases. The cell was fixed in a vacuum chamber. A precision valve outside the chamber controlled the gas pressure in the cell. The HHG was emitted just in the glass cell. As the laser propagated forwards, two 400nm thick Aluminum foils blocked it, but the highly collimated coherence harmonic radiation propagated

* The work was supported by the National Natural Science Foundation of China (Grant Nos.69925513 and 19974058) and the Chinese National Major Basic Research Project (Grant No. G1999075204).

though them. A flat-field X-ray spectrograph [9] was used to obtain the spectrum of HHG in the EUV and soft X-ray range. The spectrograph consists of a gold-coated spherical mirror, and a cylindrical mirror, a slit, a flat-field grating and a soft X-ray CCD. The harmonic spectra is spatially resolved in horizontal axis and spectrally resolved in vertical axis.

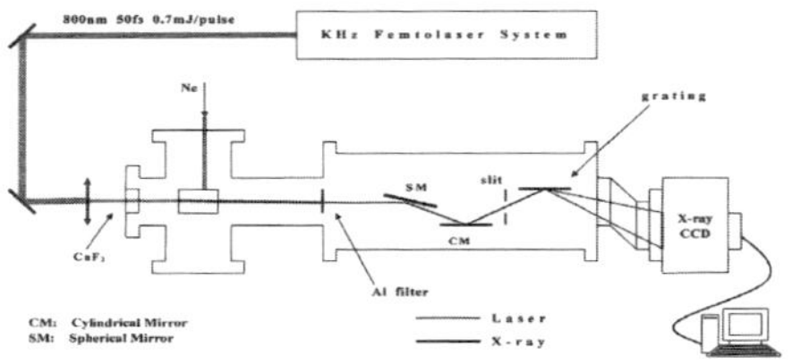

Fig. 1. Experimental setup of HHG in noble gas-filled cell.

In the experiment, the length and diameter of the cell used in the experiment is 20mm and 10mm respectively, and the intensity of the driving laser in the focus spot is about 2×10^{14} W/cm^2. The signal from 21st to 33rd order harmonics were obtained in Ar gas, and a typical spectrum of HHG in Ar are shown in Fig. 2. The highest conservation efficiency of 27th order was about 1×10^{-7} experimentally in the optimal phase-matching condition of low driving laser energy.

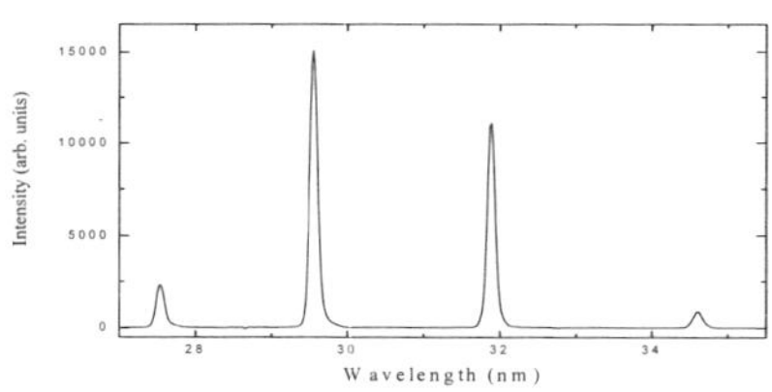

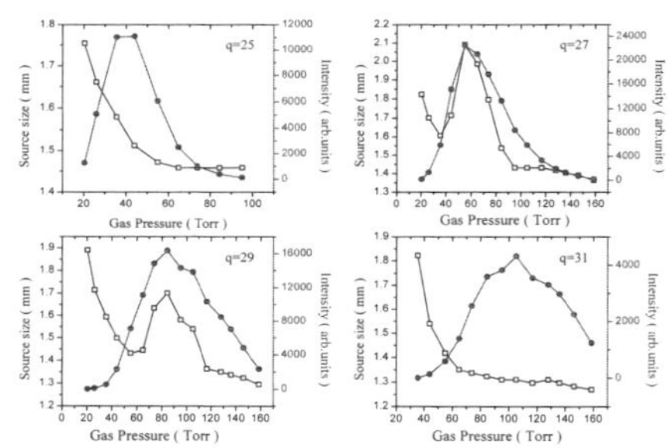

Fig. 2. Typical experimental spectrum of HHG in 44Torr Ar gas, harmonics from 23rd to 29th are shown from right to left. (left)

Fig. 3. The source size (line with hollow cubes) recorded on the CCD plate and the intensity (line with filled circles) for different order harmonics in different gas pressure conditions. (right)

As far as we know, we first observed that the source sizes (FWHW measured in the CCD plate) of HHG were enhanced under phase-matching conditions, which was quite different from the experimental results of hollow fibers (Ref. [7] as an example) and loosely focused (f/71) laser driven for static cells [8].

The source size and the intensity for different order harmonics in different gas density conditions are presented in Fig. 3. In the experiment, the measurement was one dimensional spatially resolved. The harmonic source was not sharply imagined in the CCD plate of the spectrograph, however, which would not influence the interpretation in this section. The source size was confirmed by the spatial distribution in the CCD plate. For the phase-matched 27th order harmonic and 29th order harmonic, the source sizes were augmented about 30% and 20% respectively under theirs phase-matching conditions.

In the experiment, the spatial distribution in the CCD plate is directly related to the spatial distribution of the intensity of high-order harmonic generated in the interaction region of the driving laser with Ar gas. Therefore, the variations of the source size and the spatial distribution reflect the change in the spatial distribution of HHG. The distributions of HHG mainly rely on the distributions of the driving laser intensity and the free electron density. The distribution of the laser intensity was nearly a Gauss distribution in the experiment, and the distribution of the free electron density was straight determined by that of the diving laser intensity. Under ordinary conditions, Ref. [11] as an example, the distribution of high-order harmonic in the CCD plate was nearly a Gauss distribution, like the intensity distribution of the driving laser.

However, we observed that the spatial distribution in the CCD plate was not always a Gauss distribution under the phase-matching condition in the experiment. In Fig. 4, it is shown that the spatial distribution of the intensity of phase-matched 27th order harmonic generation in 55.0Torr Ar gas is truncate, which is similar to the result of the saturated case. In fact, the counts in the CCD plate were not saturated, and the HHG was not saturated under the intensity of the driving laser in the experiment. Furthermore, the spatial distribution of the intensity of 27th order harmonic generation in 26.0Torr gas cell without phase-matching is quite different, also shown in Fig. 4.

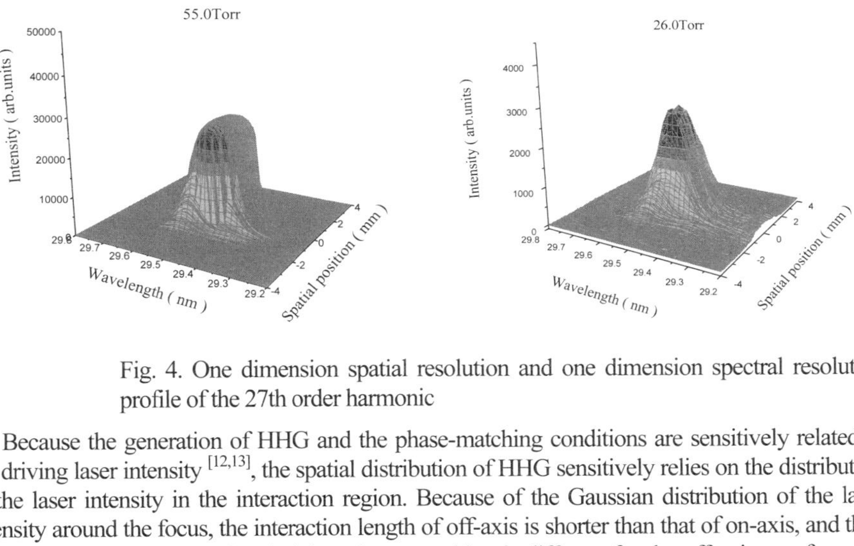

Fig. 4. One dimension spatial resolution and one dimension spectral resolution profile of the 27th order harmonic

Because the generation of HHG and the phase-matching conditions are sensitively related to the driving laser intensity [12,13], the spatial distribution of HHG sensitively relies on the distribution of the laser intensity in the interaction region. Because of the Gaussian distribution of the laser intensity around the focus, the interaction length of off-axis is shorter than that of on-axis, and then the gas density condition of optimizing phase-matching is different for the off-axis one from the on-axis one. For the shorter interaction length one, one needs shorter absorption length to achieve the optimizing phase-matching, as mentioned in Ref. [10]. Therefore, the optimized gas density for the off-axis is higher than that of the on-axis one, which is presented in the results of the experiment. As the gas density increases, the on-axis emission achieved optimizing phase-matching before the off-axis one, and there has an optimized gas density for the highest conversion efficiency. As a result, the spatial distribution of HHG will not be a Gauss distribution, but the distribution shown in Fig. 4.

The spatial distributions for the intensity of phase-matched 27th order harmonic generation in different gas density are presented in Fig. 5, which shows that the enhancement of the source size of HHG under phase-matching condition is mainly related to the intensity increase in the off-axis of the interaction region.

Moreover, in Fig. 4 and Fig. 5, it is also shown that the spatial contours of phase-matched harmonics are quite smooth and those of harmonics without phase-matching are fluctuant. We contribute the difference to the influence of the variation of dipole phases of HHG from different spatial position in the interaction region. Without phase-matching, the dipole phases of HHG from different place are quite different and will not coincide with each other after the propagation in Ar gas. The superposition of the dipoles with different phases will make the spatial contour very complex. Therefore, the spatial distributions of the phase-mismatched HHG will be fluctuant [12]. In contrast, for the phase-matched HHG, only the dipoles with the distinct phase can be enhanced by the superposition, and then the HHG with the same phase can be enlarged by the coherent superposition through the propagation in Ar gas, so the spatial distribution will be smooth. Furthermore, the phase-matching condition is sensitive to the laser intensity, so the spatial distribution has sharp edge as shown in Fig. 5.

In addition, in Fig. 3, the enhancement of source size is not observed for the 25th order harmonic. This is due to that the absorption edge of Ar gas is around the wavelength of this harmonic. On the other hand, for the higher order harmonics, the influence of absorption difference is not obvious. For the 31st order harmonic in Fig. 3, there is no enhancement in the size because this harmonic is not well phase-matched.

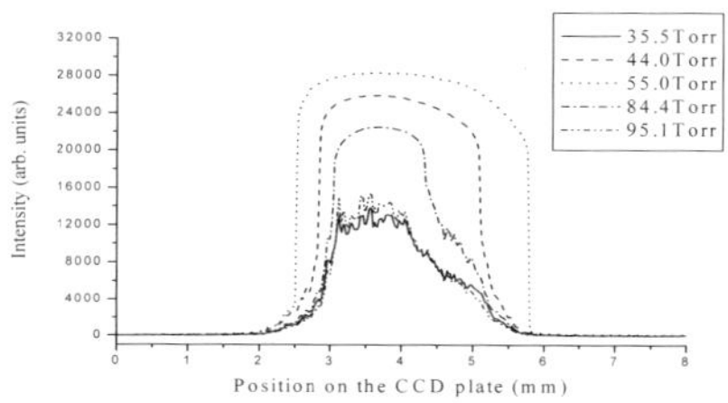

Fig. 5. The spatial distribution of the intensity of the 27th order harmonic.

In conclusion, we obtained phase-matched HHG in static Ar gas-filled cell driven by a low energy (0.55mJ/pulse) femtosecond laser pulse. Moreover, we found that the source size and the spatial distribution of the intensity of HHG is quite different under the phase-matching condition from that of the phase-mismatching case, which is mainly related to the spatial distribution of the intensity of driving laser, and lead to the flat-top output of HHG when the conversion efficiency is optimized.

References

[1] Brabec T and Krausz F 2000 Rev. Mod. Phys. 72 545
[2] Spielmann Ch et al 1997 Science 278 661
[3] Drescher M et al 2001 Science 291 1923
[4] Drescher M 2002 Nature 419 803
[5] Baltuška A et al 2003 Nature 421 611
[6] L'Huillier A et al 2003 Eur. Phys. J. D 26(1) 91
[7] Takahashi E et al 2002 Opt. Lett. 27 1920
[8] Tamaki Y et al 2000 Phys. Rev. A 62 063802
[9] Li R et al 1994 J. Optics 25 143
[10] Salières P et al 1999 Adv. At. Mol. Opt. Phys. 41 83
[11] Takahashi E et al 2003 J. Opt. Soc. Am. B 20 158
[12] Milosevic N et al 2002 Phys. Rev. Lett. 88 093905
[13] Gaarde M B et al 2002 Phys. Rev. A 65 031406

Inst. Phys. Conf. Ser. No 186
Paper presented at 9th Int. Conf. X-Ray Lasers, Beijing, China, 24–28 May 2004

Enhancement of high harmonic generation at tuned wavelengths via adaptative control

D.H. Reitze*, S. Kazamias, F. Weihe, D. Douillet, F. Augé, G. Mullot, O. Albert, Ph. Balcou**

Laboratoire d'Optique Appliquée (LOA) , CNRS-ENSTA-Ecole Polytechnique, batterie de l'Yvette, F-91128 Palaiseau cedex-FRANCE

Present address: * University of Florida, Gainesville, FL 32611 USA

** LIXAM, Université Paris XI, bat 350, F-91405 Orsay cedex- FRANCE

Abstract. We present an experimental study about the spectral tunability of the extreme UV radiation produced by high order harmonic generation in rare gases. We concentrated on the spectral region around 30 nm for which simple chirp tuning using the infrared pump beam is not sufficient. The best generation conditions were found using a genetic algorithm and playing with the spectro-temporal characteristics of the infrared laser. We demonstrated a full tunability for harmonic orders H23 to H27 generated in argon. The best conditions were obtained within a few generations only and with a reduced loss in the conversion efficiency.

1. Introduction

Compared to X ray lasers, the high order harmonic generation in rare gases provides for a smaller soft X ray source in terms of pump laser but also in terms of the number of photons produced, especially for wavelengths shorter than 15 to 20 nm. On the contrary, the present article will show that the harmonic source has the advantage to be quite easy to manipulate in order for example to produce spectrally tunable coherent extreme UV radiation.

This may be of a great importance for many applications and especially for the experimental demonstration of the amplification of an harmonic beam in an X ray laser medium. The injected harmonic wavelength must indeed be spectrally matched to the X ray laser transition line, which is spectrally quite narrow and very well defined [1].

The problem of the tunability of high harmonics is more crucial for relatively low orders because simple chirp tuning of the infrared laser is in that case not able to achieve total spectral covering. For example, the studies carried out in the KAIST laboratory and presented by C. H. Nam [2] show interesting results for harmonic orders around the 61st corresponding to the 13.4 nm spectral region. In the following, we will concentrate on the 30 nm spectral region for which high conversion efficiency (up to 3.10^{-5}) was demonstrated in argon in absorption limited conditions [3].

A very simple model can be used to define the harmonic order above which there is no need of a complex process to achieve total tunability. Let us consider as a rough approximation that the XUV radiation exactly corresponds to odd harmonics of the laser radiation. The infrared pump beam is represented by a given spectral width (FWHM) $\delta\lambda$ (typically 30 nm for Ti: Sa installations) around a central wavelength λ_0 (typically 805 nm). Each wavelength near the centre of the infrared spectrum may lead to the generation of harmonic radiation. Total tunability will be easily obtained if a certain wavelength can be reached either by a given harmonic order q generated with the "blue" part of the infrared spectrum, either by the next harmonic order q+2 generated with the "red" part of the spectrum. We then obtain the following equation: $(\lambda_0 - \delta\lambda/2)/q = (\lambda_0 + \delta\lambda/2)/(q+2)$

The condition on the order is thus: $q + 1 \geq 2\lambda_0/\delta\lambda$, this corresponds in our case to q=53 or 15 nm. This model is of course oversimplified and does not take into account the complex contribution of parameters such as ionization or quantum phase of the harmonic spectral profile but it nevertheless gives an estimation of the role of the laser spectral width on the harmonic wavelength. The present work is done with relatively low orders as compared to the former condition and that explain why we did not observe satisfactory results only by playing with the infrared laser chirp.

2. Experimental set-up

The high harmonic generation experiment was carried out on the 1 kHz laser system at the LOA ("Salle rouge" Palaiseau- France). It is a Ti:Sa multistage chirped pulse amplifier producing 6 mJ, 30 fs pulses centered at 805 nm. An acousto-optic programmable dispersive filter located before the first amplifier stage was used to change the pulse spectral phase (the "dazzler" is produced by the Fastlite company [4]). Upon compression, the pulses were first apertured with a 11 mm wide iris in order to optimize the phase-matching [5]. They were then focused via a 1m focal length lens into a 1mm long gas cell filled with 15 torr of argon. The focal point was 2 mm before the cell entrance and the estimated intensity in the medium for non-reshaped pulses was around 2.10^{14} W/cm^2. The resulting harmonics were separated from the infrared radiation using a 300 nm thick aluminium filter and sent to a VUV transmission spectrometer and CCD detector. The acquisition time for each harmonic spectrum was about 0.1s corresponding to 100 shots.

A genetic algorithm was used to search for the infrared laser spectral phase configurations to allow for the production of the desired harmonic spectrum. One spectral phase configuration was parametrized by 12 terms related to the Taylor-series coefficients. This high number of parameters allows to work with complex phase structures but remains well within the limitations of the acousto-optic filter. The so-called "cost function" or "fitness function" that was used to characterize the spectra obtained after each run was based on the Gaussian fit of the harmonic peaks in order to know their central wavelength, width and height and compare them to the final goal of the algorithm. In addition, a term was added to maximize the total harmonic output. We found that the choice of the fitness function was very important for rapid and successful optimization.

3.Experimental results

A series of optimizations were performed for a discretized set of wavelengths spanning the gap between two harmonic peaks. For each optimization, 41 individuals were tested and a transform limited pulse was included in the first generation to ensure that the algorithm began from a reasonable point in parameter space with good conversion efficiency.

Fig. 1 shows an example of an optimization to tune H27 from a nominal wavelength of 30 nm to a target wavelength of 31.2 nm. The best spectra for different generations are presented. The conversion efficiency for the optimized pulse shape is estimated to 1.4 10^{-6} The inset shows the laser pulse intensity and phase that produced the optimal output spectrum. The result is rather complicated and indicates that a simple chirp could not have led to this optimization. Finally, we demonstrated a quasi complete spectral tuning for H27 between 28.9 nm and 31.15 nm except for a 0.6 nm band just below 30nm.

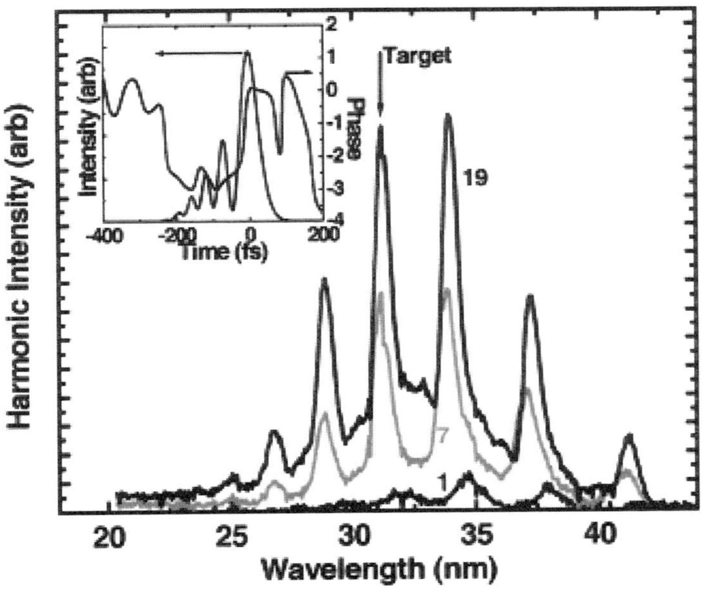

Figure 1: Evolution of the optimal harmonic spectrum for the tuning of H27 to 31.2 nm. The numbers next to the spectra indicate the generation number. The last generation was 19 and the inset shows the laser characteristics that led to the optimization. This figure is reproduced from reference [6].

4. Interpretation of the results and conclusion

A one-dimensional time dependant simulation of high harmonic generation and phase matching was performed to understand the relevant parameters for the former optimization. The temporal phase and intensity profiles of the optimized laser pulses were recomputed from the 12 parameters found by the genetic algorithm. The harmonic temporal phase could then be deduced from the intrinsic atomic phase [7] plus the propagation phase plus q times the infrared laser phase (q is the harmonic order considered). Many physical processes implied in the generation of high harmonics can lead to a time dependant harmonic phase and consequently to modified spectral characteristics of the radiation. The latter can be obtained by the Fourier transform of the temporal harmonic dipole amplitude and phase into the spectral domain.

Simulations made on the configurations that led to the demonstration of the tunability of H27 allowed to show that no parameter could be neglected to explain the results and confirmed that the optimization performed by the algorithm is rather complex and implies many processes such as ionization time profile, pump laser phase but also intrinsic atomic phase. This atomic phase plays a more important role in our experiment performed in a gas cell with a multimillijoule laser than in the experiment reported in reference [8] for which a gas filled capillary was used to generate the harmonics at a smaller laser intensity.

In conclusion, we performed an experimental demonstration of the quasi full tunability of the harmonic radiation in the 30 nm spectral range for which the conversion efficiency is high. The optimization was obtained through a genetic algorithm within a few tens of generations and the conversion efficiency was only slightly reduced (in the 10^{-6} range). The optimization conditions could not have been found using simpler configurations such as playing only with the laser chirp. More details on this experiment can be found in reference [6].

References

[1] G. Faivre, Ph. Zeitoun et al., proceeding of the ICXRL2004 "Second generation x-ray laser"

[2] C.H. Nam, H.T. Kim, I J. Kim, C.M. Kim, V. Tosa, proceeding of the ICXRL2004 "Optimization of high harmonic generation in a long gas jet in the space and time domains"

[3] S. Kazamias, D. Douillet, et al., Physical Review Letters 90, 193901 (2003).

[4] http://www.fastlite.com/

[5] S. Kazamias, F. Weihe, D. Douillet, C. Valentin, T. Planchon, S. Sebban, G. Grillon, F. Augé, D. Hulin, and Ph. Balcou, European Physical Journal D 21, 353 (2002).

[6] D. H. Reitze, S. Kazamias, F. Weihe, G. Mullot, D. Douillet, F. Augé, O. Albert, V. Ramanathan, J.P. Chambaret, D. Hulin et Ph. Balcou, Optics Letters 29, 86 (2004)

[7] Lewenstein et al., Phys. Rev. A 52, 4747 (1995)

[8] R.A. Bartels, S. Backus et al., Nature 406, 164 (2000)

Inst. Phys. Conf. Ser. No 186
Paper presented at 9th Int. Conf. X-Ray Lasers, Beijing, China, 24–28 May 2004

High-intensity attosecond high-order harmonic generation driven by synthesized laser field

Zhinan Zeng, Ruxin Li, Xinhua Xie, Zhizhan Xu

Laboratory for High Intensity Optics, Shanghai Institute of Optics and Fine Mechanics, Chinese Academy of Sciences, Shanghai 201800, China

Abstract The scheme of high-intensity attosecond high-order harmonic generation driven by a synthesized laser field is proposed. The synthesized laser field is obtained by an appropriate superposition of a few-cycle laser pulse and a relatively long pulse of several tens of femtoseconds. Calculated results show that, the intensity of the attosecond high-order harmonic pulse in helium driven by the synthesized laser field with a 0.05a.u./5fs laser pulse and a 0.1a.u./50fs laser pulse is several orders of magnitude higher than that driven by a single 0.05a.u./5fs laser pulse, and it is even stronger than that driven by a single 0.15a.u./5fs laser pulse, although the peak electric field of driving laser pulses is the same.

1. Introduction

The generation of attosecond pulses is very important and it opens a brand new chapter in ultrafast spectroscopy [1-2]. Just as femtosecond pulses allow the observation of the motion of atoms in molecules, attosecond pulses will allow us to trace electronic motion and electronic relaxation (such as inner-shell dynamics) in atom and molecules. A number of methods have been proposed for attosecond pulse generation [3-5]. High harmonic generation (HHG) is currently the experimentally furthest advanced method that can supply single as-pulses, which is essential for many potential applications. Recently M. Hentschel et al produced the first single attosecond x-ray pulse with a 7fs laser pulse as the driving laser for high order harmonic generation [2].

However, the XUV/x-ray pulses generated by HHG process are currently too weak to induce measurable nonlinearities of atomic media. As a result, although the autocorrelation scheme is conceptually simple and well understood, so far it has been impossible to implement in the x-ray regime. In the experiment of measuring the duration of the generated attosecond pulse, the cross correlated method is used in which an x-ray photon excites a bound atomic electron into a positive-energy state in the presence of a laser field and then laser-induced transitions from this state or a laser-induced shift in its energy (Stark shift) provide the nonlinearity linking the x-ray to the laser pulse that enables us to compare the x-ray pulse duration with that of the visible laser pulse. Although the x-ray pulse duration can be measured with this method, to produce an attosecond pulse with higher intensity is highly desired for other measurements and its application, e.g. the detailed temporal characterization of the x-ray pulse may need higher energy and higher intensity.

In current experiments on the attosecond pulse generation, the few cycle driving laser pulse (typically 0.7mJ/5fs) was obtained by the post amplification spectral broadening technique in which the hollow-fiber chirped-mirror high-power compressor was used [6]. However, the energy of the few-cycle laser pulse produced by this method is presently limited in the level of 1mJ due to the ionization of the gas atoms in the hollow fiber, which results in the limitation in the energy and intensity of the attosecond x-ray pulse generation by HHG.

2. Theoretically model

In this paper we proposed a scheme of generating high-intensity attosecond high-order harmonic by using a synthesized laser field. The synthesized laser field is obtained by an appropriate superposition of a few-cycle laser pulse and a relatively long pulse of several tens of femtoseconds whose energy can be increased to the level of several tens of millijoules. Calculated results show that, the intensity of attosecond high-order harmonic pulse in helium driven by the synthesized laser field with a 0.05a.u./5fs laser pulse and a 0.1a.u./50fs laser pulse is several orders higher than that driven by a single 0.05a.u./5fs laser pulse, and it is even stronger than that driven by a single 0.15a.u./5fs laser pulse, although the peak electric field of the driving laser pulses are the same.

The approach in the calculation of attosecond HHG is similar to that described in previous works [7, 8]. We calculated the dipole acceleration by solving numerically the time-dependent Schrödinger equation (TDSE). The synthesized laser pulse in the calculation is described by $E(t) = E_1(t) \cos(\omega t + \varphi_1) + E_2(t-t') \cos[\omega(t-t') + \varphi_2]$, where ω is the frequency of the laser field and φ_i the absolute phase, $E_i(t) = E_i \exp(-t^2 / \tau_i^2)$ denotes the envelope of the driving electric field, where E_i is the peak amplitude of the laser pulse and $(2\ln 2)^{1/2}\tau_i$ is its duration, $i = 1, 2$. t' is the delay between two pulses. The wavelength of the laser pulse is 800nm. The dipole acceleration is then Fourier transformed to obtain the spectrum, and a time-frequency analysis (the Gabor analysis[9]) finally provides the temporal profile of the harmonics. We calculated the HHG in helium atoms and the attosecond pulses are generated at the cutoff regime of the HHG.

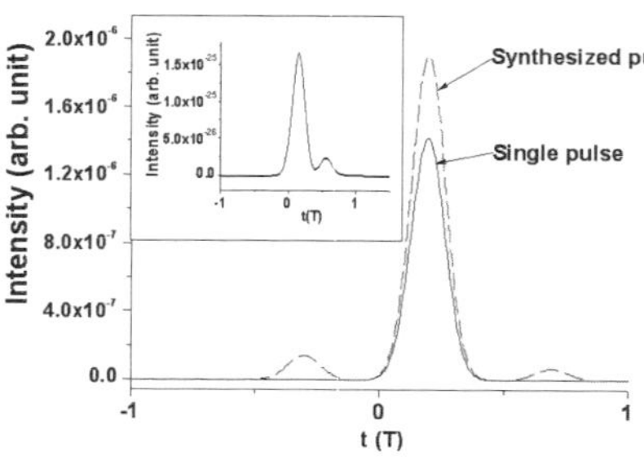

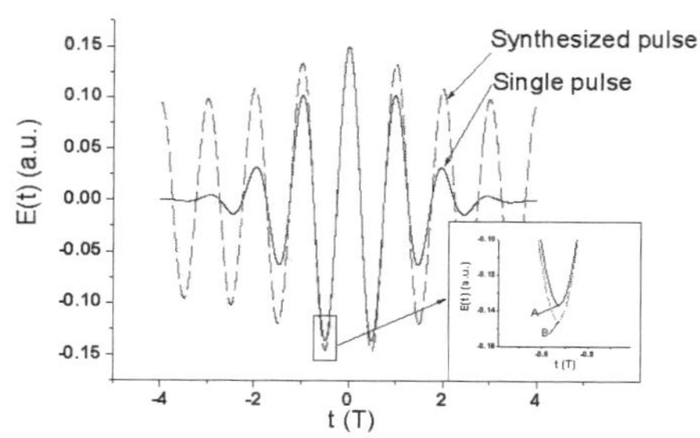

Fig. 1 The temporal profile of attosecond x-ray pulses generated by single few-cycle laser pulse (0.15a.u./5fs, solid line) and synthesized pulse scheme (0.05a.u./5fs+0.1a.u./50fs, dashed line). The inset shows the temporal profile of the attosecond x-ray pulse by the single laser pulse of 0.05a.u./5fs in the synthesized pulse.

Fig. 2 The electric field curves of the single 5fs laser pulse (solid line) and the synthesized (50fs + 5fs) pulse (dashed line) used in obtaining the results shown in Fig.1.

3. Simulation result and discussion

Fig. 1 shows the temporal profiles of the attosecond high-order harmonic pulses produced by a single 5fs laser pulse and a synthesized laser pulse of 50fs and 5fs pulse pair. The laser parameter in the calculation is 0.15a.u./5fs for the single laser pulse case and 0.05a.u. /5fs+0.1a.u./50fs for the synthesized pulse case. The carrier-envelope phase and the time delay between two pulses are zero in the calculation. The attosecond x-ray pulses are produced in the cutoff regime of HHG and the photon energies (174eV in the calculation) of the attosecond x-ray pulses are the same. From this figure one can see that, not only a single attosecond x-ray pulse can be produced with the synthesized pulse, but its intensity is even higher that generated by a single laser pulse with the same intensity. For comparison, the inset in this figure shows the attosecond x-ray pulse with the same photon energy generated by just the 0.05a.u./5fs part of the synthesized laser pulse. The intensity is lower by 19 orders of magnitude in comparison with that driven with the addition of 50fs pulse (Because there is only the noise in the calculation for the result of the single 0.05a.u./5fs laser pulse.), although the difference in the driving laser electric field is only 3 folds.

Fig. 2 shows the electric field curves of the single 5fs laser pulse (solid line) and the synthesized (50fs + 5fs) pulse (dashed line) used in obtaining the results shown in Fig.1. According to the Corkum's model of HHG [10], before the attosecond x-ray pulses are generated (as shown in Fig. 1), the electrons will be ionized at the position A (solid line, single laser pulse) and B (dashed line, synthesized pulse scheme). The electric field at the position B is a little stronger than that at the position A. Therefore in the synthesized pulse scheme, more electrons will be ionized. Then after the free motion in the laser field, more electrons will recombine to the atom to generate the attosecond x-ray pulse. This is the reason why the intensity of the attosecond x-ray pulse generated by synthesized pulse scheme is higher than that generated by a single laser pulse as shown in Fig.1, although the peak electric fields of the laser pulses of two schemes are the same.

Within this synthesized pulse scheme, the carrier envelope phase (CEP) of the laser pulse is very important. Our calculation shows that comparing to the case of zero CEP difference, the intensity of the attosecond pulse will reduce to be half if the CEP difference of two laser pulses is $\pi/8$. When the difference of the CEP of two pulses increases, the intensity of the attosecond pulse decreases very fast. But this does not mean that the scheme is difficult to realize in a realistic experiment. According to the Ref. [11], the CEP jitter emerging in the waveguide is found to be less than 50mrad. That means the difference of CEP of two laser pulses can be very little. The only thing we need to do in the experiment is to adjust the time delay of these two laser pulses. Fig. 3 shows the result of different CEPs when the two laser pulses have good time delay. It is similar to the result of single driven laser pulse with different CEPs [7]. If the CEP of the driven laser pulse is stabilized, the similar result can be obtained as that in Ref. [11].

To implement this proposal experimentally is straightforward (Fig. 4). One can use a normal chirped pulse amplification laser system to obtain 50fs laser pulse at the energy level of 22mJ and split it into two parts, and then send one part into a hollow fiber chirped mirror compressor to obtain few cycle laser pulse at the energy level of 0.5mJ[6], and then combine and synchronize the two pulses by suitable delay line before delivering the synthesized laser pulse to the laser atom interaction region for HHG. In recently reported attosecond pulse generation experiments, tight focusing is needed due to the energy of the few-cycle laser pulse produced by the hollow-fiber chirped mirror compressor is limited, and the area of the focus is very small which will limit the volume of the interaction and the energy of generated attosecond pulse. With the synthesized laser pulse scheme proposed in this work, the area of the focus can be extended, more atoms and more laser energy in the long pulse (50fs) can participate in the interaction, higher energy attosecond pulse can therefore produced, higher intensity attosecond pulse for potential nonlinear spectroscopy study is possible with suitable beam-splitter and focusing optics.

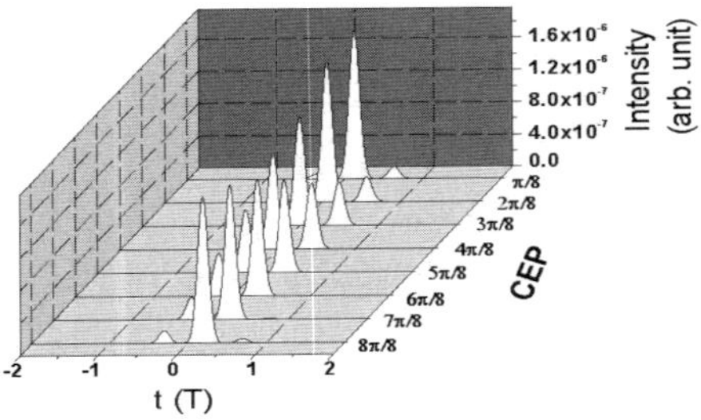

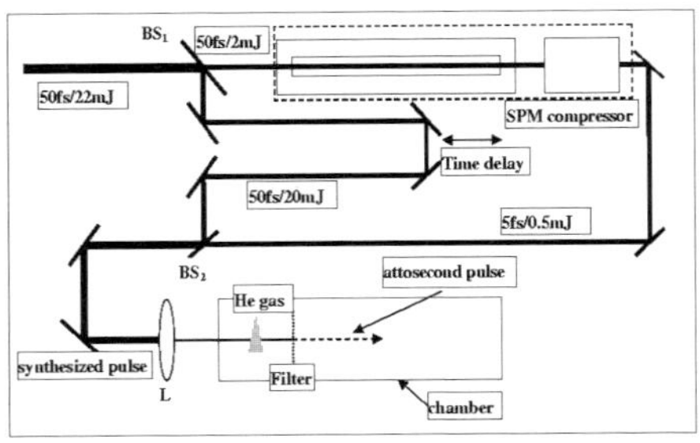

Fig. 3. The attosecond pulse changes with the CEP of the synthesized pulse. It is similar to the result of single driven laser pulse.

Fig. 4. The scheme for the synthesized pulse. The laser pulse is split into two beams by BS_1. One part is compressed with SPM compressor and a 5fs pulse is generated. Then two pulses are combined and synchronized by suitable delay line at BS_2. Within the synthesized pulse, most energy is offered by the long pulse (50fs pulse). The synthesized pulse can be focused loosely and more atoms can participate in the interaction.

4. Summary

In conclusion, we have presented a new scheme of generating high-intensity attosecond high-order harmonic by using a synthesized laser field. The synthesized laser field is obtained by an appropriate superposition of a few-cycle laser pulse and a relatively long pulse of several tens of femtoseconds whose energy can be increased to the level of several millijoules. This new scheme for attosecond pulse generation gives a way for the realization of high intensity attosecond pulse which can be used in attosecond nonlinear spectroscopy.

References

[1] Drescher M, Hentschel M, Kienberger R *et al* 2001 Science 291 1923-1926
[2] Hentschel M, Kienberger R, Spielmann Ch *et al* 2001 Nature 414 509-512
[3] Kapteyn H C and Murnane M M 1991 J. Opt. Soc. Am. B 8 1657-1662
[4] Kaplan A E 1994 Phys. Rev. Lett. 73 1243-1246
[5] Christov I P, Murnane M M and Kapteyn H C 1997 Phys. Rev. Lett. 78, 1251-1254
[6] Brabec T and Krausz F 2000 Rev. Mod. Phys. 72 545-576
[7] Z. Zeng et al 2003 Phys. Rev. A 67 013815
[8] Z. Zeng et al 2003 Chin. Phys. Lett. 19 1112
[9] Antoine P and Piraux B 1995 Phys. Rev. A 51 1750-1761
[10] Corkum P B 1993 Phys. Rev. Lett. 71 1994-1997
[11] Baltuška A, Udem Th, Hentschel M *et al* 2003 Nature 421 611-613

Inst. Phys. Conf. Ser. No 186
Paper presented at 9th Int. Conf. X-Ray Lasers, Beijing, China, 24–28 May 2004

Point-diffraction interferometry using high-order harmonic soft-x-ray

J. J. Park, J. H. Sung, G.-H. Lee, and C. H. Nam

Department of Physics and Coherent X-ray Research Center, Korea Advanced Institute of Science and Technology

Abstract:
A point-diffraction interferometer has been realized using high-order harmonic beams. High-order harmonics were generated by focusing intense Ti:sapphire laser pulses into an Ar- or Ne-filled tube. Results of double-pinhole interference showed that the generated harmonic beam possessed good spatial coherence. The wave-front phase of the harmonic beam estimated from the point-diffraction interferogram was nearly spherical.

1. Introduction

High-order harmonics generated from the coherent interaction between atoms and intense femtosecond laser pulses have been studied intensively due to the potentiality as a coherent soft-x-ray source. High-order harmonic generation has several good properties such as low divergence [1], ultrashort duration [2], and good spatial coherence [3,4]. Comparing to other EUV or soft x-ray sources, high-order harmonics have shown excellent spatial coherence even though a spatial filtering was not used. Bartels *et al.* obtained a Gabor hologram using this coherent soft-x-ray light source [3]. The spatial coherence of the high-order harmonic beam has been usually measured using double-pinhole interferometry because the beam splitter and mirror in the x-ray region are not easily available.

It is important to measure the wave-front phase of the harmonic beam from the viewpoint of improving beam quality and investigating its propagation. Point-diffraction

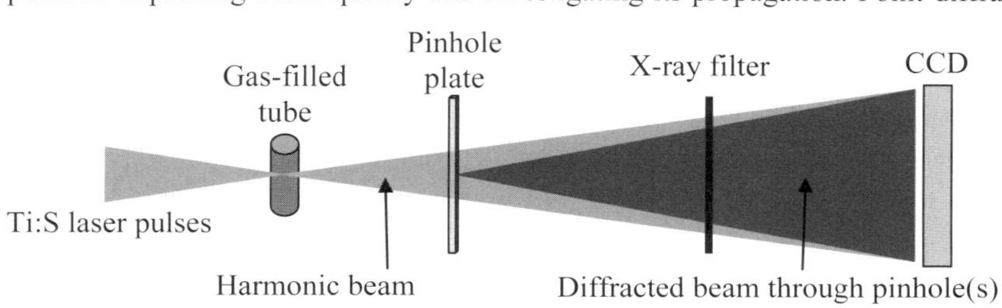

Figure 1. Optical layout to measure the spatial coherence and wave-front phase of a harmonic beam.

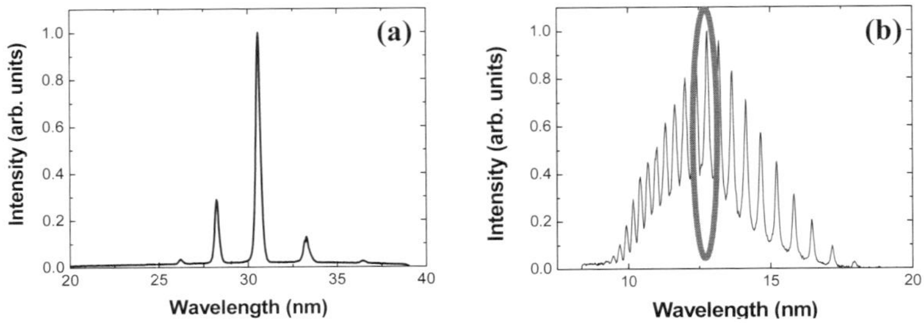

Figure 2. High-order harmonic spectra from **(a)** Ar and **(b)** Ne atoms. For the 13-nm harmonic beam in (b), only one harmonic order, inside the gray ellipse, was selected using a Mo:Si x-ray mirror.

interferometry (PDI) is a simple, but powerful technique that can be applied to the measurement of the wave-front phase of a harmonic beam. A pinhole in a semi-transparent material generates a reference wave front that interferes with the transmitted harmonic wave at the detector. Analysis of the PDI interferogram enables us to measure the wave-front phase of a harmonic beam.

Here, we report the spatial coherence measurement of 30 and 13 nm harmonic beams performed using the double-pinhole interferometry, and, then, we describe the measurement of the wave-front phase of 30-nm harmonic beam carried out using PDI.

2. Experimental setup

The spatial coherence and the wave-front phase of the harmonic x-ray beam were measured with the setup shown in Fig. 1. Ti:sapphire laser pulses with 28-fs duration and 3-mJ energy, and 1-kHz repetition rate were focused into a 10-mm gas tube filled with either Ar or Ne atoms. The gas density was controlled using a needle valve up to 100 Torr. Harmonic spectra were obtained with a flat-field soft x-ray spectrometer equipped with an x-ray CCD. Harmonic spectra from Ar and Ne atoms are shown in Fig. 2. The 27th harmonic from Ar was made dominant by controlling such experimental parameters as incident laser energy and Ar density in the gas tube. For the coherence measurement of high harmonic beam using a double-pinhole interferometry, we drilled several pinhole pairs on an 18-μm thick Al foil using femtosecond laser pulses. The pinhole diameter was 10 μm and the separation between the pinholes varied from 40 μm to 200 μm. We installed it after the target to investigate the spatial coherence of the generated harmonic beam. For the 13-nm soft x-ray, a Mo:Si x-ray mirror was additionally installed in the beam path to select only the 63rd harmonic order, which gives better coherence. Al- or Zr-filter, whose thickness was 400 nm, was used to block the fundamental laser pulses. We also drilled a 10-μm diameter hole in the center of a 1-μm thick Al filter for the measurement of the wave-front phase. The thickness and the pinhole diameter were determined so as to make intensities of transmitted and point-diffracted beam comparable at the detector.

3. Results

Double-pinhole interferograms obtained at 30 nm and 13 nm are shown in Fig. 3(a) and (b), respectively. Pinhole separations for both cases were 100 μm. Both 30- and 13-nm harmonic soft x-ray beams were found to have good spatial coherence. The visibilities of the interference patterns are over 0.9 and 0.8 at 30 and 13 nm, respectively. Harmonic x-ray beam size at the pinhole plate was 130 μm (FWHM) for the 27th and 500 μm

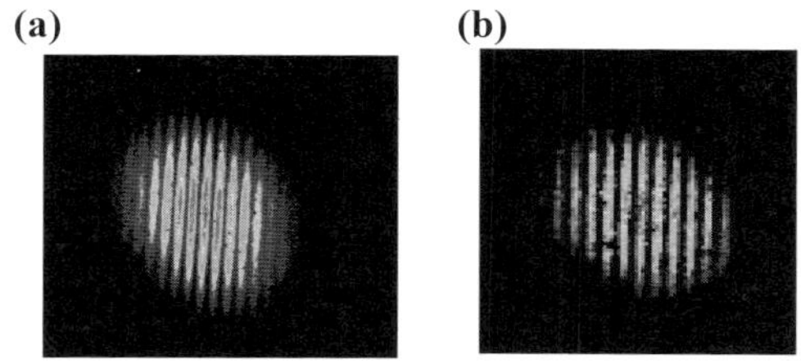

Figure 3. Double-pinhole interference fringe patterns generated by the harmonics at **(a)** 30 nm, and **(b)** 13 nm with the pinhole separation of 100 μm.

(FWHM) for the 63rd harmonic beam. Because the distance between the gas cell and the pinhole plate was 15 cm (93 cm) for the experiment of the 30-nm (13-nm) double-pinhole interferometry, the beam divergence was found to be almost 0.9 mrad (0.5 mrad). We also tried different pinhole separation. The fringe visibility was decreased moderately as the pinhole separation increased. From the measurement of the fringe visibility, we conclude that the harmonic beam has good spatial coherent over the whole beam cross-section.

We measured the wave-front phase of the 30-nm harmonic x-ray beam using point-diffraction interferometry. Measured interferogram is shown in Fig. 4. When the pinhole was moved to the center of the harmonic beam, the PDI interferogram had some concentric fringes with azimuthal symmetry. We calculated the far-field amplitudes of both transmitted and point-diffracted beam under the assumptions of Gaussian intensity distribution of the harmonic beam and the uniform illumination within the pinhole aperture. The calculated result shows good agreement with the measured interferogram (Fig. 4(b)). From the interferogram, the wave front of the harmonic beam was found to be nearly spherical with the radius of 1.19 m.

The wave-front phase deviation, calculated from the optical-path difference between the harmonic beam and a perfect spherical wave, was plotted in Fig. 4(c). It shows that

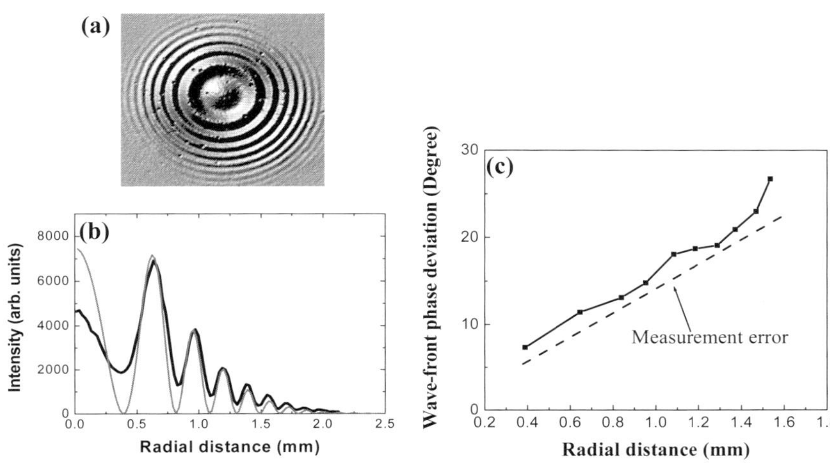

Figure 4. **(a)** Point-diffraction interferogram. **(b)** Intensity profile of the interferogram shown in (a) (thick black line) and calculated profile (thin gray line). **(c)** Wave-front phase deviation from a spherical wave with the radius of 1.19 m. Dashed line shows the measurement error due to the CCD pixel size.

the source of the deviation mainly comes from the finite CCD pixel size of 24 μm. The error was less than 10° at the central region and 30° at the edge of the interferogram. This results show that PDI is a compact and powerful tool suitable for EUV metrology.

4. Conclusions

We measured the spatial coherence of the high-order harmonic x-ray beam using double-pinhole interferometry, and found that generated harmonics had good spatial coherence. Using this good spatial coherence, we demonstrated a point-diffraction interferometry and measured the wave-front phase of the 30-nm harmonic beam. Harmonic beam has nearly spherical wave-front phase with the radius of 1.19 m, the distance from the birthplace to the detecting position of the harmonic. The error of the wave-front phase was less than 30° ($=\lambda/12$) for the whole beam region. The coherent harmonics will be valuable as a source of the x-ray microscopy, metrology, and holography.

This research was supported by the Ministry of Science and Technology of Korea through the Creative Research Initiative Program.

5. References

[1] D. G. Lee, H. T. Kim, et.al.Appl. Phys. Lett. **81**, 3726 (2002).
[2] P. M. Paul, E. S. Toma, et.al. Science **292**, 1689 (2001);
A. Baltšuka, Th. Udem, et.al.Nature **421**, 611 (2003).
[3] R. A. Bartels, A. Paul, et.al. Science **297**, 376 (2002).
[4] D. G. Lee, J. J. Park, J. H. Sung, and C. H. Nam, Opt. Lett. **28**, 480 (2003).

Inst. Phys. Conf. Ser. No 186
Paper presented at 9th Int. Conf. X-Ray Lasers, Beijing, China, 24–28 May 2004

High Harmonics Generated from Photoelectron Escape Process

Dong Sheng Guo[1,*], Jingtao Zhang[2], Wei Yu[2], Zhizhan Xu[2], Xiaofeng Li[3], Panming Fu[3], and R.R. Freeman[4]
[1]Huazhong University of Science & Technology, College of Optoelectronics & Information
[2]Laboratory for High Intensity Optics, Shanghai Institute of Optics and Fine Mechanics, CAS, P. O. Box 800-211, Shanghai 201800
[3]Laboratory of Optical Physics, Institute of Physics, CAS, Beijing 100080, China
[4]Department of Applied Science, University of California-Davis, Davis,California 95616
[*]Department of Physics, Southern University and A&M College, Baton Rouge, Louisiana 70813

Abstract. We report an unknown mechanism for high harmonic generation (HHG) in contrast to the well-known recombination mechanism. In this mechanism, the harmonics are generated when the photoelectrons exit the laser field. Our calculations show high harmonics in both odd and even orders also exhibit plateau and cut-off features similar to those with only odd orders. Comparison is made with spectra of harmonics produced from plasma and shows a good agreement. We reach a unified picture for above-threshold ionization (ATI) companied by HHG: there are two coexisting mechanisms to produce high harmonics. In the recombination mechanism, photoelectrons fall back to the original atomic bound state with an emission of odd-only harmonics. In the escape mechanism, photoelectrons escape from the laser field with an emission of both even and odd harmonics.

1. Introduction

High harmonic generation (HHG) accompanying above-threshold ionization (ATI) is one of the most interesting phenomena in strong-field laser physics. There are two different types of high harmonic spectra showed by experiments. One exhibits only odd harmonics [1] while the other exhibits both even and odd harmonics [2]. The former was showed by many ATI experiments and successfully described by recombination models [3-8], which assume that a photoelectron falls back to the atomic ground state and emits a high harmonic, with plenty discussions; while the latter was found from plasma irradiated by laser light [2] and has much less discussions [9-11]. There is an apparent theoretical difficulty in recombination models: since HHG is a phenomenon accompanying ATI why high harmonics are produced by the electrons not forming ATI peaks.

In 1989, Guo, Åberg and Crasemann (GAC) [12] published a scattering theory which treats multiphoton ionization (MPI) as a scattering process. According to GAC, the final state of the photoelectron is not a Volkov state, the state for an otherwise-free electron in a plane light wave, rather a plane wave; while the Volkov states play a role of intermediate states. With the single-mode assumption for the laser light, GAC theory results an integer condition for the ponderomotive parameter $u_p \equiv U_p/\hbar\omega$ which indicates the incompleteness of the single-mode assumption commonly adopted by most other theories and further indicates the existence of a spontaneous light emission [13].

The assertion about the final state from GAC was soon verified by an existing measurement which manifested the half Kapitza-Dirac (KD) effect performed by Bucksbaum et al. [14] in a standing-wave MPI experiment and theoretically treated by Guo and Drake [15]. The strong agreement between Bucksbaum et al.'s measurement and GAC theory in the standing wave case further strengthened the believe that the extra photon mode has to be included in a single-mode MPI.

According to GAC theory, Gao, Guo, and Wu [13] derived the formula for ATI rate with the inclusion of the spontaneously emitted photon mode. The numerical calculations in the current paper are based on the transition rate formula of Gao, Guo, and Wu. Comparison between current calculations and existing experimental measurements shows that there are two coexisting mechanisms for HHG, recombination mechanism and escape mechanism. A unified picture about ATI accompanied by HHG is given in the last part of this paper.

2. Transition rate formula with the inclusion of spontaneous emission

The transition formula derived by Gao, Guo, and Wu is the angular distribution for a given ATI peak is

$$\frac{d^4W}{d\Omega_{p_f}d\Omega_{k'}} = \frac{2e^2\omega^{9/2}}{(2m_e)^{1/2}(2\pi)^4}(j-\varepsilon_b-u_p)^{1/2}(j-u_p)^2 \times \sum_q (u_p-j+q)|\Phi_i(\vec{P}_f-q\vec{k}+\vec{k}')|^2 \, |\aleph_q(\vec{P}_f,\vec{k})|, \tag{1}$$

where j is the transferred-photon number in the ionization process and signifies the order of ATI peaks. For a specific ATI peak, the sum in Eq. 1 is performed over all possible q to add all the photoelectron rates with the emission of different harmonics;

$$\aleph_q(\vec{P},\vec{k}') \equiv \frac{1}{\omega}\aleph_{-j}(Z,\eta)\sum_{j'}\frac{\aleph_{-j'}(Z_f,\eta)}{u_p-j'}\{-[\vec{P}_f+(j-q-u_p)\vec{k}]\cdot\vec{\varepsilon}'^* X_{q-j+j'}(Z_{k'}) + e\Lambda\vec{\varepsilon}\cdot\vec{\varepsilon}'^* X_{q-j+j'+1}(Z_{k'}) + e\Lambda\vec{\varepsilon}^*\cdot\vec{\varepsilon}'^* X_{q-j+j'-1}(Z_{k'}).$$

where k' and ε' are the wave vector and polarization vector of harmonics mode.

Each term in the summation indexed by q in Eq.(1) corresponds to one harmonic light emission while the electron transits from the initial bound state to the final plan wave of the jth order ATI peak with the frequency

$$\omega' = q + (u_p - j)\omega \tag{2}$$

Here $q=n_i-n_f$ is the overall transferred photon number through the transition. This relation is obtained from the energy conversation required by the scattering theory. The quantity $q+u_p-j$ signifies a physical process as follows: the electron ionizes into the laser field by absorbing j photons, then, exits the field by re-absorbing q-j photons from the field if q>j and converting these photons into a photon with a high frequency ω' as the spontaneous-emission mode. If q<j, the photoelectron exits the field by emitting j-q photons back to the field directly without any spontaneous emission.

The spontaneous emission has a discrete energy spectrum which is the most important feature of harmonics. Thus, we can identify the spontaneous emission derived by this theory as a light harmonic. The other important features shared by the spontaneous emission derived here and the HHG will be discussed in later sections. The spacing between two neighboring harmonics in the spectrum of the spontaneous emission is exactly the laser photon energy

$$\Delta\omega' = \Delta[q + (u_p - j)\omega] = \omega$$

The order of the harmonic can be determined by q-j+[u_p]. An interesting phenomenon here is that the ponderomotive energy cannot be kept by the photoelectron when it exits the field, but turns into a part of the energy of a harmonic.

3. The spectra of spontaneous emission featured by even and odd harmonics

In our sample calculation, the wavelength of the incident laser is chosen as 1064nm and linearly polarized. The laser intensity varies from 10^{13}W/cm^2 to 2×10^{14}W/cm^2. Our calculations include angular distributions and spectral distributions of the spontaneous emission. The angular distribution calculation reveals the forward-propagation property. The harmonic emission rate in the laser propagation direction is, at least, twice stronger than that in the laser polarization plane. The spectral distribution calculation further reveals that the spontaneous emission does have all main characters of HHG in the odd-order-only case, such as the plateau and cut-off structures.

Our calculations show that the emission rate falls off rapidly at low orders, then exhibits a plateau where all the harmonics have the comparable strength, and ends up with a sharp cut-off. Plateau and cut-off are two important feature of the spectra of odd-only harmonics, now are shared with also the spectra of even-odd harmonics, as a consequence of our calculations.

Just as in the odd-only case, the plateau and the cut-off frequency of even-odd harmonics also vary with the laser intensity. At low laser intensities, the plateau and the cut-off are both not outstanding. With increasing laser intensity, a clear plateau and a sharp cut-off appear in the harmonic spectra, as shown in Figs. 1 at laser intensity 10^{14}W/cm^2. The clear plateau and the cut-off structures are similar to those obtained by recombination mechanism, but the harmonics emitted in the exit process differs from the recombination mechanism that the even order harmonics show comparable strength with the odd order harmonics. For this reason, we term the harmonics in the current study as even-odd harmonics.

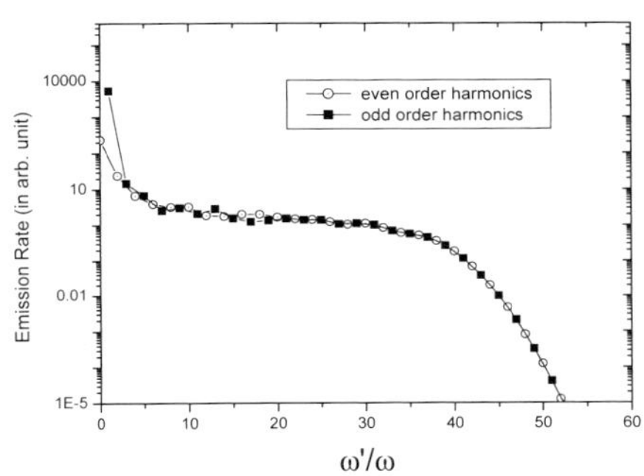

Fig.1, The calculated harmonics spectra at laser intensity 100TW/cm^2.

4. Discussion and Conclusion

From our calculations, we assert that there are two coexisting mechanisms to produce high harmonics. In the recombination mechanism, photoelectrons fall back to the original atomic bound state with an emission of odd-only harmonics. In the escape mechanism, photoelectrons escape from the laser field with an emission of both even and odd harmonics.

An immediate question to challenge this assertion is why people do not observe both even and odd harmonics in most ATI experiments? Following analysis answers this question. Figure 4 shows a spectra calculation in an even-odd case which is comparable to an existing calculation in an odd-only case [3]. In the two cases, the plateaus have the same length, say 80, but the required laser intensity for the odd-only case is little lower, about 1:1.2, than the even-odd case. Thus, at the same laser intensity, the HHG rate in the odd only case versus the rate in the even-odd case is about 1.2:1. Since the two mechanisms coexist, total rate of odd harmonics versus that of even harmonics is 2.2:1. So, the expected strength of signals of even harmonics is about 45% of the full strength of the signals of odd harmonics, less than 50%. Experimental reports show [17] that in the spectra of harmonics, all spectrum lines presenting each harmonic have certain width. The envelopes of the spectrum lines of neighboring harmonics intersect at certain heights, below which is the noise area. The ratio for signal versus noise from reference [17] is about 60-80%, at least more than 50%. This analysis answers why existing ATI experiments do not show both even and odd harmonics. We expect, in future experiments, if one can fully ionized atoms such that much more electrons escape from both ionic and laser fields with improved spectral resolution, the harmonics with both even and odd orders can be observed.

References

[1] Xiaofeng Li et al., 1989 Phys. Rev. A39, 5751.

[2] R.L. Carman et al., 1981 Phys. Rev. Lett.46, 29; R.L. Carman et al., 1981 Phys. Rev. A24, 2649.

[3] Gao Liang-hui et al, 1999 Chin. Phys. Lett. 16,502-4. .

[4] Gao Liang-hui et al, 2000 Phys. Rev. A, 61, 063407(5).

[5] Ph. Antoine et al., 1996 Phys. Rev. Lett., , 77(7) 1234-7.

[6] I. P.Christov et al., 1997 Phys. Rev. Lett., 78, 1251-4.

[7] J. Zhou, et al., 1996, Phys. Rev. Lett., 76, 752-5.

[8] M. Lewenstein, et al., 1994 Phys. Rev. A, 49(3) 2117-32.

[9] Y. I. Slamin, et al., 1997, Phys. Rev. A, 55(5) 3964-67; ibid. 1996, 54(5) 4383-95.

[10] C. Bamber, et al., 1996, Phys. Rev. Lett., 76(17) 3116-9; D. G. Lappas, et al., 1993 Phys. Rev. A, 47(2) 1327-35.

[11] E. Esarey, S. K. Ride, and P. Sprangle, 1993 Phys. Rev. E 48 3003-21.

[12] Dong-Sheng Guo, T. Aberg, and B. Crasemann, 1989 Phys. Rev. A, 40(6) 4997.

[13] J. Gao, Dong-Sheng Guo, et al., 2000 Phys. Rev. A, 61 043406(5)

[14] P.H. Bucksbaum et al., 1988 Phys. Rev. Lett. 61, 1182.

[15] Dong-Sheng Guo and G. W. Drake, 1992 Phys. Rev. A 45, 6622.

[16] M. Bashkansky, et al., 1987 Phys. Rev. Lett. 59, 274

[17] J. W. G. Tisch, et al., 1994 Phys. Rev. A, 49 R28.

Inst. Phys. Conf. Ser. No 186
Paper presented at 9th Int. Conf. X-Ray Lasers, Beijing, China, 24–28 May 2004

Efficient and powerful generation of XUV radiation with a femtosecond KrF laser system

B. Wellegehausen, C. Reinhardt, A. Rühl, B.N. Chichkov*
Institut für Quantenoptik,Universität Hannover,Welfengarten 1, D-30167 Hannover
*Laser Zentrum Hannover e. V., Hollerithallee 8, D-30419 Hannover

Abstract

Third and fifth harmonic generation with a short-pulse KrF laser system is investigated. As nonlinear material argon is used. For the third harmonic, a near three photon resonance yields a high nonlinear susceptibility and allows phase matching in the negative dispersive region. The fifth harmonic generation takes place in argon ions, generated in the leading part of the pulse at intensities above 10^{14} Wcm^{-2}. In this case, a ponderomotively shifted five photon resonance enhances the process. At optimized densities and interaction lengths, output energies of 150 μJ at 82.9 nm and of 15 μJ at 49.7 nm have been achieved, corresponding to conversion efficiencies of 2.5 % and 0.25 % respectively.

1. Introduction

For the generation of coherent radiation in the XUV by nonlinear optics, the high order harmonic process is well suited and applicable down to the water window (for a recent review see [1]). Unfortunately, the conversion efficiency is low in the plateau region, typically 10^{-6} or much less for very high orders. Only at relatively low orders, higher efficiencies have been demonstrated, with a so far highest value of 4.3×10^{-4} for the 11th harmonic of 810 nm Ti:sapphire laser radiation in xenon at 72.7 nm, and an output energy of 7 μJ [2]. The main reason for the low efficiency is the reabsorption of plateau harmonics, which limits the interaction length. Consequently, much higher conversion efficiencies should be possible for very low-order harmonics still in the pertubative regime, where absorption does not play a role. However, in this case already short wavelengths pump lasers are needed for XUV generation. Good suited for this purpose are short-pulse KrF laser systems, which are commercially available with output energies of up to 100 mJ [3]. Short-pulse KrF lasers have been used for high harmonic generation [4], for the generation of broadly tunable VUV radiation by four wave mixing [5] and for the efficient and powerful generation of third harmonic radiation at 82.8 nm in some recent experiments [6]. In this contribution we report on the further enhancement of the power and efficiency of the third harmonic generation and on first results of powerful fifth harmonic generation at 49.7 nm.

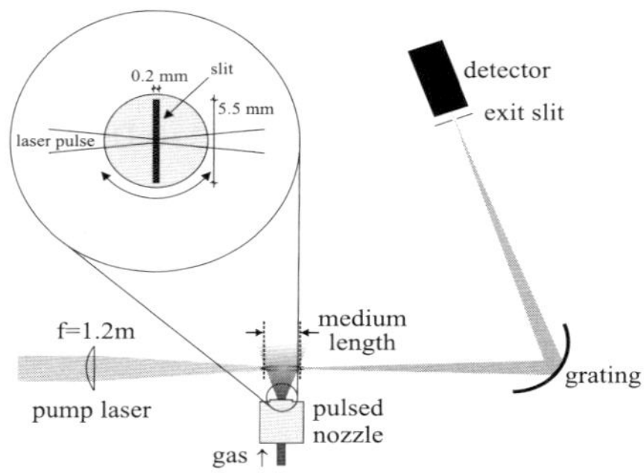

Figure 1. Experimental setup for harmonic generation.

2. Experiments

The experimental setup for 3rd and 5th harmonic generation of tunable KrF laser radiation is schematically shown in Fig.1. The gas is injected into the vacuum chamber via a pulsed rotatable slit nozzle, enabling interaction lengths between 1.2 mm and 6.7 mm at maximum densities of 2×10^{18} cm^{-3} for backing pressures of 2 bar. The maximum laser input energy was 6 mJ, giving for the focussing lens with 1.2 m focal length a focal intensity of 5×10^{14} W/cm^2 (for a pulse duration of 500 fs and a focus radius of about 30 μm). The generated radiation is spectrally selected with a toroidal grating monochromator (Seya-Namioka-mounting) and detected with an open XUV photomultiplier with a Cu-BeO photocathode, having a quantum efficiency of 12 % at 83 nm as well as at 49 nm.

For the frequency conversion argon gas is used as nonlinear material. In case of the third harmonic the conversion takes place in neutral argon around the $3p^5 4d$-three-photon-resonance (Fig.2), which allows almost perfect phase matching at high densities, as discussed in detail in [6]. For the frequency tripling the intensities in the interaction volume should remain below 3×10^{13} Wcm^{-2} to avoid ionization. In case of the fifth harmonic much higher intensities, well above 10^{14} Wcm^{-2} have to be applied, leading to almost full ionization of argon by a four photon process resonantly enhanced by the above mentioned three photon resonance. Consequently, fifth harmonic generation occurs in argon ions near the $3p^4 5s$- and $3p^4 4d$- states, enhancing the nonlinear response for this process. Additionally, in both cases ponderomotive level shifts have to be taken into account at the applied intensities. The relevant level scheme for the processes is shown in fig. 2. In both cases reabsorption is avoided, as the harmonics are generated in a medium where the photon energy is below the ionisation energy to the next ion stage. The optimum intensity for the processes can be adjusted by changing the focus position relative to the gas jet as shown in fig. 3 for a gas jet of 1.2 mm length and a backing pressure of 1 bar. Right at the focus with an intensity of 4×10^{14} Wcm^{-2} the fifth harmonic signal is maximum and strongly reduces with a shift of the focus, while the third harmonic is optimum when the focus is about 7 mm behind the gas jet, corresponding to the lower intensity of 3 x 10^{13} Wcm^{-2}.

In addition to the focussing, interaction length L and density N of the nonlinear material have to be optimized, which strongly influence the phase matching. In the pertubative regime and weak focussing limit (medium length L < confocal parameter b = $2\pi w^2/\lambda M^2 \approx 5.4$ mm for the here given wavelength $\lambda = 248$ nm, beam radius $w \approx 30$ µm,

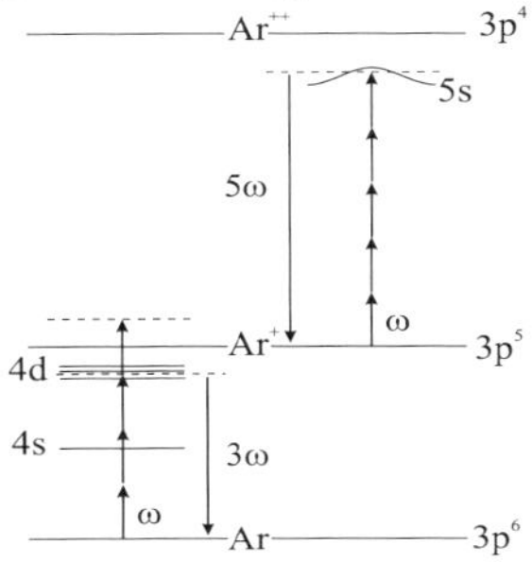

Figure 2. Reduced energy-level schema of neutral and singly ionized argon with relevant levels for third and fifth harmonic generation.

M^2=4.5), which is approximately valid here, the intensity $I_{n\omega}$ of the n- harmonic is given by $I_{n\omega}$ = const $|\chi|^2N^2L^2\ (I_\omega)^n \text{sinc}^2\Delta kL/2$, where χ is the susceptibility per atom or ion, N the corresponding density and ΔkL the total phase mismatch. The wave vector mismatch $\Delta k = k_{n\omega} - nk_\omega = \Delta k_{geom} + N_{at}\Delta k_{at} + N_{ion}\Delta k_{ion} + N_{electr}\Delta k_{electr}$ thereby consists of dispersive contributions Δk_{at}, Δk_{ion} and Δk_{electr} due to atoms, ions and electrons with corresponding densities N_{at}, N_{ion}, N_{electr}. The geometrical contribution due to the focussing is given by $\Delta k_{geom} = 2(n-1)/b$.

For third harmonic generation the dispersive contributions of ions and electrons to the phase matching can be neglected for the used maximum intensity of 3 x 10^{13} Wcm^{-2}. Δk_{geom} has a value of 7.4 cm^{-1} and Δk_{at} is about -1 x 10^{-17} cm^2 for the near resonance condition as discussed in [6]. Consequently, exact phase matching should be possible at densities around 7.4 x 10^{17} cm^{-3}, and quasi phase matching ($L < L_{coh} = \pi/\Delta k$) at even higher densities. Changing of the density and of the interaction length by rotating the slit nozzle resulted in optimum output energies at densities of 1.2 x 10^{18} cm^{-3} for the maximum possible length of 6.7 mm. Under these conditions a maximum output power of 150 µJ was measured for the here used 6 mJ input power, corresponding to a conversion efficiency of 2.5 %.

For the fifth harmonic the geometrical phase mismatch Δk_{geom} is $8/b \approx 14.8$ cm^{-1}. As at the used intensities argon is almost fully ionized to Ar^+, an atomic contribution Δk_{at} can be neglected, but now contributions Δk_{ion} and Δk_{electr} due to the ions and electrons have to be considered. The electron contribution is given by $\Delta k_{electr} = r_e(5\lambda - \lambda/5)$ with the constant r_e given as r_e = 2.181 x 10^{13} cm. For an optimum initial density of 1 x 10^{18} cm^{-3} found in the experiments and assuming almost full ionization to Ar^+, an electron contribution $\Delta k_{electr} N_e$ of about 33 cm^{-1} is obtained. For perfect phase matching $\Delta k = 0$, the ionic contribution therefore has to compensate the sum of the electron and geometrical contribution of about 47.8 cm^{-1}. For an ion density of 1 x 10^{18} cm^{-3}, this would require a Δk_{ion} -value of – 4.8 x 10^{-17} cm^2. Such a negative dispersive ionic contribution appears possible above the $3p^45s$ state of Ar^+, and in fact, measured length and density dependencies indicate quasi phase matching at the used density of 1 x 10^{18} cm^{-3} and maximum interaction length of 6.7 mm. With these parameters and a pump power of 6 mJ a so far highest energy for the fifth harmonic of 15 μJ was obtained, corresponding to a conversion efficiency of 0,25 %.

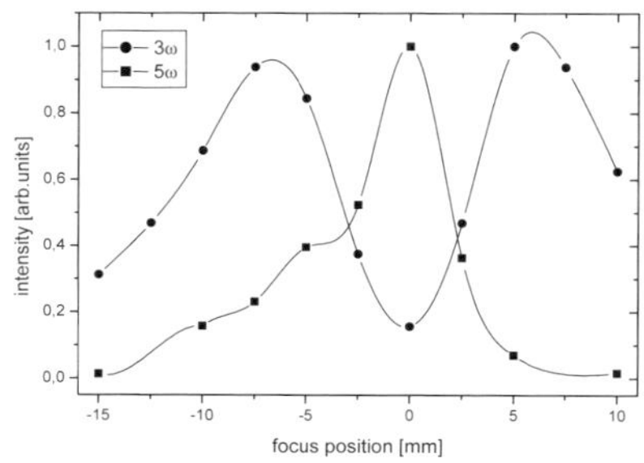

Figure 3. Normlized harmonic energy versus position of the beam focus relative to the nozzle. Positive values correspond to positions behind the nozzle.

For the absolute determination of the harmonic energies, a vacuum photodiode of known quantum efficiency was used, consisting of a polished aluminium cathode and a stainless steel grid as anode [6]. Instead of using calibrated aluminium filters to separate the harmonic photons from the pump photons, the grating was used. The grating reflectivity for both harmonics was determined in an independent step with the same aluminium cathode. Thereby, the third and fifth harmonic can be separated by the generation process itself, such as by the adjustment of the focal position, as discussed in connection with fig. 3. By measuring the signals with and without gas jet in front and behind the grating, the individual power values can finally be determined. For the third harmonic energy an error of 10 % and for the fifth harmonic of 15 % is estimated, mainly resulting from the error for the quantum efficiency of the aluminium cathode and the pump laser energy fluctuations.

3. Conclusions

Powerful harmonic generation of short-pulse KrF laser radiation in an argon gas jet has been demonstrated. Maximum output energies of 150 μJ for the third harmonic at 83 nm and of 15 μJ for the fifth harmonic at 49 nm at a pump energy of 6 mJ have been obtained by optimization of the particle density and by using an interaction length of 6.7 mm. The achieved powers are among the highest reported for nonlinear optical processes in this spectral range and are even comparable to those of x-ray lasers. With presently available up to 100 mJ short pulse KrF-lasers, an upscaling to much higher output energies appears feasible.

References

[1] D. Descamps et al.; AIP Conf. Proc. **525**, 385 (2000)
[2] K. Midorikawa, E. Takahashi, Y. Nabekawa; AIP Conf. Proc. **641**, 412 (2002)
[3] J. Bekesi, G. Marowski, S. Szatmari, P. Simon; Zeitschrift für Physikalische Chemie **215**-12, 1543 (2001)
[4] D. Yoshitomi, T. Shimizu, et al. Opt. Letters **27**-24, 2170 (2002)
A. A. Offenberger et al.; AIP Conf. Proc. **369**, 559 (1996)
[5] A. Tünnermann, C. Momma, et al. IEEE J. QE-**29,** 1233 (1993)
[6] C. Doelle, C. Reinhardt, et al. Appl. Phys. B **75**, 629 (2002),
C. Doelle, C. Reinhardt, et al. Appl. Phys. B **76**, 891 (2003)

Grazing Incidence Pumping (GRIP) X-ray lasers

Inst. Phys. Conf. Ser. No 186
Paper presented at 9th Int. Conf. X-Ray Lasers, Beijing, China, 24–28 May 2004

Grazing Incidence Pumping for Efficient X-ray Lasers

J. Dunn, R. Keenan, P.K. Patel, D.F. Price, R.F. Smith
Lawrence Livermore National Laboratory, Livermore, CA 94551, USA

V.N. Shlyaptsev
University of California, Davis, Livermore, CA 94551, USA

Abstract. We report progress in developing efficient pumping of laser-driven x-ray lasers that opens new possibilities for both high average power x-ray lasers as well as producing progressively shorter wavelength lasers. The new scheme of grazing incidence pumping (GRIP) is described. In essence, a chosen electron density region of a pre-formed plasma column, produced by a longer pulse at normal incidence onto a slab target, is selectively pumped by focusing the short pulse ~ps laser at a determined grazing incidence angle to the target. The controlled use of refraction of the pumping laser in the plasma results in several benefits: The pump laser path length is longer and there is an increase in the laser absorption in the gain region for creating a collisional Ni-like ion x-ray laser. There is also an inherent travelling wave, close to *c*, that increases the overall pumping efficiency. The scheme requires careful tailoring of the pump and plasma conditions to the specific x-ray laser under investigation but the main advantage is a 3 – 30 times reduction in the laser pump energy for mid-Z materials. We report several examples of this new x-ray laser on two different laser systems. The first demonstrates a 10 Hz x-ray laser operating at 18.9 nm pumped with a total of 150 mJ of 800 nm wavelength from a Ti:Sapphire laser. The second case is shown where the COMET laser is used both at 527 nm and 1054 nm wavelength to pump higher Z materials with the goal of extending the wavelength regime of tabletop x-ray lasers below 10 nm.

1. Introduction

Since the first demonstration of the laboratory x-ray laser (XRL) in 1984 on the LLNL Novette laser facility [1], there has been a world-wide effort to make these high brightness, short wavelength sources more efficient, more compact and to operate at a higher repetition rate. Until relatively recently, the mid-1990s, all research was conducted on very large, national class facilities requiring typically kilojoules of laser energy and at shot rates of a few per day. Some improvements, for example the pre-pulse technique, allowed approximately one order of magnitude reduction in the laser drive energy, ~150 J

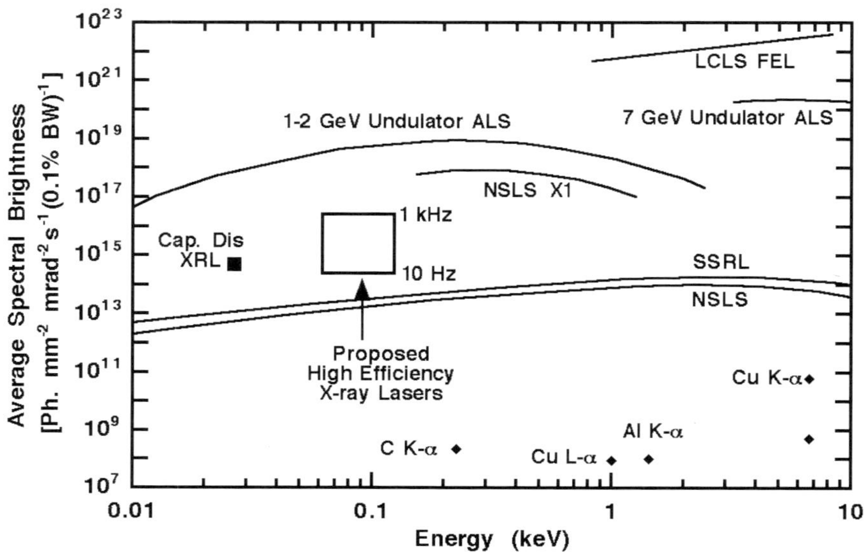

Figure 1. Comparison of average spectral brightness for various sources including present 3rd generation synchrotrons, the proposed LCLS Free Electron Laser, and K-α sources. The proposed 100 fs – 5 ps duration pumped high efficiency GRIP x-ray lasers of this work are shown as a box defined for 10 - 20 nm operation and 10 Hz – 1 kHz shot rates.

in 70 - 100 ps pulses, to drive a 13.9 nm x-ray laser in 1997 [2] and even as low as 30 J in 1999 [3].

Recently there have been a number of remarkable scientific breakthroughs that have led to the development of efficient, compact tabletop (occupying less than 10 m^2) collisional excitation x-ray lasers with high output and a higher repetition rate. The picosecond laser-driven transient inversion lasers have allowed a dramatic reduction of pump energy to 5 - 10 J [4] while achieving saturated output for 14.7 nm x-rays at a repetition rate of 1 shot/4 minutes [5]. The fast capillary discharge technique developed at Colorado State University operates at a higher rate of 7 Hz for slightly longer 46.9 nm wavelength x-rays and is the first demonstration of a high-average power, tabletop system [6]. The capillary discharge scheme still remains the highest average power tabletop system with 10 mW of power at 46.9 nm. These new short wavelength sources have ultra-high peak brightness 10^{24} - 10^{25} ph mm^{-2} mrad^{-2} s^{-1} (0.1% BW)$^{-1}$ typically 4-5 orders of magnitude higher than 3rd generation synchrotrons. However, with the exception of the capillary discharge scheme shown in Fig. 1, most of the laser-pumped x-ray lasers until very recently had low average brightness.

A high repetition rate, high-average brightness laser-driven x-ray source would be very attractive if it could operate at shorter wavelengths 8 – 20 nm. Near synchroton output levels with average brightness between the bending magnets and undulators, see Fig. 1, would be possible from a tabletop source. It should be kept in mind that this would be from a relatively inexpensive source (less than $1M) when compared to the large expensive, central facilities like the 3rd generation and planned 4th generation synchrotron sources. High repetition rate 10 Hz laser-driven x-ray lasers are beginning to be studied using the optical field ionization (OFI) process and have been demonstrated at

41.8 nm and 32.8 nm with less than 1 J pump energy/pulse [7]. It is a substantial technical and scientific challenge to operate x-ray lasers below 20 nm with the present high-repetition rate laser systems. The power required to drive an x-ray laser of wavelength, λ, typically scales $\sim\lambda^{-4} - \lambda^{-6}$. Furthermore, to operate our present picosecond-laser driven x-ray lasers [5] from ~5 J/shot to 150 mJ/shot (about 30 times reduction) for 10 Hz operation for a small high-power laser requires a revolutionary step forward in efficiency. Nevertheless, the canonical x-ray conversion efficiency numbers, x-ray laser energy out divided by laser pump energy in, are in the range of $10^{-7} - 10^{-5}$. This suggests that there is still substantial room for improvement in pumping efficiency.

This paper addresses research activities using existing laser facilities within LLNL to study a new high-efficiency x-ray laser concept based on collisional excitation of Ni-like ion schemes. The grazing incidence pumping (GRIP) uses refraction of the short laser pump in a pre-formed plasma to deliver the laser energy into a specific electron density region where the inversion takes place. The benefits in achieving an efficient high-repetition rate x-ray laser can be used for further development of shorter wavelength x-ray lasers below 10 nm with modest ~10 J facilities. It may also be possible to produce lasing close to the water-window with two orders of magnitude less energy than the 10 kJ required for the first x-ray laser cellular imaging experiment conducted on Nova [8].

2. Grazing Incidence Pumping Scheme for Mo 18.9 nm laser

The GRIP scheme is proposed to dramatically improve the efficiency of laser-pumped x-ray lasers below 20 nm for input energies of ~150 mJ. This is 30 times smaller than the energy of our picosecond laser-driven scheme that uses the well-established transverse pumping geometry for mid-Z Ni-like ions [5]. Simulations using the RADEX code indicate that while this is considerably more efficient than previous x-ray laser pumping, only a very small fraction of the absorbed laser energy, approximately 5 – 8%, is deposited in the gain region of interest and pumping the inversion process. The recently developed diagnostic tools of near-field imaging of the x-ray laser gain region and picosecond x-ray laser interferometry [9] have given new insight into the x-ray laser generation process. These results show a very narrow plasma gain region of ~20 μm width at a distance of 20 – 50 μm from the target surface at an electron density of $\sim10^{20}$ cm^{-3}. Thus, in agreement with the predictions, much of the laser energy heats regions of the plasma that do not contribute to the observed lasing. In principle, longitudinal optical pumping, where the laser beam is focused at high intensity into the end of the plasma column, is appealing since it allows efficient traveling-wave heating and therefore co-propagation with the x-ray laser pulse. Recently, a longitudinally pumped Ni-like Mo laser has been demonstrated [10]. However the considerable absorption, refraction, and

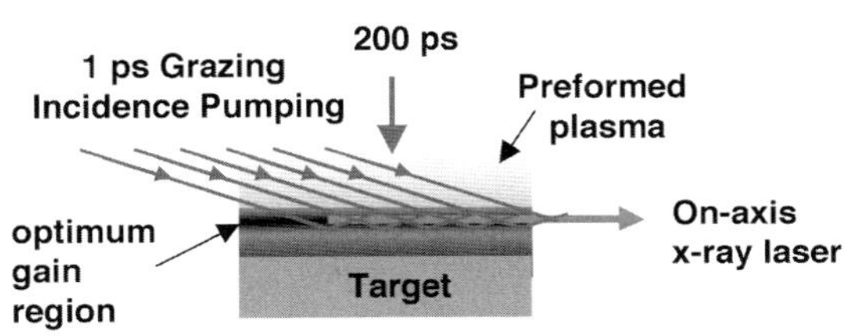

Figure 2. Schematic of grazing incidence pumping scheme. A pre-formed plasma is first produced in a flat target to generate the optimum gain region. Short pulse 1 ps, 800 nm wavelength laser is incident at a grazing angle to strongly heat this region producing efficient on-axis x-ray lasing.

relativistic self-focusing of the high intensity short pulse optical pump make the longitudinally-pumped x-ray laser progressively difficult at higher electron densities.

However, utilisation of the high efficiency, inherent travelling wave for longitudinal schemes can be combined with some small component of transverse pumping to improve laser absorption in the desired density region with reduced detrimental refraction. We will call this grazing incidence pumping as proposed by Shlyaptsev and illustrated in Fig. 2 above. The main reason for the choice of the grazing angle is to use refraction of the pumping laser in a controlled way depending only on the predicted optimum electron density and wavelength of the laser pump. In the GRIP scheme the same two stage pumping is used as in the conventional transient scheme but with a significant difference. A long pulse, 200 ps in this case, focused on a Mo slab at normal incidence preforms a plasma column to create a tailored density profile. After a certain delay when the plasma conditions, particularly the electron density profile is correct, the short pulse is incident at a grazing angle to the target surface in a line focus. The short pulse beam is refracted in the plasma at a chosen electron density, well below critical, and is strongly absorbed. The required angle of incidence of the short pulse, measured relative to the target surface, is chosen to allow the laser pump to be refracted in the plasma. This is obtained from the equation for the refraction angle, ϕ_r, simplified to $\phi_r \sim (n_{eo}/n_{ec})^{1/2}$ where n_{eo} is the maximum density within the gain region and n_{ec} is the critical density for the optical pump [11]. This has several advantages that increase the laser coupling into the gain region. The grazing angle gives a longer path length to increase absorption within a specific plasma region so that a smaller *active* volume is pumped. Refraction causes the short pulse to turn back into the gain region thus allowing the laser pump to traverse this region twice. The short pulse beam has an inherent travelling wave very close to the speed of light. Also there is no restriction in the length of the plasma to be pumped. For these reasons, a sub-20 nm x-ray laser should be able to be generated on a small, high-repetition rate laser.

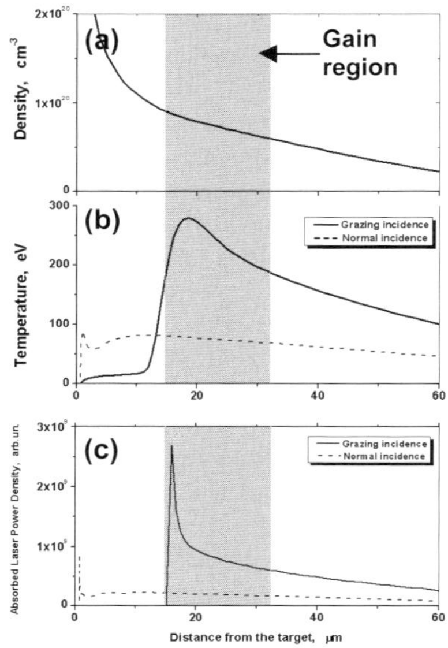

Figure 3. RADEX simulations show (a) electron density profile of x-ray laser medium for Ni-like Mo preformed plasma, (b) electron temperature and (c) absorbed laser energy of short pulse for normal incidence (dashed line) and grazing incidence (solid line) pumping.

Initial RADEX simulations of the electron density profile 500 ps after the pre-pulse is fired, as well as electron temperature and absorbed laser power just after the short pulse is fired are shown in Fig. 3 for 800 nm pumping of a Mo target [11]. The predicted gain region is shown in the shaded area. Laser energy of 75 - 120 mJ, 200 ps (FWHM) is incident at normal incidence at an intensity ~2×10^{11} W cm^{-2}. The short pulse pumping conditions are chosen at a grazing angle of 13.6° to couple strongly into 10^{20} cm^{-3} region.

The pulse duration is 4 ps (FWHM) with 64 - 100 mJ focused to ~3 – 5 × 10^{13} W cm^{-2}. Some optimisation of the focal conditions is essential in order to maximise the deposition of laser energy into the gain region in the most efficient manner. The comparison between the GRIP and the transverse schemes show higher electron temperature T_e, Fig. 3(b), and higher absorbed power, Fig. 3(c) in the preferred gain region when the short pulse is incident at a grazing angle.

3. Results for GRIP Ni-like Mo 10 Hz, 18.9 nm Laser Pumped with 150 mJ

The Ni-like Mo laser experiment is described in more detail in several places including these proceedings [12 –14]. The laser pumping conditions at a repetition rate of 10 Hz were very similar to those described above. The long 200 ps pulse contained 70 mJ, 800 nm, and was focused at an intensity of 5 × 10^{11} W cm^{-2} onto the Mo target at normal incidence in a line of 5 mm × 15 μm (FWHM). A maximum energy of 80 mJ in 1.5 ps, 800 nm, was focused, at an intensity of 4 × 10^{13} W cm^{-2}. An on-axis parabola generated a very uniform line focus 4 mm × ~25 – 30 μm (FWHM) wide at the chosen 14° angle with an intrinsic travelling wave velocity of 0.97c [14]. Careful adjustment of the pumping conditions were required to achieve the soft x-ray laser line. Reducing the laser energy by more than 10% in each beam resulted in a strong fall in x-ray laser intensity. This indicated that we were pumping close to the threshold for 18.9 nm lasing on the 4*d* – 4*p* line. The optimised pump conditions were found to be a 1.5 ps short pulse delayed by 500 ps after the peak of the long pulse. Lasing was also observed for pumping from 100 fs to 6 ps short pulse durations. High gains of ~60 cm^{-1} were determined with strong output for target lengths up to a maximum of 4 mm. The gain saturation regime was achieved based on the intensity versus length measurements [13, 14].

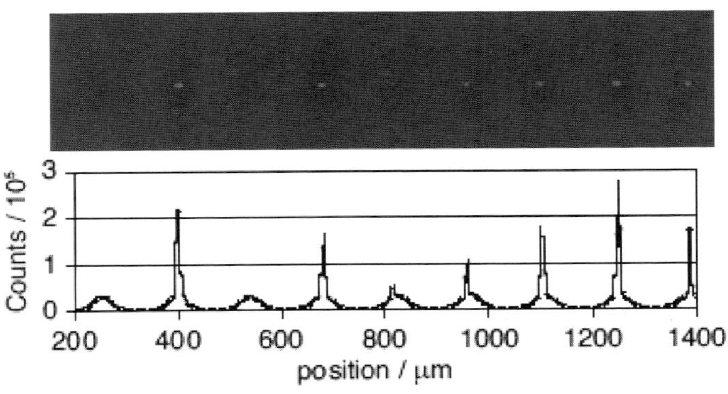

Figure 4. Near-field imaging (using multilayer optics) of the GRIP Ni-like Mo 18.9 nm x-ray laser. The x-ray laser is pumped at high repetition rate during the CCD camera readout. Lower image is lineout through near-field intensity.

An on-axis soft x-ray Mo:Si multilayer-coated mirror imaging system, consisting of a 12.5 cm focal length mirror in combination with a 45° mirror, was used to image the near-field pattern of the 18.9 nm Mo x-ray laser at 17× magnification onto a back-illuminated CCD. Many laser near-field image x-ray laser shots could be recorded in one CCD frame by synchronising the repetition rate of the laser with the CCD readout. Figure 4 shows a sequence of frames for the x-ray laser plasma imaged on-axis running at high repetition rate. A variation in the XRL output was observed which was attributed to a vibration-induced low frequency oscillation in the short pulse line focus. The overlap between the two line foci separates on some shots by no more than one width ~20 μm. As the short pulse line focus moves back towards the center of the preformed plasma, 5 images on the right hand-side, the XRL output increases to a maximum. This allows us to test the sensitivity of the line focus overlap important for high repetition sources. The near-field x-ray laser beam size was less than ~$9 \times 20\ \mu m^2$ and showed some structure present in the image. Initial estimates of the x-ray laser energy were in the range of 10 - 20 nJ but could be higher. This will be reported in a Ni-like Mo characterisation paper in detail at a later date.

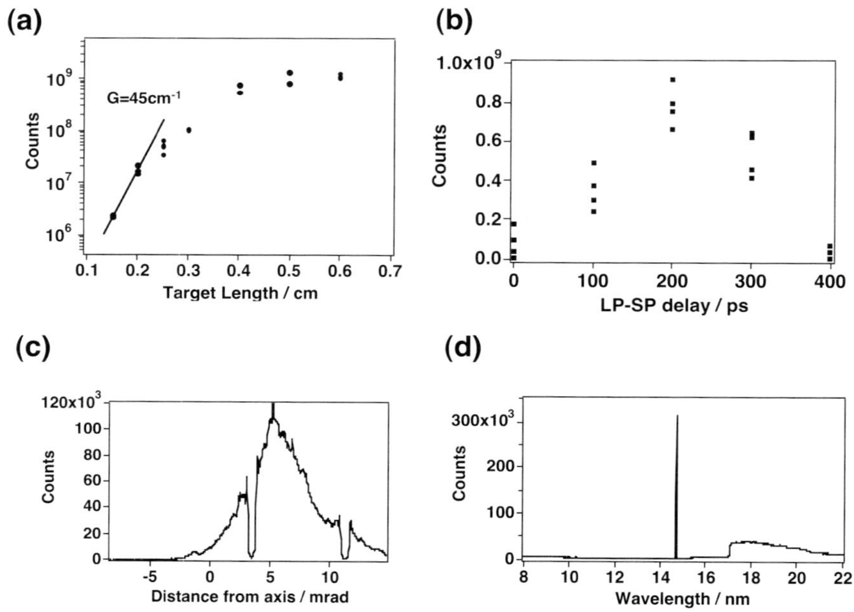

Figure 5. GRIP experiments conducted on the COMET laser using 1ω LP/2ω SP pumping on the Ni-like Pd 14.7 nm x-ray laser. (a) X-ray laser 14.7 nm output intensity plotted as a function of Pd target length indicating gains of 45 cm^{-1}. (b) Optimum delay between the laser pulses was found to be Δt of 200 ps. (c) Horizontal beam deflection for 14.7 nm x-ray laser. (d) X-ray laser spectrum filtered by 200 nm Al filter showing very strong output in the Ni-like Pd line pumped by ~2 J energy.

4. GRIP Scheme Implementation on COMET

The first Ti:Sapphire, 800 nm, 10 Hz GRIP experiments reported in the previous section were working at the limits of the available laser pump energy. More energy to produce a larger line focus, both in width and length for both beams, was required. So the GRIP experiments were transferred to COMET to test the pumping principle with more energy at both 1054 nm and 527 nm wavelengths. It is not possible to summarise all of the parameters studied here but the technique was checked and lasing was achieved for numerous mid-Z Ni-like ions as short as 12 nm with ~2 - 4 J pump energy.

One example is described where a 1ω, (1054 nm) 600 ps pulse was used in combination with a 2ω, (527 nm) 1.5 ps short pulse. Both beams were focused to ~7 – 8 mm × 40μm (FWHM). The Ni-like Pd laser was chosen for this study. The optimum electron density for the turning point was chosen to be 1.2×10^{20} cm^{-3} which required a pumping angle of 10°. The laser energy was approximately 1.2 J - 1.3 J in each beam. After some optimisation it was found that there was very strong output available in the x-ray laser. The results are shown in Fig. 5. The optimum delay was found to be 200 ps between the pulses with a wider window than the previous Mo results [13, 14], Fig. 5(b). The gain was determined to be 45 cm^{-1} and the laser goes into saturation with gL ~14 at 4mm, see Fig 5(a). The deflection angle of the 14.7 nm x-ray laser as it leaves the plasma was 5 mrad away from the target, Fig. 5 (c). Although the larger focal geometries require ~10× larger pumping energy they appear to generate more than 100× higher x-ray laser output. For targets above 4 mm in length, the output had to be attenuated using 200 nm Al filters to prevent the CCD from saturating on the on-axis spectrometer, Fig. 5(d). Initial comparisons suggest that the GRIP XRL output in a 6 mm target are comparable with a 1.2 cm target pumped transversely with ~5 J in the short pulse. This indicates that a reduction of 3 – 4× in short pulse pump energy has been achieved for the same x-ray laser output. These are only the second set of experiments and further improvements will be possible as more groups employ the grazing incidence pumping scheme.

5. Summary and Future Directions

In summary the grazing incidence x-ray laser scheme shows substantial promise for producing higher efficiency coupling of the laser pump energy into the plasma gain region. The underlying principle is the controlled use of refraction of the pump beam to selectively pump a specific electron density and gain volume. We have shown that it can work with as little as 150 mJ pump energy for a 10 Hz Ni-like Mo 18.9 nm x-ray laser. It has been driven into saturation though there is still more potential to extract higher x-ray laser output energy. Additional results demonstrating the GRIP pumping are reported by the Colorado State University group [15]. The GRIP experiments on COMET show that the scheme appears to be universally applicable to many pumping configurations and pulse lengths. It can operate on numerous higher Z materials at different angles and different pump wavelengths though in each case some optimisation is likely to be required. Increasing the laser pump energy to ~1 J/pulse gives substantial output for the mid-Z Ni-like ions x-ray lasers. There is also the prospect of extending lasing to shorter wavelengths with this technique to re-visit water-window x-ray lasers for biological applications.

Acknowledgments

The continued support and encouragement of Al Osterheld is greatly appreciated. The authors would like to thank Joe Nilsen for the use of the multilayer optics in part of this work. Work performed under the auspices of the US Department of Energy by the University of California Lawrence Livermore National Laboratory under Contract No. W-7405-Eng-48 and in part by funding from the National Science Foundation through the Center for Biophotonics, an NSF Science and Technology Center, managed by the University of California, Davis, under Cooperative Agreement No. PHY 0120999.

References

[1] D. Matthews *et al.*, *Phys. Rev. Lett.* **54**, 110 (1985).

[2] J. Nilsen *et al.*, *Phys. Rev. A* **48**, 4682 (1993); J. Zhang *et al.*, *Phys. Rev. Lett.* **78**, 3856 (1997).

[3] R. Tommasini *et al.*, *Phys. Rev. A* **59**, 1577 (1999).

[4] P. Nickles *et al.*, *Phys. Rev. Lett.* **78**, 2748 (1997); J. Dunn *et al.*, *ibid.* **80**(13), 2825 (1998).

[5] J. Dunn *et al.*, *ibid.* **84**, 4834 (2000).

[6] B.R. Benware, C.D. Macchietto, C.H. Moreno and J.J. Rocca, *Phys. Rev. Lett.* **81**, 5804, (1998).

[7] S. Sebban *et al.*, *Phys. Rev. Lett.* **86**, 3004 (2001); S. Sebban *et al.*, *Phys. Rev. Lett.* **89**, 253901 1-4 (2002).

[8] L.B. Da Silva *et al.*, Science **258**, 269 (1992).

[9] R.F. Smith *et al.*, Phys. Rev. Lett, **89**, 065004-1 (2002); R.F. Smith *et al.*, *Soft x-ray lasers and Applications V*, SPIE Int. Soc. Opt. Eng. Proc, vol. **5197**, ed. E.E. Fill and S. Suckewer, 155-167 (2003).

[10] T. Ozaki *et al.*, Phys. Rev. Lett. **89**, 253902-1 (2002).

[11] V.N. Shlyaptsev *et al.*, *Soft x-ray lasers and Applications V*, SPIE Int. Soc. Opt. Eng. Proc, vol. **5197**, ed. E.E. Fill and S. Suckewer, 221 - 228 (2003).

[12] R. Keenan *et al.*, *Soft x-ray lasers and Applications V*, *ibid.*, 213 - 220 (2003).

[13] R. Keenan *et al.*, "Grazing Incidence Pumping for Efficient X-ray Lasers", in these proceedings (2004).

[14] R. Keenan *et al.*, "High Repetition Rate Grazing Incidence Pumped X-ray Laser Operating at 18.9 nm", submitted to *Phys. Rev. Lett.* (2004).

[15] J.J. Rocca *et al.*, "Demonstration of High Repeition Rate Desk-Top and Table-Top Soft X-ray Lasers", in these proceedings (2004).

Inst. Phys. Conf. Ser. No 186
Paper presented at 9th Int. Conf. X-Ray Lasers, Beijing, China, 24–28 May 2004

Development of advanced 20 TW Ti:S laser system for x-ray laser and coherent XUV generation

H Kuroda, M Suzuki, R A Ganeev, J Zhang, M Baba,
The Institute for Solid State Physics, The University of Tokyo
5-1-5 Kashiwanoha, Kashiwa, Chiba 277-8581, Japan

T Ozaki
Institut national de la recherche scientifique
1650 boul. Lionel-Boulet Varennes (Quebec) Canada J3X 1S2

Z Y Wei, J Zhang
Laboratory of Optical Physics, Institute of Physics,
Chinese Academy of Sciences, Beijing 100080, People's Republic of China

Abstract. In order to develop the high repetition rate x-ray lasers, the longitudinal-pumped transient collisional excitation (TCE) x-ray laser is one of the most effective pumping schemes. The high directive Ni-like Mo 18.9 nm soft x-ray laser pumped by modest laser energy has already been demonstrated by using the tabletop size Ti:sapphire / Nd:glass laser system that delivering energy of 150 mJ in 475 fs at the center wavelength of 1054 nm. The total energy in the pre-pulse and the main pulse was 150 mJ, which will make possible multi-hertz operation. To pursue the high repetition rate of the laser-driven TCE x-ray laser, we have designed a new 20 TW Ti:Sapphire laser system (600 mJ, 30 fs, 10 Hz). Special attention was paid to improve the contrast ratio, control of pulse shape as well as phase by an acoustic optic programmable dispersive filter (AOPDF) and 1 kHz preamplifier. Preliminary data have shown good laser characteristics.

1. Introduction

So far, extensive studies have been devoted to develop coherent x-ray and extreme ultraviolet (XUV) laser generation, which are expected to open new fields in physics, chemistry and biophysics. Development of high repetition rate x-ray laser is an important factor for a real application experiment. Recently, capillary discharge [1] and optical-field ionization [2] schemes have achieved the high repetition rate x-ray laser generation. The longitudinal-pumped TCE x-ray laser and XUV generation from femtosecond laser produced solid surface plasma have been studied in our group [3,4]. Since the development of high directive Ni-like Mo 18.9 nm soft x-ray laser pumped longitudinally by only total 150 mJ laser, there has been strong interest in making table-top soft x-ray laser pumped by such modest energy and equipped with qualified laser characteristics [4]. In order to develop the high repetition rate of the laser-driven TCE x-ray laser, the longitudinal-pumped TCE x-ray laser is the most effective pumping scheme. In our demonstration the new aspects of pulse shape control (energy and duration of pedestal) and concave electron density are discussed and compared with simulations of x-ray laser propagation [5,6]. Recently Keenan et al reported a different scheme using traveling wave grazing incidence pumping [7]. They showed the Ni-like Mo 18.9 nm lasing pumped by total 300 mJ (main pulse 70 mJ) of which energy is 1/20 of usual transverse pumping. Nickles also reported on the success to reduce the pumping energy by a factor 1/5 – 1/6 by changing the shape of pump pulses, including its shape and amplitude of pedestals [8]. It is an important factor of the pedestal, contrast ratio, and phase of laser pulse to generate the TCE x-ray laser. On the other hand, an important factor for XUV harmonics generations from solid surface plasma is the high contrast ratio ($\sim 10^{-6}$) laser pulse is required [9,10].

In this paper, we report the investigation on the longitudinally pumping TCE x-ray lasers and high order harmonic generations from solid surface of plasmas. According to results from our investigations, we have designed new 20 TW Ti: Sapphire laser system (600 mJ, 30 fs, 10 Hz) and describe the necessity of its new technology for high repetition x-ray lasers.

2. Longitudinally pumped Ni-like Mo x-ray Lasers

2.1 Experiment Results

The details of our pumping laser and experimental setup used for the longitudinally-pumped Ni-like Mo x-ray laser experiment have been described elsewhere [11,12]. A part (~30 %) of stretched amplified long pulse (300 ps) and a short pulse (475 fs) beam from a tabletop laser are sent into the target chamber for generation of coherent x-ray radiation. The longitudinally pumping scheme is used because of its advantages, namely, perfect traveling-wave pumping, which is crucial factor for efficient collisional excitation, and the highly efficient pumping that is inherent in longitudinal geometry [13]. The long pulse is focused onto a Mo target to form a 100 μm × 2 mm line focus by using a cylindrical lens, and the short pulse propagated through the pre-plasma. The beam waist size and the confocal parameter of the femto-second radiation are 30 μm and 10 mm, respectively. The maximum intensities of the long and short pulses at the target surface are 1.5×10^{11} and 2.0×10^{16} W/cm^2, respectively. Figure 1 shows the intensity trace of a typical on-axis 18.9 nm Ni-like Mo line ($3d^9 4d^1 S_0$-$3d^9 4d^1 P_1$). At the same time, the millimeter-long jetlike structures in the visible emission from laser plasma are observed [14]. The pump laser energy in the long and short pulse is 30 and 120 mJ, respectively, and the delay between the long and short pulse is 4 ns. When the short pulse intensity is decreased, the x-ray laser can not be observed, as is shown in Fig. 2. The threshold intensity of Ni-like Mo x-ray laser generation is around 2.0×10^{16} W/cm^2, respectively. We are now performing additional experiment.

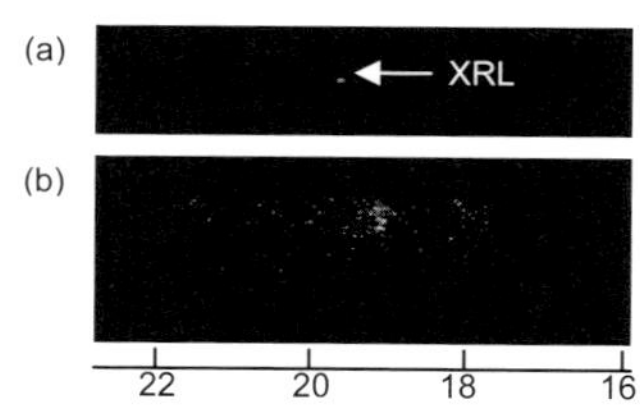

Figure 1. Spectrum image of the on-axis emission for plasma length of (a) 2.0 mm and (b) 0.5 mm. The sensitivity of the photo-cathode camera was increased by 8 times for (b).

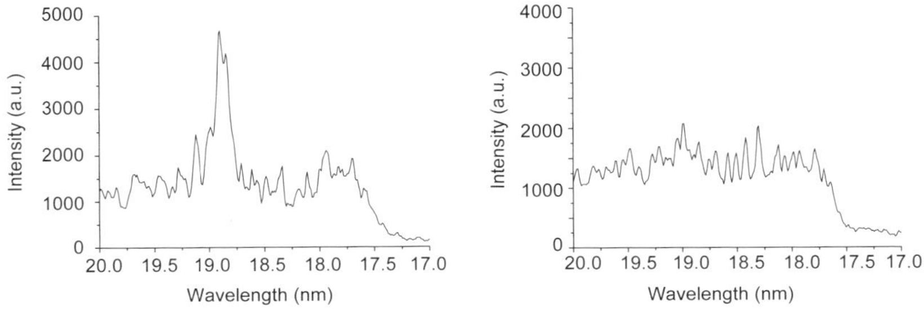

Figure 2. Intensity trace of the on-axis soft X-Ray spectrum in the wavelength range between 17 nm and 20 nm. (a) The spectrum shows 18.9 nm line at theshold conditions of lasing. (b) The spectrum of Mo plasma without longitudinal pumping by femtosecond radiation.

2.2 Pre-plasma conditions [3]

A big question arises as to why such highly directive x-ray lasers can be generated from short 2-mm-long plasma. We attribute this unique phenomenon to the presence of a pedestal in the longitudinal pump beam. We have experimentally measured the pedestal-to-peak contrast ratio of our 475 fs pump laser to be about 1×10^{-5} for the present pump conditions, using photomultiplier tube and calibrated neutral density filters. There are two outcomes of this pedestal that can greatly affect the divergence of the present x-ray laser. First is the localization of the Ni-like ion abundant region to the position of the longitudinal beam. Simulations show that the 1.5×10^{11} W cm^{-2} intensity pre-pulse is only capable of producing pre-plasma with an average ionization of 8, much less than 14 required for Ni-like molybdenum ions. However, by introducing a 300 ps duration, 1×10^{-5} contrast ratio pedestal to the main pulse, the temperature and density is increased to 90 eV and 1×10^{20} cm^{-3}, respectively. As a result, narrow Ni-like molybdenum abundant region is produced within the pre-plasma, which can generate high gain with the irradiation of the high intensity main pump. To illustrate this effect, the result of gain calculation with and without the above pedestal condition is shown in Fig.3, respectively. Here the pedestal produces a density profile within the pre-pulse-produced plasma in the x-direction normal to the target surface, which is assumed to be uniform in the z-direction parallel to the x-ray laser axis. We find that gain coefficients exceeding 70 cm^{-1} are generated at the entrance of the pre-plasma, over a length of 500 μm. A long and narrow positive gain region extends over the full length of the 2-mm-long pre-plasma, which corresponds to the region that is pumped by the pedestal. Note that gain outside of this region is strongly negative, resulting in strong absorption of the x-ray laser. In case of a longitudinal pump without pedestal, the x-ray beam is bent by a relatively large electron density gradient and no gain is observed both in the experiments and in the simulation. Figure 3 shows the simulation result.

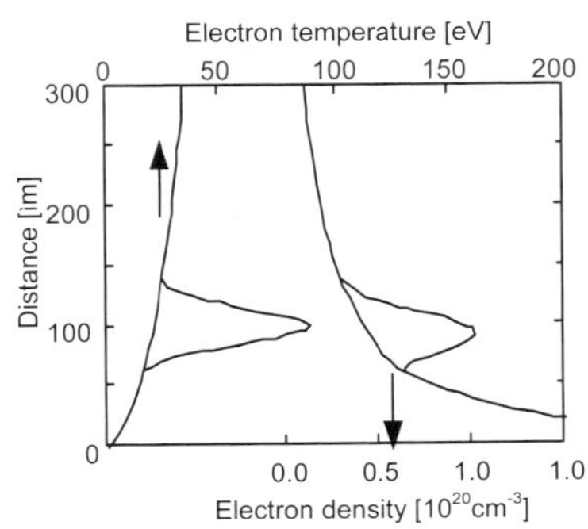

Figure 3. Effect of pedestal in pumping pulse

2.3 Pedestal conditions [3]

Another important outcome of the presence of a pedestal in the main pulse is the generation of electron density and gain distributions that have large spatial gradients. The parabolic electron density and gain profile that are produced by the pedestal is found to limit the number of transverse modes that are amplified in the gain medium. In the work by London *et al.* [15], it was numerically shown that for gain medium with such unique profiles, the efficiently guided modes that receive large amplification could be restricted to the lower order modes (See Fig. 4.). Using formulations from this paper, we can calculate the number of transverse modes that receive large amplification in our experiment. Assuming that the density and gain profile is homogeneous along the x-ray laser axis, we take the lower limit of $0.18 \times gL_{max}$ for counting the number of modes, where gL_{max} is the maximum gain-length product experienced by a single transverse mode. If the density profile vertical to the target surface was flat, the number of effectively amplified one-dimensional Hermite-Gaussian modes is 47, and in this case the geometrical dimensions of the gain region determines the divergence of the x-ray laser. However, this number is reduced to 5 for the parabolic profile predicted in our experiment, with the maximum gain obtained for the diffraction-limited transverse mode. This shows that for our experimental condition, amplification prefers transverse modes with high spatial coherence, which explains the observed small divergence.

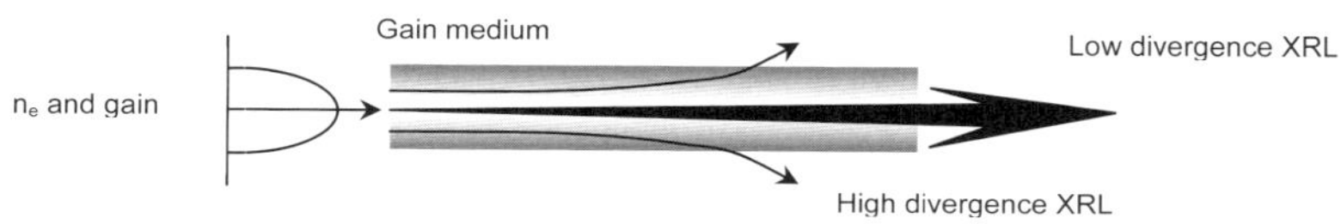

Figure 4. The parabolic electron density and gain profile that are produced by the pedestal is found to limit the number of transverse modes that are amplified in the gain medium.

3. High order harmonics generations

The availability of compact, high-intensity sub-picosecond lasers with chirped pulse amplification (CPA) has opened the new field of study of laser-matter interactions with solid targets. At the high laser intensities above 10^{16} Wcm^{-2}, the electron temperature is more than one keV and the main laser absorption is related to collisionless absorption. Collisionless absorption mechanisms include vacuum heating, J x B heating, sheath inverse Bremstrahlung, and the anomalous skin effect. Hansen et al. reported for the first time that the second harmonic is blueshifted when the laser has intensity exceeds 5 x 10^{16} Wcm^{-2} and the blueshift is due to the absorption by the anomalous skin effect [17]. However, no detailed study of the blueshifts in the other harmonic orders, nor the dependence on the laser intensity at the incidence angle of 45 degrees has yet been reported. It was commonly observed that efficiency of harmonic generation for the p-polarized laser pump was much larger than that for the s-polarized laser beam. Only a few experimental studies have been reported on the effects of polarization on the harmonic radiation. However, the cause of the disagreement between some recent theoretical and experimental results is not yet clear.

In this paper, we present blueshifts of up to about 2 nm in the second harmonic and for the first time about 5 nm in the fifth. We carry out more detailed model-calculations of the electron temperature in our experimental condition using a 1D hydrodynamic code HYADES. We find that the blueshift in our experiment can be well explained from the absorption mechanism due to the anomalous skin effect. We also show that the harmonic dependence on the pump laser polarization decreases when laser intensity exceeds 10^{17} W cm^{-2}. Figure 5 shows the dependence of the wavelength shift of the second harmonics on the incidence laser intensity. The peak frequency of the second harmonic emission is blueshifted for intensities greater than 2 x 10^{16} Wcm^{-2}. It is found that the increase of this shift energy with an increase in the laser intensity is small in comparison with the experimental results by Hansen et al. [17].

Unfortunately, we cannot detect spectral shift below 1 x 10^{16} Wcm^{-2} that the fifth harmonic light is blue shifted by 5.1 nm for p-polarization at the intensity of 1 x 10^{17} Wcm^{-2}. This shift is much larger in magnitude than the Doppler shift seen in the spectrum of the fundamental. Therefore, these results cannot be explained by the hydrodynamic motion of plasma. The calculation results by Rozmus et al. [18] show that at low temperature (Te<300 eV), the collision absorption mechanism is dominant, and at higher temperature (Te>800 eV), the collisionless absorption mechanism due to the anomalous skin effect becomes dominant.

The electronic temperature profile in our experiment condition was calculated using a 1D hydrodynamic code HYADES. From the results, it is found that the electron temperature in the skin depth from the critical density surface is much higher than 800 eV. Therefore from our experimental and calculation results, we attribute the harmonic frequency shift to the collisionless absorption mechanism, and the difference in the wavelength shift between the orders of harmonics in our experimental results is attributed to the phase shift accumulation or phase shift rotation at a certain cycle. A more detailed analysis of the experimental data has been reported elsewhere [19].

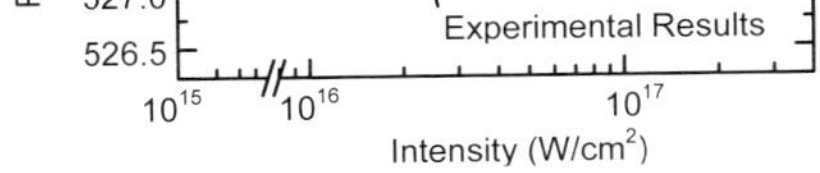

Figure 5. Location of the second harmonic peak as a function of incident intensity in our experiment.

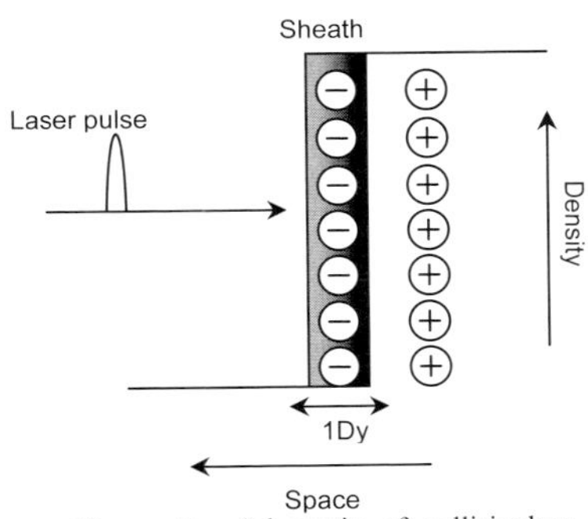

Figure 6. Schematic of collisionless absorption image on the solid surface plasma irradiated by an ultra-intense laser pulse

4. New 20 TW laser system design and performance

Layout of the 20 TW Ti: sapphire laser system (Ultra-laser Inc. Beijing) presently under costraction at our laboratory is shown in Fig.7. The whole of tabletop 20 TW Ti:Sapphire laser system consists of a femtosecond oscillator, an Offner-type stretcher, an AOPDF [20], a kHz regenerative amplifier, a 6-pass main amplifier, and a vacuum pulse compressor. Special attention was paid to improve the contrast ratio and control of pulse shape and phase by adaptive AOPDF and 1 kHz preamplifier. Preliminary data have shown good laser characteristics. The femtosecond oscillator is pumped by a compact solid-state diode-pumped, frequency-doubled Nd:Vanadate (Nd: YVO_4) laser (Verdi-V8, COHERENT Inc.) that provides single-frequency green (532 nm) output with maximum power of 8W. The oscillator with 1.1 W output power and pulse width of 30 fs pulse duration at 83 MHz repetition rate is obtained for a 6.5 W pump power of the Verdi-V8. The FWHM spectral bandwidth of 48 nm is obtained for the center wavelength of 800 nm. The short pulse from the oscillator is stretched to 300 ps by the Offner-type stretcher. In order to control the dispersion and shaping of the laser pulse, the AOPDF (800 WB-AT, FASTLITE Inc.) was installed after the Offner-type stretcher. The AOPDF can control the pedestal and shape of the laser pulse in x-ray and XUV laser experiments. Figure 8 shows the laser spectrum shape controlled by the AOPDF. To improve the stability of laser and contrast ratio, the regenerative amplifier was operated at 1 kHz repetition rate. This also allows a beam pattern with high quality. The stretched pulse is amplified to 3 W by the 1 kHz Ti:Sapphire regenerative amplifier which pumped by a diode-pumped laser providing average power in excess of 20 W at second harmonics wavelength of Nd: YLF (JADE, THALES LASER). The amplifier with ~2.3 W output power is obtained with 15 W pump power of JADE. The amplified pulse from the regenerative amplifier change the 1 kHz to the 10 Hz repetition rate by using the Pockels cell. The 10 Hz laser pulse from the Pokels cell is amplified to 1.2 J with the

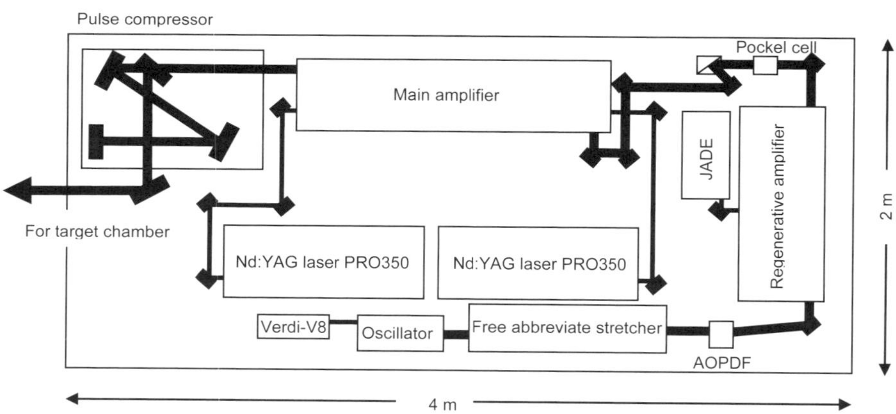

Figure 7. Layout of the 20 TW Ti:Sapphire laser system. The system consists of a femtosecond oscillator, a free abbreviate stretcher, an AOPDF, a kHz regenerative amplifier, a 6-pass main amplifier, and a vacuum pulse compressor.

6-pass Ti: Sapphire amplifier pumped by using two Nd: YAG lasers (Spectra Physics PRO350), which provides a pump laser energy of 1.4 J in 8 ns pulse at a wavelength of 532 nm. The pulse compressor is composed of two gratings of 1800 lines/mm in a vacuum chamber. In our estimations the laser would generate more than 20 TW after compression.

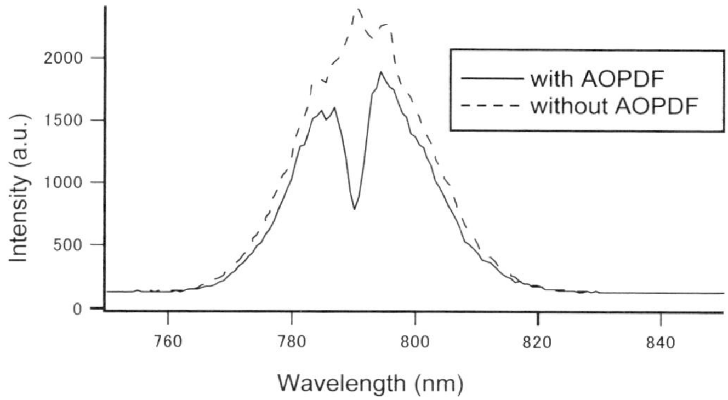

Figure 8. Spectra property of the laser pulse from AOPDF. Using the AOPDF, the spectra shape can be controlled.

5. Summary

In order to develop the high repetition rate x-ray laser, the longitudinal-pumped TCE x-ray laser is one of the most effective pumping schemes. We have described the results of longitudinally pumping TCE x-ray laser and high order harmonic generation from the solid surface plasma. The most important factor of the longitudinally pumping x-ray generations is concave density profile of pre-plasma and pedestal of the pumping laser pulse. To pursue the high repetition rate of the laser-driven TCE x-ray laser, we have designed new 20 TW Ti:Sapphire laser system (600 mJ, 30 fs, 10 Hz). Special attention was paid to improve the contrast ratio and control of pulse shape and phase by adaptive AOPDF and 1 kHz preamplifier. At present we have already obtained the laser energy of 1.2 J at 10 Hz repetition rate before the pulse compressor. Precise measurements of characteristics and soft x-ray laser experiment by this system will soon be started.

Acknowledgements

The authors would like to acknowledge A. Ishizawa for the early stage of harmonics experiments and Z. Wang, W. Lin, and P. Wang for the development of the 20 TW laser system. This work was partly supported by the Grant-in-Aid for Creative Scientific Research (14GS0206) of Japan Society for the Promotion of Science (JSPS).

References

[1] Benware B R, Seminario M,et.al. 2001 Opt. Soc. Am. B 18 1041-1045

[2] Sebban S, Mocek T, et.al. 2002 Phys. Rev. Lett. 89 253901 1-4

[3] Kuroda H, Ozaki T, et.al. 2002 J. Lumin. 100 291-300

[4] Ozaki T, Ganeev A R, et.al. , 2002 Phys. Rev. Lett. 89 253902 1-4

[5] Ozaki T, Nakano H, Kuroda H 2002 Phys. Rev. E 66 047402 1-4

[6] Ozaki T, Nakano H, Kuroda H 2002 J. Opt. Soc. Am. B 19 1335-1341

[7] Keenan R, Dunn J et al., 2003 SPIE Vol 5197 213-219

[8] Nickles P V 2003 EQEC 11

[9] Tarasevitch A, Orisch A, et.al. 2000 Phys. Rev. E. 62 023816 1-6
[10] Teubner U, Eidman K, et.al. 2004 Phys. Rev. Lett. 92 185001 1-4
[11] Itatani J, Kanai T, et.al. , 1996 Prog. Cryst. Charact. Growth Mater. 33 281-292
[12] Ganeev R A, Kanai T, et.al. 2004 Appl. Opt. 43 1396-1403
[13] Li R, Ozaki T, Kanai T, and Kuroda H 1998 Phys. Rev. E 57 7093-981
[14] Ozaki T, Yamamoto K et al., 2002 J. Phys. Soc. Jpn. 71 2963-2968
[15] London R, Strauss M et al., 1990 Phys. Rev. Lett. 65 563-566
[16] Wilks S C and Kruer W L 1997 IEEE J. Quantum. Electron. 33 1954-1968
[17] Hansen C T, Wolks S C et al., 1999 Phys. Rev. Lett. 83 5019-5022
[18] Rozmus W, Tikhonchuk V T et al., 1996 Phys. Plasmas 3 360-367
[19] Ishizawa A, Ozaki T, et.al. 2002 Phys. Rev. E. 66 026414 1-7
[20] Monmayrant A, Joffre M, et.al. 2003 Opt. Lett. 28 278-280

Inst. Phys. Conf. Ser. No 186
Paper presented at 9th Int. Conf. X-Ray Lasers, Beijing, China, 24–28 May 2004

Spatial coherence of longitudinally pumped soft x-ray lasers

T Ozaki[1], R.A. Ganeev[2], M. Suzuki[2], M. Baba[2] and H. Kuroda[2]
[1] Institut national de la recherche scientifique, Université du Québec, 1650 Lionel-Boulet, Varennes, Québec J3X 1S2 Canada.
[2] Institute for Solid State Physics, University of Tokyo, 5-1-5 Kashiwanoha, Kashiwa, Chiba 277-0882 Japan.

Abstract. Simulations on longitudinally pumped Ni-like Mo x-ray lasers are performed, aimed at clarifying the mechanism of the high spatial coherence observed in our experiments. Transverse mode calculations show that refractive antiguiding of the x-ray laser beam within the low gain region can greatly improve the transverse mode of the x-ray laser.

1. Introduction

Experimental demonstration of longitudinally pumped nickellike molybdenum soft x-ray laser at 18.9 nm have shown that soft x-ray lasers with near-diffraction limited spatial coherence can be achieved at pump energies of only a few 100 mJ [1]. This work has opened the possibility of generating a highly practical, highly coherent x-ray laser within the soft x-ray wavelength range, which should result in new applications of these unique sources. However, the detailed physics leading to the high spatial coherence and low pump energy requirement has yet to be clarified. In this paper, we perform numerical investigations to clarify some of these questions.

2. Gain Calculations with Ray Tracing

Previous simulations have shown that with the long prepulse alone, the ionization stage of molybdenum can only reach Mo^{8+} [1]. However, ionization is enhanced to the Ni-like stage if the pedestal of the main longitudinal pulse is included in our simulations. This dual ionization of the preplasma by the long prepulse and the pedestal of the main longitudinal pump is also predicted to be the source of the high spatial coherence of the longitudinally pumped Ni-like Mo x-ray laser. In the present work, we include in our calculation code the effects of non-identical plasma distributions along the major (z) axis. This will allow taking into account, for example, the bending of the longitudinal pump beam resulting from the electron density gradient in the direction vertical to the target surface (x direction).

2.1. The Code

In our simulation code, the formation and expansion of the preplasma by the transverse pump is modeled using the *HYADES* code [2]. The plasma parameters of the preplasma are calculated at the time just before the pedestal of the longitudinal beam begins to irradiate the preplasma. Due to limited calculation resources, we use a compact collisional-radiative model that includes the ground states of Ni-like, Co-like and Cu-like ions, and 7 excited states of Ni-like ions. The propagation of the longitudinal beam through the preplasma is modeled using a ray trace code, which simultaneously calculates the absorption of the laser as well as the plasma dynamics. The longitudinal pump beam has a Gaussian intensity distribution in the plane normal to the target surface, and is incident onto the preplasma at an angle parallel to the target surface. The absorption of the longitudinal pump beam by the preplasma deforms the temporal pulse shape, since the lower electron temperature at the rising edge of the pump results in a larger absorption, compared with that at the trailing edge. In our code, this change in the pulse shape is approximated by the shift in the time of the pulse peak.

2.2. Results

2.2.1 *Modeling the Experiment.* First, the exact parameters used in our experiments are adopted for simulations. The 300 ps, 0.15 TW cm^{-2} peak intensity prepulse is line-focused on to the molybdenum solid target. The delay between the prepulse and main pulse is 4 ns. The pedestal of the main pulse has a 300-ps duration with an intensity of 0.3 TW cm^{-2} at a time 4 ps before the peak. The longitudinal pump beam enters the preplasma at a position 100 μm from the target surface.

In Fig.1, we show the electron temperature and density of the preplasma at a time just

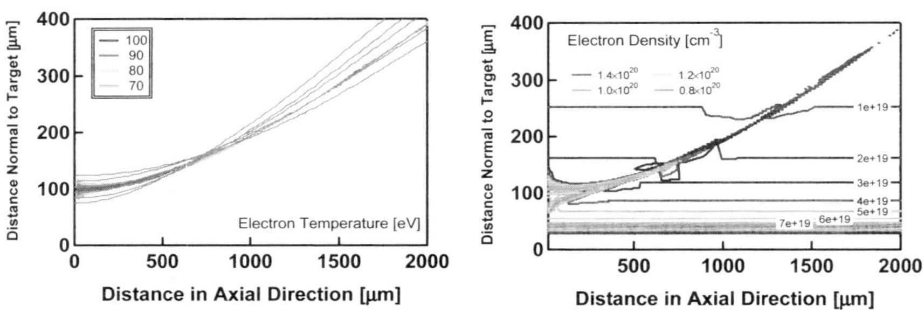

Figure 1. Two-dimensional distribution of the electron temperature and density using the experimental condition.

before the high intensity 475 fs pulse irradiates the preplasma. The longitudinal beam bends away from the target due to the electron density gradient. A relatively narrow region with increased electron temperature and density is produced at the center of the beam. The electron temperature is between 70 and 100 eV, which results in Ni-like Mo ion abundance of greater than 0.4. The above narrow ionization channel can be divided into two regions, one with a relatively high electron density near the target and the preplasma entrance, and one with a relatively low electron density, at a distance greater than 200 μm from the target surface. The maximum electron density within the channel can reach 1.4 x 10^{20} cm^{-2}, which for peak main pump intensity of 30 PW cm^{-2} will result in a maximum gain of 75 cm^{-1}.

2.2.2 *Effect of prepulse to main pulse time delay.* Experiments have shown that the maximum intensity of the Ni-like Mo x-ray laser is obtained when the prepulse to main pulse time delay Δt is 4 ns. When Δt is either reduced to 2 ns or increased to 10 ns, the x-ray laser intensity decreases by one order of magnitude and two orders of magnitude, respectively. Simulations using the ionization and ray trace code were performed to investigate the main cause of this tendency.

The variation of average ionization of the molybdenum ion within the preplasma is shown in Fig.2, for (a) $\Delta t = 2$ ns and (b) $\Delta t = 10$ ns. Results show that at Δt=2 ns, the longitudinal pump beam is more strongly refracted due to the larger electron density gradient in the x direction. This results in a spreading of the beam and a reduction in the peak pump intensity at the exit end of the preplasma. Consequently, the ionization level is far below the Ni-like stage in these regions, which will produce a highly absorbing region and thus reducing the x-ray laser intensity. On the other hand, the average ionization distribution for $\Delta t = 10$ ns is similar to that calculated for the optimum $\Delta t = 4$ ns case. The simulations does not reveal any evident cause for the large reduction in the x-ray laser intensity for $\Delta t = 10$ ns. However, one possible explanation is the change in the preplasma expansion from a cylindrical to a spherical one at large delay times. Such a change will result in increased plasma cooling and lower ionization stages, thus producing regions with strong absorption.

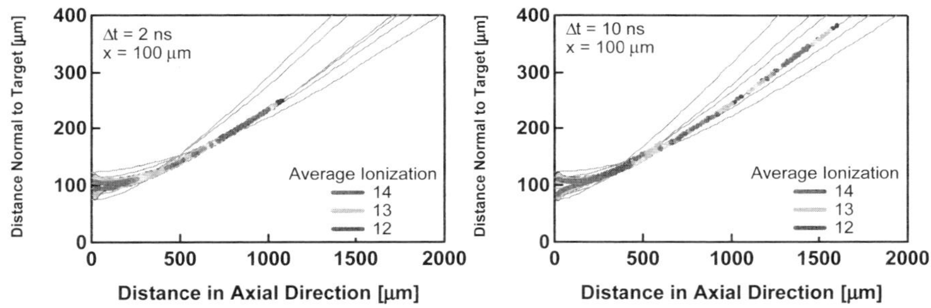

Figure 2. 2-D ionization distribution for (a) $\Delta t = 2$ ns and (b) $\Delta t = 10$ ns.

3. Far-field Transverse Mode Calculations

3.1. The Code

To investigate the spatial coherence of longitudinally pumped x-ray lasers, we developed a simulation code based on paraxial wave equation [3]:

$$\left[\frac{1}{k}\nabla_{\perp}^{2}-2i\frac{\partial}{\partial z}-h(x,y)+ig(x,y)\right]E(r)=-4\pi kP(x,y) \qquad (1)$$

Here, h is the refractive strength, defined as $h(x,y)=\omega_p^2(x,y)/kc^2$, g is the small-signal gain coefficient, and P is the spontaneous emission intensity. To include the random nature of spontaneous emission, P is modeled as a random function, and so we used the finite element method to solve the above equation. g and h are assumed to be parabolic functions with cylindrical symmetry, and the minimum gain is limited to $-5\times g_{max}$, where g_{max} is the maximum gain at the axis center. h is assumed to be constant outside of the gain region, to better model the actual gain medium.

3.2. Results

We first performed simulations to show the effects of gain-length product on the far-field transverse mode of a longitudinally pumped x-ray laser. In this simulation, we assumed that the parabolic gain profile in the direction perpendicular to the target surface is identical at all positions along the x-ray laser axis. The maximum small-signal gain coefficient is 50 cm^{-1}, with a transverse dimension of 25 μm full-width at half maximum

(FWHM). In Fig.3 we show the results of our transverse mode calculations for gain

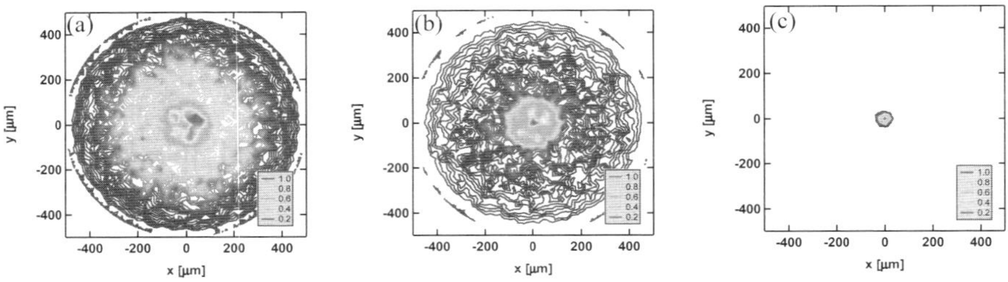

Figure 3. Transverse mode of Ni-like Mo x-ray laser with a maximum gain coefficient of 50 cm^{-1}, for gain lengths of (a) 0.5, (b) 1.0, and (c) 2.0 mm.

lengths of 0.5, 1.0 and 2.0 mm. These modes correspond to those observed at a position 100 mm from the source. We can see the amplified spontaneous emission (ASE) gradually emerging from the background spontaneous emission as the length of the gain medium is increased. For 1.0 mm length gain medium, the transverse mode size is 200 μm FWHM and still non-uniform, with the background of considerable intensity compared with the ASE. The condition completely changes for gain lengths of 2.0 mm. The ASE beam is uniform, and is confined within a small spot of 40 μm FWHM, corresponding to an x-ray laser divergence of 0.4 mrad.

Our calculations of Section 2.2.1 showed that the narrow plasma column is separated into a high gain region, close to the entrance of the preplasma, and a low gain region at the exit side. In order to model this unique gain profile, we varied the gain so that the first 1 mm length of the preplasma has a parabolic gain profile with a maximum of 50 cm^{-1}, while the last 1 mm had uniform zero gain. The results of the mode calculation revealed some interesting facts. The peak intensity of the ASE in this case decreases by a factor of $6x10^{-3}$ compared with a 2-mm-long, 50 cm^{-1} maximum gain profile, due to a reduction in the *gL* product. However, the beam still possesses a small divergence of 0.4 mrad, comparable to that observed in Fig.3(c). These results show that under our experimental conditions, the high spatial coherence of the longitudinally pumped Ni-like Mo x-ray laser is largely due to refractive antiguiding of the x-ray laser beam through the highly ionized narrow column of the preplasma.

4. Summary

We have performed simulations to model the plasma dynamics and transverse mode characteristics of longitudinally pumped Ni-like Mo x-ray lasers. Ray trace calculations have shown that due to the pedestal of the longitudinal pump beam, a narrow region with high Ni-like ion abundance is produced. Transverse mode calculations show that such a narrow gain region with high electron density gradient will induce refractive antiguiding of the x-ray laser beam, thus selecting only the high spatial coherence modes.

References

[1] Ozaki T et al. 2002 Phys. Rev. Lett. 89, 3902
[2] Larsen J T and Lane S M 1994 J. Quant. Spectrosc. Radiat. Transf. 51, 179
[3] London R A et al. 1990 Phys. Rev. Lett. 65, 653

Inst. Phys. Conf. Ser. No 186
Paper presented at 9th Int. Conf. X-Ray Lasers, Beijing, China, 24–28 May 2004

Pumping efficiency of transverse, longitudinal and grazing incidence schemes

V.N. Shlyaptsev[1], J. Dunn[2], R. Keenan[2], R. Smith[2], P. Patel[2], D.F. Price[2], J.J. Rocca[3], J. Filevich[3], B. Luther[3], Y. Wang[3], M.C. Marconi[3]
[1]Department of Applied Science, University of California Davis-Livermore, Livermore, CA 94551
[2]Lawrence Livermore National Laboratory, Livermore, CA 94551
[3]Electrical Engineering and Computer Department, Colorado State University, Fort Collins CO 80523

Abstract The grazing incidence concept of plasma pumping is described here and is shown to have properties surpassing both transverse and longitudinal pumping schemes. It has been demonstrated recently as the basis for a high-repetition rate x-ray laser (in the following papers [1, 2]). Simulations show that grazing incidence pumping and the effect of refraction boost the coupling efficiency and provide the spatial selectivity of pumping simultaneously with near speed of light traveling wave excitation.

In development of shorter wavelength lasers substantial efforts were devoted to investigating plasma physics and hydrodynamics, different kinds of atomic mechanisms for creating inversion, implementing traveling wave excitation and decreasing gradients to reduce tremendous energy requirements of first generation x-ray lasers. Many different targets types were used, starting from plane solid targets, foils, gas-puff jets to capillaries. There was a lot of optimization performed to obtain good efficiency of coupling of pumping energy and at the same time to minimize detrimental effects like self-absorption and refraction that also strongly influence lasing parameters.

Experimentally, the lasing in transient collisional excitation concept at wavelengths of 30 nm and shorter has been achieved using the *traditional* normal incidence (or transverse) pumping of flat metal targets to generate plasmas containing Ne-like or Ni-like ion charge states. An order of magnitude higher gain for transient schemes, for example, was achieved by pumping with the short pulse durations from tabletop chirped pulse amplification (CPA) systems of just a few joules of energy [3-5]. Other pumping geometries for example longitudinal pumping have been explored at longer wavelengths for low density gas targets, specifically the 41.8 nm Pd-like Xe ions and the 32.6 nm Ni-like Kr ion schemes at high repetition rates [6-8]. These show promise but are limited to gases or metal vapors operating at low pressures [9]. The attempts to create hybrid schemes using capillary discharge plasmas combined with longitudinal laser pumping have been reported [10]. The vast majority of proposed and demonstrated x-ray lasers during two decades of all varieties from photo-pumped and recombination to OFI- and collisional schemes mostly utilize either transverse or longitudinal geometry to excite the plasma. The methods of selective high efficiency pumping described here, in [11] and present Proceedings [1, 2] will hopefully be the next improvements on this route.

1. From transverse and longitudinal to grazing incidence design

Each of these laser pumping designs has their advantages and disadvantages. Transverse pumping of the active medium was a natural method of plasma creation for most of x-ray laser schemes and it has been successfully used for decades. It can pump long 1 – 6 cm targets where the target length is only limited by the available energy or loss of amplification from refraction of x-ray radiation. To deal with the latter case, some control of the plasma density profile, important for x-ray laser amplification and propagation, can be achieved by using pre-pulses or specially designed targets for example gas-puff or exploding foils [12-14]. The major disadvantage for normal incidence pumping is that the laser energy is absorbed inefficiently, i.e. at various places but mostly at dense regions up to the critical surface, n_{cr}, that typically do not contribute to the amplification. But there still exists ways to improve the absorption here so that most of laser energy will be deposited right at the optimum for laser amplification conditions [13].

The longitudinal pumping scheme on the other hand has the advantage of intrinsic traveling wave pumping of the plasma volume. The disadvantages include a strong dependence on refraction and, a situation opposite to the transverse scheme, too high absorption. The laser pump has to pass through the entire plasma column longitudinally where strong absorption and refraction processes limit the target length to a few millimeters. This requires low plasma densities substantially below 10^{20} cm^{-3}. There is also a major problem with longitudinal pumping which is potentially hard to solve to produce high efficiency x-ray lasers. This is the requirement of homogeneous pumping of active medium column with very high aspect ratios (plasma length to diameter), which can be as high as a 1000. The pump radiation focused on one end of the medium has to travel the distance of not one but tens of Rayleigh lengths which is hard to achieve without implementing wave-guiding techniques. The capillary plasma, for example, can potentially transport radiation greater than ten Rayleigh lengths [10,14,15]. The propagation and absorption along the distances of typical lengths of laser medium $L \sim 1$ cm sets a low limit on plasma density n_e. But the density has to be high to follow the major scaling laws to maintain high x-ray laser gain, g, as a function of x-ray wavelength, λ_L. Typically, the density must grow as $\sim\lambda_L^{-3.5}$ (for Ne- and Ni-like ions for example $g \sim n_e^{4.5}$ while $n_e \sim \lambda_L^{-8}$ [3]). Note, that some solutions to lower absorption here still are possible if shorter wavelengths pump lasers are used, high laser intensities (where electron quiver velocity in the laser field of intensity I, $v_q \sim eI^{1/2}/m\omega$, exceeds the electron thermal speed $v_e = (T_e/m)^{1/2}$) or to some extent overheat the plasma, if population inversion mechanism allows, which may help to fit the absorption coefficient at the level of $k \sim L^{-1} \sim 1$ cm^{-1}. But most severe difficulty and worst obstacle in longitudinal case could be caused by refraction. It is clear that since even for amplified x-rays the refraction still plays detrimental role turning x-ray photons out of the gain region, it is specifically severe for optical radiation. The grazing incidence pumping scheme in its turn addresses the pumping radiation refraction problem of longitudinal scheme and small absorption in transverse case in very elegant way.

The idea of improvement of transverse scheme utilizing pumping at grazing angles was pushed forward when different channels of dissipation of pumping laser energy were analyzed in detail. The RADEX calculations of the plasma conditions of molybdenum Ni-like x-ray laser which was demonstrated experimentally in the following paper [1] show that typically the gain region is positioned at the slope of electron density profile near the $n_f \sim 10^{20}$ cm^{-3} where gradient are much smaller than near the critical density ($n_c \sim 1.7x10^{20}$ cm^{-3} for $\lambda = 0.8$ μm in our case). But due to smaller density the absorption coefficient which is $k \sim n_e^2$, the fraction of laser energy absorbed in narrow gain region of size 15-30 microns in transverse case is very small, typically ~5%. Fig 1 clearly illustrates this showing the distribution of laser power from which it is clear that most of the pumping power pass through the amplification region not being absorbed and is dissipated uselessly near the critical density because of both transverse dimensions and absorption coefficients in active medium region are typically small.

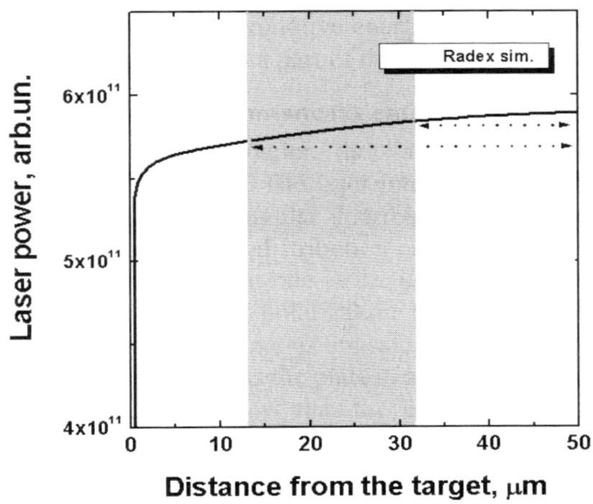

Fig. 1 Short pulse laser power distribution in transverse case (Ni-like Mo x-ray laser). Shaded area is transient gain region. Most of radiation comes through this region without absorption reaching critical density where it is mostly absorbed.

If we increase the pass length of pump laser radiation by constructing the optics system in such a way that radiation enters the plasma at grazing angles covering whole active medium (see schematics in [11]), the electron thermal energy equation [16] can be modified as

$$N\frac{dE}{dt} = qk/\cos(\theta) - Q \qquad (1)$$

where for simplicity we include in Q all kind of losses, dissipation processes and the work $\nabla(P\boldsymbol{v})$ against the thermal pressure P and leave here only the external source of energy due to absorption of laser radiation q (N is ion density, E is thermal energy of particles, $\boldsymbol{v}$ is mass speed, q is laser flux density on the surface, θ is angle of incidence). As it seen from (1) at shallow angles the improvement in absorption efficiency is clearly associated with $\cos(\theta) \rightarrow 0$. This simple picture is missing the refraction of laser radiation, which obviously will limit minimum grazing angles which can be applied.

2 Refraction of pumping radiation

As it follows from ray propagation equation derived in geometrical optics approximation [17]

$$\frac{dz}{dr} = \frac{n_0 \sin\theta_0}{\sqrt{n^2(r) - n_0^2 \sin^2\theta_0}}$$

(where z is coordinate along the axis of active medium, r is distance in transverse direction, n_0 denotes index of refraction and θ_0 is angle of incidence of laser radiation in the fixed initial

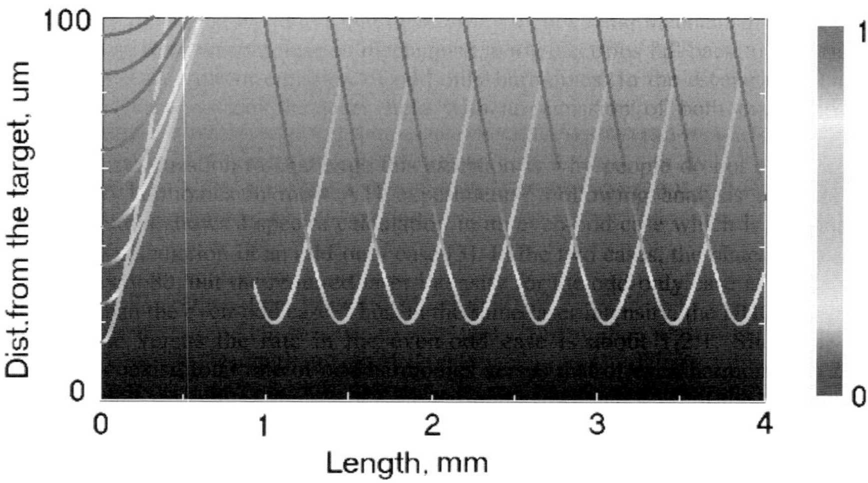

Fig. 2 RADEX calculated ray tracing of 800 nm pump beam in preformed plasma for longitudinal (laser pump comes from the left) and grazing incidence (laser pump comes from the top) cases. Grazing angle is 14°. Color scale shows pump laser pump power along the trajectory of the ray. The distances from the target of faster color change around 20 μm indicate the areas of maximum of energy deposition which is the gain region. Darker gray-scale background reflects more dense plasma near the target surface.

point, in our case for vacuum $n_0=1$) the rays of laser radiation are gradually bended in the plasma until it reach so called turning point where rays go parallel to z axis, and then are reflected back to vacuum. The shape of the ray trajectory curves in real plasmas are enough complex, but for reference, in some simplistic cases when plasma radial density profiles can be approximated with analytical fits, the shape of the rays can be also obtained analytically [18]. For example for the $n_e(r) = n_{e0}(r_0/r)^2$ the curves are hyperbolas. In general case of arbitrary density profiles the ray tracing codes which also solve radiation transport equations have to be used. Fig.2 shows code RADEX ray tracing results for both longitudinal (curves at the left) and transverse cases on the density profiled obtained from hydro calculations of recently demonstrated transient Mo x-ray laser described in more detail in [1]. From this figure it is clearly seen severe disadvantages of longitudinal design specifically the ones caused by the refraction. In this specific case the pumping length barely reach 1mm and pumping homogeneity along the active length is good only in thin layer for which the absorption has to be around 50%.

In the grazing case the presence of refraction turns into additional advantages. First, it acts as a mirror placed exactly at the boundary of active medium which is alone very valuable because it effectively doubles the laser path inside the medium. With active medium width 15-25 microns in radial direction the ray path is 500 microns, as is seen from Fig.2b, which approximately characterizes the efficiency increase by factor 20-30. To use this property the setup for the experiment has to be done such a way that turning point has to be the bottom of active medium. This task will be extremely facilitated by the another property of refraction which also follows from above equation. The position of turning point (radius or electron density) does not depend on any other plasma parameters, including plasma gradients, refraction index etc besides just grazing angle and wavelength of the pumping laser which is usually fixed. So just one external parameter defines position of the

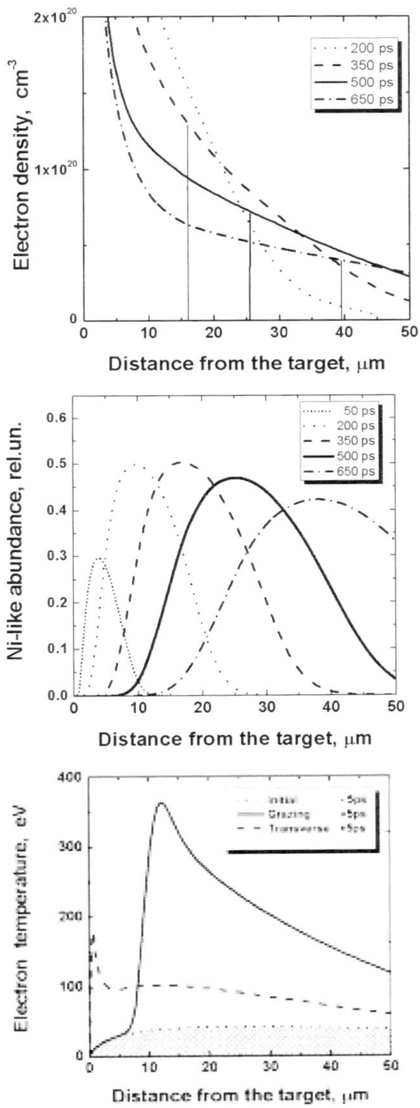

Fig. 3a,b,c The calculated electron density, relative abundance of Ni-like Mo ions and electron temperature for Mo x-ray laser. Parameters used are described in the text.

turning point and the gain region incredibly simplifying the whole design. Besides in high-Z plasma the grazing angle $\varphi_r = 90^o - \theta$ being small has simple relationship with electron density at turning point $\sin\varphi_r \approx \varphi_r = (n_e/n_c)^{1/2}$ because the index of refraction can be well approximated with the expression $n = (1 - n_e/n_c)^{1/2}$ where $n_c = \pi mc^2 / \lambda^2 e^2$ is critical density for laser radiation with wavelength λ. The density at the point is optimized for a particular x-ray laser from atomic kinetics and refraction of the x-ray signal itself. Given this selected density and the wavelength of pumping laser the angle of incidence is chosen.

Some additional properties of refractions can be used for further improvement of amplification of x-ray radiation. They depend on specifics of implementation of focusing optics which delivers pump laser radiation on the target. In our setup the angle of incidence is not actually constant at 14° and vary along the medium z axis so it deliver maximum of pump laser flux density at higher electron density around 10^{20} cm^{-3} at the beginning of amplifier and at smaller densities $5x10^{19}$ cm^{-3} at its exit due to divergence $\Delta\varphi \sim 2.8^o$ caused by finite aperture-to-focal distance ratio $\Delta\varphi = d/f$. Such setup allows to some extent to follow the refraction path of x-ray photons itself.

Finally, the grazing incidence scheme delivers one more benefit for x-ray lasers. With small angles 12 – 17° the pump wave front speed is $c / \cos(\psi)$ or just 2-5% faster than speed of light thus providing traveling wave pumping along the medium.

Numeric simulations below show efficiency improvement of GRIP compared to standard transverse pumping case. The RADEX calculations of hydrodynamics with both grazing incidence and transverse pumping cases for Ni-like Mo x-ray laser are shown in Fig.3. As in the transverse pumping designs, the pre-pulse creates initial plasma density profiles (Fig.3a) with temperature of the order of 50 eV and the required ion composition for which as little as

60 mJ or less can be sufficient at optimal conditions with tighter focusing (15-20 microns was used). When the main heating pulse of 60-80 mJ of energy is turned on (at the grazing angle 14° which was calculated to get the turning point at density around ~10^{20} cm^{-3}) the temperature in the active medium jumps to 200 – 300 eV needed for efficient pumping of Ni-like Mo ions. In the transverse case with such incident laser energy the temperature jump barely reach 50 eV (Fig3c). The optimal time delay for switching on the main short pulse depends mostly on two parameters. If delay is small the density profile is too steep as it can be seen from Fig.3a for 200ps and 350ps curves. This will cause strong refraction of amplified radiation. At too large delays the density gradients decrease but the density value drop too much decreasing absorption and hence temperature. Another key parameter is Ni-like ions relative abundance space and time history (Fig3b). At small time delays the maximum abundance of ions positioned at very steep slope of density profile (these maxima also marked with vertical bars on Fig.3a), while at large delays ions move too far from the surface where, again, the pumping laser radiation absorption will be small. The optimal timing is positioned somewhere in between of these extremes, in this case [1] around 500ps relative to prepulse intensity maximum. Both optimal conditions for ion abundances and appropriate density gradient and its value exist for around 150-200 ps only. The x-ray laser intensity vs pulse delay curve has even sharper shape since it is approximately factor $(gL)^{0.5}$ narrower, ~50ps, in accordance with the experiment [1]. The x-ray laser in experiment [2] works in somewhat different regime. The initial pumping energies and focusing parameters there differ from ones described above having several times larger energy and larger focal width for prepulse intensity, so the optimal delay window is much wider, with around 600ps FWHM for the resulting x-ray laser intensity at total ~1J in energy for both pulses. Note that transient inversion x-ray laser using Ni-like Mo demonstrated before with transverse pumping utilized almost 5 J of laser energy [19]. It can be seen from these results that with GRIP geometry, as opposed to transverse pumping, the refraction, known for its notoriously detrimental influence for x-ray lasers, plays positive role substantially increasing pumping efficiency and generally making whole design more robust and predictable. GRIP delivers the maximum pumping rate precisely near the gain region (Fig.3c) and absorption efficiency jumps by an order of magnitude to 40-50%.

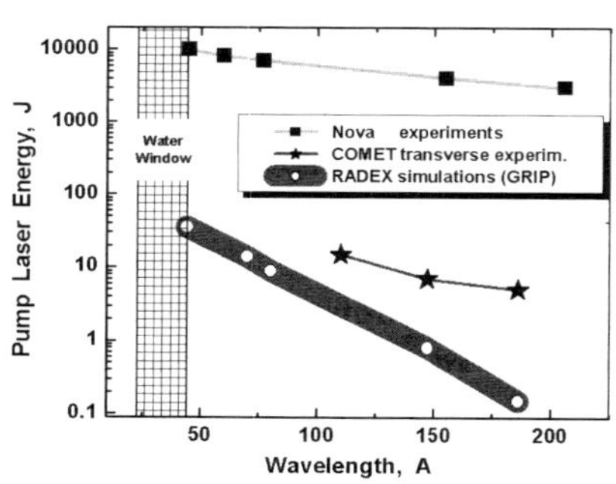

Fig. 4 Pumping energy vs wavelength for Nova experiments, Comet transverse pumping transient XRL experiments and Radex simulations for grazing incidence pumping.

Note, that the temperature of plasma with good precision can be calculated analytically from energy equation (1):

$$T_e = \left(T_{e0}^{5/2} + \beta\, q \tau \Lambda z \frac{n_e}{n_c} \right)^{2/5} \qquad (2)$$

where T_{e0} is initial temperature in eV, flux density is in W/cm^2, τ is pulse duration in seconds, z ions charge, Λ is Coulomb logarithm. The high accuracy of this formula (typically within several percents from numerical calculations) can be explained that when the laser pulse duration is shorter then characteristic times of other processes in plasma, one can neglect them so, for example, hydro, dissipative and ionization processes do not play significant role simplifying the equation (1). In grazing case $\beta = 1984/\cos(\Theta)$ (we include refraction by simplistic way as a mirror causing the rays pass the plasma twice) while for transverse and longitudinal pumping $\beta \sim 992$. In high-flux regimes the Langdon correction factor ~0.5 due electron distribution function change and other factors like electron collision frequency modification can be taken into account [20].

The pumping energy trend in scaling of this scheme to shorter wavelengths is shown in Fig.4. Here RADEX calculations for GRIP design with high-Z atomic elements shown together with older experimental results obtained previously on Nova [21] and COMET [22] lasers. As it follows from the calculations, with proper choosing of pumping laser and target parameters it could be possible to reach, for example, the long wavelength edge of "water window" 23-44Å region with as little as 30-50 J of pumping energy.

In summary, we proposed a new efficient pumping geometry that combines the best properties of previously used schemes while introducing a number of novel features. Utilization of this scheme on high-Z element will allow 3-10 times decrease in laser power requirements and facilitate the achievement of shorter wavelength. The high laser absorption efficiency, simplicity and inherent precision in pointing at needed densities, close to traveling wave conditions are among the most important of them. Experimentally GRIP allowed to reach high repetition regime with our JanuSP laser at just 150mJ per pulse. In general, this approach substantially decreases the cost of x-ray lasers and lowers the threshold of affordability of x-ray lasers for numerous applications.

The support of this research by A.Osterheld is greatly appreciated. This work was performed under the auspices of the U.S. Department of Energy by the University of California Lawrence Livermore National Laboratory under Contract No. W-7405-Eng-48 and in part by the National Science Foundation through the University of California, Davis Center for Biophotonics, Science and Technology under award number Agreement No. PHY 0120999.

References

[1] R.Keenan, J.Dunn, R.Smith et all, These Proc

[2] Y.Wang, B.M.Luther, F.Pedacci et.al, These Proc

[3] Yu.V.Afanasiev and V.N. Shlyaptsev, Sov.J.Quant.Electron, Sov. J. Quant. Electron. **19**, 1606 (1989).

[4] P.V. Nickles, V.N. Shlyaptsev, et al,Phys. Rev. Lett., **78**, 2748-2751 (1997).

[5] J. Dunn, Y. Li, et al, Phys. Rev. Lett., **84**, 4834-4837 (2000).

[6] B.E. Lemoff, G.Y.Yin, et al, Phys. Rev. Lett. **74**, 1574-1577 (1995).

[7] D. Korobkin, C.H. Nam, et al. Phys. Rev. Lett. **77**, 5206(1996)

[8] S. Sebban, et al, Phys. Rev. Lett. **86**, 3004-3007, 2001; S. Sebban, et al, Phys. Rev. Lett.
89, 253901 1-4 (2002).

[9] T. Ozaki, et al, Phys. Rev. Lett. **89**, 253902 1-4 (2002).

[10] K.A.Janulewicz, J.J.Rocca, et al, Phys.Rev.A, **63**, No.3, p.033803 (2001).

[11] V.N.Shlyaptsev et al Proc.of SPIE, 2003; R.Keenan et al, ibid

[12] J. Nilsen, B. MacGowan, et al, Phys. Rev. A, **48**, 4682-4685, 1993

[13] V.N.Shlyaptsev, A.V.Gerusov, Proc. "3 Int. Colloquium on X-ray lasers", Schliersee, Germany, in X-RAY LASERS 1992, IOP Publishing, Bristol, Vol.125, 195 (1992)

[14] H.Fiedorowicz, A.Bartnik, et al, Optics Letters, 26(18), 1403-5 (2001)
[15] Y. Ehrlich, C. Cohen, et al, Phys. Rev. Lett. **77**, 4186, (1996);
J.J. Rocca, F.G. Tomasel, et al, SPIE **3156**, 164, (1997);
M. C. Marconi, C. H. Moreno, et al, Phys.Rev.E, **62**(5), 7209 (2000).
[16] V.P.Avtonomov, Yu.V.Afanasiev, et al, J. Sov. Laser Research **10**, 1 (1989).
[17] V.L. Ginzburg, *Propagation of Electromagnetic Waves in Plasma*, Addison–Wesley, Reading, 1964
[18] R.London, Physics of Fluids, **31,** 184, (1987);
V.N.Shlyaptsev, Ph.D Thesis, P.N.Lebedev Physical Institute, 1987
[19] Y. Li, J. Dunn, et.al. , *J. Opt. Soc. Am.* **17**, 1098 (2000).
[20] A.B.Langdon, Phys. Rev. Lett. **44**, 575 (1980),
V.P.Silin, Sov.Phys.JETP 20,1510 (1964)
[21] B.MacGowan, "X-ray Laser Program", Proc of "Workshop on X-ray laser applications", San Francisco (1992)
[22] J.Dunn LLNL Preprint UCRL – JC-149104 (2003)

Inst. Phys. Conf. Ser. No 186
Paper presented at 9th Int. Conf. X-Ray Lasers, Beijing, China, 24–28 May 2004

Absorption of longitudinally pumping laser pulses in slab-plasmas

Y. J. Li[1,2], X.Lu[1] and J. Zhang[1]
[1]Laboratory of Optical Physics, Institute of Physics, Chinese Academy of Science, Beijing 100080, P.R.China
[2]Physics department, China University of Mining and Technology, Beijing 100083, P.R.China

Abstract. The absorption of longitudinally pumping laser pulses in slab plasmas is investigated. The calculations show that the intensity decreases exponentially with the transmission length of the laser pulses in the plasma. The transmission length sensitively varies with the electron density. The density gradient plays an important role to determine the maximum transmission length.

1. Introduction

The major motivation of development of the longitudinally pumping scheme is to reduce the energy requirement; and second is to overcome the problem of the smallgain length product from a transient time scale in the "traditional TCE" scheme. The gain length of TCE scheme is limited by the life of the TCE population inversion, in which the maximum length is only 5mm [1]. In order to overcome the problem, Li et al.[2] proposed the longitudinally pumping scheme in 1998. In 2002 the scheme is experimentally demonstrated in Ni-like Mo by T.Ozaki et al.[3]. Firstly, a 300 ps prepusle is line focused on a 2mm target to produce a pre-plasma. Then, after several ns delay time, an intense 475 fs main pulse is point focused along the longitudinal direction to pump the pre-plasma. They obtain a highly directional x-ray laser output. However, only 2mm length is still too short to obtain a saturated x-ray laser output. The shorter length is due to the influence of the plasma absorption and the refraction of the plasma density gradient. So it is necessary to understand the longitudinally pumping laser propagation characteristics in slab plasmas.

In this paper, the absorption and refraction of longitudinally pumping x-ray laser in slab plasmas is analytically investigated using a ray-tracing equation and the similarity equation.

2. The model

In the longitudinally pump scheme, a transverse (perpendicular to the surface of the target) prepulse is used to perform and generate a narrow and long plasma. After a delay time, a longitudinal main laser pulse with a pedestal is used to ionize the plasma to the Ne-like (or Ni-like) and rapidly heat the plasma to result in a transient x-ray laser output. The longitudinally pump scheme is able to save energy while keeping a traveling pumping. However, the intensity of the longitudinally pumping laser rapidly decreases in the pre-plasma with the influence of the plasma absorption and the refraction of the plasma density gradient.

In this paper, we assume that the laser refraction is due to the density gradient of the pre-plasma, and the laser absorption is mainly through the inverse bremsstrahlung.

2.1 The laser refraction

In order to see the characteristics of the laser rays in the gain region more clearly, as shown in Fig.1, the electron density profile is taken to be

$$n_e(x,t) = n_0(t)(1 - x / L_x) \tag{1}$$

Using the ray tracing equation,

$$\frac{\mathrm{d}}{\mathrm{ds}}(\eta \frac{\mathrm{d}\bar{\mathrm{r}}}{\mathrm{ds}}) = \nabla \eta \tag{2}$$

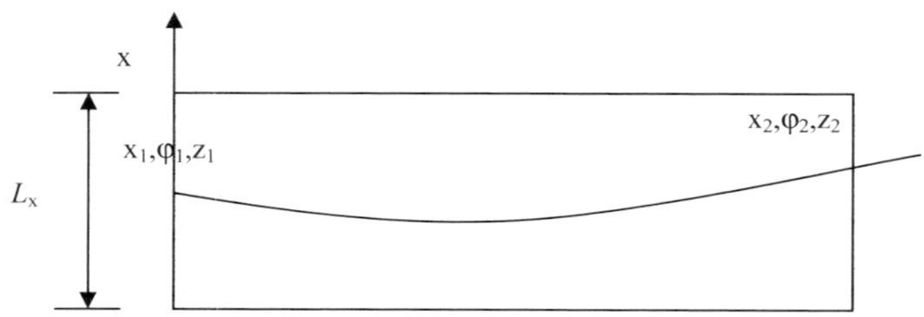

Fig1. Schematic of the rays in a linear density region

we can obtain the ray tracing function under an assumption of linear density condition,

$$x = \frac{\varphi_r}{2L_r} z^2 + \varphi_1 z - \frac{\varphi_r}{2L_r} z_2^2 - \varphi_1 z_2 + x_2 \tag{3}$$

where $\varphi_r = (n_0 / 2n_c)^{1/2}$, $L_r = L_x (2n_c / n_0)^{1/2} = L_x / \varphi_r$, φ_1 is the incidence angle of laser beam, x_2 and z_2 are the position of the ray in the end of the gain region.

2.2 The absorption

Because the pumping laser can not propagate in the plasma region of $n_e > n_c$, the longitudinally pumping laser energy is deposited in the pre-plasma region of $n_e < n_c$ mainly through inverse bremsstrahlung. The laser deposition is assumed to be the result of inverse bremsstrahlung in our derivation. The laser intensity of the longitudinal length z is determined from the absorbed fraction of laser flux:

$$I = I_0 e^{-\tau_{ib}} \tag{4}$$

where τ_{ib} is the inverse bremsstrahlung optical depth through the plasma:

$$\tau_{ib} = \int \kappa_{ib} ds \tag{5}$$

here s is the coordinate in the laser incidence direction. The opacity k_{ib} is determined by the following equation [4]:

$$\kappa_{ib} = \left(\frac{2}{\pi}\right)^{1/2} \frac{4}{3} \left(\frac{Z n_e^2 e^6 (\log \Lambda) \lambda^2}{c^3 (m_e T)^{3/2} (1 - n_e^2 / n_c^2)^{1/2}} \right) \tag{6}$$

where λ is the laser wavelength, n_e is the electron density, n_c is the critical density$(=1.1\times10^{21}cm^{-3}\lambda^{-2})$, T is the electron temperature in energy unit, and Z is the average ion charge.

Considering the assumption of the linear electron density in x coordinate and the relation formula (3) between the x and z, the laser intensity with the longitudinal length z and the incidence angle φ_1 in the gain area is obtained:

$$I = I_0 \exp[-1.3368\frac{Z\Lambda\lambda^2 n_0^2}{T^{3/2}}\left(\frac{2\varphi_r^2 z_2^3}{3L_x^3}+\frac{\varphi_1 z_2^2}{L_x^2}\right)] \tag{7}$$

where I_0 is the initial intensity of the longitudinally pumping laser pulse, n_0 is the maximum electron density in x coordinate in the gain region.

2.3 The determination of T, L_x and n_0

The values of electron temperature T, scaling length L_x and density n_0 in the gain area of the pre-plasma are gotten from the similarity solutions [5]. They are respectively

$$T = T_{1L}t_{1L}^{2/3}t^{-2/3} \qquad L = L_{1L}t_{1L}^{-5/9}t^{5/9} \qquad n_o = n_{1L}t_{1L}^{7/9}t^{-7/9} \tag{8}$$

where T_{1L}, L_{1L} and n_{1L} are the electron temperature, scale length and electron density at the laser duration t_{1L}.

3. Results and discussion

In order to understand the absorption effects of the pre-plasma from the slab to the longitudinally pumping laser for different conditions, we calculate the absorption rate of the Ni-like Mo x-ray laser plasma for different density and scaling length, respectively.

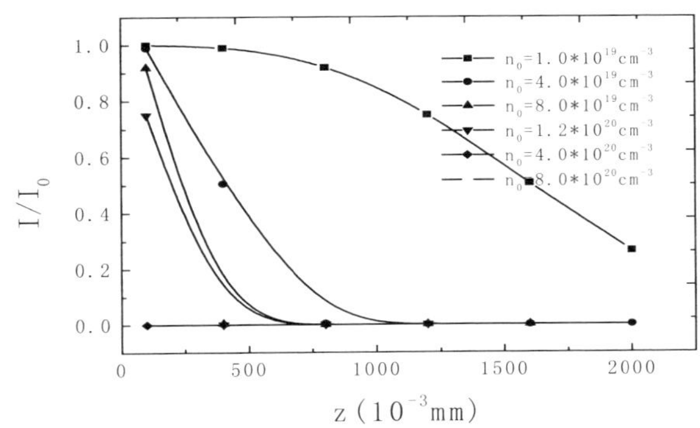

Fig2. Intensity *vs* propagation length for different density

We first calculate the change of the laser intensity with the propagation length for different electron density, as shown in Fig.2. The intensity changes from 0.9998 to 0.264 with z from 100μm to 2mm while the density is $1.0\times10^{19}cm^{-3}$ and the scaling length is 30μm. The intensity rapidly decreases to nil while z get to be 100μm and the density is $8.0\times10^{20}cm^{-3}$. So, it is clear that the propagation length sensitively changes with the electron density.

We then calculate the effects of the laser intensity with the propagation length for different scaling lengths. From Fig.3, the intensity decreases with the propagation length z for different scaling length is more slowly than one in the density. The intensity changed from 0.994 to nil with z from 100μm to 1.3mm while the scaling length is 200μm and the density is $1.5\times10^{20}cm^{-3}$. However, The intensity rapidly decreases from 0.86 to nil with z from 100μm to 200μm while the scaling length is 10μm. This means the longer scaling length is beneficial from the decrease of the absorption.

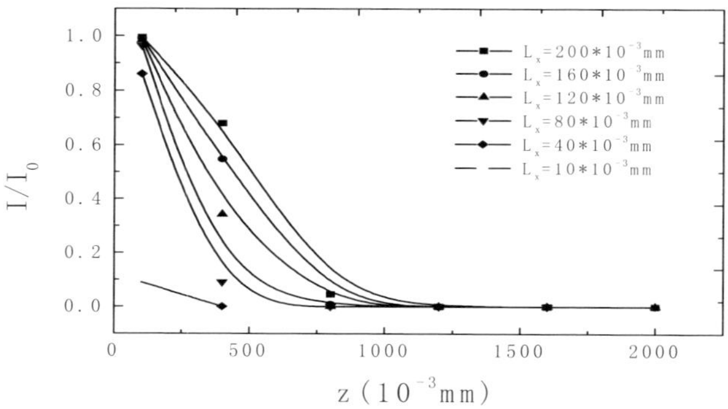

Fig3. Intensity *vs* propagation length in the different scaling length

4. Conclusions

In conclusion, we have investigated the absorption of longitudinally pumping laser pulses in slab plasmas. The results show that the propagation length sensitively changes with the electron density and the low density will reduce the absorption in the pre-plasma. The scale length (density gradient) plays an important role to generate the maximum transmission length.

5. Acknowledgement

This work was supported by the NSFC under Grant Nos. 10374114, 60321003, 10474134, 10374116 and 60128505, and the NKBRSF under Grant No. G1999075206, and the National Hi-tech ICF Program.

References

[1] P.V.Nickles et al., Phys.Rev.Lett. **78**, 2748(1997).
[2] R.Li et al., Phys.Rev.E, **57**, 7093(1998).
[3] T.Ozaki et al., Phys.Rev.Lett, **89**, 253902(2002).
[4] T.W.Johnston and W.M.Dawson, Phys.Fluids, Vol.16, 722(1973).
[5] Y.J.Li et al., Phys.Rev.E. 63, 410(2001).

Inst. Phys. Conf. Ser. No 186
Paper presented at 9th Int. Conf. X-Ray Lasers, Beijing, China, 24–28 May 2004

A 10 Hz Grazing Incidence pumped Ni-like Mo laser at 18.9 nm with 150 mJ pump energy

R. Keenan, J. Dunn, P.K. Patel, D.F. Price, R.F. Smith
Lawrence Livermore National Laboratory, Livermore, CA 94551, USA

V.N. Shlyaptsev
University of California, Davis, Livermore, CA 94551, USA

Abstract. The first demonstration of the grazing incidence pumping (GRIP) scheme for laser-driven x-ray lasers (XRLs) is described utilizing 2-pulse pumping. A long pulse is incident normal to the target to produce a plasma with a particular density profile. Then a short pulse is incident at a grazing angle, chosen to optimally couple the short pulse laser energy into the specific density region where the inversion process will occur. The short pulse is simultaneously absorbed and refracted at a maximum electron density specified by the chosen pump angle then turns back into the gain region. The increased path length gives improved absorption allowing a reduction in the drive energy required for lasing. A Ni-like Mo XRL at 18.9 nm has been demonstrated with only 150 mJ total pump energy and a repetition rate of 10 Hz. We report high gains of 60 cm^{-1} and the achievement of gain saturation for targets of 4 mm length.

1. Introduction

High power, tabletop lasers [1] can be used as a pump source for short wavelength, high repetition rate lasers. A longitudinally pumped, high repetition rate 41.8 nm laser was first reported for a low density Xe gas cell heated by a 40 fs laser using optical field ionisation (OFI) followed by collisional excitation [2]. Transient collisional excitation x-ray lasers were demonstrated where Ne- and Ni-like ion schemes were pumped with ~1 ps laser pulses with less than 10 J of energy and saturated output achieved [3, 4]. Transverse pumping geometry was utilised here where the laser beams were incident at normal incidence and the pump energy was absorbed over a wide range of plasma densities. Further progress in the OFI scheme was reported and the gain saturation regime was achieved for the Pd-like Xe laser at 41.8 nm [5]. This was extended to the shorter wavelength Ni-like Kr at 32.8 nm pumped with energy < 1 J and at a repetition rate of 10 Hz [5]. A longitudinal pumped Ni-like Mo x-ray laser at 18.9 nm was also demonstrated [6], where a short pulse pumps the inversion along a pre-formed plasma column. This laser operated with a pump energy of 150 mJ and produced a highly directional output. High repetition rate and millijoule output 46.9 nm x-ray lasers have also been achieved in the tabletop capillary discharge scheme at 4 Hz and higher [7].

The continuing trend has been to look at ways of improving the efficiency for laser-pumped x-ray lasers operating at shorter wavelengths. We report the first demonstration of a novel pumping scheme using the excitation pulse incident on the plasma column at a grazing incidence angle to create an 18.9 nm Ni-like Mo $4d - 4p$ x-ray laser at 10 Hz with a very low energy ~150 mJ laser pump [8].

2. Grazing Incidence Pumping (GRIP) Scheme Description and Results

The GRIP geometry allows improved laser coupling efficiency into a pre-determined optimum gain region of the plasma by using refraction to turn the pump laser at an electron density below the critical density. This increases the path length and absorption in this specific region of the plasma and allows a reduction of the pump energy threshold for inversion. The Ni-like Mo laser was chosen for the initial study of the GRIP scheme using a 100 fs, 800 nm Ti:Sapphire laser at the Lawrence Livermore National Laboratory. A maximum of 300 mJ/pulse uncompressed energy at 10 Hz was split with 33% into a 200 ps long pulse arm and 66% into the short pulse arm. The long pulse energy 70 mJ was focused onto the Mo target at normal incidence in a line of 5 mm long by 15 μm (FWHM), at an intensity of 5×10^{11} W cm^{-2}, see Fig. 1. The short pulse arm was compressed under vacuum where a pulse length of 125 fs to 8 ps could be produced with a maximum of 80 mJ on target. RADEX simulations determined that the turning electron density was determined to be ~10^{20} cm^{-3} from the equation for the refraction angle, ϕ_r, simplified to $\phi_r \sim (n_{eo}/n_{ec})^{1/2}$ where n_{eo} is the maximum density within the gain region and n_{ec} is the critical density for the optical pump [9, 10]. The grazing incidence

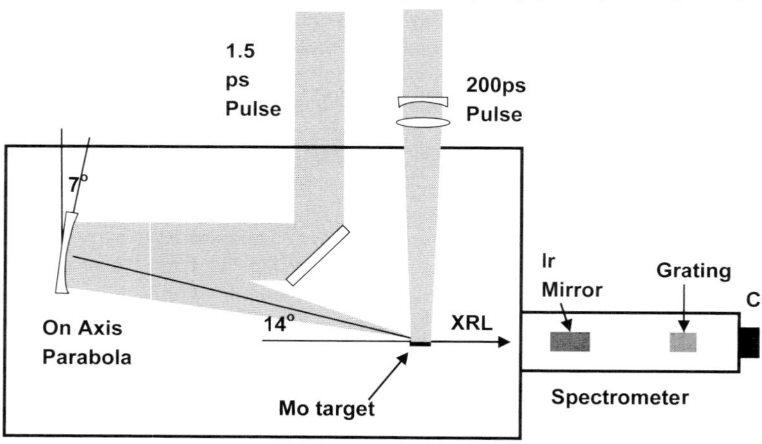

Figure 1 Experimental setup of grazing incidence pumping for Ni-like Mo x-ray laser.

angle for the short pulse beam was determined to be 14° from the critical density of 1.74×10^{21} cm^{-3} for the 800 nm wavelength. An on-axis parabola was used to generate a very uniform line focus 4 mm long by ~25 – 30 μm (FWHM) wide at this angle with an intrinsic travelling wave velocity of 0.97c. A typical short pulse was 1.5 ps at an intensity of 4×10^{13} W cm^{-2}. The long pulse arm was sent through a delay line that could adjust the separation of the two pulses from 0 – 1 ns.

Initially a longer line focus of 8 mm was tried but no lasing was observed in the 18.9 nm line even though Cu-like Mo emission lines were identified in the spectrum [10]. The line focus was reduced to 4 mm which limited the Mo target length to a maximum of 4 mm. Some optimisation of the laser pumping conditions was required to achieve output on the x-ray laser. For the results presented in this paper a 500 ps delay between the laser pulses and a 1.5 ps short pulse duration gave the highest x-ray laser output. More experimental details are reported elsewhere [8, 10]. A spectrum from the on-axis flat-field 1200 l mm^{-1} grating spectrometer with a back-thinned CCD camera is shown in Fig. 2. This is a single shot spectrum from a 4 mm long Mo target pumped by a total of 150 mJ energy. A 1 μm Al foil is used to cut down scattered laser pump light. The x-ray laser line at 18.9 nm is orders of magnitude stronger than any other line. Fiducial wires, observed at 0 and 10 mrad in Fig. 2(b), were placed in front of the spectrometer to determine the angular pointing of the x-ray laser beam. The x-ray laser shows a narrow divergence angle of 3.6 mrad (FWHM) in the horizontal direction with a deflection angle of 3.2 mrad from the target, Fig. 2(b). The most intensely peaked shots had typical average deflection and divergence angles of 3 mrad and 4.5 mrad (FWHM), respectively. RADEX simulations indicate that the measured deflection angles are consistent with optimum electron densities from 4×10^{19} cm^{-3} and up to a maximum of 10^{20} cm^{-3}.

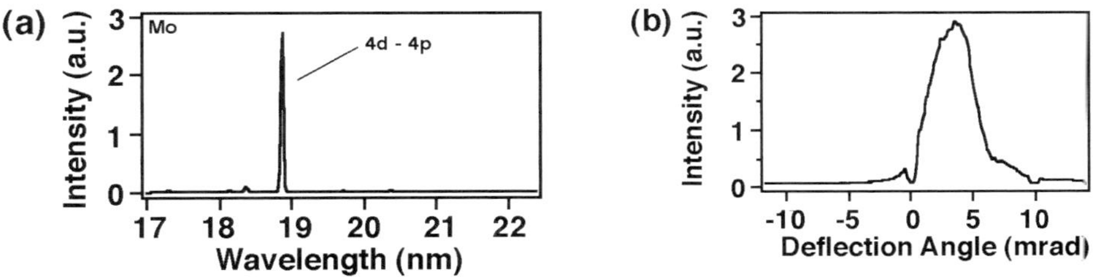

Figure 2(a) Soft x-ray spectrum from on-axis flat-field spectrometer. Strong lasing is observed on Ni-like Mo 4d – 4p transition at 18.9 nm. (b) Horizontal deflection angle showing angular pointing of x-ray laser line. Target axis is 0 mrad and increasing angle is deflected away from the target.

It is noted that if the delay between the two laser pulses was changed by ± 50 ps from the chosen 500 ps then the 18.9 nm intensity dropped substantially. This is understood to be a consequence of the low laser energy and narrow line focus of the 200 ps plasma-forming beam. The GRIP scheme short pulse pumps a specific density region and hence a spatial region of the plasma. This places more constraints on the initial conditions before short pulse pumping than for the transverse scheme where a range of plasma densities are pumped. Subsequent experiments with more energy and a wider line focus for the long pulse have shown a wider "lasing window" as a function of this delay [11].

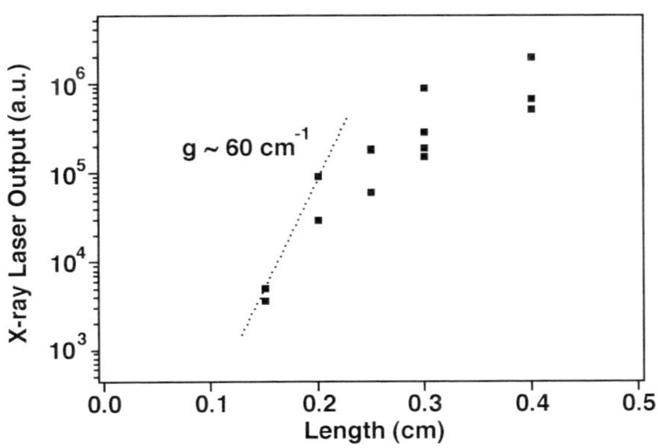

Figure 3 X-ray laser intensity as a function of length for Mo targets up to 4 mm.

Ni-like Mo lasers have shown laser action in recent years to varying degrees with similar small pump energies but with no indication of possible output as a function of length [6, 12, 13]. It was important to determine the gain for the GRIP scheme with travelling wave pumping and scaling of the amplification process with plasma length. The 18.9 nm output of single shots for targets up to 4 mm were recorded using a near-field imaging technique and on the spectrometer, Fig. 3. The XRL intensity increased by approximately three orders of magnitude from the threshold of detection for 1.5 mm targets to the maximum of 4 mm. The highest gain is determined to be 60 cm^{-1} for targets up to 2 mm with no output detected for targets below 1.5 mm. For a 2 mm target the gain length (GL) product of 12 is achieved with the output rolling over indicating a saturation-like behavior. The intensity increases by a further factor of 20 for targets up to 4 mm long which determines an upper limit of GL~15 which would be into saturation.

3. Conclusions

The new grazing incidence pumping scheme has demonstrated x-ray laser action for Ni-like Mo at 18.9 nm with only 150 mJ total pump energy on target operating at 10 Hz. The angle of incidence of the short pulse is chosen to pump the inversion efficiently at a specific density region where the gain occurs. Gain saturation regime has been achieved. We believe this is a major step forward for efficient x-ray lasers that will lead to new applications requiring high average-power x-ray sources at a high repetition rate.

Acknowledgments

The continued support and encouragement of Al Osterheld is greatly appreciated. The authors would like to thank Joe Nilsen for the use of the multilayer optics in part of this work. Work performed under the auspices of the US Department of Energy by the University of California Lawrence Livermore National Laboratory under Contract No. W-7405-Eng-48 and in part by funding from the National Science Foundation through the Center for Biophotonics, an NSF Science and Technology Center, managed by the University of California, Davis, under Cooperative Agreement No. PHY 0120999.

References

[1] D. Strickland and G. Mourou, Opt. Comm. **56**, 219 - 221 (1985)
[2] B.E. Lemoff et al., Phys. Rev. Lett. **74**, 1574 - 1577 (1995).
[3] P.V. Nickles et al., Phys. Rev. Lett. **78**(14), 2748-2751 (1997).
[4] J. Dunn et al., Phys. Rev. Lett. **84**, 4834 - 4837 (2000).
[5] S. Sebban et al., Phys. Rev. Lett. **86**, 3004 - 3007 (2001); S. Sebban et al., Phys. Rev. Lett. **89**, 253901 1-4 (2002).
[6] T. Ozaki et al., Phys. Rev. Lett. **89**, 253902 1 - 4 (2002).
[7] C. D. Macchietto, B. R. Benware, and J. J. Rocca, Opt. Lett. **24**, 1115- 1117 (1999).
[8] R. Keenan et al., "High Repetition Rate Grazing Incidence Pumped X-ray Laser Operating at 18.9 nm", submitted to Phys. Rev. Lett. (2004).
[9] V.N. Shlyaptsev et al., Soft x-ray lasers and Applications V, SPIE Int. Soc. Opt. Eng. Proc, vol. **5197**, ed. E.E. Fill and S. Suckewer, 221 - 228 (2003).
[10] R. Keenan et al., Soft x-ray lasers and Applications V, ibid., 213 - 220 (2003).
[11] J. Dunn et al., "Grazing Incidence Pumping for Efficient X-ray Lasers", in these proceedings (2004).
[12] R. Li, Z.Z. Xu et al., X-ray Lasers 2000, ed. G. Jamelot, C. Möller, and A. Klisnick, J. Phys. IV **11**, Pr27 - 34 (2001).
[13] R.Tommasini et al., Proc. SPIE Int. Soc. Opt. Eng. **4505**, 85 (2001).

Inst. Phys. Conf. Ser. No 186
Paper presented at 9th Int. Conf. X-Ray Lasers, Beijing, China, 24–28 May 2004

Investigations on longitudinal pumped Ni-like Mo X-Ray lasers by fs laser pulses

M Suzuki, R A Ganeev, J Zhang, M Baba, and H Kuroda
The Institute for Solid State Physics, The University of Tokyo
5-1-5 Kashiwanoha, Kashiwa, Chiba 277-8581, Japan

T Ozaki,
Institut national de la recherche scientifique
1650 boul. Lionel-Boulet Varennes (Quebec) Canada J3X 1S2

Abstract. In order to improve the repetition rate of X-Ray laser, the longitudinal-pumped transient-collisional excitation (TCE) is one of the most effective schemes. The latest experimental investigations on the longitudinal-pumped TCE Ni-like Mo X-Ray lasers at 18.9 nm wavelength by a Ti:sapphire laser pulses that delivered energy of 180 mJ in 130 fs at 795 nm wavelength have been prepared.

1. Introduction

Since the transient-collisional excitation (TCE) scheme proposed by Nickles et al., [1] and Dunn et al., [2], and then advances in TCE X-Ray lasers have been demonstrated by using 2-10 J of pumping laser energy. However the spatial coherence of these traditional TCE X-Ray lasers has been still low [2]. Different techniques were used to improve the spatial coherence of X-Ray lasers. It is an important subject to improve the spatial coherence of these traditional TCE X-Ray lasers.

A high coherence laser-drive Ni-like Ag X-Ray laser at 13.9 nm wavelength has been demonstrated by an oscillator-amplifier scheme with double targets [3]. However, these configurations are complicated and became a large system. Practical applications of X-Ray lasers require compact system and high pulse repetition rate. In order to solve these problems, the longitudinal-pumped TCE X-Ray laser is one of the most effective pumping schemes [5]. Ozaki and co-workers [6] first demonstrated this novel scheme by using Mo slab target irradiated by the tabletop hybrid glass laser [7] that delivering energy of 150 mJ in 475 fs at the center wavelength of 1060 nm. The divergence of the 18.9 nm X-Ray laser was less than 1 mrad. Since then, the advanced longitudinal-pumped TCE X-Ray laser schemes has been studied [8] actively.

To improve the pulse repetition rate of longitudinal-pumped TCE Ni-like Mo X-Ray lasers, the Ti:sapphire laser pulses that delivered energy of 180 mJ in 130fs at 795 nm wavelength was prepared. In this paper, we summarize our results on 18.9 nm X-Ray laser generation by using 1060 nm radiation and describe new 1.5 TW Ti:Sapphire laser system.

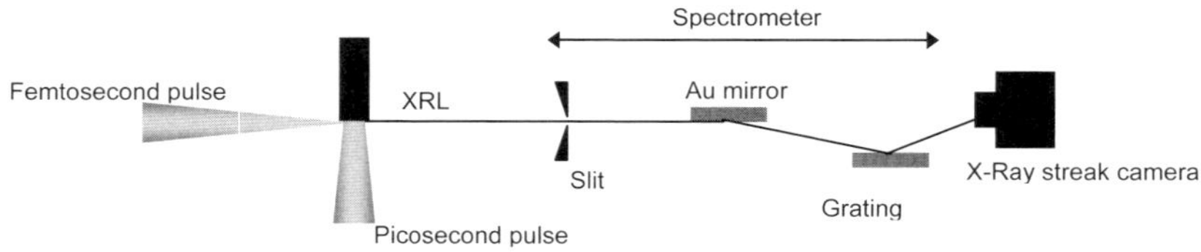

Figure 1. Experimental setup for the measurement of X-Ray laser (XRL). The femtosecond laser pulse irradiates a pre-plasma produced by picosecond laser pulse and then a XRL is generated from the plasma. The XRL is measured by a soft X-Ray spectrometer equipped with X-Ray streak camera.

2. Longitudinally pumping X-Ray laser

At first, we describe the longitudinally pumping TCE X-Ray laser results by using the hybrid glass laser system. The details of laser system were described in Ref 7. Experimental setup is shown in Fig.1. The 300 ps laser pre-pulse is line focused onto the Mo target by using the cylindrical lens. After that the femtosecond main pulse is point focused on the pre-plasma generated by 300 ps laser pulse from longitudinal direction. The line focus is with a length of 2 mm and a width of 100 μm. The delay between main pulse and pre-pulse duration were 4 ns. The pump energy of the pre-pulse and main pulse is 30 and 120 mJ, respectively. The generated X-Ray laser is measure by a Hitachi 1200 grooves/mm flat-field soft X-Ray spectrometer [9] equipped with X-Ray streak camera (Hamamatsu C1936).

Figure 2 shows the 18.9 nm Ni-like Mo ($3d^94d^1S_0$-$3d^94p^1P_1$) soft X-Ray laser spectrum. For the calibration of the spectrometer, He- and H-like spectra were used. The total energy of the pre-pulse and the main pulse are 150 mJ, which will make possible the high pulse repetition rate operation by using Ti:Sapphire laser system.

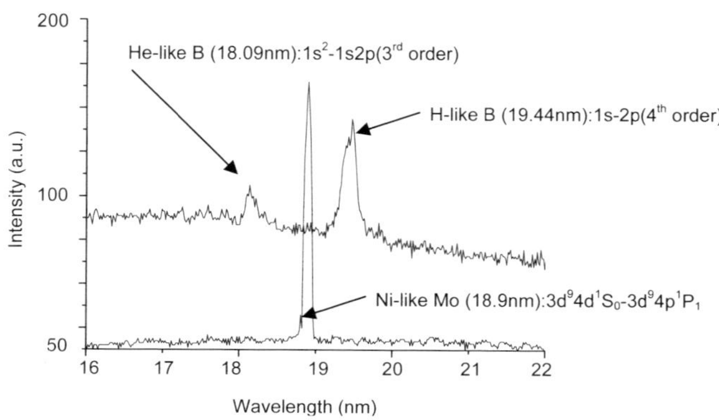

Figure 2. 18.9 nm X-Ray laser from Mo plasma was obtained by using hybrid glass laser (150 mJ, 475 fs). For the calibration of the spectrometer, He- and H-like B line spectra was used.

3. 1.5 TW laser system performance

Figure 3 shows the layout of 1.5 TW laser system. A Ti:sapphire laser consists of a femtosecond oscillator (Tsunami, 82 MHz, 100 fs), a stretcher, a regenerative amplifier, a double pass amplifier, a 6-pass main amplifier, and a pulse compressor. The stretcher and regenerative amplifier are produced by Spectra Physics. The laser pulses at 10 Hz repetition rate from the regenerative amplifier are amplified to 380 mJ with a 6-pass Ti: Sapphire amplifier pumped by two Nd:YAG lasers (Spectra Physics PRO230) that providing a laser energy of 650 mJ in 8 ns pulse at a wavelength of 532 nm. The pulse compressor is composed of two gratings of 1800 lines/mm. The output energy is ~180 mJ in 130 fs at 10 Hz repetition rate and λ=795nm. This system will be used in the longitudinally pumping TCE X-Ray laser experiments.

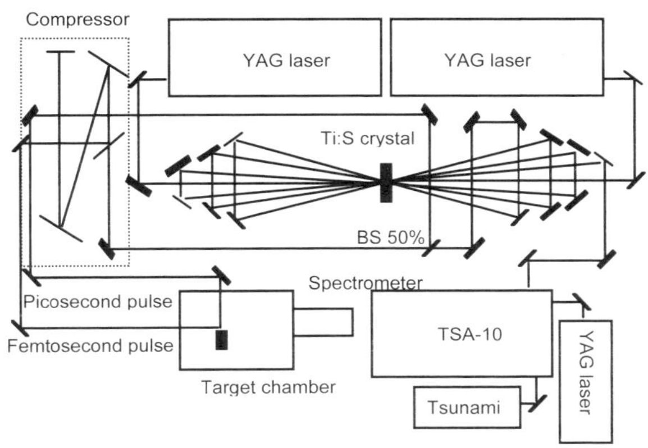

Figure 3. Ni-like Mo X-Ray laser experiment layout by using the multi-tera-watt laser system. The Ti:Sapphire laser that delivering laser energy of 200 mJ in 130 fs at the center wavelength of 795 nm.

4. Conclusion

Recently, the 18.9 nm Ni-like Mo soft X-Ray laser was observed by using a tabletop hybrid glass-Ti:Sapphire laser. In order to improve the high repetition rate soft X-Ray lasers, the experimental investigations on the longitudinal-pumped TCE Ni-like Mo X-Ray lasers at 18.9 nm wavelength by a Ti:sapphire laser system that delivered energy of 180 mJ in 130 fs at 795 nm wavelength have been preparing.

Acknowledgements

The authors would like to thank Japan Society for the Promotion of Science (JSPS) for supporting us in carrying out our project studies in Grant-in- Aid for Creative Scientific Research (14GS0206).

References

[1] Nickles P V, Shlyaptsev V N,et.al. 1997 Phys. Rev. Lett. 78, 2748-2751
[2] Dunn J, Li Y, et.al. 2000 Phys. Rev. Lett. 84, 4834-4837
[3] Tang H, Daido H, et.al. 2003 Jpn. J. Appl. Phys., Part 1 42 443-448
[4] Nishikino M, Tanaka M, et.al. 2003 Phys. Rev A 68, 061802R 01-04
[5] Li R, Ozaki T, et.al. 1998 Phys. Rev. E 57, 7093-7102
[6] Ozaki T, Ganeev R A, et.al., 2002 Phys. Rev. Lett. 89 253902 01-04
[7] Itatani J, Kanai T, et.al. 1996 Prog. Cryst. Charact. Growth Mater. 33 281-292
[8] Keenan R, Dunn J, et.al., 2003 SPIE Vol. 5197 213-219
[9] Kita T, Harada T, et.al. , 1983 Appl. Opt. 22 512-513

Theory and simulations of X-ray lasers and plasma media

Inst. Phys. Conf. Ser. No 186
Paper presented at 9th Int. Conf. X-Ray Lasers, Beijing, China, 24–28 May 2004

Copper neon-like x-ray laser modelling including hyperfine structure levels

J. Dubau*, **O. Zabaydullin**† **C. Möller*** **and D. Benredjem***

*L.I.X.A.M., Université Paris-Sud, 91405 Orsay, France

†N.F.I., Kurchatov Institute, 123182 Moscow, Russia

Abstract. Atomic data have been calculated for Ne-like copper including the first 89 fine-structure levels, i.e. $(1s^2)\ 2s^22p^6$, $2s^22p^5nl$, $2s2p^6nl$ configurations with $n = 3, 4$ (energies, oscillator strengths, and collision strengths). Plasma hydro/atomic simulations have been carried out for some experimental conditions giving important gain ($\simeq 40$) for the $J = 0 \rightarrow 1$ x-ray laser line. Hyperfine structure level energies have been calculated for the $J = 1$ lower level of the laser line. Amplification of the laser line has been studied both with fine and hyperfine levels. The gain is decreased by almost 15 %.

1. Introduction

Hyperfine structure splitting was experimentally observed [1] for the $J = 0 \rightarrow 1$ x-ray laser line of Ne-like niobium (nuclear charge Z =41), ten years ago. In Ref. [1], is written "*Hyperfine splitting can affect the gain of the laser line by effectively increasing the linewidth. Since the gain is inversely proportional to the linewidth the gain will decrease.*" It is also mentionned that '*The hyperfine-splitting reduces the gain of odd Z ions and helps to explain the anomalous behavior observed in which low Z Ne-like ions with odd Z do not lase.*" As niobium, copper is an odd Z elements, (Z =29).

Since Fermi's paper [2] and many other famous papers published between 1930 and 1935 [3, 4], the hyperfine structure (*hfs*) is well known. Although hyperfine-splitting is often too small to be measured, *hfs* can have measurable effects on forbidden-transition rates [5, 6] and on line polarization [7], for examples.

Copper has 2 abundant isotopes with nuclear spin $I = 3/2$ and nuclear magnetic moment $\mathcal{M}_N \simeq 2.2721$ (i.e. 69.17 %: 2.2233; 30.83%: 2.3817). Its nuclear spin and magnetic moment being relatively small compared to Nb ($I = 9/2$, $\mathcal{M}_N = 6.17$) one does not expect a very strong reduction of the $J = 0 \rightarrow 1$ x-ray laser gain. Nevertheless since copper is an attractive element for future x-ray laser applications, refined studied are needed for it.

2. Copper atomic data

Ne-like copper data of Blancard et al.[8] have been extended from 65 to 89 fine-structure (*fs*) levels, belonging to the ($1s^2$) $2s^22p^6$, $2s^22p^5nl$, $2s2p^6nl$ configurations with $n = 3$, 4. Energy levels and radiative decay rates have been obtained by the SUPERSTRUCTURE code [9]. Electron scattering calculations between target LS-terms have been calculated by DISWAV [10] for partial wave $l < 20$, which have been transformed to *fs* J-levels collision strengths by JAJOM [11], using Term-Coupling-Coefficients supplied by SUPERSTRUCTURE. As mentioned in Ref.[8], different term energy order can happen between SUPERSTRUCTURE and DISTWAV, for closed-shell target, such as Ne-like and Ni-like ions. To fix the problem, a program has been used to correct the wrong order for matrix elements, before data transfer from DISTWAV to JAJOM. For $l \geq 20$, extrapolations have been done using the geometric series properties of partial wave contributions. Such a procedure has been controled carefully using different methods, (Ref.[8]).

The *hfs* F angular momentum is obtained by coupling electron total angular momentum $\mathbf{J}$ with nuclear spin $\mathbf{I}$: $\mathbf{F} = \mathbf{J} + \mathbf{I}$. To a *fs* level, $J = 0$, corresponds one *hfs* level, $F = I$. To a *fs* level, $J' = 1$, corresponds three *hfs* levels $F' = I - 1$, $F' = I$, $F' = I + 1$. The $J = 0 \rightarrow J' = 1$ line is then plitted into 3 lines:

$$(J=0)\ \ F = I \left\{ \begin{array}{cc} \rightarrow & F' = I - 1 \\ \rightarrow & F' = I \\ \rightarrow & F' = I + 1 \end{array} \right\} \ \ (J' = 1)$$

It is possible to transform collision strengths from *fs* to *hfs* using a similar method to JAJOM (see D'Cruz et al. [12], formulae (1) and (2)). The Term-Coupling-Coefficients being replaced by Level-Coupling-Coefficients which can be obtained by diagonalizing the total target hamiltonian, including both the *hfs* operator, H_{hfs}, and the Breit-Pauli hamiltonian H_{BP}. In the present case, electron density is high enough ($10^{10} - 10^{11}$ cm^{-3}) to enable us to neglect forbidden transition rates which are sensitive to *hfs* . Furthermore, one can neglect *hfs* effect on wavefunctions : H_{hfs} operator has therefore only an effect on energy levels. It is then possible to prove that *hfs* level population, $N(\alpha JIF)$, can be derived directly from *fs* level population $N(\alpha J)$ [12, 13], as long as x-ray laser has no effect on population, i.e. before the saturation region:

$$N(\alpha JIF) = \frac{(2F+1)}{(2J+1)(2I+1)} N(\alpha J) \tag{1}$$

Plasma hydro/atomic simulations were performed by the EHYBRID code[14], using *fs* levels. From *fs* level populations were derived *hfs* level populations, Eq. (1).

3. Hyperfine structure splitting of energy levels

The H_{hfs} operator contains *magnetic dipole* and *electric quadrupole* terms [15]. Only magnetic dipole moment data being available for Cu, it was not possible to calculate the electric quadrupole contribution, which is usually smaller than the magnetic dipole contribution. SUPERSTRUCTURE wavefunctions being non-relativistic, the non-relativistic expression of H_{hfs} has been used [16, 17]

$$H_{hfs} = 2\,\alpha^2\,\mu_B\,\mu_N\,g_I\,\mathbf{I}\,.\left\{\sum_{i=1}^{10}\left[\frac{8\pi}{3}\delta(\mathbf{r}_i)\mathbf{s}_i + \frac{1}{r_i^3}\mathbf{l}_i - \frac{\sqrt{10}}{r_i^3}(s^{(1)}.C^{(2)})_i^{(1)}\right]\right\} \tag{2}$$

Table 1. Hyper-fine level energy separations in a.u. (1 u.a. = 27.2 eV), $\epsilon(-1) = E(F = I - 1) - E(F = I)$ and $\epsilon(+1) = E(F = I + 1) - E(F = I)$, of the $1s^22s^22p^53s\ ^3P_1$ level, for Cu (Z=29) and Nb (Z=41).

Z	I	$\mathcal{M}_N$	$\epsilon(-1)$	$\epsilon(+1)$
29	3/2	2.30	-5.3432(-5)	8.9052(-5)
41	9/2	6.17	-4.7429(-4)	5.7965(-4)

In the Appendix, are given the H_{hfs} matrix element expressions.

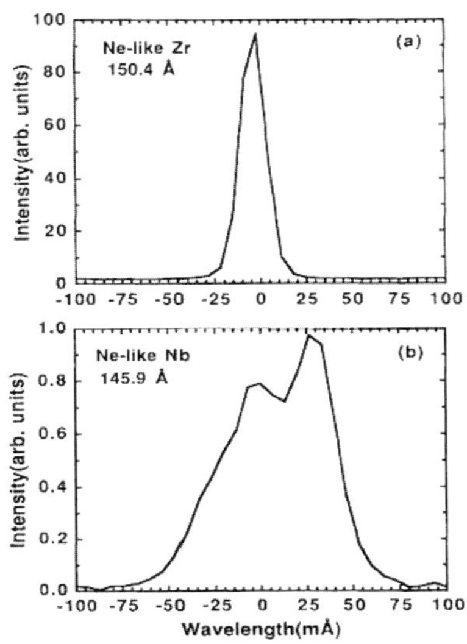

Figure 1. Measured intensity vs wavelength around the line center for the $J = 0 \rightarrow 1$ of Ne-like Zr (a), $Z = 40$, and Ne-like Nb (b), $Z = 41$ Ref.[1]

Before studying copper, it is worth analyzing the Ne-like niobium results obtained in Ref.[1]. In Fig 1. is presented the "*measured intensity vs wavelength around the line center for the* $J = 0 \to 1$ *laser line of Ne-like Nb (b) at 145.9 Å and in Ne-like Zr (a) at 150.4 Å. A 28 mÅ splitting of* $F = 11/2$ *and* $F = 9/2$ *components is observed in Nb*". Experimental conditions for atomic elements were similar. On the contrary, niobium measured intensity profile is considerably broader, with 2 peaks instead of one. On the opposite, the intensity scale is considerably smaller.

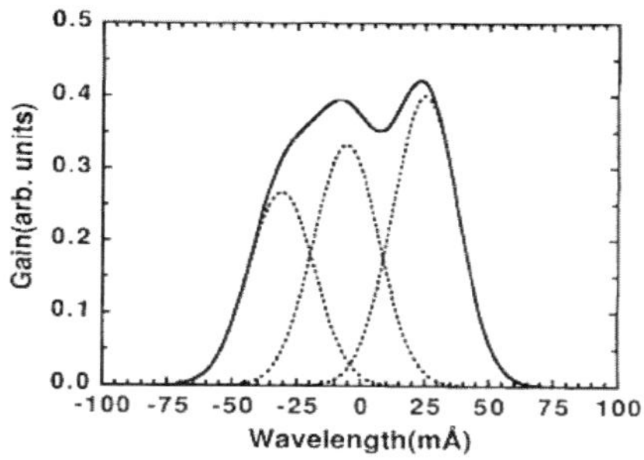

Figure 2. Gain vs. wavelength around the line center for the $J = 0 \to 1$ of Ne-like niobium, Ref.[1]

In Fig. 2, is presented the theoretical gain obtained in Ref.[1], for Nb. We see clearly 3 line profiles corresponding to $F' = 7/2, 9/2, 11/2$. The line profiles are identical except magnitures which are proportional to the statistical weight of F' level belonging to the $J' = 1$ level, i.e. $\omega(-1) = 8/30$, $\omega(0) = 10/30$, $\omega(+1) = 12/30$

4. X-ray laser simulations without hyperfine structure

Plasma hydro/atomic simulations have been carried out for some experimental conditions (2 and 3 pumping-laser pulses) giving important gain ($\simeq$ 40) using EHYBRID [14]. Effective charge $< Z >$ =19.6 and electron density $N_e = 9.10^{20}$ cm^{-3} were obtained. The amplification up to saturation of the x-ray laser has been calculated by the Bloch-Maxwell code of Holden and Sureau [18]. This code solves simultaneously M_J sublevel population equations and radiative transfer equations of different lines including the x-ray laser line. In Fig 3, are given the fractional sublevel populations of the lasing line along the amplifying plasma for J=0 and J=1. The quantization axis z used to define the sublevels is the x-ray laser direction. It can be noticed that the population are nearly constant up to $z = 0.45$ cm, after which saturation effect takes place.

The radiative transfer equation of a lasing line between levels J and J' has the following form [19]

$$\frac{d\,I(z,\nu)}{d\,z} = f(z)\,\phi(\nu - \nu(0)) + G(z)\,\phi(\nu - \nu(0))\,I(z,\nu) \tag{3}$$

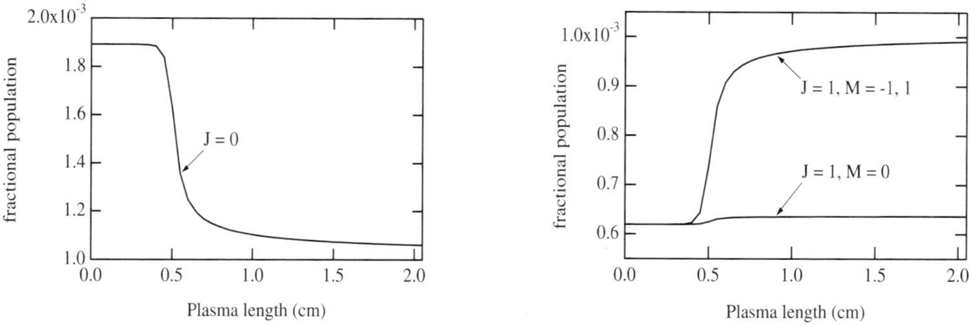

Figure 3. *Fractional sublevels population of $M_J = 0$ of the upper level $J = 0$, and frational sublevels population of $M_J = 0, \pm 1$ of the lower level $J = 1$, relative to copper population.*

where $\phi(\nu - \nu(0))$ is a Doppler-Lorentz profile obtained from a plasma temperature, T=206 eV, derived from EHYBRID, and the sum of initial and final level life-times of the lasing transition. The emissivity, $f(z)$, is proportional to $M_J = 0$ sublevel population of the upper level $J = 0$. The gain, $G(z)$, is proportional to the difference of sublevel populations, M_J and M'_J, of the levels $J = 0$ and $J' = 1$. $\nu(0)$ is the frequency corresponding to the line center. The x-ray path length is $z = ct$.

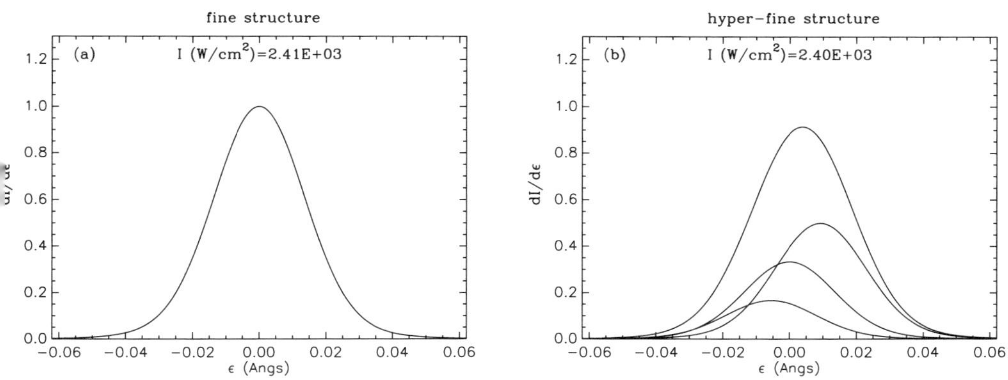

Figure 4. *Emissivity profile of the lasing line integrated over 1cm : (a) fine-structure; (b) hyperfine structure.*

The emissivity profiles integrated over 1 cm, are shown in Fig 4 : (a) without *hfs* , (b) with *hfs* . The integrated value over the profile is the same but the width is different. Hyperfine levels give three lines centered at $\nu(-1), \nu(0)$ and $\nu(+1)$. From Eq.(1), we get that emissivity and gain are statistically distributed over the 3 lines.

$$\frac{d\,I(z,\nu)}{d\,z} = f(z)\,\Big(\omega(-1)\,\phi(\nu-\nu(-1)) + \omega(0)\,\phi(\nu-\nu(0)) + \omega(+1)\,\phi(\nu-\nu(+1))\Big)$$

$$+G(z)\,\Big(\omega(-1)\,\phi(\nu-\nu(-1)) + \omega(0)\,\phi(\nu-\nu(0)) + \omega(+1)\,\phi(\nu-\nu(+1))\Big)\,I(z,\nu)) \qquad (4)$$

For $z < 0.45$ cm, $f(z) = f(0)$ and $G(z) = G(0)$. Eqs. (3) and (4) can therefore be solved analytically.

$$I(z,\nu) = \frac{f(0)}{G(0)}\left\{\exp\Big[G(0)\,\phi(\nu-\nu(0))\,z\Big] - 1\right\} \qquad (3')$$

$$I(z,\nu) = \frac{f(0)}{G(0)}\left\{\exp\Big[G(0)\,\Big(\omega(-1)\,\phi(\nu-\nu(-1))+\omega(0)\,\phi(\nu-\nu(0))+\omega(+1)\,\phi(\nu-\nu(+1))\Big)\,z\Big]-1\right\} \qquad (4')$$

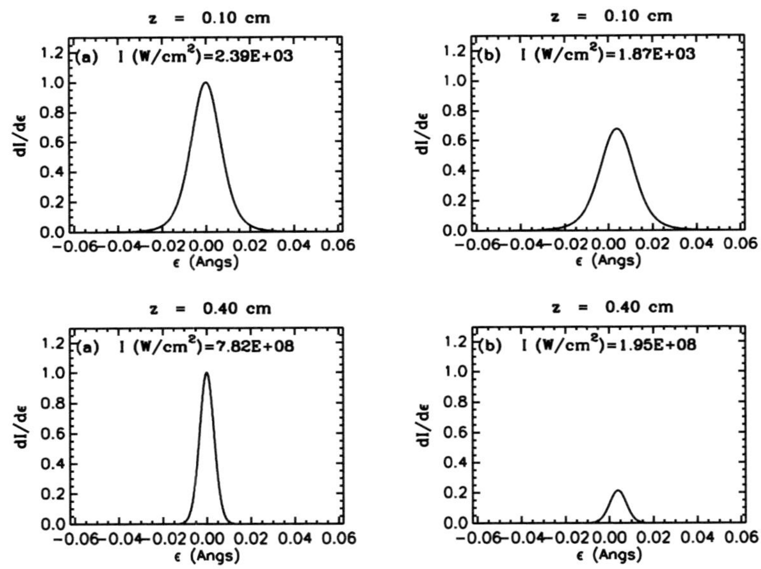

Figure 5. *Intensity profile of the lasing line : (a) fine-structure Eq.(3'); hfs Eq.(4') (for given z, same normalisation in (a) and (b)).*

In Fig (5), are plotted, for $z = 0.1$ cm and $z = 0.4$ cm, the intensity profile without *hfs* (Eq.(3')) and with *hfs* (Eq.(4')). One can notice that the ratio of *hfs* and *fs* intensities integrated over the profile rapidly decrease with z.

The total intensities integrated over the line profile are shown in Fig. 6(a) for the *fs* level, with (Eq.(3)) and without (Eq.(3')) saturation effect. In fig 6(b), are plotted using natural log scale, the total intensities obtained without (continuous curve) and with (dotted line) *hfs* levels. The dashed curve, represents the same quantity in the extreme case where the 3 *hfs* levels are completely separated. From the slope of the

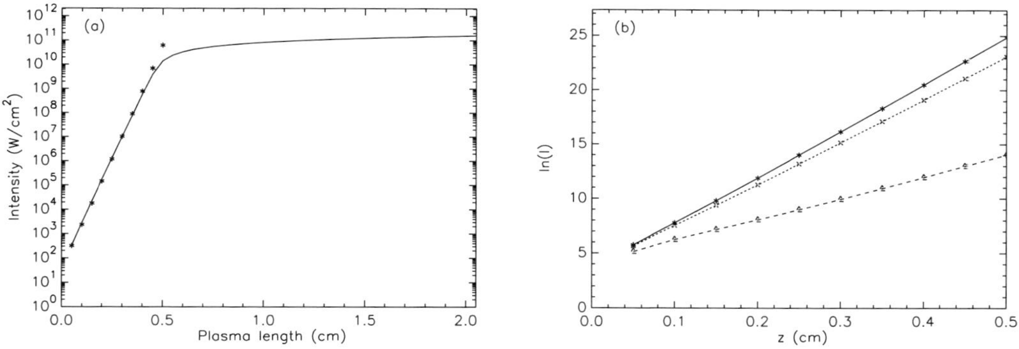

Figure 6. *Intensity of the lasing line: (a) fine-structure, continuous curve Eq.(3) point Eq.(3'); (b) continuous curve Eq.(3'), dotted line Eq.(4'), dashed line hfs-levels completely separated.*

lines of Fig 6(b), one can get the gain in the 3 cases : the gain is decreased by 15 % from *fs* to *hfs*. In the extreme case, it is decreased by 50 %.

We did not study saturation for *hfs* levels but we expect that saturation intensity would be the same as before. Then, the *hfs* effect would only be to increase the plasma length z necessary to reach saturation.

References

[1] Nilsen J, Koch J A, Scofield J H, Mc Gowan B J, Moreno J C and Da Silva L B 1993 *Phys. Rev. Lett.* **70** 3715-3717

[2] Fermi E 1930 *Z. Phys.* **60**, 320

[3] Condon E U and Shortley G H 1935 (1991) *The Theory of Atomic Spectra* (Cambridge University Press)

[4] Bethe H A and Salpeter EE 1933 (1957) *Quantum Mechanics of One and Two electron atoms* (Springer-Verlag Berlin)

[5] Gould H, Marrus R and Mohr P J 1974 *Phys. Rev. Lett.* **33** 676-680

[6] Indelicato P, Birkett B B, Briand J-P, Charles P, Dietrich D D, Marrus R and Simionivici A 1992 *Phys. Rev. Lett.* **68** 1307-1310

[7] Henderson J R, Beiersdorfer P, Bennett C L, Chantrenne S, Knapp D A, Marrs R E, Schneider M B, Wong K L, Doschek G A, Seely J F, Brown C M, La Villa R E, Dubau J and Levine M A 1990 *Phys. Rev. Lett.* **65** 705-708

[8] Blancard C, Cornille M and Dubau J 2001 *Phys. Scr.* **64** 452-454

[9] Eissner W, Jones M and Nussbaumer H 1974 2001 *Comput. Phys. Commun.* **8** 270-306

[10] Eissner W 1998 *Comput. Phys. Commun.* **114** 295-341

[11] Saraph H E 1972 *Comput. Phys. Commun.* **15** 247-258

[12] D'Cruz N L, Sarazin C L and Dubau J 1998 *Astrophys. J* **501** 414-424

[13] Sampson D H and Zhang H L 1997 *J. Phys. B: At. Mol. Opt. Phys.* **30** 1449-1456

[14] Pert G 1983 *J. Fluid Mech.* **131** 401

[15] Bransden B H and Joachain C J 1983 *Physics of Atoms and Molecules* (Longman London)

[16] Trees R E 1953 *Phys. Rev.* **92** 308-314

[17] Inal M K, Sampson D H, Zhang H L and Dubau J 1997 *Phys. Scr.* **64** 452-454

[18] Sureau A and Holden P B 1995 *Phys. Rev.* **A52** 3110-3124

[19] Benredjem D, Sureau A, Cornille M and Dubau J 2000 *J. Quant. Spectrosc. Radiat. Transfer* **65** 751-767

Appendix. H_{hfs} matrix elements between *hfs* levels of the $1s^2 2s^2 2p^5 3s$ configuration

$$< 2p^5 3s\, L\, S\, J\, I\, F\, M_F\, |H_{hfs}| 2p^5 3s\, L\, S'\, J'\, I\, F'\, M'_F >= \delta_{F\,F'}\,\delta_{M_F\,M'_F}$$
$$\times \frac{g_I\,\alpha^2}{2}\,(-1)^{F+I+J'} \begin{Bmatrix} J' & I & F \\ I & J & 1 \end{Bmatrix} \sqrt{I\,(I+1)\,(2I+1)}\, \times$$
$$< 2p^5 3s\, L\, S\, J|| \sum_{i=1}^{10} \left[\frac{8\pi}{3}\delta(\mathbf{r}_i)\mathbf{s}_i + \frac{1}{r_i^3}\mathbf{l}_i - \frac{\sqrt{10}}{r_i^3}(s^{(1)}.C^{(2)})_i^{(1)} \right] ||2p^5 3s\, L\, S'\, J' > \qquad (A1)$$

Fermi contact term (interaction with 3s orbital electron):

$$< 2p^5 3s\, L\, S\, J\, || \sum_{i=1}^{10} \delta(\mathbf{r}_i)\mathbf{s}_i || 2p^5 3s\, L\, S'\, J' >= (-1)^{J+S+S'}\, 4\,\pi\, |R_{3s}(0)|^2 \qquad (A2)$$
$$\times \sqrt{\frac{3}{2}}\, \sqrt{(2J+1)(2J'+1)(2S+1)(2S'+1)} \begin{Bmatrix} S & 1 & S' \\ \frac{1}{2} & \frac{1}{2} & \frac{1}{2} \end{Bmatrix} \begin{Bmatrix} S' & J' & 1 \\ J & S & 1 \end{Bmatrix}$$

Orbital term (interaction with 2p orbital electrons):

$$< 2p^5 3s\, L\, S\, J\, || \sum_{i=1}^{10} \mathbf{l}_i || 2p^5 3s\, L\, S'\, J' >= (-1)^{J'+S}\, \delta_{S\,S'}\, < r^{-3} >_{2p} \qquad (A3)$$
$$\times \sqrt{6}\, \sqrt{(2J+1)(2J'+1)} \begin{Bmatrix} 1 & 1 & 1 \\ S & J & J' \end{Bmatrix}$$

Spin dipole term (interaction with 2p orbital electrons):

$$< 2p^5 3s\, L\, S\, J\, || \sum_{i=1}^{10} (s^{(1)}.C^{(2)})_i^{(1)} || 2p^5 3s\, L\, S'\, J' >= (-1)^{S'}\, < r^{-3} >_{2p}$$
$$\times \sqrt{\frac{27}{5}} \sqrt{(2J+1)(2J'+1)(2S+1)(2S'+1)} \begin{Bmatrix} S & 1 & S' \\ \frac{1}{2} & \frac{1}{2} & \frac{1}{2} \end{Bmatrix} \begin{Bmatrix} 1 & S' & J' \\ 2 & 1 & 1 \\ 1 & S & J \end{Bmatrix} \qquad (A4)$$

Inst. Phys. Conf. Ser. No 186
Paper presented at 9th Int. Conf. X-Ray Lasers, Beijing, China, 24–28 May 2004

Photopumping of XUV lasers by XFEL-Radiation

Ke Lan[1], Ernst Fill, and Jürgen Meyer-ter-Vehn
Max-Planck-Institut für Quantenoptik, D-85748 Garching, Germany

Abstract. Within the next few years X-ray Free-Electron Lasers (XFEL´s) now under construction are expected to generate highly collimated XUV pulses with 10^{13} photons and a duration of 100 fs. Focusing this radiation to a spot a few μm in diameter generates intensities of up to 10^{16} W/cm^2. Such pump intensities make feasible the investigation of photo-pumped XUV lasers using this radiation. We present simulations taking into account two different mechanisms generating the gain: 1. Photopumping with subsequent three-body recombination which takes advantage of the monochromaticity of the pump radiation to generate very cold electrons. 2. Innershell photo-ionization in which transient inversion is obtained by generating a hole in an otherwise completely filled shell. The simulations show that under appropriate conditions both mechanisms generate very high gain. However, a number of further issues must be considered, such as the propagation of the pump pulse in the medium to be pumped and refraction of the generated XUV radiation in a plasma with a high electron density gradient. These problems and possible solutions will be discussed as well as the motivation for proposing a laser-pumped XUV laser.

1. Introduction

Though photopumped x-ray lasers have been investigated by many researchers, an experimental demonstration of this pumping method is still lacking. Now the high brilliance expected from the X-ray Free-Electron Laser (XFEL) under construction at DESY/Hamburg makes this proposal again feasible. In this paper, we investigate the possibility of using radiation from the TESLA Test Facility at DESY/Hamburg to demonstrate a photopumped XUV laser. The simulations apply to two different mechanisms for generating the gain: 1. photoinization with subsequent three-body recombination [1] and 2. innershell photo-ionization [2]. In particular we investigate photo-ionization pumping with three-body recombination for a He-α x-ray laser ($\lambda = 58.4$ nm, He I) and a Ly-α laser (30.4 nm, He II) in helium, and for an innershell photo-ionization He-α laser (19.9 nm, Li II) and a Ly-α laser (13.5 nm, Li III) in lithium under the pump parameter conditions to be provided by XFEL at DESY/Hamburg. We will introduce the model and code used in our simulation in Sec. 2. We present simulations of photo-ionization with three-body recombination He I and

---

[1] On leave from the Institute of Applied Physics and Computational Mathematics. P.O. Box 8009-12 Beijing, 100088

He II lasers in Sec. 3, and innershell Li II and Li III lasers in Sec. 4. To attain a high gain length product the pump pulse must propagate through the gain medium for a sufficiently long distance, and thus, absorption of the pump pulse in the medium is a critical issue. We therefore simulate the propagation process of the pump pulse in medium for He I laser, and the results will be discussed in Sec.5. The experimental setup is introduced in Sec.6. Finally, conclusion and outlook will be given in Sec.7.

2. Model and code

The simulations are performed by using the CRMHA code [3]. Because of the very short XFEL pulse, we assume that ions keep cold during pumping. The electron velocity distribution is Maxwellian. In addition, hydrodynamics is neglected, as is justified for such a short pumping time. Hence, the equation for the electron temperature T_e and the coupled rate equations of the atomic states are solved. To simulate the propagation process, the photon equation is also solved. The atomic energy levels in [4] are used, and the population of the sublevels is assumed to be in statistical equilibrium.

The gain is very sensitive to the width of the x-ray laser line. For helium, the Doppler broadening can be neglected because the time for collisional energy exchange between ions and electrons is much longer than the duration of pumping, and ions stay at a very low temperature. However, the Doppler broadening must be taken into account for lithium because we are going to use the lithium vapor in the experiments and the vapor temperature is around 0.12 eV. The dominant contribution to the line width comes from Stark broadening. In this paper, the Stark line-broadening widths are taken from Refs. [5] and [6]. The life time broadening is also considered in the simulation.

To fit the pulse shape of DESY/XFEL, the pump intensity $I(t)$ is taken as

$$I(t) = \begin{cases} I_p \sin^2(\pi t / 2\tau), & 0 \le t \le 2\tau \\ 0, & t > 2\tau \end{cases}, \tag{1}$$

where t is time, τ is the Full Width at Half Maximum pump intensity (FWHM) and I_p is the peak value of I at $t = \tau$. According to the experimental parameters of DESY, the XFEL spectral bandwidth is 0.6% FWHM and the pulse duration $\tau = 100$ fs.

3. Photoionization with three-body recombination lasers in He I and He II

This mechanism takes advantage of the monochromaticity of the pump radiation to generate very cold electrons, a prerequisite for rapid three-body recombination. Shown in fig.1 are the level schemes for photopumped XUV He I and He II recombination lasers. Detailed simulations of He I and He II lasers were presented in [7,8]. Here, we only give the main results for the two lasers.

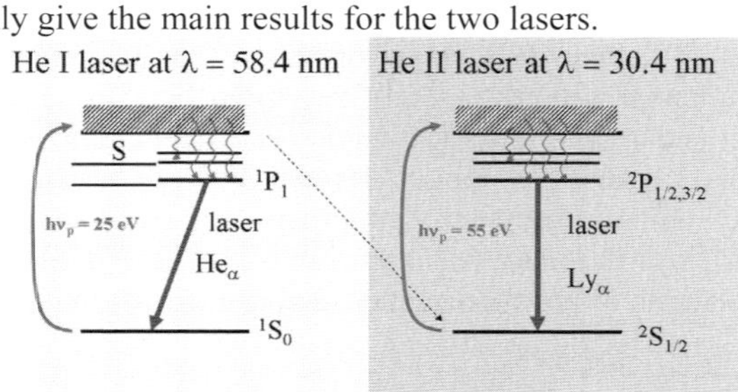

Figure 1. Level schemes for photopumped XUV recombination He I and He II lasers

3.1. *He I (58.4 nm) laser*

One step of photo-ionization is needed to generate the parent ions. Pump photons of 25 eV are used (just above the ionization potential of 24.6 eV). The main results are:

a. Initial T_e is near zero just after pumping, and later it is mainly increased by three-body recombination and collisional de-excitation.
b. The gain duration decreases with increasing gas density.
c. A pump intensity window exists, which is $10^{13} < I_p < 10^{17}$ W/cm^2. At lower intensity not enough atoms are ionized, while at higher intensity heating and ionization from upper laser level decrease gain.
d. An optimum pump intensity around 10^{14} W/cm^2 generates 100 cm^{-1} of gain with 300 fs duration at gas density of 10^{20} cm^{-3}.

3.2. *He II (30.4 nm) laser*

Pump photons of 55 eV are used (just above the ionization energy of H-like He). In contrast to the He I laser, two steps of photo-ionization are needed in the He II laser to get the parent ions, and thus a relatively high T_e is obtained. The main results are:

a. Electron temperature starts at 20 eV due to excess energy of pump photons, while second ionization reduces temperature.
b. Due to relatively high T_e, $N_0 \geq 10^{19}$ cm^{-3} is required to provide rapid three-body recombination.
c. There is also a pump intensity window, between 5×10^{14} W/cm^2 and 10^{18} W/cm^2.
d. The optimum pump intensity is 10^{15} W/cm^2, a factor of 10 higher than that for He I.
e. At this intensity the gain is comparable to the He I gain.

4. Innershell photo-ionization lasers in lithium

In this section, we study two innershell photo-ionization lasers in lithium. Interest in innershell photo-ionization lasers derives from the fact that the states forming the upper laser level are already occupied and, thus, very high gain should be generated. In general, innershell photo-ionization lasers suffer from a high rate of loss of inversion due to the Auger effect. However, the problem of the Auger transition is not pertinent if the atom contains only a single outer electron, as in Li and Na. Ionization of a 1s electron of neutral lithium generates gain at the He-α line of Li-II with a wavelength of 19.9 nm, and further ionization of the 1s electron of Li-II generates gain at the Ly-α line of Li-III at 13.5 nm. However, transfer of the 2s electron of the neutral to the 2p level is required. This is possible by means of a visible laser tuned to the 2s-2p resonance line of atomic lithium at λ = 671 nm at the beginning of the pump. The principle of an innershell photo-ionization laser in lithium is shown in fig.2.

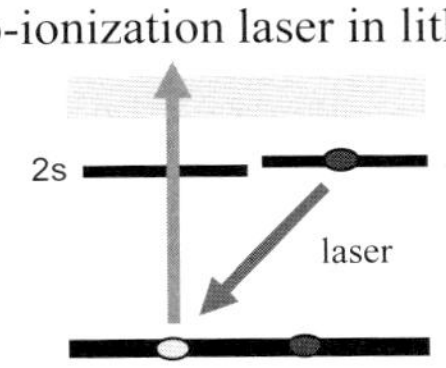

Figure 2. Principle of innershell photo-ionization laser in lithium.

4.1. Li II (19.9 nm) laser

In the process of producing Li II laser, the whole ionization phase and the gain are determined by photo-ionization if the pump intensity is so strong that all other atomic processes can be neglected. A pump photon energy of 68 eV, above ionization energy from atomic lithium $1^2 2^1$ to $1^1 2^1$ ion, is used. When population of $1^1 2^1$ ions dominates the ion distribution, the maximum gain is generated. The gain lasts until $1^1 2^1$ is strongly photo-ionized to $1^1 2^0$. We use P_g^{Li} and $P_1^{He-like}$ to express the fractional populations of $1^2 2^1$ and $1^1 2^1$ ion, F_g^{Li} and $F_{1,n=2}^{He-like}$ to express the photo-ionization rates from $1^2 2^1$ to $1^1 2^1$ and from $1^1 2^1$ to $1^1 2^0$. Only considering the photo-ionization process, we have:

$$P_g^{Li}(t) = \exp(-\int_0^t F_g^{Li} dt'), \tag{2}$$

$$P_1^{He-like}(t) = [1 - P_g^{Li}(t)]\exp(-\int_0^t F_{1,n=2}^{He-like} dt') \tag{3}$$

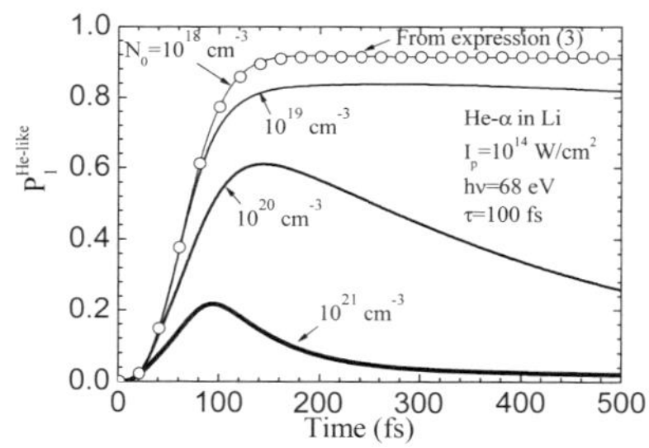

Figure 3. Time dependent $P_1^{He-like}$ of Li II laser for different gas densities.

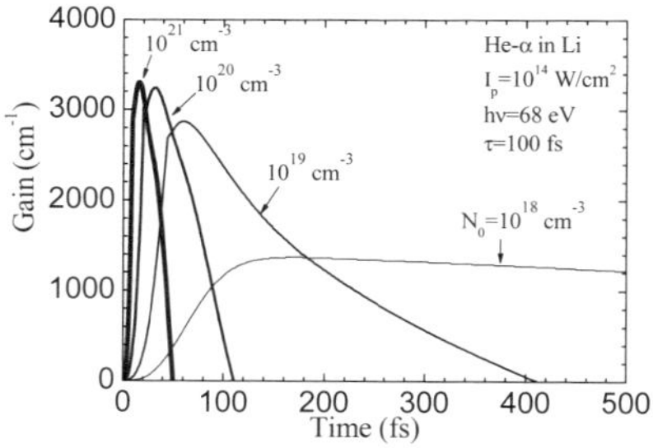

Figure 4. Time dependent gain of Li II laser for different gas densities.

From the photo-ionization rate expression [1], we get $F_{1,n=2}^{He-like} \approx (1/110) F_g^{Li}$. Defining $\eta = \int_0^t F_g^{Li} dt'$ and taking expression of $I(t)$ (given in Sec.2) into η, we conclude that $I_p > 3.5 \times 10^{13}$ W/cm^2 is required for getting the maximum $P_1^{He-like}$ before the pump ends. At $\eta \approx 5$, $P_1^{He-like} \approx 95\%$ and $\gamma \approx 1$. Here γ is inversion factor Thus, the gain produced at $\eta \approx 5$ is mainly determined by the plasma density and the line broadening.

In the high density plasmas, collisional processes can not be neglected. They may seriously decrease the life time of $1^1 2^1$ and influence the gain. The temporal evolutions of $P_1^{He-like}$ and gain at different N_0 are shown in fig. 3 and fig. 4, respectively. At a higher N_0, $P_1^{He-like}$ is smaller, and the gain is produced earlier with a shorter duration. We note that the gain shape is quite different from the $P_1^{He-like}$ shape, this being due to line broadening. In addition, the gain is produced earlier with a shorter duration at a higher I_p, while the maximum gain is almost the same.

4.2. Li III (13.5 nm) laser

For the Li III laser, pump photons of 108 eV are used. There are two steps of photo-ionization in getting the upper laser state. Electron temperature starts at 28 eV due to excess energy of pump photons. Because of relatively high T_e, a low density is required to obtain a long life time of the upper laser level. At $N_0 < 10^{19}$ cm^{-3}, the maximum value of P_1^{H-like} is about 80%. There is also a low limit of I_p for getting the maximum P_1^{H-like} before the end of the pump, which is 8.7×10^{14} W/cm^2. The variations of the gain with N_0 and I_p are shown in fig. 5 and fig. 6, respectively.

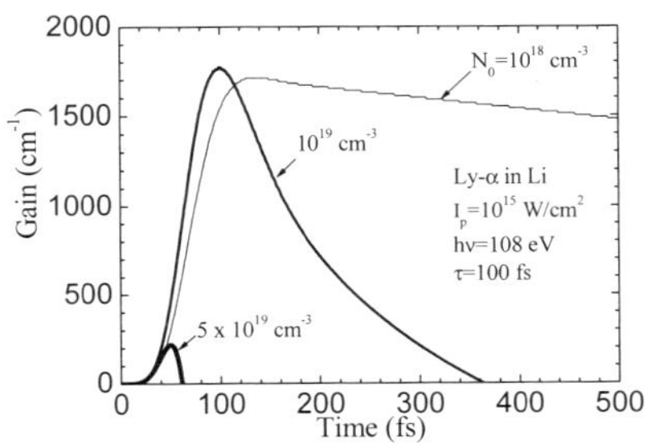

Figure 5. Temporal evolutions of Li III laser gain for different densities.

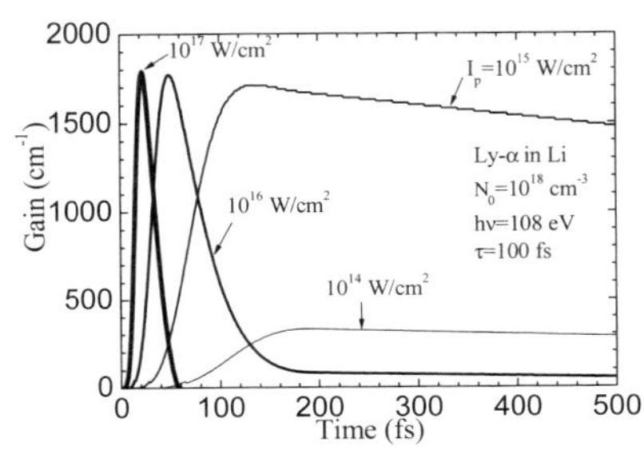

Figure 6. Temporal evolutions of Li III laser gain for different I_p.

It appears that very high gain is indeed generated for the innershell photo-ionization laser. The duration of the gain is limited by ionization of the electron in the n=2 level. For the Li II laser, the gain duration is 60 fs at 10^{20} cm^{-3} and only 30 fs at 10^{21} cm^{-3}. These conditions would thus lead to considerable shortening of the laser pulse compared with the pump pulse, a feature most welcome for many experiments.

5. Propagation of pump pulse in helium gas for He I laser

As shown in section 3.1, the pump parameters obtainable by the DESY XFEL are sufficient to generate He-α laser gain of ~ 100 cm^{-1}. However, saturation requires $gL > 15$. Hence, the gain must be generated over a length of a few mm in order to achieve gain saturation. In this section, we study the problem of propagation of the pump radiation through the medium.

Shown in fig. 7 are contour plots of the pump intensity at $I_p = 10^{14}$ W/cm^2 under different gas densities. There is high absorption in the pump pulse front, and so the pulse becomes shorter with a faster rising front as it propagates, clearly seen in Fig. 7(b). Moreover, the propagation length is longer at a higher pump intensity I_p or in a lower helium density N_0. After the pump, there is an obvious increase in T_e and P_g^{He} due to three-body recombination, as shown in fig. 8 and fig. 9. Here, P_g^{He} is the fractional population of atomic He in the ground state. Because the pump pulse becomes shorter with a faster rising front as its propagation, the maximum gain may be generated deeper inside the medium. In fig. 10, we give contour plots of the gain under three different conditions. As shown in fig. 10(a), the pump propagates a distance of almost 2 mm into the medium in spite of a high absorption coefficient. This is near the optimum case. In fig. 10(b), the propagation length is too short because of too high gas density. In fig. 10(c), a high pump intensity results in a very low gain. The optimum case is $I_p = 2\times10^{14}$ W/cm^2 and $N_0 = 10^{19}$ cm^{-3}, which generates $g > 90$ cm^{-1} with $L > 0.4$ cm. To have $gL > 15$, I_p should be between 10^{14} and 10^{15} W/cm^2 at $N_0 = 10^{19}$ cm^{-3} and N_0 should be lower than 3×10^{19} cm^{-3} at $I_p = 2\times10^{14}$ W/cm^2.

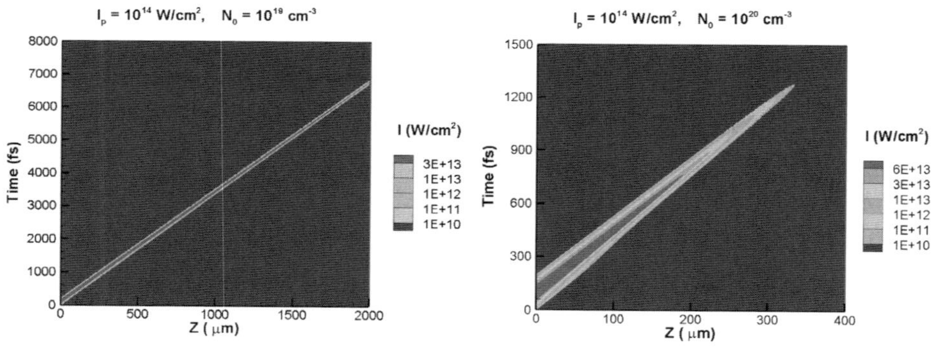

(a) **(b)**

Figure 7. Contour plots of the pump intensity at $I_p = 10^{14}$ W/cm^2. (a) $N_0 = 10^{19}$ cm^{-3}, (b) $N_0 = 10^{20}$ cm^{-3}.

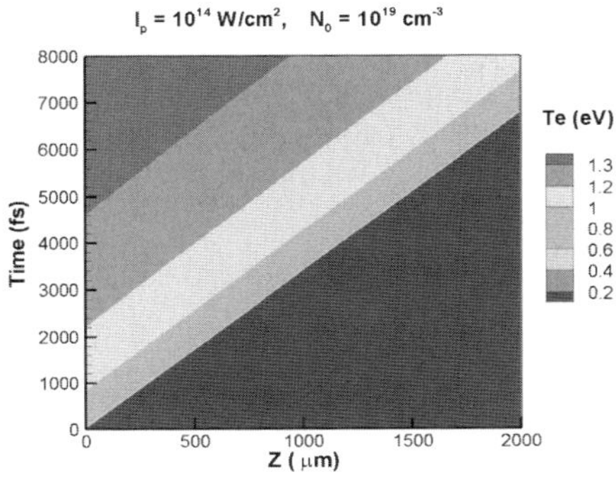

Figure 8. Contour plot of T_e at $I_p = 10^{14}$ W/cm^2 and $N_0 = 10^{19}$ cm^{-3}.

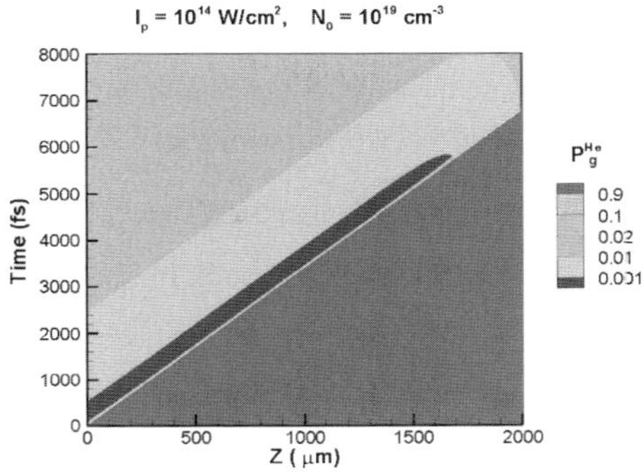

Figure 9. Contour plot of P_g^{He} at $I_p = 10^{14}$ W/cm^2 and $N_0 = 10^{19}$ cm^{-3}.

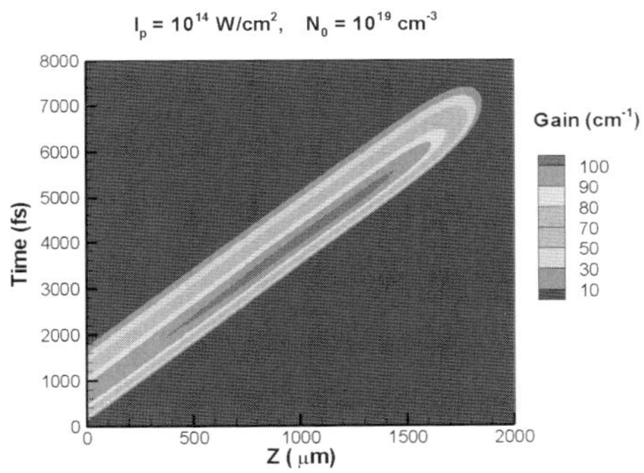

(a)

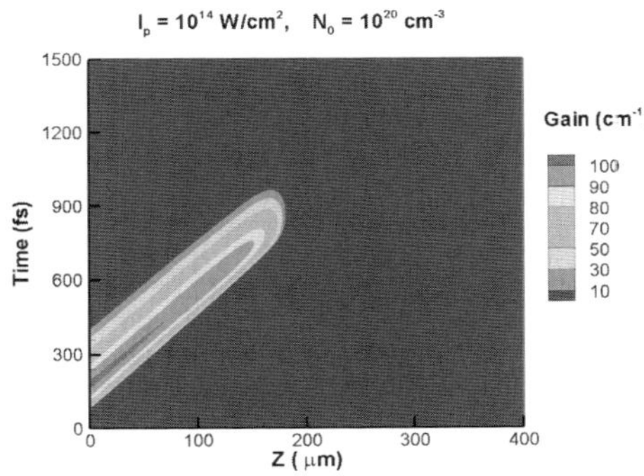

(b)

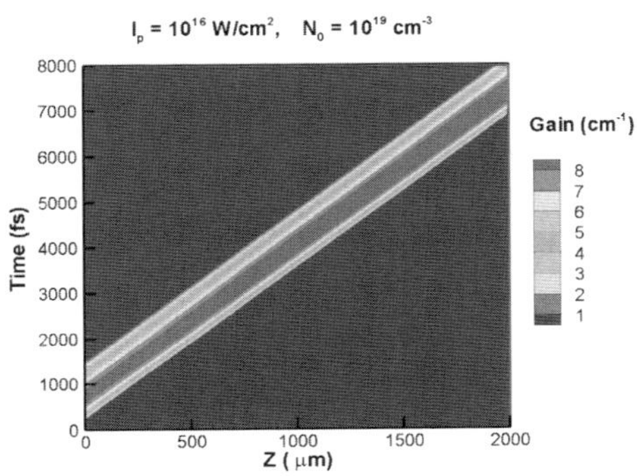

(c)

Figure 10. Contour plots of gain.
(a) $I_p = 10^{14}$ W/cm^2 and $N_0 = 10^{19}$ cm^{-3},
(b) $I_p = 10^{14}$ W/cm^2 and $N_0 = 10^{20}$ cm^{-3},
(c) $I_p = 10^{16}$ W/cm^2 and $N_0 = 10^{19}$ cm^{-3}

6. Experimental setup

Shown in fig.11 is a layout of photopump experiment using XUV-FEL radiation. The beam is focused to a 10 μm spot by a toroidal mirror M. A nozzle N generates a gas jet with appropriate density. The emitted radiation is dispersed by a transmission grating TG and detected by means of a soft X-ray CCD.

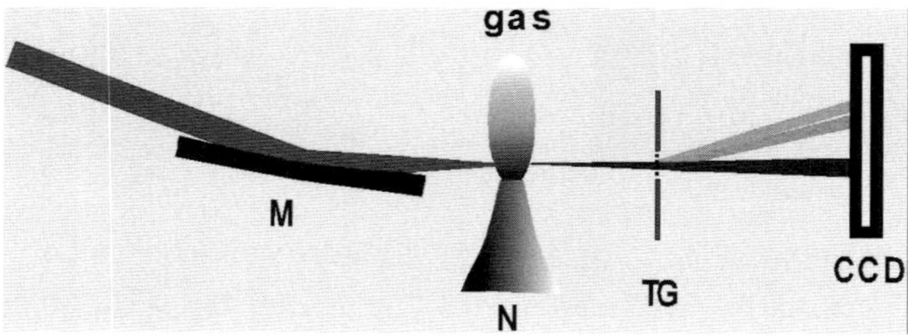

Figure 11. Layout of the experiment.

7. Conclusion and outlook

Our investigation showed that photoionization-recombination He I and He II lasers and innershell photo-ionization Li II and Li III lasers can be generated by using radiation from the TESLA Test Facility at DESY/Hamburg. The main motivation for this study is the hope that experiments with a well-defined pump source will increase our understanding of photopumping and may subsequently lead to table top X-ray lasers at shorter wavelengths. Additional arguments are that for innershell lasers the gain duration is mainly limited by ionization of the electron in the n=2 level, which would lead to considerable shortening of the laser pulse compared with the pump pulse. Moreover, by carefully designing the geometry of the laser and by an appropriate choice of the plasma parameters, the photo-pumped laser could yield pulse parameters superior to those provided by the pump laser.

Acknowledgement

This work is supported by an Alexander von Humboldt Fellowship.

References

[1] D. G. Goodwin and E. E. Fill 1988 J. Appl. Phys. **64**, 1005-1014.
[2] M. A. Duguay and P. M. Rentzepis 1967 Appl. Phys. Lett. **10**, 350-352.
[3] K. Lan and Y. Zhang 2002 Eur. Phys. J. AP **19**, 103-112.
[4] Raymond L. Kelly 1987 Journal of Physical and Chemical Reference Data 16 Supplement No.1.
[5] Hans R. Griem 1964 Plasma Spectroscopy (McGraw-Hill, New York).
[6] Hans R. Griem 1974 Spectral line broadening by Plasmas (Academic Press, New York and London).
[7] K. Lan, E. E. Fill et al., in Proceedings of SPIE Vol. **5197** *Soft X-Ray Lasers and Applications V,* edited by E. E. Fill and S. Suckewer (SPIE, Bellingham, WA, 2003) 253-257.
[8] K. Lan, E. E. Fill et al., 2003 Europhys. Lett. **64**, 454-460.

Inst. Phys. Conf. Ser. No 186
Paper presented at 9th Int. Conf. X-Ray Lasers, Beijing, China, 24–28 May 2004

Modeling of Ni-like Ag laser. Time dependence of the output and saturation regime

D. Benredjem[1], J. Dubau[1], C. Möller[1], J. Kuba[2] and O. Zabaydullin[3]

[1] Laboratoire d'Interaction du Rayonnement X avec la Matière (LIXAM), Université Paris-Sud, Orsay, France

[2] Lawrence Livermore National Laboratory, California, USA

[3] Institute of Nuclear Fusion, Kurchatov Institute, Moscow, Russia

Abstract. In this contribution we present recent results on the lasing transition at 13.9 nm in Ni-like silver. We have modeled an experiment done at the Rutherford Laboratory. Our calculations involve a 3D ray-trace code which is a post-processor of the hydro-atomic code EHYBRID. This first step gives the output intensity as a function of time and space. In a second step we have modeled the saturation effect on populations. To this end, we have used the Maxwell-Bloch formalism where the interaction between the X-ray laser field and the amplifying medium is taken into account properly. Gain, level populations and integrated intensity values are obtained consistently.

1. Introduction

We study the amplification of the X-ray beam by using two methods. In the first one, a ray-trace code accounting for refraction in the directions perpendicular and parallel to the target surface is coupled to the hydro/atomic package EHYBRID [1]. Saturation effect is taken into account in a simple manner. We obtain the integrated intensity at the exit of the plasma. In a second approach we use the Maxwell-Bloch formalism [2] which describes more accurately the interaction between the X-ray laser (XRL) field and the amplifying medium. Refraction is also taken into account.

Our calculations model the experiment done at the Rutherford Laboratory [3], in the transient collisional excitation scheme. In this experiment, a slab target was irradiated by a driving laser consisting of a pulse of 300-ps duration (FWHM), followed by a short pulse of 1-ps duration. The long pulse is of low intensity (6×10^{11} W/cm^2) and the second pulse, arriving 200 ps later, is of high intensity (5.6×10^{14} W/cm^2).

2. Ray-tracing

We use the system of coordinates shown in Fig. 1.

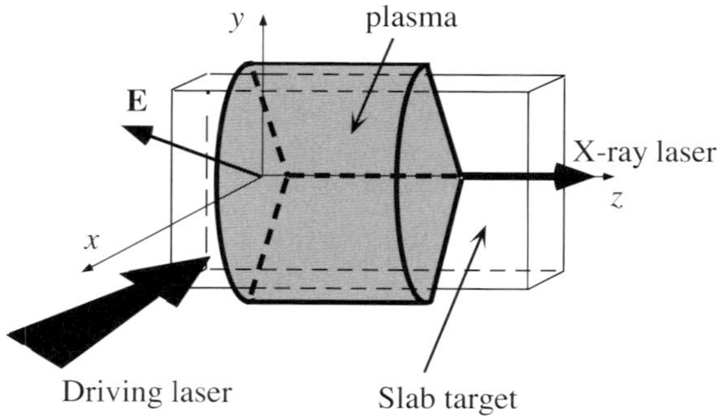

Figure 1. Geometry of the target irradiation

The beam path in a medium with an electron density gradient is described by the ray equation:

$$\frac{d}{ds}\left(n\frac{d}{ds}\mathbf{r}\right) = \nabla n(\mathbf{r}), \tag{1}$$

where n is the index of refraction and $ds = \sqrt{dx^2 + dy^2 + dz^2}$. In the beam approximation, one can derive the following relation from Maxwell's equations:

$$n = \sqrt{1 - \frac{n_e}{n_c}}, \tag{2}$$

where n_e is the electron density and n_c the critical density$\approx 1.1 \times 10^{21} /(\lambda(\mu m))^2$. z being the coordinate with the largest variation, the paraxial approximation allows us to write $d/ds \approx d/dz$, and the ray equation reduces to the following differential equations:

$$\begin{aligned} \frac{d\alpha}{dz} &= \frac{d^2x}{dz^2} = -\frac{1}{2n_c}\frac{\partial}{\partial x}n_e(x,y) \\ \frac{d\beta}{dz} &= \frac{d^2y}{dz^2} = -\frac{1}{2n_c}\frac{\partial}{\partial y}n_e(x,y), \end{aligned} \tag{3}$$

where α (β) is the angle between the z axis and the projection of the wave vector on the xz (yz) plane.

The hydro/atomic code EHYBRID gives the electron density, local gain and saturation intensity as functions of time and distance (x) to the target surface. It involves an atomic data base consisting of 272 Ni-like levels. The collisional excitation/deexcitation rates are given by the code HULLAC [4] for all transitions between the $n = 4$ levels. Transitions involving more excited levels are handled by means of empirical formulae. For the variation of the electron density with y, we assume a Gaussian profile:

$$n_e(x,y,t) = n_{e0}(x,t)\exp\left[-4\ln(2)\frac{y^2}{\delta^2}\right], \tag{4}$$

where n_{e0} is the electron density given by EHYBRID, δ the focal line half-width. Similarly, the electron and ion temperatures as well as the ion density (even though we do not use them in the ray-trace calculations) would be expressed in the same way in this approximation. The electron density gradient in the direction y is then

$$\frac{\partial n_e}{\partial y} = -\frac{8\ln(2)n_{e0}}{\delta^2}\, y\exp\left[-4\ln(2)\frac{y^2}{\delta^2}\right]. \tag{5}$$

For a set of initial time, angles and coordinates (t_0, α_0, β_0, x_0, y_0) the ray-trace code gives for each z value the electron density, i.e., $n_e(x,y,z,t)$. In order to take into account the saturation effect on the gain, we have used the well-known relation $G(s) = G_{ss}/\left[1 + I(s)/I_{sat}\right]$, where $I(s)$ is the integrated intensity. G_{ss} is the small-signal gain and I_{sat} the saturation intensity. Both quantities are given by EHYBRID. The intensity is given by the following relation:

$$I(s) = \frac{\sigma}{G}\left\{\exp\left[G\Delta s - 1\right]\right\} + I(s-\Delta s)\exp\left[G\Delta s\right], \tag{6}$$

where σ is the emissivity, $\sigma = N_u h\nu_0 r^2 A_{ul}/(4\pi L^2)$. The emissivity depends on the radius r of the output region, the length L of the plasma column, and the Einstein coefficient for spontaneous emission between the two lasing levels u and l, N_u being the population density of the upper level.

We have calculated the intensity as a function of time. Figure 2 (next page) shows the intensity obtained when 10^6 rays are launched 1 ps before the time of peak gain (a), at the time of peak gain (b), and 1 ps after the time of peak gain (c). It is clear that the X-ray beam exits the plasma 45 μm away from target. The largest intensity, I_{max}, is obtained when the rays are launched at the time of peak gain.

When the intensity of the XRL beam is large, absorption and induced emission have an important effect on the populations of the lasing levels. The collisional-radiative equations should then account for these two processes. To this end, we have developed a satisfactory approach described below.

3. Maxwell-Bloch formalism

The propagation of a radiation through globally neutral plasmas obeys the Maxwell wave equation

$$\Delta\mathbf{E} - \frac{1}{c^2}\frac{\partial^2}{\partial t^2}\mathbf{E} - \frac{\omega_{pe}^2}{c^2}\mathbf{E} = \frac{1}{\varepsilon_0 c^2}\frac{\partial^2}{\partial t^2}\mathbf{P}, \tag{7}$$

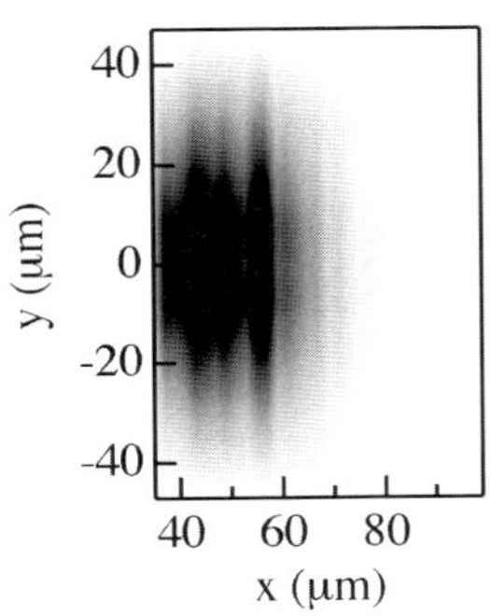

(a)

$I_{max} = 2.8 \times 10^{11}$ W/cm^2

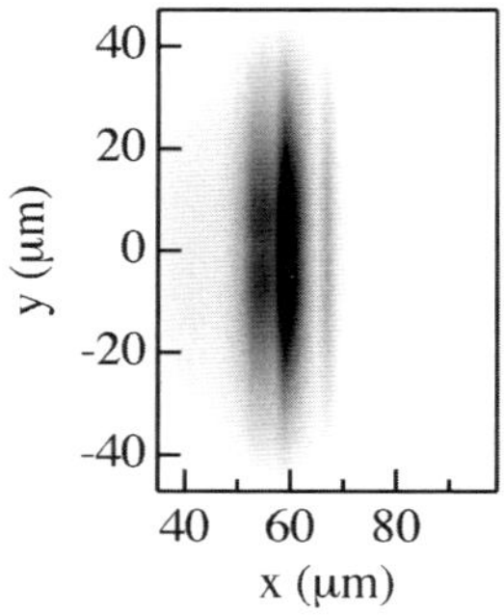

(b)
$I_{max} = 1.4 \times 10^{13}$ W/cm^2

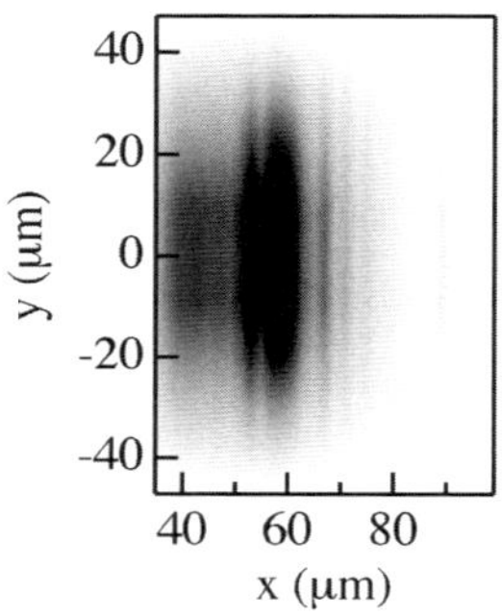

(c)
$I_{max} = 3.2 \times 10^{12}$ W/cm^2

Figure 2. Integrated intensity at the exit from plasma. The propagation length is 5 mm. The rays are launched at three different times around the time of peak gain predicted by the code EHYBRID.

where **E** is the electric field associated with the amplified beam, **P** the polarization vector of the lasing ions, and ω_{pe} the electron plasma frequency. The polarization vector is given by

P=Tr(ρ**d**), where ρ is the density matrix operator and **d** the electric dipole of the emitter. At t=0, the coherences (off-diagonal elements of ρ) are equal to zero, and ρ may hence be written $\rho = \sum_{JM} |JM\rangle\langle JM| n_{JM}$, where $n_{JM} = \langle JM|\rho|JM\rangle$ is the fractional population density of the Zeeman sublevel $|JM\rangle$. There is a simple relation between the fractional populations and the actual population densities N_{JM}. We then consider the Bloch relation

$$i\eta\partial\rho/\partial t = [H,\rho], \tag{8}$$

where $H=H_A$-**d**.**E**, with H_A designating the Hamiltonian of the lasing ion without radiation field, and -**d**·**E** the interaction between the radiation and the medium, in the dipole approximation.

The combination of Eqs. 7 and 8 gives the radiative transfer equation written below for an isolated transition between an upper lasing level u and a lower lasing level l:

$$\frac{\mathbf{k}}{k}\cdot\nabla I_{ul}(\nu,\mathbf{r},t) = \sigma(\nu,\mathbf{r},t) + G_{ul}(\nu,\mathbf{r},t)I_{ul}(\nu,\mathbf{r},t) \tag{9}$$

and a set of population rate equations for the Zeeman sublevels i that belong to the two levels u and l:

$$\frac{\partial}{\partial t} N_i(\mathbf{r},t) = -N_i(\mathbf{r},t)\Lambda_i(\mathbf{r},t) + \sum_{j \neq i} \lambda_{ji}(\mathbf{r},t) N_j(\mathbf{r},t), \tag{10}$$

where Λ and λ are respectively the total decay rate and a rate of population increase of the level i. These rates account for absorption and induced emission.

Equations 9, 10 and a set of rate equations for non lasing levels are treated in a consistent calculation that gives the intensity and gain during amplification.

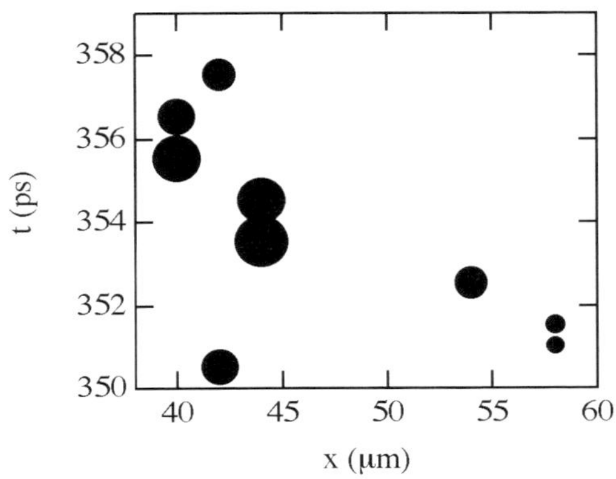

Figure 3. Integrated intensity calculated for a propagation length of 5 mm. Disk area is proportional to the intensity. I_{max} is equal to 1.15×10^{10} W/cm^2.

In this approach, the highest small-signal gain is much smaller than the value predicted by EHYBRID (140 instead of 800 cm^{-1}). Figure 3 shows the integrated intensity at the exit from plasma. The highest intensity is obtained at a distance from target approximately equal to 40 μm. The values of the intensity are more realistic than those obtained with the ray-trace code with saturation described by Eq. 6, typically two to three orders of magnitude smaller.

4. Concluding remarks

We have calculated the output intensity of the 13.9 nm Ni-like silver x-ray laser line by a Maxwell-Bloch approach suitable for the treatment of the saturation regime. The required hydrodynamic quantities are obtained by a ray-trace code working as a post-processor of EHYBRID.

Acknowlegment

We are indebted to G. J. Pert (York University) for providing the code EHYBRID.

References

[1] Pert G J 1980 J. Comput. Phys. 39 251
[2] Bar-Shalom A et al 2001 J. of Quantit. Spectrosc. and Radiat. Transf. 71 169
[3] Klisnick A et al 2002 Phys. Rev. A 65 033810
[4] Bar-Shalom A, Kalpisch M and Oreg J 2001 J. of Quantit. Spectrosc. and Radiat. Transf. 71 169

Inst. Phys. Conf. Ser. No 186
Paper presented at 9th Int. Conf. X-Ray Lasers, Beijing, China, 24–28 May 2004

Numerical investigation of a transient Ni-like Ag x-ray laser at 13.9nm

F. Yan, J. Zhang, X. Lu
Laboratory of Optical Physics, Institute of Physics, Chinese Academy of Science, Beijing 100080, People's Republic of China

J.Y. Zhong
National Astronomical Observatories, Chinese Academy of Sciences, Beijing 100012, People's Republic of China

Abstract. The Ni-like Ag x-ray laser at 13.9nm (4d→4p, J=0→1) is numerically studied, using one-dimensional hydrodynamic code coupled with the Cowan atomic data package. High gain (>150cm^{-1}) is predicted using a 1ps, 10^{15}w/cm^2 main drive pulse proceeded by a 2ns, 6×10^{11}W/cm^2 prepulse with a temporal separation of 1.0ns between these two pulses. The characteristics of the x-ray laser, including the evolution of gain in time and space, the electron temperature and the electron density obtained using the optimized drive configuration is investigated. The effects of refraction and saturation on the propagation of the x-ray laser in the plasma medium are also investigated by using a two-dimensional ray tracing code. We predict that a saturated Ni-like Ag soft x-ray laser at 13.9 nm could possibly be generated using an energy of several joules under the optimized drive configuration. These results contribute to the further understanding of Ni-like x-ray lasers.

1. Introduction

In Ni-like x-ray lasers, the upper laser level $3d^9 4d\ ^1S_0$ is populated through the monopole collisional excitation from the ground state $3d^{10}\ ^1S_0$. The strong lasing lines are dominated by the $3d^9 4d\ ^1S_0 \rightarrow 3d^9 4d\ ^1P_1$ and $3d^9 4d\ ^1S_0 \rightarrow 3d^9 4d\ ^3D_1$ transitions. The energy level interval between 4d and 4p levels for the Ni-like ions is larger than that between 3p and 3s levels for the Ne-like ions while the bound energy of the ground state for Ni-like and Ne-like ions is similar. This results in a higher quantum efficiency for Ni-like x-ray lasers than Ne-like x-ray lasers. So Ni-like x-ray lasers have a more favorable scaling of laser wavelength with drive energy than Ne-like x-ray lasers [1]. However, the population kinetics of Ni-like ions is much more sensitive to plasma conditions than that of Ne-like ions, which results in difficulties in providing a saturation output for Ni-like x-ray lasers. So the theoretical simulation and optimization of parameters for experiments are necessary.

In this paper, the Ni-like Ag x-ray laser at 13.9 nm (4d→4p, J=0→1) is simulated to predict laser-plasma interaction and gain coefficient, using a one-dimensional Lagrangian hydrodynamic code MEDUSA [2], coupled with an atomic physics code. A two-dimensional ray tracing code is developed to calculate the x-ray laser output intensity, as a postprocessor to the hydrodynamic code MEDUSA. In the calculation, the saturation and refraction effects are taken into account. Based on these simulations, we can further understand the plasma dynamics of x-ray lasers and optimize the parameters for the experiments.

2. Optimization of the drive pulse configuration

Atomic data including the energy levels, radiative transition probability (Einstein coefficient A) and the electron collisional excitation rates used in these simulations are calculated using the code RCN/RCG developed by Cowan [3], to complete the insufficient published data of Ag XX ions. A detailed description of the atomic data calculation for predicting x-ray lasers by code RCN/RCG has been given in [4].

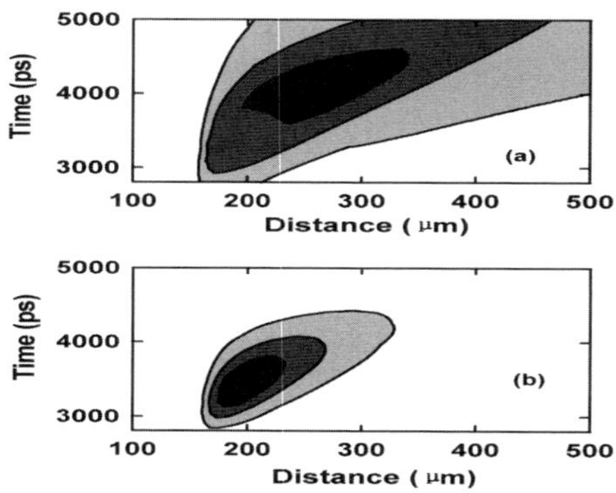

Figure 1 Contours of ion abundance vs. space and time for (a) Ni-like Ag ions and (b) Co-like Ag ions.

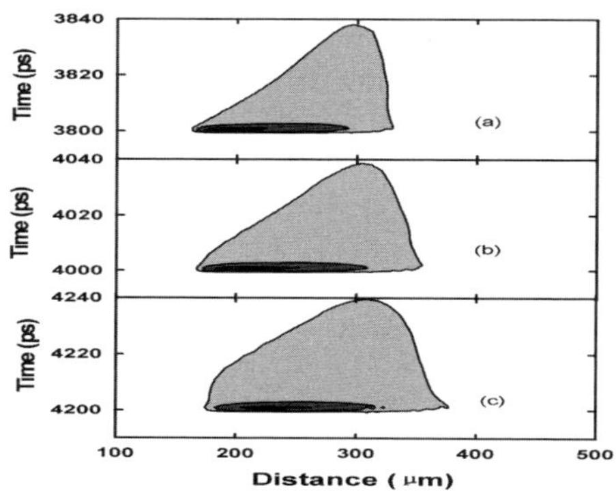

Figure 2 Gain contours vs. space and time for the delay time of (a) 0.8ns, (b) 1.0ns and (c) 1.2ns.

In the simulation we use Gaussian pulses at 1.053μm and 100-μm-thick slab targets. The pump geometry is set to be the standard line focus on slab targets. A nanosecond prepulse followed by a picosecond main pulse is used. An optimum preplasma condition can be achieved with large Ni-like Ag ion abundance when the prepulse was set at 6×10^{11}W/cm^2 in 2 ns. Fig.1 gives the contours of ion abundance vs space and time for (a) Ni-like Ag ions and (b) Co-like Ag ions, respectively. The target surface is located at 100 μm and the drive pulse reaches its peak at 3ns. The ion abundance of the Ni-like Ag ions reaches 70% while the ion abundance of the Co-like Ag ions only reaches 10%. This is the optimum condition for high gain operation. The delay time between the prepulse and the main pulse is optimized for high gain operation. In order to investigate the effect of delay time between the two pulses on gain coefficient, we set the peak intensity of the main pulse at 1×10^{15}W/cm^2 while varying the delay time. Fig.2 shows the gain contours at (a) 3.8ns, (b) 4ns, (c) 4.2ns for the delay time of (a) 0.8ns, (b) 1.0ns and (c) 1.2ns, respectively. High gain over 150cm^{-1} appears just before the peak of the main pulse. It is apparent that 1.0ns delay gives a better condition for high gain operation, due to the high Ni-like ion abundance and the low Co-like ion abundance at 4ns, as shown in Fig. 1. The gain area is enlarged with the delay time, but the peak gain coefficient is decreased. This is due to the larger scale length of the plasma with larger delay time. Lower population of Ni-like Ag ions causes the decrease of gain.

The main pulse duration and intensity are also optimized. The results show that for 1ps and 5ps duration, the gain reaches 150cm^{-1}. There is a larger region of high gain

(>150cm^{-1}) in space and time for a 1ps duration. Moreover, the drive laser energy for 1ps duration is also less than the case for 5ps duration. So the main pulse duration is set at 1ps in our simulation. The gain can be over 150cm^{-1} for an intensity of 1, 2×10^{15}W/cm^2. Although the region for high gain (>150cm^{-1}) is larger in space and time for 2×10^{15}W/cm^2 than that for 1×10^{15}W/cm^2, the main pulse with intensity of 1×10^{15}W/cm^2 is used for a more efficient drive.

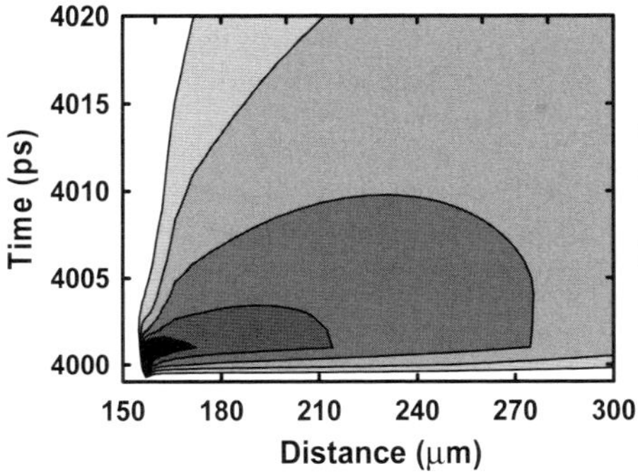

Figure 3 Contours of electron temperature vs. space and time.

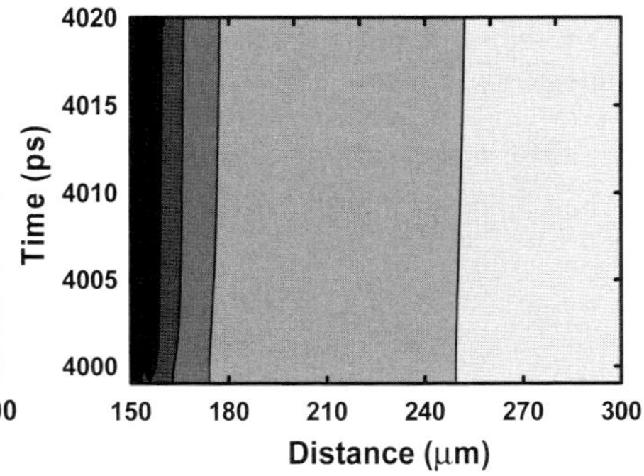

Figure 4 Contours of electron density vs. space and time.

4. Plasma conditions

In this section we analyze the plasma conditions obtained using the above optimized drive configuration. Fig. 2 (b) shows the gain contours vs space and time. The main drive pulse reaches its peak intensity at 4.0ns. The high gain region (>150cm^{-1}) has a large dimension in space over 50 μm and a small dimension in time of less than 2ps. As the plasma further expands, gain (<150cm^{-1}) with a larger region in space and time can be obtained. The distribution of electron temperature is similar to that of the gain profile, as shown in Fig. 3. The electron temperature is about 500-600eV in high gain (>150cm^{-1}) region and electron temperature reaches 700eV at the critical density surface. The refraction of x-ray lasers propagation in the plasma strongly affects the laser output. A smaller density gradient in the gain region is favorable for the amplification and propagation of x-ray lasers. Fig. 4 shows the contours of electron density. The electron density in the high gain region is about $1\text{-}2\times10^{20}$cm^{-3}, which is far away from the critical density region (1×10^{21}cm^{-3} for 1.06 μm laser driver). Moreover, the density gradient is small in the high gain region. This is certainly favorable for x-ray laser propagation while avoiding refraction.

5. Refraction and saturation effects

In order to investigate the effects of refraction and saturation on the propagation of x-ray lasers in the plasma medium, a two-dimensional ray tracing code is used to calculate the x-ray laser output intensity, as a postprocessor to the hydrodynamic code MEDUSA. The electron density, local gain and energy level population are given by the code MEDUSA, coupled with the Cowan atomic physics code. A traveling wave excitation is used to keep the spatial profiles of gain and electron density unchanged along the line focus. The total laser output as a function of the plasma length is calculated.

Fig. 5 shows the distance from the target surface of x-rays passing through a 1.5cm long plasma medium for various initial incidence angles and location x_0. The most important refraction is observed for rays incident at $x_0\approx170\sim200$ μm. The reason is that in these regions the electron density is close to critical density and the electron density gradient is large. When the x rays are incident with negative angles, that is, the rays propagate towards the target, the refraction effect is less important than that when the x rays propagate away from the target. Fig. 6 shows the profile of spatially integrated laser output intensity as a function of the plasma length. Saturation effect is taken into account by a simplified approach described by Benredjem [5]. The solid line describes the x-ray laser output intensity when the refraction is taken into account. In fact, some rays initially directed away from the target, rapidly reach a small signal gain region or propagate out of the gain region due to the refraction effect. This is the reason why the plasma length for saturation output is larger than that without considering refraction effect (dash line).

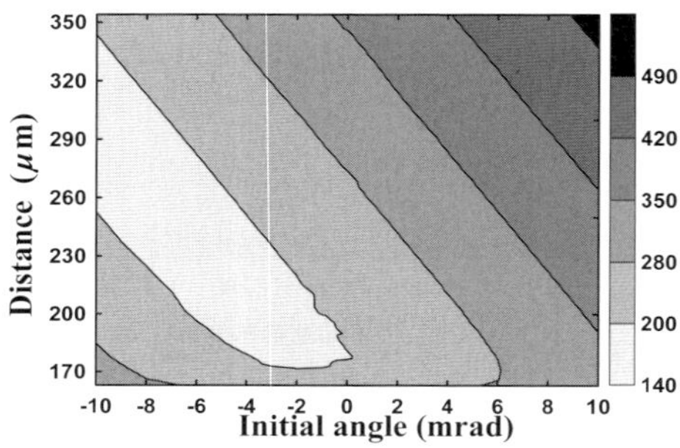

Figure 5 Distance from the target surface of x rays passing through a 1.5cm long plasma medium for various initial launching angles and location x_0.

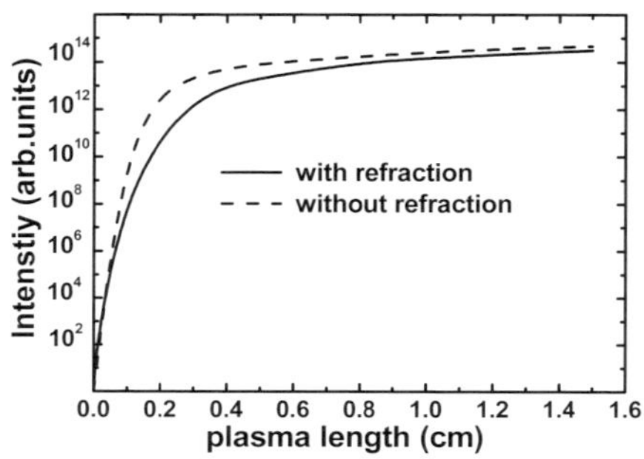

Figure 6 Spatially integrated laser output intensity as a function of the plasma length.

6. Conclusions

The Ni-like Ag x-ray laser at 13.9nm (4d→4p, J=0→1) is numerically studied, using one-dimensional hydrodynamic code coupled with the Cowan atomic data package. High gain (>150cm^{-1}) is predicted using a 1ps, 10^{15}W/cm^2 main drive pulse proceeded by a 2 ns, 6×10^{11}W/cm^2 prepulse with a temporal separation of 1.0 ns between these two pulses. The characteristics of the x-ray laser, including the evolution of gain in time and space, the electron temperature and the electron density obtained using the optimized drive configuration is investigated. The effects of refraction and saturation on the propagation of the x-ray laser in the plasma medium are also investigated by using a two-dimensional ray tracing code. We predict that a saturated Ni-like Ag soft x-ray laser at 13.9 nm could possibly be generated using energy of several joules under the optimized drive configuration. These results contribute to the further understanding of Ni-like x-ray lasers.

Acknowledgements

This work was supported by the NSFC under Grant Nos. 10374114, 60321003, 10474134 and 60128505, and the NKBRSF under Grant No. G1999075206, and the National Hi-tech ICF Program.

References

[1] Zhang J et al., 1996 Phys. Rev. A **54** R4653
[2] Christiansen J P, Ashby D E T et al., 1974 Comput. Phys. Commun. **7** 271
[3] Cowan R D, 1981 The Theory of Atomic Structure and Spectra (University of Califormia Press, Berkeley, CA)
[4] Lu X, Zhong J Y, Li Y J and Zhang J, 2003 Physics of Plasma **10**, 2978
[5] Benredjem D, Kuba J, Möller C et al. 2003 Phys. Rev. A **67,** 043816

Inst. Phys. Conf. Ser. No 186
Paper presented at 9th Int. Conf. X-Ray Lasers, Beijing, China, 24–28 May 2004

Theoretical research on Ni-like Ag X-ray lasers at 13.9nm

Guoping Zhang, Tianxin Zhang and Wudi Zheng
Institute of Applied Physics and Computational Mathematics P. O. Box 8009, Beijing 100088, China

Abstract: A lot of experiments were done on Ni-like Ag X-ray Lasers at 13.9nm by some laboratory in the world. Simulation was done on these experiments by our series codes. The simulation results were compared to the experiments. According to simulation, the electronic density of the gain region for Ni-like Ag was about $7\text{-}2\times10^{19}cm^{-3}$, using lower prepulse as 1% and longer delay time as 3-4 ns between pre-main pulse, it was be got higher gain.

1. Introduction

The first saturated Ni-like x-ray laser (XRL) has been demonstrated by J. Zhang et al in RAL, at 13.9nm on Ni-like Ag [1,2]. Double targets opposite coupling with 150J driving laser energy was adopted, the output XRL energy was 90μJ. Optimization to prepulse intensity, a saturated XRL at 13.9nm on Ni-like Ag has been got by S. Sebban et al in ILE. Double targets opposite coupling with 280J driving laser energy was adopted, the output XRL energy was near 300μJ. Only using 30J driving laser energy and single 23mm flat target, a saturated XRL has also been got by R. Tommasini and Ch. Siegel et al. at 13.9 and 14.7nm on Ni-like Ag and Pd, respectively in IAP [4-6].

Simulation was done on these experiments by our series codes. The simulation results were consistent with the experiments. A system analysis on simulation results was given.

2. Simulation on experiment in RAL

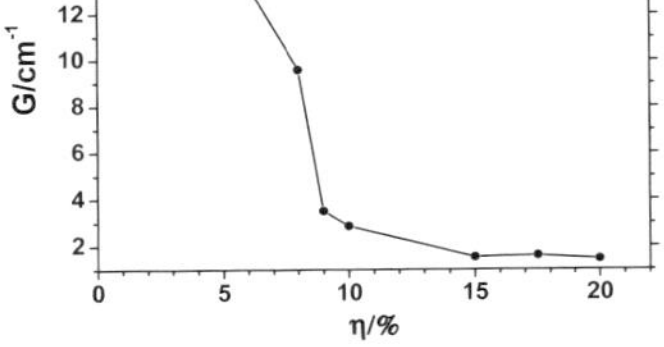

Fig. 1. Optimization to η.

In order to consider plasma aging, in simulation by XRL transport and amplify program, after XRL photon went 1ps, the plasma condition was be changed. Driving laser condition: wavelength is 1.053μm, duration τ is 75ps, line focus is 2.5cm long 200μm wide, intensity is $2.0\times10^{13}W/cm^2$, time interval between pre-main pulse $T_{1,2}$ is 2.2ns, prepulse intensity ratio η is 15-20%. The optimization to η have been done by our series codes, for 18mm flat target the

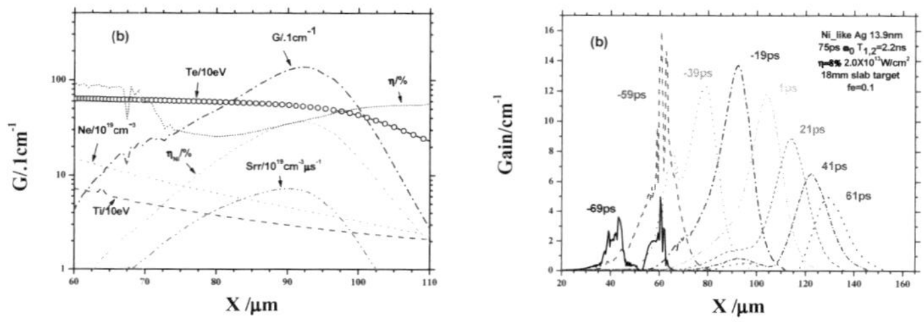

Fig. 2. (a) Status of gain region, (b) Evolvement of gain for model A7

effective gain G vs. η put in Fig. 1. The flux limit factor of electronic thermal conduct fe is 0.1. When η is 1-3%, G>16cm^{-1}, then G is decreasing as η increasing, as η >6%, decreasing very fast. When η is 8%, G is 9.6cm^{-1}, it is very near experiment results G=7.2±0.4cm^{-1}. When η is 15-20%, G is less then 2cm^{-1}. For η is 8% with model A7, status of gain region at 18ps before the peak of main pulse and evolvement of gain put in Fig. 2. At this time, XRL laser photo in output peak intensity of time distributing goes to the middle of 18mm target. The maximum of gain (G) Gmax is 14cm^{-1} at 92μm from target surface, gain region is 84-98μm, width of gain region wg is 14μm. Electronic and ionic temperature Te, Ti is 530, 28 eV, respectively, electronic density Ne is 4.3 × 10^{19} cm^{-3}, abundance of Ni-like ions η_{Ni} is 37%, at 92μm. According simulation, as η is 1-3%, G>16cm^{-1}, using single flat target saturated XRL could be got.

Driving laser energy was mainly deposited on the plasma with high density close to critical surface. In the gain region with low density the absorption to driving laser was weak; the electronic temperature was increased mainly by electronic thermal conduction. The radiation thermal conduction was faster than electronic thermal conduction. If the side radiation thermal conduction was not considered, it was resulted to over-ionization with low electronic temperature, to cause no gain. The electronic thermal conduction was gone slowly, the ionization of plasma was fast, so that the gain region was narrow, inside the plasma was over-ionization, and outside was under-ionization. For each ion, the gain lasted about 10ps. Gain region was moving outside. At total, the gain last about 80ps, the gain region was extended about 70μm. The moving angle for gain region θy was about 2.2mrad/cm. Due to the low gradation of electronic density -1.54×10^{22}cm^{-4}, the refraction angle ration of X-ray laser line dθ/dz was 1.32mrad/cm. In half of target, refracted angle θr is 1.2mrad, refracted distance is 5.3μm. In total the refracted angle θ=θy+θr is 3.4mrad, it is near 3.1mrad which given by series codes, in experiment θe is 4.5mrad. The X-ray laser line could be transported more than 2cm in the gain region and be got high amplification.

If the intensity of driving laser was not uniform in length direction of line focus, the gain region was not matched in time and in space, the amplification of X-ray laser line was be reduced. In two targets opposite coupling, the gain region was fast passed though each other, so that the efficiency of two targets opposite coupling was lower to that in the same direction coupling. Due to the movement outside of the gain region with big range, in the two targets opposite coupling, the output intensity of X-ray laser was not very sensitivity to vertical distance between two targets and to the time match for two driving laser beams. Rotated suitability angle as 4.4mrad for second target could be got higher output.

3. Simulation on experiment in ILE

Driving laser condition: wavelength is 1.053μm, τ is 100ps, line focus is 2.5cm long and 100μm wide, intensity is $5.0\times10^{13}W/cm^2$, $T_{1,2}$ is 3.0ns, η is 1-10%. The optimization to $T_{1,2}$ and η had been done by series codes, with 20mm flat target. fe is 0.03, The results were been put in Fig3. For 3ns $T_{1,2}$, there are two peaks, if η could precisely be controlled, just select 1%, otherwise select 5% as in ILE experiment. If 1% was been selected, $T_{1,2}$ 3.5-4.5ns was optimal. If 5% was been selected, $T_{1,2}$ 2.5ns was optimal.

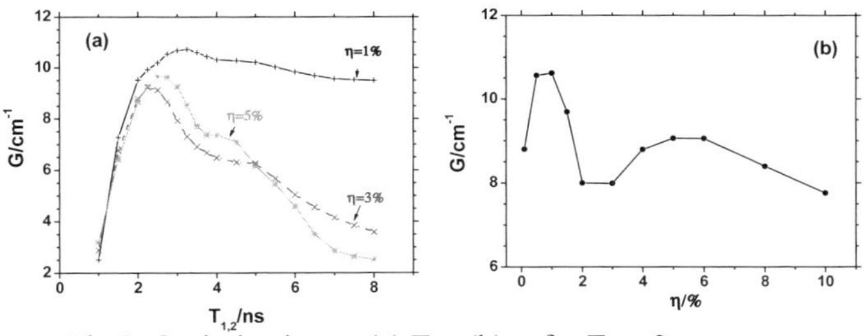

Fig 3. Optimization to (a) $T_{1,2}$ (b) η for $T_{1,2}$=3ns.

For model B1, η is 5% and $T_{1,2}$ is 3ns, at 89ps after peak of main pulse, status of gain region and evolvement of gain put in Fig. 4. Gmax is $10.6cm^{-1}$ at 116μm from target surface, gain region is 106-120μm, wg is 14μm; Te, Ti is 540, 32 eV, η_{Ni} is 48%, Ne is 2.1×10^{19} cm^{-3}, at 116μm. At total, the gain last about 120ps, the gain region was extended about 70μm. The moving angle for gain region θy was about 1.2mrad/cm. Gradation of Ne is $-1.60\times10^{22}cm^{-4}$, dθ/dz was 1.4mrad/cm, θr is 1.4mrad, refracted distance is 7.0μm. In total θ=θy+θr is 2.6mrad, it is near 2.1mrad which given by series codes, θe is ~5mrad. The X-ray laser line could be transported more than 2cm in the gain region and be got high amplification.

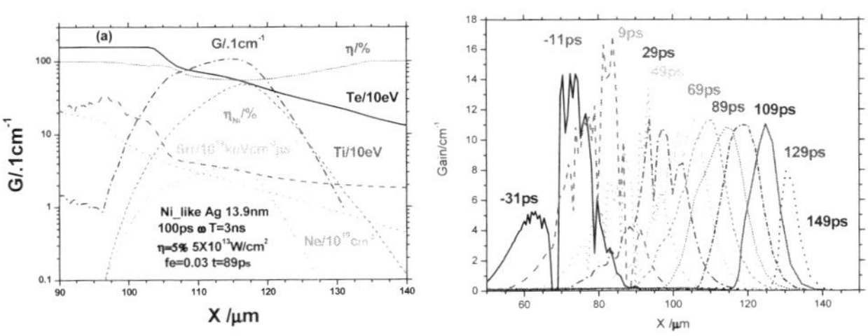

Fig. 4. (a) Status of gain region, (b) Evolvement of gain for

4. Simulation on experiment in IAPBE

Driving laser condition: wavelength is 1.053μm, τ is 100ps, line focus is 2.5cm long and 100μm wide, intensity is $1.2\times10^{13}W/cm^2$, $T_{1,2}$ is 1.0ns, η is 8%. The optimization to $T_{1,2}$ and η had been done by series codes, with 23mm flat target, fe is 0.3, The results were been put in Fig5. In

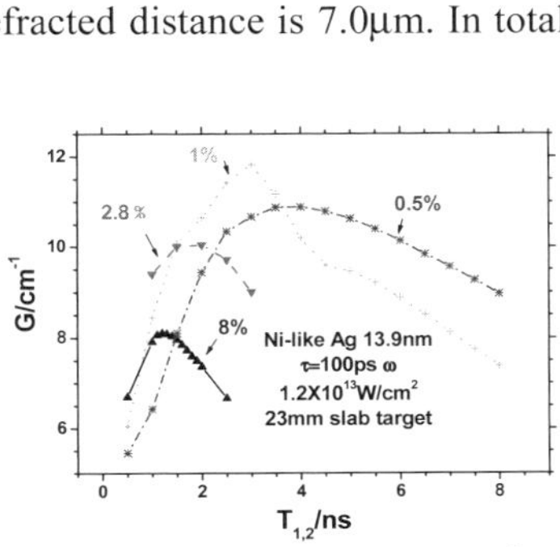

Fig 5.Optimization to $T_{1,2}$ and η.

simulation, for 8% η, optimal $T_{1,2}$ is 1.3ns, in experiment, optimal $T_{1,2}$ is 1.0ns with model C1. At 29ps after peak of main pulse, status of gain region and evolvement of gain put in Fig. 6. Gmax is 11cm^{-1} at 76μm from target surface, gain regin is 65-84μm, wg is 19μm; Te, Ti is 460, 36eV, Ne is 3.0×10^{19} cm^{-3}, η_{Ni} is 42%, at 76μm. At total, the gain last ~80ps,the gain region was extended about 70μm. θy was about 1.2mrad/cm. Gradation of Ne is -1.35×10^{22}cm^{-4}, dθ/dz was 1.2mrad/cm, θr is 1.6mrad, refracted distance is 7.7μm. In total θ=θy+θr is 2.8mrad, it is near 2.9mrad which given by series codes, θe is 1-2mrad. The X-ray laser line could be transported more than 2.4cm in the gain region and be got high amplification. The small signal G is 7.9 and 7.3cm^{-1}; small signal GL is 18.2 and 16.8 for simulation and experiment, respectively. In simulation, small signal output XRL energy is 24μJ.

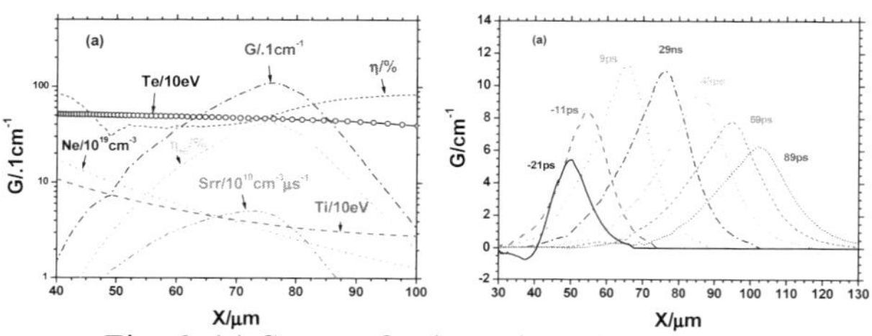

Fig. 6. (a) Status of gain region, (b) evolvement of gain for model C1

Only using single flat target and 30J driving energy, saturated has been got. If η is 1% and $T_{1,2}$ is 3ns, small signal G and GL with 11.8cm^{-1} and 27.1, respectively could be got. In this case, a deep saturated will be got.

In conclusion, Ne of the gain region for Ni-like Ag was about $7\text{-}2\times10^{19}$cm^{-3}, using lower η as 1% and longer $T_{1,2}$ as 3-4 ns, higher gain can be got. The gain region was fast moving outside, last 80-120ps, extend 70μm. The XRL refraction is low. XRL laser can be got high amplify in flat target with 2-3cm length. Following coupling can be got high amplify than opposite coupling. The match time of two pumping laser was not serious. Usually the output XRL beam was off axial, its shape and intensity was be changed shot to shot, optimization $T_{1,2}$, η and other factors go by near or far field. Driving laser intensity from 12 to 70TW/cm^2, effect gain bigger than 8cm^{-1} also can be got. High intensity results in little fe, wide line focus width affects side radiation thermal conduction; these are not good for XRL amplify. Adopt suitability intensity of driving laser and increasing in length of target could be got higher gain length product GL and higher output energy of X-ray laser. Of course, when length of target longer than 3cm, quasi traveling wave driving will be need.

This work was supported by Chinese national 863 high technology plan ICF field (grant No. AA847010) and national natural science fund (grant No. 69989801). This work was directed by Prof. Yu Min.

References

[1] J. Zhang, A.G. MacPhee, J. Nilsen, et al., Phys. Rev. Lett. 1997, **78**: 3856.
[2] J. Nilsen, J. Zhang, A.G. MacPhee,et al., Phys. Rev. A 1997,**56**: 56.
[3] S. Sebban, H. Daido, N. Sakaya, et al., Phys. Rev. A 2000, **61**: 3810
[4] R. Tommasini, F. Loewenthal, and J.E. Balmer, J. Opt. Soc. Am. B 1999, **16**: 1664.
[5] R. Tommasini, F. Loewenthal, and J.E. Balmer, Phys. Rev. 1999, **59**: 1577.
[6] Ch. Siegel, M. Braud, J.E. Balmer, et al., Opt. Commu.2002, **210**: 305.

Inst. Phys. Conf. Ser. No 186
Paper presented at 9th Int. Conf. X-Ray Lasers, Beijing, China, 24–28 May 2004

The Theoretical Research on the Ni-like Ta X-ray laser

Guoping Zhang, Tianxin Zhang and Wudi Zheng
Institute of Applied Physics and Computational Mathematics P.O.Box8009, Beijing, 100088, China

Abstract: To obtain saturated gain at spectral 'water window' is one of dreams in X-ray laser research, but now, its largest gain length product GL obtained is still 8 only. To Ni-like Ta and W, electron density should be around $2\times10^{21}cm^{-3}$, double frequency 2ω of driven laser is needed. In this paper, a new idea is proposed; using two driven laser beams combined one beam in prime frequency ω with another beam in 2ω. A design of long distance four targets series coupling is also presented; the X-ray laser produced by former two targets is collected by a pair of relay mirrors with grazing incidence, and then amplified by latter two targets. The theoretical research predicts that, taking advantage of this method, saturated gain of X-ray laser at spectral 'water window' can be got by Shenguang □facility. In 2002, the Ni-like Ta experiment was done in the Shenguang II facility. When the target was 2-3mm, the gain G at 4.4nm line was ~ 15.5±3.5 cm^{-1}. As length of target was increasing, the gain was decreasing. The maximum gain length product GL was 5.2. Different from simulation, the refraction of the X-ray laser line caused the gain decreasing.

1. Introduction

Until now at spectral 'water window', its largest gain length product GL obtained was still only 8 and 7 by LLNL to Ni-like Ta and W, respectively, 2ω driving exploding target[1]. In 1998 and 2000 Chinese and Japanese scientists in Gekko XII using 100ps duration pre-main driving slab target do same experiment. In 1998, using ω, very weak Ni-like Ta and Ni-like W XRL has been recorded[2]. In 2000, using 2ω, it is need 6 lines focus from 6 beams put together form one line focus, due to line focus badly quality and other problems, no laser line be recorded[3].

In this paper, a new idea is proposed; using two driven laser beams combined one beam in prime frequency with another beam in double frequency. A design of long distance four targets series coupling is also presented[4]. The simulation of the scheme was done by series codes. In 2002, the Ni-like Ta experiment was done in the Shenguang II facility. Its results are also being introduced.

2. Two driven laser beams combined one beam in prime frequency with another beam in double frequency

When the prime ω and double frequency 2ω combined driving target, 2ω let the plasma near ω critical density with low gradient Ne, increasing the absorption for ω, the energy of ω mainly deposit near its critical density. ω increase Te for plasma with Ne less then $10^{21}cm^{-3}$, 2ω increase Te for plasma with Ne bigger than $10^{21}cm^{-3}$, it is favorable for thermal electronic conduct transport ω energy to gain region. Avoid energy for ω beam loss for 2ω, in the end, Te increase about 30% to two beams with 2ω, so does G.

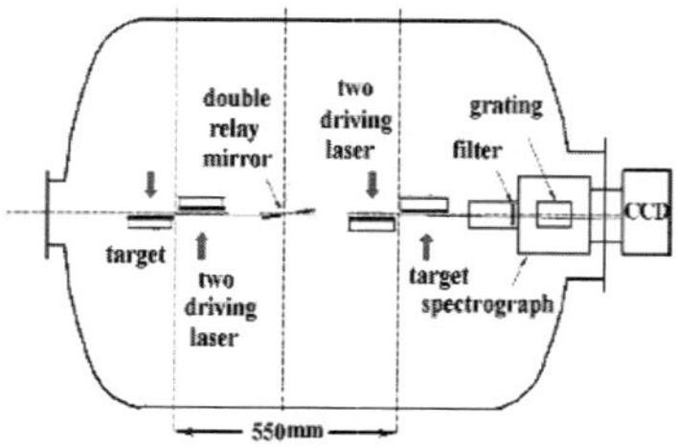

Fig. 1. Scheme diagram of four targets series coupling. (planform)

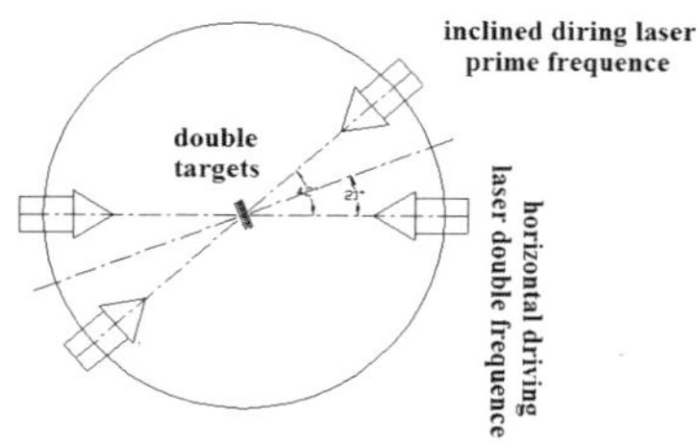

Fig.2. Scheme diagram of four targets series coupling. (side elevation)

3. Long-distance four targets series coupling experiment scheme

The experiment scheme put in Fig.1 and 2. Two laser beams drive one target, each beam shots on target oblique with 21° to the normal of target. Double targets opposite coupling are put 3mm away. Two double targets are put ~537mm away. Each target is 5mm long; four targets are 20mm long; line focus is 6mm long 100μm wide. The driving laser duration was 100ps. Output driving laser energy for each beam is 100J in ω and 50 J in 2ω, respectively. In the total, the power intensity on the target surface was about $2.5\times10^{14}W/cm^2$. Avoid attenuation 500-1000 from long distance coupling, the double relay mirrors focus XRL from double target to second.

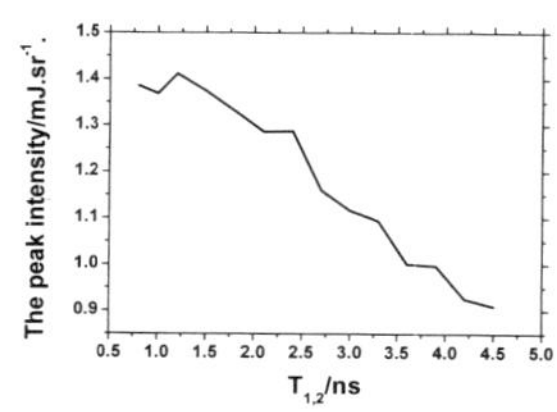

Fig.3 optimization on $T_{1,2}$ in single target.

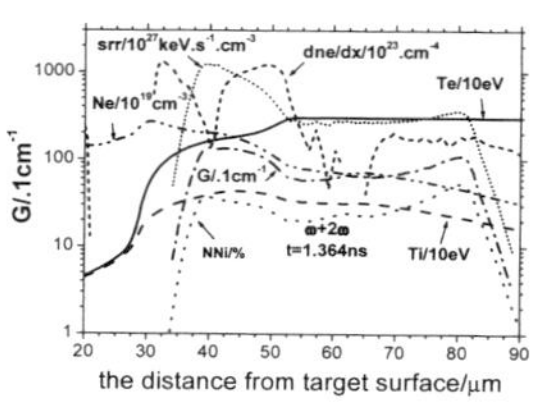

Fig.4 status of gain region

4. Optimization on experiment scheme by series codes

At first do optimize on single target, with prepulse 5%, the output peak intensity vs. pre-main relay time $T_{1,2}$ put in Fig.3. $T_{1,2}$ 1.2ns is best. The status of gain region put in Fig.4. In gain region, Ne is $(4.5\text{-}20)\times10^{20}cm^{-3}$.Te and Ti are 1.5-3 and 0.2-0.4 keV, respectively. η_{Ni} is~20-50%. Gain region from initial target surface is 38-82μm, with 44μm width. Gain evolvement put in Fig.6

About 10ps plasma in middle of gain region has been over-ionization, G has been decreased, and the peak of G has been moved to two directions. G_{max} ~15cm^{-1}. In the simulation, every 1ps the plasma condition has been changed, in this way the plasma aging was been simulated [4,5]. The output laser intensity in single, double and four targets are put in Fig.7. For single target refracted angle θ is 2.0mrad, divergence angle ϕ is 9.4mrad, peak intensity Ip is 1.4mJ/sr., output energy E_{out} is 0.10μJ. In double targets

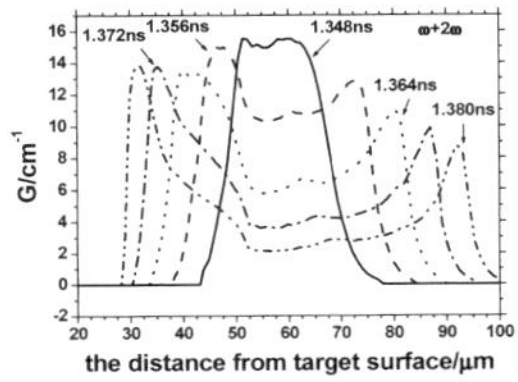

Fig. 6 gain evolvement.

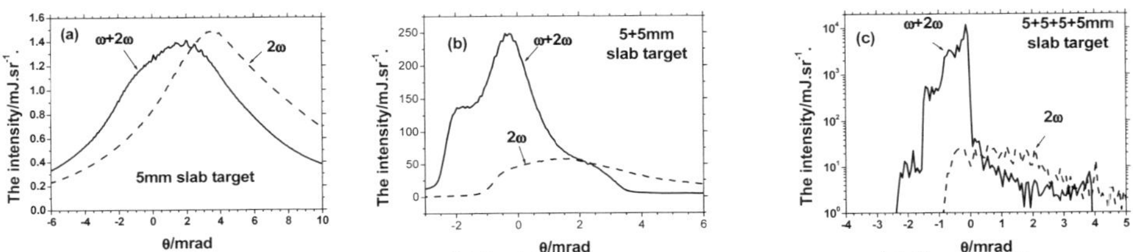

(a) single target (b)double targets (c) four targets

Fig.7. the intensity of X-ray laser vs. the refracted angle

opposite coupling, two target perpendicular distance $Y_{1,2}$ is 105μm, the angle between two targets $\theta_{1,2}$ is -3.75mrad, G is 10.9cm^{-1}, Ip has been amplify 176 to single target, E_{out} is 4.4μJ. In long distance four targets series coupling, $Y_{3,4}$ is 105μm and $\theta_{3,4}$ is -3.75mrad, $Y_{2,3}$ is -20μm, $\theta_{2,3}$ is -4.2mrad, GL is 22.0, E_{out} is 0.55mJ. For 8 beam are 2ω, the results are put in Fig.7 (dash line); in this case, $Y_{2,3}$ is 700μm, $\theta_{2,3}$ is -3.75mrad, GL is 15.2, E_{out} is 9.6μJ; in gain region, Te is down 30%, so is G, but its spontaneous radiation rate is high, in single target, Ip is a little higher than that of ω and 2ω joint driving. We suppose when GL≥16 and XRL energy transition efficiency≥10^{-6}, the XRL line has got saturated gain, four targets long distance coupling, in joint driving, XRL has got saturated gain, but in 2ω driving, XRL has not got saturated gain. If the efficiency of double relay mirrors can get 20-40 times amplification, a near saturated gain can be got in 2ω driving.

5. Experiment results

In 2002, the Ni-like Ta experiment was done in the Shenguang II facility. I vs. θ is put in Fig.8. When the target was 2-3mm, G at 4.4nm line was ~ 15.5±3.5 cm^{-1}. As length of target was increasing to 4, 5, 6, 8 and 10mm, the gain was decreasing to 11±2, 8±1, 7.4±1, 6.5±0.8and 4.8±0.5cm^{-1}, respectively. The maximum gain length product GL was 5.2. Different from simulation, θ was large. In most experiment, θ was beyond the region (±15mrad), which recorded by spectrograph. In experiment, to alter below condition: prepulse and main laser intensity, time interval between ω to 2ω and between targets, $T_{1,2}$ and $Y_{1,2}$; to improve uniformity of driving laser intensity along the line focus, the θ and ϕ of XRL is not be reduced. The refraction of the X-ray laser line caused the gain decreasing. When GL is only 5.2 for double targets coupling, long distance four targets

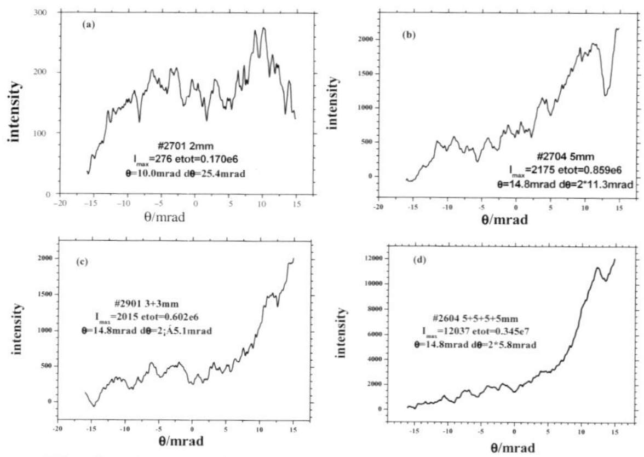

Fig.8. I vs. θ (a) L=2mm, (b) L=5mm, (c) L=3+3mm, (d) L=5+5+5+5mm

coupling is difficult to farther increasing output laser intensity than double targets coupling, even if the double relay mirrors have high efficiency.

In conclusion, the experiment on Ni-like Ta was done by ω and 2ω normal driving flat target, respectively. In despite of output XRL intensity at 4.48nm was weak, very large refracted angle was not been recorded in the experiment. So it was possible that non-symmetry in two driven laser beams caused non-symmetry distribution of Ne in direction of line focus width, resulted to X-ray laser beam gone out of gain region, gain was decreasing. In this case, if two driving laser beams normal irradiated the target and the reflected line focus system was adopted, refracted angle could be reduced and the higher GL could be got.

This work was supported by Chinese national 863 high technology plan ICF field (grant No. AA847010). This work was directed by Prof. Yu Min. Prof. Wang Shijie, Gu Yuan, Wang Guanlong and many personnel in experiment and shooting range from Shenguang II facility.

Reference

[1] MacGowan B J, Da Silva L B, Fields D J, et al., Phys. Fluids B **4**,2326(1992)

[2] Daido H, Sebban S, Sakaya T et al., Inst. Phys. Conf. Ser. No **159**,83

[3] Tang H, Daido H, Suzuki M, et al., Annual progress Report 2000, Insititue of laser Engineering, Osaka University, Japan, 23

[4] Zhang G, Zhang T and Wu J, Chinese Journal of Lasers, **A29,** Supplement, 26(2002)

[5] Zhang G, Zhang T and Wu J, High Power Laser and Particle Beams, **10**, 352 (1998)

Inst. Phys. Conf. Ser. No 186
Paper presented at 9th Int. Conf. X-Ray Lasers, Beijing, China, 24–28 May 2004

The Theoretical Research on the Ne-like Ge X-ray Laser

Guoping Zhang, Tianxin Zhang and Wudi Zheng
Institute of Applied Physics and Computational Mathematics
P.O.Box8009, Beijing, 100088, China

Abstract: The simulation of the Ne-like Ge experiments completed by the Institute of Shanghai Laser Plasma, MPQ, RAL and LLNL has been also completed by ours series codes. The results of simulation compared with experiments were also introduced. The optimization of Ne-like Ge X-ray laser at 19.6nm with prepulse in 100ps duration has done by ours series codes. The curved target and the double targets opposite coupling were also done.

1. Introduction

With duration ~100ps multi-pulses pumping Ni-like and Ne-like ions, the X-ray laser (XRL) with high gain, short duration, small divergence and high brightness was be produced [1-3]. It was suitable for XRL source of diagnostic laser plasma.

Series codes which can simulated all process that involved in XRL have been developed in our institute [4,5]. This series codes are buildup by one dimension non-equilibrium radiation hydrodynamics program JB19, one dimension reversion dynamics on Ne-like and Ni-like collisional exciting scheme program ALPHA and two dimension XRL propagation and small signal amplification in geometry optics paraxial approximate program XBPA. In order to test series codes, simulations of some experiments of Ne-like Ge XRL were completed, including domestic experiments driven with long-pulse (1ns) and international experiments driven with multiple-pulse [6]. Systemic optimum designs to Ne-like Ge XRL driven by pre-main short pulses are conducted by our series codes, the two targets opposite coupling and the curved target are also be researched [7].

2. Simulation compare to data from ISLP experiment with 1ns duration [8,9]

The driven laser condition is below: wavelength is 1.053μm, line focus is 100μm wice, and power density is 10^{13}W/cm^2. At 23.2 and 23.6nm two laser lines, for double target following coupling with 14mm length for each target, the optimal angle between two targets is 4.5mrad. According to our simulation, at 19.6, 23.2 and 23.6nm three laser lines, the optimal angle is 6.0, 7.0 and 6.0mrad, respectively. In this condition, output peak intensity is increasing to 6, 11 and 6% than that of 4.5mrad, respectively. At these three lines, Gain (G) is 2.82, 3.48 and 3.47/cm, respectively. In three targets opposite coupling with 14mm length for each target, the optimal perpendicular distance between second and third target is 250μm in the experiment. At these three lines, the optimal distances are 220, 250 and 260μm, respectively. For 22mm single target, at 23.6nm line the refracted and divergence angle are same as 11mrad, according our simulation there are 11.6 and 11.3mrad, respectively.

According to simulation, with long duration irradiated slab target a stable ablation is formed, gain region can be supply continuous from high density, the width of gain region is bigger than 100μm, gain can be last 800ps, it is suitable to multiple targets series coupling and using multiplayer mirror to get double amplification, ununiformity in driven laser intensity also has a little influence on the XRL laser line amplification.

3. Simulation compare to data from RAL experiment [1]

We simulated 15 models on Ne-like Ge scheme doing by RAL. The output XRL energy at 19.6nm laser line is put in Table 1 from experiment and simulation. E_{xrl}, E_{xrlu} and E_{xrll} are XRL output energy for expected value, upper limit and lower limit in the experiment, respectively. E_{xrls} is the XRL output energy for simulation. For 6 of 15 models, $E_{xrll}<E_{xrls}<E_{xrlu}$, for 5 models, $E_{xrll}/3.5<E_{xrls}<E_{xrll}$, for 2 model, $E_{xrlu}<E_{xrls}<3\times E_{xrlu}$, for 2 models, $E_{xrls}\approx E_{xrlu}/11$. Last two models also concerned with seriously over-ionized recombination plasmas.

Using pre-pulses with short duration, output energy at 19.6 nm laser line is exceeding that at 23.2nm laser line; latter exceed that at 23.6nm laser line. In these sides, both results from experiment and simulation are same.

Table 1. XRL Output energy at 19.6nm laser line in experiment at RAL.

No	R1	R2	R3	R4	R5	R6	R7	R8	R9	R10	R11	R12	R13	R14	R15
N_p	1	1	2	2	2	4	2	4	2	2	2	2	2	2	2
E_l(J)	125	337	320	100	234	146	134	140	131.2	131.2	131.2	126.3	137.5	187.5	312.5
τ(ps)	100	600	600	100.	100	100	100	100	100	100	100	100	100	100	100
dt_{12}(ps)			3200	400	800	800	2000	400	400	800	2000	800	800	800	800
dt_{23}(ps)						1200		1600							
dt_{34}(ps)						800		400							
η_1(%)			2	75	150	5	10	2.5	5	5	5	1	10	50	150
η_2(%)						5		5							
η_3(%)						100		50							
E_{xrlu}(nJ)	4.2	30	290	3000	5300	6500	18000	70000	38	340	1800	275	980	3400	4200
E_{xrl}(nJ)	3.0	20	180	1700	3600	3500	11000	40000	25	220	1050	225	850	2800	3600
E_{xrll}(nJ)	2.2	8.0	70	800	1800	480	4200	10000	13	110	450	175	660	2300	2800
E_{xrls}(nJ)	3.51	85.6	212	73.7	527	527	14500	902	20.8	205	2621	67.1	448	916	1423

*18mm flat target, No. is number of experiment, N_p is number of pulses, E_l is driving laser energy on the target surface,τ is driving laser duration. dt_{12}, dt_{23} and dt_{34} are delay time for first pulse to second one and so on, η_n is driving laser intensity for the *n*th pulse (the last one is 100%).

4. Simulation compare to data from MPQ experiment [2]

The driving laser condition is below: wavelength is 1.315μm, duration is 450ps, delay time is 5.23ns, power density is 16TW/cm^2. Target is 28mm long. At 19.6nm laser line the relative output XRL energy E_{out} be put in Table 2. $E_{out,exp}$ and $E_{out,sim}$ are relative output XRL energy in experiment and simulation at 19.6nm, respectively. η is driving laser intensity for the prepulse (main pulse is 100%). Both experiment and simulation choose E_{out} from single pulse driving flat target results as 1.2×10^4. From Table 2, $E_{out,exp}/E_{out,sim}\approx$ 5 to 8 for M1 models, $E_{out,sim}/E_{out,exp}\approx5.6$ for M4 model with 5mrad/cm. In others, $E_{out,exp}/4<E_{out,sim}<4\times E_{out,exp}$. As η and the bending of curved target increasing, E_{out} also

No		M1	M2	M3	M4	M5
η(%)		15	1.65	0.165	0.087	No
	flat target	5.2×10^6	6×10^5	1.4×10^4	9×10^3	1.2×10^4
$E_{out,exp(a.u)}$	5mrad/cm	8.0×10^6	5×10^5	2.5×10^5	3.3×10^4	9×10^3
	10mrad/cm	1.3×10^7	4×10^6	2.4×10^6	5.5×10^5	7.5×10^4
	flat target	8.47×10^5	2.47×10^5	4.26×10^4	3.22×10^4	1.20×10^4
$E_{out,sim(a.u)}$	5mrad/cm	1.43×10^6	7.00×10^5	2.08×10^5	1.84×10^5	2.07×10^4
	10mrad/cm	1.61×10^6	1.27×10^6	6.10×10^5	5.59×10^5	5.70×10^4

Table 2. Output XRL energy at 19.6nm laser line

increasing, it is same for experiment and simulation.

5. Simulation compare to data from LLNL experiment [2]

The driving laser condition is below: power intensity is 110TW/cm², wavelength is 0.53μm, duration is 100ps, three pulses are used, each delays 400ps. Target is 25.2mm long. The near field distribution for two and three pulses at 19.6nm is measured in experiment. The peak in the near field is at 140-160μm and 130-200μm from target surface for second and third pulse, respectively. The intensity from third pulse is 10 times as that of second one. Simulation from LLNL indicated that the peak of gain is 50-60μm from target surface. Due to refraction the peak move to 150μm [2]. In our simulation the peak of gain is also at 50-60μm. Due to refraction the peak move to 160μm. Output XRL energy in the third pulse is only 30% that in the second one. The plasma is seriously over ionization in the second pulse. For seriously over-ionized recombination plasmas, the recombination in simulation is slow. In the third pulse plasma is also over ionization, the abundance of Ne-like ions is low, so does the gain.

6. Optimization on Ne-like Ge

There are 8 beams in Shenguang II. At 100ps duration, output laser energy is 100J for each beam. The line focus is 100μm wide and 18mm long. The flat target is 16mm long. On the target surface, the intensity is 4×10^{13}W/cm^2. At first fix η, we do optimization to $T_{1,2}$. For 16mm flat target, in the small signal, the output peak intensity in time integral I vs. $T_{1,2}$ is given in Fig. 1(a). Then fix $T_{1,2}$, we do optimization to η, The I vs. η is given in Fig. 1(b). M1 model with η of 3% and $T_{1,2}$ of 7.5ns is best. The status of gain region is given in Fig. 2(a) at 22ps after the peak of main pulse. At this time XRL photon goes on the middle of target for output peak intensity. Ne is electronic density. η_{Ne} is abundance of Ne-like ions Te, Ti is electronic and ionic temperature, respectively. Srr is spontaneous radiation rate at 19.6nm. G decreases as η_{Ne} does. For high Ne, plasma is Over ionization. For low Ne, plasma is under ionization. The evolvement of gain at 19.6nm laser line is given in Fig. 2(b). The -18ps means 18ps before the peak of main pulse. The gain region moves outside continuous. The peak of gain bigger then 10/cm at -18

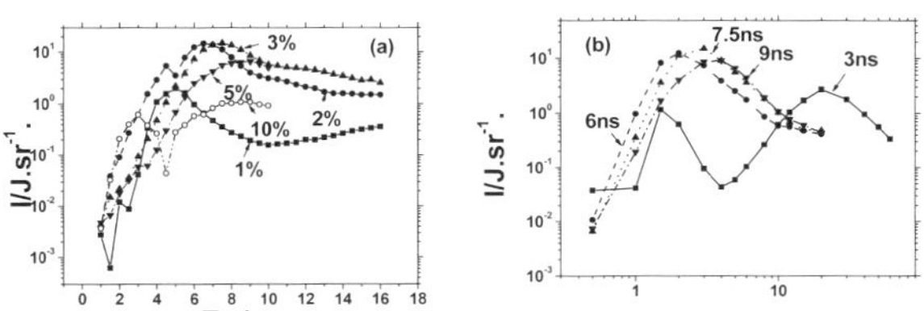

Fig.1. Optimization for Ne-like Ge at 19.6 nm laser line.

to 72ps, it lasts 90ps. The wide of gain region wider than 40μm at 2 to 72ps, it lasts 80ps. For 16mm single target, the small signal G is 11.8/cm at 19.6 nm laser line; the small signal GL is 18.9. Saturated gain can be got only using single target. In double target opposite coupling, the optimal perpendicular distance between two targets is 200μm. The small signal GL is 38.4. Deeply saturated gain can be got. For 16mm curved target, when bending is 6mrad/cm, the output peak intensity is strongest, it is 3.76 times to that of flat target. The small signal G is 13.3/cm. The small signal GL is 21.3.

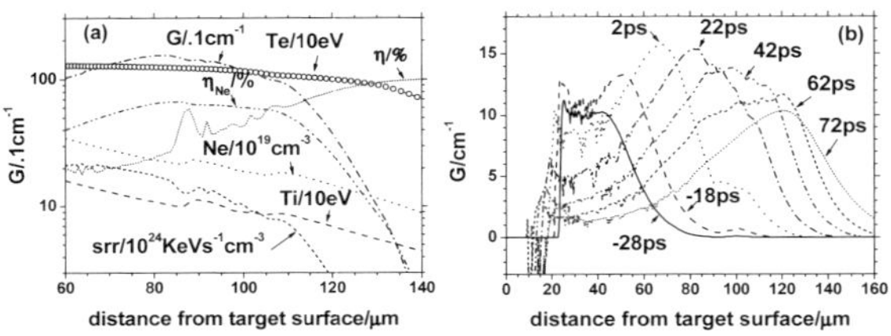

Fig.2. (a) The Status of Gain Region and (b) The Evolvement of Gain

In summary, in the case of long-pulse, at three lines of 19.6, 23.2 and 23.6nm, theoretical results are mainly consistent with experimental data. And in the case of multiple-pulse, theoretical simulations are also consistent with experiments except for the experiments being concerned with seriously over-ionized recombination plasmas. Adopting driven conditions of 2~3% pre-pulse, 6~8ns pre-main pulse interval and 4×10^{13}W/cm^2 power intensity, a gain area of 19.6nm laser line beyond 60μm and gain duration of 90ps can be obtained. Furthermore, with 16mm flat target, gain gets to 11.8/cm, and if 6mrad/cm curved target is adopted, gain reaches to 13.3/cm, and small signal gain-length product GL of single target is about 21.3, so saturated gain can be realized by a single target. With double targets opposite coupling, GL can reach to 38.4, deep saturation can be gotten, and XRL source required by demonstration of application is out of question.

This work was supported by Chinese national 863 high technology plan ICF field (grant No. AA847010) and national natural science fund (grant No. 69989801) This work was directed by Prof. Yu Min.

References

[1] A. G. MacPhee, A. Behjat, G. F. Cairns et al., Inst. Phys. Conf. Ser. No **151**: Section 7 Paper presented at X-ray Lasers 1996 conf., Lund, Sweden: 250

[2] J. Nilsen , J. C. Moreno, L. B. Da Silva et al., Inst. Phys. Conf. Ser. No **151**: Section 7 Paper presented at X-ray Lasers 1996 conf., Lund, Sweden: 240

[3] Zhang J, A. G. MacPhee, J. Lin, et al., *Science,* 1997,**276**:1097

[4] Zhang G P, Zhang T X and Wu J Z, Inst. Phys. Conf. Ser. No **159**: Section 6 Paper presented at X-ray Lasers 1998 conf., Kyoto, Japan: 391

[5] Zhang G P, Sheng J T, Yang M L, et al., High Power Laser and Particle Beams, 1992,**4**:521

[6] Zhang G P, Zhang T X and Zheng W D, High Power Laser and Particle Beams, 2004,**16:**35

[7] Zhang G P, Zhang T X and Zheng W D, High Power Laser and Particle Beams, 2004,**16:**171

[8] Wang S J, Gu Y, Fu S Z, et al., Science in China(series A), 1991,**34**:1388

[9] Wang S J, Gu Y, Zhou G L, et al., High Power Laser and Particle Beams, 1992,**4**:165

Inst. Phys. Conf. Ser. No 186
Paper presented at 9th Int. Conf. X-Ray Lasers, Beijing, China, 24–28 May 2004

Energy levels and radiative rates for transitions in Fe XVII

K M Aggarwal, F P Keenan and R Kisielius*
Department of Pure and Applied Physics, Queen's University, Belfast BT7 1NN, UK

Abstract. Energies of the lowest 157 levels belonging to the ($1s^2$)$2s^22p^6$, $2s^22p^53l$, $2s^22p^54l$, $2s^22p^55l$, $2s2p^63l$, $2s2p^64l$ and $2s2p^65l$ configurations of Fe XVII have been calculated using GRASP code [1]. Additionally, radiative rates, oscillator strengths, and line strengths are calculated for all electric dipole (E1), magnetic dipole (M1), electric quadrupole (E2), and magnetic quadrupole (M2) transitions among these levels. Results for energy levels, radiative rates (for transitions to the ground state), and lifetimes are reported here.

1. Introduction

Emission lines of Ne-like Fe XVII have been observed in a variety of plasmas including astrophysical, laser produced, magnetically confined, Z-pinch and EBIT. To interpret the vast amount of observational data, atomic parameters such as energy levels, radiative rates, and collision strengths are required. Since there is a paucity of experimental data for these parameters, theoretical results are of vital importance. Therefore, in this work we report our results for energy levels and radiative rates for transitions among the lowest 157 levels of the ($1s^2$)$2s^22p^6$, $2s^22p^53l$, $2s^22p^54l$, $2s^22p^55l$, $2s2p^63l$, $2s2p^64l$ and $2s2p^65l$ configurations of Fe XVII. This is an extension of our earlier work [2], which was confined to 89 levels among the $2s^22p^6$, $2s^22p^53l$, $2s^22p^54l$, $2s2p^63l$, and $2s2p^64l$ configurations. To obtain results, we have adopted the GRASP code [1], as in our earlier work.

2. Results

Our calculated energy levels are listed in Table 1. Detailed comparisons with other available theoretical and experimental results have already been made in our earlier papers [2,3]. Experimental energies for all levels are not available, but the agreement is better than 1% for common levels. Furthermore, level orderings are the same in both theoretical and experimental results, which is highly satisfying. Similarly, apart from the additional 68 levels included in the present work, the agreement with the previous results [2] is better than 1%.

---

* Associate research fellow, Vilnius University Research Institute of Theoretical Physics and Astronomy, A. Goštauto 12, 01108 Vilnius, Lithuania

Table 1. Energy levels, radiative rates, and lifetimes for Fe XVII. ($a \pm b \equiv a \cdot 10^{\pm b}$)

Level	E (Ry)	A (s^{-1})	τ (s)	Level	E (Ry)	A (s^{-1})	τ (s)
$2s^22p^6\ ^1S_0$	0.000			$2s^22p^54p\ ^3P_0$	72.591	0.00+00	2.75-12
$2s^22p^53s\ ^3P_2$	53.170	2.22+05	4.50-06	$2s^22p^54s\ ^3P_1$	72.615	2.47+11	1.62-12
$2s^22p^53s\ ^1P_2$	53.314	9.52+11	1.05-12	$2s^22p^54p\ ^3D_2$	72.653	1.98+08	2.96-12
$2s^22p^53s\ ^3P_0$	54.099	0.00+00	6.35-05	$2s^22p^54p\ ^3D_3$	72.072	0.00+00	3.10-12
$2s^22p^53s\ ^3P_1$	54.190	8.36+11	1.20-12	$2s^22p^54p\ ^1P_1$	72.732	3.08+00	2.96-12
$2s^22p^53p\ ^3S_1$	55.395	1.80+05	2.84-10	$2s^22p^54p\ ^3P_2$	72.776	1.91+08	3.18-12
$2s^22p^53p\ ^3D_2$	55.664	5.15+08	1.77-10	$2s^22p^54p\ ^3P_0$	73.063	0.00+00	3.35-12
$2s^22p^53p\ ^3D_3$	55.780	0.00+00	1.56-10	$2s^22p^54p\ ^3D_1$	73.565	4.60+00	2.89-12
$2s^22p^53p\ ^1P_1$	55.868	7.13+03	1.65-10	$2s^22p^54p\ ^3P_1$	73.652	7.68+04	3.14-12
$2s^22p^53p\ ^3P_2$	55.999	5.57+08	1.19-10	$2s^22p^54p\ ^1D_2$	73.668	2.32+08	3.11-12
$2s^22p^53p\ ^3P_0$	56.410	0.00+00	1.10-10	$2s^22p^54d\ ^3P_0$	73.813	0.00+00	1.77-12
$2s^22p^53p\ ^3D_1$	56.553	4.17+03	2.01-10	$2s^22p^54d\ ^3P_1$	73.845	5.66+10	1.61-12
$2s^22p^53p\ ^3P_1$	56.789	2.05+05	1.36-10	$2s^22p^54d\ ^3F_4$	73.885	0.00+00	1.77-12
$2s^22p^53p\ ^1D_2$	56.817	6.72+08	1.34-10	$2s^22p^54d\ ^3P_2$	73.898	3.01+06	1.79-12
$2s^22p^53p\ ^1S_0$	57.942	0.00+00	3.46-11	$2s^22p^54d\ ^3F_3$	73.908	0.00+00	1.80-12
$2s^22p^53d\ ^3P_0$	58.776	0.00+00	8.50-11	$2s^22p^54p\ ^1S_0$	73.915	0.00+00	3.40-12
$2s^22p^53d\ ^3P_1$	58.847	9.23+10	9.61-12	$2s^22p^54d\ ^1D_2$	73.954	8.54+05	1.82-12
$2s^22p^53d\ ^3P_2$	58.984	6.12+06	8.67-11	$2s^22p^54d\ ^3D_3$	73.982	0.00+00	1.81-12
$2s^22p^53d\ ^3F_4$	58.991	0.00+00	8.86-11	$2s^22p^54d\ ^3D_1$	74.177	5.75+12	1.59-13
$2s^22p^53d\ ^3F_3$	59.052	0.00+00	7.65-11	$2s^22p^54f\ ^3D_1$	74.462	2.33+02	7.40-13
$2s^22p^53d\ ^1D_2$	59.182	1.12+06	7.31-11	$2s^22p^54f\ ^1G_4$	74.463	0.00+00	7.94-13
$2s^22p^53d\ ^3D_3$	59.263	0.00+00	7.85-11	$2s^22p^54f\ ^3G_5$	74.464	0.00+00	7.91-13
$2s^22p^53d\ ^3D_1$	59.613	5.89+12	1.69-13	$2s^22p^54f\ ^3D_2$	74.473	7.43+08	7.55-13
$2s^22p^53d\ ^3F_2$	59.978	4.45+05	7.41-11	$2s^22p^54f\ ^3F_3$	74.498	0.00+00	7.70-13
$2s^22p^53d\ ^3D_2$	60.037	2.67+05	8.01-11	$2s^22p^54f\ ^1D_2$	74.502	1.86+09	7.97-13
$2s^22p^53d\ ^1F_3$	60.078	0.00+00	8.24-11	$2s^22p^54f\ ^1F_3$	74.504	0.00+00	7.92-13
$2s^22p^53d\ ^1P_1$	60.637	2.48+13	4.04-14	$2s^22p^54f\ ^3F_4$	74.510	0.00+00	7.96-13
$2s2p^63s\ ^3S_1$	63.212	1.94+04	6.25-12	$2s^22p^54d\ ^3F_2$	74.829	1.29+05	1.80-12
$2s2p^63s\ ^1S_0$	63.699	0.00+00	9.17-12	$2s^22p^54d\ ^3D_2$	74.848	4.47+05	1.79-12
$2s2p^63p\ ^3P_0$	65.634	0.00+00	6.79-12	$2s^22p^54d\ ^1F_3$	74.871	0.00+00	1.79-12
$2s2p^63p\ ^3P_1$	65.667	4.07+11	1.80-12	$2s^22p^54d\ ^1P_1$	75.036	7.00+12	1.33-13
$2s2p^63p\ ^3P_2$	65.837	8.48+05	6.67-12	$2s^22p^54f\ ^3G_3$	75.409	0.00+00	7.91-13
$2s2p^63p\ ^1P_1$	65.980	3.32+12	2.88-13	$2s^22p^54f\ ^3G_4$	75.417	0.00+00	7.95-13
$2s2p^63d\ ^3D_1$	68.920	2.10+03	8.56-12	$2s^22p^54f\ ^3D_3$	75.426	0.00+00	7.69-13
$2s2p^63d\ ^3D_2$	68.930	1.87+07	8.38-12	$2s^22p^54f\ ^3F_2$	75.427	1.27+09	7.87-13
$2s2p^63d\ ^3D_3$	68.949	0.00+00	8.19-12	$2s^22p^55s\ ^3P_2$	79.605	1.04+05	3.56-12
$2s2p^63d\ ^1D_2$	69.325	1.09+10	7.03-12	$2s^22p^55s\ ^1P_1$	79.629	2.11+11	1.97-12
$2s^22p^54s\ ^3P_2$	71.660	1.15+05	2.76-12	$2s^22p^55p\ ^3S_1$	80.063	5.95+04	3.69-12
$2s^22p^54s\ ^1P_1$	71.707	3.49+11	1.39-12	$2s^22p^55p\ ^3D_2$	80.099	3.82+08	3.69-12
$2s^22p^54s\ ^3S_1$	72.587	1.22+05	3.05-12	$2s^22p^55p\ ^3D_3$	80.123	0.00+00	3.91-12

Table 1. (continued)

Level	E (Ry)	A (s^{-1})	τ (s)
$2s^22p^55p\ ^1P_1$	80.138	2.16+02	3.52-12
$2s^22p^55p\ ^3P_2$	80.159	3.46+08	3.94-12
$2s^22p^55p\ ^1S_0$	80.325	0.00+00	3.94-12
$2s^22p^55s\ ^3P_0$	80.536	0.00+00	3.54-12
$2s^22p^55s\ ^3P_1$	80.547	2.49+11	1.83-12
$2s^22p^55d\ ^3P_0$	80.665	0.00+00	2.39-12
$2s^22p^55d\ ^3P_1$	80.681	2.19+10	2.27-12
$2s^22p^55d\ ^3F_4$	80.699	0.00+00	2.42-12
$2s^22p^55d\ ^3P_2$	80.707	1.73+06	2.39-12
$2s^22p^55d\ ^3F_3$	80.709	0.00+00	2.41-12
$2s^22p^55d\ ^1D_2$	80.731	5.73+05	2.40-12
$2s^22p^55d\ ^3D_3$	80.745	0.00+00	2.39-12
$2s^22p^55d\ ^1P_1$	80.855	4.15+12	2.17-13
$2s^22p^55f\ ^3D_1$	80.976	3.70+02	1.46-12
$2s^22p^55f\ ^3D_2$	80.984	3.66+08	1.45-12
$2s^22p^55f\ ^3G_5$	80.987	0.00+00	1.49-12
$2s^22p^55f\ ^1G_4$	80.987	0.00+00	1.49-12
$2s^22p^55f\ ^3F_3$	81.000	0.00+00	1.45-12
$2s^22p^55f\ ^1D_2$	81.006	1.10+09	1.48-12
$2s^22p^55f\ ^1F_3$	81.007	0.00+00	1.47-12
$2s^22p^55f\ ^3G_4$	81.010	0.00+00	1.47-12
$2s^22p^55g\ ^3F_2$	81.010	3.97+00	2.81-12
$2s^22p^55g\ ^3F_3$	81.012	0.00+00	2.82-12
$2s^22p^55g\ ^1H_5$	81.019	0.00+00	2.80-12
$2s^22p^55g\ ^3H_6$	81.021	0.00+00	2.80-12
$2s^22p^55g\ ^3G_3$	81.024	0.00+00	2.81-12
$2s^22p^55g\ ^3G_4$	81.026	0.00+00	2.81-12
$2s^22p^55g\ ^1G_4$	81.030	0.00+00	2.80-12
$2s^22p^55g\ ^3G_5$	81.032	0.00+00	2.80-12
$2s^22p^55p\ ^3D_1$	81.022	5.39+03	3.43-12
$2s^22p^55p\ ^3P_1$	81.054	3.20+04	3.72-12
$2s^22p^55p\ ^1D_2$	81.073	3.71+08	3.85-12
$2s^22p^55p\ ^3P_0$	81.164	0.00+00	3.82-12
$2s2p^64s\ ^3S_1$	81.593	1.67+04	2.06-12
$2s^22p^55d\ ^3F_2$	81.636	5.36+04	2.42-12
$2s^22p^55d\ ^3D_2$	81.644	3.64+05	2.38-12
$2s^22p^55d\ ^1F_3$	81.657	0.00+00	2.40-12
$2s^22p^55d\ ^3D_1$	81.729	3.44+12	2.57-13
$2s2p^64s\ ^1S_0$	81.746	0.00+00	2.01-12
$2s^22p^55f\ ^3G_3$	81.926	0.00+00	1.48-12
$2s^22p^55f\ ^3D_3$	81.930	0.00+00	1.45-12
$2s^22p^55f\ ^3G_4$	81.931	0.00+00	1.49-12
$2s^22p^55f\ ^3F_2$	81.934	7.07+08	1.47-12
$2s^22p^55g\ ^1F_3$	81.954	0.00+00	2.82-12
$2s^22p^55g\ ^3F_4$	81.955	0.00+00	2.81-12
$2s^22p^55g\ ^3H_4$	81.957	0.00+00	2.80-12
$2s^22p^55g\ ^3H_5$	81.959	0.00+00	2.80-12
$2s2p^64p\ ^3P_0$	82.552	0.00+00	2.15-12
$2s2p^64p\ ^3P_1$	82.564	2.39+11	1.42-12
$2s2p^64p\ ^3P_2$	82.632	6.44+05	2.22-12
$2s2p^64p\ ^1P_1$	82.677	1.65+12	4.75-13
$2s2p^64d\ ^3D_1$	83.788	1.12+03	1.44-12
$2s2p^64d\ ^3D_2$	83.792	8.25+06	1.43-12
$2s2p^64d\ ^3D_3$	83.801	0.00+00	1.43-12
$2s2p^64d\ ^1D_2$	83.930	2.99+09	1.47-12
$2s2p^64f\ ^3F_2$	84.352	2.52-01	7.19-13
$2s2p^64f\ ^3F_3$	84.354	0.00+00	7.19-13
$2s2p^64f\ ^3F_4$	84.358	0.00+00	7.19-13
$2s2p^64f\ ^1F_3$	84.368	0.00+00	7.31-13
$2s2p^65s\ ^3S_1$	89.493	5.09+03	2.32-12
$2s2p^65s\ ^1S_0$	89.574	0.00+00	2.35-12
$2s2p^65p\ ^3P_0$	89.979	0.00+00	2.46-12
$2s2p^65p\ ^3P_1$	89.985	1.86+11	1.68-12
$2s2p^65p\ ^3P_2$	90.019	5.29+05	2.53-12
$2s2p^65p\ ^1P_1$	90.041	1.17+12	6.25-13
$2s2p^65d\ ^3D_1$	90.580	3.38+02	1.81-12
$2s2p^65d\ ^3D_2$	90.583	8.03+06	1.80-12
$2s2p^65d\ ^3D_3$	90.587	0.00+00	1.79-12
$2s2p^65d\ ^1D_2$	90.650	2.42+09	1.76-12
$2s2p^65f\ ^3F_2$	90.862	1.97+00	1.25-12
$2s2p^65f\ ^3F_3$	90.863	0.00+00	1.25-12
$2s2p^65f\ ^3F_4$	90.864	0.00+00	1.25-12
$2s2p^65f\ ^1F_3$	90.872	0.00+00	1.26-12
$2s2p^65g\ ^3G_3$	90.891	0.00+00	2.08-12
$2s2p^65g\ ^3G_4$	90.891	0.00+00	2.08-12
$2s2p^65g\ ^3G_5$	90.893	0.00+00	2.08-12
$2s2p^65g\ ^1G_4$	90.893	0.00+00	2.08-12

In the present work, we have calculated transition energies, radiative rates, oscillator strengths, and line strengths for all 3401 electric dipole (E1), 3334 magnetic dipole (M1), 4439 electric quadrupole (E2), and 4503 magnetic quadrupole (M2) transitions among the 157 levels of Fe XVII. The detailed results for all transitions, and comparisons with earlier work, are available from our recent publication [3]. Here in Table 1, we report results for only transitions to the ground state. It may be noted that E1 (allowed and intercombination) transitions are the most important for applications, but data for other transitions are also required for modelling calculations. Additionally, the radiative rates listed in Table 1 are the sum over all types of transitions, although for a majority of them only one type of transitions dominates.

A comparison made between our earlier [2] and present results for transitions among common levels, shows an agreement of better than 20% for all strong transitions with $f \geq 0.01$, and the only exception is the $2s^22p^54p\ ^3P_2$ – $2s^22p^54d\ ^1P_1$ (47–71) intercombination transition, for which the present results are lower by 30%. This highly satisfactory agreement between the two sets of results confirms that A-values have converged within the n = 5 configurations, and the present results are accurate to better than 20% for a majority of strong transitions. However, for weaker transitions the results may be comparatively less accurate.

Also listed in Table 1 are lifetimes for all excited levels of Fe XVII. No other similar theoretical or experimental results are available to compare with, but we hope the presently listed lifetimes will be useful for future comparisons. Similarly, calculations for collision strengths (Ω) and excitation rate coefficients are in progress, but will take a considerably long time for completion, because we have employed the fully relativistic Dirac Atomic R-matrix Code (DARC) of Norrington and Grant [4] (*http://www.am.qub.ac.uk/DARC*). Because of the inclusion of fine-structure in the definition of channel coupling, the matrix size of the Hamiltonian increases manyfold, making the calculations computationally challenging. Apart from this, in order to account for resonances, calculations for Ω need to be performed at several thousand energy points (in a fine energy grid) in thresholds region. However, we hope to conclude our calculations in near future, which in combination with the present results may resolve some of the discrepancies between theory and experiments, especially for the prominent X-ray lines of Fe XVII, namely 3C, 3D, 3E, 3F, 3G and 3H at wavelengths of ~ 15.0 Å and ~ 17.0 Å.

Acknowledgment

This work has been financed by the EPSRC and PPARC of UK, and FPK is grateful to AWE Aldermaston for the award of a William Penny Fellowship.

References

[1] Dyall K G, Grant I P, Johnson C T, Parpia F A and Plummer E P 1989, Comput. Phys. Commun., 55, 424

[2] Aggarwal K M, Keenan F P and Msezane A Z 2003, ApJS, 144, 169

[3] Aggarwal K M, Keenan F P and Kisielius R 2004, A&A, 420, 783

[4] Ait-Tahar S, Grant I P and Norrington P H 1996, Phys. Rev. A 54, 3984

Inst. Phys. Conf. Ser. No 186
Paper presented at 9th Int. Conf. X-Ray Lasers, Beijing, China, 24–28 May 2004

Atomic data for H-like aluminium

K M Aggarwal[1], F P Keenan[1] and S J Rose[2]
[1]Department of Pure and Applied Physics, Queen's University, Belfast BT7 1NN, UK
[2]Department of Physics, Clarendon Laboratory, Parks Road, Oxford OX1 3PU, UK

Abstract. Energy levels, radiative rates, collision strengths, and excitation rates for all transitions among n = 5 levels of Al XIII have been computed in *jj*-coupling scheme including relativistic effects. Resonances have been resolved in a fine energy grid in the threshold region, and the importance of including relativistic effects is discussed in comparison with the earlier available non-relativistic results.

1. Introduction

Atomic data (namely, energy levels, radiative rates, and excitation rate coefficients) for hydrogenic ions are required for the modelling of astrophysical plasmas, as well as for the study of power loss in fusion reactors. Such data for transitions in Al XIII are of additional interest, because this ion is employed to pump Fe XXIV in x-ray lasing plasmas. Therefore, here we present our results for above parameters for transitions among the $n \leq 5$ levels of Al XIII.

The earlier available atomic data for Al XIII are from our own calculations [1], which reported collision strengths (Ω), and effective collision strengths (Y), but for transitions in *LS*-coupling scheme. As a result of this, desired atomic data for fine-structure transitions are not available. Therefore, in the present work we are performing a fully relativistic calculation in order to report results for fine-structure transitions. Additionally, we report radiative rates, which are required in any modelling or diagnostic analysis and are not yet available in the literature.

The $n \leq 5$ configurations of Al XIII give rise to 25 fine-structure levels. To calculate energy levels and radiative rates, we have used the GRASP code [2]. For the computations of collision strengths Ω, we have adopted the Dirac Atomic *R*-matrix Code (DARC) of Norrington and Grant [3]. In our calculations the *R*-matrix radius is 7.36 au, and 55 continuum orbitals have been included for each channel angular momentum for the expansion of the wavefunction. This allows us to compute Ω up to an energy of 300 Ry, which will easily permit us to calculate excitation rate coefficients up to a temperature of 10^7 K, sufficient for applications to a wide variety of plasmas. Furthermore, all partial waves with angular momentum $J \leq 60$ have been included, in order to ensure convergence of Ω at all energies, and for all transitions.

Table 1. Threshold energies, radiative rates, and life-times for Al XIII. ($a \pm b \equiv a \cdot 10^{\pm b}$)

Index	Config.	Level	NIST (Ry)	GRASP (Ry)	A (s^{-1})	τ (s)
1	1s	$^2S_{1/2}$	0.000000	0.000000		
2	2s	$^2S_{1/2}$	126.986288	126.988960	3.474+05	2.879-06
3	2p	$^2P_{1/2}$	126.982560	126.985252	1.792+13	5.580-14
4	2p	$^2P_{3/2}$	127.078367	127.081062	1.787+13	5.597-14
5	3p	$^2P_{1/2}$	150.531340	150.534500	4.773+12	1.846-13
6	3s	$^2S_{1/2}$	150.532450	150.535599	1.542+05	5.443-12
7	3d	$^2D_{3/2}$	150.559680	150.562851	2.860+09	5.387-13
8	3p	$^2P_{3/2}$	150.559700	150.562897	4.775+12	1.847-13
9	3d	$^2D_{5/2}$	150.569110	150.572266	2.847+09	5.406-13
10	4p	$^2P_{1/2}$	158.769510	158.772842	1.948+12	4.302-13
11	4s	$^2S_{1/2}$	158.769980	158.773300	7.372+04	7.858-12
12	4d	$^2D_{3/2}$	158.781460	158.784805	1.574+09	1.258-12
13	4p	$^2P_{3/2}$	158.781490	158.784821	1.947+12	4.310-13
14	4d	$^2D_{5/2}$	158.785450	158.788773	1.567+09	1.263-12
15	4f	$^2F_{5/2}$	158.785440	158.788773	3.785-02	2.533-12
16	4f	$^2F_{7/2}$	158.787420	158.790756	0.000+00	2.536-12
17	5s	$^2S_{1/2}$	162.581180	162.584579	3.988+04	1.215-11
18	5p	$^2P_{1/2}$	162.580950	162.584351	9.795+11	8.336-13
19	5d	$^2D_{3/2}$	162.587060	162.590469	8.886+08	2.422-12
20	5p	$^2P_{3/2}$	162.587070	162.590485	9.815+11	8.331-13
21	5f	$^2F_{5/2}$	162.589000	162.592499	3.308-02	4.899-12
22	5d	$^2D_{5/2}$	162.589000	162.592499	8.852+08	2.433-12
23	5g	$^2G_{7/2}$	162.590100	162.593521	0.000+00	8.218-12
24	5f	$^2F_{7/2}$	162.590100	162.593521	0.000+00	4.905-12
25	5g	$^2G_{9/2}$	162.590710	162.594131	0.000+00	8.222-12

2. Results

Our calculated energy levels, along with those of NIST, are listed in Table 1. The agreement between the theoretical and experimental results is excellent, both in magnitude and ordering. Our radiative rates for transitions from the ground state are also listed in Table 1. These radiative rates are the sum over four types of transitions, namely electric dipole (E1), electric quadrupole (E2), magnetic dipole (M1), and magnetic quadrupole (M2), although in practice, only one type of transitions generally dominate. Additionally, the calculated life-times for all excited levels are also listed in Table 1. No other similar results are available to compare with, but the accuracy of our results is assessed to be better than 5% for all transitions.

In Table 2 we list our results of Ω, at three energies of 200, 250 and 300 Ry, for transitions from the ground state. Similar results for Y, which are simply related to the excitation and de-excitation rate coefficients, are also listed in Table 1, at three electron temperatures of 10^5, 10^6 and 10^7 K. While calculating Y, values of Ω have been integrated over a Maxwellian distribution of electron velocities, over a wide energy range (up to 300 Ry), and in a fine energy mesh in the thresholds region. This is because

Table 2. Collision strengths (Ω) and effective collision strengths (Y) for transitions in Al XIII. ($a \pm b \equiv a \cdot 10^{\pm b}$)

Transition		Ω			Y		
I	J	200 Ry	250 Ry	300 Ry	10^5 K	10^6 K	10^7 K
1	2	4.285-3	4.416-3	4.539-3	4.180-3	4.546-3	4.501-3
1	3	7.390-3	8.630-3	9.807-3	5.757-3	6.204-3	7.372-3
1	4	1.471-2	1.720-2	1.956-2	1.101-2	1.228-2	1.464-2
1	5	1.376-3	1.536-3	1.710-3	1.263-3	1.403-3	1.461-3
1	6	8.659-4	8.712-4	8.855-4	9.535-4	1.061-3	9.081-4
1	7	1.877-4	1.679-4	1.701-4	2.991-4	4.589-4	2.440-4
1	8	2.746-3	3.072-3	3.424-3	2.417-3	2.723-3	2.895-3
1	9	2.792-4	2.505-4	2.543-4	4.425-4	6.642-4	3.585-4
1	10	5.164-4	5.598-4	6.179-4	5.413-4	5.332-4	5.459-4
1	11	3.301-4	3.248-4	3.250-4	4.654-4	4.096-4	3.386-4
1	12	9.964-5	8.383-5	8.209-5	2.306-4	1.937-4	1.114-4
1	13	1.031-3	1.120-3	1.238-3	1.041-3	1.036-3	1.085-3
1	14	1.481-4	1.250-4	1.227-4	3.334-4	2.741-4	1.630-4
1	15	4.961-6	3.112-6	2.499-6	8.320-5	5.699-5	1.162-5
1	16	6.554-6	4.121-6	3.317-6	1.095-4	7.214-5	1.487-5
1	17	1.668-4	1.599-4	1.579-4	2.171-4	1.954-4	1.656-4
1	18	2.550-4	2.711-4	2.963-4	2.435-4	2.502-4	2.548-4
1	19	5.597-5	4.550-5	4.379-5	9.119-5	8.044-5	5.753-5
1	20	5.093-4	5.428-4	5.942-4	4.803-4	4.980-4	5.344-4
1	21	4.177-6	2.536-6	1.984-6	1.223-5	8.910-6	4.461-6
1	22	8.320-5	6.781-5	6.542-5	1.374-4	1.199-4	8.609-5
1	23	9.911-8	4.377-8	2.685-8	7.213-7	4.262-7	1.390-7
1	24	5.522-6	3.357-6	2.633-6	1.634-5	1.182-5	5.902-6
1	25	1.231-7	5.437-8	3.341-8	9.023-7	5.310-7	1.729-7

the threshold region is dominated by numerous resonances, as shown in Fig. 1 for only one transition, namely 3 – 4 (2p $^2P_{1/2}$ – 2p $^2P_{3/2}$). Clearly such resonances have a dominating effect, over the background values of Ω, in determining the results for Y, especially towards the lower end of the temperature range.

Our detailed results for all parameters and for all transitions, along with comparisons, will soon be reported in a regular journal. However, on the basis of our earlier non-relativistic results [1], the accuracy of our results for Ω and Y is assessed to be ~ 20%. Furthermore, the contribution of relativistic effects is found to be very important, because inclusion of fine-structure in the definition of channel coupling gives rise to additional resonances, which increase the value of Y up to a factor of three, for some of the transitions, particularly at lower temperatures.

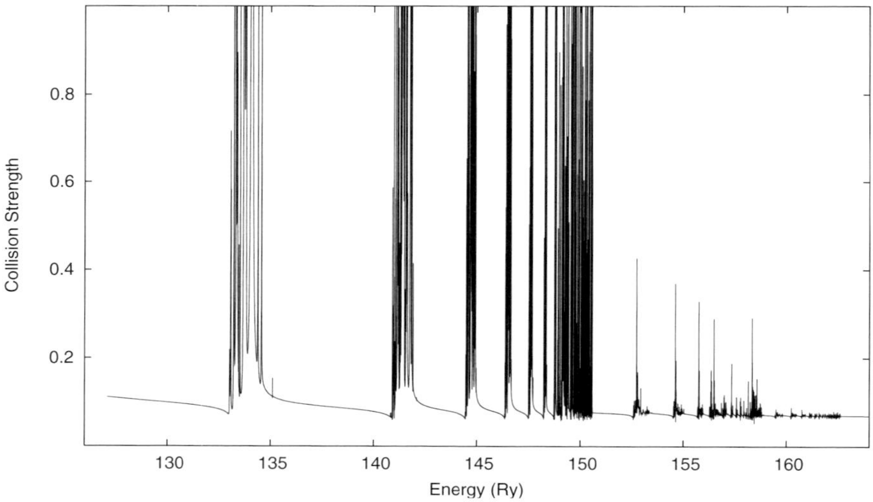

Figure 1. Collision strengths for the 3 – 4 transition in Al XIII.

Acknowledgment

This work has been financed by EPSRC, PPARC and AWE Aldermaston of UK, and we wish to thank Dr. Patrick Norrington for providing his code available to us prior to publication.

References

[1] Aggarwal K M and Keenan F P 2001, Phys. Scr. 63, 95.

[2] Dyall K G, Grant I P, Johnson C T, Parpia F A and Plummer E P 1989, Comput. Phys. Commun., 55, 424

[3] Norrington P H and Grant I P 2004, Comput. Phys. Commun. - in preparation

Inst. Phys. Conf. Ser. No 186
Paper presented at 9th Int. Conf. X-Ray Lasers, Beijing, China, 24–28 May 2004

Electron impact collision strengths of Ti XVIII

J. Y. Zhong[1,2], J. Zhang[1], J. L. Zeng[2], and G. Zhao[2]

[1]Laboratory of Optical Physics, Institute of Physics, Chinese Academy of Science, Beijing 100080, People's Republic of China

2 National Astronomical Observatories, Chinese Academy of Sciences, Beijing 100012,People's Republic of China

Abstract. We present results of a Dirac R-matrix calculation for the electron impact excitation of Boron-like Titanium ions. The lowest 125 fine-structure levels belonging to the $(1s^2)$ $2s^22p$, $2s2p^2$, $2p^3$, $2s^23l$, $2s2p3l$ and $2p^23l$ (l=s, p, d) configurations are included in the calculations. The target model space encompassed the first 15 lowest levels, and all 105 Δn =0 transitions together with 40 partial waves are included in the calculation. For comparison, the calculations are also performed using the relativistic distorted-wave approximation.

1. Introduction

Atomic data for highly charged B-like ions is of considerable interest in modelling high temperature astrophysical and laboratory plasmas. As emphasized in [1], the spectra from ions in the Boron isoelectronic sequence are observed from a variety of astrophysical objects such as novae, planetary nebulae, Seyfert galaxies, the interstellar medium and the Sun. The atomic data of B-like ions are also required in the research of high-temperature plasmas. Earlier work by Luo and Pradhan [2] and by Blum and Pradhan [3] dealt with the calculation of collision strengths and rate coefficients for the first three members of the sequence: CII, NIII, and OIV. Then most recently the work by Zhang and Pradhan [4] calculated the other ions in the B sequence: Ne VI, Mg VIII, Al IX, Si X, S XII, Ar XIV, CaXVI, and FeXXII. However, most previous work on electron-impact excitation of heavily charged ions involved some kinds of limitation. Such as the resonance and coupling effects on the collision rates were not included (RDW) and relativistic effects were only included in an approximation method through the term coupling coefficients (TCC's) [5]. To avoid these kinds of limitation, the Dirac atomic R-matrix code (DARC) of Norrington and Grant [6] just meet the needs.

In this paper the collision strengths B-like Ti ions including first 15 lowest levels, and all 105 $\Delta n = 0$ transitions are calculated using relativistic distorted waves and Dirac R-matrix theory. In Sec. II the theoretical background to the calculations and present some details of the models for the target states and collision strengths are outlined. The results and the discussions were given in Sec. III. Section IV is devoted to the conclusions.

2. Atomic calculations

2.1. Target Calculations

The wave functions and energies of the B-like Ti ions are obtained by the GRASP multiconfiguration Dirac-Fock (MCDF) code [10]. The configuration interaction method is used to obtain accurate wave functions in this calculation. Firstly the 1s, 2s, 2p orbitals are initially chosen for the ground configuration $1s^2 2s^2 2p$ in an average level (AL) calculation. The other orbitals, 3s, 3p and 3d, are generated in a fixed core. Finally, the results are combined in a CI calculation, where $(1s^2)$ $2s^2 2p$, $2s2p^2$, $2p^3$, $2s^2 3l$, $2s2p3l$ and $2p^2 3l$ (l=s, p, d) configurations are included. The energy results of MCDF are listed when the transverse Breit interaction and QED contributions are included. Also, the energy levels for these configurations of B-like Ti ions are calculated using the Flexible Atomic Code (FAC) [14]. It is a fully relativistic approach based on the Dirac equation. The FAC code integrates various atomic processes within a unified theoretical framework, ensures the self-consistency between different parts and provides a uniform, flexible and user-friendly interface for implementing all computational tasks. Configuration interactions between the configurations mentioned above are also included in the present calculations.

2.2. Scattering Calculations

The theory for present calculation is mainly based on the Dirac R-matrix way, which can be found elsewhere [11-14]. The DARC code is similar in structure to the R-matrix package. The MCDF results were used as input for the threshold energies. The radial Dirac equations were solved for 80 continuum angular momentum corresponding to the relativistic quantum number $\kappa = \pm1, \pm2, \pm3, \ldots, \pm40$, and all possible waves in the range l=0-40 were included. Given the range of values of the target angular momentum, this allows for a maximum value of J=40 for the total angular momentum of the system.

For the electron impact strengths, we also perform the calculation using distorted wave (DW) methods. The present DW implementation in FAC is used and the detailed theory can be found in reference. The factorized formula of the collision strength is used in FAC, which separates the angular integration from the radial integrals, and permits efficient interpolation of the radial integrals. Partial collision strengths for all other orbital angular momenta are obtained by interpolation. Very large angular momentum contributions to the allowed transitions are taken into account by the rapid Coulomb-Bethe approximation [14].

3. Discussions and results

3.1. Resonance and coupling effects

The most important features of the close-coupling calculations are the inclusion of the strong coupling between the low-lying states of atoms and ions and the autoionizing resonances that generally enhance the effective scattering cross sections by a large amount [1]. We discuss transition from the ground level to the three low-lying levels in B-like ions, which represent three types of transitions: forbidden, intercombination, and optically allowed.

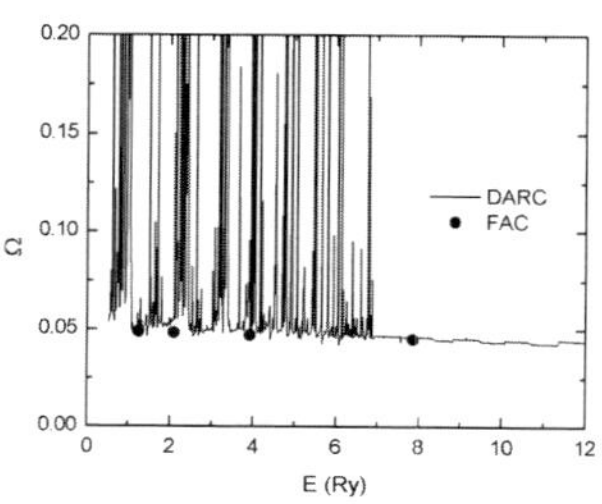

Figure 1. Collision strength for the forbidden transition (1-2) in Ti XVIII in the resonance-energy region. The solid line indicates the DARC results and the circle give the FAC values.

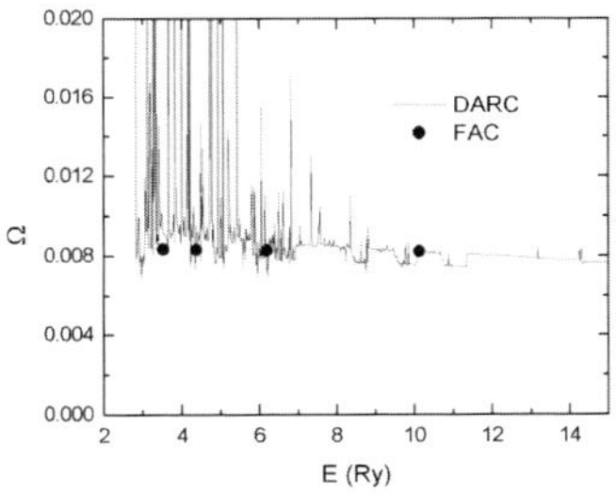

Figure 2. Collision strength for the intercombination transition(1-3) in Ti XVIII in the resonance-energy region. The solid line indicates the DARC results and the circle give the FAC values.

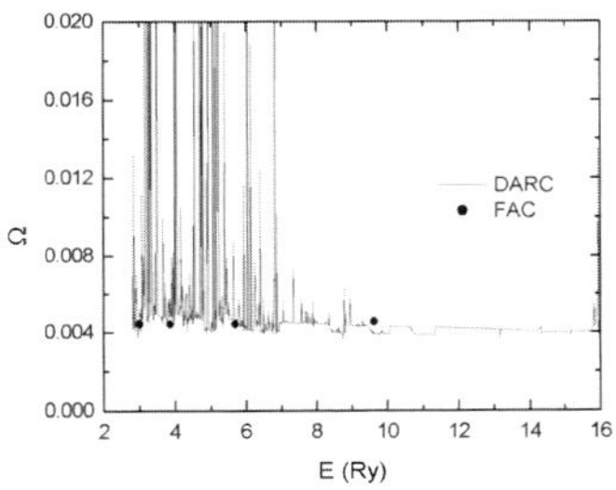

Figure 3. Collision strength for the optically allowed transition (2-3) in Ti XVIII in the resonance-energy region. The solid line indicates the DARC results and the circle give the FAC values

3.2. Convergence of partial waves expansion

The convergence of the partial waves expansion is usually rapid for forbidden transitions, but it can be quite slow for allowed transitions. Also, the number of partial waves needed for convergence increases with increasing energy of the free electron. In Figure 4 we show the variation of Ω with angular momentum J at three energies of 15, 60 and 130 Ry, for the 1-2 transitions. The values of Ω are fully converged below J =7.0. However for a few transition and particularly towards the higher end our energy range, even such a large set of partial waves is not sufficient for the convergence. We demonstrate this in Figure 5 for the 2-3 transitions.

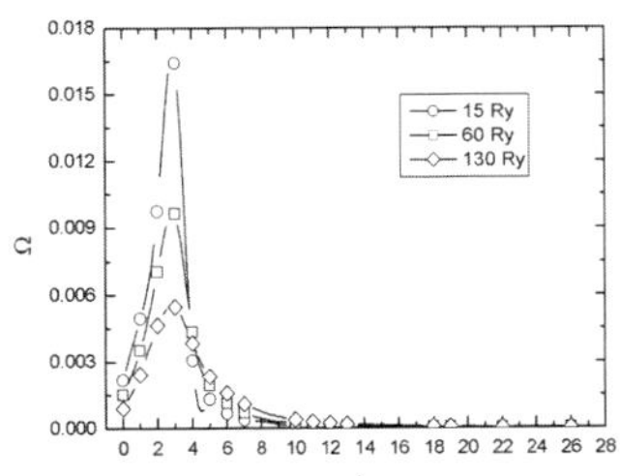

Figure 4. Partial collision strengths for the 1-2 transition of B-like Ti, at three energies 15, 60, 130 Ry

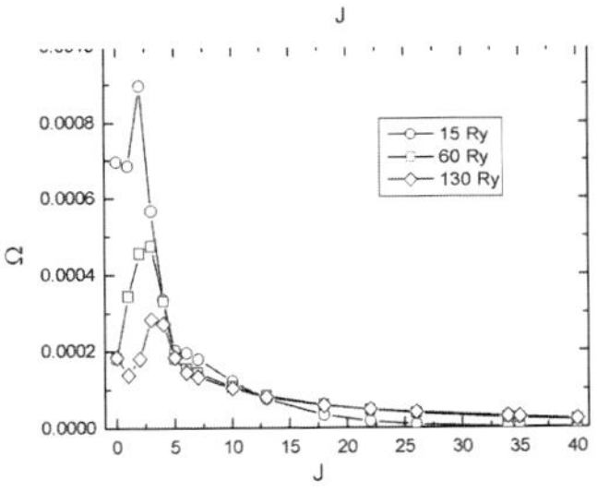

Figure 5. Partial collision strengths for the 2-3 transition of B-like Ti, at three energies 15, 60, 130 Ry

3.Conclusions

In conclusion, two independent calculations are performed for collision strengths of boron-like titanium by using DARC and FAC code. 125 fine-structure levels belonging to the ($1s^2$) $2s^22p$, $2s2p^2$, $2p^3$, $2s^23l$, $2s2p3l$ and $2p^23l$ (l=s, p, d) configurations are included in the calculations. Some important properties are considered in the close-coupling calculations.

Acknowledgments

This work is jointly supported by the NSFC under Grant Nos. 10176034, 10374114, 60321003, 10474134 and the NKBRSF under Grant No. G1999075206, No. G19999075406 and the National Hi-tech ICF program.

References

[1] H. L. Zhang, M. Graziani, et al. Astron. Astrophys. 283, 319(1994)
[2] D. Luo and A. K. Pradhan, Phys. Rev. A. 41, 165(1990)
[3] R. D. Blum and A. K. Pradhan, Astrophys. J. Suppl. Ser. 80 425(1992)
[4] H. L. Zhang, and A. K. Pradhan, Phys. Rev. A 50, 3105(1994)
[5] S. Ait-Tahar and I. P. Grant, Phys. Rev. A 54, 3984(1996)
[6] X. Lu, Y. J. Li, et al. Phys. Rev. A. 67, 013810 (2003)
[7] S. B. Utter, P. Beiersdorfer, et al. Phys. Rev. A. 67, 012508 (2003)
[8] J. F. Peng, and A. K. Pradhan et al. Astrophysical Journal 432, L123(1994)
[9] K. G. Dyall, I. P. Grant et al. Comput. Phys. Commun. 55, 425 (1989)
[10] J. J. Chang, J. Phys. B 8, 2327; 10, 3335 (1975).
[11] P. H. Norrington and I. P. Grant, J. Phys. B 20, 4869 (1987)
[12] W. P. Wijesundera, F. A. Parpia et al. J. Phys. B, 1803 (1991)
[13] W. P. Wijesundera, I. P. Grant et al. J. Phys. B, 1017 (1991)
[14] M. F. Gu, Astrophys. J. 582, 1241 (2003)

Inst. Phys. Conf. Ser. No 186
Paper presented at 9th Int. Conf. X-Ray Lasers, Beijing, China, 24–28 May 2004

Collisional radiative modelling of the Ne-like and F-like line emission from Ne-like Titanium X-ray laser

E Hajiyev[1] A Demir[1,*] G J Tallents[2]
[1]University of Kocaeli, Department of Physics, 41200 Kocaeli, TURKEY
* Laser Technologies Research and Application Center, University of Kocaeli, 41200, TURKEY
[2]University of York, Department of Physics, York,YO10 5DD, U.K.

Abstract. We describe a collisional radiative code to simulate the Ne-like and F-like resonance lines emitted from Ne-like Titanium X-ray laser media. The code calculates intensities of Ne-like $1s^2\,2s^22p^6 \rightarrow 1s^22s^22p^5\,nl$ and F-like $1s^2\,2s^22p^5 \rightarrow 1s^22s^22p^4\,nl$ resonance lines emitted from titanium. Intensities of photons emitted between Ne-like $1s^2\,2s^22p^5n'l' \rightarrow 1s^22s^22p^5\,nl$ and F-like $1s^2\,2s^22p^4\,n'l' \rightarrow 1s^22s^22p^4\,nl$ excited levels are also calculated. The optimum electron temperature and density for J=0-1 lasing at 32.6nm is evaluated. The ratios of F-like to Ne-like 3s-2p resonance lines are calculated with the aim that such ratios can serve as a diagnostic to measure electron temperatures.

1. Introduction

Saturated soft X-ray lasing in collisionally excited Ne-like ions has been demonstrated in highly ionised plasmas in several elements ranging from Ti (Z=22) to Ag (Z=47) [1]. To produce collisional-excited x-ray lasing, several methods such as the employment of double targets [2], irradiation with single long pulses [3], pre-pulses [4], multi-pulses [5] and superimposed short pulses on long pulses [6] have been used.

An understanding of the excitation and ionization behavior of Ne-like and F-like ions is of fundamental importance for x-ray laser development using Ne-like ions. Codes such as Ehybrid [7] and Lasnex [8] have been used to simulate x-ray laser media by combining fluid modeling of the plasma expansion with collisional-radiative modeling of the ionization and excitation. The codes can predict gain and are useful for predicting the line emission from Ne-like and F-like ions [9]. In this paper, we describe a stand-alone collisional-radiative code, NeF, that simulates line emission from Ne-like and F-like ions for given electron temperature and density. We report the result of the collisional radiative code NeF in steady state condition for diagnostic purposes relevant to Ne-like titanium x-ray laser media. Ne-like titanium lasing at 32.6nm is a useful x-ray laser to study with small pumping laser systems now becoming available in University and small research laboratories. We have modelled the Ne-like $1s^2\,2s^22p^6 \rightarrow 1s^22s^22p^5\,nl$ and F-like $1s^2\,2s^22p^5 \rightarrow 1s^22s^22p^4\,nl$ resonance lines emitted from titanium plasma. Also we calculated emission between excited levels and determined the lasing line gain coefficient. Resonance lines ratio can be used to determine the electron temperature and electron density in experiments, so we have calculated the ratio of F-like resonance lines to Ne-like resonance lines using the collisional-radiative model for electron density between $n_e=10^{18}$ and 10^{21} cm^{-3} and electron temperature $T_e = 100$ and 600 eV.

2. Model for calculation

The ionization in the plasma is obtained by solving rate equations. In the steady state condition, for an initial ground state of density N^i in an ionization stage $i+$ ($i = 1,2,3,\ldots, Z-1, Z$) where the $Z+$ ion is completely stripped, the equations takes the form

$$N_e\left\{N^{i-1}I^{i-1} + N^{i+1}\left[R_{rr}^{i+1} + N_e R_{cr}^{i+1} + R_{de}^{i+1}\right] - N^i\left[I^i + R_{rr}^i + R_{cr}^i + R_{de}^i\right]\right\} = 0, \quad (1)$$

Here I and R represent ionization and recombination rate coefficients. The subscripts rr, cr and de stand for radiative, collisional and dielectronic recombination, respectively. The ionization rate, recombination rate coeffients have been calculated using the equations in references [10,11]. In the steady state solution, a general form of rate equation for a specific excited level m is given by [12]

$$\sum_k \left\{n_k\left[n_e C_{km} + A_{km} + B_{km}u(\lambda_{km})\right] - n_m n_e\left(C_{mk} + B_{mk}u(\lambda_{km})\right)\right\} \\ + n_e\left\{n^{i-1}I^{i-1} - n_i I^i + n^{i+1}\left[R_{rr} + n_e R_{cr} + R_{de}\right]\right\} = 0 \quad (2)$$

where $N_e C_{km}$ and A_{km} represent respective electron collisional excitation and spontaneous radiative decay rates from an arbitrary level k to a lower specific level m, B_{km} and B_{mk} represent photo excitation and photo de-excitation respectively; $u(\lambda_{km})$ is the radiation energy density. The required atomic data for the rate coefficients have been calculated using the Cowan code [13]. The collisional rate is calculated using equation in reference [14]. After calculating the excited level population density, line intensities are calculated by multiplying the population densities, the transition probability and an escape factor obtained assuming the Holstein approximation [15].

3. Results and discussion

To produce an x-ray laser from Ne-like ions, there must be a considerable fraction of Ne-like and F-like ions in the plasma. Figure 1 shows the ionic fraction and the Ne-like 3p-3s J=0-1 lasing line gain coefficient in titanium plasma as a function of electron temperature. The optimum gain in the steady state occurs at 200 eV electron temperature. The gain is calculated by keeping the electron density constant at 5×10^{20} cm^{-3} close to the critical density of an Nd:YAG laser. At 200 eV electron temperature, Ne-like and F-like ions dominate the titanium plasma.

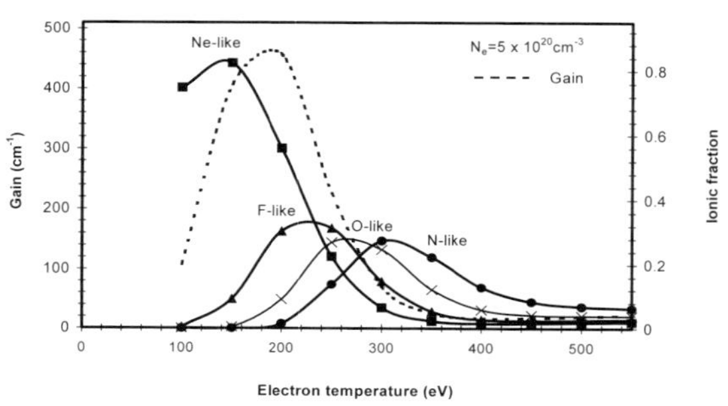

Figure 1. The ionic fraction and the Ne-like J=0-1 lasing line gain coefficient of titanium as a function of electron temperature at electron density 5×10^{20} cm^{-3}.

Figure 2 (a) shows the modeled titanium spectrum emitted from Ne-like and F-like excited levels around the lasing line wavelength and the intensity of the lasing line assuming an approximate radiative transfer along a plasma of length 1 cm. The electron density is $1x10^{19}$ cm^{-3}. The lasing line intensity is not strong and does not dominate the spectrum. However, at electron density $5x10^{19}$ cm^{-3}, emission from the lasing line dominates the spectrum (Figure 2(b)).

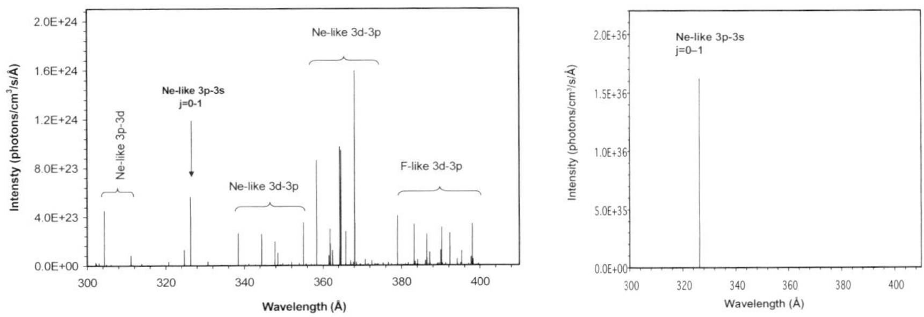

Figure 2. The Ne-like and F-like line spectrum emitted around the J=0-1 lasing line, (a) at 220 eV, $1x10^{19}$ cm^{-3} and (b) at 200 eV, $5x10^{19}$ cm^{-3}. The lasing line intensity is calculated for a plasma length of 1 cm

Resonance line emission from x-ray laser media is recorded using crystal or grating spectrometers viewing the transverse direction. Figure 3 shows the modelled Ne-like and F-like resonance line intensities emitted from a titanium plasma.

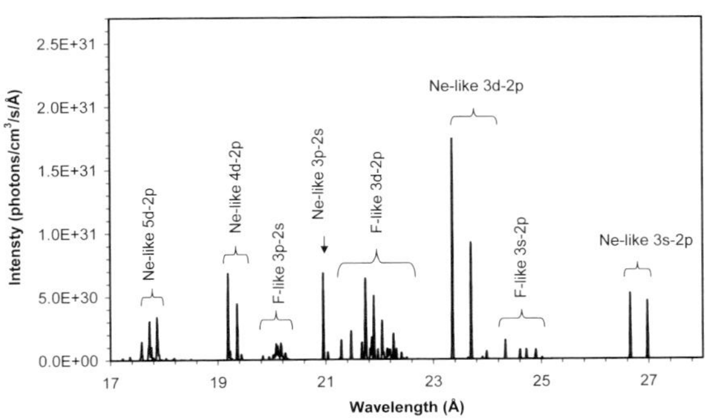

Figure 3. The Ne-like and F-like resonance line spectrum emitted at 200 eV electron temperature and $5x10^{20}$ cm^{-3} electron density.

The excited levels in Ne-like collisional x-ray laser media are populated mainly by electron collisions. Therefore, electron temperatures can be obtained using resonance line intensity ratios. Figure 4 (a) shows the F-like 3s-2p lines to Ne-like 3s-2p resonance lines as a function of electron density at different temperatures. The line ratios do not strongly depend on electron density. Figure 4 (b) shows the line ratios as a function of electron temperature. The lasing line gain coefficient can be approximated by measuring the resonance line ratio (see figure 4(b)).

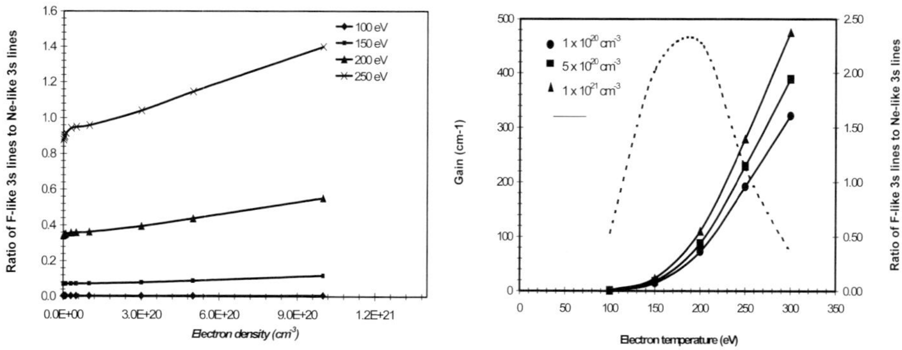

Figure 4. (a) The F-like 3s-2p to the Ne-like 3s-2p resonance lines as a function of (a) electron density and (b) electron temperature. The predicted gain coefficient is also superimposed on (b).

4. Conclusions

The Ne-like and F-like titanium resonance lines have been modeled by solving collisional-radiative equations assuming steady state. The optimum gain is obtained at 200 eV electron temperature. The F-like 3s-2p to Ne-like 3s-2p resonance line intensity ratio is suggested as a diagnostic to measure electron temperatures since the line ratios do not depend strongly on electron density.

Acknowledgements:

This work is supported by the Research Found of University of Kocaeli under the contract 2002/41.

References

[1] Carillon A et al. 1992 Phys. Rev. Lett. 68 2917; M.P. Kalachnikov et al 1998 Phys. Rev.A 57 4778
[2] Zhang J et al. Phys. Rev. A 54 6 4653.
[3] Cairns G F et al. 1996 Optics Commun. 123 777.
[4] J.Y. Lin et al. 1998 J.App. Phys. 83 1863
[5] Behjat et al. 1997 Optics Commun. 135 54
[6] Warwick P J et al. 1998 J. Opt. Soc. of Am. B: Opt. Phys. 15 6 1808.
[7] Pert G J J. 1983 Fluid. Mech. 131 401.
[8] Zimmermann G B 1975 Comments Plasma Physics Controlled Fusion 11 51.
[9] Abou-Ali Y. et al. 2003 J.Phys. B: Atom. Mol. and Opt. Phys. 30 20 4097.
[10] Lotz W. 1970 Zh. Phys. 232 101.
[11] Hutchinson I H 1987 Principles of Plasma Diagnostics (Chambridge Univ. Press).
[12] Elton RC 1990 X-ray Lasers (Academic Press).
[13] Cowan R D 1968 J. Opt. Soc. Am. 58 808.
[14] Wymgaarden et al. 1979 Phys. Rev. A 1409.
[15] Holstein T. 1947 Phys. Rev. 72 12 1212.

Inst. Phys. Conf. Ser. No 186
Paper presented at 9th Int. Conf. X-Ray Lasers, Beijing, China, 24–28 May 2004

Modelling of the Ni-like tin x-ray laser driven by 1.2 ps short pulse and 280 ps background pulse configuration

A Demir[1,*] E Hajiyev[1] H. Goktas[2] and G J Tallents[3]
[1]University of Kocaeli, Department of Physics, 41200 Kocaeli, TURKEY
* Laser Technologies Research and Application Center, University of Kocaeli, 41200, TURKEY
[2]Ankara Nucleer Research and Training Center, Ankara 06105, TURKEY
[3]University of York, Department of Physics, York,YO10 5DD, U.K.

Abstract. Detailed simulations of a Ni-like Sn x-ray laser are undertaken using the EHYBRID code. The atomic physics data are obtained using the Cowan code. The optimum value is obtained for the conditions of a Nd:glass laser with 1.2 ps pulse at 6.7 x 10^{14} W cm^{-2} irradiance pumping a plasma pre-formed by a 280 ps duration pulse at 5x10^{12} W cm^{-2} with peak-to-peak pulse separation set at 200 ps. X-ray resonance lines between 12 Å and 25 Å emitted from tin plasmas have been simulated. Free-free and free-bound emission from the Zn-, Cu-, Ni-, Co- and Fe-like ions is calculated in the simulation.

1. Introduction

Nickel-like x-ray lasing in tin is in fact the first proposed nickel-like ion laser [1], and gain production in an imploding gas puff was discussed in 1985 [1]. Collisional nickel-like ions x-ray lasers have been developed using a wide range of elements with saturated output down to 5.9 nm now achieved [2-5]. Lasing in many nickel-like lanthanide ions has been demonstrated by Daido *et al.* [6], and by Nilsen and Moreno [7]. In one of the experiments [5], lasing at wavelengths from 6 to 8 nm was obtained with a drive energy as low as 200 J. One of the efficient ways of producing lasing at shorter wavelengths is provided by nickel-like ions [6-9] with which lasing at 3.6 nm has been demonstrated in gold. A modified ASE gain model is used to characterise the saturated Sn x-ray laser output pumped by 75 ps pulse or 3 ps short pulse [10]. Experimentally, it has been previously investigated using an exploding foil target [11], in which an indication of amplification on one of the J=0-1, $4d$-$4p$ transitions at 11.91 nm was reported. In particular, hydrodynamic simulations [12] have shown that using short ~ps driving laser pulses superimposed on long pre-pulses reduces the required driving energy and increases the lasing output of collisional x-ray lasers. The initial low intensity pulse produces Ne-like or Ni-like ions over a large plasma volume and the high intensity second pulse creates the population inversions.

In this paper, theoretical results of Ni-like Sn x-ray laser gain and resonance line emission with a short pulse (1.2ps) superimposed on a long background pulse (280 ps) pumping is presented. Simulated results of keV emission due to the Ni-like, Co-like Sn resonance lines and continuum emission due to the free-free and free-bound emission from the Zn-, Cu-, Ni-, Co- and Fe ions are also presented.

2. Method of simulation

Laser pulses of 1.064μm wavelength focused on 100 μm wide and 1cm long slab targets have been simulated. Tin has been assumed pumped with a double pulse configuration in which a 1.2 ps short pulse is superimposed on a 280 ps long background pulse. Peak intensities assumed are respectively $5x10^{12}$ W/cm^2 and $6.7x10^{14}$ W/cm^2 for the background and the short pulse respectively. The peak to peak time between the short pulse and background pulse is 200 ps.

Line intensities are calculated in a post-processor to EHYBRID using the simulated excited level population densities. The original EHYBRID code [13] was modified to calculate the spontaneous transition rates from the absorption oscillator strengths used in the evaluation of the ionization balance. The absorption oscillator strengths and excited level energy and statistical weights for all levels are obtained from the Cowan code [14]. 212 Ni-like $1s^22s^22p^63s^23p^63d^{10}$ - $1s^22s^22p^63s^23p^63d^9$ nl and 537 Co-like $1s^22s^22p^63s^23p^63d^9$ - $1s^22s^22p^63s^23p^63d^8$ nl resonance line intensities have been calculated for the simulation of the spectral emission from the tin plasma.

In addition to the Ni-like and Co-like resonance line intensities, the emission due to free-free Bremsstrahlung and free-bound continuum emission is obtained from EHYBRID deduced plasma parameters in our post-processor. Bremsstrahlung and free-bound emission [15] between 1 and 30 Å, arising from recombination of free electrons with Zn-, Cu-, Ni-, Co- and Fe –like ions have been considered.

3. Results

To produce an x-ray laser from Ni-like ions, there must be a considerable fraction of Ni-like and Co-like ions in the plasma. Since collisional excitation rates strongly depend on electron temepratures, gain coefficient in collisional x-ray laser follows electron temperature time variation. Figure 1 shows the gain coefficient of the Ni-like Sn lasing line at 12 nm and the electron temperature as a function of time from the start of the short pulse interaction time. The peak gain occurs at peak of short pulse interaction time with plasma. Figure 2 shows the gain coefficient of the Ni-like Sn lasing line at 12 nm and the electron temperature as a function of distance at the time of peak gain.

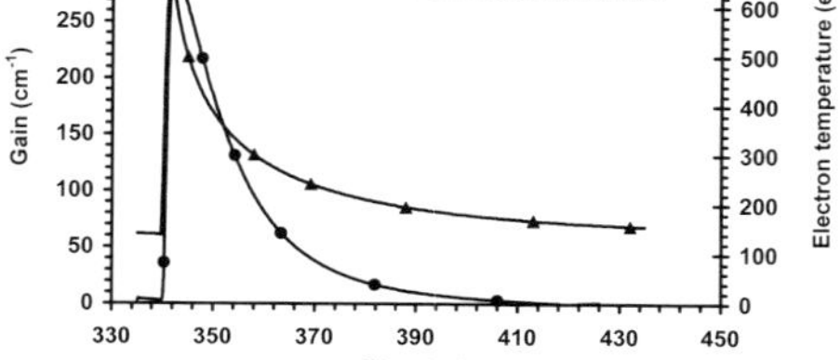

Figure 1. The Ni-like Sn lasing line gain coefficient and electron temperature as a function time

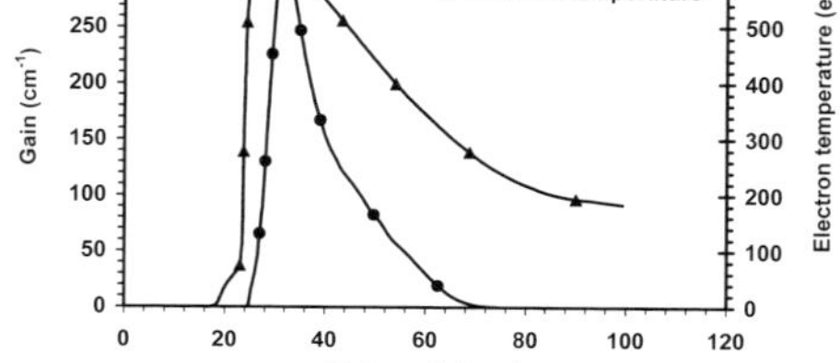

Figure 2. The Ni-like Sn lasing line gain coefficient and electron temperature as a function of distance.

In experiments, ionization balance is often checked using a spectrometer transversely viewing the plasma (at an angle to the line focus) and recording keV photon energy resonance lines emitted by transitions from the excited levels to the ground level. Figure 3 shows the simulated time-integrated Ni-like and Co-like Sn resonance line and continuum spectrum for intensity 6.7×10^{14} W/cm^2. The Co-like resonance lines are more intense than the Ni-like resonance lines.

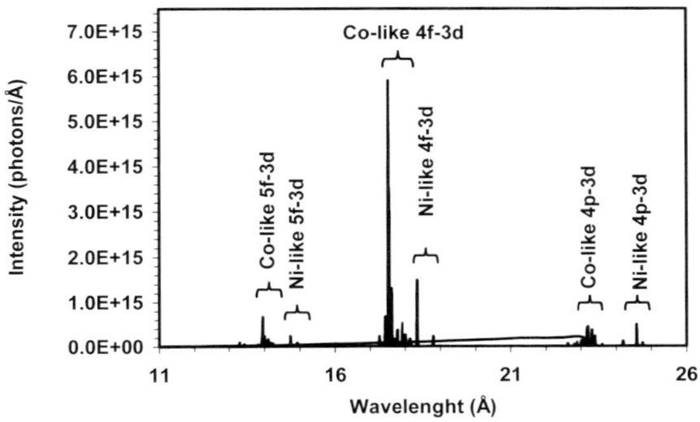

Figure 3. The time-integrated Ni-like and Co-like resonance line and continuum spectrum emitted from tin plasma.

Figure 4 shows the Ni-like and Co-like resonance lines intensity and the gain coefficient of the Ni-like Sn lasing line at 12 nm as a function of time from the start of the short pulse interaction time. The FWHM of the gain coefficient duration is ~11ps. Figure 5 shows the gain coefficient of the Ni-like Sn lasing line at 12 nm and the Ni-like and Co-like resonance lines intensity of distance at the time of peak gain. The FWHM of the gain size is ~14μm.

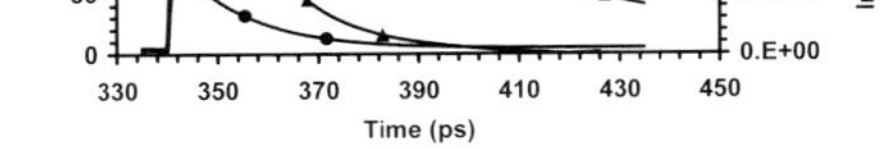

Figure 4. The time variation of the Ni-like, Co-like resonance lines and gain coefficient. ▲ indicates the Ni-like, ■ represents Co-like lines and •indicates gain coefficient.

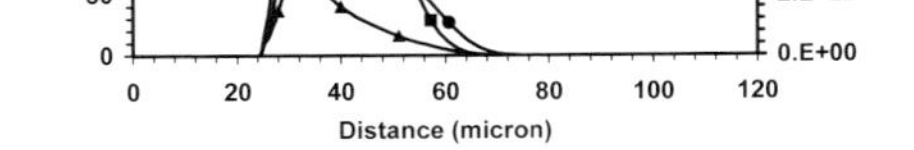

Figure 5. The time variation of the Ni-like, Co-like resonance lines and gain coefficient. ▲ indicates the Ni-like, ■ represents Co-like lines and •indicates gain coefficient.

4. Conclusions

X-ray laser output at 12 nm and the intensity of resonance lines and continuum emission between 0.1-3.2 nm emitted from a tin plasma has been simulated using the EHYBRID fluid and atomic physics code. A post processor code coupled to the EHYBRID code has been developed to model the transverse emission of resonance lines and continuum emission in the keV spectral region. The simulated duration of gain coefficient (11 ps) is obtained. The temporal output of continuum and resonance lines has been found to be (24 ps).

Acknowledgements:

This work is supported by the Prime Ministry State Planning Organization of Turkey under the contract no: 2004-K120710.

References

[1] S. Maxon, P. Hagelstein, K. Reed, and J. Scofield, J. Appl. Phys. 57, 971 (1985).
[2] J Zhang et al Phys. Rev. Lett. 78, 3856 (1997).
[3] R. Smith, et al Phys.Rev. A, 59,R47 (1999)
[4] MacGowan B J et al. 1988 J. Opt.Soc. Am. B 5 1858
[5] Balmer J E et al. 2002 AIP Conference proceedings 641 1 147.
[6] H. Daido, Y. Et al. Phys. Rev. Lett. 75, 1074 [1995].
[7] Nilsen J et al. 1995 Opt. Lett. 20 1386.
[8] Li Y et al. 1996 Phys. Rev. A 53 2 562.
[9] MacGowan B J et al. 1987 Phys. Rev. Lett. 59 2157
[10] J.Y.Lin et al J.App.Phys., 85,672 (1999) ; J.Y. Lin et al. Opt. Comm. 166, 211 (1999)
[11] G. D. Enright et al. X-Ray Lasers 1992, edited by E. E. Fill ~Institute of Physics, Bristol, (1993), p. 45.
[12] A. Demir et al. Czechoslovak Journal of Physics, Vol. 54 (2004), Suppl. C
[13] Pert G J J. 1983 Fluid. Mech. 131 401.
[14] Cowan R D 1968 J. Opt. Soc. Am. 58 808.
[15] Abou-Ali Y. et al. 2003 J.Phys. B: Atom. Mol. and Opt. Phys. 30 20 4097.

Inst. Phys. Conf. Ser. No 186
Paper presented at 9th Int. Conf. X-Ray Lasers, Beijing, China, 24–28 May 2004

Gain coefficients for possible 4d-4p x-ray lasers of Ni-like Nd, Sm, Eu, Gd, Ta and W ions

Jiaolong Zeng[1], Jiayong Zhong[1,2], Jie Zhang[2], and Gang Zhao[1]
[1] National Astronomical Observatories, Chinese Academy of Sciences, 20A Datun Road, Chaoyang District, Beijing 100012, PR China
[2] Laboratory of Optical Physics, Institute of Physics, Chinese Academy of Sciences, Beijing 100080, PR China

Abstract. The populations of the 249 levels belonging to the ground configuration $3s^23p^63d^{10}$ and those of single excitations of a 3d, 3p and 3s electron to the orbitals of n=4 and 5 have been calculated for Nickel-like Nd, Sm, Eu, Gd, Ta and W ions for temperatures from 300 to 2000 eV and densities of 1.0×10^{21} and 1.0×10^{22} cm^{-3}. The gain coefficients for possible x-ray laser lines have also been calculated at these plasma conditions. Estimation is made on the optimum plasma condition for x-ray laser gain.

1. Introduction

Researchers have been trying to realize a shorter wavelength x-ray laser for holography of biological living cell specimens within the water window. Electron collisionally pumped x-ray laser in Ni-like ions due to 4d-4p transitions in laser plasmas of high atomic number elements have been demonstrated by various workers[1-4]. Take Ni-like Sm as an example. Zhang *et al.*[1] reported a saturated laser beam output at a wavelength of 7.3 nm with an output energy of 0.3 millijoule in 50-picosecond pulses. The laser line at at a wavelength of 6.8 nm was also visible. Lin *et al.*[2] carried out a systematic study of double pulse pumping at 7.3 nm. To model x-ray lasers and their gains, a better understanding of atomic physics and plasma kinetics is essential.

For x-ray lasers, Ni-like Nd (Z=60), Sm (Z=62), Eu (Z=63), Gd (Z=64), Ta (Z=73), and W (Z=74) ions are of particular interest. In the present work we first study the population inversion between the levels of possible 4d-4p x-ray lasers by solving the rate equation and then calculate the gain coefficients of the possible x-ray lasers for these ions at different temperatures and densities.

2. Theoretical method

Level populations are obtained by solving the set of coupled steady-state rate equations

$$N_j[\sum_{i<j} A_{ji} + n_e(\sum_{i<j} C^d_{ji} + \sum_{i>j} C^e_{ji})] = \sum_{i>j} N_i A_{ij} + n_e(\sum_{i<j} N_i C^e_{ij} + \sum_{i>j} N_i C^d_{ij}) \qquad (1)$$

where N_j is the number density of level j, A_{ji} is the spontaneous decay rate, n_e is electron density and the superscript e and d refer to electron excitation and de-excitation, respectively. The populations are calculated for each of the 249 levels belonging to the configurations of $3s^23p^63d^{10}$, $3s^23p^63d^9nl$, $3s^23p^53d^{10}nl$, and

$3s3p^6 3d^{10}nl$ (n=4, 5; l=0, 1, ..., n-1) of Ni-like Nd, Sm, Eu, , Gd, Ta and W ions. The atomic data required are calculated using the FAC Code developed by Gu[5].

The gain coefficient G for a Doppler-broadened laser transition, dominant in high-temperature plasmas, is given by

$$G = \frac{\lambda^3}{8\pi} A_{ul} (\frac{M}{2\pi kT})(N_u - \frac{N_l g_u}{g_l}) \qquad (2)$$

where λ is the wavelength of the transition between the upper and lower levels, A_{ul} is the rate of spontaneous emission from the upper level u to lower level l, N_j is the ion population density in the level j having statistical weight g_j, k is the Boltzmann constant, M is the ion mass, and T_i is the ion temperature. It is assumed to be equal to the electron temperature in the present work.

3. Results and discussions

Table 1 shows the five levels which may be connected with the x-ray lasers of Ni-like Nd, Sm, Eu, Gd, Ta and W ions on the $3d^{-1}4d$ J=0 → $3d^{-1}4p$ J=1 transitions. $3d^{-1}$ means a hole in the 3d sub-shell. Fig. 1 shows the variations of the reduced population fractions n_j / g_j with the temperature at an electron density of 1.0×10^{21} cm^{-3} for the six Ni-like ions. The numbers labeled in the plots refer to the No. given in table 1. As can be seen from Fig. 1, the population inversion can be occurred between the upper level of $(3d_{3/2}^{-1}4d_{3/2})_0$ (No. 5) and all those lower levels of $(3d_{3/2}^{-1}4p_{1/2})_1$ (No. 1), $(3d_{5/2}^{-1}4p_{3/2})_1$ (No. 2), and $(3d_{3/2}^{-1}4p_{3/2})_1$ (No. 3) for all six ions. The most favorable population inversion occurs between levels of Nos. 5 and 2. Most of the experiments which have demonstrated x-ray lasers observed laser line of this transition. For example, this laser line for Ni-like Sm corresponds to wavelength of 7.3 nm. It has been experimentally demonstrated by Zhang *et al.*[1] and Lin *et al.*[2]. For the population inversion of level Nos. 4 and those belonging to the configuration of $3d^{-1}4p$ whose angular momentum is 1, the situation is different. For Ni-like Nd, Sm, Eu, and Gd ions, population inversion can only be occurred between levels of Nos. 4 and 2 at the given electron density. However, population inversion can be occurred between levels of Nos. 5 and 1, 2, and 3. Note that the population difference from No. 4 is much smaller than 5.

Table 1. The energy levels (in eV) related to the 4d-4p x-ray lasers of Ni-like ions

No	Designation	$J\pi$	Nd	Sm	Eu	Gd	Ta	W
1	$3d_{3/2}^{-1}4p_{1/2}$	1°	932.15	1032.57	1084.50	1137.56	1664.28	1728.05
2	$3d_{5/2}^{-1}4p_{3/2}$	1°	942.45	1045.06	1098.23	1152.65	1697.56	1764.17
3	$3d_{3/2}^{-1}4p_{3/2}$	1°	964.44	1071.22	1126.67	1183.52	1757.87	1828.72
4	$3d_{5/2}^{-1}4d_{5/2}$	0^e	1059.23	1170.64	1228.33	1287.35	1877.96	1950.17
5	$3d_{3/2}^{-1}4d_{3/2}$	0^e	1106.13	1220.35	1279.50	1340.03	1946.89	2021.30

Figure 2 shows variation of gain coefficients for the possible x-ray lasers of six Ni-like ions with the temperature at the same electron density as in Fig. 1. As can be seen from the figure, the variation trend of gains is basically the same for Ni-like Nd, Sm, Eu and Gd ions. For these four ions, the laser line 5-2 has the largest gain among all possible laser lines, while the gains of 5-1 and 5-3 laser lines are much smaller than that of 5-2. For the strongest laser line 5-2, the gain of all four ions increases with temperature and then decreases. The turning point of temperature occurrs at 700-800 eV for the four ions. This behavior of variation means that there is an optimized plasma condition for x-ray laser gain at a given electron density. On the other hand, at a given temperature, the gain decreases with the atomic number Z. For other two ions, i.e., Ni-like Ta and W, the trend of gain variation with the temperature is different from the four ions. As mentioned above, there are six possible x-ray laser lines for these two ions, while there are only four lines for Ni-like Nd, Sm, Eu and Gd. In Fig. 2, we only give the gains of the four lines whose gains are relatively large.

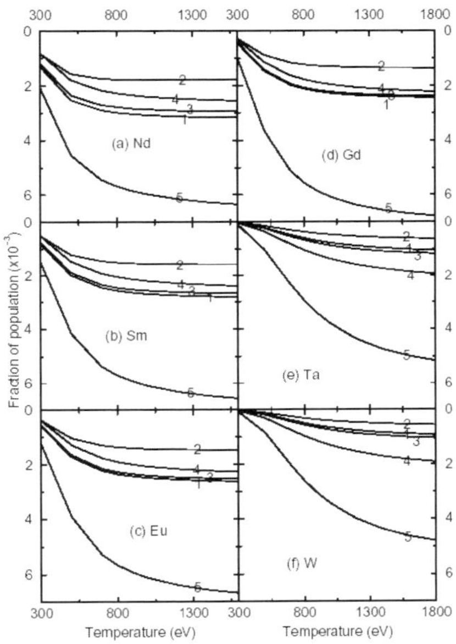

Figure 1. Variation of level population fractions with temperature at 1.0×10^{21} cm^{-3}

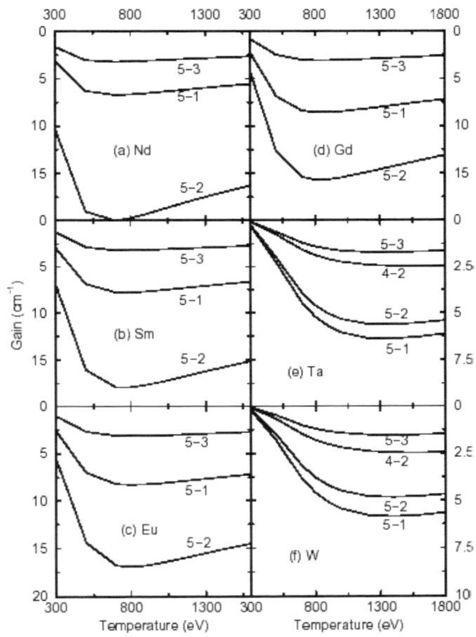

Figure 2. Gain coefficients for possible x-ray lasers at 1.0×10^{21} cm^{-3}

Practical calculations show that the gains are too small to demonstrate efficient x-ray lasers near the electron density of 1.0×10^{20} cm^{-3}. With the increase of electron density, the gains become larger. Fig. 3 shows the gain coefficients at an electron density of 1.0×10^{22} cm^{-3}. At this density, the behavior of the gains is different from that of at an electron density of 1.0×10^{21} cm^{-3}. Firstly, the gains for Ni-like Ta and W are much larger than in Fig. 2, while the gains for Ni-like Nd, Sm, Eu and Gd do not increase so much. Such a result shows that saturated output has been obtained at some density between 1.0×10^{21} cm^{-3} and 1.0×10^{22} cm^{-3}. Secondly, the trend of the gain variation with the atomic number Z is on the opposite at the electron densities of 1.0×10^{21} cm^{-3} and 1.0×10^{22} cm^{-3}. Thirdly, for Ni-like Ta and W, the gains for laser lines of 5-1 and 5-2 are very close at the density of 1.0×10^{22} cm^{-3}, which is not the case at the density of 1.0×10^{21} cm^{-3}.

In conclusion, level populations for Ni-like Nd, Sm, Eu, Gd, Ta and W ions have been calculated for plasma conditions of variant temperatures and densities. Population inversion and gain coefficient have been calculated for possible x-ray laser lines of transitions 4d-4p J=0-J=1. For Ni-like Nd, Sm, Eu, Gd, the x-ray laser line which has the largest gain is due to the transition from level $(3d_{3/2}^{-1}4d_{3/2})_0$ to $(3d_{5/2}^{-1}4p_{3/2})_1$. For

Ni-like Ta and W, however, the x-ray laser line of the largest gain may be the transition from $(3d_{3/2}^{-1}4d_{3/2})_0$ to $(3d_{5/2}^{-1}4p_{3/2})_1$ or to $(3d_{3/2}^{-1}4p_{1/2})_1$, depending on the electron density of the plasma. At higher electron density, the gain is very close for these two laser lines. The favorable plasma condition for the output of x-ray laser can be determined.

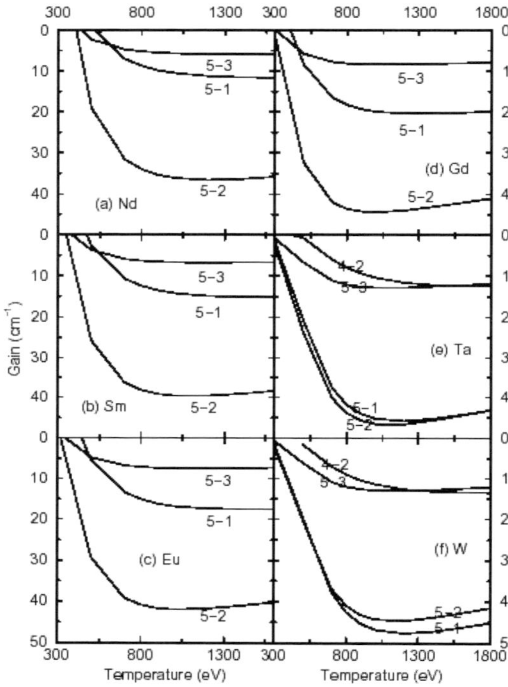

Figure 3 Gain coefficients for possible Ni-like x-ray lasers at 1.0×10^{22} cm^{-3}

Acknowledgments. This work was supported by the National Natural Science Foundation of China under Grant No. 10204024 and No. 10373014.

References

[1] Zhang J. *et al.*, 1997 Science **276**, 1097
[2] Lin J. Y. *et al.*, 1998 J. Appl. Phys. 85, 672
[3] Zhang J. *et al.*, 1996, Phys. Rev. A 54, R4653
[4] Basu S. *et al.*, 1993 Appl. Phys. B 57, 303
[5] Gu M.F., 2003, Astrophys. J. 582, 1241

Inst. Phys. Conf. Ser. No 186
Paper presented at 9th Int. Conf. X-Ray Lasers, Beijing, China, 24–28 May 2004

A Nickel-like Yttrium x-ray laser at 23.4 nm

J. Zhao, F. Yan, JY Zhong, J. Zhang[@]
Laboratory of Optical Physics, Institute of Physics, Chinese Academy of Sciences, Beijing 100080, China

Abstract: A nickel-like yttrium collisional-excitation x-ray laser at 23.4 nm pumped by a 1ns prepulse followed by a 1ps main pulse is simulated using a one-dimensional Lagrangian hydrodynamic code MED103 coupled with an atomic physics data package. The optimized conditions of preplasmas are achieved by optimizing the intensity of the prepulse and the time delay between the prepulse and the main pulse. By using a two-dimensional ray tracing code, the output intensity and the propagation properties of the x-ray laser are simulated with different parameters of the main pulse. The results show that Ni-like yttrium ions are very sensitive to the intensity and the duration of the main pulse. We predict a saturated output with an effective gain product length of 14 can be generated by the optimized drive configuration with an energy about 1.0J.

1. Introduction

A major objective in the development of x-ray lasers is to significantly reduce the drive energy and hence to make these coherent soft-x-ray sources widely available. It is therefore important to improve the efficiency and reduce the drive energy of soft-x-ray lasers.

As one of the electron collisional excitation schemes, Ni-like x-ray lasers have higher quantum efficiency than Ne-like x-ray lasers, especially at shorter wavelengths.[1] In the Ni-like scheme, the upper laser level $3d^9 4d\ ^1S_0$ is populated due to the monopole collisional excitation from the ground state $3d^{10}\ ^1S_0$. The strong lasing lines are dominated by the population inversion of $3d^9 4d\ ^1S_0 \rightarrow 3d^9 4p\ ^1P_1$ and $3d^9 4d\ ^1S_0 \rightarrow 3d^9 4p\ ^3D_1$ transions. Principally, lasing in Ni-like ions is more favorable to scale down to a short wavelength because of the inherently higher quantum efficiency. But on the other hand, it is more difficult for the Ni-like sequence to provide saturated output. Values of the gain-length product $gL \geq 14$ are typically required to reach saturation in laser-driven soft-X-ray lasers.[2]

In recent years, using the prepulse or multi-pulse technique, saturated output of Ni-like x-ray lasers has been obtained at wavelengths close to the water window. Furthermore, the drive energy requirements were recently dramatically reduced to a few joules by application of the transient collisional excitation scheme.[3-4] The initial ~1 ns low-intensity pulse creates a pre-plasma with a large abundance of the desired ion species (i.e Ni-like or Ne-like). The ~ ps high intensity main pulse then rapidly heats up the plasma while the ion temperature remains low, thus a transient population inversion can be created.

[@] Corresponding author: jzhang@aphy.iphy.ac.cn

In this paper, we simulate a nickel-like yttrium x-ray laser at 23.4 nm. The five main states of Ni-like Yttrium ion are $3d^{10}\ ^1S_0$, $3d^94p\ ^3D_1$, $3d^94p\ ^1P_1$, $3d^94d\ ^1S_0$, $3d^94f\ ^1P_1$. A modified version of the one-dimensional Lagrangian hydrodynamic code MED103 coupled with an atomic physics data package is used to predict the time evolution of laser-plasma interactions and gain coefficients. The atomic physics data including the energy levels, radiative transition probability (Einstein coefficient A) and the electron collisional excitation rates are calculated using the code FAC developed by Gu.[5] The calculated wavelength of $3d^94d\ ^1S_0 \rightarrow 3d^94p\ ^1P_1$ is 23.4 nm. The driven pulse configuration is a ~1 ns prepulse followed by a ~1ps ultra-short main pulse. The optimization calculations are performed through terms of the peak intensity of the prepulse, the delay time between the two pulses and the peak intensity of the main pulse. A 2D ray-tracing code is used to calculate the output and the averaged gain-length product.

2. Optimization of a preplasma

In our simulations the wavelength of the pump laser is 1.05μm. The pump laser is set to be the standard line focus on a slab target. We use 100μm –thick slab target. Using the prepulse technique, prepulses produce a properly ionized preplasma with large scale length while the main pulse is used to rapidly heat the preplasma and make a sudden jump of electron temperature. The functions of the two pulses can be independent of each other, so they can be optimized separately. We simulate various prepulses with different peak intensities but same full width at half maximum of 1ns. Fig.1 give the contours of Ni-like Y ion fraction vs space and time in the plasma generated by a 1ns prepulse with peak intensities of (a) 0.8×10^{11} W/cm^2, (b) 0.9×10^{11} W/cm^2, (c) 1.0×10^{11} W/cm^2, respectively. The fraction in the plot is from 20 % (white), 40 % (gray), to 70 % (black). We find that the ionization state of the preplasma is very sensitive to the intensity of the prepulse. From Fig. 1 we can see the prepulse at 0.8×10^{11} W/cm^2 can not ionize the preplasma sufficiently. While we can find an over-ionized region near 150μm for the pulse with 1.0×10^{11} W/cm^2. Therefore the prepulse with an intensity of 0.9×10^{11} W/cm^2 can produce a high fraction of Ni-like ions without over-ionization.

3. Optimization of the delay time

In order to increase the efficiency of the pump pulse, a rich population of Ni-like ions and large scale length is required before the main pulse. The delay time between the two pulses also operates the density gradient which is critical for the propagation of the

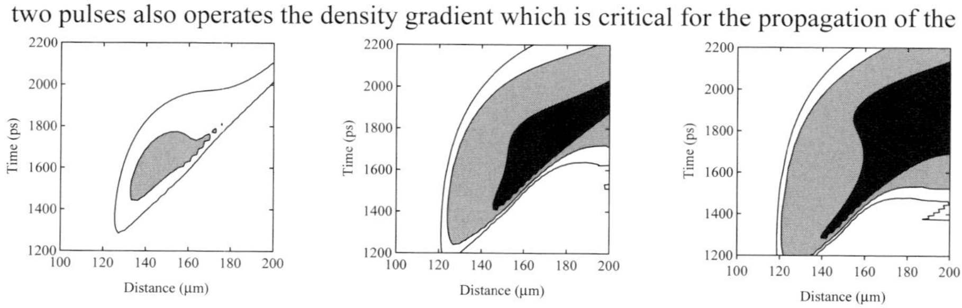

Figure 1. The contours of Ni-like Y ion fraction vs space and time in the plasma generated by 1 ns prepulse with intensities of (a) 0.8×10^{11} W/cm2, (b) 0.9×10^{11} W/cm2, (c) 1.0×10^{11} W/cm2. The fraction in plot is from 20 % (white), 40 % (gray), to 70 % (black).

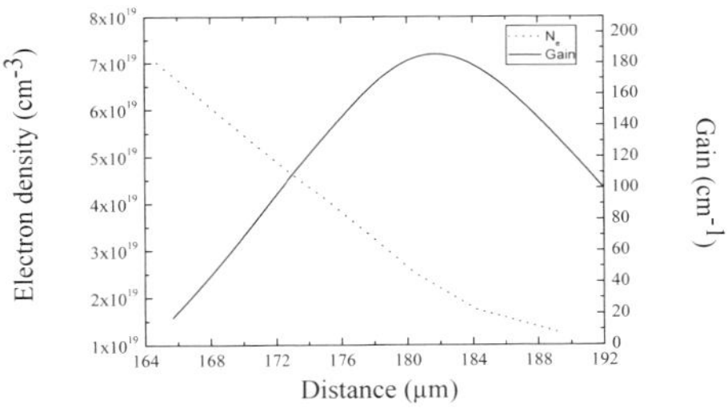

Figure 2. The spatial profiles of the local gain coefficient and electron density at the peak gain time for delay time of 0.6 ns

x-ray laser in the plasma. So we compare the electron density gradient under different delay time to get an optimum separation time.

According to the contours of the nickel-like Y fraction, the time region at 1.7-1.9 ns is found to be a good time for the peak intensity of the main pulse. The spatial profiles of the local gain coefficient and electron density at the peak gain time are similar at different delay time of 0.5 ns, 0.6 ns and 0.7 ns. The longer delay time can result in a lower density gradient. With a lower density gradient x-ray laser can propagate a longer distance in the plasma. However a too long delay time will reduce the peak gain coefficient due to a lower Ni-like ion population and a reduced electron temperature. We choose 0.6 ns to be the optimized delay time between the prepulse and the main pulse because of high fraction of nickel-like yttrium, low density gradient, high electron temperature. Fig.2 gives the spatial profiles of the local gain coefficient and electron density at the peak gain time for delay time of 0.6 ns.

4. Optimization of the main pulse

With the optimized preplasma, main pulse then heats the plasma and creates a transient population inversion. In this section we use a two-dimensional ray tracing code, which was developed in our previous work[1] to simulate the output intensity for different parameters of the main pulse with the optimum delay time of 0.6 ns.

Table 1 lists several examples of peak gain produced by the main pulse with different intensities and durations. From Table Ⅰ, the gain coefficient is listed for the main pulse with the same energy, but different duration and intensity.

To investigate whether the x-ray laser can be saturated, we simulate the profile of spatially integrated laser-output intensity as a function of the plasma length. By fitting the results with the Linford formula, we obtain the small signal gain coefficient.[6] Generally, the x-ray laser output increases exponentially with the target length when the effective gain length product is smaller than ~15. Once the gain length product is great than 15, the output laser is in the saturation region.

TABLE 1. Peak gain produced by different main pulses

Peak intensity of Main pulse ($\times 10^{15}$ W/cm^2)	Duration of main pulse (ps)	Peak gain (cm^{-1})
1.0	1.0	192
0.8	1.0	159
0.6	1.0	116
1.0	0.8	160
1.0	0.6	117

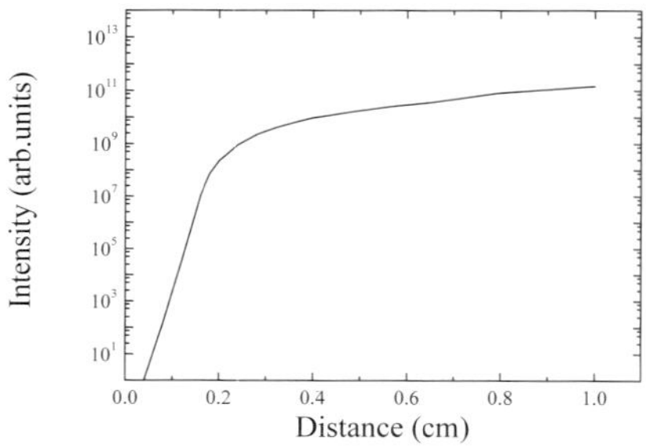

Figure 3. The profile of spatially integrated laser-output intensity produced by main pulse with peak intensity of 0.8×10^{15} W/cm^2

When we change the peak intensity from 1.0×10^{15} W/cm^2 to 0.6×10^{15} W/cm^2 but with the same duration of 1ps the profile changes a little. Fig.3 gives the profile of spatially integrated laser-output intensity produced by the main pulse with the peak intensity of 0.8×10^{15} W/cm^2. Using the configuration, the x-ray laser output intensity is saturated for a plasma length of about 0.23cm. The effective gain coefficient is about 61.86 ± 1.48cm^{-1} and the gain-length product achieves 14.23 which approaches the critical required value 14. For this situation, if the line focus is 50 μm wide the total required pump energy is about 1.02J. That means it is possible to get a saturated x-ray laser output with an energy of less than 1J.

5. Conclusions

We investigate a nickel-like yttrium laser at 23.4 nm numerically by a one-dimensional Lagrangian hydrodynamic code MED103 coupled with an atomic physics data package. With the optimized conditions of plasmas achieved by optimizing the intensity of the prepulse and the delay time between the two pulses, we find that the x-ray laser output is very sensitive to the energy of the main pulse. The results show that a saturated output laser can be achieved using a 1ns, 0.9×10^{11} W/cm^2 prepulse followed by a 1ps, 0.8×10^{15} W/cm^2 main pulse. We predict that a saturated output with an effective gain product length about 14 can be generated by the optimized drive configuration with an energy of about 1 J.

Acknowledgements

This work was supported by the NSFC under Grant Nos. 10374114, 60321003, 10474134 and 60128505, and the NKBRSF under Grant No. G1999075206, and the National Hi-tech ICF Program.

References

[1] X. Lu. and J. Y. Zhong. et al, 2003 Phys. Plasmas.**10**, 2978
[2] Yariv A. Quantum Electronics 2nd ed Wiley, p.282
[3] P. J. Warwick C. L. S. Lewis, M. P. Kalachnikov, P. V. Nickles, et al, 1998 J.Opt. Soc. Am. B **15**, 1808
[4] M. P. Kalachnikov, P. V. Nickles, et al, 1998 Phys. Rev. A **57**, 4778
[5] R. D. Cowan, 1981 The Theory of Atomic Structure and Spectra (University of California Press, Berkeley, CA)
[6] G. J. Linford, E. P. Poressini, W. R. Sooym and M. L. Spaeth, 1974 Appl. Opt. **13**
[7] F. Yan and J. Zhang. et al., 2004 Phys. Plasmas. – in preparation
[8] J. Nilsen. et al. 2000 J. Opt. Soc. Am. B.**17**, 1084
[9] J. Zhang, et al., 1996 Phys. Rev. A.**53**, 3640

Inst. Phys. Conf. Ser. No 186
Paper presented at 9th Int. Conf. X-Ray Lasers, Beijing, China, 24–28 May 2004

Effective collision strengths for Ni-like Tungsten ions

J. Y. Zhong[1,2], J. Zhang[2], J. L. Zeng[1], and G. Zhao[1]

[1] National Astronomical Observatories, Chinese Academy of Sciences, Beijing 100012,People's Republic of China

[2] Laboratory of Optical Physics, Institute of Physics, Chinese Academy of Science, Beijing 100080, People's Republic of China

Abstract. The energy levels, spontaneous radiative decay rates, and electron impact collision strengths are calculated for Ni-like tungsten ions. The data refer to 107 fine-structure levels belonging to the configurations $(1s^22s^22p^6)\ 3s^23p^63d^{10}$, $3s^23p^63d^94l$, $3s^23p^53d^{10}4l$, and $3s3p^63d^{10}4l$ (l = *s, p, d, f*). The collision strengths are calculated for the six scattering electron energy at 20 scattered electron energies (10-20000 eV) by using distorted-wave approximation. Effective collision strengths, obtained after integrating the collision strengths over a Maxwellian distribution of electron energies, have also been calculated for electron temperature in the range (100-2500 eV). Some calculations are compared with other theoretical results and a good agreement is found.

1. Introduction

Recently, for the requirement of simulation in Ni-like x-ray lasers the atomic data of Ni-like ions are required urgently. An important objective in the development of x-ray lasers is to deliver a coherent, saturated output at wavelengths towards the water window [1]. Such saturated x-ray lasers are required for holography [2] and microscopy [3] of biological specimens and for deflectometry [4], interferometry [5], and radiography [6] of dense plasmas relevant to inertial confinement fusion and laboratory astrophysics [5]. Ni-like x-ray lasers, in principle, have a more favourable scaling of laser wavelength with drive-laser energy [7,8]. For high-Z ions, although there are a few studies on the energy levels, oscillator strengths and spontaneous radiative decay rates [9-13], much less work are devoted to the electron impact excitation. Hagelstein et al [9] calculated the lowest 107 fine-structure levels of Ni-like Gd ion, belonging to the $(1s^22s^22p^6)\ 3s^23p^63d^{10}$, $3s^23p^63d^94l$, $3s^23p^53d^{10}4l$, and $3s3p^63d^{10}4l$ (l = *s, p, d, f*) configurations, and the radiative rates between the levels using a relativistic atomic structure code called YODA. Aggarwal et al [10] reported the energy levels and radiative rates for the allowed transitions of the Ni-like Nd, Sm, Eu, Ta, and W ions using the multi-configuration Dirac-Fock method implemented by the GRASP code. They also calculated the electron collision strengths and excitation rates for Gd XXXVII as a test case by using the Dirac atomic R-matrix code (DARC). Considering up to the n=5 orbital, Zhang et al [11], calculated the electron impact collision strengths of the ions with $60 \leq z \leq 92$ between the ground state to the 249 energetically lowest levels. No data are reported for the collision strengths between the excited levels. However, not only the collision strengths from the ground sate, but also

from the excited states are needed in the simulations of the x-ray lasers.
In this paper we report the calculations of the energy levels, oscillator strengths, radiative rates for dipole allowed transitions and collision strengths between the all 107 levels of W XXXXVII. Effective collision strengths, obtained after integrating the collision strengths over a Maxwellian distribution of electron energies, have also been calculated for electron temperature in the range (100-2500 eV).

2. Energy levels and radiative rates

The energy levels and spontaneous radiative decay rates for the configurations of ($1s^22s^22p^6$) $3s^23p^63d^{10}$, $3s^23p^63d^94l$, $3s^23p^53d^{10}4l$, and $3s3p^63d^{10}4l$ (l = *s, p, d, f*) of Ni-like Tungsten ions are calculated using the Flexible Atomic Code (FAC) [8]. It is a fully relativistic approach that is based on the Dirac equation. The configurations mentioned above are included in the present calculations. The resulting energy levels of Ni-like Tungsten ion are given in table I. For comparison, the theoretical results obtained using the Grasp [3] and Dirac-Fock-Slater (DFS) code [11] are also given in table I. (ftp://159.226.36.237/Ni-likeW/Table1) From the table one can see a good agreement among these calculations.

The Spontaneous decay rates are calculated in the single multipole approximation. The detailed theory can be found in the reference [8]. In table II (ftp://159.226.36.237/Ni-likeW/Table2) we list the oscillator strengths and all dipole allowed spontaneous radiative decay rates for transitions.

3. Collision strengths and Effective collision strengths

For calculating the collision strengths, we adopt the first order Born approximation which free orbitals are calculated in a more realistic potential taken into account the electronic structure of the target ion. This method is called distorted wave approximation. In this paper we use the FAC code to calculate the collision strengths. In order to save the computation time we choose the quasi-relativistic approximation, in which the two wave functions with the same i but different j are treated in the same way. The results are very close to the fully relativistic results for the nuclear charge as large as 74 and collision energy as high as 30 keV [14]. We perform he code for 6 representative energies following Zhang et al.[11], E'=0.002, 0.01, 0.03, 0.07, 0.15 and 0.28, where E' in units of Z_{eff}^2 rydbergs with $Z_{eff}=51.5$. This range of energies is expected to cover those ordinarily needed for determining the collision rates for applications to very high temperature plasmas. In table III (ftp://159.226.36.237/Ni-likeW/table3), we list all the collision strengths of all excited levels of the Ni-like Tungsten ion. Excitation rate coefficients (in cm^3s^{-1}) for a transition I to f was calculated using the following formula,

$$C(i \to f) = \frac{8.629\times10^{-6}}{g_i\sqrt{kT_e}}\exp(-\Delta E_{if}/kT_e)\gamma(i \to f), \tag{1}$$

Where γ is the effective collision strength. Using the results of FAC 20 collision strengths are averaged over a Maxwellian velocity distribution to obtain effective collision strengths as a function of electron temperature. It can be defined as,

$$\gamma(i \to f) = \int_0^\infty \Omega_{i\to f}\exp(-\varepsilon_f/kT_e)d(\varepsilon_f/kT_e). \tag{2}$$

Here ε_f is the energy of the electron after the collision, g_i the statistical weight of the level i, ΔE_{if} the excitation energy, $\Omega_{i\to f}$ the collision strength. The temperature is Kelvin. In table IV the effective collision strengths are listed for electron temperature in the range (100-2500 eV), which can be used in the simulation of x-ray laser. The deexcitation rates are obtained by detailed balancing. The monopole collision excitation results of YODA code can be obtained by the fitting formula [15],

$$C(i \to f) = 1.58 \times 10^{-5} \frac{np(b)\exp(-b)}{\Delta E_{i \to f} \sqrt{kT_e}} \tag{3}$$

Where the notation $p(b)$ was described as,

$$\ln p(b) = A_0 + A_1 \ln b + A_2 (\ln b)^2 + A_3 (\ln b)^3 \tag{4}$$

b was $\Delta E_{if} / kT_e$, n the multiplier, all fitting parameters can be found in the reference [15]. In FIG.1, as an example, we show the electron excitation rate coefficient as a function of electron temperature for excitation of 1→36 obtained by our calculation and that of Daido et al [15]. It can be seen that good agreement is found between the two results.

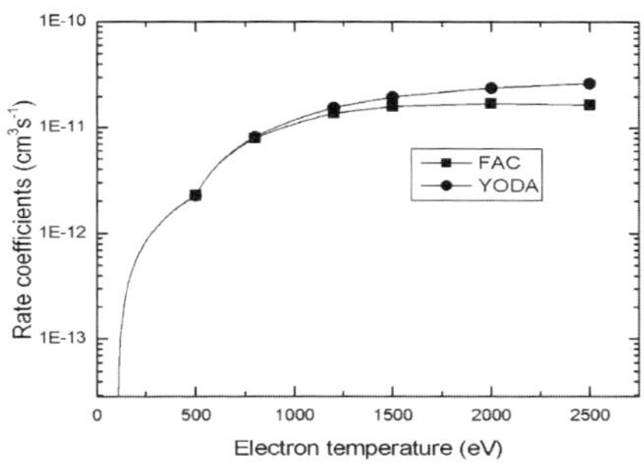

FIG.1. Collisional excitation rates as a function of electron temperature for transitions from the ground level 1 to the upper energy level 36. The solid curve with square is calculated by the FAC code and the solid curve with circle from the YODA code.

3. Conclusions

In conclusion, the energy levels, spontaneous radiative decay rates, and electron impact collision strengths are calculated for Ni-like Tungsten ions using FAC code. These data can be used for simulations of Ni-like Tungsten x-ray lasers. Effective collision strengths, obtained after integrating the collision strengths over a Maxwellian distribution of electron energies, have also been calculated for electron temperature in the range of 100-2500 eV. Some calculations are compared with other theoretical results and a good agreement is found.

AcknowledgmentsThis work is jointly supported by the NSFC under Grant Nos. 10176034, 10374114, 60321003, 10474134 and the NKBRSF under Grant No.G1999075206, No.G19999075406 and the National Hi-tech ICF program.

References

[1] R. C. Elton, X-Ray Lasers (Academic Press, New York, 1990)
[2] J. E.Trebes et al., Science 238, 517 (1987)
[3] L. B. Da Silva et al.,Science. 258, 269 (1992)

[4] D. Rees et al., Science. 265, 514 (1994)
[5] R. Cauble et al., Science. 273, 1093 (1996).
[6] D. H. Kalantar et al.,Phys. Rev. Lett. 76, 3574 (1996)
[7] J. P. Christiansen, D. E. T. F. Ashby, and K. V. Roberts, Computer Phys. Commun., 7, 271 (1974).
[8] M. F. Gu, Astrophys. J. 582, 1241 (2003)
[9] P. L. Hagelstein, Phys. Rev. A 34, 874 (1986)
[10] K. M. Aggarwal, P. H. Norrington et al., ATOMIC DATA AND NUCLEAR DATA TABLES 74, 157 (2000)
[11] H. L. Zhang, D. H. Sampson et al. ATOMIC DATA AND NUCLEAR DATA TABLES 48, 91 (1991)
[12] U. I. Safronova, W. R. Johnson et al., Phys. Rev. A 62(052502), 2000
[13] C. Z. Dong, S. Fritzsche et al., Journal of Quantitative Spectroscopy and Radiative Transfer, 6, 447(2003)
[14] H. L. Zhang, D. H. Sampson et al., Phys. Rev. A 40, 616(1989)
[15] H. Daido, S. Sebba et al., J. Opt. Soc. Am. B, 16, 2295 (1999).
[16] J. P. Christiansen, et al., Computer Phys. Commun., 7, 271 (1974).

Inst. Phys. Conf. Ser. No 186
Paper presented at 9th Int. Conf. X-Ray Lasers, Beijing, China, 24–28 May 2004

Atomic Data for Li-like Fe XXIV

K McKeown, K M Aggarwal, F P Keenan and S J Rose†

Department of Pure and Applied Physics, Queen's University, Belfast BT7 1NN, UK

† Department of Physics, Clarendon Laboratory, Oxford OX1 3PU, UK

Abstract. Energy levels are presented for transitions among the lowest 24 fine-structure levels of the $1s^2\ nl$ ($n \leq 5$) configurations of Li-like Fe XXIV. These were calculated using the fully relativistic GRASP code. Comparisons are made with other available results. The energy levels are assessed to be accurate to better than 1%. Also presented are the resonant transition radiative rates and the lifetimes for the excited levels.

1. Introduction

Li-like ions are of considerable importance in astrophysical and magnetically confined plasmas. Emission lines of Fe XXIV belonging to transitions among the $n \leq 5$ levels have been observed in solar flares, particularly in the X-ray region (below 50Å), and have been exclusively listed by Dere et al [1]. Also, radiation from the 1s $^2S_{1/2}$ – 2p $^2P^0_{3/2}$ transition of H-like aluminium is used to pump the 5p $^2P^0_{1/2}$ and $^2P^0_{3/2}$ levels of Fe XXIV in lasing plasmas [2,3], which populates the other $n = 5$ levels, making it possible to have gains for the 4p-5d, 4d-5f and 4f-5g transitions. The gains for these transitions are larger than that for the 4s-5p transitions [2]. Therefore, atomic data for transitions among the $n \leq 5$ levels of Fe XXIV are particularly important for a variety of applications.

For the reliable modeling and diagnostics of plasmas, accurate theoretical atomic data (such as energy levels, radiative rates and excitation rate coefficients) are required, which are generally not available from measurements. This paper presents our energy levels and radiative rates for resonance transitions in Fe XXIV, which is abundant in solar and fusion plasmas.

2. Calculations

The General purpose Relativistic Atomic Structure Package (GRASP) code of Dyall et al [4], is used to calculate energy levels and radiative rates. GRASP is a fully relativistic code based on the *jj* coupling scheme. We have used the option of extended average level (EAL), in which a weighted trace of the Hamiltonian matrix is minimized, to produce a compromise set of orbitals describing closely lying states with moderate accuracy. In addition, configuration interaction (CI), and Breit and Quantum Electrodynamics (QED) corrections have been included in the calculations.

3. Results

3.1 Energy Levels

Our calculated energy levels, along with the available experimental results of NIST (http://physics.nist.gov/PhysRefData) and the theoretical results of Whiteford et al [5] and Nahar and Pradhan [7], who used the Autostructure (AS) code of Badnell [6] and the R-matrix (RM) code of Berrington et al [8], respectively, are listed in Table 1.

Discrepancies among different sets of energy levels are within 1%. Furthermore, level orderings are generally consistent, with some minor discrepancies amongst the highest three levels. The RM results are slightly higher in general, and so are comparatively less accurate than the other results because two-body relativistic operators are not included in the RM code [8]. The GRASP energies are the lowest, except for the lowest three levels, are and in better agreement with the experimental results.

3.2 Radiative Rates

Radiative rates are calculated for electric dipole (E1), electric quadrupole (E2), magnetic dipole (M1), and magnetic quadrupole (M2) transitions, although E1 transitions are the dominant ones. However the other types of transitions are also required in plasma modeling and diagnostic calculations. In Table 1 we list our radiative rates for the resonance transitions alone. These radiative rates are the sum over all types of transitions, although in practice only one type of transition dominates. However, detailed results of radiative rates, oscillator strengths and line strengths for all types of transitions, and comparisons for E1 transitions, are given in our recent publication [9], and here we summarize below the main conclusions.

In general, the agreement between our GRASP and the AS results is within 20% for a majority of transitions and the greatest disagreement is a factor of two for the 2-16 transition. However, the RM results differ by up to a factor of 5 for 6 transitions, namely, 5-7, 6-7, 6-8, 13-14, 20-21 and 22-23 –see Table1 for level indices. All of these transitions belong to the degenerate levels of a state, for which the energy differences are very small and arise due to the inclusion of relativistic effects. The RM results are comparatively less accurate due to the omission of two-body relativistic operators. Finally, the AS radiative rates for the 5f $^2F^o_{7/2}$ – 5g $^2G_{7/2}$ (22-23) transition is in error (by several orders of magnitude), because it disagrees with all other known results – see Ref. [9] for further details and comparisons. Based on the comparisons made among different sets of results, and between the length and velocity forms of the radiative rates, we can state with confidence that the presently calculated GRASP data are the most accurate available to date for transitions in Fe XXIV.

Also included in Table 1 are the lifetimes (τ_j), which are obtained by the following simple expression:

$$\tau_j = \frac{1}{\sum_i A_{ji}}.$$

At present, we do not have any available theoretical or experimental results to compare with, but believe that the present results will be useful for future comparisons.

Table 1. Energy levels (Ry.), radiative rates (s^{-1}) and lifetimes (s) for transitions in Fe XXIV.
($a \pm b \equiv a \times 10^{\pm b}$).

Index	Configuration	NIST	GRASP	AS[5]	RM[7]	A_{ji} (s^{-1})	τ_j (s)
1	$1s^2 2s\ ^2S_{1/2}$	0.000	0.000	0.000	0.000	0.000-00	0.000-00
2	$1s^2 2p\ ^2P^o_{1/2}$	3.572	3.583	3.578	3.578	1.832+09	5.459-10
3	$1s^2 2p\ ^2P^o_{3/2}$	4.745	4.797	4.739	4.838	4.346+09	2.301-10
4	$1s^2 3s\ ^2S_{1/2}$	84.497	84.510	84.532	84.525	1.553+05	3.581-13
5	$1s^2 3p\ ^2P^o_{1/2}$	85.459	85.504	85.520	84.501	7.499+12	1.333-13
6	$1s^2 3p\ ^2P^o_{3/2}$	85.814	85.863	85.858	85.891	7.252+12	1.378-14
7	$1s^2 3d\ ^2D_{1/2}$	86.197	86.244	86.258	86.575	1.189+10	4.538-14
8	$1s^2 3d\ ^2D_{1/2}$	86.316	86.354	86.367	86.696	1.195+10	4.591-14
9	$1s^2 4s\ ^2S_{1/2}$	113.581	113.610	113.632	113.960	1.308+05	5.273-13
10	$1s^2 4p\ ^2P^o_{1/2}$	114.010	114.020	114.038	114.365	3.285+12	2.268-13
11	$1s^2 4p\ ^2P^o_{3/2}$	114.155	114.171	114.178	114.530	3.212+12	2.331-13
12	$1s^2 4d\ ^2D_{3/2}$	114.283	114.329	114.343	114.686	1.590+09	1.053-13
13	$1s^2 4d\ ^2D_{5/2}$	114.328	114.376	114.389	114.737	1.627+09	1.066-13
14	$1s^2 4f\ ^2F^o_{5/2}$	...	114.384	114.398	114.743	3.641-00	2.170-13
15	$1s^2 4f\ ^2F^o_{7/2}$	...	114.407	114.422	114.768	0.000-00	2.179-13
16	$1s^2 5s\ ^2S_{1/2}$	...	126.965	126.986	127.308	8.381+04	8.269-13
17	$1s^2 5p\ ^2P^o_{1/2}$	127.113	127.172	127.191	127.513	1.701+12	3.834-13
18	$1s^2 5p\ ^2P^o_{3/2}$	127.113	127.249	127.261	127.597	1.669+12	3.932-13
19	$1s^2 5d\ ^2D_{3/2}$	127.223	127.330	127.345	127.676	4.047+08	2.019-13
20	$1s^2 5d\ ^2D_{5/2}$	127.259	127.336	127.368	127.702	4.237+08	2.048-13
21	$1s^2 5f\ ^2F^o_{5/2}$	...	127.358	127.373	127.706	3.205-00	4.192-13
22	$1s^2 5f\ ^2F^o_{7/2}$	...	127.370	127.385	127.719	0.000-00	4.209-13
23	$1s^2 5g\ ^2G_{7/2}$	...	127.370	127.385	127.719	0.000-00	7.058-13
24	$1s^2 5g\ ^2G_{9/2}$	...	127.377	127.392	127.727	0.000-00	7.070-13

4. Further Work

We are currently calculating collision strengths (Ω) and excitation rate coefficients for all transitions among the 24 fine-structure levels of Fe XXIV using the Dirac Atomic R-matrix Code (DARC) of Norrington and Grant [10]. The DARC program includes relativistic effects in a systematic way.

In our calculations, the R-matrix has been taken to be 4.0 atomic units, and 80 continuum orbitals have been retained for the expansion of the wavefunction for each channel angular momentum. This permits us to calculate Ω over a wide range of energies below 2000 Ryd. All partial waves with angular momentum $J \leq 60$ are included to achieve convergence in Ω for all transitions and at all energies. Furthermore, we are resolving resonances in a fine energy mesh in order to account for their contribution, which is often significant, particularly for forbidden transitions, and at lower temperatures. These calculations are in progress, and will take some time to complete. We hope to report our results for the collisional data in the near future.

Acknowledgements

Financial support from AWE Aldermaston, DENI, EPSRC and PPARC of the United Kingdom are gratefully acknowledged. FPK and SJR are grateful to AWE Aldermaston for the award of William Penney Fellowships.

References

[1] Dere K P, Landi E, Young P R and Del Zanna G 2001 Astrophys. J. Suppl. 134 331
[2] Lee Y T, Howard W M and Nash J K 1990 J. Quant. Spectros. Radiat. Transfer 71 335
[3] Al'Miev I R, Rose S J, and Wark J S 2001 J. Quant. Spectros. Radiat. Transfer 71 129
[4] Dyall K G, Grant I P, Johnson C T, Parpia F A and Plummer E P1989 Comput. Phys. Comm. 55 425
[5] Whiteford A D, Badnell N R, Ballance C P, Loch S D, O'Mullane M G and Summers H P 2002 J. Phys. B 35 3729
[6] Badnell N R 1997 J. Phys. B 30 1
[7] Nahar S N and Pradhan A K 1999 Astron. Astrophys. Suppl. 135 347
[8] KA Berrington, W Eissner and PH Norrington 1995 Comput. Phys. Comm. 92 290
[9] McKeown K, Aggarwal K M, Keenan F P and Rose S J 2004 Phys. Scr. 70 - in press
[10] Norrington P H and Grant I P 2004 Comput. Phys. Comm. – in preparation

Inst. Phys. Conf. Ser. No 186
Paper presented at 9th Int. Conf. X-Ray Lasers, Beijing, China, 24–28 May 2004

Modelling of seeded X-ray laser

G. Faivre, Ph. Zeitoun
Laboratoire d'Optique Appliquée, chemin de la Hunière, 91761 Palaiseau, France

S. le Pape
Laboratoire pour l'Utilisation des Lasers Intenses, Ecole Polytechnique, 91128 Palaiseau, France

Abstract
X-ray lasers (XRL) and High order Harmonic Generation (HHG) are two complementary sources: XRL is energetic, but unpolarized and weakly coherent, whereas HHG are coherent, polarized, but produce low energy beam. By seeding XRL medium with HHG one may obtain an energetic beam with very good optical qualities. Among the different XRL schemes, solid target pumped by an intense laser seems to be the most favourable scheme for reaching the 0.1 mJ level for the amplified beam. But the problem for this kind of target is the refraction of the medium, which can deflect the seeded beam and make it leave the gain zone. It is well known that plasmas created from solid target exhibit strong density gradients that may degrade the amplified beam. To overcome this deleterious effect, HHG has to be seeded in a precise way. In this aim, we have developed a numerical code, called SHADOX describing the HHG seeding and amplification in the XRL plasma. Here we will show several examples of the abilities of this code, and the effect of different density gradient.

The main goal of this work was to simulate the amplification of an harmonic beam in a solid target plasma, and to check the effect of the refraction, so of the density profile of the plasma. In a solid target the electron density is much higher than for gas, for example Optical Field Ionisation (OFI) XRL and thus the expected output energy is higher than the output energy obtained with OFI XRL (P. Zeitoun et al., this Proceedings). However, the refraction becomes also more significant, so it is important to numerically investigate such system.

SHADOX is a numerical ray trace code originally developed at LIXAM (France) by S. Le Pape et al. [1], able to simulate the amplification and propagation of an x-ray laser in a gain medium. It is a post processor of EHYBRID, a 1D1/2 hydrocode describing the plasma evolution [2]. SHADOX solves the eikonal equation (propagation of rays in an inhomogeneous medium) according the paraxial approximation. More than 50 millions of rays can be sent, providing a better precision for the output beam. The position of each ray is calculated by taking into account the refraction of the medium. The code is fully coupled with EHYBRID, and has the possibility to update the data of the plasma at each time step. So it can simulate for example a travelling wave. Special attention has been taken for to the calculation of the amplification and saturation in order to estimate the absolute output energy.

This code was recently modified by including the propagation and amplification of a seeded beam (fig 1). In order to do that the laser is described as a gaussian beam in the following form:

$$I(\rho,z)=I_0\times\left[\frac{W_0}{W(z)}\right]^2\times\exp\left[-\frac{2\rho^2}{W^2(z)}\right] \tag{1}$$

where z is the axial distance, ρ is radial distance, and W0 is the beam radius in the waist.This beam can be focus independently in both direction X and Y, using the same equations as for lenses. The magnification of the "virtual lens" is calculated:

$$M=\frac{\left|\frac{f}{z-f}\right|}{\sqrt{1+\frac{z_0}{z-f'}}} \tag{2}$$

Thanks to this magnification the code calculates the new properties of the focused beam (new waist position, divergence, depth of focus..).

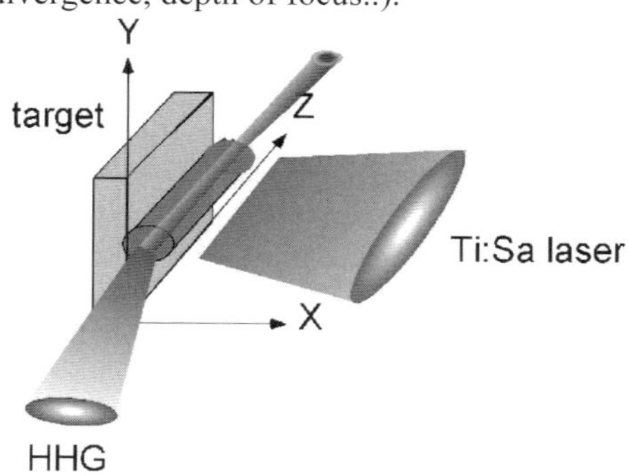

Figure 1 : **Geometry of the injection. The driving laser hits the target perpendicularly to the surface while harmonics are sent in the plasma longitudinally.**

An angle can also be assigned to the beam, to try to compensate the refraction effect. Many output parameters are available: intensity map at near and far field, dimensions of the beam, pointing angle and wave vector.

We performed several simulations for different cases relevant to the future "LASERIX" facility. Here we will show an example of raytracing simulation in a solid target plasma. Plasma was created by 3 pulses having the following parameters: 1/ Δt = 300 ps FWHM, 0.6 J, λ = 400 nm, 2/ 300 ps, 6 J, 400 nm, 1.5 ps after the first pulse, 3/ 10 ps, 20 J, 800 nm, 2.1 ps after the 2nd pulse. The plasma is 2 mm long and 200 μm wide. The seeded soft X-ray laser at the entry of the plasma has a gaussian intensity profile, and is focused in the Y direction. The gain (fig. 2a) and density (fig. 2b) profiles are calculated with EHYBRID.

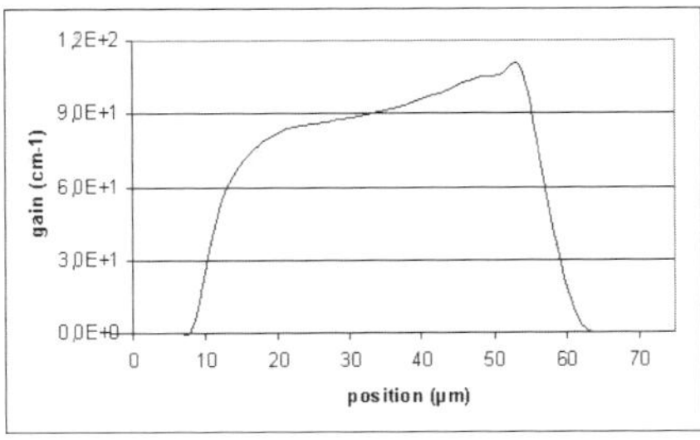

(a)

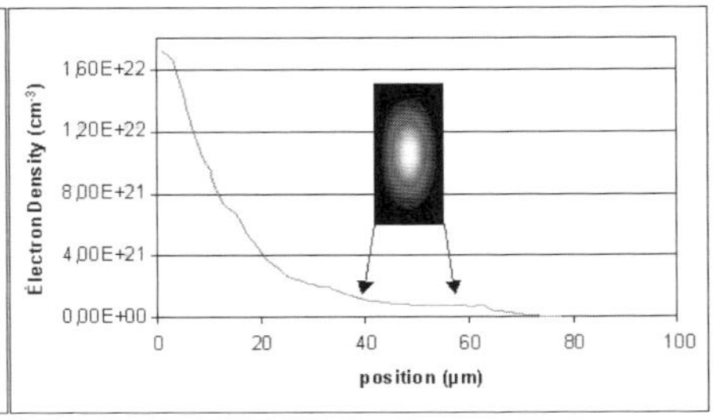

(b)

Figure 2: **gain (a) and density (b) profiles as coming from EHYBRID calculation at the time of seeding. The focal spot of the injected beam is about 14*90μm.**

In this simulations, we can see that the initially elliptic beam with a gaussian intensity profile begins to be distorted after only 1 mm propagation (fig 3a). After 2 mm it breaks up into several beamlets (fig 3b) although the density profile of the plasma seemed to be smooth.

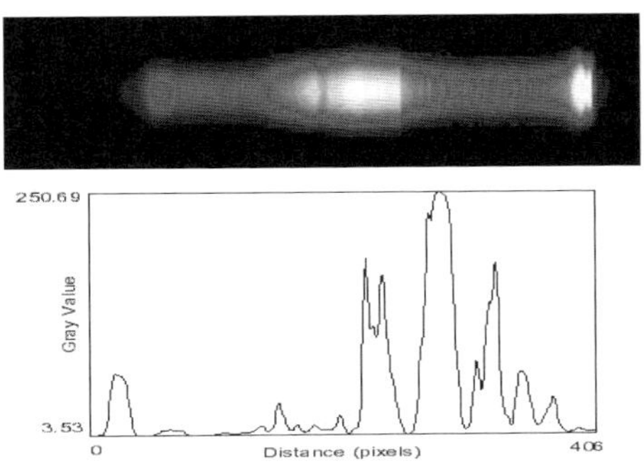

Figure 3a : **Shape and density profile of the beam after 1 mm propagation in the plasma**

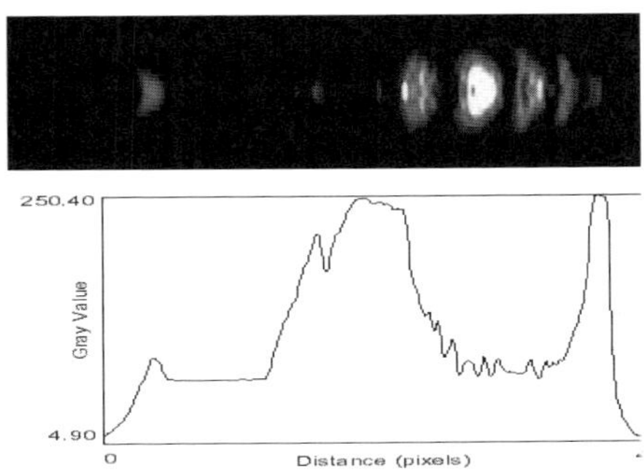

Figure 3b : **Shape and density profile of the beam at the end of the plasma (i. e. a 2 mm propagation)**

We have tried to estimate the influence of several plasma shapes on the beam degradation. In this aim we fitted the density profile with equation of different orders. The case with an equation of 0th order, i. e. a constant, is not interesting. For a linear density profile (equation of 1st order), there is no change in the beam shape while we observed a change in the beam pointing as expected. When we used an equation of second order to model the density profile (fig 4a) the refraction changes, so the beam propagation is not the same. In the middle of the plasma (fig 4b) the beam begins to break up, and at the end (fig 4c), after 2 mm propagation, the beam is separated in 2 main beamlets.

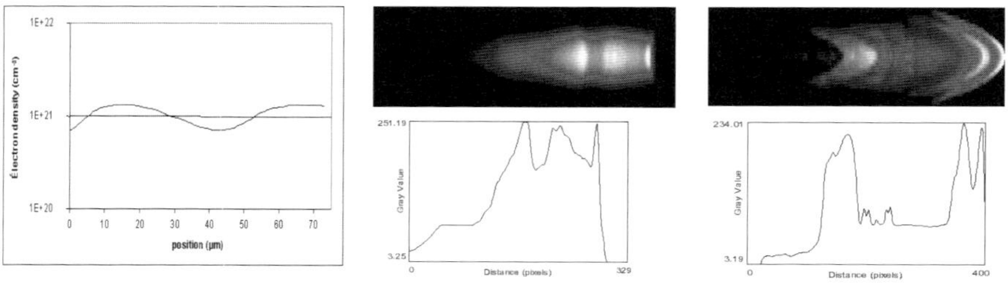

(a) (b) (c)

Figure 4 : **from the right to the left: a/ density profile of the plasma b/ beam in the middle of the plasma c/ beam at the end of the plasma**.

This example demonstrates the major role of the refraction in the propagation of a beam in a plasma. Even a small change in the density can lead to a strong refraction and so disturb the propagating beam. Since the density depends on the hydrodynamic of the plasma, it has to be modelled with an accurate hydrocode.

In this aim a new hydrocode, ARWEN [3], developed in Instituto Polytecnico de Madrid (Spain) will soon be coupled with SHADOX. This new code will allow us to describe more precisely the plasma. ARWEN is a fluid dynamic simulation code for 2D systems. It works under Adaptive Mesh Refinement (AMR), a very optimized scheme of calculations. Many physical features of the system are solved: fluid motion and heat exchange through both electron conduction and radiation transport, and energy deposition by several sources. It is also able to calculate temperature, velocity and density maps and can work with various kinds of target (slab, curved…).

The main goal of all these simulations is to prepare a future seeding in solid target. We have seen that the density gradient of the target, *i. e.* the amplifier, plays a major role in the propagation of the beam in a solid target. A minor change in this profile can lead to a complete distortion of the beam. Thus we have to simulate more accurately the evolution of plasma. In this aim we plan to make ARWEN fully coupled with SHADOX. The final goal is to define one or several types of plasma able to preserve the qualities of seeded beam after amplification.

References

1 S. Le Pape, Ph Zeitoun, Opt. Com. 219 (2003), 323
2 G. J. Pert, J. Fluid Mech. 131 (1983), 401
3 F. Ogando, P. Velarde, JQSRT 71 (2001), 541

Free-electron lasers and accelerator-based X-ray sources

Inst. Phys. Conf. Ser. No 186
Paper presented at 9th Int. Conf. X-Ray Lasers, Beijing, China, 24–28 May 2004

Experimental Results Towards X-Ray Free-Electron Lasers

S. Reiche
University of California, Los Angeles

Abstract. The Free-Electron Laser exploits the collective instability of the radiation self-interaction of an electron beam, coupled by the periodic motion within an undulator or wiggler. The beam current is modulated and emits coherently on the resonant wavelength of the FEL. Because the wavelength is determined by the energy of the driving electron beam, there are no fundamental limitations to achieving radiation in the X-ray regime. The Self-Amplified Spontaneous Emission (SASE) Free-Electron Lasers avoids the problem of an external radiation source for seeding, by starting from the spontaneous undulator radiation within the bandwidth of the FEL. Several experiments have successfully proven this mechanism down to the VUV regime, supporting the case for X-ray FELs, planned or under construction, worldwide. This presentation gives the status of several milestone FEL experiments, achieving shorter and shorter wavelengths and discusses the performance and limitations of X-ray Free-Electron lasers.

1. Introduction

In free-electron lasers [1] a relativistic electron beam is injected into a periodic magnetic field. Radiation is emitted by the enforced transverse oscillation, which then acts back on the electron beam. Under the right condition the radiation can be amplified up to multiple gigawatts of radiation power. The advantage of an FEL is that the radiation wavelength is freely tuneable by the electron beam energy. It also overcomes any restrictions in the achievable wavelength by optical elements when saturation is reached within a single pass.

Earlier FEL experiments [2] showed only a gain of a few percents and relied on an optical cavity to accumulate the gain and to reach saturation. However with the ongoing improvement of the quality of the driving electron beam single-pass experiments became feasible [3]. Aiming for a possible application in the X-ray regime the recent results have shown a steady decrease in the wavelength down to 80 nm Based on that Free-Electron Lasers in the VUV and X-ray regime are currently under construction [4,5].

2. Working Principle of Free-Electron Lasers

The underlying principle of the Free-Electron Laser is the emission of radiation in an undulator [6] - a magnet with a periodic transverse magnetic field. The radiation adds up constructively for the resonant wavelength

$$\lambda = \frac{\lambda_u}{2\gamma^2}(1 + a_u^2)$$

where λ_u is the period of the magnetic field, $a_u = e\lambda_u B_{rms}/4\pi mc$ is the undulator parameter, B_{rms} is the rms on-axis magnetic field and γ is the electron beam energy in units of its rest mass mc^2. The radiation field advances by one radiation wavelength with respect to the electron over one undulator period.

Because the undulator field enforces a transverse velocity component, the electrons can couple with the co-propagating radiation field. Depending on its relative position to the radiation field, electrons are loosing or gaining energy and a sinusoidal energy modulation is imprinted into the electron beam. On the microscopic scale of the radiation wavelength faster electrons are catching up and slower ones are falling behind, resulting in an periodic accumulation of the electrons with the spacing of the radiation wavelength. This current modulation causes a stronger emission of radiation because the degree of coherence is increased.

This "run-away" process can yield a collective instability with an exponential growth of the radiation field [7] as well as the energy and current modulation. The characteristic scale of this process is the gain length

$$L_g = \frac{\lambda_u}{2\pi\sqrt{3}\rho}$$

defining the ρ-parameter [8]

$$\rho = \left(\frac{a_u}{4\gamma}\frac{\omega_p}{\omega_u}\right)^{\frac{2}{3}}$$

with $\omega_u = 2\pi c/\lambda_u$, the plasma frequency $\omega_p = (4\pi r_e 2 n_0/\gamma)^{1/}$, the electron density n_0 and the classical electron radius r_e. Typical values for ρ are between 10^{-2} - 10^{-3}.

The amplification, which is started either by a seeding radiation field or a modulation in current or energy, comes to a halt when maximum bunching is achieved and the electrons slips out of the FEL bandwidth. At that point the electron beam has transferred about ρ times its energy to the radiation field. The saturation power is therefore

$$P_{sat} = \rho P_b = \rho\frac{\gamma mc^2 I_p}{e}$$

where P_b is the beam power and I_p is the peak current.

Analyzing the motion in the longitudinal phase space the FEL process rotates the electron distribution about 90 degree within the relative FEL bandwidth of 2ρ. This defines the requirements for the beam quality. The relative energy spread has to be much smaller than the bandwidth ($\sigma_\gamma/\gamma \ll \rho$). Focusing, which strength is described by the β-function

[9], couples the transverse emittance ε_n to the longitudinal motion similar to an effective energy spread [10], which has to fulfil the requirement as well. It yields the constraint $\varepsilon_n \ll (4\lambda\beta\gamma/\lambda_u)\rho$.

FEL amplifiers, which rely on an external seeding field, generate a fully transverse and longitudinal coherent output signal with a Fourier-limited bandwidth in the spectrum, but they are limited to the visible and near UV regime. Towards shorter wavelength the FEL can start from its own incoherent undulator radiation and thus overcoming the wavelength limitation. These Self-Amplified Spontaneous Radiation (SASE) FELs [11] have different output characteristics. The field information propagates one radiation wavelength per undulator period. Thus, the longitudinal coherence is limited to the cooperation length $L_c=\lambda/4\pi\rho$ [12], resulting in spikes in the radiation profile, when the coherence length is shorter than the bunch length. The same number of spikes is also observed in the spectrum, filling up the full FEL bandwidth of 2ρ. As the FEL amplifier the SASE FEL provides full transverse coherence.

3. Experimental Results

This section discusses the results of two single pass-high gain free-electron lasers, both operating as SASE Free-Electron Lasers. Their main parameters are listed in Tab.1 together with other SASE FELs.

3.1. The TESLA Test Facility FEL at the Deutsches Elektronen Synchotron

The beamline of the Tesla Test Facility serves two purposes. First, it tests the feasibility of a linear collider -- TESLA -- based on superconducting technology. Eight nine-cell cavities, made out of Niobium, are combined to a single accelerating module. Due to its low losses a single L-band klystron at 1.3 GHz can operate two accelerating modules, delivering an rf pulse of several hundreds of microseconds. An exception is the normal conducting 1.6 cell rf photo-electron gun, because the strong solenoid field for a low emittance electron beam excludes superconducting technology.

The second purpose of the Tesla Test Facility is to provide a high-brightness electron beam to drive a SASE Free-Electron Laser in the UV regime. In the initial design of the future X-ray FEL it was foreseen that the operation of the linear collider would be interlaced with generating bunch trains, serving multiple undulators, tuned to different wavelengths. Recently the two projects have been separated.

Table 1: Parameters of single pass high-gain free-electron laser experiments.

	Units	LANL [13]	UCLA [14]	LEUTL [15]	TTF [16]	VISA [17]	DUVFEL [18]
Energy	MeV	17	18	255	250	71	177
Spread	%	0.33	0.25	0.4	0.06	0.1	0.15
Emittance	μm	7	5.3	8.5	6	2.3	5
Current	kA	0.25	0.17	0.63	1.3	0.25	0.3
Period	Cm	2	2.05	3.3	2.73	1.8	3.9
Parameter	-	0.62	0.74	2.1	0.82	0.89	0.78
Length	m	1	2	25	15	4	10
Wavelength	μm	15.3	12.0	0.385	0.095	0.84	0.266

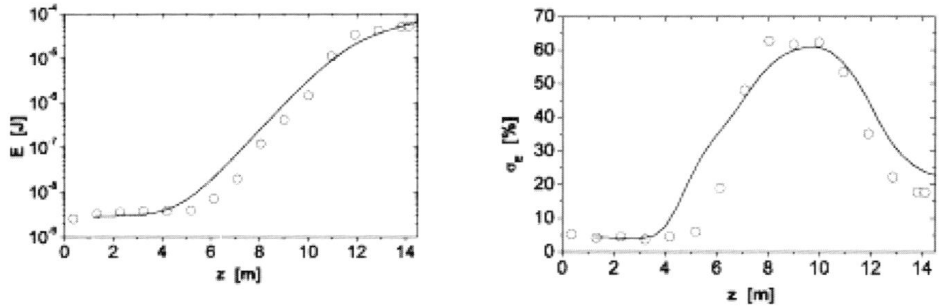

Figure 1 - Energy and fluctuation of the radiation pulse (left and right plot, respectively) as a function along the undulator. The circles are measurements and the solid line results from the start-end simulation.

The TTF FEL undulator consists out of three modules with 5 m length each. Between each module a diagnostic section allows for bunch position profiles measurements, using wire scanners. Additional beam position monitors and corrector magnets are integrated into the vacuum chamber. The corrector magnets can excite a large betatron oscillation so that any further FEL amplification is inhibited. With that the FEL properties along the undulator can be measured. A strong focusing quadrupole lattice reduces the beam size within the undulator to enhance the FEL performance. Because the quadrupoles are integrated into the undulator modules the alignment tolerances of the quadrupoles are directly carried over to those of the modules.

The beam energy of the TTF Free-electron laser in its first phase was 250 MeV and the bunch length was designed to be compressed down to 50 micrometer rms. However beam diagnostic and simulations of the beam line reproduce a different setting for the best performance of the FEL. The initial bunch length was longer than planned by design. The magnetic bunch compressor over-compressed the electron bunch resulting in a current spike of more than 2.5 kA and a FWHM length of less than 10 micrometer. The majority of the bunch charge of up to 3 nC was in the trailing tail in the bunch profile with low current values and spoiled emittances.

Saturation was achieved within 14 m [16] for a wavelength down to 80 nm (see Fig.1). The shot-shot fluctuation in the FEL signal had its maximum of 60% in the linear regime of the FEL before saturation. In the post-saturation regime the fluctuation was reduced to 25%. The fluctuation and the single-shot measurement of the spectrum indicated the presence of 2.6 spikes in average. The experimental results have been reproduced in great detail by start-end simulations [19] where the results of the electron beam diagnostic (beam emittance, energy spread, bunch length) had been matched to the numerical predictions. The resulting electron distribution then was used for the FEL simulation, verifying all observed FEL parameters such as central wavelength, radiation divergence, gain length, shot-shot fluctuation and average spectrum (see Fig.2).

Phase II of the Tesla Test Facility FEL is currently under commissioning. More accelerating structures were added to reach beam energies of 1 GeV. Three more undulator modules increases the total undulator length to 30 m. Due to a more stringent alignment tolerance the quadrupoles are separated from the undulator modules and placed in between. The nominal radiation wavelength will be 6.4 nm and the FEL will serve after initial measurement as a user facility.

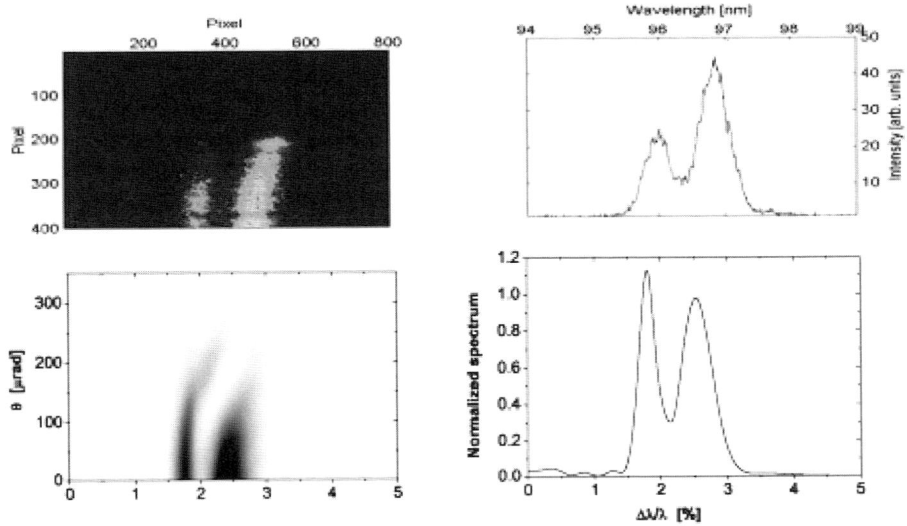

Figure 2 - Measurement and prediction by simulation for a single-shot spectrum (top and bottom plots, respectively).

3.2. The VISA FEL at Brookhaven National Laboratory

The Visible-Infrared Self-Amplified (VISA) Free-Electron Laser is located at the Accelerator Test Facility at the Brookhaven National Lab. The SASE FEL operates at a wavelength of 840 nm, corresponding to a beam energy of about 70 MeV.

The FEL reached saturation [17] for rf phases, significantly deviating from on-crest acceleration in the accelerator. This coincides with a maximum signal from the Coherent Transition Radiation (CTR) measurement, indicating a maximum compression of the electron bunch. It occurred in the dogleg of the beam transport line from the acceleration section to the undulator. At optimum setting the current was enhanced by a factor 5 to 250 A. Modelling the beamline by start-end simulations indicates that a big contribution arises from non-linear compression, namely the second order terms in the beam energy and emittance. The presence of a scraper in the dogleg introduced a correlation between rf phase and bunch charge. This eliminated the machine jitter in the FEL performance by binning the results with the charge.

The VISA undulator consists out of 4 modules of 1 m length each. The modules are placed close together to exclude field taper at the entrance and exit to avoid kicks in the electron orbit. An alternating gradient quadrupole lattice is superimposed onto the undulator lattice to increase the electron density. With the resulting beam size the FEL operation is limited by diffraction, where the FEL performance depends rather on the beam current and further increase in the electron density would not improve the performance. The modules are put into vacuum so that a vacuum chamber does not limit the available space for intra-undulator diagnostic. Pop-ins with OTR screens and plain mirrors allow the alignment of the beam trajectory and the measurement of the z-dependence of FEL parameters, such as power, fluctuation and spectrum.

Exponential growth of the radiation field has been observed with a gain length of about 18 cm [17]. The FEL reached saturation after 3.5 m with an output pulse energy of 20 microjoule. The central wavelength was 840 nm with a bandwidth of 2%. The strong shot-shot fluctuation as well as the observation of a single spike in the spectrum were in agreement with a short electron bunch length due to the compression in the transport line. The hollow angular distribution in the far field zone indicated that higher modes were excited parasitically in the FEL amplification. It has been reproduced by start-end simulations (see Fig.3) and arises from a correlation between energy and transverse position. An additional source is the asymmetric transverse bunch profile due to the scraper in the transport line.

A second series of measurement investigated the FEL performance with a strong energy chirp of the electron beam. Although the aperture of the beam transport limits the chirp to 2%, the measured spectral width was up to 10%. It is caused by the coupling of higher modes with stronger Guoy phase shifts, resulting in a significant broadening of the FEL spectrum.

It is planned to repeat the experiment of chirped electron pulses with a linearized beam transport. This suppresses the coupling of higher modes and allows for advanced measurements such as FROG traces or to slice the pulse with a monochromator. The generation of a chirped FEL pulse is essential for proposed schemed of X-ray FELs to achieved radiation pulse length below 10 fs.

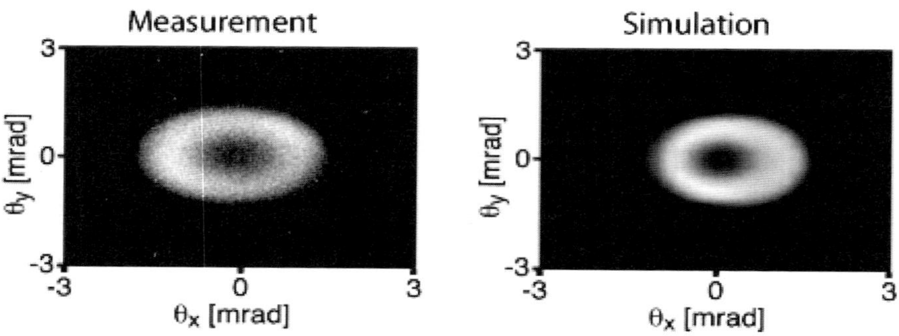

Figure 3 - Measured and predicted angular distribution of the VISA FEL in the far-filed zone (left and right plot, respectively).

4. Outlook

Based on the successful proof-of-principle experiments, listed in the previous section, several proposals for Free-Electron Lasers in the VUV and X-ray regime have been granted. One is the energy upgrade of the Tesla Test Facility FEL. With a beam energy of 1 GeV the wavelength range of the TTF FEL is extended down to 6.4 nm. The commissioning starts in the summer of 2004. Two Free-Electron Lasers in the Ångstroem regimes - the Linac Coherent Light Source [4] at the Stanford Linear Accelerator Laboratory and the TESLA Free-Electron Laser [5] at the Deutsches Elektronen Synchrotron - are currently under construction and be expected to generate the first light in 2008. Although the proof for a successful operation of an FEL below 10 nm has still to be shown this next generation of FELs are designed as a radiation source user facility. The FEL beam can be directed to multiple user station or in the case of the TESLA FEL several electron bunches can serve different undulators.

Acknowledgement

I would like to thank S. Milton, L.H Yu, J. Rossbach, M. Yurkov, A. Murokh and M. Hogan for their help to provide us with the information for the experimental results, presented in this paper.

References

[1] Madey J M J 1971 J. Appl. Phys. 42 1906

[2] Elias R L et al 1976 Phys. Rev. Lett. 36 717

[3] to be filled

[4] Linac Coherent Light Source (LCLS) 1998 SLAC-R-521 UC-414

[5] TESLA – Technical Design Report 2001 DESY 2001-011 TESLA Report 2001-23 TESLA-FEL 2001-05

[6] Motz H 1951 J. Appl. Phys. 22 527

[7] Kondradenko A M and Saldin E L 1980 Par. Acc. 10 207

[8] Bonifacio R, Pellegrini C and Marducci L M 1984 Opt. Comm. 50 373

[9] Edwards DA and Syphers M J 1993 *An Introduction to the Physics of High Energy Accelerators* John Wiley and Sons, New York

[10] Smith T I and Maday J M J 1982 Appl. Phys. A381 545

[11] Derbenev Y S, Kondradenko A M, et al., 1982 Nucl. Inst. & Meth. A193 415

[12] Bonifacio R et al 1994 Phys. Rev. Lett. 73 70

[13] Nguyen D C et al 1998 Phys. Rev. Lett. 81 810

[14] Hogan M et al 1998 Phys. Rev. Lett. 81 4867

[15] Milton S V et al 2001 Science 292 2037

[16] Ayvazyan V et al 2002 Phys. Rev. Lett. 88 104802

[17] Murokh A et al 2003 Phys Rev E67 0666501

[18] DiMauro L et al 2003 Nucl. Inst. & Meth. A507 15

[19] Saldin E L et al, 2003 Proc. Of the 25th Free-Electron Laser Conference 2003, Tsukuba

Inst. Phys. Conf. Ser. No 186
Paper presented at 9th Int. Conf. X-Ray Lasers, Beijing, China, 24–28 May 2004

The Analysis of Laser Electron X-ray Generator based on Thomson Scattering

E.G. Bessonov, R.M.Fechtchenko, M.V. Gorbunkov, A.V. Vinogradov*
P.N. Lebedev Physical Institute, 119991 Russia, Moscow Leninskii Prospect 53

V.I. Shvedunov,
Nuclear Physics Institute of Moscow State University, 119899 Russia, Moscow, Vorobyevy Gory

Abstract. A source of medical x-rays based on a 50 Mev storage ring and a quasi-continues picosecond laser is considered. It is shown that such generator produces useful X-ray flux with higher average power and higher efficiency than that of conventional X-ray tubes. The main energy losses are related to coherent synchrotron radiation. Taking this into account the required parameters of the storage ring and injector are determined.

1. Introduction

The experiments with synchrotron radiation sources confirmed that the use of narrowband X-ray beams extends substantially diagnostics capabilities in mammography, bronchography, angiography, computer tomography etc [1-3]. Apart from the image contrast improvement there are several other advantages of this method: possibility for subtraction of images taken at the wavelengths before and after absorption K-edge of the contrast element (usually iodine), reduction of radiation dose and amount of the contrast material, intravenous (catheterless) angiography of coronary arteries, etc. Actually new attractive methods of X-ray diagnostics are emerging.

However further development and practical applications of these methods depend on the availability of a compact (as compared with synchrotrons) narrowband X-ray source suitable for installation in medical centers and hospitals. As a rule it should work in the energy range 20 -50 kev that includes K absorption edges of the elements I and Gd used in contrast compounds.

As a solution of the problem of narrowband compact X-ray source it was proposed to generate X-rays in collisions of a laser beam with an electron beam moving in opposite direction. In such an approach the electron beam energy decreases from several Gev (as for the systems based on synchrotrons with magnetic undulators) to tens Mev and the dimension of the device changes from tens of meters down to several meters. To produce electron beams for

* Corresponding author: vinograd@sci.lebedev.ru

this purpose both storage rings [4,5] and linear accelerators [6] can be utilized. In this paper the energy losses and efficiency of X-ray generator based on a storage ring and repetitive laser is considered.

2. Laser-electron X-ray generator (LEXG)

The X-ray photon flux required for coronary arteries subtraction angiography is estimated as:

$$\Phi \approx 2*10^{14}\ \mathrm{s}^{-1} \tag{1}$$

at the surface of a human body [7,8]. The time necessary to obtain a single image must be 2-4 ms, and the frame rate – 20 – 30 s^{-1} [9].

In the electron rest frame the process of laser photon scattering by an electron is the classical Thompson scattering. The energy of the scattered (X-ray) photon in the laboratory frame is:

$$\hbar\omega = 4\gamma^2 \hbar\omega_L, \tag{2}$$

where $\hbar\omega_L$ – laser photon energy and $\gamma = E/mc^2$, E – electron energy. The average X-ray photon flux caused by Thomson scattering is:

$$\Phi = N_e \frac{\sigma_T}{s} \Phi_L (1-r)^{-1}, \quad s = s_e + s_L, \tag{3}$$

where N_e is the number of electrons in the bunch, $\sigma_T = 6.6 \cdot 10^{-25}\ \mathrm{cm}^2$ – Thompson cross-section, Φ_L – the laser photon flux, $(1-r)^{-1}$ – the finesse of the optical cavity, which is used to enhance the power of the laser beam in the interaction region, s_e and s_L – the cross section areas of the electron and laser beams respectively. We assume that both laser and electron beams consist of bunches following with equal repetition rate and having equal cross-section areas $s_e = s_L = s/2$ (see Appendix A).

Using (2), (3) it is easy to show that X-ray flux required by (1) can be attained with a system consisting of a storage ring, a laser and an optical cavity (see Fig.1),

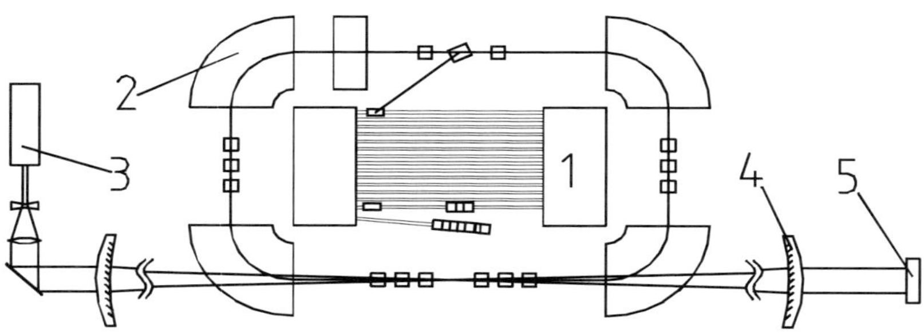

Fig 1. The scheme of laser-electron X-ray generator: 1 - injector, 2 – storage ring, 3 - laser, 4 – optical cavity, 5 – a damp for laser beam

which parameters are given in Table I. The listed parameters either have been already achieved or close to the achieved ones in existing setups [10].

The electron bunch length in Table I was chosen equal to double Raleigh length of laser beam: $l_e = 2z_L = 4s_L/\lambda = 4s_e/\lambda$ assuming the Gaussian electron density distribution in transverse direction (see Appendix A). Then the storage ring emittance is ~0.1 mm·mrad. In the next sections we estimate the electron beam energy losses and the X-ray generator efficiency.

Table I. The parameters of the laser electron X-ray generator. E – the electron energy, R – the bending radius, C – the orbit circumference, T – the revolution period, J – current, l_e – the bunch length, f – repetition frequency, ε_L – the laser pulse energy, P_L – the average laser power, τ_L – the laser pulse duration, ε_r – the energy of the laser pulse in the cavity.

Storage ring							
E, Mev	γ	***R*, m**	***C*, m**	T, ns	N_e	J, A	l_e, cm
46	91	0.5	6	20	$1.25 \cdot 10^{10}$	0.1	1.2

The source of optical radiation							
Laser						Cavity	
$\hbar\omega_L$, ev	F, MHz	s_L cm^2	ε_L, μJ	$P_L = f \cdot \varepsilon_L$, W	τ_L, ps	$(1-r)^{-1}$	ε_r, mJ
1	50	$3 \cdot 10^{-5}$	1.2	60	<40	$4 \cdot 10^3$	4,8

3. The synchrotron radiation losses

The main source of electron energy losses in our storage ring (see Table I) is coherent synchrotron radiation (CSR), which spectrum lies in the microwave region. The CSR power for a bunch with Gaussian longitudinal charge density distribution can be estimated as (see Appendix A):

$$P = \frac{3^{1/6}\Gamma^2(2/3) r_e c N^2}{2\pi R^{2/3} l^{4/3}(1+\mu)} mc^2 \tag{4}$$

where r_e=2.8·10^{-13} cm, $1+\mu = C/2\pi R$, C – is the perimeter of the electron orbit. Substituting the values from Table I for the LESG into (4) and also supposing $\mu = 1$ we obtain:

$$P = 1.1\ \text{kW}, \qquad \Delta E = 11\ \text{kev}, \tag{5}$$

where ΔE is the energy lost by an electron in one revolution. This energy loss is compensated by the storage ring radio frequency (RF) cavity. The accelerating voltage there should not be less than $U=250$ *kV* in order to provide the stability of the electron bunch (see Appendix B). In the next section the storage ring power consumption is estimated.

4. Storage ring and injector

As an injector for the storage ring it is reasonable to take a compact highly efficient race-track microtron with photo-cathode electron gun [11]. To simplify storage ring operation the injection energy should be close to the final energy of the stored electrons. To store ~1 nC charge in a single ring bunch multi-turn injection can be used with the injector operating in single bunch mode with 100-150 pC bunch charge and bunch repetition frequency dependent on the ring damping time. The use of rare-earth permanent magnets in the injector and storage ring design will essentially decrease energy consumption.

In the storage ring with permanent magnets the required electric power is defined mainly by an RF power P_{RF} necessary to produce the accelerating structure voltage U inside the accelerating structure (single RF cavity or a chain of coupled cavities):

$$P_{RF} = \frac{U^2}{ZL}, \tag{7}$$

where Z – is the effective shunt impedance and L – is the structure's electrical length. Because of $Z \propto \sqrt{f_{RF}}$, where f_{RF} is operating frequency, for a fixed structure length the RF power required decreases with frequency increase. Some aspects of storage ring beam dynamics are also favorable to higher frequency. Other considerations, in particular the separatrix size, the cavity beam hole aperture favors lower frequency. The final choice hasn't been made yet, so two possibilities were considered: 2856 MHz – equal to the injector operating frequency [11], and 571.2 MHz – the fifth sub-harmonic. Consider elliptical form cavity, which has 1.5-2 times lower shunt impedance as compared with the optimized cavity, but is more stable with respect to multipactor discharge and parasitic modes excitation. With such a cavity one obtains $Z \approx 30$ MΩ/m and $P_{RF} \approx 8$ kW for $L = 0.2625$ m (five half wavelength coupled cavities chain) at 2856 MHz, and $Z \approx 13.4$ MΩ/m, $P_{RF} \approx 18$ kW for the single cavity of the same length at 571.2 MHz. The RF power dissipated per unit length in the considered cases is 30 kW/m and 69 kW/m, respectively. These thermal loads will not change essentially structure parameters if the cooling channels are properly constructed. Taking into account beam loading and RF amplifier efficiency, the full power consumption from the socket will be about 20 and 40 kW, respectively.

5. The source of the optical radiation and its efficiency.

Power consumption of optical unit of LEXG can be an order of magnitude less than that required for RF system if a solid state laser with diode pumping is used. Such lasers reached tens of percent efficiency during last decade. For example, in [12] the efficiency 28% is reported for Nd:YAG laser having quasi-continuous output with average power more than 300 W. The active element was 6 mm diameter and 50 mm length cylinder pumped by six high power diode bars.

Besides of efficiency, the stability of laser radiation is one more important parameter for LEXG. The enhancement of laser light power inside high finesse optical resonator implies both equal repetition rates and careful phase matching of laser and resonator pulses. In other words the most part of every laser mode energy must be concentrated inside the bandwidth $\Delta\nu_r$ of the corresponding resonator mode, $\Delta\nu_r = c(1-r)/2\pi \cdot r^{1/2}L_c$, where L_c is the length of the cavity. From this point of view optically pumped solid state lasers generating subpicosecond- and femtosecond pulses again have evident advantages [13]. For example, crystalline active mediums doped with ytterbium are successfully utilized for continues sequences of powerful subpicosecond pulses production. In [14] the average power of 60 W was obtained in a Yb:YAG laser with the duration of a single pulse 0.81 ps, and optical power of a linier diode assembly being 370 W. Estimated efficiency is not less than 5%.

A significant decrease of the optical source average power can be achieved with the transition from quasi-continues ultrashort pulse sequence to generation of pulse trains with train duration 2-4 ms corresponding to accumulation time of one X-ray image and with train repetition rate being 20-30 Hz. Such mode of operation can be reached using external negative feed-back loop [10].

Thus from the above discussion it follows that the power necessary to feed the optical unit of LEXG is less than 2-3 kW.

6. Discussion

Estimations given in sections 4 and 5 show that for production of ~1 W (see (1)) of X-ray radiation with photon energy $\hbar\omega_L = 33$ kev a laser-electron X-ray generator with electrical power consumption 25-45 kW is required. In respect of efficiency (the ratio of useful X-ray flux to electric power consumption) LEXG surpasses X-ray tubes. Note that medical X-ray tubes utilized in angiography have power up to 100 kW [9,15,16]. No doubts that if physical and technical problems connected with the development of LEXG are resolved such X-ray sources will find wide application in medicine and other fields.

Except the scheme shown in figure 1 there are other possibilities to increase the coupling of electron and laser beams in order to provide necessary X-ray flux. For instance high-energy laser pulse circulation obtained by switching of a Pockel's cell inside the optical resonator may be used [17]. The practical optical scheme should satisfy several conditions: resistance of the mirrors and other optical elements to laser and X-ray radiation loading, synchronization of laser pulses with electron bunches, time and phase matching of optical resonator with the laser beam etc. This will require special research.

7. Conclusion

A new X-ray generator for medical applications can be designed on the basis of a storage ring with electron energy ~50 Mev and a picosecond laser. If high finesse optical resonator is used the plug efficiency of such a generator can be higher than that of conventional X-ray tubes. Other advantages of LEXG as compared to X-ray tubes are: narrowband X-ray beam, the absence of high energy "tail" in photon spectrum, the possibility of fine tuning of X-ray photons energy by changing the electron energy, image quality improvement by subtraction of images taken at different wavelengths near iodine K absorption edge, reduction of the necessary quantity of contrast materials and radiation dose received by a patient and medical staff, opportunity of catheterless angiography introduction into the wide medical practice.

Thus, a compact laser-electron X-ray generator can be considered as a promising direction of accelerator, laser and optical technology development.

Appendix A. On the electron energy losses connected with coherent synchrotron radiation.

Derivation of formula (3) assumed Gaussian radial density distribution of electrons and photons:

$$\rho(x) = \frac{1}{\sqrt{2\pi}\sigma} e^{-x^2/2\sigma^2}, \tag{A1}$$

where σ is the mean square transversal beam size connected with the effective area of the cross section as $s = 2\pi\sigma^2$. In the longitudinal direction the mean square transversal laser beam size changes in accordance with formula:

$$\sigma_L(z) = \sigma_L \sqrt{1 + \frac{z^2}{z_L^2}}, \tag{A2}$$

where $z_L = 2s_L / \lambda = 4\pi\sigma_L^2 / \lambda$ is the Raleigh length.

The mean square transversal size of the electron beam is governed by the same law. The role of Raleigh length in this case plays the β-function at the central point of interaction of laser and electron beams. The focal beam size is determined by the expression $\sigma_e = (\varepsilon\beta)^{1/2}$, where ε is the emittance of the electron beam.

Formula (4) for coherent synchrotron radiation losses was obtained by L.Schiff in 1946 [18]. Similar formulas were derived in [19-21]. However numerical coefficients there are respectively $2^{1/6} \approx 1.12$, $27^{/3} \approx 5.04$ and $2^{8/3} \approx 6.35$ times more than in (4) . The analysis made by E.G. Bessonov and R.M. Fechtchenko [22] supported the results of L.Schiff.

Appendix B. The effect of coherent synchrotron radiation on the storage ring beam dynamics.

Balance of energy gained by an electron in the RF cavity and lost due to synchrotron radiation and Thomson scattering defines an equation for electron phase oscillations in the storage ring [23,24]:

$$\frac{d^2\varphi}{dt^2} + \frac{k_i P_{noncoh}^{rad}}{E}\frac{d\varphi}{dt} - \frac{he\Omega^2}{2\pi E}\frac{\alpha\gamma^2 - 1}{\gamma^2 - 1} V(\varphi) = 0, \tag{B1}$$

$$V(\varphi) = \frac{2\pi}{\Omega e}\left[P_{coh}^{rad}(\varphi) - P_{coh}^{rad}(\varphi_s)\right] + U(\cos\varphi - \cos\varphi_s), \tag{B2}$$

where k_i is the exponent in the energy dependence of incoherent losses, P_{noncoh}^{rad} - the total energy losses due to incoherent synchrotron radiation and Thomson scattering, $P_{coh}^{rad}(\varphi)$ - the energy losses for coherent synchrotron radiation, $\Omega = 2\pi R/(1+\mu)c$ – the revolution frequency (which is supposed to be equal to the laser repetition rate), α - the momentum compaction factor, φ_s – the particle phase in the center of the bunch, the rest parameters were defined above. Equations (B1), (B2) determine conditions for bunch stability in the storage ring.

Incoherent synchrotron radiation and Thomson scattering provide slow damping of the phase oscillations and can be neglected in the first approximation. Coherent synchrotron radiation force acting on the electrons enlarges the bunch length [20]. The stability criteria in terms of equation (B1) are provided by relation $\frac{dV(\varphi)}{d\varphi} < 0$ kept in the vicinity of phase φ_s, occupied by bunch electrons. Using equation for coherent synchrotron radiation power $P_{coh}^{rad}(\varphi)$ [20] we obtain:

$$U > U_m = \frac{3^{1/6}2^{2/3}(1+\mu)\Gamma^2(2/3)}{\sqrt{e_n}\sin\varphi_s}\frac{NeR^{4/3}}{hl_e^{7/3}}, \quad e_n = 2{,}718... \tag{B3}$$

Under this condition electron trajectories stay stable in spite of destabilizing role of coherent synchrotron radiation.

Substituting h =10, $\sin\varphi_s \approx 1$ and other parameters from the Table I into (B3) we get U > 120 kV. In order the coherent synchrotron radiation would not change essentially equilibrium bunch dimensions this inequality must be fulfilled with some spare. More details for electron beam dynamics in the coherent synchrotron radiation field can be found in [22].

References

[1] B.Jacobsen, Acta Radiol., **39**, 437 (1953).
[2] E.Rubinstein et.al., Proc. SPIE, **314**, 42 (1981).
[3] M.Ando and C.Uyama (Eds.), Medical application of synchrotron radiation (Springer-Verlag, 1998)
[4] R. Tatchyn, Remarks on the role of multilayer optics and short period insertion devices for medical imaging sources and application (see[3], pp139-148).
[5] Zhirong Huang and R.D.Ruth, Phys. Rev. Lett., **80**, 976 (1998).
[6] F.E.Carrol, AJR, **187**, 1197 (2003).
[7] T.Dill, Synchrotron Radiation News, **11**, N2, 12 (1998).
[8] Y. Sugishita, Medical activities of synchrotron radiation in Japan – Intravenous coronary angiography, (see [3], pp 15-21).
[9] A.M.Babunachvili, V.A.Ivanov, S.A. Biryukov and Stenting of Coronary Arteries, ACB-Publishing, Moscow, 1996 (in Russian).
[10] E.G.Bessonov, et al, Physics-Uspekhi, **46**, 872 (2003).
[11] A.S. Alimov, et al, NIM B, **139**, 511 (1998).
[12] S. Fujikawa, K. Furuta, and K. Yasui, Optics Letters, **26**, 602 (2001).
[13] P.G.Kryukov, Quantum Electronics, **31**, 95 (2001).
[14] E.Innerhofer, et.al., Optics Letters, **28**, 367 (2003).
[15] N.N.Blinov (ed.), Principles of X-ray Diagnostics Technology (Medicine, Moscow, 2002) (in Russian).
[16] A.M.Babunachvili, I.Kh.Rabkin and V.A.Ivanov, Coronary Angioplasty (ACB-Publishing, Moscow, 1996) (in Russian).
[17] M.Uesaka, A.Fukasawa, K.Dobashi, H.Iijima, J.Urakawa, T.Higo, M.Akimoto and H.Hayano, Proc. XXI Int. Linac Conf., Gyuongju, Korea, Aug 19-23, 2002
[18] L.Schiff, Rev. Sci. Instr., **17**, 7 (1946).
[19] J.S.Nodvic and D.S.Saxon, Phys. Rev., **96**, 180 (1954).
[20] L.V.Ioganson and M.S.Rabinovich, JETP, **37**, 118, (1959) (in Russian).
[21] E.L.Saldin, E.A.Schneidmiller and M.V.Yurkov, TESLA FEL Report 96-14, November 1996.
[22] E.G.Bessonov, R.M.Fetchenko and V.I.Shvedunov, http://xxx.lanl.gov/abs/physics/0404141
[23] A.A.Kolomensky and A.N.Lebedev, Theory of Cyclic Accelerators (North Holland Publishing, 1966).
[24] H Wiedemann, Particle Accelerator Physics (Springer Verlag, New York, 1993).

Inst. Phys. Conf. Ser. No 186
Paper presented at 9th Int. Conf. X-Ray Lasers, Beijing, China, 24–28 May 2004

Electron-based EUV and ultrashort hard-x-ray sources

B N Chichkov, A Egbert, and U Hinze
Laser Zentrum Hannover e.V., Hollerithallee 8, 30419 Hannover, Germany

Abstract. A brief review of our progress in the realization of femtosecond laser-driven ultrashort hard-x-ray sources is given. New results on the development of a compact electron-based EUV source for at-wavelength metrology and microscopy are presented.

1. Introduction

In this paper, electron-induced characteristic emission from solid targets is used for the development of compact EUV and ultrashort hard-x-ray sources.

EUV source. At present, synchrotron radiation is widely used for EUV and soft-x-ray metrology and microscopy. Synchrotrons are high-quality sources which are ideal for scientific research but are very complex and expensive for applications. Therefore, the development of compact and cheap EUV and soft-x-ray radiation sources optimized for potential users is of great practical importance. There are many efforts aimed at the development of laser-produced plasmas and discharge-produced plasmas as radiation sources for next generation lithography, metrology, and microscopy. In spite that these sources can generate high radiation powers there are many problems with their long-term stability and with the generation of debris.

Often overlooked, there is a straight-forward method to generate EUV and soft-x-ray radiation directly from solid targets avoiding all these plasma-related drawbacks [1]. This can be done by using electron-induced characteristic emission from solids. This type of EUV/soft-x-ray source can be considered as an analog to a conventional x-ray tube. Electrons are generated by a tungsten filament (cathode), accelerated in a high-voltage electric field, and focused onto a solid target (anode) which allows the generation of characteristic emission in the EUV/soft-x-ray spectral range. This radiation is produced due to electron-impact ionization (excitation) of atomic inner shells, which is followed by radiative decay. There are different elements emitting characteristic radiation in the wavelength region from 1 nm to 20 nm. For example the L-shell emission of silicon ($Z = 14$) is located at 13.5 nm which if of high practical importance for EUV metrology-related applications.

Ultrashort hard-x-ray source. The development of reliable, compact, high-repetition rate ultrashort hard-x-ray sources is required for many applications in science and technology, including time-resolved diffraction studies, control of chemical reactions, time-gated medical imaging, etc. Ultrashort hard-x-ray pulses are presently produced with synchrotrons or by focusing high-power femtosecond laser radiation onto solid targets. These sources are very complex and expensive for practical applications. In contrast with laser-plasma x-ray sources which require high-power laser systems in the x-ray diode one can use low-energy, compact, high-repetition rate, commercially available femtosecond lasers. The generated hard-x-rays are synchronized with the femtosecond laser pulses which is important for time-resolved applications. The shortest hard x-ray pulse generated with our x-ray diode had a pulse duration of 3 ps.

2. Compact EUV source

A modified commercial microfocus x-ray tube (phoenix|x-ray Systems + Services GmbH) is used for the generation of electron-induced EUV and soft-x-ray radiation (see Fig. 1). The source is equipped with an electro-magnetic focusing unit which enables the variation of the electron spot size at the target surface. The open design of this source allows one to use different target materials and access a broad wavelength range. The accelerating voltage can be adjusted in the range of 5 to 20 kV. The maximum electrical power which can be applied to the target is 50 W. The tube is evacuated to a pressure below 10^{-4} mbar by a turbo molecular pump. All parameters of the EUV tube are computer-controlled.

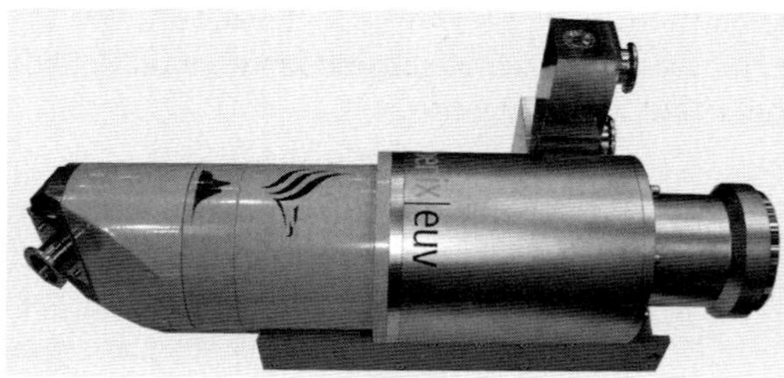

Fig. 1. Photograph of the electron-based EUV tube.

Fig. 2 illustrates a typical silicon EUV spectrum which is recorded at an accelerating voltage of 10 kV using a silicon wafer target.

The absolute conversion efficiencies from electrons into EUV radiation are measured for silicon targets using a calibrated EUV spectrograph. This efficiency is defined as the emitted EUV power at 13.5 nm (in 2% bandwidth and a solid angle of 2π sr) per applied target current. The conversion efficiency exhibits a maximum in the range of 7 to 10 kV and reaches a value of approximately 16 μW/mA. Therefore, a total EUV power of more than 30 μW (2π sr, 2% bw) can be obtained with the present version of the EUV tube at the maximum target current. This corresponds to an in-band EUV flux of 2×10^{12} photons/s. The total EUV power emitted from the tube in a spectral range from 12.4 nm to 15 nm is >100 μW in 2π sr. An upgraded version of the EUV tube is presently under development.

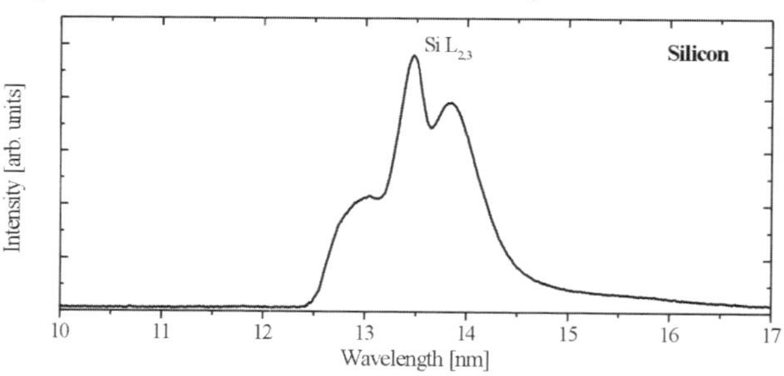

Fig. 2. Silicon EUV spectrum an accelerating voltage of 10 kV.

The electron optics used in the EUV tube allows one to achieve source sizes well below 20 μm in diameter. Therefore, the brightness (power per area) obtained with the electron-based EUV tube can be higher than that of plasma-based EUV metrology sources.

A great benefit compared to plasma-based EUV sources is the excellent temporal long-term stability. It has been demonstrated that the intensity fluctuations of the EUV tube are below ±0.5% over a total run time of nearly 70 hours. Together with a completely

debris-free operation, these features make the EUV tube an ideal source for metrological applications with synchrotron-like precision.
In one application example, the transmission of a double multilayer mirror device (multilayer tool) developed and fabricated by JENOPTIK Mikrotechnik GmbH is measured using the EUV tube. This device reflects and filters the incoming EUV radiation and is used (in combination with an EUV diode) as an energy monitor for the characterization of EUV lithography sources. A detailed description of this tool can be found elsewhere [2]. In our experiments the reflectivity of this multilayer tool as a function of wavelength is determined. In order to perform this measurements the multilayer tool was directly mounted to the entrance flange of the EUV spectrograph and connected to the EUV tube.

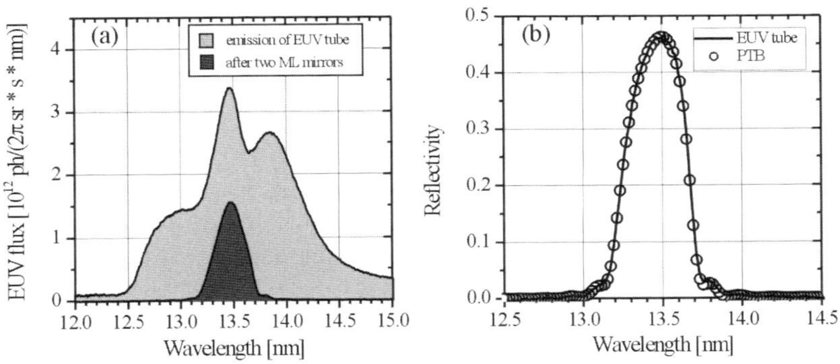

Fig. 3. (a) EUV spectra as emitted from the EUV tube (light gray spectrum) and after reflection from the two mirrors of the multilayer tool and (b) reflectivity curve of the double multilayer mirror tool measured with the EUV tube (solid line) and the corresponding data from a PTB measurement (open circles).

Fig. 3a illustrates the silicon EUV spectrum as it is emitted from the EUV tube (light gray spectrum) and the spectrum after reflection from the two mirrors of the multilayer tool (dark gray spectrum). These two spectra are recorded using identical EUV tube parameters and are obtained after a 4-minute illumination of the spectrograph CCD camera. The spectral response of the double multilayer mirror device is determined by the division of the reflected spectrum by the original source spectrum. This reflectivity curve is illustrated in Fig. 3b. The solid line represents the multilayer tool reflectivity measured with the EUV tube. As a reference Fig. 3b also illustrates the corresponding data from the Physikalisch Technische Bundesanstalt (PTB). It can be seen that both reflection curves are in excellent agreement. The observed high accuracy originates from the very smooth shape of the silicon EUV spectrum. Note, that such a high precision cannot be achieved by EUV plasma sources with emission lines in the relevant wavelength range. In this case a noise-like structure appears in the center of the measured curve [2].

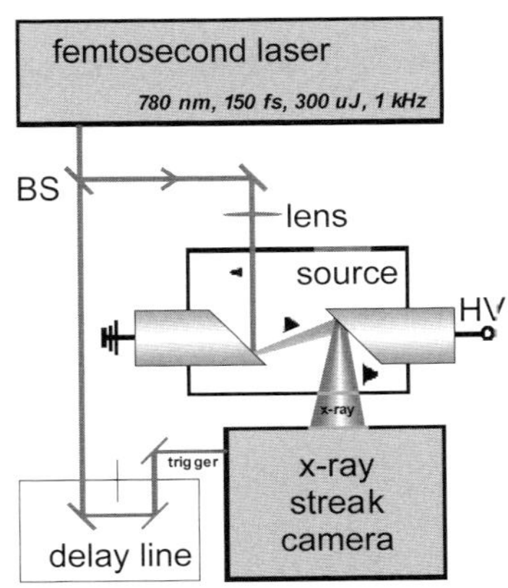

Fig. 4: Experimental setup for hard x-ray pulse length measurement. BS: beam splitter, HV: high voltage

3. Ultrashort hard x-ray source

We report on our recent progress in the temporal characterization of ultrashort x-ray pulses generated in a fs-laser driven x-ray diode. The experimental setup (Fig. 4) consists of three main parts: a fs-laser system, the x-ray diode, and an x-ray streak camera for the detection and characterization of the x-ray pulses. Laser pulses from a 1 kHZ Ti:sapphire femtosecond laser system (150 fs, 780 nm, 300 μJ) are split into two parts by a beam splitter: 80% are directed to the streak camera's optical trigger input, and the remaining part of the energy is used for the generation of ultrashort electron bunches and hard x-rays. The x-ray diode consists of a vacuum chamber (pressure $<10^{-5}$ mbar) containing two polished 45° copper-electrodes at a distance of about 10 mm. The anode is connected to a high voltage power supply allowing the application of voltages from 0 to 100 kV whereas the cathode is set to ground potential. Femtosecond laser pulses induce photoemission of electrons from the photocathode. The electron bunch is then accelerated toward the high-Z anode by the external high-voltage electric field. Hard-x-rays are produced by bremsstrahlung of accelerated electrons and characteristic emission inside the target The diode current is measured with a pico-amperemeter. A detailed characterization of the x-ray diode in terms of spectral properties, conversion efficiency, and x-ray flux can be found in Ref. [3].

The measured values of the x-ray pulse duration as a function of the acceleration voltage are shown in Fig 5 for different laser pulse energies. As can be seen, the duration of the x-ray pulses measured in our experiments varies from about 3 ps to 21 ps. The x-ray pulse duration becomes shorter when the laser pulse energy (i.e. the number of photoelectrons) is decreased or when the acceleration voltage is increased. In the first case the space-charge broadening effects and in the second case the acceleration time are reduced.

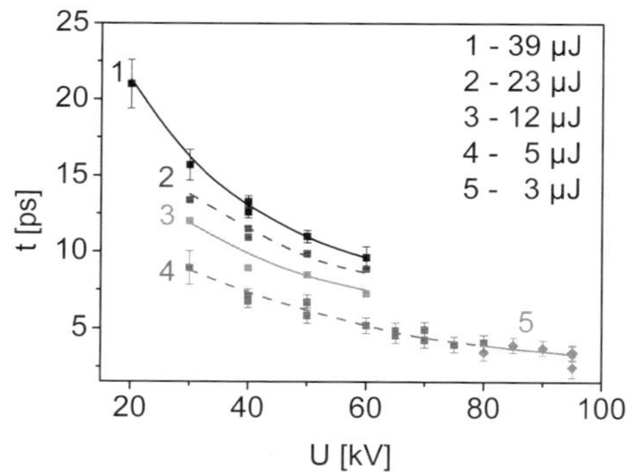

Fig. 5: Dependence of the measured hard x-ray pulse duration on the acceleration voltage.

To the best of our knowledge, the experiments presented here demonstrate for the first time that 3 ps Cu K_α x-ray pulses can be generated in an x-ray diode driven by femtosecond laser pulses. Certainly, these results do not represent any physical or technical limits. By further improvement of the source setup, e.g. by reducing distance between the electrodes, optimizing their geometry, using the third harmonic of Ti:sapphire laser radiation, etc. one can reduce the x-ray pulse duration down to 400 fs by still providing more than 10^4 hard-x-ray photons per pulse [4].

References

[1] A. Egbert, B. Mader, et al., J. Microlith., Microfab., Microsyst. **2**, 136 (2003).
[2] M.C. Schürmann, T. Missalla, et al., Proc. SPIE **5037**, 378 (2003).
[3] A. Egbert and B. N. Chichkov, Proc. SPIE **4978**, 92 (2003)
[4] U. Hinze, A. Egbert, et al., accepted for publication in Opt. Lett **79**,(2004)

Inst. Phys. Conf. Ser. No 186
Paper presented at 9th Int. Conf. X-Ray Lasers, Beijing, China, 24–28 May 2004

X-ray radiation in supercontinuum generated by intense laser-plasma interactions

Qiu-Ju Zhang, Zheng-Ming Sheng, Jie Zhang
Laboratory of Optical Physics, Institute of Physics, Chinese Academy of Sciences, Beijing 100080, China

Abstract. A supercontinuum (SC) spectrum is produced by the nonlinear interaction through the strong coupling of the stimulated Raman scattering (SRS), the relativistic self-modulation instability and the cascading SRS process. The SC spectra can extend to so high frequency as well as X-ray wave band. The relation between the extent of the SC spectra and the width of merged unstable region in the wave-vector space is discussed.

1. Introduction

There have been a lot of works about the supercontinuum (SC) generation in fibers, air and water over the last a few years, but there are few studies on the SC generation in plasmas up to now [1-6]. The electron parametric instabilities have been studied in one-dimensional (1D) theory [7-10], however, there are few numerical simulations and experimental studies on this phenomenon. In this work, using 1D particle-in-cell (PIC) simulations, we observed the SC generated by intense laser-plasma interaction, which extend to the X-ray wave band. The results may be useful for SC generation in the laser-plasma experiments and for the diagnosis of laser plasma parameters.

2. One-dimensional dispersion relation for cold plasmas

We consider a laser pulse propagating in homogeneous cold plasma, with a vector potential of $\mathrm{a} = (a_0/2)[\,\mathrm{e}_0 \exp(ik_0 x - i\omega_0 t) + c.c.]$, where a is normalized by mc^2/e, $\mathrm{e}_0 = (\mathrm{e}_y \pm i\,\mathrm{e}_z)/\sqrt{2}$, e, m, c, k_0, and ω_0 are the electron charge, electron mass, the speed of light in vacuum, the wave vector of the laser pulse and its frequency, respectively. In 1D case, the dispersion relation for electron parametric instabilities can be written as $D_+D_- - (D_+ + D_-)S = 0$, where $D_\pm = \omega^2 - c^2k^2 \pm 2(\omega_0\omega - c^2k_0k)$ with ω and k the frequency and wave vector of the involved plasma wave, respectively, and $S = a_0^2\omega_p^2[k^2c^2/(\omega^2 - \omega_p^2/\gamma_0) - 1]/4\gamma_0^3$ [8]. We solve this dispersion relation and compare it with the spectrum of the laser pulse obtained from 1D PIC simulations in next section.

3. Generation of supercontinuum from laser-plasma interaction

Using the 1D PIC code, we study the propagation of a linearly polarized laser pulse with duration $T = 400T_0$ in a plasma slab, where T_0 is the period of laser. The plasma slab is initially uniform and with a thickness of 100 laser wavelengths.

Figure 1(a) shows three unstable regions for relativistic modulation instability (RMI), Raman forward scattering (F-SRS) and Raman backward scattering (B-SRS) for increasing k, where $a_0 = 0.1$ and $n_0 = 0.1n_c$. Figs. 1(b), obtained in PIC simulations,

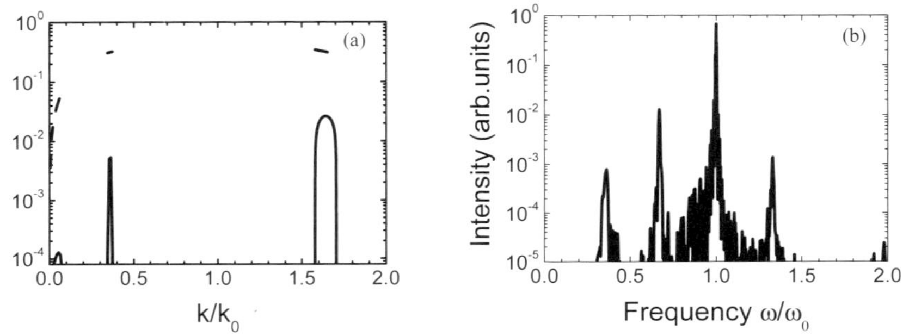

Figure 1. (a) The growth rate (solid line) (normalized by ω_0) and real part (dashed line) (normalized by ω_0) of frequency; (b) Frequency spectrum of transmitted laser. Here $a_0 = 0.1$, $n_0 = 0.1n_c$.

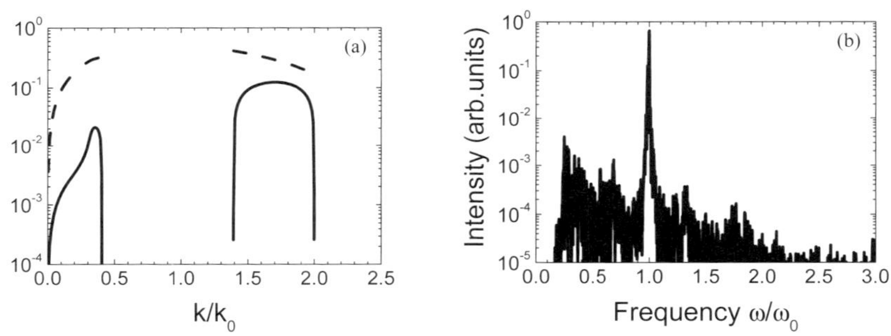

Figure 2. (a) The growth rate and the real part of frequency; (b) Frequency spectrum of transmitted laser pulse. Here $a_0 = 0.5$, $n_0 = 0.1n_c$.

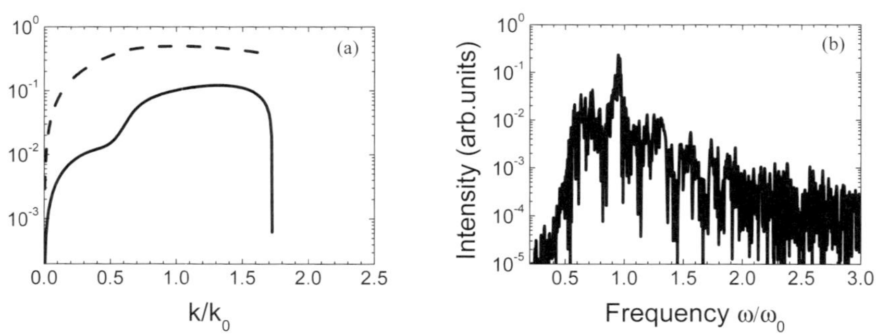

Figure 3. (a) The growth rate and the real part of frequency; (b) Transmitted frequency spectrum. Here $a_0 = 0.5$, $n_0 = 0.3n_c$.

shows the transmitted light spectra for the same parameters. We can see clearly the first Stokes, anti-Stokes components, and the second Stokes line, respectively. This can be explained as due to a secondary B-SRS.

When $a_0 = 0.5$ and $n_0 = 0.1n_c$, as shown in Fig. 2(a), the unstable regions of RMI and F-SRS merge into an enlarged RMI/ F-SRS region. The spectrum of RMI and F-SRS

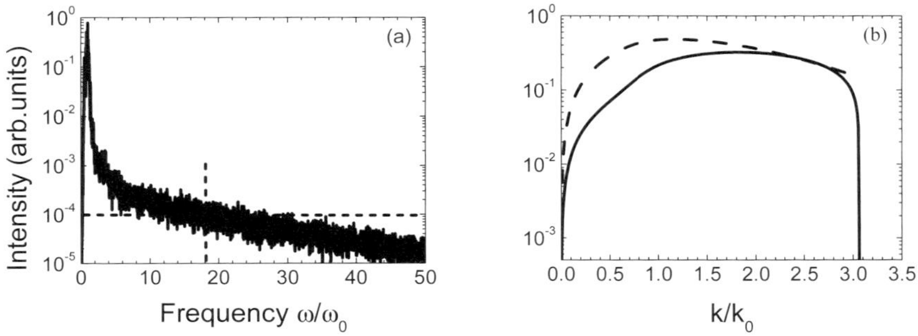

Figure 4. (a) Transmitted frequency spectrum; (b) The growth rate and the real part of frequency. Here $a_0 = 1.6$ and $n_0 = 0.7n_c$

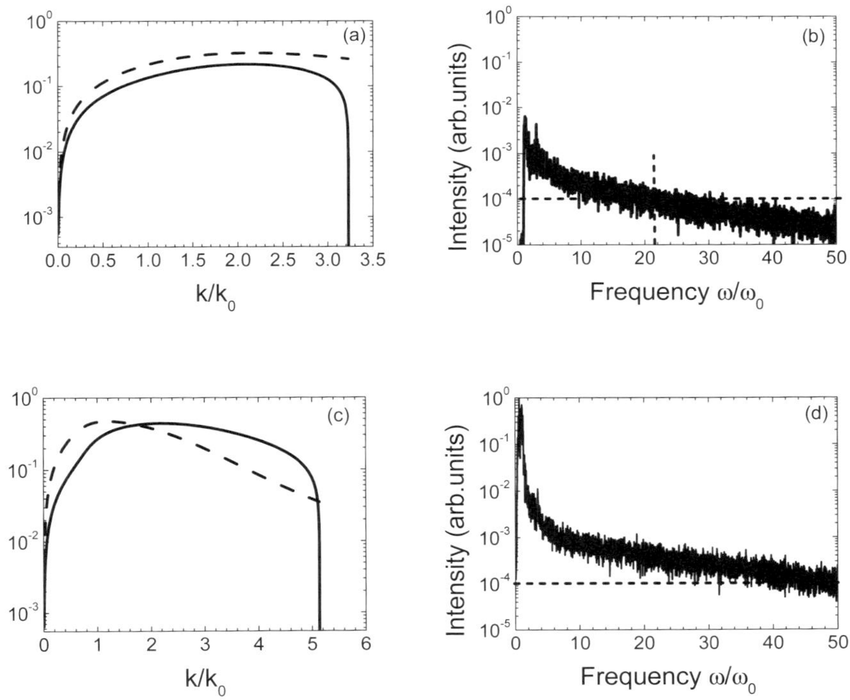

Figure 5. (a) The growth rate and the real part of frequency; (b) Transmitted frequency spectrum. Here $a_0 = 1.6$, $n_0 = 1.2n_c$. (c) The growth rate and the real part of frequency; (d) Transmitted frequency spectrum. Here $a_0 = 3.0$, $n_0 = 1.2n_c$.

are mixed in Fig. 2(b). What important is that the spectrum generated has basically changed into a continuous one. This is caused through the coupling of the backward and forward moving wave through the parametric instabilities.

When $n_0 = 0.3n_c$ and $a_0 = 0.1$, we cannot obtain a broaden spectrum. However, when $a_0 = 0.5$ in Fig. 3(a), the unstable region becomes significantly enlarged, and the spectrum broadens obviously as shown in Fig. 3(b).

Figure 4(a) shows a SC obtained for $a_0 = 1.6$ and $n_0 = 0.7n_c$. Only the part between $\omega/\omega_0 = 0$ and $\omega/\omega_0 = 50$ of SC spectrum is shown. The corresponding unstable region shown in Fig. 4(b) illustrates that B-SRS, F-SRS, and RMI have merged into a single unstable region. The SC generation can be associated with the strong coupling of SRS and RMI. Obviously, cascading SRS may also play an important role.

As referring to the extents of SC spectra thereinafter, we mean the spectrum with intensity above 10^{-4}. When $a_0 = 1.6$ and $n_0 = 1.2n_c$, the unstable region gets broaden, as shown in Fig. 5(a). Compared with the SC extent from $\omega/\omega_0 = 0$ to $\omega/\omega_0 = 18$ in Fig. 4(a), the SC extent in Fig. 5(b) is extended to $\omega/\omega_0 = 22$. It appears that more energy near the fundamental frequency has been transferred to the high-frequency components. When $a_0 = 3.0$ and $n_0 = 1.2n_c$, the unstable region can also get broaden, as shown in Fig. 5(c), and the corresponding SC extent can reach up to $\omega/\omega_0 = 50$, as displayed in Fig. 5(d). However, at higher plasma densities, high harmonics generation can significantly reduce the extent of SC generation, even though the unstable regions are quite broad.

4. Summary

We have investigated the SC generation, which can reach to X-ray wave band, in the regime of nonlinear interaction of a relativistic intense laser pulse with plasmas. The SC generation scheme will be investigated in the present experiments of short-pulse laser-plasma interactions.

Acknowledgments

This work was supported by the NSFC under Grant Nos. 10105014, 10425416, 60321003, and 60128505, and the NKBRSF under Grant No. G1999075206, G1999075203 and the National Hi-tech ICF Program.

References

[1] Alfano R R 1989 The Supercontinuum Laser Source (New York: Springr-Verlag)
[2] Coen S, Chau A H L et al., 2002 J. Opt. Soc. Am. B 19 753
[3] Husakou A V and Herrmann J 2002 J. Opt. Soc. Am. B 19 2171
[4] Sandhu A S, Banerjee S et al., 2000 Opt. Commun. 181 101
[5] Kolesik M, Moloney J V et al., 2002 Phys.Rev. Lett. 89 283902
[6] Ting A, Krushelnick K et al., 1996 Opt. Lett. 21 1096
[7] Sakharov A S, Kirsanov V I, 1994 Phys. Rev. E 49 3274
[8] Guerin S, Laval G et al., 1995 Phys. Plasmas 2 2807
[9] Decker C D, Mori WB et al., 1996 Phys. Plasmas 3 2047
[10] Sheng Z M, Mima K et al., 2000 Phys. Rev. E 61 4362

Inst. Phys. Conf. Ser. No 186
Paper presented at 9th Int. Conf. X-Ray Lasers, Beijing, China, 24–28 May 2004

Parameter studies of attosecond x-ray pulses generated by nonlinear Thomson scattering

Jun Zheng, Zheng-Ming Sheng, Jie Zhang, Zhi-Yi Wei
Laboratory of optical Physics, Institute of Physics, CAS, Beijing 100080, China

Wei Yu
Shanghai Institute of Optics and Fine Mechanics, CAS, Shanghai 201800, China

Abstract. The attosecond x-ray pulse trains emitted by a single electron subjected to the electromagnetic field of an intense ultra-short laser pulse in vacuum have been investigated theoretically and numerically by use of the classical theory of Thomson scattering. The temporal and spatial characteristics of the pulse trains are presented for different laser parameters (including intensities, pulse durations, initial phases and polarizations) and initial states of the electron (including initial momenta and positions). Usually, with the increase of the incident laser intensity, the radiation power and its central frequency increase, while the emission duration and angular spread reduce. The radiation produced under a linearly polarized laser has a higher amplitude than under a circularly laser pulse at the same intensity. In both cases, the polarization state of the radiation is very complicated, which depends on the observation direction as well. Whatever its initial energy and moving direction are, the radiation produced by a relativistic energetic electron tends to be predominantly along the moving direction of the electron.

1. Introduction

The development of the ultra-intense lasers has led to significant renewed interests in the classic problem of Thomson scattering both for fundamental studies and applications [1-4]. An important application of the scattering processes is the generation of x rays with ultrashort pulse lasers. Recently, Ta Phuoc *et al.* [3] have observed x-ray radiation from the nonlinear Thomson scattering of an intense femtosecond laser on relativistic electrons in Helium plasma. Lee *et al.*[4] have studied the Thomson scattering of a free electron in an ultrashort intense laser pulse, where the electron is initially at rest. In this work, the nonlinear Thomson scattering of an electron with certain initial momenta is investigated. Some new features associated with the scattered radiation are found.

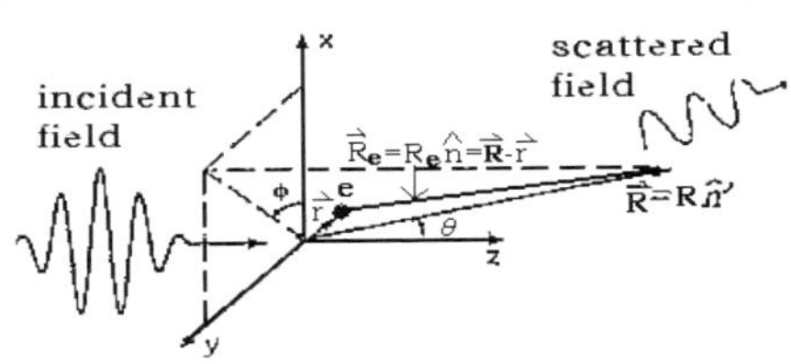

FIG. 1 Schematic diagram showing the Thomson scattering of an intense laser field from a free electron.

The schematic diagram showing the Thomson scattering of a laser field from a free electron is given in Fig. 1. Neglecting the radiation reaction, the equation of motion for an electron in an arbitrarily intense, elliptically polarized, plane electromagnetic wave in vacuum can be solved exactly[5]. So the angular distribution of the radiated power detected far away by observers in the direction $\hat{n}$ at a time t' is calculated as[6]

$$\frac{dp(t')}{d\Omega} = \frac{e^2}{4\pi c} \frac{\left| \hat{n} \times [(\hat{n} - \vec{\beta}(t)) \times \dot{\vec{\beta}}(t)] \right|^2}{\left(1 - \hat{n} \cdot \vec{\beta}(t)\right)^6} \tag{1}$$

where $\vec{\beta} = \vec{v}/c$, and t is the electron's time or retarded time, which is related to t' by $t' = t + \frac{[\vec{R} - \vec{r}(t)] \cdot \hat{n}}{c} \cong t + \frac{R - \vec{r}(t) \cdot \hat{n}}{c}$. The energy radiated per unit solid angle per unit frequency interval can be obtained by the Fourier transform as[6]

$$\frac{d^2 I}{d\omega d\Omega} = 2 \left| \sqrt{\frac{e^2}{8\pi^2 c}} \int_{-\infty}^{\infty} \left[\frac{\hat{n} \times [(\hat{n} - \vec{\beta}(t)) \times \dot{\vec{\beta}}(t)]}{\left(1 - \hat{n} \cdot \vec{\beta}(t)\right)^3} \right]_{ret} e^{i\omega t} dt \right|^2 \tag{2}$$

2. The radiation for different laser parameters: intensities, pulse durations, initial phases and polarizations

As the laser intensity increases, the total radiated energy and radiation flux are enhanced, the radiation's central frequency becomes higher, its duration becomes shorter, and its angular distribution becomes smaller, as shown in Fig. 2. Note that the observer receives two radiated peaks during every laser cycle (Fig. 2(a-d)). However, only a single peak is found every laser cycle for a circularly polarized laser [Fig. 3]. This is associated with the electron's parallel motion. The radiation produced by the electron under a linearly polarized laser has a higher amplitude than under a circularly polarized laser pulse at the same intensity. The time interval between the double pulses is only $\tau_0/40$ (τ_0 is the incident laser period) when a_0=3.0 and $\tau_0/120$ when a_0=6.8. If the incident laser's period is 3.3fs, the time intervals mentioned above are 83as and 28as, respectively. Fig. 2(c) and 2(f) show the modulations with an interval of about 40 and 120, respectively.

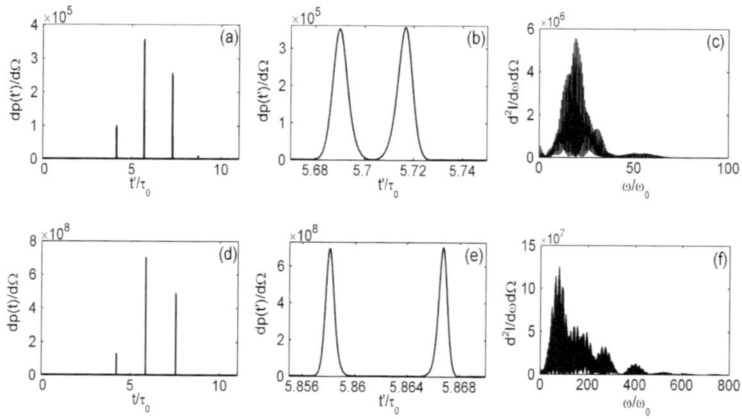

FIG. 2 The radiation from the initially stationary electron for various incident laser intensities. The laser is linearly polarized and its duration is 10 laser

periods. (a, d) the time history of the radiated power[in unit of $e^2/(4\pi c\tau_0^2)$] for the laser amplitude parameter a_0=3, 6.8, respectively, at the direction of maximum radiation($\phi=0^\circ$, $\theta=\theta_M$). The highest peak in (a) and (d) are magnified in (b) and (e), respectively. (c) and (f) are the energy radiated per unit solid angle per unit frequency interval [in unit of $e^2/(4\pi c\tau_0^2)$] in the same conditions as in (a) and (d), respectively.

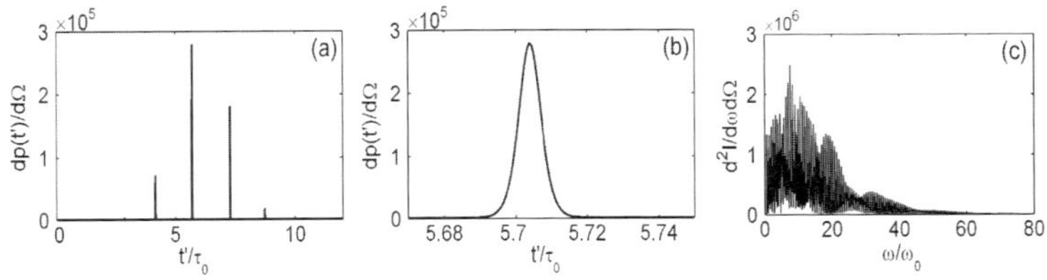

FIG. 3 The time history of the radiated power for a circularly polarized laser. The parameters are same as in Fig.2(a). The highest peak in (a) is magnified in (b). (c) The energy radiated per unit solid angle per unit frequency interval.

The initial phase of the laser pulse changes the symmetry of the electron's motion. As a result , it changes the symmetry of the radiation emitted by the electron (Fig. 5).

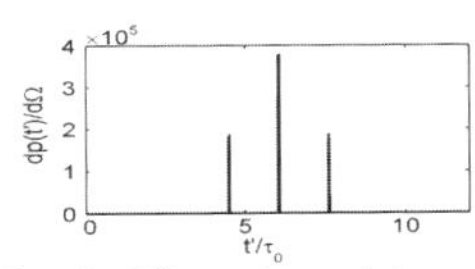

FIG. 5 The time history of the radiated power for the laser's initial phase $\phi_0=90^\circ$. The other parameters are same as in Fig. 2(a).

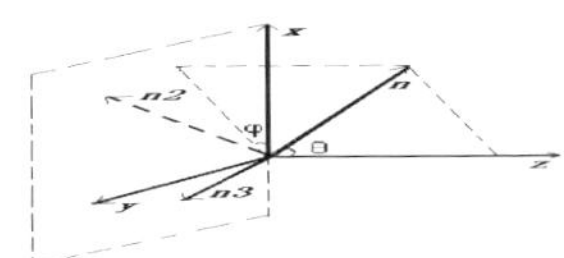

FIG. 6 Schematic diagram of the vectors of the radiated electric field. **n**=($sin\theta cos\varphi$, $sin\theta sin\varphi$, $cos\theta$), **n2**=($sin(\theta+\pi/2)cos\varphi$, $sin(\theta+\pi/2)sin\varphi$, $cos(\theta+\pi/2)$), **n3**=($sin(\pi/2)cos(\varphi+\pi/2)$, $sin(\pi/2)sin(\varphi+\pi/2)$, $cos(\pi/2)$).

3. The polarization states of the scattered fields

The polarization state of the radiation is very complicated, which depends on the laser's polarization states and the observation direction as well. Assume that the radiation direction is vector **n,** the vectors of the radiated electric field are showed in Fig. 6. Fig. 7 shows the relationship between the projections of the electric field along **n2** and **n3** for a linearly polarized laser and a circularly polarized.

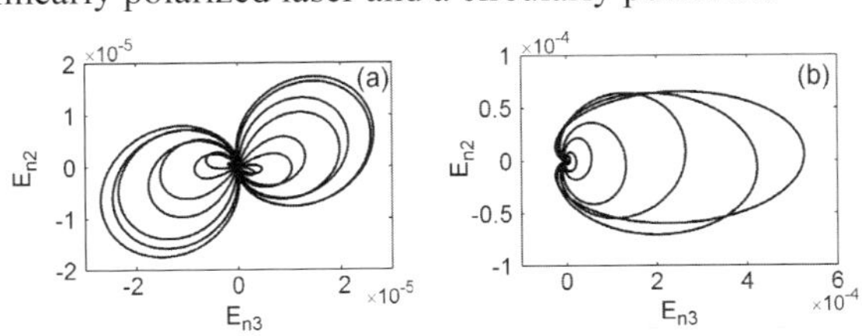

FIG. 7 The components of the radiation fields along **n2** and **n3** directions for (a) a linearly polarized laser and (b) a circularly polarized laser.

4. The radiation for different initial velocities of the electron

The initial velocity of the electron can significantly change the angular distribution of the radiated energy. Whatever the initial energy and moving direction of the electron as well as the laser polarizations are, the radiation generated by a relativistic energetic electron tends to be predominantly along the moving direction of the electron. If the initial moving direction of the electron directed to –z, ± x or ± y axis, with the increase of initial energy, the radiated pulse duration becomes small. Fig. 8 shows the angular distribution of the radiated energy for various initial velocities of the electron.

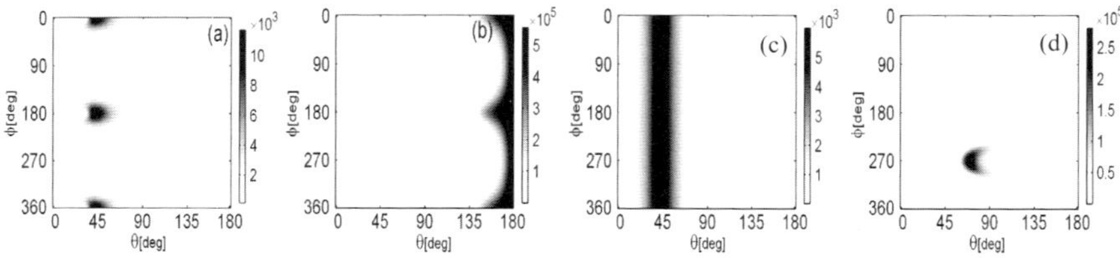

FIG. 8 Angular distributions of the radiated energy [in unit of $e^2/(4\pi c\tau_0)$] for various initial velocities of the electron. The incident lasers are linearly polarized in figure (a-b) and circularly polarized in (c-d). The electron's initial velocity directions and energies are: (a, c) 0MeV, (b) axis -z, 1Mev, (d) axis –y, 1MeV. The other parameters are same as in Fig. 2(a).

5. Conclusions

The attosecond x-ray pulse trains emitted by a single electron under the ultra-short and ultra-intensity laser pulse in vacuum have been investigated theoretically and numerically. The radiation reaction to the electron has been neglected in the calculations. The temporal and spatial characteristics of the pulse trains are presented for different laser parameters and initial states of the electron. Usually, as the laser intensity increases, the radiation power and its central frequency increase, while the emission duration and angular spread reduce. The radiation produced by the electron under a linearly polarized laser has a higher amplitude than under a circularly laser pulse at the same intensity. In both cases, the polarization state of the radiation is very complicated, which depends on the observation direction as well. The electron's initial states effect the radiation too. Whatever its initial energy and moving direction are, the radiation generated by a relativistic energetic electron tends to be predominantly along the moving direction of the electron.

Acknowledgement

This work was supported by the NSFC under Grant Nos. 60321003, 10474134 and 60128505, and the NKBRSF under Grant No. G1999075203, G1999075206, and the National Hi-tech ICF Program.

References

[1] Sarachik E S and Schappert G T, Phys. Rev. D **1**, 2738 (1970).
[2] Yu W, Yu M Y, Ma J X, Xu Z, Phys. Plasmas **5**, 406 (1998).
[3] Phuoc K Ta, Rousse A et al., Phys. Rev. Lett. **91**, 195001(2003).
[4] Lee K and Cha Y H, et al., Phys. Rev. E **67**, 026502 (2003).
[5] Meyer-ter-Vehn, Pukhov A, and Sheng Z M. , "Relativistic laser-plasma interaction" in Atoms, Solids, and Plasmas in Super-Intense Laser Fields, Ed. By D. Batani et al., Kluwer Academic/Plenum Publishers, 2001, pp 167-192.
[6] Jackson John David, Classical Electrodynamics , 3rd ed. (Wiley, New York, 1998) Chap. 14.

Alternative pumping schemes for X-ray lasers

Inst. Phys. Conf. Ser. No 186
Paper presented at 9th Int. Conf. X-Ray Lasers, Beijing, China, 24–28 May 2004

Generation and applications of 10-mJ XRL pulses near 20 nm using long scale length plasmas

B Rus, M Kozlová, J Polan, A R Präg, M Stupka, T Mocek
Dept of X-ray Lasers, Institute of Physics /PALS Centre, Na Slovance 2, 18221 Prague 8, Czech Republic
M Bittner, L Juha
Department of Laser Plasmas, Institute of Physics, Na Slovance 2, 18221 Prague 8, Czech Republic
M Begusová, V Štísová
Nuclear Physics Institute, Academy of Sciences of the Czech Republic, Na Truhlárce 39/64, 18086 Prague 8, Czech Republic
D Ros, K Cassou, S Kazamias, A Klisnick, G Jamelot
Laboratoire d'Interaction du rayonnement X avec la Matiere, Université Paris-Sud, 91405 Orsay, France
G Eot, E Sage
Institut Curie, Bâtiment 110-112, Université Paris-Sud, 91405 Orsay Cedex, France

Abstract. We report recent achievements in development and applications of deeply saturated Ne-like zinc soft X-ray laser, employing a sequence of multiple-100-ns IR pump pulses. The population inversion is produced in the regime of long scale-length density plasma, which is achieved by very large separation of the prepulse (<10 J) and the main pump pulse (~600 J), of up to 50 ns. An extremely bright and narrowly collimated double-pass XRL beam was obtained, providing ~10 mJ pulses and ~100 MW of peak power, which is the most powerful soft X-ray laser yet demonstrated. The applications include pilot radiobiology studies of DNA damage in the soft X-ray region, material ablation relevant to microfabrication technologies, and soft X-ray interferometry. A concomitant topic is focusing the XRL beam down to a narrow spot, with the goal of achieving ~10^{13} Wcm^{-2}.

1. Introduction

The recent X-ray laser program pursued at PALS has been aimed at investigation of novel pumping techniques of QSS systems, and at building a platform for derivative "user" applications of these devices. The model X-ray laser investigated is Ne-like zinc employing slab target. This laser was previously found very robust when pumped by a weak ~1.6 J prepulse and main ~500 J pulse separated by 10 ns [1], delivering up to ~4 mJ at 21.2 nm (J=0-1 transition) in a high quality, narrowly collimated (4×6 mrad) beam [2,4].

A particular objective of this work was to investigate whether amplification in long scale-length plasmas, produced from a slab target by a sequence of prepulse and main pulse, separated by 50 ns, can be achieved. The motivation for this effort is a prospect to reduce refraction of the XRL beam in the perpendicular direction, while simultaneously increasing the size of the gain region potentially leading to strong lasing action.

2. Double-pass Ne-like zinc XRL using 50-ns prepulse-to-main-pulse delay

The experimental arrangement, represented in Figure 1, is similar to the previous experiments at PALS [1-3]. The pump laser, operating at 1.315 μm, produces 300-450 ps pulses, which are delivered to the target in line focus of 3 cm long. The pump sequence consists of a prepulse, deliberately defocused to generate ~600-800 μm wide "buffer plasma", followed by the main pulse of ~500 J tightly focused to ~130 μm wide line.

The maximum ASE single-pass signal for the prepulse-to-main pulse delay of 50 ns was found for the prepulse energy of ~6 J. This value is nearly 4-times larger than that optimising the ASE single-pass beam for the prepulse-to-main pulse delay of 10 ns [1,2]. Furthermore, in stark contrast to the 10-ns configuration, the ASE intensity strongly depends on the position of the main focus with respect to the target; detuning the lens-target distance by less than 500 μm leads to a drop of the emitted intensity by one order of magnitude. At present, it is not clear whether this behaviour is caused by a complex character of the pump beam focus due to the used matrix optics [5], or whether it reflects a more fundamental aspect of the laser-plasma interaction under the given conditions.

The near-field image ("footprint") of the ASE signal is shown in Figure 2(a). As seen from the lineout through this footprint pattern in Figure 3(a), the ASE exhibits a horizontal divergence of nearly 10 mrad, which is a value about 3.5× larger than the divergence of the ASE beam generated for the prepulse delay of 10 ns [1,2], see Figure 4(a). In contrast, the refraction angle of the beam of ~4.2 mrad is only marginally smaller than the ASE refraction angle of ~5 mrad for the prepulse delay of 10 ns.

With the half-cavity mirror aligned to reinject to the plasma along the reverse trajectory the rays emerging under an angle of 5 mrad, an extremely intense beam, highly collimated in the horizontal direction, is generated. Its footprint image and the corresponding horizontal profile are shown in Figure 2(b) and 3(b), respectively. The beam delivers about 3× larger signal than the in the corresponding double-pass configuration driven with the prepulse and main pulse 10 ns apart, cf. Figure 4(b). While the horizontal beam divergence of 2.2 mrad is fairly smaller than that in the 10-ns configuration, the most striking difference

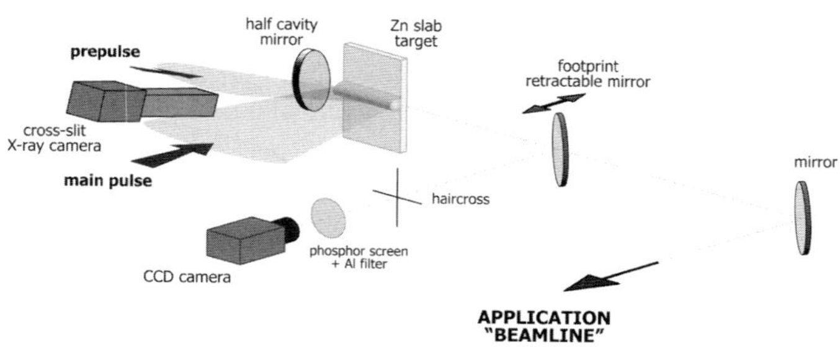

Figure 1. Generic arrangement for X-ray laser experiments at PALS: the amplifying plasma 3 cm long is produced by separately delivered and loosely focused prepulse, and by tightly focused main pump pulse. The half-cavity Mo:Si mirror is positioned typically at 8.5 mm from the plasma end.

is observed for the mutual shape of the single- and double-pass beams in the individual configurations, as evident from Figs 3 and 4. Specifically, the significant narrowing of the horizontal profile of the double-pass beam in the 50-ns configuration points at markedly distinct character of gain region morphology and/or possibly gain time history compared to plasma produced with the prepulse and main pulse 10 ns apart.

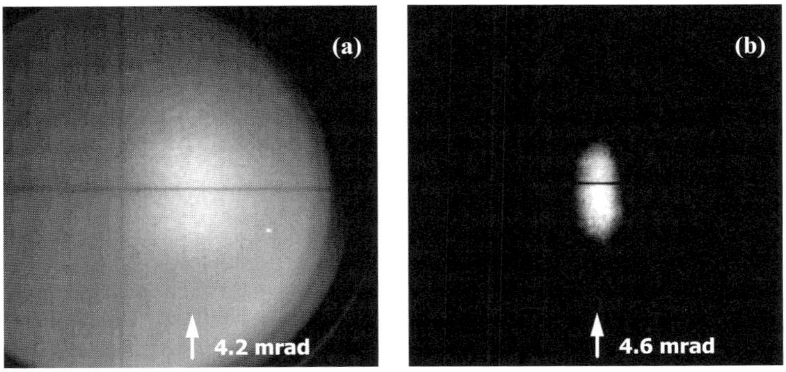

Figure 2. Footprints of zinc X-ray laser beam produced for prepulse-to-main-pulse delay of 50 ns: (a) ASE single-pass beam, (b) double-pass beam produced by the half cavity. The detector sensitivity in (b) is reduced by a factor of 16.

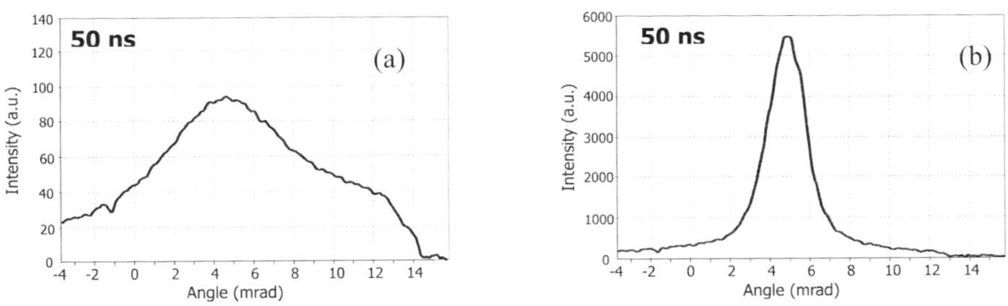

Figure 3. Horizontal lineout of the XRL beams generated for 50 ns prepulse-to-main-pulse delay, as shown in Fig.2: (a) ASE single pass, (b) double-pass beam. The beam shapes are qualitatively very different from those produced with the prepulse and main pulse 10 ns apart, as seen from Fig.4 below.

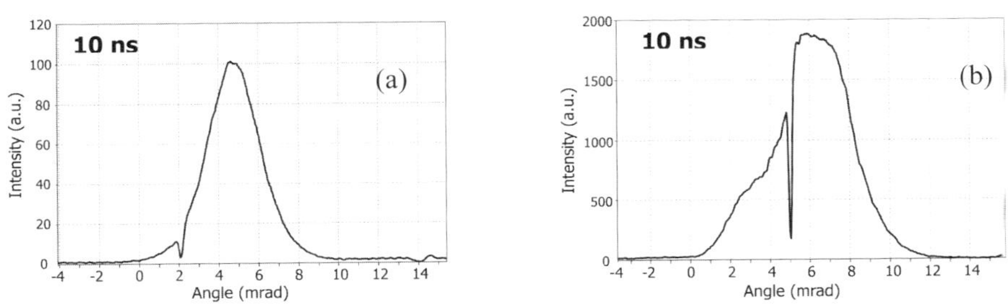

Figure 4. Comparative lineouts of XRL beam emitted for the prepulse-to-main-pulse delay of 10 ns; (a) ASE single-pass, (b) double-pass beam.

3. Applications

The developed half-cavity zinc X-ray laser delivering ~4 mJ in the prepulse configuration of 10 ns, and ~10 mJ in the prepulse configuration of 50 ns, has been employed in the applications outlined below. For the radiobiology studies and material ablation, the X-ray laser was employed as a tool by specialists in disciplines remote to laser plasma interactions.

3.1. Pilot studies of DNA damage by soft X-ray radiation

In molecular radiobiology, there is a lack of information on biological effects of low energy radiation, including soft X-rays, to biomolecular object such as DNA and proteins, notably concerning double and multiple ionisation events [6]. The unique potential of X-ray lasers consists in a possibility to deposit a large radiation dose to the objects under study in one single pulse, thereby alleviating problems associated with a long exposition of the samples to vacuum conditions, and to produce a significant yield of higher-order ionisation events; these studies are unattainable with other soft X-ray sources such as synchrotron radiation.

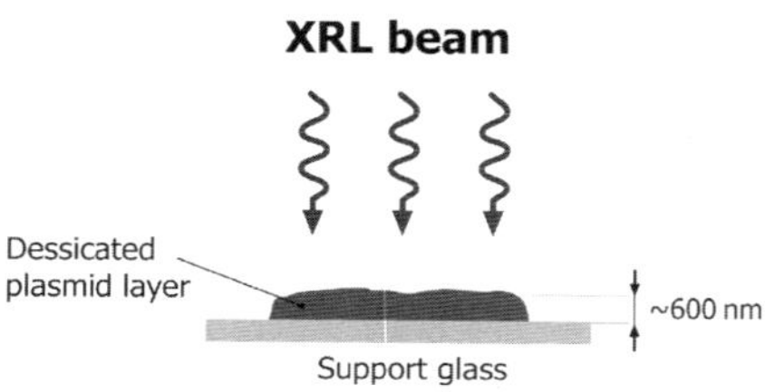

Figure 5. Damage to DNA plasmids by XRL: schematic of the samples.

The study was carried out using DNA plasmid pBS 189 as a model object; a thin dried plasmid layer was placed to the path of the X-ray laser beam reflected off a multilayer mirror and appropriately filtered. Each irradiated set included a control sample not exposed to vacuum and another control sample inserted to vacuum but not irradiated by the X-ray laser. After irradiation, the samples were dissolved in water and analysed by electrophoresis.

Plasmid is extrachromosomal form of DNA and in normal state it is twisted ("supercoiled", see Fig.6a). The damaging radiation may either nick one or both strands of the helix, relaxing the superhelical tension and changing respectively the spatial form of the plasmid to a floppy circle (in case of single strand break, SSB), or to a linear chain (in case of double strand break, occurring at the same place, DSB).

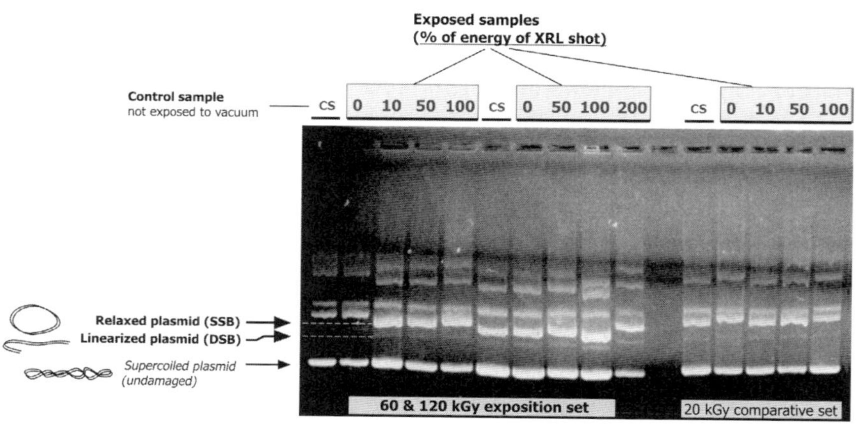

Figure 6(a). Agarose gel electrophoresis data of irradiated pBS 189 (Blue Script) samples; the exposed dose is adjusted by 0.4 and 1.5 μm Al filters attenuating the XRL beam, and by performing two shots on a single sample. Peaks due to the plasmids cut on one (SSB) or both (DSB) strands of the DNA helix are apparent on the samples exposed up to 120 kGy (left), and are absent on the "20 kGy" sample.

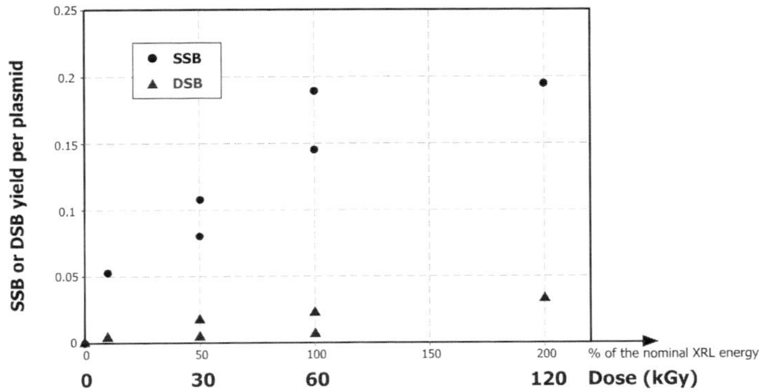

Figure 6(b) Yield of SSB and DSB damage as a function of the absorbed XRL energy, from two sets of irradiated samples.

The results obtained are shown in Figure 6(a). Due to the different migration speed in the gel, the individual spatial forms of plasmid appear as different peaks in the raw image. Figure 6(b) shows quantitatively analysed data from Fig.6(a) as a dependence of the yield of SSB and DSB damage on the total deposited dose induced by the 21.2-nm radiation. The data indicate that while the damage yields increase with dose, the dependence is probably complex and not necessarily linear; further analysis including comparison with radiation damage due to conventional sources as ^{60}Co60 is currently under progress.

3.2. Spot focusing of the 21-nm XRL beam

A number of potentially unique applications of X-ray lasers include configuration where the XRL beam is focused down to a narrow spot, producing irradiance of 10^{13} Wcm^{-2} or higher. An arrangement used at PALS for investigation of the focused half-cavity XRL beam at 21.2 nm is shown in Figure 7; the design is a compromise between simplicity, throughput, and acceptable optical aberrations. The focusing element is a spherical mirror with f=300 mm; the focused beam is further reflected off the edge of a flat mirror at 45 degrees. The position of the focus is identified using a simple miniature Hartmann plate placed in the XRL beam before it arrives at the focusing system.

The obtained focus pattern, obtained with the X-ray laser operating with the prepulse-to-main-pulse delay of 10 ns, is displayed in Figure 8(a), showing a 100×100 μm imaged region of the focal plane. Figure 8(b) shows a lineout through the raw image along the horizontal direction. By taking into account the instrumental function of the used detector (spatial resolution of ~10 μm), the size of the produced XRL focus is ~50 μm.

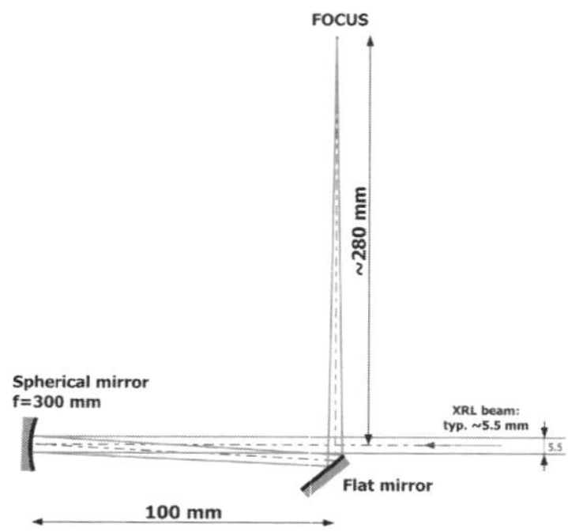

Although both the focus position and its size are very well reproducible from shot to shot, the measured focus size is larger than expected, leading to the focused intensity of ~5×10^{10} Wcm^{-2}. It is however to be noted that only few shots were taken to investigate more systematically this setup and the properties of the focused XRL beam.

Figure 7. Nominal setup for XRL point focusing. A Tb:doped fine phosphor screen and a CCD equipped with a macro objective is used to visualise the focus pattern.

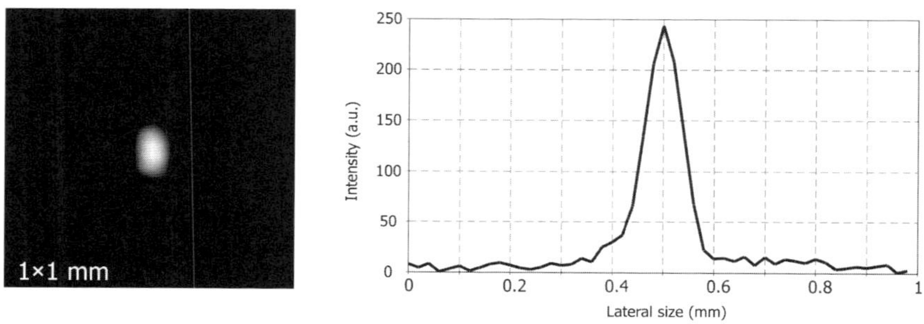

Figure 8. (a) Zinc double–pass XRL focus spot as viewed in a 1×1 mm window; (b) horizontal lineout of the raw image.

3.3. Ablation of materials (PMMA, Si, amorphous C) by focused X-ray laser beam

The motivation for studies of ablation of materials consists both in the possibility to obtain pilot, first-ever, data about the ablation rate and the processes involved in ablation in the soft X-ray region, and in prospects for applications where large energy fluence is involved. These include microfabrication technologies, design of coatings of optical elements, and choice of the "first wall" materials for future fusion devices. The QSS half-cavity zinc laser offers a unique platform for single-shot ablation experiments due to its pulse energy and peak power, besides making it possible to carry out measurements on the currently shortest soft X-ray wavelength (21.2 nm).

Samples of PMMA, polycrystalline Si, and amorphous C were irradiated by the XRL beam focused down to a spot with size of typically ~100 μm, both in a setup involving a single f=300 mm spherical mirror, and in the setup displayed in Figure 7; thin Al filters were employed to appropriately adjust the intensity of the focused XRL beam. Some of the obtained results are shown in Figures 9 and 10. The first AFM scan in Fig. 9 shows a ~150-nm deep crater ablated in PMMA by ~100 μJ of focused XRL energy, exhibiting net ablation features (sharp edges, absence of material deposited in the vicinity of the ablated pattern). In contrast, the second image in Fig. 9 displays a crater produced by the XRL beam in polycrystalline Si, showing that the rim of the crater was formed by melted matter ejected out from the XRL impact region, and indicating that the irradiance was below the ablation threshold in this case.

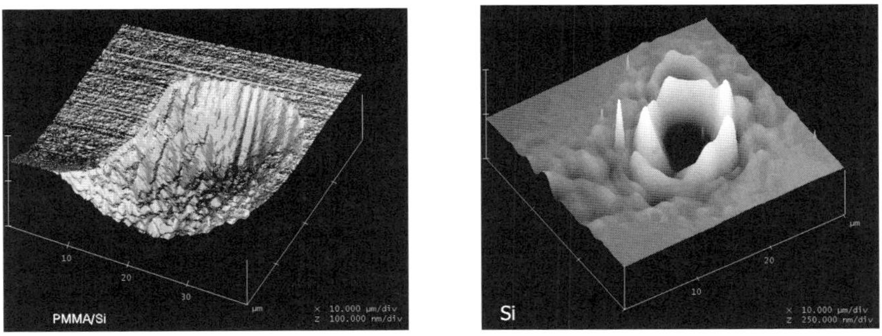

Figure 9. AFM (Atomic Force Microscopy) surface scans of the craters produced in a single shot by appropriately attenuated XRL in PMMA (100 µJ) and Si (200 µJ).

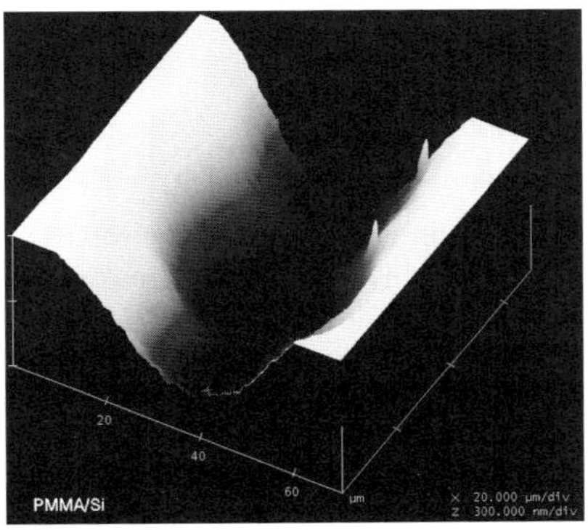

Another result of PMMA ablation is shown in Figure 10, demonstrating the possibility to obtain high aspect ratio structures using the 21-nm XRL radiation. While the results obtained in the recent XRL campaign are still analysed, future experiments might be aimed to imprint a simple lateral structure to the PMMA resists assessing the potential of the single-shot ablation in the soft X-ray region.

Figure 10. Crater ~600 nm deep produced in PMMA by three cumulative XRL shots (net focused energy ~200 µJ per pulse).

3.4 Development of adjustable soft X-ray interferometer using double Lloyd's mirror

Interferometric probing of surfaces using 21.2-nm zinc X-ray laser was successfully performed at PALS in 2001 [5], using double mirror Fresnel interferometer developed at Université Paris-Sud [7]. An effort has recently been made at PALS to implement versatile wavefront-splitting interferometer making it possible to adjust the fringe period and to offer a versatile platform for reflection surface mapping. The basis of the device, schematised in Figure 11(a), is a double Lloyd's mirror (arrangement parallel to double Fresnel mirror but functioning for both parallel and divergent beam), in which one reflective surface provides the reference wave while the probed object constitutes the second "mirror". The possibility to adjust the mutual angle of the mirror No 1 and mirror No 2 / specimen holder makes it possible for the XRL beam to be incident to the mirrors under a large angle (thereby maximising the surface altitude resolution), and to adjust the fringe period to fit the detector resolution. The interference pattern is recorded by a camera inclined under a small angle to augment the viewed fringe period.

The first test of the device was carried with a phosphor-coated CCD camera Photonic Science, equipped with optical fibre demagnificator. The Lloyd's mirrors consisted of optically polished Zerodur blocks 15×25×100 mm.

An example of the test interference fringes obtained is shown in Fig. 11(b). The fringes exhibit significant local intensity perturbations due to the uneven surface of the camera phosphor layer; this issue is expected to be resolved by a new detection system in which phosphor is replace by Ce:YAG fluorescent crystal. On the other hand, the fringe period tunability was successfully demonstrated in this experiment.

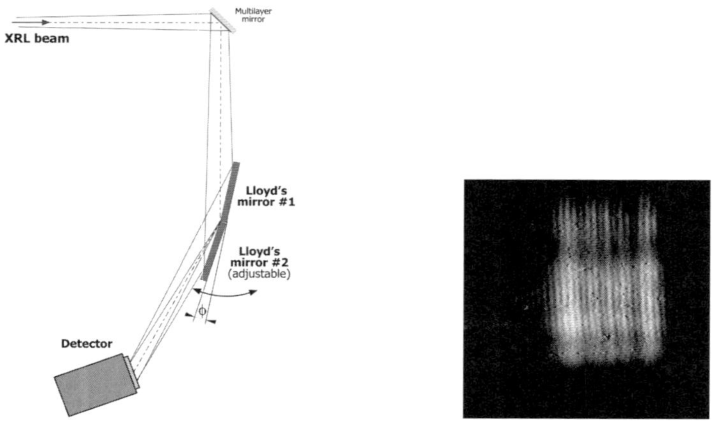

Figure 11. (a) Setup of the double Lloyd's mirror interferometer, top view; (b) early test interferogram with fringe period of 62 μm.

4. Conclusions

The potential of a novel pumping regime of the QSS systems, exploiting a prepulse applied 50 ns ahead of the main pulse, was demonstrated. This approach makes it possible in the case of Ne-like Zn to generate an extremely collimated XRL beam delivering about 10 mJ of energy, equivalent to $\sim 1.2\times10^{15}$ photons. Further research of this approach, including both numerical modelling and experimental characterisation of the amplifying plasma, will be carried out in near future. On the other hand, the program of derivative applications will be pursued, both in the fields described above and in other topics including laboratory astrophysics and basic processes involved in the interaction of focused soft X-ray radiation with solid matter.

References

[1] Rus B, Mocek T, et.al., 2002 Phys. Rev. A 66, 063806-12
[2] Mocek T, Rus B, et.al. 2003, J. Opt. Soc. Am. B 20, 1386–1391
[3] Rus B, Mocek T, et.al. 2001 Journal de Physique IV 11, 589-596
[4] Präg AR, Mocek T, et.al.2003 Eur. Phys. J. D 22, 31–40
[5] Rus B, Mocek T, et.al. 2002 Plasma Phys. Control. Fusion 44, B207-B223
[6] Sonntag C 1987 Chemical basis of radiation biology (Taylor & Francis, London)
[7] Svatos J, Joyeux D, et.al.1993 Opt.Lett. 18, 1367-1369

Inst. Phys. Conf. Ser. No 186
Paper presented at 9th Int. Conf. X-Ray Lasers, Beijing, China, 24–28 May 2004

M-L-K-Shell x-Ray Laser with 10 Femtosecond Laser Driven Proton Blocks

Wei Hong[a], Hansheng Peng[a], Yuqiu Gu[a], Heinrich Hora[b], George Miley[c], Jan Badziak[d], Frederick Osman[e]

[a]Research Center of Laser Fusion, China Academy of Engineering Physics, Mianyang 621900, China
[b]University of New South Wales, Sydney 2052, Australia
[c]University of Illinois, Urbana IL 61801, USA
[d]Inst.Plasma Physics and Laser Microfusion, 00-908 Warsaw, Poland
[e]University of Western Sydney, Penrith 1797, Australia

Abstract. The use of CPA laser pulses of less than 50 fs duration and power above 30 TW may permit conditions for x-ray lasers where in solid state dense heavy nuclei materials K-shell electrons are emitted by irradiation of 5 MeV protons with subsequent transition of L-shell electrons into the K-shell level. Alternatively, if the L-shell emission dominates, M-L-x-ray laser operation may be interesting. The CPA laser pulses are used to produce plane plasma blocks of several wave length thickness where space charge neutral 5 MeV proton beams of 10^{11} Amps/cm^2 current densities are moving co-axially into the heavy nuclei material.

1. Introduction

The L-K-shell x-ray laser was discussed before under several aspects [1-5]. There may be the possibility that ions with energy above 5 MeV per nucleon penetrating targets are emitting electrons from the K-shell of the target nuclei with higher probability than other interactions including bremsstrahlung generation. After this K-shell ionisation, there is preference that an electron from the L-shell of the target is filling up the K-shell under emission of precise line x-rays in the range of above keV photons. If these photons could be led to stimulated emission, x-ray lasers with a wave lengths of 10 Angstrom or less would be possible.

The spontaneous L-K-shell transition was estimated to be 100 fs [1-5]. The aim is then to work under conditions shorter than this time. This can well be done by several longer duration schemes for which examples are well known [4,5]. Further possibilities are available now by using laser pulses with chirped pulse amplification CPA [6] where laser pulses up to Petawatt (10^{15} W) are available. The interaction of these pulses led to never seen phenomena as the generation of new isotopes from the ps very intense 10 MeV gamma pulses [7,8] including the transmutation of long lived radioactive waste [9], electron acceleration to several 100 MeV energy [10]. The possibilities for Exawatt and Zetawatt pulses are discussed [11] where pair-production in vacuum may be envisaged [12]. In the case of preferential electron emission from L-shells the before mentioned considerations may apply for an M-L-transition laser. Without discussing what preference is to be considered, the following study is about consequences which are provided now by the application of the CPA laser pulses of less than 50 fs duration to conditions of a M-L-K-shell x-ray laser, where the generation of proton beams of current densities of 10^{11} Amps/cm^2 in space charge neutralized plasmas may be applied [13-17].

2. Acceleration of plane plasma blocks

One of the new phenomena observed is the laser acceleration of blocks of plasmas with measured acceleration above 10^{17} g (g = gravitational acceleration) where ps or shorter laser pulses of TW or higher power were used. After the first observation of these accelerations spectroscopically from the Doppler shift in the back scattered light [13], the measurement of a drastic decrease of the maximum ion energy form targets irradiated by ps-TW pulses was a problem, where Cu^{+13} ions did not reach the 22 MeV energy as known from experiments after relativistic self focusing, but only about 0.5 MeV [14]. The number of fast ions did not change when the laser intensity was changed by a factor 30. The explanation was that the suppression of a laser prepulse by a factor of 10^8 (contrast ratio) avoided relativistic self-focusing and there was only a plane geometry interaction within the fixed skin depth volume by the nonlinear (ponderomotive) force [15].

The acceleration of the plasma blocks is due to the nonlinear force which is given by the plasma frequency ω_p, the laser radian frequency ω, and the dielectrically modified amplitudes of the electrical and magnetic laser fields E and H respectively by [16,17]

$$\mathbf{f}_{NL} = -(\omega_p/\omega)^2\,\nabla\mathbf{E}^2/(8\pi) \tag{1}$$

$$\mathbf{f}_{NL} = -(\partial/\partial x)\,(\mathbf{E}^2+\mathbf{H}^2)/(8\pi) \tag{2}$$

Where the first expression is the ponderomotive force into which the general nonlinear force [17] is reduced, and the second expression is the algebraically identical with the first one but better demonstrated how the force densities are gradients of energy densities.

The generation of two very collinear plasma blocks, one against the laser light and one into the plasma was measured [16] with ion current densities of at least 10^{10} A/cm^2 in agreement with the theoretical expectation [15,17].

The advantage of this block acceleration for use to the L-K-shell laser consist in the fact that the blocks are parallel beams of space charge neutralized ions whose lateral expansion is given only by the rather low thermal motion at the basically non-thermalising nonlinear force interaction. Even if the lateral thermal motion has to be compensated, the target surface in the focus may be a little bit spherically bent such that a fully parallel beam of 5 MeV protons may be produced with a current density above 10^{11} A/cm^2. It should be noted that the alternatively generated 5 MeV proton beams [18] after relativistic self-focusing [19] are essentially divergent. The relativistic self focusing channel reduces the plasma density there such that the electrons are accelerated by the direct Lorentz force - as confirmed by comparison with experiments [10] – and are accelerated into a cone with subsequent acceleration of the ions.

In contrast to this, the plasma blocks following the conditions just described have the high cut-off density. Indeed to achieve our 5 MeV plane intense blocks of proton beams, the laser intensity for Ti:sapphire laser has to be close to 6×10^{19} W/cm^2 or less by the factor of maximum dielectric swelling [19]. What is important is that the Debye length for these blocks with protons at the neodymium glass laser wave length of

$2.3x10^{-5}$ cm is more than ten times smaller than the thickness of the blocks if only the vacuum wave length is used in case that there is no dielectric swelling at all with usual radiation pressure interaction. In the case of swelling with several vacuum wavelength thick blocks the relation to the Debye length is even less effective.

Since the energy transfer to the blocks as a kind of collisionless nonlinear absorption is well known and even turning out as one of the rare analytical solutions of an integral equation [20] which method was proposed by Shank [21] for measuring the pulse lengths and energy transfer of sub-picosecond laser pulses. This all has to be taken into account when we are now looking into a special x-ray laser scheme.

3. Application to an L-K-shell or M-L-shell laser mechanism

The x-ray laser scheme is based on the studies of F.P. Schäfer [1] and Harry Paul et al [2] where the option was discussed [3,4] that there may be a very high cross section of K-shell ionisation for ions with energies above 5 MeV/nucleon compared to Coulomb collisions of the MeV ions with electrons or lighter bound electrons than those in K-shells. There is also a relation to the gas lasers with neutron pumping from nuclear reactors [22]. The high ionisation probability of atoms by irradiation with MeV ions was confirmed also by Ulrich et al [23]. If the probabilities are better for an M-L-shell laser this may be considered as an analogy.

The laser mechanisms may first hand be based on the transition of an L-shell electron into the K-shell which was emptied by the 5 MeV protons [2]. The spontaneous transition of this kind was estimated by Schäfer to be within 10^{-13} seconds [5]. While a 5 MeV proton produces a trace in a cloud chamber of few cm, we have to use heavy atoms in a metal of about 10 g/cm^3 density for our purpose, such that about 10^4 ionisations along a trace of 10 μm will be generated corresponding to an inversion

$$n^* = 2x10^{21} \text{ excited states/cm}^3$$

This is sufficient for a laser in the range of keV photons. The only condition is along the 10 μm trace that instead of super-radiance recombination radiation the one going into the direction of the trace passes this length in a shorter time than the spontaneous recombination time. This condition is fulfilled just by a factor 10 and it can be expected that from all the 10^{11} A/cm^2 5 MeV protons in the nonlinear force driven plasma block hitting the heavy atom density metal behind the skin layer, at least 1% may have produced a laser emission along the traces. The laser property of this x-rays is confirmed by the directivity of the x-ray beam. A further proof and an access to study the details of this process is given by application of a magnetic field in the range of Tesla perpendicular to the traces bending the otherwise straight traces. Since the photons at the stimulated emission cannot follow the bent curvature, the x-ray emission will then become isotropic and not longer directed.

This method of magnetic field bending of traces e.g. of alpha particles or reaction products in low energy nuclear reactions in solids [1] could be used as proof whether the emission of x-rays in the keV range from solid density alpha emitters or similar cases is laser emission or not. However in these cases the direction of the traces is random, while the ion beams for the skin layer generated blocks for MeV ion excited x-ray lasers are primarily leading to parallel traces and a collinear laser beam.

References

[1] Schäfer F.P. 1986 Appl. Phys. B39, 1

[2] Paul H. and Müller J. 1986 Phys. Reports 135, 47

[3] Hora H. 1987 *Soft x-Ray optics and Technology*, EE. Koch and G. Schmahl eds, SPIE Proceed. 733, 47

[4] Hora H. 1988 Estimations for Optical or MeV Ion Impact K-Shell Ionization X-ray Laser Schemes,*Short Wave Length Lasers*, C. Yamanaka ed. (Springer, Heidelberg) 87

[5] Schäfer F.P. 1986 Physikalische Blätter 42, 283

[6] Mourou G. and Umstadter R. 2002 Scientific American 286 No.5, 81

[7] Ledingham K.W.D. et al 2000 Phys. Rev. Letters 84, 899

[8] Roth M., Cowan T.E., Key M.H., et al. 2001 Phys. Rev. Lett. 86, 436

[9] Magill I., Schwoerer H., et al., 2003, Appl. Phys. B77, 387

[10] Hora H., 1988 Nature 333 337; Hora H., Hoels M., et al., 2000 Laser and Particle Beams 18, 135

[11] Mourou G. and Tajima T. 2001 Inertial Fusion Science and Applications, Tanaka K.A., Meyerhofer D.D. and Meyer-ter-Vehn J. eds. 2002, (Elsevier, Paris) 831

[12] Hora H., Osman F. et al 2002 Laser and Particle Beams 20, 79

[13] Sauerbrey R. 1996 Physics. of Plasmas 3, 4712

[14] Badzziak J., Kozlov A.A., et al., 1999 Laser and Particle Beams, 17, 323

[15] Hora H., Badziak J. et al. 2002, Opt. Commun. 207, 333

[16] Badziak J. et al 2003 Phys. Letters A315, 452

[17] Miley G.H. et al. 2004 *High Power Laser Ablation, Taos April*, Phipps C. ed., SPIE Proceedings 5449, 138; Hora H., Peng Hansheng, et al., 2002 SPIE Proceedings 4914, 37

[18] Patel P.K., Mackinnon A.J., et al., 2003 Phys. Rev. Lett. 91, 125004; Wilks S.C., Langdon A.B., et al. 2001 Phys. Plasmas 8, 542

[19] Hora H. 1991, *Plasmas at High Energy and Density*, (Springer, Heidelberg)

[20] Batchelor M.T. and Stening R.J. 1985 Laser and Particle Beams 3, 189

[21] Shank C.V., 1985 private communication, Baltimore

[22] Miley G.H. in *Laser Interaction and Related Plasma Phenomena*, Schwarz H. et al eds., 1977 Plenum Press News York Vol. 4A, 181; De Young R., Wells W.E., et al., 1976 Appl. Phys. Lett. 28, 519; Helmick H.H., Fuller J.L. et al., 1975 Appl. Phys. Lett. 26, 327

[23] Ulrich A., Knorer H.J., et al, 1987, J. Appl. Phys. 62, 357

Inst. Phys. Conf. Ser. No 186
Paper presented at 9th Int. Conf. X-Ray Lasers, Beijing, China, 24–28 May 2004

Charge Transfer as Alternative Pumping Scheme for XUV Lasers

V.A. Vorontsov, I.F. Shaikhislamov*, M. Born, B.N. Chichkov, and B. Wellegehausen**

Institut für Quantenoptik, Universität Hannover, Welfengarten 1, D - 30167 Hannover, Germany
*Institute of Laser Physics, pr. Lavrentyeva 13/3, Novosibirsk, 630090 Russia
**Laser Zentrum Hannover e.V., Hollerithallee 8, D-30419 Hannover, Germany

Abtract: Investigations on a novel set up for charge exchange pumping of ions in femtosecond laser-produced plasmas colliding with a pulsed gas jet are presented. Strong selective excitation of XUV ionic transitions is observed for a number of reactions. The experimental data obtained are in a good agreement with an analytical model. Analysis of experimental data provides evidence that efficient charge exchange pumping is realized at densities of reagents well in excess of 10^{16} cm^{-3}, which is essential for the realization of XUV lasers. In preliminary investigations on the reaction C^{6+} + H $\rightarrow$ C^{5+} (n=3,4) + H^+ a strong increase of line intensities at 13.5 and 18.2 nm is observed.

1. Introduction

Among the alternative XRL pumping schemes charge exchange [1] is one of the most promising candidates. Thanks to its quasi-resonant character, this scheme could provide better conversion efficiency than currently operating collisional and recombination soft-x-ray lasers. Population inversion produced by charge exchange pumping was observed already in the 70's [2] in experiments with laser-produced plasmas expanding into a uniform background gas. However, the charge-transfer pumping efficiency was very low, not more than 10^{-3} of the maximum possible value for the given ion and background gas densities. The reasons of this have been clarified in later experiments and numerical simulations discussed in [3]. When nanosecond laser pulses with energy of a few joules are used to produce plasma, the plasma radiation and hot electrons can ionize the background gas in a region of up to several millimeters from the target surface. At densities of the surrounding pre-ionized gas larger than ~10^{16} cm^{-3}, plasma expansion generates a shock wave which further ionizes gas particles and prevents a direct interaction.

To realize an intermixing of different ion species which are initially separated, several other schemes were proposed and investigated. An ion-ion charge-exchange process was suggested in [4], and in experiments with hot and cold colliding plasmas an increase in intensity of the C^{5+} ($3d$-$2p$) line at λ=18.2 nm was reported and explained as an effect of the reaction $C^{6+} + C^{2+} \rightarrow C^{5+}(n=3) + C^{3+}$ [5], but no improvement has been made in further experiments with colliding plasmas.

In [3] a compact gas cloud created by laser ablation was used as a source of neutrals flowing towards the laser produced ions. Sharp density gradients of the gas cloud and a controllable gap between the plasma target and the gas front allowed to realize efficient and direct charge-exchange interaction at densities of neutrals $>10^{16}$ cm^{-3} and of ions at about 10^{15} cm^{-3}. In this approach, however, a further increase of the ion density in the interaction region seems to be problematic since it is difficult to position the ablation gas cloud close enough to the plasma target.

An analytical model of charge-exchange interaction of dense interpenetrating flows was developed in [6]. It follows from this model, that for an efficient charge-transfer pumping at relatively high densities of reagents it is crucial to have, besides a compact gas cloud, a well-localized plasma jet with a sharp ion density gradient. At desired particle densities ($>10^{16}$ cm^{-3}), the characteristic length of charge exchange which determines the interpenetration of ions into a gas medium is less than 1 mm. Thus, effective pumping can only be realized on a spatial scale of the order of 1 mm or less, and the widths of the plasma and gas fronts should be correspondingly small. To ensure these conditions, in a preceding publication [7] a novel experimental scenario for charge exchange pumping, consisting of a pulsed gas jet interacting with plasma produced by ultrashort laser pulses was introduced and tested. With this technique efficient charge exchange pumping was demonstrated for the reactions $C^{4+} + H \rightarrow C^{3+}(n=3) + H^{+}$ and $O^{3+} + H \rightarrow O^{2+}(n=3) + H^{+}$ [7, 8]. The achieved results clearly demonstrate the pumping of specific levels of the newly created ions in qualitative agreement with the charge exchange interaction. In this contribution we present new experimental data and dependencies based on time resolved measurements which allowed us to perform quantitative analysis of the results obtained by comparison with a kinetic model of charge exchange pumping. This provides additional information on the optimum conditions and proves that charge exchange interaction at densities of both reagents exceeding 10^{16} cm^{-3} was realized for the first time. Towards XUV gain experiments, first results on the reaction $C^{6+} + H \rightarrow C^{5+}(n=3,4) + H^{+}$, which is promising for lasing on the 4-2 and 3-2 transitions at 13.5 and 18.2 nm, are given.

2. Experimental set up and diagnostics

In Fig. 1, the basic experimental arrangement as developed earlier [7] is shown. The plasma is produced by focusing of 250 fs, 70 mJ Ti:sapphire laser pulses with a f = 250 mm lens on a flat rotating target. The angle of incidence is normal to the target surface and the size of the focal spot is approximately 200 μm. XUV radiation of the produced plasma is analyzed either by a single shot spectrometer (McPherson instrument 248/310G) equipped with a MCP-phosphor-CCD time- integrating detection unit or by Jobin Yvon Monochromator equipped with MCP, allowing us to perform time of flight measurements. To obtain a spatial resolution in the direction normal to the target, a 200 µm slit is placed at a 2 cm distance from the focal spot.

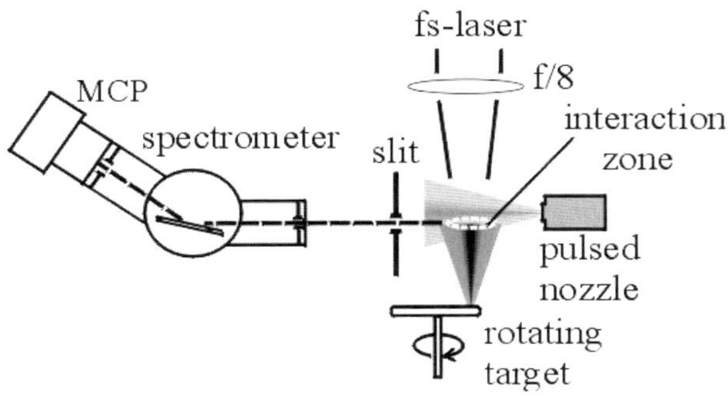

Figure 1. Experimental set up.

The gas is injected into the vacuum chamber (evacuated to below 10^{-4} mbar) by a pulsed nozzle (General Valve Corporation) located close to the target surface near the focal spot. The pulsed nozzle has an opening time of about 1 ms and is operated at backing pressures p of up to 3 bar. In most measurements the nozzle axis coincides with the optical axis defined by the slits. The position of the focal spot on the target surface relative to the nozzle opening (diameter $\approx$1 mm) can be easily varied by moving the target along its normal. The particle density n at the nozzle exit is determined from interferometric measurements and can be approximated by $n = 7\cdot10^{17}\cdot p$, where n is measured in cm^{-3} and p in bar.

3. Experimental results and discussion.

For basic studies, the well known charge-exchange reaction between C^{4+} ions and Hydrogen gas was used [9]. In the reaction $C^{4+} + H \rightarrow C^{3+}(n=3) + H^{+}$, which has a cross-section of about 2.5×10^{-15} cm^2, the quantum level n=3 of the C^{3+} ion is selectively pumped with a probability of over 90 %. This process can be observed through radiative transitions $3d$-$2p$ (38.4 nm), $3p$-$2s$ (31.2 nm), and $3s$-$2p$ (42 nm), and in our previous work we reported, that by applying the H_2 gas jet all charge exchange influenced C^{3+} lines (from the n=3 level) are strongly enhanced in comparison to the case of expansion into vacuum and the lines which cannot be pumped by charge exchange (from the n=4 and higher) show even a decrease [7]. Such strong effect occurs at distances of 1-2 mm from target surface, hydrogen backing pressure of 100 mbar, and at time delay between the valve opening and laser pulse of 600 ms.

Most informative data of the process are obtained from time resolved measurements. In Fig. 2(a), line intensities for two distances - close (0.5 mm) and far (1.5 mm) from the target surface are presented. The position of the gas nozzle relative to the line of sight of the monochromator was kept the same. These data demonstrate, that at relatively large distances between the target and the gas jet the lines from the level n=3 show an increase with respect to the vacuum case during the whole time when the plasma streams through the gas jet. At smaller distances, where the density of the plasma is comparable to or higher than the density of neutral particles, the lines show dramatic increase only at the plasma front. This increase of intensity at the plasma front exhibits a linear dependence with the gas jet pressure up to about 100 mbar, exactly as expected for

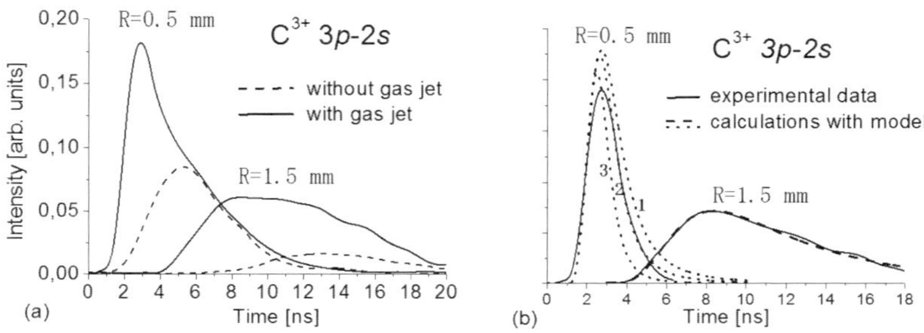

Figure 2. (a) Time resolved intensity of C^{3+} $3p$-$2s$ line at two distances R from the target surface with gas jet (solid lines) and in vacuum (dashed lines). (b) Solid curves: time dependent charge-transfer induced part of the C^{3+} $3p$-$2s$ line at distances R=1.5 and 0.5 mm (calculated as difference between the solid and dashed curves shown in (a)). Dashed curve: calculation results for a distance R=1.5 mm with the mean velocity $V_m = 1.27\times10^7$ cm/s obtained using Eqs. (1-2). Dotted curves: calculation results for a distance R=0.5 mm with $V_m = 1.27\times10^7$ cm/s and three different values of a total number of ions N_{tot}: 1: 3×10^{12}, 2: 4×10^{12} and 3: 6×10^{12} Sr^{-1}.

charge exchange pumping as the rate of charge transfer pumping is proportional to the density of neutrals.

In order to better understand the behavior of the lines, the experimental curve is analyzed with the existing model of charge exchange interaction. The pumping rate of charge exchange pumping J can be approximately expressed by the equations [6]

$$J(R,t)=n_i^o(R,t)\,V_i\,n_a^o(R)\,\sigma_i\,\kappa, \quad \text{where} \quad \kappa=\frac{e^{-N_i\sigma_i+N_a\sigma_i}}{[1+e^{-N_i\sigma_i}(e^{N_a\sigma_i}-1)]^2}, \qquad (1)$$

which provide exact solutions for the case of planar geometry. Here V_i is the ion velocity, n_i^o, n_a^o are the densities of species at a distance R and time t taken without interaction, σ_i is the charge exchange cross-section. The efficiency of interaction is described by a factor κ; which depends on the integral densities N_a and N_i: $N_a=\int^R n_a^o dR$ is the total number of neutrals along the path from the target surface to the observation point R, $N_i=\int n_i^o(R,t)V_i dt$ is the total number of ions passed through the unit square (located at the observation point) to the moment of observation t.

When the ion density is small and $N_i \ll 1/\sigma_i$ is fulfilled, the efficiency coefficient κ is constant and can be approximated by $\kappa \approx exp(-N_a\,\sigma_i)$. In this case the line intensity, which is determined by the factor J, is proportional to the ion flux n_iV_i, see Eq. (1). In the opposite case of large ion density, it follows that $\kappa \approx exp(-N_i\,\sigma_i)$. Because N_i is the time integrated ion flux, the pumping rate exponentially decreases with time. In other terms, the pumping rate decreases because the number of neutral particles available for interaction decreases with time due to the charge exchange process. Thus, from Eq. (1) it follows that most effective interaction can be realized only if the number of neutrals and ions in front of the gas cloud and
plasma flow, respectively, are approximately equal to each other, e.i. $N_i=N_a$. It is also desirable to keep this values not too large, $N_i,N_a\sim 1/\sigma_i$. Otherwise the length of charge exchange is too small and interaction takes place only in a thin layer. N_a can be calculated as $N_a=n_a\times L_a$, where L_a is the width of the jet boundary. For our experimental parameters, the estimation $N_a\sigma_i\sim 1$ is valid.

To calculate the time behavior of signals, a model of plasma expansion should be added to the solution of charge transfer equations. Plasma produced by a short laser pulse is described by a self-similar expansion of ions with Maxwell velocity distribution

$$n_i(R,t) = (4/\sqrt{\pi})N_{tot}exp(-(R/V_mt)^2)/(V_mt)^3 \quad (2)$$

where N_{tot} is the total number of particles of interest (in our case C^{4+} ions) per unit solid angle, and V_m is their mean velocity. The velocity of each plasma element is given by $V=R/t$. The value of V_m can be found by fitting of the calculated curve to the experimental one. By the solid curves in Fig. 2(b), the difference between the line intensities measured with a gas jet and in vacuum are presented for two distances R=0.5 and 1.5 mm (see Fig. 2(a)). These curves represent the signals only due to charge exchange pumping. For the curve at R=1.5 mm, a best calculated fit is shown by the dashed line, which corresponds to $V_m=1.27\times10^7$ cm/s and agrees very well with the experimental measurements. Because only relative line intensities were measured in these experiments, the value of N_{tot} cannot be derived from comparison with calculations as long as the ion density is small and the efficiency coefficient κ is independent of it. However, near the target the behavior of signals get dependent on the total number of ions involved into this process. By dotted lines in Fig. 2(b), results of calculations are shown for the distance R=0.5 mm for three values of N_{tot} at the same mean velocity V_m found earlier. A best fit with $N_{tot}=4\times10^{12}$ Sr^{-1}

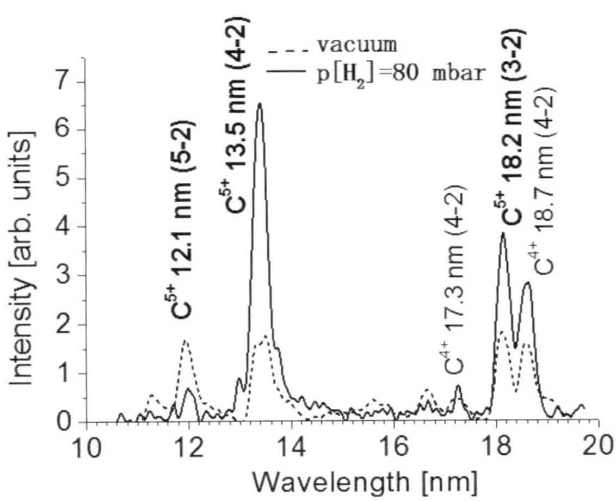

Fig. 3 Carbon spectra recorded at a distance of 1.5 mm from the target. Dashed line: in vacuum, solid line: at 80 mbar H_2 backing pressure.

corresponds to an ion density of 2.8×10^{16} cm^{-3} at the signal maximum located at t = 2.7 ns, as follows from Eq. (2). This analysis confirms that highly selective charge exchange pumping has been realized at densities of both reagents exceeding 2×10^{16} cm^{-3}.

At the optimized experimental conditions, preliminary investigations on the reaction $C^{6+} + H \rightarrow C^{5+}$ (n=3,4) + H^+ were conducted. For the 3→2 transition of C^{5+} ions at 18.2 nm gain was already demonstrated in recombination type scheme. For the

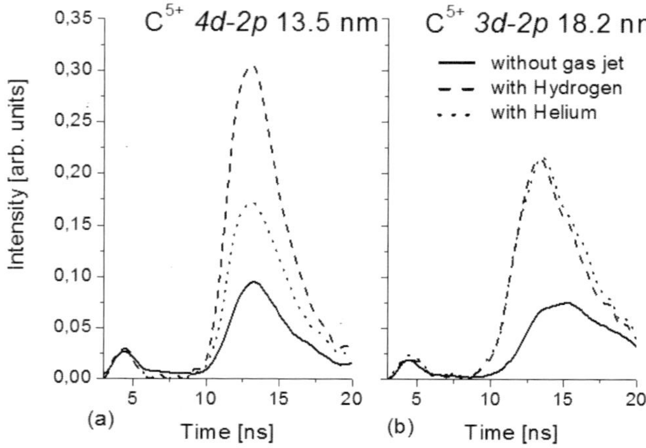

Fig.4 Time resolved measurements for C^{5+} 3*d*-2*p* and C^{5+} 4*d*-2*p* lines at a distance of 2.5 mm from the target: solid lines – expansion in vacuum, dashed lines – with hydrogen gas jet, dotted lines – with Helium gas jet.

production of bare Carbon ions from a solid carbon target, 200 mJ of uncompressed 100 ps-pulses from the Ti:Sapphire laser system have been used. A corresponding intensity at the focal spot was 10^{13} W/cm^2. In Fig. 3 examples of spectra recorded in vacuum and with a Hydrogen gas jet at a distance of 1.5 mm from the target surface are shown. One can see substantial intensity increase of the lines that should be pumped by the charge-transfer reaction, especially of the 4-2 line at λ=13.5 nm wavelength. The time of flight measurements also show that there is a dramatic increase in the front of the signal for both lines with hydrogen gas jet [Fig. 4, dashed lines] in comparison with the lines induced in vacuum. Unlike the results of experiments with the plasma expanding into background gas [2], a clear dependence of measured line intensities on the sort of gas was observed. Charge-exchange, being a resonant process, depends mainly on the ionization potential of the collision partners. In case of C^{5+} ions, the level $n=4$, for instance, can better be pumped with hydrogen gas, but not with Helium which has higher ionization potential, and this is confirmed in our experiment [Fig. 4(a)], where the best increase is for the hydrogen gas jet. The level $n=3$, however, has the better decrement of energy with Helium gas, although the cross section is larger for the hydrogen gas, which explains the equal effect of these gases [Fig. 4(b)].

4. Conclusions

Interaction of ions produced by short laser pulses with atoms from a pulsed gas jet has been studied by means of time and space resolved spectroscopy in the XUV range. New and previously obtained dependencies on the gas pressure and gas sort are explained by direct and highly selective charge exchange pumping. Experimental results are in a good agreement with an analytical model and confirm that charge exchange pumping has been realized at densities of both reagents well in excess of 10^{16} cm^{-3}, which is at least an order of magnitude higher than in previous experiments. At optimum conditions, a strong increase in the intensities of several potential lasing lines in the XUV range was observed in the presence of the gas jet. These observations provide new prospects for the realizations of XUV lasers with higher efficiency than in presently demonstrated schemes. For the 18.2 nm line of the C^{5+} ion, a gain coefficient exceeding 10 cm^{-1} is expected for a length of medium of 1 cm with the densities of reagents of 10^{16} cm^{-3}. In preliminary investigations towards XUV gain experiments, it has been found that about 2.5 J of energy in a ps-laser pulse is needed to produce sufficient amount of C^{6+} ions in a line focus of about 1 cm in length and 0.07 cm in width.

References

1. Vinogradov A.V. and Sobel'man I.I. 1973 *Sov. Phys. JETP* **36** 1115
2. Dixon R.H. and Elton R.C. 1977 *Phys. Rev. Letters* **38** 1072
3. Ponomarenko A.G., Shaikhislamov I.F., et.al. 1998 *J. Phys. D* **31** 2117
4. Seely J.F. and McKnight 1977 *J. Applied Phys.* **48** 3691
5. Ruhl F., Aschke L., Kunze H.-J. 1997 *Physics Letters A* **225** 107
6. Shaikhislamov I.F. 2000 *Russian J. of Applied Mech. And Tech. Phys.* **41** 212
7. V.A. Vorontsov, M. Born, et.al. 2003 *J. Phys. B: At. Mol. Opt. Phys.* **36** 3865
8. V.A. Vorontsov, M. Born, et.al. 2004 *Appl. Phys. B* **78** 979
9. Hoekstra R., De Heer F.J. and Morgenstern R. 1991 *J. Phys. B* **21** 81 supplement

Inst. Phys. Conf. Ser. No 186
Paper presented at 9th Int. Conf. X-Ray Lasers, Beijing, China, 24–28 May 2004

Coherence and other characters of supercontinuum generated in air by ultrashort laser pulses

K. Li, J. Zhang, M. H. Xu, and Y. T. Li
Laboratory of Optical Physics, Institute of Physics, Chinese Academy of Sciences, Beijing, 100080, China

Abstract. The combined spectrum of white light pulses generated by fs pulses in air is measured here. This demonstrates that in spite of the high nonlinear process in the generation of white light pulses, the white light pulses are well phase-locked to the main pulses. The temporal coherence length of white light pulses, being the same to that of main pulses, is independent of the wavelength. Two spatially separated filaments are also generated from the same main pulses. The interference pattern is recorded, which gives another proof of the non-degradation of the coherence of the white light pulse, compared to that of the main pulses. Other characters of white light pulses, such as the pulse width and divergence angle, are measured too.

1. Introduction

Discovered in 1970 (Alfano and Shapiro, 1970a, 1970b), continuum generation is a ubiquitous response of transparent materials (liquids, solids, and gases) to high-power, ultrashort pulse radiation [1]. The physical origin of this phenomenon lies in the fact that the strong field of a laser beam induces an appreciable intensity-dependent refractive index change in the medium, which results in the phase modulation of the pump laser pulses and is call self-phase-modulation. The supercontinua generated in different kinds of medium has found utilities in many fields, such as broadband absorption and excitation spectroscopy, light clock, the generation of shorter laser pulses and so on. The character of SC generated in liquid, solid [2][3], and fibres [4] has been studied extensively, while the character of supercontinuum (SC) generated in air is still to be investigated. So it is our aim here to investigate of the character of SC generated in air by femtosecond laser pulses.

Experiments

Ultrashort laser pulses from the 20TW JG-II Ti:sapphire chirped-pulse-amplification laser system at the institute of Physics, CAS were focused by a 4m positive lens in air. The beam diameter was about 40mm (FWHM), the energy of the each pulse was about 40mJ. The emitted white light pulses were directed into a stable Michelson interferometer with 50/50 beam splitters ranging from 650nm to 900nm to get balanced white light pulses. The interferometer arm displacements were controlled by means of high-precision motor. The two arms had to be carefully controlled to collinear the two white light pulses thus to observe clear interference fringes. In one experiment, a spectrometer, based on a 2048-pixel photodiode array, ranging from 200nm to 1200nm was placed behind the Michelson interferometer to measure the combined spectrum of white light generated here. In another experiment, the spectrometer is replaced by a BBO crystal followed by a photodiode as an intensity autocorrelator to measure the pulse width of both the main pulses and the generated white light pulses. In other experiments, double slits with different gaps were positioned along the SC to measure the transverse coherence length of white light pulses. And different interference filters centred at 635nm, 770nm, 805nm, 850nm and 880nm with a bandwidth of 6nm were used to choose different part of the spectrum of SC to measure their divergence angle. The power of the main pulse is carefully decreased to achieve a single filament. A digital camera is used to record the image of the light on a white screen.

The pump laser pulses were also directed into the same Michelson interferometer followed by a positive lens to generate two spatially separated filaments. The interference fringes of the white light pulses generated from the two filaments were recorded by a CCD camera.

2 Results and discussion

In measuring the combined spectrum of the pulse from two arms, the spectrum is modulated. Theoretically in the ideal case, the output pulses from the two arms are phase-locked with electric fields $E_1(t)$ and $E_1(t+\tau)$, so their combined spectrum is easily found to be

$$I_t(\omega,\tau) = I_1(\omega) + I_2(\omega) + 2\sqrt{I_1(\omega)I_2(\omega)}\cos(\omega\tau) \qquad (1)$$

By changing the delay τ between the SC pulses, the periods of the spectral fringes are changed, and it is simply found from (2) to be

$$\Delta\lambda = \frac{\lambda^2}{c\tau} \qquad (2)$$

τ is the temporal delay of the two supercontinua. So the combined spectrum should

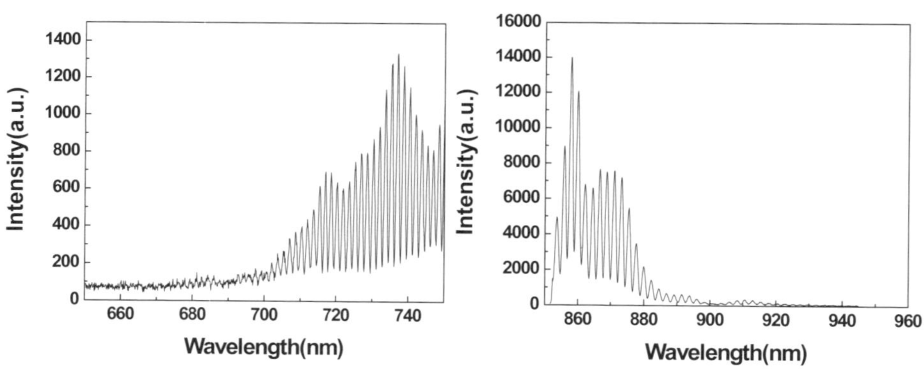

Fig.1 The combined spectrum of the white light pulses in the IR region and the visible region.

shows sinusoidal modulations with fringe period with the inverse of the time separation between the two pulses. The measured combined spectrum of the white light pulses in the IR region and in the visible region is shown in Fig.1. As the time delay of the two pulses increase, the visibility of the combined spectrum decreases. There are two possible reasons for the degradation of the visibility of the combined spectrum. One is obviously connected with the finite resolution of the spectrometer; another is from the loss of the mutual coherence of the two pulses. Because the visibility of the combined spectrum of the white light diminished almost at the same time delay to that of the main pulses, the second reason does not exist. Compared to the analysis of ref.3, the most reasonable degradation of the visibility degradation is from the finite resolution of the spectrometer. And the coherence of main laser is kept in the white light. This is contrast to our assumption at first for the high nonlinear interaction process in the generation of while light. Our explanation is that the perturbation is much slower than that of the process of the generation of supercontinua, so the phase coherence is kept during this high nonlinear process. From the combined spectrum at different time delay we can also get the temporal coherence length of the white light pulses if it is defined as the time span across which the visibility of the combined spectrum diminishes. The length of both the main pulses and white light pulses is about 2mm and is independent of the wavelength.

The divergence angle of the supercontinuum at different wavelengths is shown in Fig. 2. So the stokes part of SC has a larger divergence angle than that of the pump pulse, and the anti-stokes part propagates collinearly with the pump pulses. This is consistent with the spatially SPM theory, in which conically radiation is reasoned. The pulse width of the white light is shown in Fig.3. The white light pulses is no longer gauss but temporally modulated. The temporal pulse width ranges from 120 to 260fs, much longer than that of the pump pulses which is about 50fs. The main reason of the prolongation of the white light pulse is the velocity mismatch of the phase velocity of the white light and the group velocity of the pump pulse. So the white light pulse must be chirped.

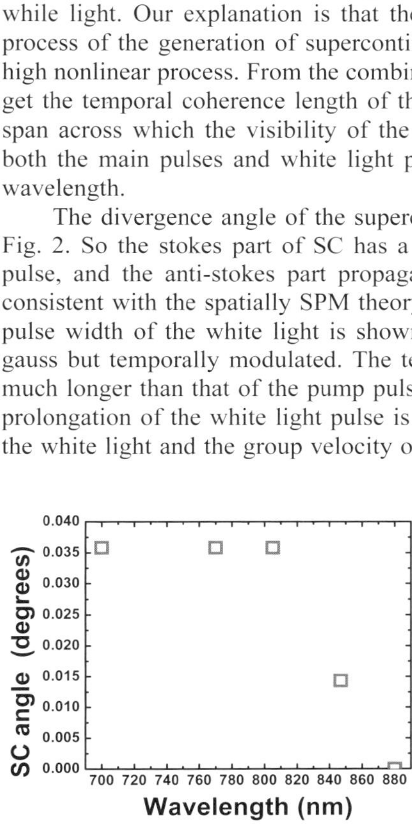

Fig.2 Divergence angle of white light pulses at different wavelength.

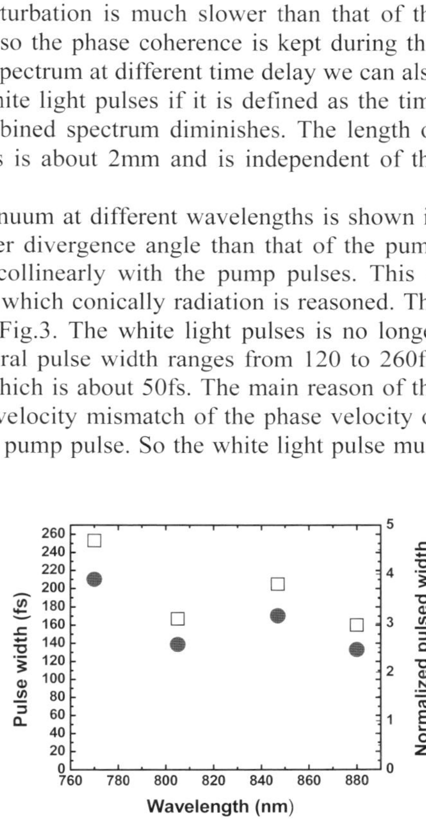

Fig.3 The pulse width of the white light pulses and the results normalized to the pulse width of

The conical radiation from a single filament is recorded by a digital camera, as is shown in Fig. 4. The single filament is achieved by diminishing the power of the energy of the pump pulse and turning the direction of the lens. Radiation with shorter wavelength is shown to be in the outer side. It is consistent to Fig. 2.

In another experiment, pump pulses were sent into a Michelson interferometer. The two output pulses were focused by a lens with a focal lens of 10 cm and white light pulses radiated from the two separated filaments interfere on a white screen. The interference pattern is shown in Fig.5. In this figure, the narrow and clear fringes are deemed to origin from the interference of the white light pulses of the filaments, while the unclear and sparse fringes are the results of the diffraction of white lights in a single filament. So the coherence of mian pulse is kept in different filaments. If the time delay of the two pulses is increased, the visibility of the interference pattern diminishes gradually. At almost the same time delay, the visibility of the interference pattern of all wavelength including that of the TH harmonics generated here diminishes, so the coherence length of white light pulses is independent of the wavelength which is consistent with our former conclusion.

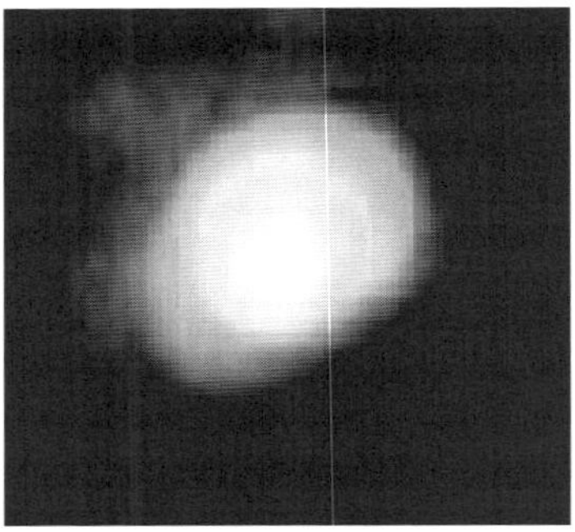

Fig.4 The radiation of white light pulses from a single filament.

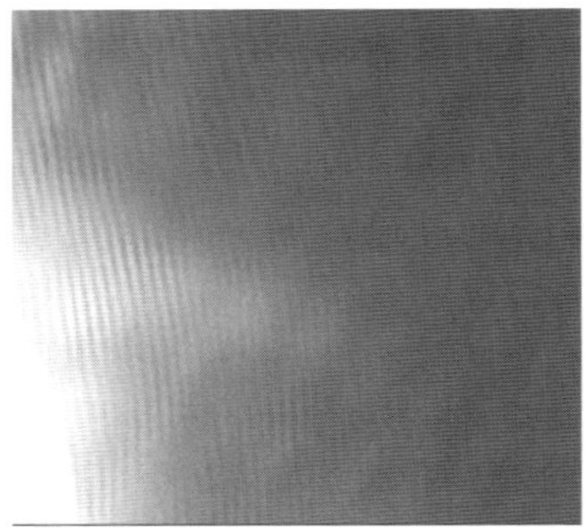

Fig.5 The interference pattern of white light pulses from two spatially separated filaments.

3. Conclusion:

We have verified that the white light pulses generated from air by femtosecond laser pulses are phase-locked to the pump pulses, even in such a high nonlinear process. Furthermore, the coherence at all wavelengths of the white light pulses is identical, being about 2mm (0.67ps). The coherence length is almost the same to that of the main pulses. The pulse width of the white light pulses ranges from 120fs to 260fs compared to 50fs of the main pulses. The anti-Stokes side of SC propagates collinearly with the pump pulse, thus being called chirped ultrashort laser pulses while the Stokes side propagates conically with colourful rings.

Acknowledgements

This work was supported by the NSFC under Grant Nos. 60478047, 60321003, 10390160, 10474134 and 60128505, and the NKBRSF under Grant No. G1999075206, and the National Hi-tech ICF Program.

Reference

[1] Alfano, The Supercontinuum Laser Source, pp.318， Springer-Verlag
[2] Marco Bellini, Theodor W. Hansch, Opt.Lett. 25, 1049 (2000)
[3] C. Corsi, A. Tortora, M. Bellini, App. Phys. B, 77, 285 (2003)
[4] John M. Dudley, Stephane Coen, Opt. Lett. 27, 1180 (2002)

Inst. Phys. Conf. Ser. No 186
Paper presented at 9th Int. Conf. X-Ray Lasers, Beijing, China, 24–28 May 2004

Charge Exchange Recombination X-Ray Laser in a Cluster Plasma Produced by an Ultra-Short High Intensity Pulse Laser

S. Namba, K. Takiyama
Graduate school of Engineering, Hiroshima University,
Kagamiyama 1-4-1, Higashi-Hiroshima, Hiroshima, 739-8527, JAPAN

N. Hasegawa, M. Kishimoto, T. Kawachi, K. Nagashima
Advanced Photon Research Center, Kansai Research Establishment,
Japan Atomic Energy Research Institute,
Umemidai 8-1, Kizu-cyo, Sourakugun, Kyoto, 619-0215, JAPAN

Abstract. Cluster plasma produced by an interaction of an ultra-short pulse laser and nitrogen clusters has been investigated as active medium of charge exchange recombination X-ray laser. A non-equilibrium charge state distribution could be easily created in the mixture of cluster and atomic gas. The possibility of the oscillation of X-ray laser due to the charge exchange was briefly discussed.

1. Introduction

For the purpose of realization of X-ray lasers, a variety of schemes, such as transient collisional-excitation [1][2] and recombination [3], have been extensively studied. Charge exchange recombination (CXR) process has been also expected to be one of the promising approaches, since electrons are captured into selective electronic states with the large cross section, followed by a creation of population inversion [4][5][6]. In order to realize CXR X-ray laser, the interaction between highly-charged ions and neutral atoms in plasma is required. Therefore, it is essential to create an extreme non-equilibrium plasma, in which highly-charged ions and neutral atoms coexist simultaneously. Although the observation of population inversion due to CXR was reported in conventional laser produced carbon plasma, the gain coefficient was not enough for the laser oscillation so far [4]. The question we have to ask here is how to produce the extreme non-equilibrium plasma.

To satisfy the above requirement, we have employed laser produced plasma that is generated by an interaction of an ultra-short, high-intensity pulse laser and clusters in gas atmosphere. Since the atomic density inside the cluster is quite high, $\sim 10^{23}cm^{-3}$, the incident laser energy is deposited efficiently due to collisional heating (inverse bemsstrahlung) and resonance absorption [7]. As a result, electron collisional ionization plays a dominant role to produce highly-charged ions. On the other hand, atoms surrounding clusters are primarily ionized by optical field ionization (OFI) [8]. It should be noted that atoms are still in neutral under laser irradiation with less than threshold intensity that may be determined by barrier suppression ionization (BSI). For helium atom, this value corresponds to be ~2×10^{15} W/cm^2. By choosing appropriate laser intensity, non-equilibrium charge state distribution could be created in cluster plasma. This allows generation of favourable conditions for efficient charge exchange pumping. Particularly, in this study, we have focused attention to the following CXR reaction with the cross section of the order in ~10^{-15}cm^2 [9], N^{7+} + He->N^{6+} (n=4) + He^+. Lasing actions with λ=1.9 nm for transition of N^{6+} (1s-4p) and λ=9.9 nm for 2p-4d are expected.

In this work, population densities of excited states involved with lasing actions as well as the characteristics of cluster plasma were investigated by spectroscopic observations. Nitrogen clusters created in a supersonic free jet was irradiated with ultra-short, high-intensity pulse laser. The possibility of the oscillation of X-ray laser due to CXR process will be briefly discussed.

2. Numerical calculation of gain coefficients

In order to investigate plasma parameters required for realization of X-ray laser, population densities of excited levels of N^{6+} ion were calculated by using a modified collisional-radiative model, in which time evolution of the excited states and the CXR were taken into accounts. In this calculation, we assumed that plasma consisted of fully ionized N^{7+} ions, electrons and helium atoms at t=0.

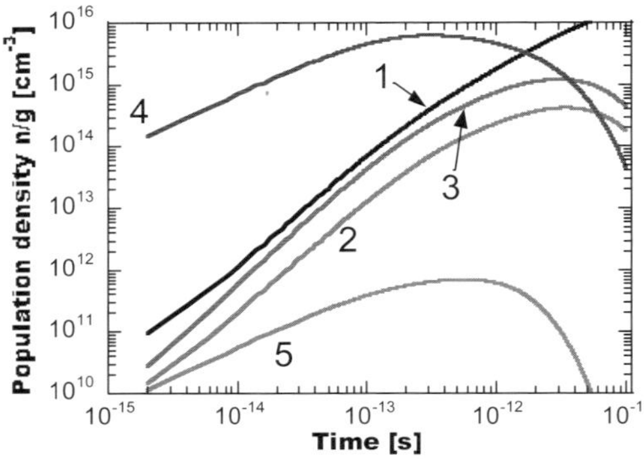

Figure 1. Population densities of N^{6+} ions.

Figure 1 shows the temporal evolution of population densities under the recombining plasma conditions of T_e=40 eV, $n(N^{7+})=n_e=1\times10^{19}$ cm^{-3}, n(He)=1×10^{18} cm^{-3}, collision velocity $v=1\times10^5$ m/s. Population density of n=4 was increased rapidly at initial stage (≦2ps) due to CXR process. Assuming that the spectral profile was Gaussian shape and angular momentum quantum number l distribution on n=4 level associated with CXR was accordance with the statistical weight, we obtained the temporal behaviour of gain coefficients, as shown in Fig. 2. Gain coefficients g=6 and 28 were obtained for 1s-4p and 2p-4d transition, respectively.

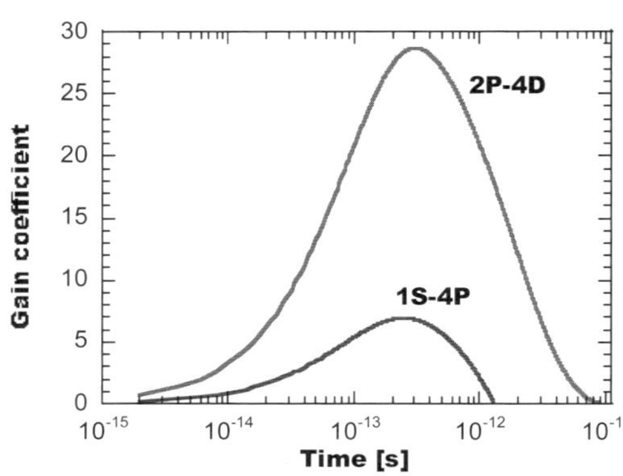

Figure 2. Gain coefficients of 1s-4p and 2p-4d transitions of H-like ions.

3. Experimental setup

The experiment was carried out with a 6 TW Ti:S laser (λ=800 nm, 60 fs, 3×10^{16} W/cm^2). In order to investigate the characteristics of cluster plasmas, the clusters were irradiated with higher intensity than the threshold intensity of helium atom (~$2x10^{15}$ W/cm^2). When high density gas was injected into vacuum, atoms bonded together by van der Waals force were produced due to cooling associated with an adiabatic expansion [10]. In this study, two types of nozzle (0.8 mmϕ circular nozzle and 0.2 mmϕ throat supersonic conical nozzle) were used. It is well-known that the supersonic nozzle is suitable for the creation of larger clusters. Moreover, a seeding technique, that is, nitrogen gas was diluted with helium, was used to enhance production efficiency of larger cluster. The average cluster size estimated by Hagena empirical law was more than 10^3 at the backing pressure of ≧10 atm with circular nozzle.

Soft X-ray emission spectra from cluster plasmas were measured by using a grazing incidence spectrometer with a flat field grating in a direction perpendicular to the laser propagation. The detector was a back-illuminated charged coupled device (CCD).

4. Results and discussion

Figure 3 shows the spectrum obtained by use of the conical nozzle under the backing pressure 23 atm (N_2 30%, He 70%). Gas density adjacent to nozzle exit was estimated to be ~$10^{17}cm^{-3}$ by measurements of a fast ionization gauge. In such a low gas density, atoms and ions are dominantly ionized due to OFI process. According to BSI model, however, the threshold intensity to generate H-like ions due to OFI is estimated to be

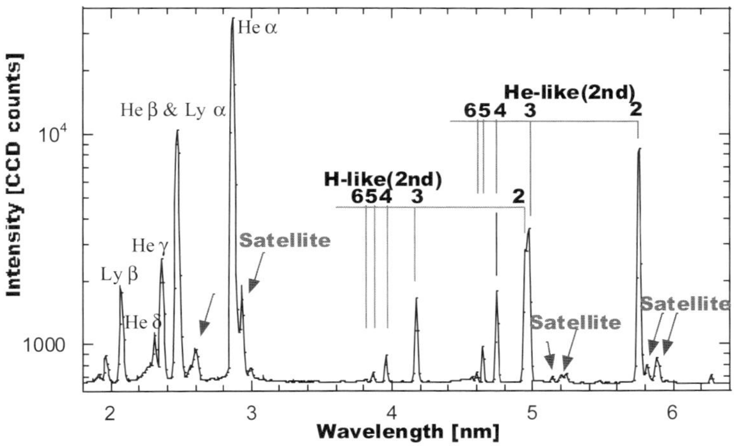

Figure 3. H-like and He-like nitrogen ion spectrum.

~$1x10^{19}$W/cm^2. The fact that H-like ion spectra as well as He-like ones were observed suggests that the large cluster must be formed in gas jet, leading to the creation of high density and high temperature plasmas due to the efficient laser energy deposition within the clusters. Further evidence of the existence of cluster can be seen in the dependence of He-α intensity on the partial pressure of helium gas as shown in Fig. 4. In spite of that total amount of nitrogen atoms decreased under the fraction of N_2 30%, the seeding technique made it possible for the intensity of He-α to increase. On the other hand, with the circular nozzle, He-like ion spectra could be still observed. The intensity of He-α, however, decreased by one order of magnitude compared with that of conical nozzle. Moreover, the dependence on partial pressure was reverse relation. This implied that even with the seeding the clustering efficiency was quite low with the circular nozzle.

Population densities of H-like and He-like ions were determined by the corresponding second order spectra. Figure 5 shows the relative population densities of

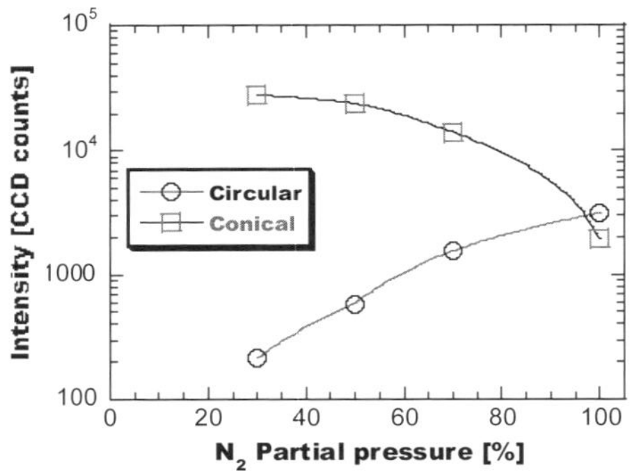

Figure 4. Dependences of the partial pressures of nitrogen gas.

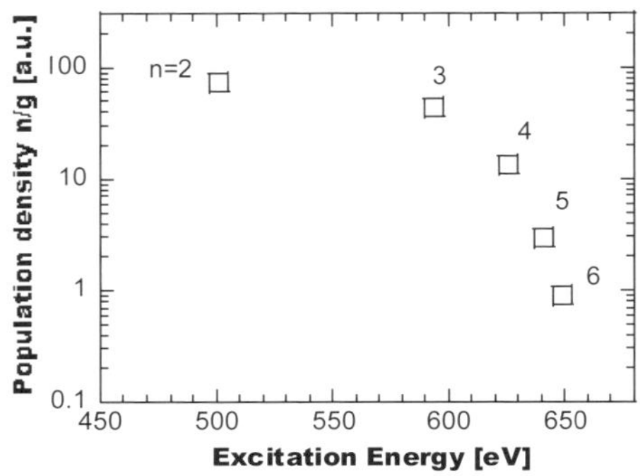

Figure 5. Boltzmann plot of H-like nitrogen ions.

H-like nitrogen ions. In this study, the population inversion between n=3 and n=4 in H-like ions was not obtained. The main reason may be that almost all of the helium atoms, subjected to laser irradiation more than the threshold intensity, are likely to be ionized. Also, we have to consider the effect of conventional three-body recombination process that is dominant process in low temperature, high density plasmas. In this process, population flow into n=4 proceeded relatively slowly, because of including cascade process, such as collisional de-excitation. It should be also noted that the electron capture into selective states due to CXR was considered to take place immediately after the laser irradiation (~1ps). Therefore, the contribution from the CXR process might be overwhelmed by strong three-body recombination process, since we observed time-integrated spectra. In order to clarify whether the CXR process takes place in the initial stage of cluster expansion, we will try to measure time-resolved spectra under irradiation with low intensity by using X-ray streak camera.

5. Conclusion

It has been shown that the cluster plasma produced by fs laser is the promising medium of X-ray laser oscillation due to CXR process, where the non-equilibrium charge state distribution can be easily created. The calculation of the collisional-radiative model showed that the gain coefficient g=6 for 1s-3p transition of N^{6+} ions was obtained under the appropriate condition. In the experiments, we could not observe the population inversion associated with CXR process, which might be overwhelmed by three-body recombination. In future, by measuring temporal evolutions of X-ray spectra, we will clarify the possibility of realization of the CXR X-ray laser.

References

[1] Dunn J, *et al.* 2000 Phys. Rev. Lett. 84, 4834
[2] Kawachi T, *et al.* 2004 Phys. Rev. A 69, 033805
[3] Zhang J., *et al.* 1995 Phys. Rev. Lett. 74, 1335
[4] Dixon R.H. and Elton R.C. 1977 Phys. Rev. Lett. **38** 1072
[5] Fuhl F., Aschke L. and Kunze H-J. 1997 Phys. Lett. A **225** 107
[6] Chichkov B. N., *et al.*, 1999 Jpn. J. Appl. Phys. 38, 1975
[7] Ditmire T. *et al.*, 1996 Phys. Rev. A 53, 3379
[8] Agostini P. 1979 Phys. Rev. Lett. 42, 1127
[9] Kimura M., *et al.*, 1984 J. Phys. Soc. Jpn. **53** 2224
[10] Hagena O.F. and Obert W. 1972 J. Chem. Phys. **56** 1739

Inst. Phys. Conf. Ser. No 186
Paper presented at 9th Int. Conf. X-Ray Lasers, Beijing, China, 24–28 May 2004

Kinetics on Argon Ionic Excimer Pumped by Electron Beam

Yongpeng Zhao, Qi Wang, Yuanli Cheng, Jincheng Liu
National Key Laboratory of Tunable Laser Technology, Harbin Institute of Technology, Harbin 150001

Abstract. The kinetics model of argon ionic excimer pumped by electron beam is built. There are 39 possible reactions in the cavity and their rate constants to be found. The reaction rate equations are built. The equations are calculated numerically and the relations between number density of argon ionic excimer and time are obtained. With the calculation results of number density, the small signal gain is calculated Experimental results are compared with the theoretical results. And it is proved that the 240nm spectrum in the third continuum originates from the transition of Ar_2^+ ionic excimer.

1. Introduction

The third continuum of argon from 160nm to 300nm has two centers which are at 185nm and 240nm respectively. In theory, the definite interpretation about origin of the continuum has not been obtained. In 1981, G.Klein et.al supposed that the third continuum should originate from $Ar_2^{+*}(^2\Sigma_u^+)$ transition[1]. Many experimental results were interpreted by this hypothesis. But in 1988, H.Langhoff et.al suggested that the transition of Ar_2^{2+} ionic excimer should be responsible for the spectrum centred at 185nm in the third continuum of argon. And they proposed that the laser oscillation should be realized with Ar_2^{2+} ionic excimer[2]. Because of this hypothesis, the studies on the continuum become more interesting.

In 1992, C.Cachoncinlle et.al studied Ar_2^{2+} and Ar_2^+ ionic excimer completely in theory [3, 4] and reported that the transition of Ar_2^{2+} can only generate the spectrum centred at 185nm, but not the spectrum between 200nm and 300nm. In the third continuum, the spectrum between 200nm and 300nm can only be interpreted by the transition of Ar_2^+ ionic excimer. Therefore, there are different hypothesis on the origination of the spectrum between 200nm to 300nm[1-5]. However, in 1994, M.Schumann and H.Langhoff analyzed the kinetics of Ar_2^{2+} ionic excimer and wanted to know the possibility in realizing laser oscillation with it[6]. They supposed that the Ar_2^{2+} ionic excimer should be quenched so fast that it seems impossible to increase number density and realize optical gain, under the electron beam pumping. In their paper, the kinetics on Ar_2^+ ionic excimer was not calculated and discussed. Therefore the kinetics analysis on the argon ionic excimer is not integrated. It should be one of important reason why the origination of the spectrum between 200nm and 300nm can't be concluded. Consequently in this paper the kinetics about argon ionic excimer is studied roundly. The reactions of as many as possible are considered to build an integrated kinetics model. In order to verify the validity of the model, the small signal gain of the spectrum around 240nm is measured.

2. The kinetics model of argon ionic excimer

In order to build the kinetics model, there are 39 possible reactions in the cavity and their rate constants to be found. In these reactions, the ionized and excited processes on ground state argon pumped by high-energy electron are related to the energy deposition of electron beam in the gas. In other words, the electron can deposited his energy through the impact reactions which can only be realized by high-energy electron. The other reactions are considered as independent of energy deposition of electron beam. In these reactions, the deposited energy is transferred among the matter participated in the reactions.

Firstly, the energy deposited reactions are discussed. There are three main reactions:

$Ar+e \rightarrow Ar^{*}+e$ Direct excitation of electron (1)

$Ar+e \rightarrow Ar^{+}+2e$ Direct ionization of electron (2)

$Ar+e \rightarrow Ar^{2+}+3e$ Double ionization of electron (3)

In order to build the reaction rate equations, the rate constants must be calculated. The calculation can be express as follows:

$$k = \int_{0}^{E_p} v\sigma(u)\frac{f(u,t)}{n_e}du \tag{4}$$

k is the rate constant. E_p is electron energy of electron beam. v is velocity of electron whose energy is u. $\sigma(u)$ is cross section. n_e and $f(u,t)$ are number density and distribution function of high-energy electron. Differential ionized cross section is selected as reference [7]. When the excited cross sections are selected, 21 allowed transitions and 19 forbidden transitions are included[7]. With the Boltzmann equation as reference [8], we calculate the electron distribution function $f(u,t)$.The number density of high-energy electron can be calculated by integrated the distribution function over energy. Because the three reactions are corresponding to different electron energy range, the number density of electron is different. On the other hand when the reaction rate equations are built, the forming terms and the quenching terms of these reactions should multiply the number density. Therefore for convenience, the rate constants given by us have been multiplied the number density and their units are s^{-1}. Furthermore, the calculating results on distribution function of high-energy electron show that the distribution functions can reach a stationary distribution within 1ns after the electron beam begins to work. And after the electron beam stops to work, most high-energy electron will be exhausted within 1ns. Because the waveforms of our electron beam is almost square and their duration is 36ns, we can suppose that the rate constants of the three reactions are changeless during the electron beam working and will become zero after the electron beam stop to work. If E_p is 466keV, current density of electron beam is $178Acm^{-2}$ and the pressure is 0.3MPa, the rate constants of reaction (1), (2) and (3) are $2.2\times10^{4}s^{-1}$, $5.5\times10^{3}s^{-1}$ and $2.4\times10^{2}s^{-1}$.

Three-body reactions are the main formation channels of ionic excimer and neutral excimer. Among the three-body, two are argon atom. One argon atom acts with species to form excimer and the other bring the surplus energy away. Because the argon number density is very large in high-pressure gas, these reactions can arise easily. In our experiment, the pressure is 0.3MPa, these reactions are very important formation channel of ionic excimer.

The number density development of every species with time can be calculated from rate equations. In these equations, the varied rate of number density equal to forming terms minus quenching terms. In the calculation, the parameters are determined by experimental condition. The current waveform is approximately dealt with as square with 36ns duration. Its peak value is 19.6kA. The pressure in gas cavity is 0.3MPa.

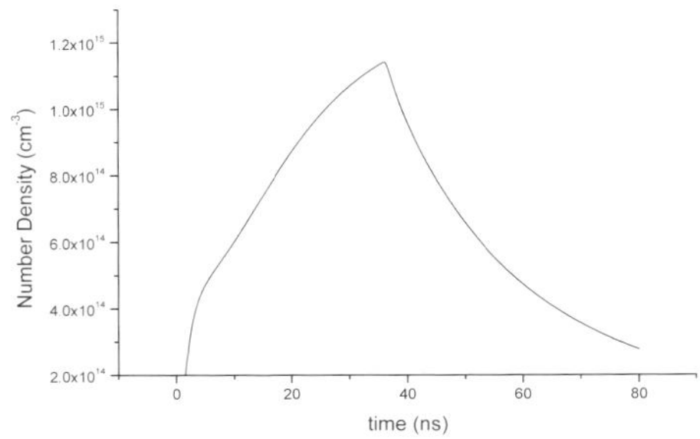

Fig. 1 Ar_2^+ number density versus time

3. Theoretical results

In figure 1, the temporal behavior of Ar_2^+ ionic excimer is shown. We will mainly discuss the peak value of number density and the temporal behavior of number density. From the figure, it can be seen that the maximum of Ar_2^+ number density is$1.142\times10^{15}cm^{-3}$. Compared with Ar_2^+ ionic excimer, the temporal behavior of Ar_2^{2+} ionic excimer is evidently different. The behavior of Ar_2^{2+} is shown in figure 2. According to the figure, the maximum of Ar_2^{2+} number density is $2.138\times10^{13}cm^{-3}$. It is less two orders of magnitude than he maximum of Ar_2^+. On the other hand, the temporal behaviors of the two ionic excimers are very different. In figure 1, during the electron beam working, the number density increases with time. But when the electron beam stopped to work, the number density begins to decrease gradually. On the contrary, the number density of Ar_2^{2+} ionic excimer increases and decreases more quickly. Especially, the number density of Ar_2^{2+} quickly decreases to zero almost at the same time when electron beam stops to work. The results can be used to interpret the time-resolved spectroscopy of the third continuum of argon. One kind of results shows that the time-resolved spectroscopy is larger than pumping pulse[15-18]. This time-resolved spectroscopy is consistent with temporal behavior of Ar_2^+ ionic excimer. In addition, the other time-resolved spectroscopy is the same as pumping pulse[5, 19-21]. This time-resolved spectroscopy is consistent with temporal behavior of Ar_2^{2+} ionic excimer. Hence the inconsistent experimental results can be interpreted with our theoretical results. The difference between the experimental results should result from the different species transitions. In order to verify our calculation, our results are compared with the results reported in reference [6] by M.Schumann and H.Langhoff. In the reference, if the number density of argon is $8\times10^{19}cm^{-3}$, the number density of Ar_2^{2+} is $1\times10^{13}cm^{-3}$. This result is in good agreement with our result. Thus our results should be right. Furthermore, our calculation results can accord with experimental

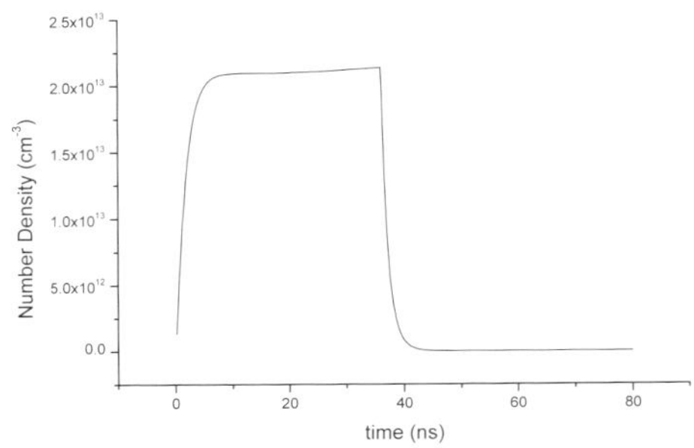

Fig. 2 Ar_2^{2+} number density versus time

results which will be introduced in the next section.

According to the number density, the gain can be calculated by multiply the cross section for stimulated emission with number density. The cross section for stimulated emission can be written as follows:

$$\sigma_{se} = \frac{\lambda^2}{8\pi\tau_s \Delta\nu} \quad (5)$$

λ and ν are the center wavelength and frequency of spectrum emitted by ionic excimer. τ_s is lifetime of upper state. $\Delta\nu$ is the FWHM of the spectrum. Based on the calculation, the cross sections for stimulated emission of Ar_2^+ and Ar_2^{2+} are $2.186\times10^{-17}cm^2$ and $3.032\times10^{-17}cm^2$ respectively. Thus we can calculate that the maximum gain of Ar_2^+ is $0.025cm^{-1}$. This value is close to experimental results. the maximum gain of Ar_2^{2+} is $6.48\times10^{-4}cm^{-1}$. The gain is so small that it is impossible to realize laser oscillation with Ar_2^{2+} ionic excimer. This conclusion is in concordance with M.Schumann and H.Langhoff. Therefore the study on Ar_2^+ is more meaningful than Ar_2^{2+}.

In order to verify our theoretical results with experiment, the small signal gain of the third continuum around 240nm are measured. According to the intensity of the continuum with and without rear mirror, the small signal gain coefficients have been measured in reference [9]. And the parameters of the electron beam are similar as figure 2 and 3 and the pressure of argon is 0.3MPa. The curve of the small signal gain coefficients versus wavelength is shown in figure 4. According to figure 4, two maxima can be found at 230nm and 244nm. At 230nm, $g_0=0.029cm^{-1}$ and $g_0l=0.78$. At 244 nm, $g_0=0.030cm^{-1}$ and $g_0l=0.81$.

4. Conclusion

According to the theory, there is not enough number density of Ar_2^{2+} ionic excimer to produce gain. At the same time, Ar_2^+ ionic excimer can produce gain and the maximum of gain is consistent with experimental results. Therefore, we suggest that the third continuum of argon around 240nm should originate from the transition of Ar_2^+ ionic excimer. This hypothesis is the same as the opinion of C.Cachoncinlle. In addition, we measured the small signal gain of the continuum. Therefore the laser oscillation is likely realized with the transition of Ar_2^+ ionic excimer.

Acknowledge

This work is supported by National Natural Science Foundation of China (No. 60178005).

References

1 Klein G and Carvalho M J 1981 J. Phys. B: At Mol. Phys. 14 1283-1290
2 Langhoff H, 1988 Opt.Commun. 68 31-34
3 Cachoncinlle C, Pouvesle J M et al 1992 J. Chem. Phys. 96 6085-6092
4 Cachoncinlle C, Pouvesle J M et al 1992 J. Chem. Phys. 96 6093-6103
5 Robert E, Khacef A et al 1995 Opt .Commun. 117 179-188
6 Schumann M and Langhoff H 1994 J. Chem. Phys. 101 4769-4777
7 Bretagne J, Callede G et al 1986 J. Phys. D: Appl. Phys. 19 761-778
8 Bretagne J, Delouya G et al 1981 J. Phys. D: Appl. Phys. 14 1225-1239
9 Yongpeng Zhao, Qi Wang et al 2000 Chinese Journal of Lasers A27 123-126 in Chinese

Applications of X-ray lasers and other bright X-ray sources

Inst. Phys. Conf. Ser. No 186
Paper presented at 9th Int. Conf. X-Ray Lasers, Beijing, China, 24–28 May 2004

Applications of X-ray lasers utilizing plasmas that are only a few times ionised

Joseph Nilsen and James H. Scofield
Lawrence Livermore National Laboratory, Livermore, CA 94550

Abstract. With the advent of tabletop X-ray lasers that operate at high repetition rate more emphasis is being put on finding useful applications for these lasers. The 14.7 nm Ni-like Pd X-ray laser at Lawrence Livermore National Laboratory is being used to do many interferometer experiments. As detailed quantitative comparisons are done between experiments and code simulations it is clear that some of the assumptions used to analyse the experiments need to be modified as one explores plasmas that are only a few times ionised. In the case of aluminium plasmas that have been analysed with interferometers there has been some unusual behaviour where the fringe lines bend the wrong way. In this work we will discuss how the index of refraction for aluminium is far more complicated than generally assumed because there are significant contributions to the index from the continuum and line structure of the bound electrons that can dominate the free electron contribution and even cause the index to be greater than one. We will also discuss some potential applications of the high repetition rate Ne-like Ar X-ray laser at 46.9 nm. In particular we will present modelling that shows how the Ar laser could be used to modify the absorption coefficient of a helium plasma and allow one to study the kinetics of plasmas with very low temperatures of a few eV. We will also discuss frequency doubling of the 46.9 nm laser.

1. Introduction

With the availability of table-top X-ray lasers that operate at high repetition rate more emphasis is being put on using these lasers in applications. A particularly useful application that many laboratories are pursuing is the use of X-ray laser based interferometers to study plasmas and other materials. At Lawrence Livermore National Laboratory, the 14.7 nm Ni-like Pd X-ray laser is being used to do many interferometer experiments. As detailed quantitative comparisons are done between experiments and code simulations it is clear that some of the assumptions used to analyse the interferometer experiments need to be modified as one explores plasmas that are only a few times ionised. In the case of aluminium plasmas that have been analysed with interferometers there has been some unusual behaviour where the fringe lines bend the wrong way. The basic assumption since the earliest days of x-ray laser interferometers using the Ne-like Y laser at 15.5 nm at NOVA is that the index of refraction can be calculated from the free electron density and that the bound electrons do not contribute to the index of refraction.

In this work we will discuss how the index of refraction for aluminium is far more complicated than generally assumed because there are significant contributions to the index from the continuum and line structure of the bound electrons that can dominate the free electron contribution and even change the sign of the gradient of the index with respect to the plasma density. We will also show how Al is not unique and how the bound electrons can have significant contributions for any material.

We will also discuss some potential applications of the high repetition rate Ne-like Ar X-ray laser at 46.9 nm. In particular we will present modelling that shows how the Ar laser could be used to modify the absorption coefficient of a helium plasma and allow one to study the kinetics of plasmas with very low temperatures of a few eV. We will also discuss frequency doubling of the 46.9 nm laser.

2. Index of refraction for Al plasmas

Recent experiments [1] at the COMET laser facility at LLNL and the Advanced Photon Research Center [2] at JAERI have observed anomalous behaviour of fringe lines in interferometer experiments of Al plasmas observed at late time when the plasma is cooling. These experiments used the 14.7 nm Ni-like Pd laser and 13.9 nm Ni-like Ag laser, respectively. The basic assumption for many years has been that the index of refraction can be calculated from the free electron density and that the bound electrons do not contribute to the index of refraction. The formula is $n = (1 - N_{elec} / N_{crit})^{1/2}$ where N_{elec} is the electron density of the plasma and N_{crit} is the plasma critical density. In the Al plasmas the fringes were observed to bend the wrong way, suggesting that the electron density is negative or the index of refraction is greater than one. This was observed at late time when the plasma is cooling and may be only a few times ionised. It was recognized that neutral Al has a negative contribution to the index of refraction due to the L3 absorption edge at 73.2 eV.

For neutral materials the complex index of refraction $n = 1 - \delta - i\,\beta$. β is related to the absorption coefficient while $1 - \delta$ is the real part of the index of refraction that is relevant for interferometers. The coefficient $\delta = 0.5 \times f1 \times (N_{ion} / N_{crit})$ where N_{ion} is the ion density and N_{crit} is the plasma critical density. For an ionised plasma the contribution to the index of refraction due to free electrons uses the same formula for δ but f1 is replaced by the number of free electrons per ion. For most neutral materials f1 > 0 but for Al at 14.7 nm and 13.9 nm f1 = -1.4 and –0.8, respectively. This raised the interesting question of whether Al was unique and what the index is for singly ionised Al and other nearby charge states.

The other interesting question is whether the X-ray laser can propagate through the partially ionised Al plasma. To address this question we used a Hartree-Slater code to calculate the absorption coefficient f2 for singly and double ionised Al and compared this with the published data [3] for neutral Al in Fig. 1. Keep in mind that $\beta = 0.5 \times f2 \times (N_{ion} / N_{crit})$. If one considers a plasma with an ion density of 10^{20} cm^{-3}, then the mean free path for 14.7 nm X-rays is only 24 μm for neutral Al but jumps to 512 and 1060 μm for singly and doubly ionised Al. This is because the absorption edge for Al moves from 73.2 eV for neutral to 91.7 and 105.4 eV for singly and doubly ionised Al. The result is that the X-ray laser goes from being highly absorbed by neutral Al to being on the less absorptive side of the edge for the partially ionised Al. This suggests that neutral Al does not likely play a large role in the anomaly because the plasma would be very absorptive. However the question then becomes what is the index of refraction for the partially ionised Al.

Table 1. Contributions to optical constant f1 at photon energy 84.46 eV for Al.

Ion	L3 edge (eV)	Line	Continuum	Free	F1-total
+0	73.1	2.61	-3.46	0.00	-0.85
+1	92.36	-0.70	-4.48	1.00	-4.19
+2	105.4	-2.91	-2.63	2.00	-3.54
+3	119.99	-3.16	-1.64	3.00	-1.80
+4	153.71	-2.49	-0.66	4.00	0.84
+5	190.47	-1.17	-0.29	5.00	3.54
+6	241.43	-0.57	-0.13	6.00	5.30
+7	284.59	-0.20	-0.06	7.00	6.73
+8	330.21	0.03	-0.03	8.00	8.00
+9	398.57	0.19	-0.02	9.00	9.18

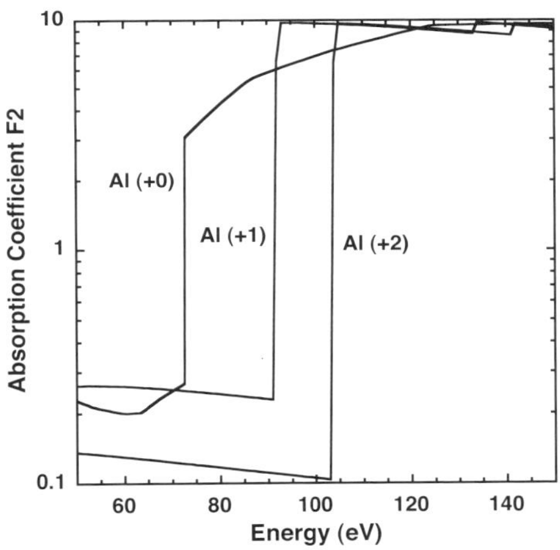

Fig. 1. Absorption coefficient f2 vs photon energy for neutral Al(+0), Al(+1), and Al(+2).

To address this we calculated the continuum absorption cross sections for each ionisation stage using a Hartree Slater code. The energies were adjusted to make sure the L3 edges agree with the experimentally measured edges for singly, doubly, and triply ionised Al. We then add the absorption for the lines below the L3 edges to the continuum absorption. For Al (+1) and Al (+2) we used the measured line positions and oscillator strengths from Refs. 4 and 5. For Al (+3) the n=2 to n=3 and 2p to 4s line positions and line strengths were used from Ref. 6. Using the absorption coefficient f2, the real part of the index, f1, is then derived using the Kramers-Kronig dispersion relation. This part of the index is the bound electron contribution so we then add the number of free electrons to determine a total value for f1. Ref. 3 has a description of the Kramers-Kronig dispersion relation. Keep in mind that the line contributions are extremely sensitive to the line positions and should include the effects of line broadening. The calculation uses the best data we have available but still has a lot of uncertainty associated with it. Table 1 shows our best calculation of the partial and total f1 value for each ionisation stage of Al for a 14.7 nm X-ray.

From Table 1 one can see immediately that there can be large contribution to f1 from both the absorption lines and from the continuum absorption edges. If one looks at the ratio of f1-total to the number of free electrons this is the same as the ratio of the measured electron density to the actual electron density when we use the usual analysis that only considers the free electron contribution. One can greatly underestimate the electron density for low ionisation states. In fact one needs to reach Al (+7) before the contribution from the bound electrons becomes less than a 10% effect.

To understand how important the line contribution can be let us look at Al (+2), which has a large line contribution. Looking at the data in Ref. 4, there is a strong line measured at 89.89 eV with oscillator strength, f_{osc}, of 0.377. The incremental contribution to f1 from a single line, $\Delta f1(E) = f_{osc} / [(E / E_0)^2 - 1]$, where E_0 is the energy of the line. For our X-ray laser at 14.7 nm (84.46 eV), $\Delta f1 = -3.22$. For the Ag laser at 13.9 nm (89.25 eV), $\Delta f1 = -26.6$. The value of $\Delta f1$ is very sensitive to the exact position of the line and its line width as you approach resonance.

One needs to remember that this problem is very complex and these results are only approximate. For all these calculations we assume the plasma is only in the ground state of the particular ionic stage while a real plasma will have a distribution of population in excited states.

3. Index of refraction for a Ti plasma

Having shown that the bound electrons can have a significant contribution to the index of refraction for an Al plasma the question remains as to whether this issue is important in other materials that do not have neutral absorption edges near the X-ray laser energy. If one looks at the ionisation energies across the periodic chart a typical material has its first ionisation about 5 to 10 eV. As it continues to ionise there will almost always be an edge near the X-ray laser energy that needs to be considered.

For Ti we decided to calculate f1 for Ti (+4), Ti (+12), Ti (+13) and compare these with neutral Ti. For each ionisation stage we calculated the photoionization cross-sections with our Hartree-Slater code. Using the Grant code [7] we then calculated oscillator strengths for the strongest lines and added these to the absorption cross section. We then estimated the term f1 by using the Kramers-Kronig dispersion relation and then added the free electrons to calculate a total value for f1.

In Fig. 2 we compare our estimated f1 for Ti (+4) with neutral Ti. Ti (+4) is interesting because it is a stable sequence with closed 3s and 3p subshells. It has an ionisation potential of 98 eV. From Fig. 2 one sees that at 84.46 eV, f1 = 10.3 for neutral Ti and still has a value of 11.0 for Ti (+4). For Ti (+4) we expect the total f1 = 4 so in an actual experiment we would overestimate the electron density by 11/4 or 175% when you take the ratio of these numbers.

If we now look at Ti (+12) where the ionisation potential is 788 eV one would expect all the absorption edges and lines to be at much higher energy than the X-ray laser energy and therefore the bound electrons should make a very small contribution. That is what we estimate in the calculation shown in Fig. 3. Now if we do the same calculation of f1 for Ti (+13) we discover that there are 2 weak lines 2s-2p lines at 96 and 102 eV that perturb the value of f1, as shown in Fig. 3. At 84.46 eV the effect is small but as you approach the resonances the effect from these small lines can be significant.

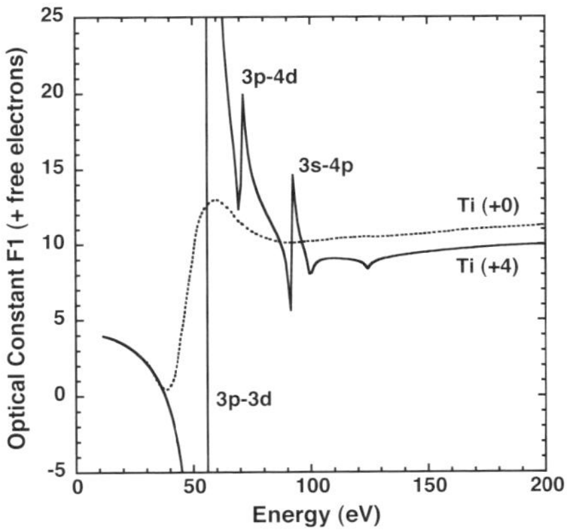

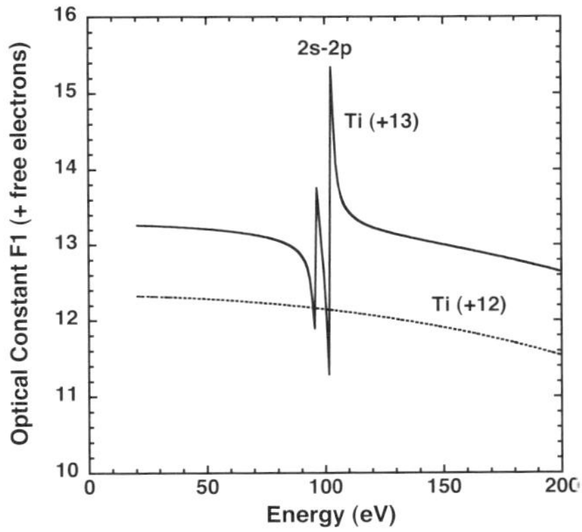

Figs. 2 (left) and 3 (right). Optical constant f1 versus photon energy for neutral Ti(+0), Ti(+4), Ti(+12), and Ti(+13).

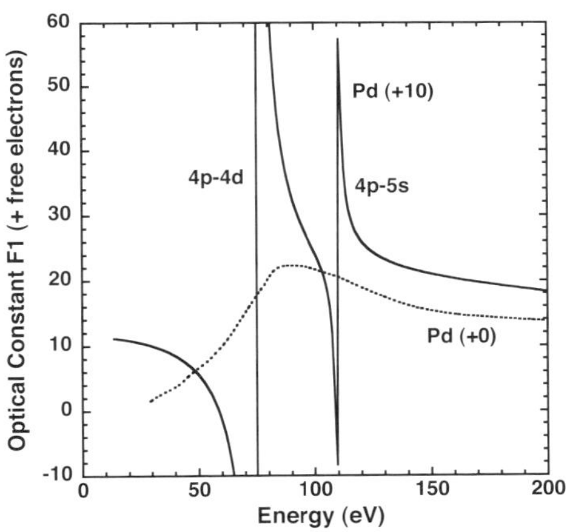

Fig. 4. Optical constant f1 versus photon energy for neutral Pd(+0) and Pd(+10).

4. Index of refraction for a Pd plasma

In order to better understand the physics of the Ni-like Pd X-ray laser, experiments have been done to look at the electron density of the Pd as prepared by the prepulse. Taking Pd (+10) as an example, we calculate the optical constants and compare with neutral Pd using the same procedure as described for Ti above. Pd (+10) has an ionisation potential of 239 eV so we might expect a significant abundance of this ion for plasmas below 100 eV. Figure 4 shows the total value of f1 including free electrons for neutral Pd (+0) and Pd (+10). At 84.46 eV, neutral Pd has f1 = 22 while Pd (+10) has f1 = 43. In both cases we would greatly overestimate the electron density of the plasma. For the Pd (+10) one can see that the strong lines, especially the 4p – 4d line estimated to be at 75 eV with oscillator strength of 7, have a large contribution to f1.

5. Studying kinetics of He plasma with Ar XRL

The Ar X-ray laser at 46.9 nm (26.46 eV) has been available for many years and offers a high repetition source with 1 mJ output. This laser has been used in many interferometer experiments but we would like to propose several other potential applications. Since neutral He has its photoionization edge at 24.59 eV we first looked at whether we could use the Ar X-ray laser to ionise neutral He and drive a laser in neutral He on either the 3d – 2p line at 667.8 nm or the 2p – 1s line at 58.4 nm as shown in Fig. 5. If we look at driving a He gas with ion density of 10^{18} cm^{-3} with an Ar X-ray laser with output of 10^{10} W cm^{-2} in a 1 ns pulse the calculations show that we quickly make a 2 eV plasma of He. The calculations do not show any gain on these lines because the plasma is too hot and the upper states tend to collisionally ionise very rapidly. There may be other conditions where gain could be achieved. However we can easily ionise the neutral He plasma which suggested that we could study the kinetics of the plasma recombining or even watch the absorption change as a function of time.

In Fig. 6 we show the absorption coefficient of He versus energy at several different times for the plasma created by driving a He gas with ion density of 10^{18} cm^{-3} with an Ar X-ray laser with output of 10^{10} W cm^{-2} in a 1 ns pulse. For simplicity we assume a square pulse that turns on at time 0 and turns off at 1 ns. In Fig. 6 one observes that the K edge at 24.59 eV is quickly bleached away by a factor of 100 as we almost completely ionise the neutral He with the Ar X-ray laser. At 1 ns the singly ionised K edge at 54.4 eV dominates the absorption. After 10 ns the He has recombined sufficiently that the absorption from the neutrals start returning. It would be very interesting to do experiments to probe a He plasma at different times and measure the absorption coefficient as a function of time. Alternatively one could watch the emission of the Kα line at 58.4 nm versus time to study the recombination process in the He plasma. Figure 7 shows the absorption coefficient (dotted line) at 46.9 nm and the Kα emission at 58.4 nm versus time for the plasma conditions described above. One could also do interesting experiments to bleach a channel in the He gas using the Ar X-ray laser.

6. Frequency doubling the Ar XRL

The Ar X-ray laser at 46.9 nm (26.447 eV) looks like a promising candidate for trying some two-photon absorption experiments. As shown in the last section it should be fairly easy to ionise a He gas. In singly ionised He there is a very good two-photon absorption resonance from the 1s ground state to the 6d excited state as shown in Fig. 8. The 6d state is nearly degenerate with the 6s and 6p states. We estimate the two-photon absorption is off resonance by only –0.011 eV for the 1s – 6d transition. For the two-photon absorption process the 2p state is used as the intermediate state. However the intermediate state is far off resonance so this does reduce the absorption rate significantly. Using the oscillator strengths for the 1s – 2p – 6d transition we estimate the two photon absorption rate W = 10^{-24} x I^2 sec^{-1} per ion where I is the laser intensity of the Ar X-ray laser in W cm^{-2}. For I = 10^{10} W cm^{-2} and an ion density of 10^{18} cm^{-3} the two photon absorption rate would be 10^5 per nsec per cm^3. This assumes a line width of 10^{-7} eV for the two-photon resonance. This would be followed by the 6p – 1s spontaneous emission at 23.4 nm. Other longer wavelength lines would also be emitted. This calculations are quite preliminary but do suggest that some interesting two photon effects might be observed using the Ar X-ray laser with a singly ionised He plasma, especially if we had a higher intensity X-ray laser.

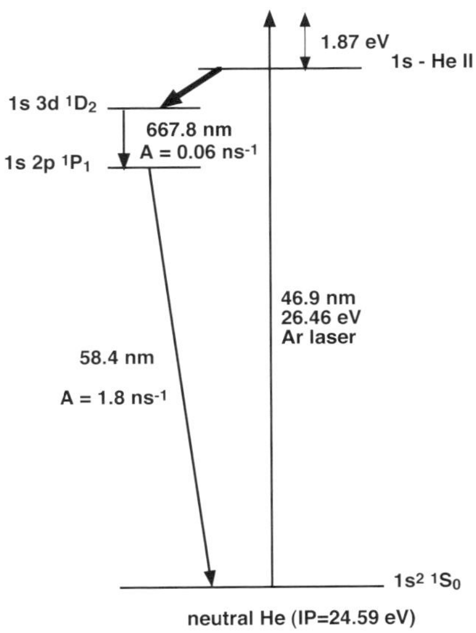

Fig. 5. Energy levels of He I showing potential laser lines.

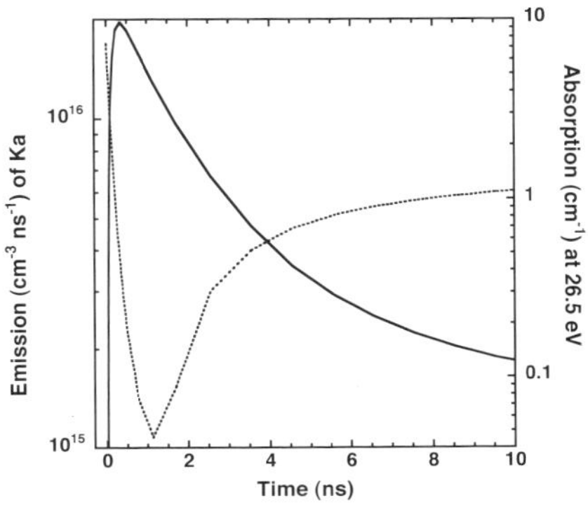

Fig. 7. Emission of He K-α line (solid) and absorption of Ar X-ray laser line (dotted) in He plasma versus time.

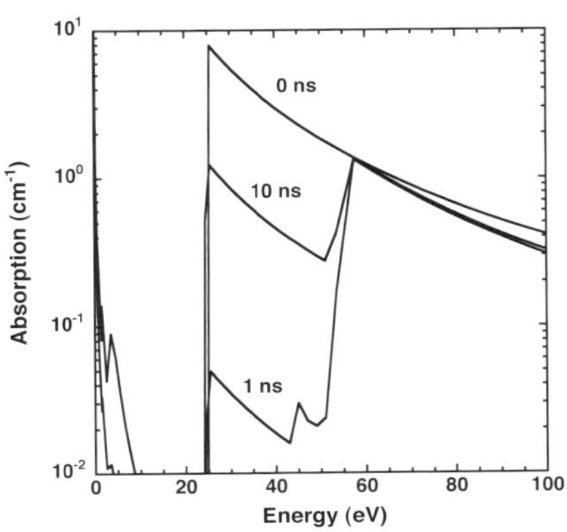

Fig. 6. Absorption spectra of He at different times before and after illumination by 1 nsec duration Ar X-ray laser.

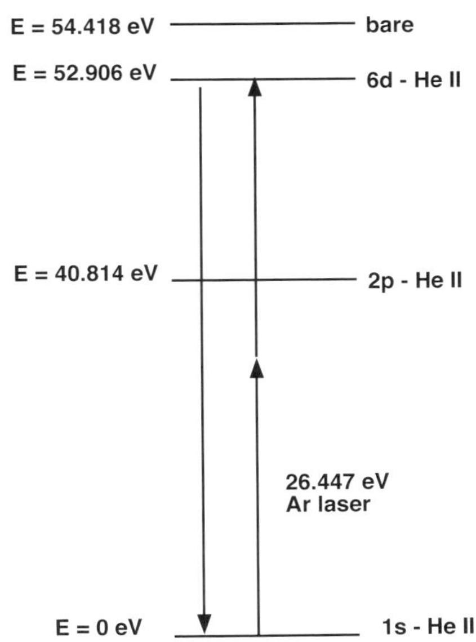

Fig. 8. Energy level diagram of He II showing two-photon absorption of Ar X-ray laser.

Acknowledgements

The authors would like to thank K. T. Cheng, Walter Johnson, John Castor, Lynn Kissel, Dave Attwood, Alexander Ryabtsev, John West and other colleagues who made valuable suggestions in the process of preparing this work. This work was performed under the auspices of the US Department of Energy by the University of California Lawrence Livermore National Laboratory under Contract No. W-7405-Eng-48.

References

[1] J. Filevich, J. J. Rocca, et.al. "Evidence of bound electron contribution to soft x-ray laser interferograms of dense plasmas," in this proceedings.

[2] H. Tang, O. Guilbaud, et.al., Appl. Phys. B **78**, 975 – 977 (2004).

[3] B. L. Henke, E. M. Gullikson, et.al., ADNDT **54**, 181 - 342 (1993).

[4] A. Aguilar, J. B. West, et.al., Phys Rev A **67**, 012701 (2003).

[5] J. B. West, J. Phys. B **34**, R45 – R91 (2001).

[6] I. M. Savukov, J Phys B **36**, 4789 - 4797, (2003).

[7] I. P. Grant, B. J. McKenzie, et.al., Comput. Phys. Comm. **21**, 207 – 231 (1980).

Inst. Phys. Conf. Ser. No 186
Paper presented at 9th Int. Conf. X-Ray Lasers, Beijing, China, 24–28 May 2004

DNA irradiation and plasma diagnostic using X-ray lasers. Recent progress and perspectives at LIXAM.

D. Ros, G. Jamelot, A. Klisnick, O. Guilbaud, H. Tang, S. Kazamias, K. Cassou
LIXAM, "X-ray Lasers and Applications group", Université Paris-Sud, Orsay, France

D. Joyeux, D. Phalippou, Y. Lechantre
LCF-Institut d'Optique, Orsay, France

H. Daido, M. Kado, K. Nagashima, M. Nishikino, M. Kishimoto, K. Sukegawa, M. Ishino
Advanced Photon Research Center, Kansai Research Establishment, JAERI, 8-1 Umemidai, Kizu-cho, Souraku-gun, Kyoto 619-0215, Japan

B. Rus, M. Kozlová, J. Polan, A. R. Präg, M. Stupka
Institute of Physics, Department of X-Ray Lasers, Na Slovance 2, CZ-182 21 Praha 8 (Czech Republic)

G. Eot-Houiller, E. Sage
Laboratoire de Génotoxicologie et cycle cellulaire, Institut Curie, CNRS UMR 2027, France

M. Begusova, V. Stísová
Nuclear Physics Institute, Academy of Sciences of the Czech Republic, Na Truhlárce 39/64, 18086 Prague 8, Czech Republic

M-A. Hervé du Penhoat, A.Touati, A. Chétioui
Groupe de Physique des solides, Universités Paris VI-VII, CNRS UMR 7588, CEA CRC, France

Abstract. In this paper we will review our experimental activity carried out since the 8th ICXRL about the utilization of X-ray lasers for applications. This work concerns both QSS and transient X-ray lasers. Besides, it involves a number of French and international collaborations and was carried out with different pump laser facilities: at JAERI (in Japan), at PALS (in the Czech Republic) and at LULI (in France). We will first present the result of diagnostic of laser-induced plasma, based on bi-mirror interference microscopy and performed at JAERI with a transient soft X-ray laser at 14 nm. We will then describe experiments performed at PALS about DNA damages using QSS soft X-ray laser at 21.2 nm. We will also discuss about the possibility of focussing these X-ray lasers, checked at PALS and LULI.

1. Introduction : X-ray lasers and applications

Development of collisional X-Ray Lasers (XRL) and their use for various applications is currently a researches subject where several approaches are being pursued in parallel [1]. The same goes for the X-ray laser group at LIXAM (Orsay, France), that is working both on optimisation of the X-ray laser source, [2] and its use for different applications, like DNA irradiation [3]. Besides, we are also involved in the built of our future LASERIX facility [4]. This paper describes more specifically the recent progress that we performed in the applications area. This work is a natural continuation of our previous ones [5]. Thus, we tried to use XRL as a source either for irradiation, either for imagery.

In the second section, we will present the results that we obtained at JAERI (Japan) about bi-mirror interferometry of laser produced plasma. This experiment was already checked, but unsuccessfully. Besides, it was the first time that we tried to use a transient X-ray laser for applications. In the third section we will present the first successful investigations that we obtained concerning DNA irradiation at PALS (Prague, Czech Republic). In Section 4 we will give some indications about experiments performed about the focalisation of the X-ray laser beam.

It is interesting to remark that in the section 2 we will present results using a transient XRL (Ni-like silver, emitting at 13.9 nm) whereas in the section 3, as we will explain later, it was necessary to consider a XRL pumped in the quasi-steady state (QSS) regime (Ne-like zinc, emitting at 21.2 nm). On the other hand, in section 4, investigations were done for the two different pumping regimes. This consideration indicates that we can realise different XRL, including different properties (spatially and temporally), leading to different applications, and increasing thus their potential number.

2. Bi-mirror interferometry of laser produced plasma

As a widely used diagnosing tool in the research of laser-produced plasma, interferometry has many advantages in the accurate measurement of the plasma electron density because it directly gives refraction index mapping from the interference pattern processing. Indeed, due to the high intensity and short wavelength, X-ray lasers applied as a probing beam for the interferometry provides better penetration into high density plasma (10^{19}–10^{22} electrons/cm^3) with small refraction, which is a main limitation of UV light interferometry in high-density gradient plasma diagnosis. In comparison with X-ray lasers working in the QSS scheme, which use a pulse duration ranging from 100 picoseconds to nanoseconds, recently developed transient X-ray lasers [6, 7] with a pulse duration of picoseconds or subpicoseconds will reduce the effects of plasma motion blurring and obtain a snapshot of the density profile with the temporal resolution of picosecond. R.F. Smith and coworkers recently demonstrated a picosecond X-ray interferometry to laser plasma with a Mach–Zehnder grating interferometer and a transient Ni-like Pd 14.7 nm X-ray laser as probing beam [8, 9]. In this section, we present the results of an X-ray interference microscopy diagnosis for a laser-induced plasma using a wavefront-division imaging Fresnel bi-mirror interferometer [10] and a transient picosecond Ni-like Ag X-ray laser [6]. One of advantages of the bi-mirror configuration is that, the utilization of the fragile or expensive X-ray beam splitters can be avoided. X-ray beam splitters worked as the key parts of the Mach–Zehnder X-ray interferometers, which are the most commonly adopted scheme in most performed X-ray interferometry for laser plasma [8, 9]. Moreover, in our configuration, the use of an ellipsoidal multilayer X-ray mirror aspherized by ion etching [11] gives a higher spatial resolution and more rigid and compact beam line.

The configuration of the experimental setup is shown in Fig. 1a. The X-ray laser illuminated the sample plasma and the bi-mirror separated the wavefront after the plasma and produced an enlarged overlapped interference region. The diffraction limited ellipsoidal imaging mirror gave a magnified image of the sample plasma onto the thinned backlighted X-ray CCD. The spatial resolution in the sample plasma plane was ~ 1.75 μm. That is not the ultimate resolution of the apparatus (~ 1 μm). The reduced spatial resolution was due to the limited space in the experimental room that reduced the image magnification.

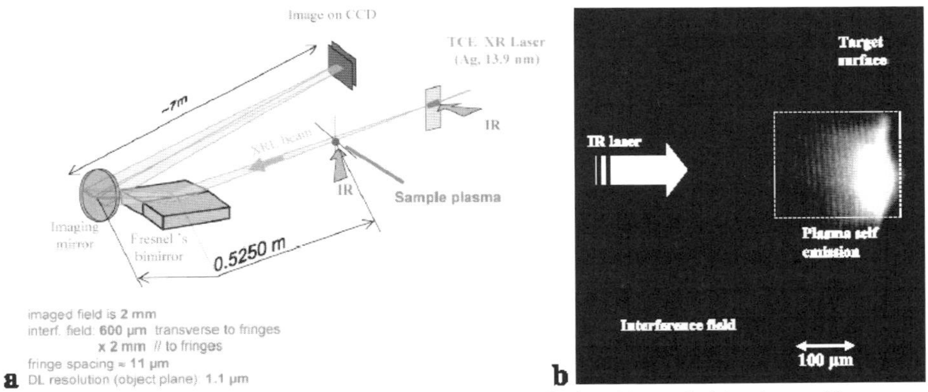

Figure 1. a) Experimental set-up of the Fresnel bi-mirror X-ray interference microscopy. b) Typical interference pattern obtained with Al plasma under the pumping laser energy of 1J.

Experiments were performed on the hybrid Nd-glass laser system of the Advanced Photon Research Center (APRC), JAERI. A 13.9 nm Ni-like Ag TCE (several picoseconds of pulse duration) X-ray laser beam was produced on the system and used as the probing beam [12].

The diagnosed plasma was formed by irradiating an Al slab targetwith 1.2 ns IR (1053 nm) pulse. A line focus system produced a plasma column of ~ 100 μm in width and 1mm in length (intensity within the line focus: 10^{10-11} Wcm^{-2}). The X-ray laser probed the plasma along the column axis (defined as X direction). Under such configuration, the electron density of the plasma can be calculated from the fringe shift by a simple linear equation without Abel inversion. A typical interference pattern of an Al laser plasma is shown in Fig. 1b. The plasma column was produced under the pumping energy of 1 J (intensity: ~ 1.4×10^{11} Wcm^{-2}). The peak-to-peak fringe separation out of the plasma region is ~ 10 μm horizontally (defined as Y direction). Pumping laser irradiates from the left side (opposite to Y direction) and the X-ray probe beam propagates along the plasma column (X direction), which is normal to the paper plane. Intense plasma self-emission can be observed close to the target surface and mixed with the interference fringes. Because the information of electron density is only carried by the fringe shift, some numerical processing should be adopted to extract the interference fringe from the self-emission and other CCD noises, under the assumption that the self-emission and other noises are linearly overlapped with the interference fringes. Figure 2a shows the temporal evolution of the phase of the Al plasmas probed by the X-ray laser at times comprised between 0 and 3 nanoseconds after the peak of the plasma heating pulse. The phase card and electron density are connected to the fringes pattern. Thus, the electron density, n_e, is related to the fringe shifts as $n_e \approx 1.583\times10^{20} N_{\rm fringe}$ [cm-3], for the plasma length of 1mm. Due to the limitation of the CCD resolution under this interferometry configuration, the minimum detectable limit of the electron density is ~ 3×10^{19} cm^{-3}.

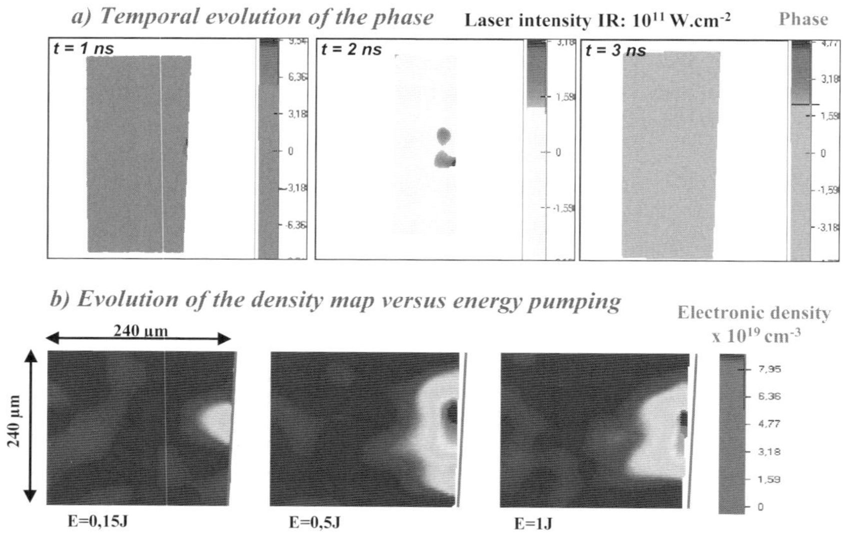

Figure 2. a) Temporal evolution of the phase. b) Evolution of the density map versus energy pumping.

The 2D electron density profile is showned in Fig. 2b, deduced from the fringe pattern. The peak electron density measured is 10^{20} cm^{-3} at the position 10 ~ 15 µm away from the target surface under the pumping intensity of ~ 1.4×10^{11} cm^{-3}Wcm^{-2}. Density fluctuations along the vertical direction (Z direction) are also observed. They are possibly due to the non-uniformity of plasma pumping intensity within the line focus; indeed, these fluctuations correspond to the vertical (Z direction) intensity profile of the self-emission. As shown in Fig. 2a, for the Al plasma diagnosed 2 and 3 ns after the peak of the heating pulse, the phase is negative, corresponding the fringe shifts bend to the opposite direction of the target position. Different explanations for this behavior are discussed [13, 14].

These results will contribute to the validation of the 1D and 2D hydrodynamic codes and a better understanding of the physics of laser-produced plasma. We just mention one possible explanation, based on the contribution of bound electrons of ions to the refraction index. A candidate could be a $2p$-$3d$ transition of Al^{2+} ions, the wavelength of which being 13.883 nm [15]. This wavelength is very close to the X-ray laser one (13.887 nm) [12]. At 2 ~ 3 nanoseconds after the heating pulse, the plasma temperature falls-off and the highly ionized plasma starts to recombine, leading to a reduced free electron density and a larger Al^{2+} ion density, which both increase the refractive index.

3. Plasmid DNA irradiation

This experiment was realized at PALS, using the QSS Ne-like Zn XRL emitting at 21.2 nm [17, 18]. The double-pass beam is deeply saturated and provides around 1mJ of energy in one pulse, corresponding to peak power in excess of 10 MW. The repetition rate give by the Asterix laser system is around 20 minutes. This experiment enrolls as addition to a previous scientific collaboration investigated at LULI (École Polytechnique, France) [19], with the purpose of identifying the physical mechanisms underlying the formation of DNA lesions induced by ionising radiation. Soft x-ray photons can be used as dosimetry tool for specific types of radiation damage and biological response. We focus on the strand breaks damage in the DNA plasmids in function of the dose. The model K predicts the break event in DNA molecule, to the K or L innershell ionisation of component DNA atoms [20, 21]. In this study, we use monochromatic XRL radiation at the energy 58.5 eV, just above the theoretical minimum energies required to produce single-strand and double-strand breaks in DNA are about 17 eV and 50 eV respectively [22].

The plasmid layer deposited on a glass support, typically 600 nm in thickness, was irradiated in vacuum by one, two or three soft X-ray laser pulses, corresponding to absorbed doses up to 10 kGy.

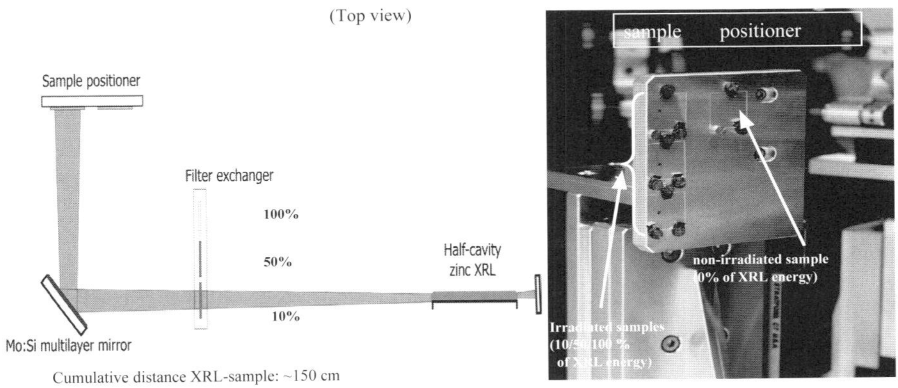

Figure 3. Top view of the experimental set-up (left picture) of the DNA irradiation at PALS and image of sample positioner (right picture) .

The beam XRL is sending in the target chamber through a filter wheel reflects by a 45° multilayer Mo:Si mirror to the sample mount, as shown in figure 3. 4 samples can be deposited on the mount holder. We have different way to play with the level of the dose, the filters, which are aluminium filters of 1%, 5%, 50% transmission and the number of shot on the sample. We are limit by the repetition rate of twenty minutes of the IR pump laser of the XRL and the lifetime of the biological sample under vacuum evaluated at one hour. The radius of the dry plasmid is around 1.5 mm. According to the XRL beam divergence the beam size is 4.5 x 7.5 mm in the ground of the sample. We have two control samples: the first one doesn't go in vacuum to determine the damage coming from the handling, and the second one put in the chamber under protection of the radiations is for estimating damage induced by the vacuum. We vary the dose from 1 shot to 3 shots. Two plasmids were irradiated : the pSB having a high degree of coiling and the pSP189 having a weaker degree of coiling, more sensitive to the radiation.

After the shot sequence, the DNA samples were then dissolved in water and were subsequently analysed by agars gel electrophoresis, making possible the accurate measurement of single and double strand breaks induced by the monochromatic XRL radiation. The electrophoresis is a technique for separating charged molecules in a buffer solution based on different mobilities in the electric field generated by a direct current. Substances for separation are allowed to move through a porous medium, agar. The separated substances occur in bands in the agar and are identified by comparison.
The geometry of DNA molecule changes with the number of break in the double helix. The DNA is supercoiled when the strands are undamaged. The SSB shows a circular shap (relax shap) and the DSB shows a linear shap. We obtain an image of the electrophoresis using an UV lamp, as presented in figure 4.

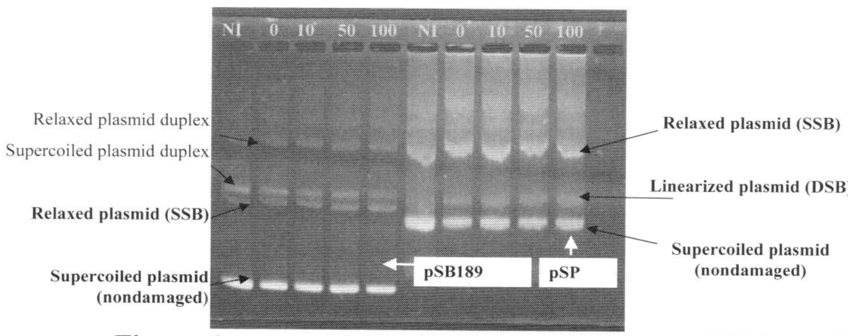

Figure 4. Image of the electrophoresis using an UV lamp for pSB and pSP plasmids. The values 0, 10, 50 and 100 indicate the XRL intensity (arb.U.). NI corresponds to the plasmid Non Inside the chamber.

The analysis of this electrophoresis shows only SSB in the pBS irradiated plasmid and SSB and DSB in the pSP189 irradiated plasmid. The variation of the dose was obtained irradiating the plasmid sample, one, two and three times. The dose is given by the following formula: $D_O(Gy)=1.6\times10^{22}\times(\mu/\rho)\times E_0\times(N/S)$, where (μ/ρ) is an absorption coefficient depending on the kind of plasmid, E_0 is the photon energy, and (N/S) the number of photons per surface unity.

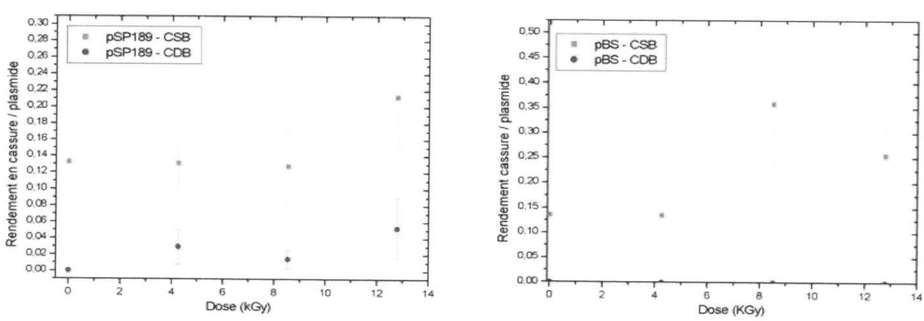

Figure 5. Ratio of single and double breaks per plasmid versus on dose (kGy) for pSP and pBS plasmids.

Even if, we don't have enough data to conclude about the process involved in the strand breaking in the DNA molecule, we observed for the first time, on figure 5, the effect of XRL dosimetry on DNA. However, the fraction of the SSB and DSB are weak, probably indicating that we are at the threshold of breaks.
This preliminary study encourage us to go thoroughly the model of K-innershell events,

and we put forward the irradiation XRL of plasmid with a new range of parameters as, flow dose effect, the effect of the number of photons and wavelength. The plamid laying has to be improved to get a quite monolayer plasmids laying.

4. Focalisation investigations

In parallel of the fundamental development of the QSS and transient X-ray lasers [2], we are checking the possibility of focussing the beam the smallest possible. Indeed, it could be very interesting in the optic of applications, which need an high photons intensity, like for example DNA irradiation.That's why we used a focusing system, at PALS using the QQS 21.2 nm X-ray laser and at LULI using the transient 13.9 nm X-ray laser.
At LULI, we performed the experiment in a separate vacuum chamber placed 3 meters after the X-Ray laser generation medium. The beam was incident on a spherical mirror with a 1.5 ° off axis. This value is a compromise between the spherical aberrations that increase very rapidly with the incidence angle and destroy the focus quality, and the geometrical aspects: a strictly normal incidence would not allow us to perform an image of the focus or to place in the future a target for applications without intercepting the incoming beam. The focal length of the mirror is also the result of the same considerations and was taken to 300 mm, the mirror treatment is a Mo:Si multilayer coating performed by the IOTA(Orsay-France) in order to optimize its reflectivity at 13,9 nm. The geometrical ratio between the focal length and the distance to the X-Ray laser source is 1/10 so that a significant reduction of the beam size could be expected. After being reflected by the spherical mirror the beam is intercepted by a coated plane mirror used with a 45° incidence angle that forwards the beam in the perpendicular direction where we can place an imagery system. This mirror was placed 17 mm after the spherical mirror so that it does not block the incident beam.
The image of the focus is performed using a Ce doped YAG crystal that emits fluorescent light in the visible range when being irradiated by the X-Ray laser beam. A magnification factor around 30 is then simple to achieve in the visible range by using a double-lens system. We performed the alignment of the system using a Nickel grid, composed of 40 μm wire separated by 350 μm. The spatial resolution in our experiment was assumed to be 6 μm. The CCD used was a 1024×1024 Princeton camera with 19μm pixels.
A typical image that we obtained is presented on figure 6. Our first investigations demonstrated, as shown on figure 6a, that we can obtain an uniform focused beam.

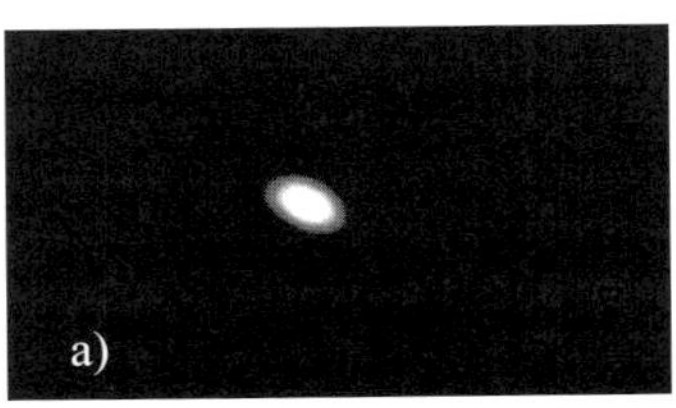

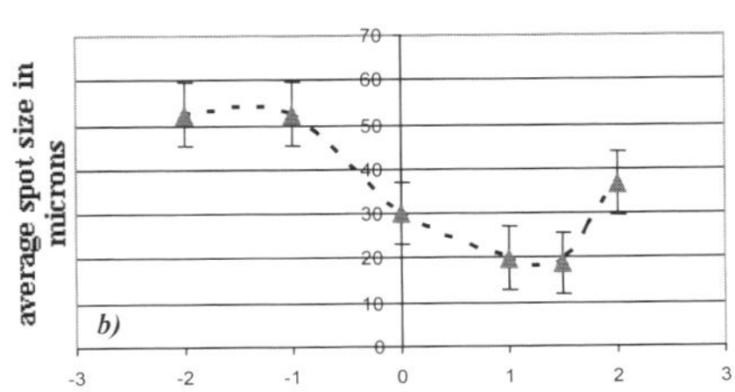

Figure 6. a) Typical image of focused XRL beam. b) Spot size versus on cristal position (mm).

As shown in figure 6b, we optimize the size of the focused spot versus on the cristal position. In the best case we found a 18×25μm^2 spot size.
At PALS, some preliminary investigations on the QSS Ne-like Zn gave a 40×50μm^2 spot size. These first promising results have to be enhanced in a close future.

5. Conclusion and perspectives

In conclusion, we have reported two different experiments: diagnostic of dense plasmas and DNA irradiation, demonstrating the potential of XRL for applications. On the one hand, we have presented the results of an X-ray interference microscopy of a dense laser-induced Al plasma using a wavefront division bi-mirror interferometer and a transient picosecond Ni-like Ag X-ray 13.9nm, leading to Electron densities measurments from the interference fringe shifts. For the aluminum plasma diagnosed 2 and 3 nanoseconds after the heating pulse, opposite fringe shifts were observed. On the other hand, we presented the first demonstration of single and double breaks on DNA using XRL source. Moreover the possibility to focus the beam, leading to high intensity and enhancing the quality of the homogeneity of the focused spot will significantly increase the possibilities of XRL use for applications.

In the future, we will need, above all, to enhance the repetition rate of XRL. It is the aim that we are now purchasing in the built of a LASERIX user facility, devoted to applications [4].

Acknowledgement:. We would like to thank the technical staff at JAERI, PALS and LULI for their support in conducting these experiments. We would like to thank Ph. Zeitoun and S. Sebban for their respective loan of focusing chamber and CCD and D. Douillet for his help in the installation of the focusing chamber.

References

[1] X-Ray Lasers 2002, Proceedings of the 8th International Conference on X-ray Lasers, edited by J.J. Rocca,J. Dunn and S. Suckewer, AIP 641
[2] Klisnick A. et al., this Conference
[3] Cassou K. et al., this Conference
[4] Jamelot G. et al., this Conference
[5] Ros D. et al., 2002 L.P.B.J. 20, pp. 23-30
[6] A. Klisnick, et al., Phys. Rev. A **65**, 033 810 (2002)
[7] T. Kawachi, et al., Phys. Rev. A. **66**, 033 815 (2002)
[8] Smith R.F. et al., Phys. Rev. Lett. **89**, 6, 065 004-1 (2002)
[9] Smith R.F. et al., 2003 Opt. Lett. 28 1
[10] Joyeux D. et al., J. Phys IV France **11**, Pr2, 511 (2001)
[11] Mercier R. et al., SPIE Proc. **3739**, 155 (1999)
[12] Kawachi T. et al ., Phys. Rev. A. **66**, 033 815 (2002)
[13] Tang H., Appl. Phys. B 00, 1–3 (2004)
[14] Nilsen J. et al., this Conference
[15] NIST Atomic Spectra Database (http://physics.nist.gov/cgi-bin/AtData/main_asd)
[16] Ros D., Thèse de l'Université Paris-Sud XI, Orsay (1998) (in French)
[17] Rus B. et al, Phys. Rev. A . **66**, 063806 (2002)
[18] Rus B. et al., this Conference
[19] Ros D. et. al, LULI report 2001, C8.
[20] Chétioui A. et. al, Int. J. Radiat. Biol., 65, 511-522, 1994.
[21] Hervé du Penhoat M.A et. al, Rad. Res.,151, 649-658,1999.
[22] Folkard et. al, Int. J. Radiat. Biol. **76,** 881-90,1995.

Inst. Phys. Conf. Ser. No 186
Paper presented at 9th Int. Conf. X-Ray Lasers, Beijing, China, 24–28 May 2004

Characterization of material properties using x-ray laser as a coherent x-ray source

R.Z. Tai [1,3], K. Namikawa [2,3], M. Kishimoto[3], M. Tanaka[3], M. Kado[3], K. Sukegawa [3], N. Hasegawa [3], T. Kawachi [3], P. Lu [3,4], M. Nishikino[3], Y. Ochi [3], K. Nagashima[3], H. Daido[3], A.Sawada [5], H. Maruyama [6], M. Ando [7] and Y. Kato [3]

[1] Shanghai Synchrotron Radiation Facility, Shanghai Institute of Applied Physics, Chinese Academy of Sciences, P.O. Box 800-204, Shanghai 201800 China

[2] Department of Physics, Tokyo Gakugei University, 4-1-1 Nukui-Kita Machi, Koganei-Shi, Tokyo 184-8501,Japan

[3] Advanced Photon Research Center, Japan Atomic Energy Research Institute, 8-1, Umemidai, Kizu, Kyoto, 619-0215, Japan

[4]State Key Laboratory of Laser Technology, Huazhong University of Science and Technology, Wuhan 430074,China

[5]Department of Physics, Okayama University,3-1-1 Tsushimanaka, Okayama, 700-8530 Japan

[6] Graduate School of Science, Hiroshima University, 1-3-1 Kagamiyama, Higashi-hiroshima, 739-8526 Japan

[7]High Energy Accelerator Research Organization,1-1 Oho,Tsukuba-shi,305-0801,Japan

Abstract. The transient ferroelectric domain structures of barium titanate were characterized near its critical temperature T_c by means of picosecond x-ray laser speckles. The results show that the correlated domain size decreases in a power law as temperature heating toward T_c. Compared to this quasi-static polarization domains existing below T_c, the dynamic polarization clusters were also confirmed to exist in this crystal above T_c and characterized for the first time by same technique.

1. Introduction

The development of coherent x-ray sources has made it possible to observe x-ray speckles [1, 2]. X-ray speckle is an important technique for study of the dynamics, with an almost atomic-scale spatial resolution, for a disordered physical system. In principle, the information such as temporal correlation and spatial correlation, related respectively to the temporal and the spatial distributions of those local scatterers within the specimen, can be obtained from the speckles. The conventional method as so called x-ray photon correlation spectroscopy is a case to measure the temporal correlation of a physical system [3]. To observe the local spatial distribution of a system, the information of spatial correlation is needed, which can be extracted straightforwardly from deconvolution of the speckle pattern diffracted from the system.

X-ray lasers operated in a transient collisional excitation scheme have advantage such as very short pulse duration, within which about 10^{12} photons assemble [4]. Since one pulse of x-ray laser can provide sufficient detectable coherent x-ray photons, the measurement of transient x-ray speckles hence becomes feasible. We chose $BaTiO_3$ as the sample because there still exist a lot of controversies on the basic problem such as the mechanism of phase transition, though it has been applied extensively and studied intensively [5]. As a first step, the transient ferroelectric domains and their evolution with respect to temperatures directly under T_c were successfully characterized by means of the picosecond x-ray laser speckles. Furthermore, the speckles patterns observed above T_c were also confirmed to originated from those dynamic polarization clusters, which had been long predicted to exist, but never been had directly observed due to the shortness of experimental techniques. The cluster measurement gives information such as dynamic correlation during phase transition, which is complementary to those obtained with other techniques currently.

In this context, we will briefly introduce this experiment. Some latest results will also be discussed. Details can be referred elsewhere [6-8].

2. Experiment

The experimental setup is schematically shown in Figure 1. The x-ray laser media here is a silver plasma produced by a chirped-pulse-amplification Nd:glass pump laser light focusing on a silver target[5]. The lasing line is the transition of $3d^94d \rightarrow 3d^94p$, with a wavelength of 13.9 nm and a bandwidth of $\Delta\lambda/\lambda \sim 10^{-4}$. The x-ray laser pulse duration is estimated to be 7 ps. The x-ray laser beam has been characterized to have a source size of 50 μm and a divergence angle of 10 mrad. The total photons are about 10^{12} per pulse.

.

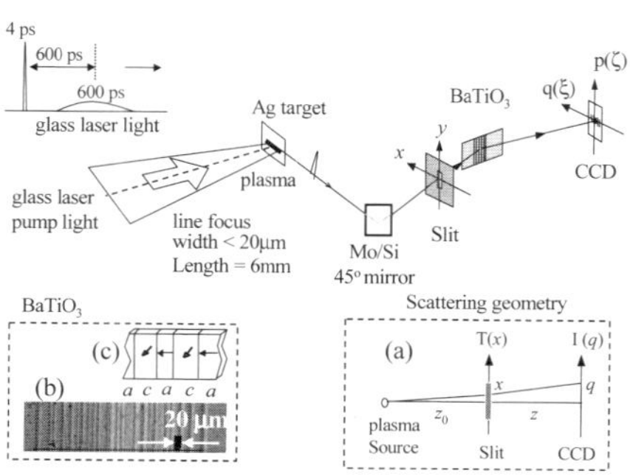

Figure 1. Experiment setup.

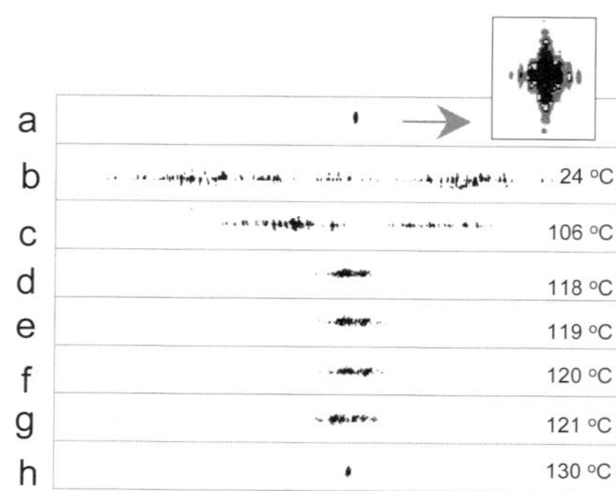

Figure 2. Speckles patterns from domains.

As shown in figure 1, a 45-degree incidence Mo/Si multilayer was used to produce a vertically polarized x-ray probing beam. A small aperture located at 1.9 m from the plasma source was used to produce a fully coherent x-ray photon beam. The sample used in the experiment was flux-grown ferroelectrics $BaTiO_3$ single crystal with alternative *a/c* domains aligned in parallel at room temperature. The Curie temperature was 122 °C. The sample was set to locate close to the slit, with its domain boundaries (walls) directed along the vertical direction. The temperature of the sample was controlled precisely. The grazing angle was 10 degrees. The diffracted speckles patterns were recorded by a soft x-ray charge-coupled device (CCD).

Figure 2 shows those patterns below T_c, where the aperture was set to be 80 μm (horizontal) × 200 μm (vertical) and the CCD was 0.5 m far from the slit. The transverse resolution (horizontal) of this setup is estimated to be around 1 μm. For comparison the pattern of the probing beam is shown as figure 2a. Figure 2b-2h show a series of the speckles diffracted from a region of multiple *a/c* domains when heating the sample from 24 °C to 130 °C. These obvious two groups in Figure 2b and 2c indicate that, at lower temperatures, *a* domains and *c* domains constitute respectively two groups of planes with a normal angle determined by the temperature-dependent *a/c* twin angle. The speckles in each of the two groups originated from the interference of x-ray photons scattered by the same type of domains (for instance, *a* domains), where the other type of domains (*c* domains) were invisible. As temperature increased toward the Curie temperature T_c, the pattern apparently evolved into one group, indicating the adjacent *a/c* twin angle is getting smaller and more irregular even at a same temperature. Above T_c, the speckles turned into one spot, as shown in Figure 2h, indicating the disappearance of the domains..

Above T_c, we still found that there exist some small structures near the main peak. To study this phenomenon, the setup has been a little modified to enhance the performance. The distance of CCD was shortened to be 0.2 m. the aperture was set to be 200 μm (horizontal) × 80 μm (vertical). An external electric field (E) could be injected normal to the sample surface. We focussed on the study of the vertical tail structure of the patterns. The results show that this kind of tail depends strongly on the temperature and the external E. The tail increased as temperature cooling down toward T_c,, while dramatically decreased if E was applied. Another interesting feature is that this tail fluctuates strongly shot by shot even at a same temperature. We have interpreted that this tail originated from the dynamic polarization clusters. Figure 3 shows an example of the temperature dependence. The black central lines show the slit images, in the surrounding some diffuse-like tail structures exist.

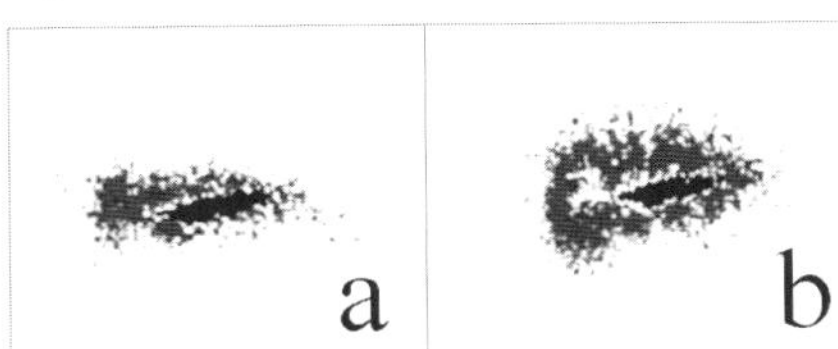

Figure 3. Speckles pattern from clusters above T_c. a: 140 °C, b: 130 °C.

3. Analysis

We can generally use a complex transmittance $T(x) = T_0(x) \exp[i\phi(x)]$ to describe domains or clusters projected on the slit, where the real part $T_0(x)$ stands for the x-ray amplitude transmittance and the $\phi(x)$ for the x-ray phase retardation respectively from the specimen. The x is the horizontal coordinate along the x-axis as shown in Figure 1. The $T_0(x)$ is of course zero if the x falls beyond the slit, that is $|x| > D/2$, where D is the horizontal width of the slit. The complex transmittance $T(x)$ carries the information of the spatial distribution of those scatterers. The matter correlation function may be defined as

$$\gamma(x)=\frac{\int_{-\infty}^{\infty} T(x')T^*(x'+x)\,\mathrm{d}x'}{\int_{-\infty}^{\infty} T(x')T^*(x')\,\mathrm{d}x'}. \tag{1}$$

From a point-source approximation, as shown in the inset (a) of Figure 1, the intensity scattered into a direction of (**k**+**q**) can be written as follows according to the Fresnel-Kirchhoff diffraction integral,

$$I(q)=\int_{-\infty}^{\infty}\int_{-\infty}^{\infty}\mathrm{d}x\,\mathrm{d}x'T(x)T^*(x')\times e^{ik(\frac{1}{2z_o}+\frac{1}{2z})(x^2-x'^2)-iq(x-x')}, \tag{2}$$

where, z_0, z is the distance of the slit from the source and the CCD respectively, the **k** is the wave vector of incident x-ray photons, and the **q** is the scattering vector defined as $\mathbf{q}=k\xi/z\mathbf{e}_\xi$, where the ξ is the coordinate on CCD along the diffraction direction with an unit vector $\mathbf{e}_\xi$. The high-order terms of x have been omitted due to their negligible contribution to the integration, considering z_0, $z >> D$ in practice. The diffraction pattern is a convolution of the slit diffraction and the domain structure diffraction.

Considering the randomness of domain size or cluster size, the Fourier transform of Eq. (2) can be factorized simply as

$$\Gamma(x)=\mathrm{F}\{I(q)\}=\gamma(x)\times\Gamma_0(x), \tag{3}$$

where the $\Gamma_0(x)$, defined as follows, is the apparatus function.

$$\Gamma_0(x)=\int_{-D/2}^{D/2-x} e^{ik(\frac{1}{2Z_0}+\frac{1}{2Z})x'^2-ik(\frac{1}{2Z_0}+\frac{1}{2Z})(x+x')^2}\mathrm{d}x'. \tag{4}$$

As shown in Eq. (4), the $\Gamma_0(x)$ is determined only by the geometric parameters of experimental setup, such as the slit width, the distances from the source and the CCD, and more, the properties of the source. It can be obtained from the Fourier transform of the specular reflection $I_{\mathrm{Sp}}(q)$ in practice. The matter correlation function can be further expressed in a compact form as $\gamma(x)=\mathrm{F}\{I(q)\}/\mathrm{F}\{I_{\mathrm{Sp}}(q)\}$ from an experimental viewpoint.

4.Discussions

The matter correlation function $\gamma(x)$ contains some important information of scatterers. The averaged size σ_s, as so called correlation length, and the averaged distance d can be estimated as the half width and the full width from the autocorrelation part of the $\gamma(x)$. From the definition of matter correlation function, we can also obtain the spatial power spectrum of those scatterers from Fourier transform of the $\gamma(x)$. From this consideration, the correlated domain size directly under T_c has been extracted, showing a feature of power-law decrease as function of temperature [6]. A dramatic structure change was also found to occur at a temperature of 2 ^{o}C below T_c. Above T_c, the local polarization clusters have been first quantitatively characterized. The polarization of the clusters was also estimated from the consideration of phase modulation to the incident x-ray photon beam [8].

5. Conclusions

A picosecond x-ray laser speckle has been successfully conducted to study a disordered structure as a function of temperature. In particular, the observation those paraelectric clusters should be attributed to the short time duration but high brilliance nature of the

TCE soft x-ray laser. The characterization the relaxation time of those clusters, estimated to be an order of nanosecond, will be a coming theme of this study.

References

[1] Sutton M, et al. 1991 Nature 352, 608-610
[2] Cai Z H et al.1994 Phys. Rev. Lett. 73, 82–85
[3] Mochrie S G et al 1997 Phys. Rev. Lett. 78, 1275–1278
[4] Takahasi H 1961 J. Phys. Soc. Jpn. 16, 1685-1689
[5] Kawachi T, et al. 2002 Phys. Rev. A **66**, 033815
[6] Tai R Z et al. 2002 Phys. Rev. Lett. 89, 257602
[7] Tai R Z et al. 2004 AIP Conference Proceeding 705, 1122-1125
[8] Tai R Z et.al. 2004 submitted to Phys. Rev. Lett.

Inst. Phys. Conf. Ser. No 186
Paper presented at 9th Int. Conf. X-Ray Lasers, Beijing, China, 24–28 May 2004

Towards the Fourier transform limit: can picosecond duration x-ray lasers be used for real applications?

G J Tallents[1], Y Abou-Ali[1], M H Edwards[1], P Mistry[1], O.Guilbaud[2], A. Klisnick[2] and D. Ros[2]
[1]Department of Physics, University of York, York YO10 5DD, U.K.
[2]LIXAM, Université Paris Sud, Orsay, France.

Abstract X-ray lasers pumped by picosecond laser pulses are bright, moderately uniform in beam profile and emit over an extremely narrow frequency range ($\nu/\Delta\nu \sim 50000$) for short durations ($\Delta t \approx 2$ - 3 ps). The high brightness is useful for applications as it can be used, for example, to 'outshine' a probed high temperature plasma so that the probing laser signal is much greater than the background plasma emission noise. The narrow frequency range of x-ray laser emission can be utilised depending on the application. In principle, high spectral resolution can be achieved with a low resolution detection instrument. The short duration of emission enables 'snap shots' of rapidly evolving events to be probed. A fundamental limit on the pulse duration Δt has recently become apparent as x-ray laser pulses are close to the Fourier transform limit with $\Delta\nu\,\Delta t \sim 1$. These issues relevant to the application of short duration x-ray laser pulses will be explored. It is shown that the Fourier transform limit has not been reached yet and that x-ray laser pulses as short as 0.5 ps should be realisable.

1. Introduction

The generation of saturated x-ray lasing at wavelengths ~ 10 nm has been achieved in many laboratories [1]. Furthermore, the pumping efficiency has increased dramatically since the first observations of x-ray lasing [2]. More improvements in efficiency now seem assured with the demonstration of grazing-incidence pumping reported at this conference [3, 4] after the initial idea was discussed at a previous SPIE meeting (see Keenan et al [5] and Schlyaptsev [6]). The efficiencies likely to be achieved will mean that small high repetition laser systems can be used to pump x-ray lasers into saturation. With the likely development of high repetition rate x-ray lasers using comparatively cheap pumping lasers systems, we can envisage the development of applications using the x-ray lasers in Universities and small research laboratories. Plasma-based x-ray lasers will offer a cheaper, smaller scale alternative for applications compared to the expensive, large scale 4th generation free-electron lasers now under construction [7].

A feature of plasma-based x-ray lasers useful for applications is obviously their short wavelength. Short wavelength means, for example, that high resolution spatial imaging is possible. Less obviously, x-ray lasers are bright, producing peak brightnesses of typically ~ 10^{24} photons s^{-1} mm^{-2} $mrad^{-2}$ which is brighter by 6 orders-of-magnitude than the instantaneous brightness of 3rd generation synchrotron light sources over 0.1% frequency bandwidth (see Turcu and Dance [8]). Even the average brightness of x-ray lasers pumped with, say, a 10 Hz repetition rate laser should be higher at an estimated 10^{14} photons s^{-1} mm^{-2} $mrad^{-2}$ (0.1% bandwidth) than 3rd generation sources [5]. High instantaneous brightness means, for example, that x-ray lasers can 'outshine' emission from a probed plasma. Plasma-based x-ray lasers also have a very narrow frequency range of emission (typically $\nu/\Delta\nu > 10^4$). A spectrally narrow output suggests that high spectral resolution using low resolution instruments can be obtained in applications experiments. It also means that the brightness remains constant as the bandwidth is reduced from the value typically employed (0.1% bandwidth or $\nu/\Delta\nu = 10^3$). Broad bandwidth sources such as synchrotrons and harmonic emission decrease linearly in brightness if the bandwidth is narrowed by frequency selecting optics.

The x-ray laser output is of short duration ($\Delta t > 2$ ps) signifying that 'snapshots' of rapidly changing events can be recorded [9, 10]. X-ray lasers also have coherence and demonstration interferometry experiments have been made [10]. 'Criticisms' of x-ray lasers that can be made are that the wavelength is really extreme ultra-violet and too soft for many applications, that the pulse durations with $\Delta t > 2$ ps are too long for many applications and that the output, though spectrally narrow, is not tunable. Though cheaper by several orders-of-magnitude, x-ray lasers are also likely to be several orders-of-magnitude less bright than 4th generation free electron laser light sources. With the aim of suggesting ways to develop x-ray lasers for applications, some of these issues are explored in this paper.

2. X-ray lasers can 'outshine' a probed plasma

Undertaking interferometry, shadowgraphy or some other probing measurement of a plasma with x-ray lasers requires that the recorded self-emission from the plasma is significantly less than the recorded x-ray laser flux. X-ray lasers can readily have an irradiance comparable or greater than the saturation irradiance of typically 3×10^{10} Wcm^{-2} [1] on output from the lasing plasma. In principle, this output can be imaged onto a plasma and the image of the x-ray laser after passing through the plasma sample can be imaged onto a detector. Parameters that are important to ensure that the x-ray laser irradiance reaching the detector is significantly greater than the plasma self-emission are (i) the spectral bandwidth of the imaging or detector, (ii) the relative detected times of emission of the x-ray laser and the plasma self-emission and (iii) the relative solid angles of the detection system and the x-ray laser beam emerging from the probed plasma.

The results of a feasibility study of the minimum transmission of a near-saturated x-ray laser imaged onto a plasma with irradiance 3×10^9 Wcm^{-2} is given in figure 1. This irradiance value is reduced from saturation by an order-of-magnitude to allow for beam magnification or losses. It is assumed that the imaging or detector has a spectral bandwidth centred on the x-ray laser frequency such that $\nu/\Delta\nu = 10$, detection is time integrated and the background emission lasts for a time 10× the x-ray laser emission and that the detection solid angle is only over the x-ray laser solid angle emerging from the probed plasma. Figure 1 shows the transmission able to be probed such that the x-ray laser flux is10× the plasma emission as a function of the plasma temperature. The highest minimum transmission curve is calculated assuming that the probed plasma emits background radiation as a black body. However, assuming that the plasma emission is black-body is almost always a severe overestimate of the actual plasma emission. For example, also shown on figure 1 is the transmission possible if the plasma is emitting only free-free continuum radiation (calculated as discussed by Abou-Ali et al [11]) from a 0.05 μm layer of solid density iron. The two curves on figure 1 represent the extreme transmission ranges to be expected.

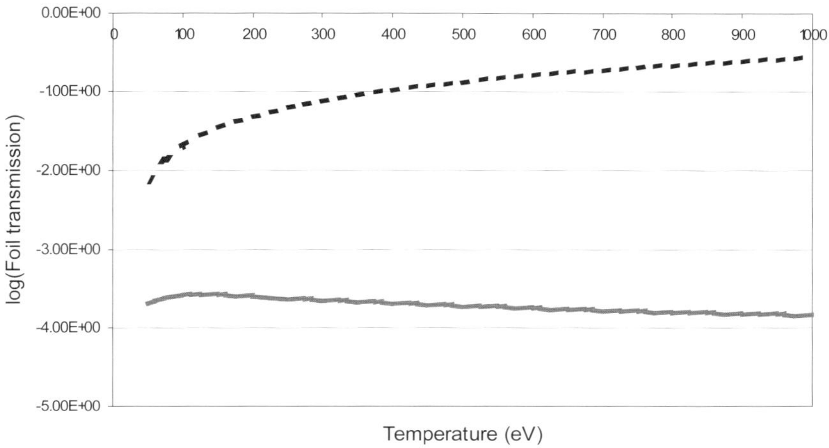

Figure 1 Minimum transmission of a probed plasma to an x-ray laser of photon energy 60 eV (21 nm) as a function of the plasma temperature such that the transmission is ten times more intense than background plasma emission assuming free-free bremsstrahlung emission from a solid density iron plasma of 0.5 μm thickness (light solid curve) or black-body emission (broken curve). Other assumptions are discussed in the text.

It is apparent from figure 1 that a plasma can have moderate opacity or be relatively thick and still be able to be probed by an x-ray laser. To indicate the typical material thicknesses that may be able to be probed with an x-ray laser, the transmission of several solid foil materials as a function of foil thickness are plotted on figure 2. Probing of plasmas produced from foils of thickness up to ≈ 1 μm of low *Z* material such as aluminium or plastic (C_8H_8) or thicknesses up to 0.05 μm of moderate *Z* material such as iron is possible up to temperatures approaching ≈ 1 keV (compare figure 2 with figure 1). The transmission will increase as the foils are heated and ionized and start to expand, so we can use the solid unheated transmissions (figure 2) as lower limits of the transmission.

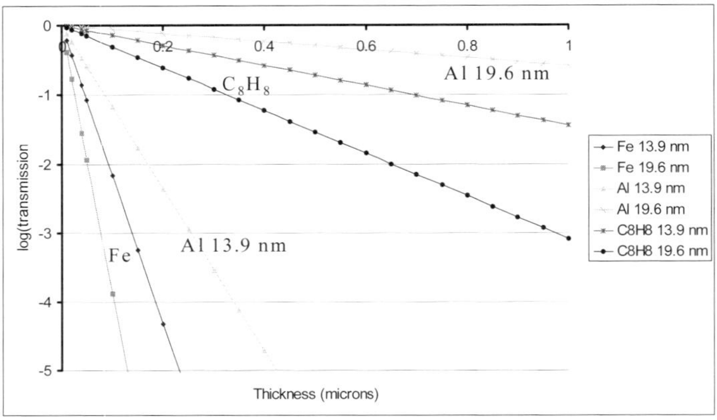

Figure 2 The transmission of solid unheated foils of iron, aluminium and plastic (C_8H_8) as a function of thickness for x-ray laser wavelengths 13.9 nm and 19.6 nm (as labelled).

3. The Fourier transform limit

Figure 4 shows the temporal variation of the Ni-like silver lasing line at 13.9 nm recorded with a fast streak camera for an experiment using a short ($\approx$ 1 ps) pumping pulse superimposed on a pre-pulse of 280 ps duration [12]. A pulse duration $\Delta t \approx 2 - 4$ ps can be measured in agreement with numerical ray tracing of the output of the Ehybrid fluid and atomic physics code [1, 13] and a simple model of the x-ray laser output assuming gain with a step rise and exponential decay [14] (see figure 3). The double pulse structure seen experimentally is reproduced in the simulations and arises due to two different gain regions in the plasma (see figure 4). There is a high gain region of narrow spatial extent close to the target surface where refraction is large and a lower gain region of broader spatial extent further from the target surface with less refraction.

It is known from interferometric measurements [15] that the x-ray laser gain bandwidth is such that $\nu/\Delta\nu > 10^4$. This means that $\Delta\nu\Delta t$ approaches unity and so implies that the pulse durations are close to the Fourier transform limit. We can examine the values of $\Delta\nu\Delta t$ associated with the Fourier transform limit assuming homogeneous Gaussian gain profiles $\Phi(\nu)$. The intensity $I(\nu)$ of the x-ray laser output is calculated using the two equations

$$\Phi(\nu) = \exp\left(-4\ln 2\left(\frac{\nu}{\Delta\nu}\right)^2\right)$$

$$I(\nu) = I_o\left[\exp(Gl\Phi(\nu)) - 1\right]$$

where $\Delta\nu$ is the frequency full width at half maximum of the gain profile, Gl is the gain-length product and I_0 is a 'spontaneous emission' parameter (see [14]). By calculating the Fourier transform of the electric field corresponding to values of $I(\nu)$ evaluated using the above two equations, the value of $\Delta\nu\Delta t$ corresponding to the Fourier transform limit is found (see figure 5). Our calculation neglects saturation effects on $I(\nu)$, so the values for figure 5 are only extended to gain length $Gl = 14$. Saturation can be expected to be important for $Gl > 14$. Doppler broadening of medium Z ions such as germanium and nickel will result in $\nu/\Delta\nu \approx 10^4$ at ion temperatures of 130 – 180 eV giving $\Delta\nu \approx 2 \times 10^{12}$ Hz. At moderate gain length product ($Gl \approx 9$), we see from figure 5 that Fourier transform limited pulses will have temporal duration $\Delta t \approx 0.5$ ps. The temporal duration of output so-far observed (e.g. figure 3) is clearly not Fourier transform limited (though it is getting close to this limit) and there is the possibility of producing shorter duration pulses. The temporal pulse shape that could be expected is not too far from Gaussian for Gaussian gain profiles despite clear gain narrowing (see e.g. figure 6).

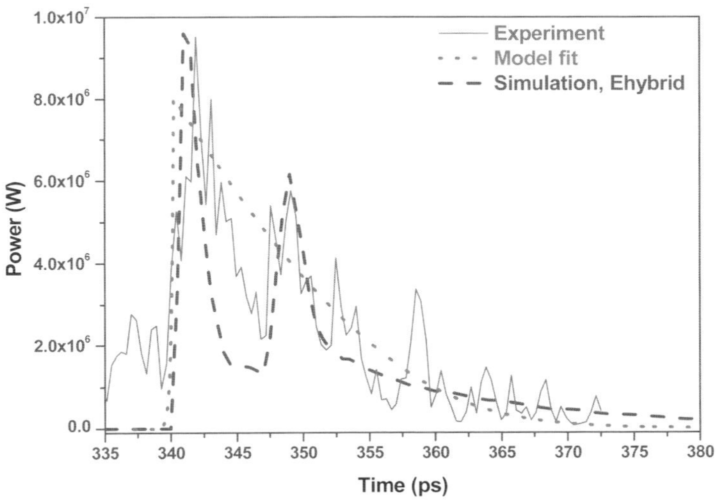

Figure 3 The measured output of lasing at 13.9 nm with a superimposed model fit and Ehybrid and ray trace simulation.

The experimentally observed temporal variations of x-ray laser output that have been recorded using fast streak cameras (see [12] and an example temporal profile on figure 3) are dominated by geometric optics. The Raytrace simulation of figure 3, for example, fits the experimentally recorded temporal variation. The recorded temporal duration of output recently recorded [16] using picosecond pumping with the LULI laser (figure 7) shows an increase in pulse duration with increasing target length as the x-ray laser is driven into saturation. Such an increase can only occur if the approach to the Fourier transform limit does not dominate the determination of the pulse duration.

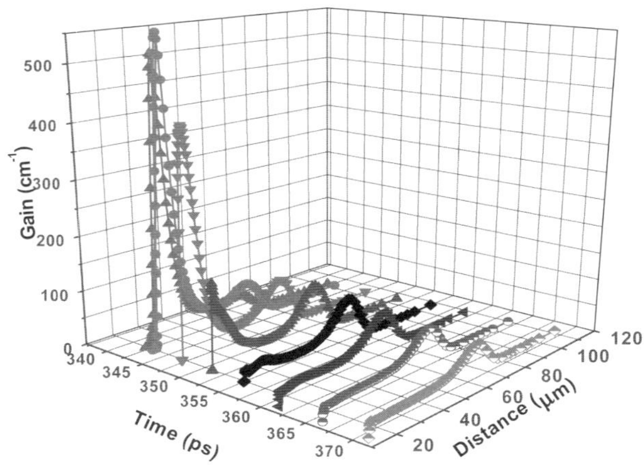

Figure 4 The gain coefficient for Ni-like silver lasing at 13.9 nm as a function of time (from the onset of the pre-pulse) and distance (from the target) predicted by the Ehybrid code.

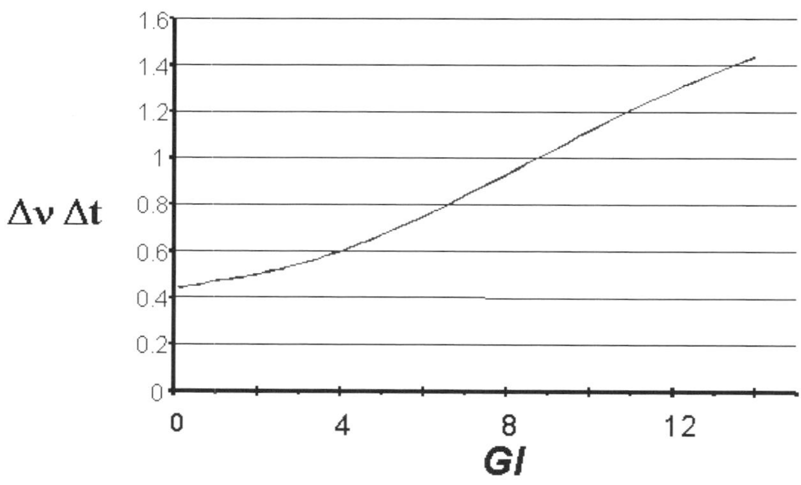

Figure 5 The full width at half maximum $\Delta\nu$ of the gain profile width in frequency units multiplied by the full width at half maximum Δt of the temporal duration of an amplified spontaneously emitted pulse at the Fourier transform limit as a function of the gain length. A Gaussian gain profile is assumed.

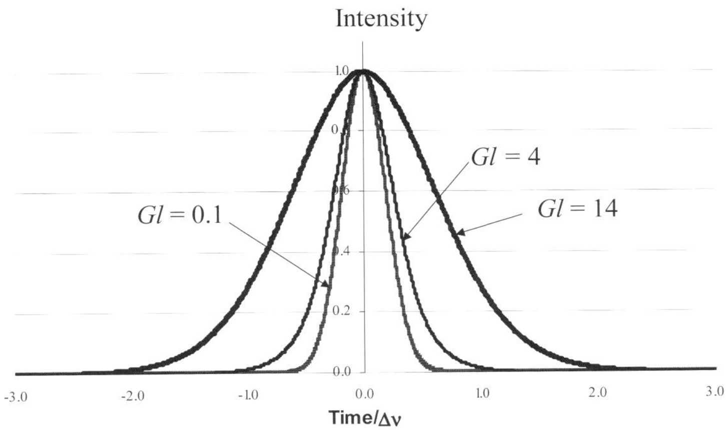

Figure 6 The temporal output expected at the Fourier transform limit for different gain length product Gl (as labeled). A Gaussian gain profile of full width at half maximum $\Delta\nu$ is assumed.

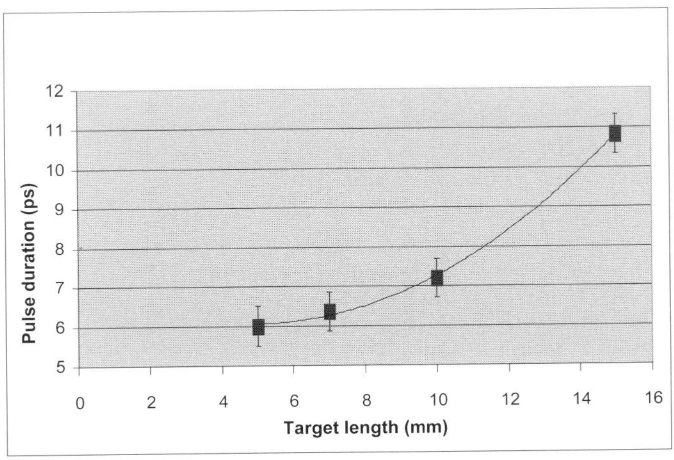

Figure 7 The pulse duration measured as a function of target length for an experiment using the LULI laser to pump x-ray lasing in Ni-like silver at 13.9 nm (see [16]). Saturation of the x-ray laser output occurs at lengths around 5 mm and the effect of saturation re-broadening as the target length increases can be clearly seen.

4. Conclusion

Issues relevant to the application of short duration x-ray laser pulses have been explored. The use of x-ray lasers to probe hot plasmas has been examined by calculating the transmitted x-ray laser flux compared to background plasma emission. A minimum transmission as a function of the probed plasma temperature so as to give a strong x-ray laser flux compared to the emission from the plasma is given for conditions that could be readily obtained in an experiment where a plasma is probed by an x-ray laser. In order to understand the temporal resolution that can be obtained with x-ray lasers, the values of the gain profile spectral width multiplied by the temporal duration for x-ray laser pulses at the Fourier transform limit are evaluated. It can be seen that x-ray laser pulse durations as short as 0.5 ps will be obtained at the Fourier transform limit. Such a temporal limit is considerably shorter than the pulse durations so-far measured with a fast streak camera which are shown to be, for example, 2 – 4 ps using the Vulcan laser at the Rutherford Appleton Laboratory and 6 ps using the LULI laser at Ecole Polytechnique.

Acknowledgements

We gratefully acknowledge financial support from the U.K. EPSRC and the EU TMR and large scale facilities access schemes.

References

[1] Tallents G J 2003 J Phys. D**36**, R259.
[2] Shlaptsev V N *et al* 2004 *these proceedings.*
[3] Rocca J *et al* 2004 *these proceedings.*
[4] Dunn J *et al* 2004 *these proceedings.*
[5] Keenan R, Dunn J, et.al., 2003 SPIE **5197**,213.
[6] Shlaptsev V N *et al*, 2003 SPIE **5197**, 221.
[7] Milton S V *et al* 2001 Science **292**, 2037.
[8] Turcu I C E, Dance J B 1999 X-rays from laser plasmas (Chichester: Wiley) p249.
[9] Wolfrum E *et al* 1998 Phys. Plasmas **5**, 227.
[10] Smith R *et al* 2002 Phys. Rev. Lett. **89**, 065004.
[11] Abou-Ali Y, Demir A, et.al., 2003 J Phys. B**36**, 4097.
[12] Abou-Ali Y *et al* 2003 Optics Commun. **215**, 397.
[13] Holden P B *et al* 1994 J Phys. B**27**, 341.
[14] Strati F and Tallents G J 2001 Phys. Rev. A**64**, 013807.
[15] Dunn J, Smith R F *et al 2003* SPIE **5197,** 43.
[16] A Klisnick *et al* 2004 *these proceedings.*

Inst. Phys. Conf. Ser. No 186
Paper presented at 9th Int. Conf. X-Ray Lasers, Beijing, China, 24–28 May 2004

Nanopatterning with a high repetition rate λ= 46.9 nm capillary discharge table-top laser

M. G. Capeluto[a], G. Vaschenko[b], M. Grisham[b], M. C. Marconi[a,b], S. Ludueña[c], L. Pietrasanta[c], Y. Lu[d], B. Parkinson[d], C. S. Menoni[b], and J. J. Rocca[b]

[a]Departamento de Física, [c]Centro de Microscopias Avanzadas, Facultad de Ciencias Exactas, Universidad de Buenos Aires, Argentina

[b]NSF ERC for Extreme Ultraviolet Science & Technology and Department of Electrical and Computing Engineering, Colorado State University, USA

[d]Department of Chemistry, Colorado State University, USA

Abstract. We have exploited the high average power and high spatial coherence of a compact high repetition rate λ = 46.9 nm capillary discharge laser to demonstrate nanometer-scale patterning in poly-methyl methacrylate. Interference patterns were generated through wave-front division interferometry using a Lloyd's mirror. Gratings with periods as small as 55 nm were imprinted on PMMA. These results, that constitute the first demonstration of nanopatterning with a soft x-ray laser, illustrate the potential of high repetition soft x-ray lasers in nanotechnology applications.

1. Introduction

The miniaturization of devices has pushed the limits of feature size fabrication to the nanometer scale. Light sources with increasingly short wavelength, down to the 157 nm corresponding to the F_2 laser, have been used to print progressively smaller features. In this path, compact high repetition rate soft x-ray lasers could provide a way to achieve the recording of nanoscale features, in particular when the pattern can be created by interference effects that take advantage of the coherence of this type of compact soft x-ray source. In these proceedings we report the imprinting of nanometer patterns in poly-methyl methacrylate (PMMA) using the combination of a 46.9 nm table-top discharge pumped laser and a Lloyd's mirror interferometer.

2. Experimental set up and procedure

To generate nanometer gratings by interferometric lithography we employed the same type of Lloyd's mirror interferometer used previously for PMMA patterning with synchrotron radiation at 13 nm [1] and for plasma diagnostics with 46.9 nm radiation

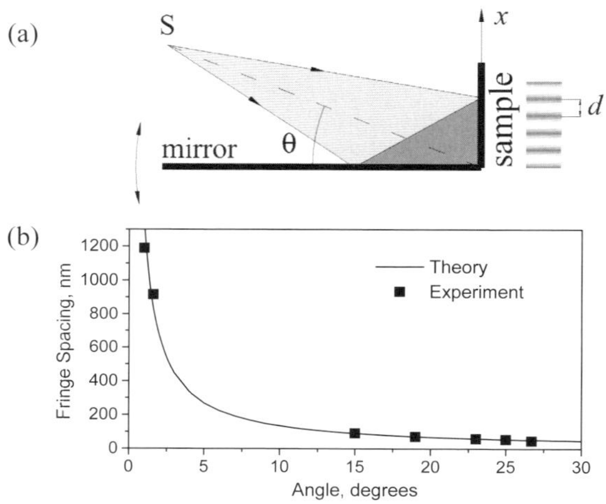

Fig. 1. (a) Schematic diagram of the interferometric lithography set-up utilizing a Lloyd's mirror; (b) variation of the grating period as a function of Lloyd's mirror angle. The points indicate the measured values.

from a capillary discharge laser [2]. At soft x-ray wavelength the Lloyd's mirror is the simplest possible interferometer configuration. In this wavefront division interferometer part of the beam impinges at grazing incidence on the mirror and is reflected to interfere with the remaining portion of the beam, as shown in Fig. 1(a). Beam interference gives rise to a sinusoidal pattern in which the period d is determined by the wavelength of the light λ and the incidence angle θ according to:

$$d \approx \lambda / 2\sin(\theta)\,. \tag{1}$$

The modulation depth of the fringes generated in this way is a function of the intensity of the interference beams I_1 and I_2, and the coherence properties of the source, and can be expressed as [3]:

$$M(x) \sim I_{\max} - I_{\min} = 2\sqrt{I_1(x)}\sqrt{I_2(x)}\left|\mu_{12}\right|, \tag{2}$$

where μ_{12} is the mutual coherence factor.

It is clear from Eq. (1) that a soft x-ray source is capable of printing nanometer gratings. For this application such light source needs to have a good degree of coherence and high average output power. All of these requirements are met by the 46.9 nm capillary discharge laser [4,5]. The spatial coherence of this laser was measured in previous experiments. It was found that for capillary lengths of 36 cm the laser output can reach essentially full spatial coherence [5]. The laser output has a divergence of ~ 4.5 mrad, and is also highly monochromatic with $\Delta\lambda/\lambda \sim 5\times10^{-5}$. Average output powers of a few mW can be obtained when the source is operating at the repetition rate of several Hz. The present experiments were performed with the laser operating at 1-2 Hz and a pulse energy of ~ 0.2 mJ.

Interference patterns were created using a gold coated mirror that was placed at 1.5 m from the source at selected grazing incidence angles θ ranging from 2 to 26 degrees. For these angles the grating spacing varied between 1200 and 50 nm respectively, as shown in Fig. 1(b). The sample, consisting of Si wafer spin-coated with ~ 500 nm PMMA layer (MicroChem 950,000 molecular weight) was placed at the farthest end of the mirror as shown in Fig. 1. Samples were illuminated with doses typically consisting of 100 laser shots, whereas the single shot fluence was ~ 0.15 mJ/cm^2. The exposed resist was then developed in 1:3 methyl isobutyl ketone/isopropanol (MIBK/IPA) solution for ~ 30 seconds and rinsed in IPA. The resulting patterns were read with an atomic force microscope (AFM).

3. Results and Discussion

Figure 2 shows the AFM images of the interferometric patterns recorded on PMMA for angles of 15, 19 and 22 degrees corresponding to imprinting gratings of periods d ~ 92, 72 and 63 nm respectively. As can be seen the 92 nm pattern consists of lines of well defined edges and an average modulation depth of 12.5 nm. For smaller line periods the modulation depth decreases and the line edge roughness increases significantly.

The decrease in the modulation depth for grating periods above 60 nm is mainly due to the decrease of the mirror reflectivity R with angle. This is shown in Fig. 3, where we have plotted Eq. (2) with the intensity of the reflected portion of the beam expressed as $I_2 = R(\theta)I_1$. It is evident from Fig. 3 that for gratings periods above 60 nm the measured modulation depth closely follows the values predicted by considering the decrease in mirror reflectivity. However, for grating periods below 60 nm the measured data abruptly departs from the predicted curve. This indicates that other factors are responsible for the rapid degradation of the modulation depth for periods smaller than ~ 60 nm. Several factors can play a role in degrading the modulation depth, among them the decreased penetration of light with angle, the increased contribution of mechanical

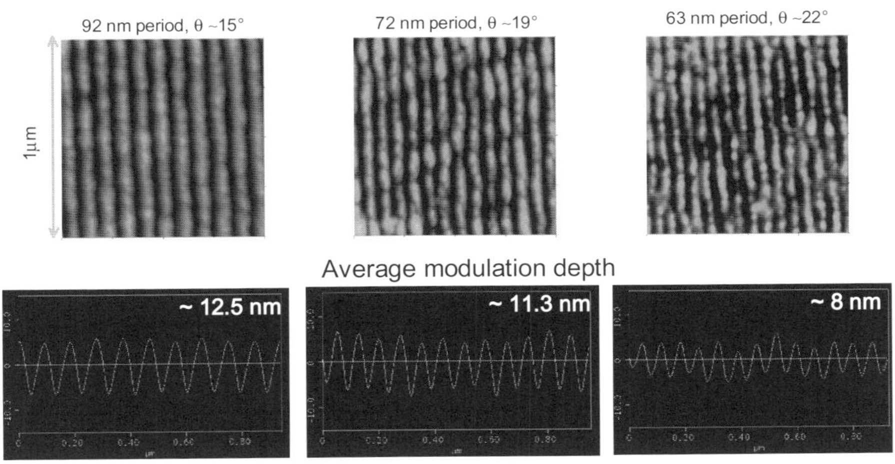

Fig. 2. Top: AFM micrographs of the grating patterns written with the 46.9 nm capillary discharge laser on PMMA. The angle of incidence and the grating period corresponding to each case are indicated. Bottom: averaged profiles corresponding to each of the images shown above. The average modulation depths are indicated in the top right corners.

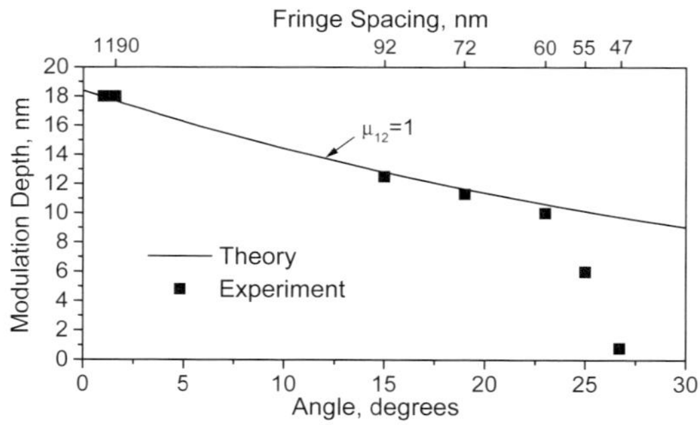

Fig. 3. Calculated (solid line) and measured modulation depth of the fringes on PMMA surface shown as a function of the beam angle.

vibrations during exposure, the comparable size of the AFM probe tip, scattering of the mirror roughness, and the fact that at this line spacing one is likely to approach the limits of the photoresist. Resist resolution is also likely to be responsible for the increase in line edge roughness as the grating period decreases.

4. Summary

We have demonstrated for the first time nanometer-scale patterning using interferometric lithography with a table-top soft x-ray laser. Dense line patterns with a period as small as 55 nm were imprinted in PMMA using a 46.9 nm capillary discharge laser and a Lloyd's mirror. The modulation depth of the groves changed with the fringe separation and was limited to a certain extent by the change in the mirror reflectivity.

Acknowledgments

This work was supported by the NSF Center for Extreme Ultraviolet Science and Technology under NSF Award Number EEC-0310717 and by CONICET. The support of the W. M. Keck Foundation is also gratefully acknowledged.

References

1. H. H. Solak, D. He, et al., Appl. Phys. Lett. **75**, 2328 (1999)
2. J. J. Rocca, C. H. Moreno, et al., Opt. Lett. **24,** 420 (1999)
3. M. Born and E. Wolf 1999 Principles of optics (Cambridge: Cambridge University Press)
4. B. R. Benware, C. D. Macchietto, et al., Phys. Rev. Lett., **81** 5804 (1998)
5. Y. Liu, M. Seminario, et al., Phys. Rev. A, **63** 033802 (2001)

Inst. Phys. Conf. Ser. No 186
Paper presented at 9th Int. Conf. X-Ray Lasers, Beijing, China, 24–28 May 2004

Evidence of bound electron contribution to soft x-ray laser interferograms of dense plasmas.

J. Filevich, J.J. Rocca, M.C. Marconi
NSF ERC for Extreme Ultraviolet Science and Technology and Department of Electrical and Computer Engineering, Colorado State University, Fort Collins, Colorado 80523

J. Nilsen, J. Scofield, S.J. Moon, J. Dunn, R.F. Smith, R. Keenan, J.R. Hunter
Lawrence Livermore National Laboratory, Livermore, California 94550

V.N. Shlyaptsev
Department of Applied Science, University of California Davis-Livermore, Livermore, California 94550

Abstract. We present clear experimental evidence showing that the index of refraction of laser created plasmas at soft x-ray wavelengths can be dominated by the contribution of bound electrons. We report picosecond resolution soft x-ray laser interferograms of Al laser-created plasmas that late in their evolution display negative fringe shifts in the plasma periphery. The comparison of measured and simulated interferograms show that this results from the dominant contribution of low charge ions to the index of refraction. The observed ion contribution to the index of refraction has broad significance, as it affects the propagation of soft x-ray radiation in plasmas and the interferometric diagnostics of plasmas for many elements.

1. Introduction

In this paper we present soft x-ray laser interferometry data obtained with a picosecond 14.7 nm (84.4 eV) laser probe that shows clear evidence of the dominant contribution of bound electrons to the index of refraction in the late stages of the evolution of an Al plasma created by a high power laser. The interferograms show that the fringes, late in the plasma evolution, in the periphery of the plasma region irradiated by the heating laser bend toward the target (negative fringe shifts), as opposed to away from the target (positive fringe shifts), as occurs at earlier times. Analysis of our data done with the assistance of hydrodynamic model simulations indicates that late in the plasma evolution the contribution of bound electrons dominates the index of refraction, causing the observed negative fringe shifts in the periphery regions of the plasma. The significance of the result goes beyond the particular case of aluminium, as this effect is computed to significantly affect the index of refraction of many materials at soft x-ray laser wavelengths.

2. Experiment results and simulations

The experiment was performed using a transient 14.7 nm Ni-like Pd soft x-ray laser [1] combined with an amplitude division diffraction grating interferometer (DGI)[2,3]. The interferometer is set in a skewed Mach-Zehnder configuration, with the principal distinguishing characteristic being the use of diffraction gratings as beam splitters. The DGI design has proven to be robust and capable of producing interferograms with very good visibility over a large field of view, as first demonstrated with a 46.9 nm capillary discharge laser [2]. The Ni-like Pd soft x-ray laser used in this experiment produced laser pulses of few 10's of μJ with a typical duration in the range of 4.5 - 5.2 ps [4]. This short pulse duration permits the acquisition of "snap-shots" of rapidly evolving plasmas, overcoming the blurring of interference fringes that occurs when the electron density profile changes significantly during the duration of the probe pulse.

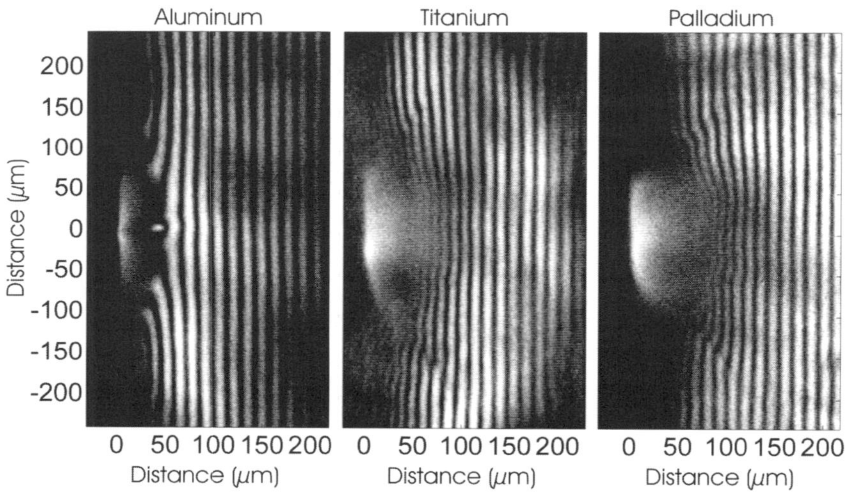

Figure 1. Comparison of interferograms obtained 3 ns after the peak of the heating laser pulse for Al, Ti and Pd targets. Fringes that bend toward the target (negative fringe shifts) are observed in the Al plasma, but not in the Ti and Pd plasmas.

Figure 1 shows interferograms of expanding Al, Ti, and Pd laser created plasmas. The plasmas were created by focusing a 600 ps, 3 J, 1054 nm laser into a 3.1 mm long x 12 μm wide line focus on a 1 mm long flat Al target. The target was positioned for the plasma to intercept one of the arms of the interferometer, and the timing between the peak of the heating beam and the probe beams was measured using a fast photodiode. These late frames (3 ns) in the evolution of the plasma show a region close to the target that presents increased absorption, an indication that the plasma is cooling. Close to this same region we observe, in the first frame, corresponding to the Al plasma, fringes that bend toward the target. Similar negative fringe shifts were also recently observed in an independently realized Al laser-created plasma soft x-ray laser interferometry experiment at 13.9 nm, but the results were not analysed in detail [5]. This is in striking contrast to the interferograms we obtained of Ti and Pd plasmas that only present fringes bending away from the target (see Fig 1.). The difference between these elements can be found by examining the index of refraction of the neutral atoms at the probe beam's wavelength. In phase shift interferometry the sign of $1-\eta$, where η is the index of refraction, determines the direction of the fringes: $N_f=1/\lambda\int(1-\eta)dl$. The real part of the index of refraction, obtained from Henke et al. [6] is larger than one in the case of Al, but smaller for all other elements with Z < 53. This suggests that a large concentration of neutral atoms is responsible for the observed negative fringe shifts. However, such a concentration, required to account for the observed fringe shifts, would completely absorb the probe beam. We conclude that the neutral density of this plasma should be negligible and that the negative fringe shifts should be caused by the ions instead (the free electron contribution to the index of refraction is equal for all elements, and always smaller than one: $\eta=(1-n_e/n_c)^{1/2}$ where $n_{crit}=5 \times 10^{24}$ cm^{-3} is the critical density for λ=14.7 nm).

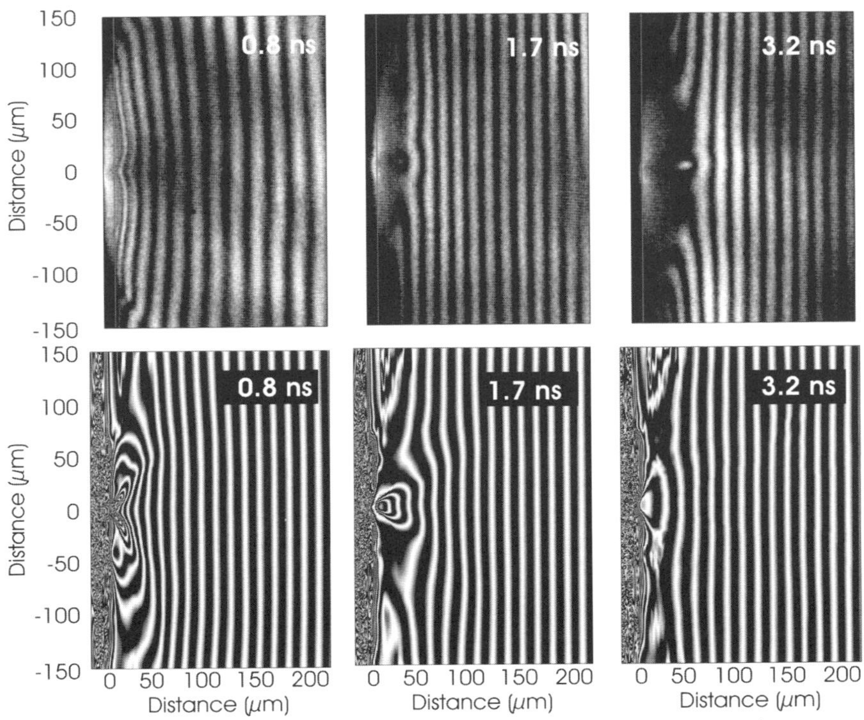

Figure 2. Comparison between the measured and synthesized interferograms.

To verify this we computed the real component of the index of refraction f_{1Z}^0 for the different ions that can be present in the plasma [7]. We define the "effective" f^*_{1Z} value as $f_{1Z}^{bound} + f_{1Z}^{free}$ that includes the separate contributions of free and bound electrons, where f_{1Z}^{free} is the number of free electrons corresponding to the ion of charge Z. We calculated the values for Al I – Al IX to be: Al I = -0.85, Al II = -4.19, Al III = -3.54, Al IV = -1.80, Al V = 0.84, Al VI = 3.54, Al VII = 5.30, Al VIII = 6.73, Al IX = 8. When the f^*_{1Z} values differ from Z (the number of free electrons corresponding to that ion) the contribution of the bound electrons is significant, and this was the case for all ions with charge up to Z=5.

To confirm this interpretation of the data, the LASNEX hydrocode [8] was used to compute the ion and electron density distributions of the measured plasmas. The simulations were conducted using the temporal and spatial profiles of the laser beam used to heat the plasma. The electron density distributions obtained from the early time interferograms, including the lateral expansion and on-axis depression, are well reproduced by the code. This indicates that the assumption that the free electrons dominate the contribution to the index of refraction is valid at early times, as is expected from the fact that at those times the plasma is hot and only highly ionised species are present. At later times, the simulations show that the degree of ionisation in the plasma periphery decreases as the plasma cools, resulting in significant concentrations of Al II-Al V ions present. The simulation results were used together with the calculated f^*_Z to synthesize the interferograms shown in Figure 2, together with the experimental data for comparison. The synthesized interferogram corresponding to 0.8 ns delay shows the central density depression and the lateral expansion observed in the measurements, and the late interferograms at 3.2 ns shows the observed negative fringe shifts in the periphery of the plasma, in good agreement with the experiment.

3. Conclusions

In conclusion, we have demonstrated that bound electrons can contribute significantly to the index of refraction of multiply ionised plasmas at soft x-ray wavelengths in the vicinity of absorption edges. In the case of an Al plasma the effect causes a dramatic change of sign of the fringe shifts in interferograms acquired using λ= 14.7 nm radiation. It should be expected that interferograms of the late stages of the evolution of many other laser-created plasmas involving other materials would also be significantly affected by the contribution of bound electrons. However, the relative contribution of bound electron scattering is not always clearly evident in soft x-ray interferograms because at these wavelengths the free and bound electrons often contribute with the same sign to the phase delay. In these cases, neglecting the contribution of bound electrons to the index of refraction could constitute a significant systematic error in the determination of the electron density. Moreover, the importance of the bound electron contribution is not always limited to plasmas with a low mean ion charge. For example, ten times ionised Pd atoms are computed to have an f_1^* value that is several times that of the corresponding number of free electrons, with the precise value of f_1^* depending strongly on the position of the resonance line[7].

Acknowledgments

This research was sponsored by the National Nuclear Security Administration under the Stewardship Science Academic Alliances program through DOE Research Grant # DE-FG03-02NA00062. Part of this work was performed under the auspices of the U.S. Dept. of Energy by the University of California, Lawrence Livermore National Laboratory through the Institute of Laser Science and Application, under contract No. W-7405-Eng-48. The CSU researchers also gratefully acknowledge the partial support of the NSF ERC Center for Extreme Ultraviolet Science and Technology, award number EEC-0310717 and the W.M. Keck Foundation.

References

[1] J.Dunn, Y.Li, et al., Phys. Rev. Lett. 84, 4834 (2000).
[2] J. Filevich, K.Kanizay, et al., Optics Letters 25, 356 (2000).
[3] J. Filevich, et al., Applied Optics 43 (2004).

[4] J.Dunn et al., in SPIE Int. Soc. Opt. Eng. Proc, edited by E. Fill (2003), vol. 5197, pp 51-59.
[5] H. Tang et al., Applied Physics B 78, 975 (2004).
[6] B. Henke, E.M. Gullikson, et al., Atomic Data and Nuclear Data Tables 54, 181 (1993).
[7] J. Nilsen and J.H. Scofield. University of California report UCRL-PROC-204652, Livermore, CA.
[8]G. Zimmerman and W. Kruer, Comments Plasma Phys. Contr. Fusion 2, 51 (1975).

Inst. Phys. Conf. Ser. No 186
Paper presented at 9th Int. Conf. X-Ray Lasers, Beijing, China, 24–28 May 2004

Experimental study on hard X-ray diffraction enhanced imaging

Hongyi Gao*, Jianwen Chen, Honglan Xie, Huafeng Zhu, Ruxin Li and Zhizhan Xu

Shanghai Institute of Optics and Fine Mechanics, Chinese Academy of Sciences, P. O. Box 800-211, Shanghai 201800, China

Peiping Zhu, Qingxi Yuan, Yulian Tian, Wanxia Huang, Junyue Wang, Xin Zheng and Dingchang Xian

Institute of High Energy Physics, Chinese Academy of Sciences, Beijing 100080, China

Abstract: Some experimental results for X-ray diffraction enhanced imaging (DEI) are presented, which demonstrates the DEI technique as a promising tool for the application in clinical, biological and industrial settings.

1. Introduction

Conventional X-ray radiography is based on the absorption of the radiation, the image contrast results from variation of the absorption in the object of varying composition, thickness or density. An alternative approach is phase contrast imaging, which has recently been studied actively in hard X-ray region due to its high sensitivity to biological soft tissues and polymeric materials consisting of light elements that poorly generate absorption contrast.

As we known, the behaviors of the X-ray passing through an object can be described by its complex index of refraction, which can be written as: $n=1-\delta-i\beta$, where β describes the absorption of X-rays and δ is the phase shift term at an order of 10^{-6}, this means the refraction index deviates only slightly from unity. Nevertheless, for the elements that atomic number lower than 15, their interaction cross-section of phase shift can be a few order of magnitude larger than that of the absorption.

As phase contrast depends strongly on the refraction variation of X-rays rather than absorption, imaging can be taken at higher energy levels and the absorbed dose of radiation is decreased, thereby the radiation damage to tissues is reduced.

Up till now, a lot of research groups are exploring ways of exploiting phase information as a means of imaging, including X-ray interferometer [1-2], diffraction-enhanced imaging (abbreviated DEI) [3-4] and in-line imaging techniques [5-6], associated with the measurements of the phase difference φ, the phase gradient $\nabla\varphi$, and the second derivative of the phase $\nabla^2\varphi$ of the X-rays that passing through the specimen. The possible applications may be the detection of tiny structure in soft tissues, the investigation of advanced materials such as ceramics and fiber composites, etc.

In this paper, some experimental results for DEI carried out at the topography station in Beijing Synchrotron Radiation Facility (BSRF) are presented. The 0.87 Å wavelength from the bending magnet was selected by a Si monochromator. The specimen used in the experiment is a fresh aquarium fish with 3~4 mm thickness and 5~6mm width.

The X-ray DEI images are recorded at Kodak high resolution films and readout by optical microscope with a maximum magnification of 40. From the magnified pictures, the vertebrae, ribs, fins and the bladders are observed.

2. Principle of DEI

In the DEI experiment, a parallel of polychromatic X-rays emitting from the Synchrotron Radiation is monochromatized by a crystal monochromator. When the x-rays traverse the specimen placed between the monochromator and the analyzer, they are absorbed, scattered coherently or incoherently at an angle of milliradians or more, or refracted at even smaller angles of microradians due to tiny variations in the refractive index. X-rays emerging from the specimen and hitting the analyzer crystal will satisfy the condition of the Bragg diffraction only in a very narrow window of incident angles, typically in an order of a few μrad, those scattered in the specimen are fall outside this window and won't be reflected. The X-rays dropped in the window or to be reflected to the analyser are strongly depends on the incident angle, the dependence is called rocking curve. When the analyzer is perfectly aligned with the monochromator, X-rays those scattered or refracted over a few μrad will be filtered. This is the reason why diffraction enhanced imaging method had a high ratio of signal-to-noise.

When the analyser is at the shoulder of the rocking curve, tiny difference of the refraction angle forms the intensity variation of the X-rays (ΔI) by the diffraction of the analyser crystal, this gives the contract of the refraction which depends on the angular width and the position on the rocking curve of the analyser.

The extremely small angular width of the analyser's rocking curve equals the Darwin width of perfect crystal diffraction. Darwin width for a symmetric diffraction is calculated from [7]:

$$\theta_D = 2\lambda/\pi\Lambda_s \sin 2\theta_B$$

where θ_B is the Bragg angle, λ is the X-ray wavelength. Λ_s is the symmetric extinction length given by

$$\Lambda_s = V_c / r_e \lambda K F_H$$

where V_c is the volume of the unit cell, $r_e = e^2/m_e c^2 = 2.818\times10^{-15}$ m is the classical electron radius, K is the polarization factor, and F_H the structure factor. For the hard X-rays transmitted through thick objects, the Darwin width of crystalline silicon is around 1-10 μrad depending on the energy and the order of crystal diffraction.

3. Experimental setup

The experiments are carried out at the topography station in Beijing Synchrotron Radiation Facility (BSRF). In general DEI setup, the obtainable image area is less than 20×4 mm^2 while the cross section of the incident 'white' X-ray beam is 20×10 mm^2 at the topographic beamline, apparently this DEI setup was deficient for large specimens as shown in figure1.

New setup has been designed and applied in DEI experiments as shown in Fig. 2. Its image area is the same as the incident "white" X-ray beam. Polychromatic synchrotron radiation with an effective source size of 800 μm in vertical direction and 2 mm horizontally, falls at one of double crystal at a distance of 43 m from the bending magnet. The double-crystal Bragg scheme contains a monochromator and analyzer. The crystals were prepared from perfect Si and installed vertically on a vibration-free setting that can be rotated and moved electrically of multi dimensions. The specimen under test was placed upstream of the analyzer in the pseudo-plane wave produced by the monochromator. The low divergence synchrotron radiation white beam was restricted in width and height by the double crystal system.

The 0.87 Å radiations are selected by the Si monochromator. The images are recorded using Kodak high resolution films and CCD.

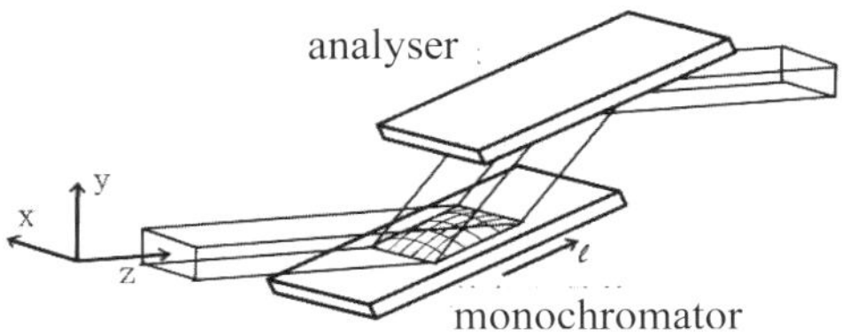

Figure 1. Schematic diagram of monochromatic crystal and analyser in general DEI setup

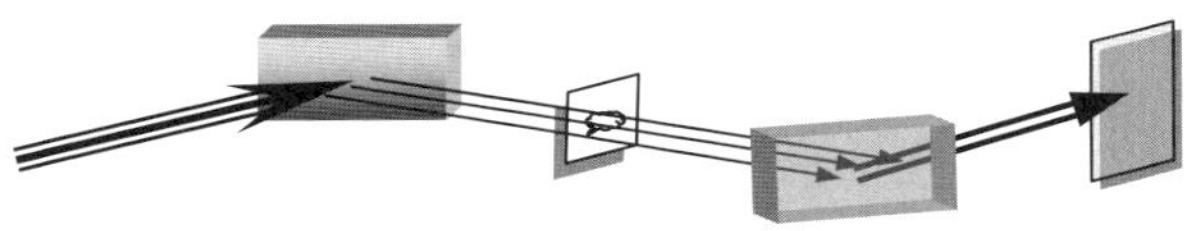

Figure 2. Schematic setup for hard X-ray diffraction enhanced imaging system

4. Experimental results

When the specimen under test was highly transparent for instance when were biological ones, the pronounced features invisible in the absorption images appeared in the DEI. For example, Fig.3 demonstrates an absorption image of an aquarium fish of maximum 5 mm thickness in which the ribs are observed faintly.

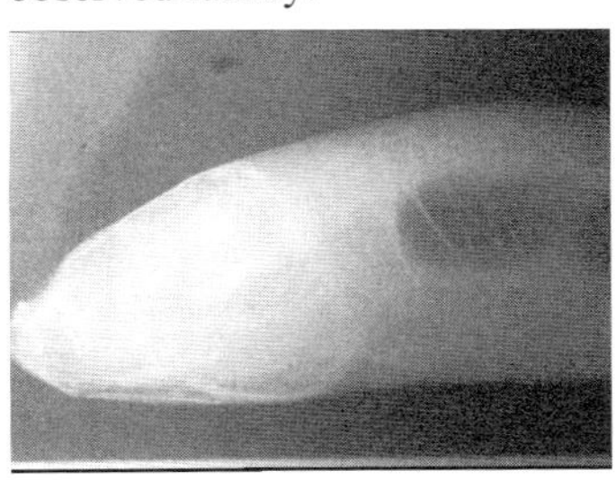

Figure 3. X-ray absorption image of a fresh aquarium fish

Figure 4(a) shows the DEI image of the same sample as figure 3 under the same condition. A local magnified image of a part of figure 4(a) is shown in figure 4 (b). When the analyser was in the Bragg position, it is immediately apparent that the DEI technique well revealed the tissue details not identified in the previous image, such as the vertebrae, the ribs, the air bladders, eyes and fins. The air bladder walls (adjacent to the spinal column) are also identified clearly.

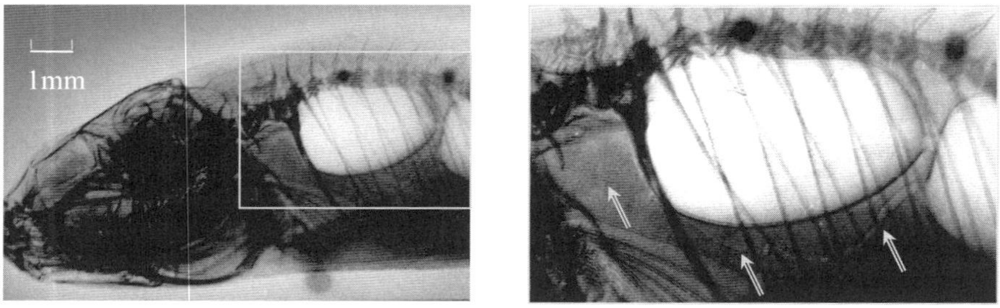

Figure 4. (a) Hard X-ray DEI image taken at the Synchrotron radiation facility. The sample is an aquarium fish with 5~6mm wide and 3~4mm thick, recorded in the double-crystal scheme; (b) Local magnified image from figure 4 (a). The arrows indicate the trace of the bladder tube.

It is interesting that we observed a soft tube that connects the air bladder to the head part of the fish, when it goes to the head it becomes slighter and slighter with its diameter varied in the range of 30 to 100 micrometers. And the resolution is estimated about 10 microns. This bladder tube is likely to serve as an air transport pipe that transfer the gas between the bladder and outside to help the fish controlling its floating and sinking.

Except the aquarium fish, other specimens such as spider, ladybug and section of human kidney are also adopted in the experiment. All the results showed the 10-micron resolution in optimum condition.

5. Conclusions

The presented experimental results allow one to conclude that: The DEI technique reveals the interior structure of tissues especially in boundary regions. It is currently most effective approach among three of phase contrast imaging; The analyser crystal improves the S/N of the DEI images effectively, the resolution goes up to 10 microns in our experiment; It is of great importance to combine the DEI techniques with CT to satisfy the actual demands.

6. Acknowledgements

This work was supported in part by National Nature Science Foundation of China (No. 60278030) and Shanghai Optical Science and Technique Foundation (No. 036105028).

References

[1] U. Bonse, M. Hart, 1965, Appl. Phys. Lett., 6, 155-156.
[2] A. Momose, 1995, Nucl. Instrum. Meth. A352, 622-628.
[3] V. N. Ingal et al., 1995, J. Phys. D: Appl. Phys. 28, 2314-2317.
[4] T. J. Davis et al., 1995, Nature, 373, 595-598.
[5] D. Chapman et al., 1997, Phys. Med. Biol., 42, 2015-2025.
[6] F. A. Dilmanian et al., 2000, Phys. Med. Biol., 45, 933-946.
[7] W. H. Zachariasenn, 1945, Theory of X-ray Diffraction in Crystals, Wiley New York.

Inst. Phys. Conf. Ser. No 186
Paper presented at 9th Int. Conf. X-Ray Lasers, Beijing, China, 24–28 May 2004

Damaging DNA using Ne-Like zinc soft x-ray laser at 21.2nm

K. Cassou[1], D. Ros[1], S. Kazamias[1], O. Guilbaud[1], A. Klisnick[1] and G. Jamelot[1], B. Rus[2], M. Kozlová[2], J. Polan[2], A. R. Präg[2], M. Stupka[2] G. Eot-Houllier[3], E. Sage[3], M.-A. H. du Penhoat[4], A.Touati[4], A. Chétioui[4], M. Begusová[5], V. Stísová[5]

[1]Laboratoire d'interaction du rayonnement X avec la matière, Bâtiment 350, Université Paris sud, 91405 Orsay Cedex, France.
[2]Department of x-ray laser/PALS Centre, Institute of Physics, Academy of Sciences of the Czech Republic, Na Slovance 2, 18221 Prague 8, Czech Republic.
[3]Laboratoire de Génotoxicologie et cycle cellulaire, CNRS UMR 2027, Institut Curie, centre Universitaire, 91405 Orsay Cedex, France
[4]Groupe de Physique des solides, Universités Paris VI-VII, CNRS UMR 7588, CEA.
[5]Nuclear Physics Institute, Academy of Sciences of the Czech Republic, Na Truhlárce 39/64, 18086 Prague 8, Czech Republic

Abstract: We report the preliminary result of the application of the collisional Ne-like soft x-ray laser as radiation source to induce DNA damage. The goal of this experiment was a test bed study of the damage yields induced by soft x-ray radiation in dried plasmid DNA sample. The saturated Ne-like soft x-ray laser available at the PALS facility, delivering several millijoules in a single 100 ps pulse at 21.2nm was used to irradiate two different plasmids: pSP 189 and pBS. The study is centered on the dose effect.

1. Introduction

Since the beginning of their development, collisional soft x-ray lasers (XRLs) have been used for various applications: they are tool for plasma interferometry, imaging and microscopy [1], and also sources for the irradiation of matter [2]. The present work enrolls as addition to a previous scientific collaboration aiming at identifying the physical mechanisms underlying the formation of DNA lesions induced by ionising radiation. Soft x-ray photons can be used to probe the dosimetry for induction of specific types of radiation damage and biological response, and so provide much needed data for understanding and modelling the processes involved. We focus on the strand breaks damage in plasmid DNA as a function of the dose. The model K, allots the break event in DNA molecule, to the K or L innershell ionisation of component DNA atoms [3, 4]. In this study, we use monochromatic XRL radiation at the energy 58.5 eV delivered at the PALS facility [5]. According to theoretical studies, it is assumed that the minimal energies required to produce single-strand and double-strand breaks in DNA are about 17 eV and 50 eV respectively [6].

The plasmid layer deposited on a glass support, typically 600 nm in thickness, was irradiated in vacuum by one, two or three soft X-ray laser pulses, corresponding to absorbed doses up to 10 kGy. The irradiated plasmid DNA samples were then dissolved in water and were subsequently analysed by agarose gel electrophoresis, making possible the accurate measurement of single and double strand breaks induced by the monochromatic XRL radiation. The main asset of those experiments consists in better understanding the mechanisms of the DNA damage formation relevant to radiotherapy.

2. Soft X-ray laser developed at PALS

The deeply saturated operation of a double-pass Ne-like zinc soft x-ray laser emitting at a wavelength of 21.2nm is currently running by DXRL in PALS, with a well optimization of its beam parameters and coherence. The active medium is a 3 cm long plasma column produced by a separately delivered prepulse and a main pulse with duration of ~450 ps, supplied by a 1.315 mm laser (Asterix laser). To generate plasma with reduced lateral density gradients, the ~130μm wide main pulse focus, producing a pump irradiance of $\sim 2.810^{13}$ $W.cm^{-2}$, is placed on top of a much broader prepulse focus. The x-ray laser emission emerges as a narrowly collimated beam possessing high spatial quality and coherence. The large gain-length product attained through half-cavity operation allows for an efficient extraction of the energy stored in the active medium. With a small-signal gain of 7 (±0.5) cm^{-1}, the double-pass beam is deeply saturated and provides around 1mJ of energy in one pulse, corresponding to peak power in excess of 10 MW. The repetition rate given by the Asterix laser system is around 20 minutes.

3. Experimental setup and method

The beam XRL is sent in the target chamber through a filter wheel, reflected by a 45° multilayer Mo:Si mirror to the sample mount. It is possible to put 4 samples on the mount holder. We have different ways to play with the level of the dose, the filters, which are aluminium filters of 1%, 5%, 50% transmission, and the number of shots on the sample. We are limited by the repetition rate of twenty minutes of the IR pump laser of the XRL and the lifetime of the biological sample under vacuum evaluated at one hour.

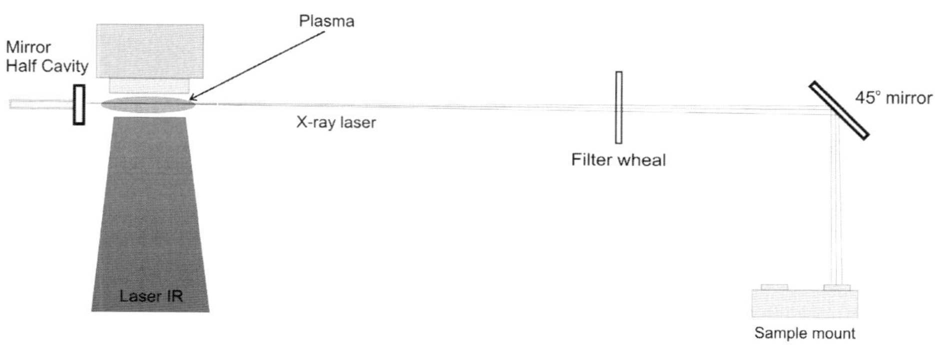

Figure 1 - Experimental Setup of DNA irradiation by X-ray laser

We deposited 4 μL of solution containing 10 μg of plasmid in the center of the glass support. The plasmid solution dries in air during 10-15 minutes. We obtain a laying which show the following radial profile (Figure 2).
The radius of the dry plasmid deposit is around 1.5 mm. According to the XRL beam divergence the beam size is 4.5 x 7.5 mm at the level of the sample. The manipulation of the plasmid could lead to strand damage in the DNA helix. We have two control samples. The first control sample is not exposed to vacuum and to determine the damage coming from the handling, and the second control sample is settled behind the sample mount for damage induced by the vacuum. We vary the dose from 1 shot to 3 shots.

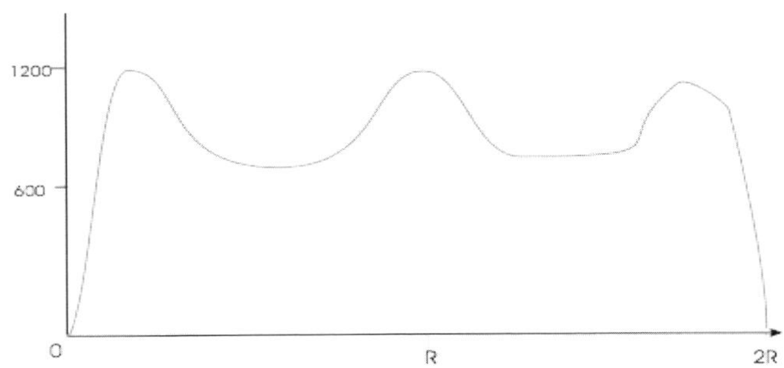

Figure 2 - Radial profile hand scheme of the dry plasmid laying

Two different plasmids were irradiated, the pSB (2961 base pairs) which has a high degree of supercoiling, and the pSP189 (4690 pairs of base), which has a weaker degree of coiling and is more sensitive to the radiation [7].
After the shot sequence, DNA samples are redissolved in distilled water and analyzed by electrophoresis on agarose gels. The electrophoresis is a technique for separating charged molecules in a buffered solution, based on different mobilities in the electric field generated by a direct current. the gel is stained with ethydium bromide, a dye which intercales into DNA and fluoresces. The separated forms of plasmid appear as bands in the gel and are then quantified by image analysis.

4. Results and discussion

The geometry of DNA molecule changes with the number of break in the double helix. The plasmid DNA is supercoiled when undamaged. A single strand break (SSB) relaxes the molecule (open circular shape) and a double strand break (DSB) generates a linear shape of plamsid. A typical image of the gel after migration obtained under a UV lamp is shown in Figure 3.

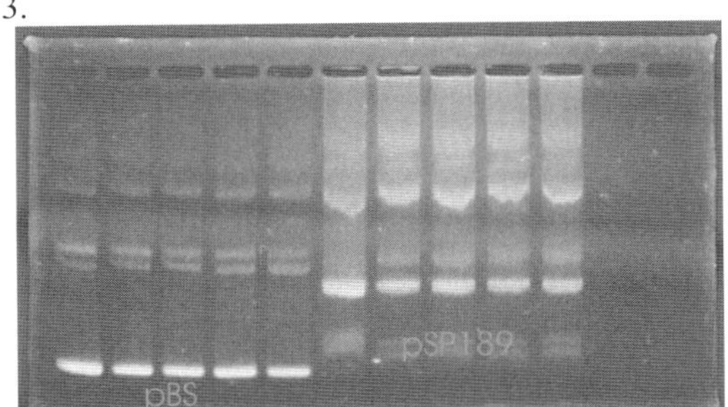

Figure 3 – Electrophoresis image

The analysis of this electrophoresis shows only SSB induction in the pBS irradiated plasmid (increasing amount of relaxed molecules, 2nd band from the bottom) and SSB and DSB in the pSP189 irradiated plasmid.

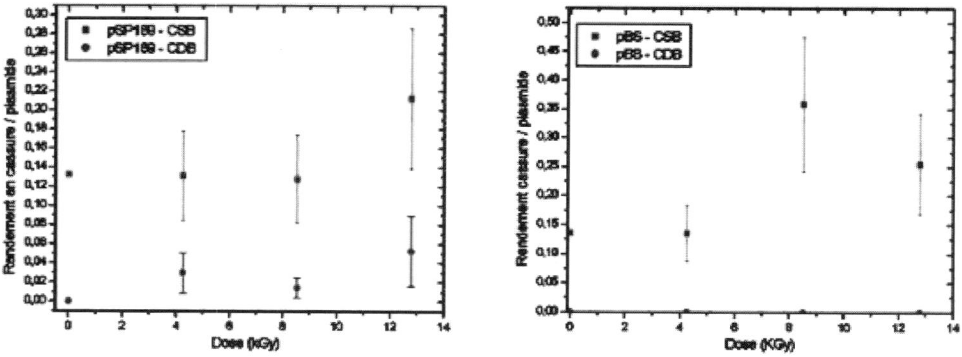

Figure 4 - Yields breaks per plasmid

The variation of the dose was obtained irradiating the plasmid sample, one, two and three times. The present data are too preliminary to conclude about the process involved in the strand breaking in the DNA molecule. We can notice a dose effect for the higher dose. The fraction of the SSB and DSB are too weak. This may partly be due to the fact that the absorption of the X-UV photons occurs within the first 100 nm in DNA layer, whereas the thickness of the DNA sample is more than 600 nm. Consequently a large fraction of the DNA molecules does not receive photons.

5. Conclusion and outlook

This preliminary study encourages us to continue the study of yields damage with higher dose, and we put forward the irradiation XRL of plasmid with a new range of parameters as, flow dose effect, the effect of the number of photons and wavelength variation. The plasmid laying has to be improved to get a thin plasmid monolayer. We only focus on the radiation induced strand breaks, however the base of DNA are also sensitive to radiation and could be revealed by action of specific enzyme. The ability of DNA damage to be repaired could also be studied versus the XRL dose.

References

[1] A. Klisnick, et. al, Meas. Sci. Technol., 12, 1813-1817, 2001.
[2] D. Ros, et. al, LULI report 2001, C8.
[3] A. Chétioui, et. al, Int. J. Radiat. Biol., 65, 511-522, 1994.
[4] M.A. Hervé du Penhouat, et. al, Rad. Res.,151, 649-658,1999.
[5] B. Rus, et. al, Phys. Rev. A, 66, 063806, 2002.
[6] Folkard, et. al, Int. J. Radiat. Biol. **76,** 881-90,1995.
[7] Giustranti, et al, J. Chim. Phys. **96**, 132-137, 1999.
A. Eschenbrenner, unpublished work, 2002.

Inst. Phys. Conf. Ser. No 186
Paper presented at 9th Int. Conf. X-Ray Lasers, Beijing, China, 24–28 May 2004

Hard X-rays holography with atomic resolution

Jianwen Chen*, Hongyi Gao, Honglan Xie, Huafeng Zhu, Ruxin Li and Zhizhan Xu

Shanghai Institute of Optics and Fine Mechanics, Chinese Academy of Sciences, P.O. Box 800-211, Shanghai 201800, China

Abstract. In this work, we discuss angular range and incident energy dependence on spatial resolution of atomic images . By means of digital simulation of a Fe single crystal of body-centred lattice , an X-ray fluorescence hologram is recorded with the high incident energy and wide multiple-energy in a 4π solid angle, the atomic images can be reconstructed from the hologram with a high spatial resolution.

1. Introduction

The basic idea of holography is to obtain both amplitude and phase information from the waves scattered by an object by means of recording its intensity distribution. The most important advantage of holography compared to other imaging techniques is that it stores the 3D information on a 2D plane without optical elements, and this information can be retrieved easily on demand. This would be of great benefit to the study on the spatial order of microscopic objects like molecules or atoms. However, its spatial resolution is limited by the wavelength used and another two factors--which are even more important--the sizes of the source and the detector element, which prevented the researchers from observing 3D inner structure of specimens at atomic scale.

In 1986, Abraham Szöke [1] proposed the concept of X-ray fluorescence holography (XFH). In this idea, a fluorescence hologram is formed by the interference between the direct waves from the photoexcited atoms and the waves scattered by the neighbouring atoms. It is called “inside source holography”. According to the optical reciprocity theorem, the optical beam paths can be reversed and the positions of the source and the detector can then be exchanged. It is regarded as the inverse of XFH and called “inside detector holography”. It is worthy of noting that different names are used for the same methods in some literatures concerning with atomic resolution holography. For short, one calls the inside source and the inside detector holography ‘normal’ and ‘inverse’ holography, respectively. Sometimes one refers to ‘inverse’ holography as ‘multiple energy X-ray holography’ (MEXH).

2. Principle of inside source holography

An atom excited by external radiation emits X-ray fluorescence radiation propagating in the form of a spherical wave. The undisturbed wave reaching the detector directly serves as a reference wave, while the portion of the radiation elastically scattered by the neighbouring atoms constitutes the object wave. The hologram is formed by the interference between the reference and the object waves, as shown in figure1

Unfortunately, making a hologram with a single atomic source is very difficult if not impossible. First of all, we have to excite only one atom in the specimen. A more severe problem is that the intensity of a single atomic source can only be very low and the recording time for a hologram with sufficient statistics becomes prohibitive.

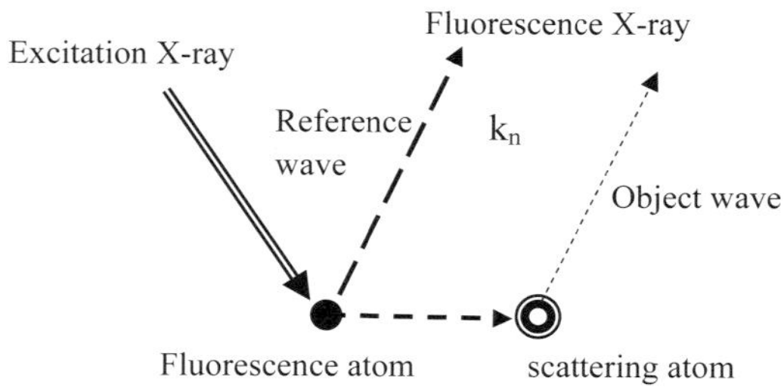

Figure 1 Principle of inside source holography

A solution to this problem is to use a large number of identical sources in identical surroundings instead of one single source. If the spacing between them is much smaller than the interference fringes and the radiation emitted from the atoms is incoherent, then the holograms created by each source are identical and their intensities simply add up at the detected area. Obviously, all single crystals fulfil these criteria.

Tegze and Faigel performed the first X-ray fluorescence holography experiment in 1995[2]. The first successful experiment using the inside detector concept was performed in 1996 at National Synchrotron Light Source of Brookhaven National Laboratory by Gog et al [3].

3. Numerical simulation of X-ray fluorescence hologram

Based on the principle of inside source holography, we numerically simulated the fluorescence hologram of Fe single crystal and reconstructed its atomic images. Fe atoms have a structure of body-centered-cubic with a lattice constant a = 0.29nm. We chose the wave vector of the incident X-ray source k to be $80nm^{-1}$.

The specimen was controlled to rotate in the range of azimuth angle 0~360° with 1° step size and the polar angle 0~180° with 5° step size. Fig. 2(a) is a projection of the spherical coordinates on the k_x-k_y plane. From the picture, we can see the normalized intensity of hologram is of the order of 10^{-3}. This is because small scattering cross-section of X-rays results in feeble holographic oscillations relative to the background. The atomic scattering length $F_j(\Theta_{\bar{r}_j}^{\bar{k}})$ is of the order of Zr_0, where Z is the atomic number of the scattered and $r_0 = 2.8\times10^{-6}$nm is the classical electron radius. For an element with a medium atomic number, it gives $F/r \cong 10^{-3}$. This means the holographic oscillations from a single atomic scattered is near the order of 10^{-3} relative to the background. The total oscillations from a number of scattering atoms may add up to a higher signal level than a single atom. There are eight scattering atoms in our theoretical model, so the holographic oscillation will be of the order of 10^{-3} relative to the background. Indeed the holographic oscillation is so feeble in the angular distribution of the fluorescence intensity that it is difficult to recognize the fine structure of the hologram from the raw data. So, many researchers adopted low-pass filters in the process of reconstruction to extract hologram from the noisy data.

From the simulated hologram of Fe atoms, the real space image is reconstructed by means of the Helmholtz-Kirchhoff theorem. Fig.2(b) is the reconstruction image of Fe atoms in the plane parallel to the (010) crystallographic plane at y = 0.145nm. Like other traditional optics holography, the two methods for a single energy also suffer from the classical problem of twin images, which dates from the original work of D. Gabor.

Barton proposed that properly phase combinations of images reconstructed from holograms recorded at several incident energies suppress twin image effects [4-5].

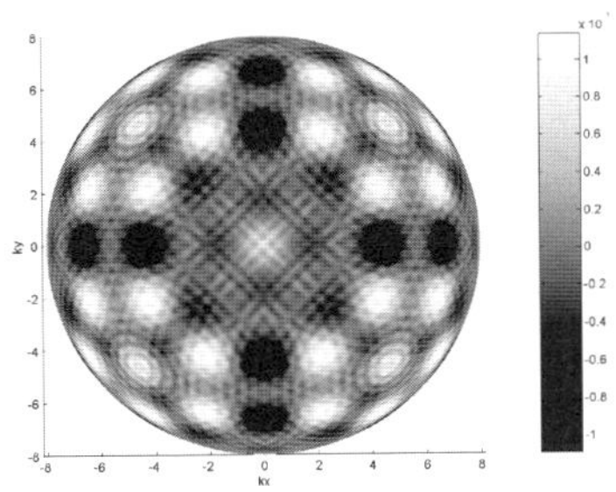

Figure 2(a) Simulated X-ray fluorescence hologram of a Fe atom in single crystal, a projection of the spherical coordinates on the k_x-k_y plane.

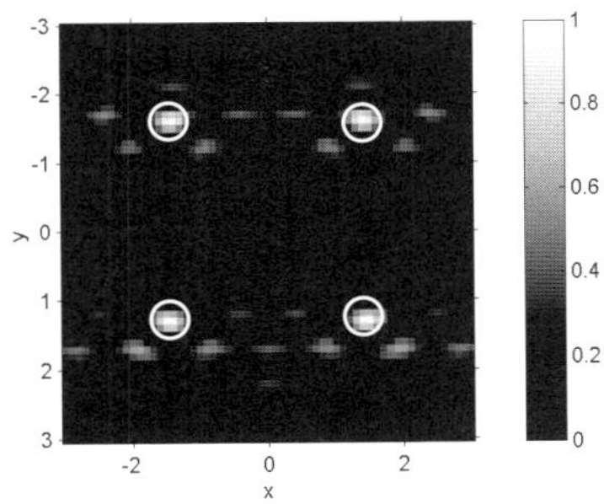

Figure 2(b) Holographic reconstruction of Fe atoms in the plane parallel to the (001) crystallographic plane at z = 0.145nm.

Discussions of early theoretical simulations about comparison of atomic images in single energy and multiple energy x-ray holography (MEXH) have been given in Reference [6] for two atoms and in Reference [7] for a cluster of bcc iron respectively. These simulated results have proved that MEXH can suppress the twin images effects. However, the influence of the magnitude and the range of incident energies on reconstructed atomic images in MEXH has not been discussed. Moreover, in the present state, the experimental results [8] are not accurate enough to discuss the actual atomic positions, due to the measurement technique of the weak holographic signal. For example, the deviation of the intensity maximal from the expected positions is less than 0.1Å in reference [9] and ± 0.2 Å in reference [8]. In the work, the influences of the range and the magnitude of incident energies on reconstructed atomic images are discussed with a single Fe atom and a Fe single crystal used as models for calculations respectively. The results show that with the increase of incident energy, the atomic images are reconstructed with an increasing resolution. At the same time, the difference between their positions and that of the actual atoms decreases. It can be reduced to 10^{-2}Å with holograms recorded at high and wide-range incident energy.

We first simulate a hologram of a single scattering Fe atom at $z = 2.90$ Å with the fluorescing atom at the origin, recorded at a single energy $k = 8.0$ Å^{-1} (E=15.8keV). Based on the Helmholtz-Kirchhoff theorem, the real-space image of the Fe atom can be reconstructed from the hologram and shown in Fig. 3 (a). Obviously, there are two equal intensity maximal at $z = 2.90$ Å and $z = -2.90$ Å respectively. The atomic image at $z = 2.90$ Å is the real image and the other one is its twin image. If holograms are recorded at widely distributed incident energies, twin image effects can be suppressed, as well as other aberrations and artefacts in reconstructed images.

In order to eliminate the twin image, we calculate the atomic image reconstructed from holograms recorded at multiple energies. Fig. 3 (b) shows the atomic image reconstructed from eleven holograms recorded with the initial incident energy $k = 28.0$ Å^{-1} and the step 0.5Å^{-1}. In comparison with image in Fig. 3 (a), the background noise almost totally disappears and the position of the atomic image is closer to that of the actual atom. The resolution is improved to as high as 0.10Å. This result shows that the increase of the incident energy in MEXH will increase the resolution of atomic image and decrease the difference between positions of reconstructed intensity peak and the actual atom.

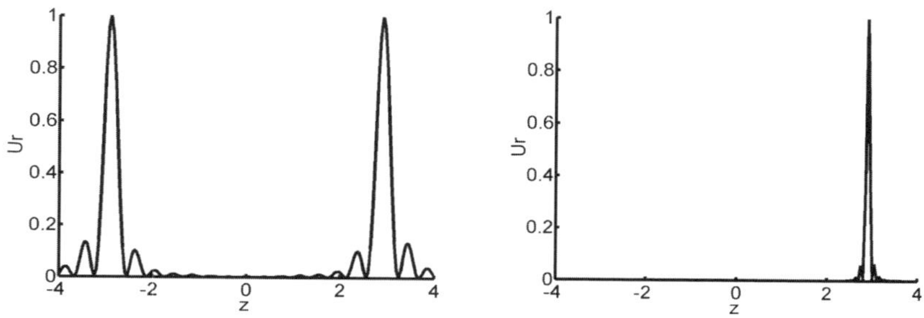

3(a) 3(b)

Figure 3. The reconstructed atomic image of a Fe single atom at $z = 2.90$ Å
(a) The atomic image reconstructed from the hologram recorded at $k = 8.0\,Å^{-1}$
(b) The atomic image reconstructed from eleven holograms recorded at $k = 28.0 - 33.0\,Å^{-1}$.

From above simulations, it is found that the wider the range of incident energies, the better the effect of removing twin images. Moreover, with the increase of incident energy, the resolution of reconstructed atomic images increases and the difference between positions of calculated intensity peak and actual atom decreases. This difference can be reduced to the order of 10^{-2}Å. In summary, with holograms recorded at high and wide-range incident energy, MEXH can provide 3D atomic images of single crystals with a high resolution and high accuracy.

References

[1] A. Szöke, 1986, "In short wavelength coherent radiation: Generation and application", edited by Atwood, T. and Boker, J., AIP Conference Proceedings, 147, 361-367.
[2] Tegze, M and Faigel, G., 1996. Nature, 380, 49-51.
[3] T. Gog, P.M. Len et al., 1996 ,Phys. Rev. Lett., 76, 3132 –3135.
[4] J. J. Barton, 1991, Phys. Rev. Lett., 61, 1356-1359.
[5] J. J. Barton, 1991, Phys. Rev. Lett., 67, 3106-3109.
[6] B. Adams, D. V. Novikov et al, 1998, Phys. Rev. B, 57, 7526-7534.
[7] P. M. Len, T. Gog, et al., 1997, Phys. Rev. B, 55, R3323-R3327.
[8] T. Hiort, D. V. Novikov, E. Kossel et al, 2000, Phys. Rev. B, 61, R830-R833.
[9] M. Tegze, G. Faigel, S. Marchesini et al., 1999, Phys. Rev. Lett., 82: 4847-4850. 3

Inst. Phys. Conf. Ser. No 186
Paper presented at 9th Int. Conf. X-Ray Lasers, Beijing, China, 24–28 May 2004

Picosecond x-ray laser photoelectron spectroscopy of room temperature and heated materials

J. Dunn, A.J. Nelson, T. van Buuren, and J.R. Hunter
Lawrence Livermore National Laboratory, Livermore, CA 94551, USA

Abstract. An 84.5 eV Ni-like Pd ion 4d – 4p x-ray laser source generated by the LLNL Compact Multipulse Terawatt (COMET) tabletop system has been used to probe the electronic structure of various metals and semiconductors. In addition to the ~4 – 5 ps time resolution, the probe provides the necessary high photon flux (>10^{12}/pulse), narrow line width ($\Delta E/E \sim 2 \times 10^{-5}$) and coherence for studying valence band and shallow core electronic structure levels in a single shot. We show some preliminary results of room temperature and heated thin foil samples consisting of 50 nm Cu coated on a 20 nm C substrate. A 527 nm wavelength 400 fs laser pulse containing 0.1 – 2.5 mJ laser energy is focused in a large 500 × 700 μm^2 (FWHM) spot to create heated conditions of 0.07 – 1.8 × 10^{12} W cm^{-2} intensity.

1. Introduction

X-ray and EUV induced photoemission spectroscopy (PES) is a powerful surface analysis tool for measuring the electronic structure of matter. The radiation photoionizes the material and the ejected electrons are subsequently detected in a photoelectron spectrometer. Different sources have been used over the years ranging from synchrotrons [1, 2], laser-produced plasmas [3, 4], laser-produced higher order harmonics (HOH) [5 – 7] to x-ray lasers [8]. The main goal has been to develop sources that have a sufficiently large photon number, a mono-energetic line and in a short pulse duration. The major advantage of using the laser-driven sources, particularly the HOH and x-ray lasers, is that the soft x-ray pulse duration can be in the ultrafast regime from below 100 fs to a few ps. This short pulse duration is required to conduct pump-probe experiments at room temperature or materials that have been excited or heated on fast timescales. For example, an optical laser pulse can be used to excite the material while the x-ray beam probes the induced changes. The time resolution in the technique is determined by the intrinsic x-ray pulse duration and over a series of shots the probe can be delayed to study the excited matter evolution [5, 6].

Our objective in recent years has been to achieve improved single-shot PES data based on a pulsed Ni-like Pd ion, $4d$ – $4p$ 85 eV x-ray laser used in conjunction with electron time-of-flight (eTOF) spectroscopy. The soft x-ray laser has a high photon flux (>10^{12}/pulse), 4 – 5 ps pulse duration and narrow line width ($\Delta E/E \sim 2 \times 10^{-5}$). This photon energy has a large photo-ionisation cross-section for valence band and shallow core level electrons with binding energies up to ~40 eV [9]. We have recently performed pump-probe experiments where the focus now is to extend the technique into dynamic studies of materials. We report on our first results x-ray laser PES data of room temperature and heated thin Cu foils.

2. Experimental Methods and Results

The experimental station at the end of the x-ray laser beam line was recently re-built [10] and a new eTOF spectrometer, designed and fabricated by the University of Nevada Reno group, installed [11]. Figure 1 shows a schematic of the x-ray laser photoelectron

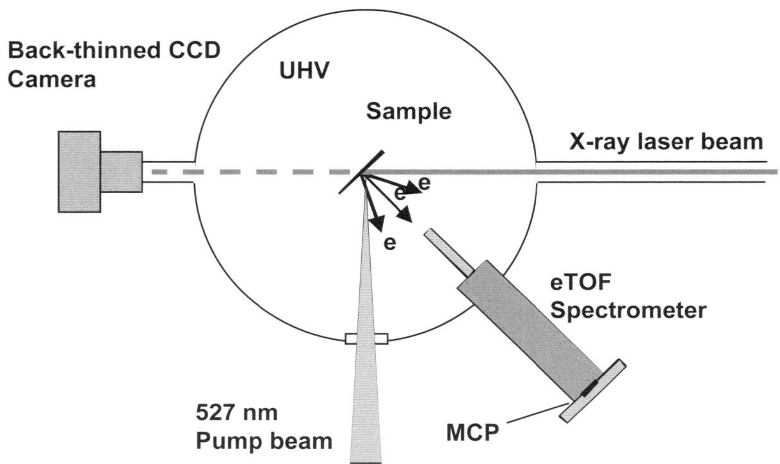

Figure 1. Schematic of x-ray laser photoelectron spectroscopy experimental setup.

spectroscopy experimental setup. The x-ray laser probe beam is collimated by a normal incidence Mo:Si multilayer spherical mirror and relayed along the vacuum beam line by a 45° Mo:Si multilayer flat mirror. The narrow reflectivity window of both mirrors selects the x-ray laser wavelength. A beam block in the x-ray laser chamber also minimizes other plasma x-rays reaching the sample. A pinhole isolates the beam line from the experimental ultra-high vacuum (UHV) chamber, which typically operates below 5×10^{-8} mbar pressure. The x-ray laser is aligned using a back-thinned CCD camera to a cross wire fiducial on the sample manipulator. A fast 400 fs, 527 nm wavelength laser pump beam can be focused and aligned similarly to overlap the two beams onto the sample. The optical pump laser is timed relative to the x-ray laser beam at the sample to within ~5 ps. The pump beam produces rapid heating and depending on the intensity can excite bound electrons, induce phase changes or ionise the sample. The x-ray laser beam can probe the material at various times generating the photoelectrons from the sample surface. These low energy photoelectrons, with K.E. < 84.5 eV, emitted from the valence band and shallow core levels are detected by the e-TOF spectrometer, the operation of which is described in detail elsewhere [11]. The signal is acquired and digitized using a fast 3 GHz oscilloscope. Due to the surface sensitivity of the technique, where the electron escape depth is typically in the range of 0.5 – 2 nm, surface oxidation and contamination are removed by UV ozone cleaning prior to analysis.

The ejected photoelectrons have discrete kinetic energy $E_k = h\nu - E_b - \phi$, where $h\nu$ is the energy of the incident x-ray laser photon (84.5 eV), E_b is the binding energy of the photoelectron relative to the Fermi level. The work function, ϕ, varies for each material and is 4.1, 4.6 and 4.8 eV for Ta, Cu and Ge, respectively (all materials that we have studied). By definition the Fermi level is used as the zero binding energy level. Each photoelectron has a discrete energy representative of the element from which it was emitted thus allowing one to identify the atomic species present. Furthermore, the binding energy of an electron is influenced by its chemical environment and thus can be used to identify the chemical state of a given atom in the sample [4]. Oxide layers in particular are prevalent for many materials. In some cases a scattered x-ray peak may be observed tens of nanoseconds before the photoelectrons and this prompt signal can be used as a timing fiducial. The valence band electrons have the highest kinetic energy and are the first arrival photoelectron signals in the time-of-flight technique. The shallow core levels arrive after the valence band electrons. Another common feature observed is a broad continuum background that originates from secondary electrons from deeper within the material that undergo multiple scattering and generate low energy electrons.

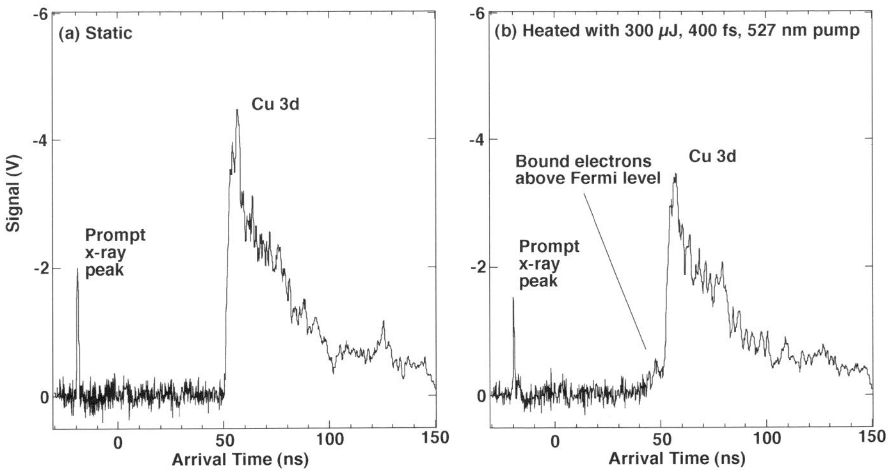

Figure 2.(a) X-ray laser generated photoelectron spectrum of 50 nm Cu on 20 nm C substrate thin foil at room temperature (static). (b) Photoelectron spectrum of same thin foil heated by 300 μJ, 400 fs, 527 nm pump beam focused at 2×10^{11} W cm^{-2} intensity.

Figure 2 is an example of very recent data that was taken from a thin foil sample consisting of 50 nm Cu coated on a 20 nm C substrate. This sample is sufficiently thin to observe the transmitted x-ray laser beam simultaneously with the photoelectron signal. The room temperature or static sample, Fig. 2(a), is shown without the optical heating beam. The peak at -20 ns is from scattered x-ray laser photons being detected by the microchannel plates at the bottom of the drift tube of eTOF spectrometer. The abrupt onset of the first photoelectron signal is recorded at ~50 ns. The peak associated with the Cu 3*d* level is evident in the valence band. The experiment was repeated again with the 400 fs, 527 nm optical pump beam focused onto a large $500 \times 700\ \mu m^2$ (FWHM) spot on the same sample. The x-ray laser probe and optical pump were timed to coincide on the sample. Approximately 300 μJ of laser energy corresponding to 2×10^{11} W cm^{-2} intensity was used on this shot. It is noted that the x-ray laser beam overfilled the optical beam on the sample so that there are detected photoelectron signals from heated and unheated regions of the foil. It is observed that the $3d$ peak intensity has been lowered and there exists a small foot just before the strong onset at ~50 ns. The latter feature is associated with bound electrons above the Fermi level which are being excited by the pump beam. Increasing the pump energy by a further factor of ten generates a strong electron signal before 50 ns that indicates the sample is in an ionized state. These experiments can be repeated by changing the pumping parameters. For example, by controlling the pump energy different states of matter can be studied. Further experiments are ongoing to study the relaxation processes at different time delays between the probe and the pump.

3. Conclusions

In summary, we have shown that picosecond x-ray lasers can be used as a surface analysis probe for metals and semiconductors. The pulsed x-ray source photoionizes the valence band and shallow core level electrons that are subsequently detected in a high resolution electron time-of-flight spectrometer. The first pump-probe experiments have been carried out where an optical laser heats a thin Cu foil sample and excites electrons above the fermi level.

Acknowledgments

This work was performed under the auspices of the U.S. Dept. of Energy by the University of California Lawrence Livermore National Laboratory under Contract No. W-7405-Eng-48.

References

[1] R. Z.Bachrach et al., Jn. Vac. Sci. Technol. **12**, 309 – 312 (1975).
[2] T.E. Glover et al., Phys. Rev. Lett. **90**, 236102 (2003).
[3] H. Kondo et al., Appl. Phys. Lett. **69**, 182 (1996).
[4] H. Kondo et al., Appl. Phys. Lett. **72**, 2688 (1998).
[5] A. Rettenberger and R. Haight, Phys. Rev. Lett. **76**, 1912-1915 (1996).
[6] M. Bauer et al., Phys. Rev. Lett. **87**, 025501-1 (2001).
[7] T. Munakata et al, Surface Science **532–535**, 1140 – 1144 (2003).
[8] J. Dunn et al., X-ray Lasers 2002, AIP Conf. Proc. **CP641**, ed. J.J. Rocca, J. Dunn, and S. Suckewer 481 – 488 (2002).
[9] J.J. Yeh and I. Lindau, At. Dat. Nucl. Data Tabl. **32**, 1 (1985)
[10] A.J. Nelson et al., Soft x-ray lasers and Applications V, SPIE Int. Soc. Opt. Eng. Proc, vol. **5197**, ed. E.E. Fill and S. Suckewer, 168 – 173 (2003).
[11] O. Hemmers et al., Rev. Sci. Instrum. **69**, 3809 - 3807 (1998).

Inst. Phys. Conf. Ser. No 186
Paper presented at 9th Int. Conf. X-Ray Lasers, Beijing, China, 24–28 May 2004

Propagation property of a femtosecond pulse train in a two-level medium

Bingxin Liu, Xiaohong Song, Shangqing Gong, Ruxin Li and Zhizhan Xu
Laboratory for High Intensity OpticsShanghai Institute of Optics and Fine MechanicsShanghai 201800, China

Abstract. The spectra of a femtosecond pulse train propagating in a two-level medium are investigated by solving the full Maxwell-Bloch equations. It is found that the higher spectral components depend crucially on both the relative shift Φ and the delay time τ between the successive pulses of the femtosecond pulse train.

1. Introduction

With the development of ultrashort optical pulses, a number of fascinating nonlinear optical propagation effects have been found. Ziolkowski *et al* [1] and Hughes [2] investigated the propagation properties of ultrashort pulses. It was demonstrated that for small area pulses, such as 2π and 4π pulses, the self-induced transparency (SIT) results are still recovered. While for large area pulses, Hughes proposed the area under individual carriers might themselves cause Rabi oscillation (RO), which can lead to carrier-wave reshaping and significantly higher spectral components [2] and even soft-X-ray generation [3]. Moreover, when a pulse with a few optical cycles is considered, the carrier-envelope phase (the absolute phase) will dramatically affect the temporal shape of electric field and the spectra [4-6]. Because intense laser-matter interactions depend on the electric field of the pulse, the absolute phase is important for a number of nonlinear processes, such as high harmonic generation [7–10], final molecular state population [11,12], etc. Furthermore, we have studied the spectra effects for two-color ultrashort pulses in a TLA medium. It is demonstrated that the interference effect of the two pulses can substantially modify the behavior of the spectra: the higher spectral components can occur even for two 4π pulses; Moreover, the resulting spectral feature depends crucially on the relative phase of the two pulses [13].

In the recent years, the research of ultrashort pulse trains has attracted much interest. Theoretical and experimental methods of designing the pulse trains have been investigated. Moreover, femtosecond pulse trains [14,15] and attosecond pulse trains [16,17] have been demonstrated, produced and applied theoretically and experimentally. If the ultrashort pulses are mutually coherent, they lead to interference effect that can lead to a large modification of the total field when the elementary pluses of the pulse train overlap while propagating [18], and can consequently result in many spectral effects. The interference effect can be affected by the relative phase Φ and the delay time τ between the successive pulses. Recently, it has been demonstrated theoretically [19] and experimentally [15] that dispersion of a femtosecond pulse train in a resonant medium depends crucially on the relative phase shift Φ between the pulses; by controlling the τ between the successive pulses, attosecond pulse trains have been used as natural tools for controlling strong field processes and the manipulation of the time-frequency characteristics of high harmonics at single atom has been demonstrated [20]. Here we will show that the spectral effects for a femtosecond pulse train propagating in a resonant TLA medium, depend crucially on both the relative shift Φ and the delay time τ between the successive pulses of the train.

2. Model and equations

The propagation property of laser light field in a resonant TLA medium with an atomic density N can be modeled using the M-B equations:

$$\partial_t H_y = -\frac{1}{\mu_0}\partial_z E_x, \tag{1}$$

$$\partial_t E_x = -\frac{1}{\varepsilon_0}\partial_z H_y - \frac{1}{\varepsilon_0}\partial_t P_x, \tag{2}$$

$$\partial_t u = -\gamma_2 u - \omega_0 v, \tag{3}$$

$$\partial_t v = -\gamma_2 v + \omega_0 u + 2\Omega w, \tag{4}$$

$$\partial_t w = -\gamma_1 (w - w_0) - 2\Omega v, \tag{5}$$

Here E_x and H_y are the electric and magnetic fields, γ_1 and γ_2 are the population and polarization relaxation constants, respectively. $\Omega = dE_x/\hbar$ is the Rabi frequency, and ω_0 is the transition frequency of the TLA medium. The macroscopic nonlinear polarization $P_x = Ndu$ is related to the off-diagonal density matrix element $\rho_{12} = (u + i\upsilon)$ and the population difference $w = \rho_{22} - \rho_{11}$ between the upper and lower states; d is the dipole moment. The refractive index is determined by the real part of ρ_{12} and the gain coefficient is proportional to the imaginary part of ρ_{12}. We employ a standard finite-difference time-domain approach for solving the full-wave Maxwell equations, and predictor-corrector method to solve the Bloch equations. The time and space increments, Δt and Δz, are chosen to ensure $c\Delta t \le \Delta z$ [1].

The electric field of the pulse train is $E_x(z,t)$ with $E_p(z,t)$, the amplitude of the total field that can be written as [19]:

$$E_x(z,t) = \sum_{p=0}^{M} E_p(z,t), \tag{6}$$

with

$$E_p(z,t) = E_0(z,t - p\tau)e^{ip\Phi}. \tag{7}$$

In this expression $E_p(z,t)$ is the amplitude of pulse (p, $p \le M$), here we suppose that $M = 24$; τ and Φ are, the time delay and the phase shift respectively between two successive pluses of the train. The expression of the total field $E_x(z,t)$ through the optically medium depends on the field $E_0(z,t)$ corresponding to $\Omega_0 = dE_0/\hbar$. The condition for a single laser pulse is

$$\Omega_0(t=0,z) = \Omega_m \cos[\omega(z+z_0)/c]\ \mathrm{sec}h[1.76(z/c + z_0/c)/\tau_0]\ ,$$

where the frequency ω is the fundamental laser field frequency, Ω_m is the maximum Rabi frequency of the fundamental field and $\tau_1 = 2ar\cosh(1/\sqrt{0.5})\tau_0$ is the full width at half-maximum (FWHM) of the pulse intensity envelope of the single pulse laser. The single pulse area is $A = \Omega_m \tau_1 \pi / 1.76$, $\Omega_m = 1 fs^{-1}$ corresponds to the electric field of $E_0 = 2.5 \times 10^7 V/cm$ or an intensity of $I = 1.6 \times 10^{12} W/cm^2$.

In the following numerical analysis, all the material and laser parameters we adopt are based on Refs. [2,3]: $\tau_1 = 18 fs$, $\omega_0 = 1.2\times10^{15} rads^{-1}$, $\omega = 1.2\times10^{15} rads^{-1}$, $z_0 = 20\mu m$, $d = 2.65e\mathring{A}$, $N = 4\times10^{18} cm^{-3}$, $\gamma_1^{-1} = \gamma_2^{-1} = 1ns$. The medium is initialized with $u = \upsilon = 0$, $\omega_0 = -1$ at $t = 0$, and the choice of z_0 ensures that the pulse penetrates negligibly into the medium at $t = 0$. For such laser intensity, recent experiments on semiconductors have shown that a description in terms of two-level systems has been able to reproduce the experimental results amazingly well [21,22]. Moreover, a few important publications that study this model exist [2,3,23–26].

3. Spectra and explanation of a 2π pulse train with different Φ

When the pulse train is considered, the interference effect between the pulses can lead to a large modification of the total field and influence the spectra accordingly. The relative phase and the delay time play great roles on the interference effect, all these can lead to interesting results.

Figures 1(a)-1(b) present the spectra produced by the femtosecond pulse train (a single $\Omega_m = 0.344 fs^{-1}$ corresponding to $A = 2\pi$, with the same delay time $\tau = 5 fs$) at the propagation distance of $z = 60\mu m$ with the different initial relative phases $\Phi = 0$ and $\pi/5$ of the pulse train, respectively. All these figures clearly show the formation of higher spectral components, and oscillatory features can be found apparently. It is also shown that when the relative phases vary from 0 to $\pi/5$, the higher spectral components become fewer and the intensities get smaller gradually. The trends continue even if the relative phases go on varying from $\pi/5$ to π. Furthermore, similar results can be achieved for other propagation distances.

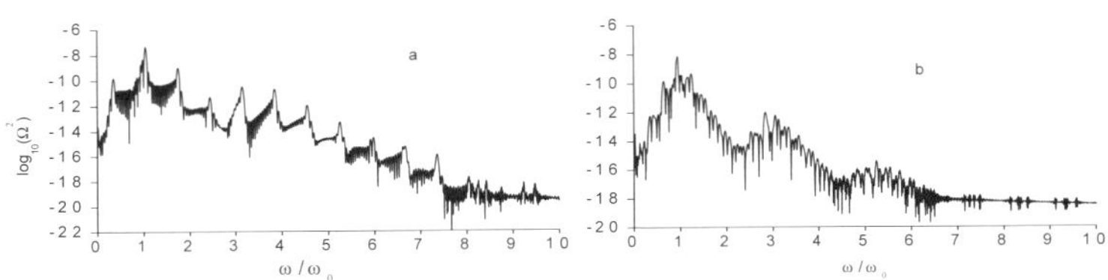

Fig.1

Figures 1(a)-1(b) present the spectra produced by the femtosecond pulse train at the propagation distance of $z = 60\mu m$ with the different initial relative phases $\Phi = 0$ and $\pi/5$ of the pulse train, respectively.

4. Spectra of a 2π pulse train with different τ

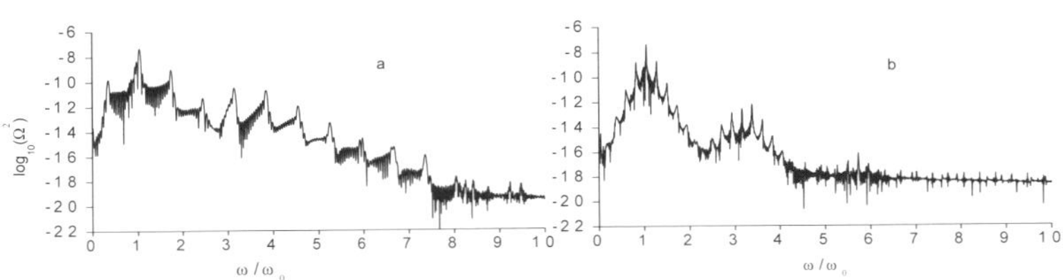

Fig.2

Figures 2(a)-2(b) present the spectra produced by the femtosecond pulse train with the delay time $\tau = 5 fs$ and $15 fs$, respectively, with the same phase $\Phi = 0$ between the successive pulses.

As well, the spectral effects depend crucially on the delay time τ. Figures 2(a)-2(b) present the spectra produced by the femtosecond pulse train (a single $\Omega_m = 0.344 fs^{-1}$ corresponding to $A = 2\pi$, with the same phase $\Phi = 0$) with the delay time $\tau = 5fs$ and $15fs$, respectively, between the successive pulses. It is clear in Fig.3 that there are shark spectra peaks in higher intensities with $\tau = 5fs$. The delay time τ varying from $5fs$ to $15fs$, the higher spectral components grow fewer and the intensities become smaller.

5. Conclusions

In this paper, we investigated the spectra for a femtosecond pulse train propagating in a resonant two-level medium by solving the full Maxwell-Bloch (M-B) equations .We demonstrate that the interference effect of the pulse train can substantially modify the behavior of the spectra: higher spectral components can occur even for a 2π femtosecond pulse train. Moreover, the features of these higher spectral components depend crucially on the relative phase Φ and the delay time τ between the successive pulses.

Acknowledgements

The work is supported by the National Key Basic Research Special Foundation of China (NKBRSFC) (Grant No. G1999075200) and the National Natural Science Key Foundation of China (Grant No. 10234030).

References

[1] Ziolkowski R W *et al.* 1995 Phys.Rev.A. 52 3082.
[2] Hughes S 1998 Phys. Rev. Lett. **81** 3363.
[3] Hughes S 2000 Phys. Rev. A. **62** 055401.
[4] Bohan A D et al.1998 Phy.Rev.Lett. 81(9) 1837-1840.
[5] Jones D J 2000 Science. **288** 635.
[6] Paulus G G *et al.* 2001 Nature **414** 182(London).
[7] Tempea G, Geissler M, and Brabec T 1999 J. Opt. Soc. Am. B. **16** 669.
[8] Potvliege R M, Kylstra N J, and Joachin C J 2000 J. Phys. B .**33** (L) 7436.
[9] Salieres P *et al.* 1998 Phys. Rev. Lett. **81** 5544.
[10] Mehendale M *et al.* 2000 Opt.Lett. **25** 1672.
[11] Brown A and Meath W J 1998 J. Chem. Phys. **109** 9351.
[12] Brown A, Meath W J, and Kondo A E 2002 Phys.Rew.A. **65** 060702.
[13] Song X H, Gong S H, Jin S Q and Xu Z Z 2004 Phys.Rew.A. **69** 015801.
[14] Assion A, Baumert T et al 1998 Scinece 282 919.
[15] Jacquey M, Bonhommeau S, and Bouchene M A 2003 Optics Lett. 28 14.
[16] Wu Jian and Zeng Heping 2003 Phy. Rew. A. **68** 063823.
[17] Paul P M *et al.* 2001 Science. **292** 1689.
[18] Spano F, Haner M, and Warren W S 1987 Chem.Phys.Lett. **135** 97.
[19] Mohamed.Aziz Bouchene 2002 Phys.Rev.A. 66 065801.
[20] Kenneth J Schafer and Mette B Gaarde 2004 Phy.Rev.Lett. 92 023003.
[21] Mucke O D *et al.* 2002 Phys. Rev. Lett. **89** 127401.
[22] Tritschler T *et al.* 2003 Phys. Rev. Lett. **90** 217404.
[23] Kalosha V P and Herrmann J 1999 Phys. Rev. Lett. **83** 544.
[24] Tritschler T, Mucke O D, and Wegener M 2003 Phys. Rev. A. **68** 033404.
[25] V.P. Kalosha, M. Muller, and J. Herrmann 1999 J. Opt. Soc. Am. B. **16** 323.
[26] Xiao J , Wang Z, and Xu Z 2002 Phys. Rev. A. **65** (R) 031402.

Inst. Phys. Conf. Ser. No 186
Paper presented at 9th Int. Conf. X-Ray Lasers, Beijing, China, 24–28 May 2004

Pressure Gradient Effects On Two-Dimensional Plasma Expansion

S. Moon, R.F. Smith, J. Dunn, R. Keenan, J. Nilsen, J. R. Hunter
Lawrence Livermore National Laboratory, Livermore, CA 94551

J. Filevich, J.J. Rocca, M.C. Marconi
NSF ERC for Extreme Ultraviolet Science and Technology and Department of Electrical and Computer Engineering, Colorado State University, Fort Collins, Colorado 80523

V.N. Shlyaptsev
University of California Davis-Livermore, Livermore, CA 94551

Abstract. Recent advances in interferometry has allowed for the characterization of the electron density expansion within a laser produced plasma to within 10 μm of the target surface and over picosecond timescales. This technique employs the high brightness output of the transient gain Ni-like Pd collisional x-ray laser at 14.7 nm to construct an effective moving picture of the two-dimensional (2-D) expansion of the plasma. We present experimentally measured density profiles of an expanding Al plasma generated through laser irradiation in a 14μm line focus geometry. Significant lateral expansion was observed at all times as well as a pronounced on-axis electron density dip. Detailed modelling with a 2-D plasma physics code gives good agreement to experimental observations. Large pressure gradients associated with the tight focal spot conditions are calculated to dominate in shaping the plasma density profile.

1. Introduction

Understanding the underlying physical processes driving the expansion of laser produced plasmas is of considerable practical and fundamental interest. Combining well characterized experimental observables with detailed simulations can give a quantitative measure of conditions within the plasma. Interferometry is a powerful tool for accurately diagnosing the two-dimensional (2-D) evolution of dense laser-produced plasmas. The recently developed technique of picosecond x-ray laser interferometry combines the short wavelength of the 14.7 nm Ni-like Pd laser with diffraction grating interferometer (DGI) instrumentation to enable the direct measurement of the electron density profile within a laser-produced plasma to within 10 μm of the target surface [1]. The picosecond duration of the x-ray laser probe pulse allows for an effective snapshot of the electron density profile while reducing the effects of plasma motion blurring at the ablation front. Also, the short wavelength, λ, and high brightness of the probe beam allows for characterization of large plasmas at high density with less deleterious effects from free-free absorption, $\propto \lambda^3(1-\exp[-hc/\lambda kT])$, and refraction ($\propto \lambda^2$) [1]. This technique produces a series of high quality 2-D density measurements providing unambiguous characterization of the time evolution in a fast evolving plasma for validation of existing 2-D hydrodynamic codes.

2. Experiment

The electron density evolution of a laser-heated Al plasma is measured using a diffraction grating interferometer (DGI) [2] at different times, relative to the peak of a 600ps plasma forming pulse. This gives an effective moving picture of how the plasma evolves in space and in time. The experimental results show pronounced two dimensional effects such as significant lateral transport and an on-axis density dip. We use 2-D hydrodynamic simulations in order to calculate the contributions of various physical mechanisms driving the plasma expansion. Density depressions have been observed before for high intensities laser drivers (pondermotive force [3]) and for lower intensities were the pumping pulse lasts for several nanoseconds (coronal x-ray ray heating [4]). For the plasma conditions explored here, simulations suggest that the on–axis density dip can be explained by rapid movement of material laterally due to the large pressure gradients set up by the tightly focused laser beam.

The Ni-like Pd 14.7 nm x-ray laser probe beam and the plasma to be studied were generated using three laser beams at 1054 nm wavelength from the COMET facility at LLNL [1]. Single pass saturated x-ray laser output of a few 10's of μJs was achieved with an optical pumping combination of a 600 ps long pulse (2 J, 2×10^{11} W cm^{-2}) and a 13 ps (5 J, 3.5×10^{13} W cm^{-2}) main heating pulse. The x-ray laser output was imaged and routed into a diffraction grating interferometer for plasma probing experiments. Recent temporal measurements [5] of the output of the Ni-like Pd x-ray laser have shown a ~4 ± 0.5 ps output for the pumping conditions detailed above. For interferometry this sets the time period over which the electron density is sampled. For a detailed description of the interferometer instrumentation see ref [2]. A plasma heated by up to 3J of energy in a 600 ps, 1054 nm pulse, corresponding to a maximum intensity of 1×10^{12} W cm^{-2}, was produced in one arm of the interferometer. A 3.2 mm long line focus with a 14 μm focal width, was generated on a polished 1mm long Al slab target using a combination of a cylindrical lens, f = -200 cm, and an off-axis paraboloid, f = 30 cm. The temporal profile of the 600ps plasma forming beam along with a cross section of the laser line focus are used as inputs into 2-D Lasnex simulations. The relative delay between the arrival of the x-ray laser probe pulse, to the peak of the plasma forming beam was measured to within 100 ps with a fast diode. The x-ray laser could probe the plasma in the temporal range of –1 ns to +2 ns relative to the peak of the 600 ps plasma forming pulse by adjusting a delay arm in the plasma laser beam. The line focus plasma was probed longitudinally by the x-ray laser, thereby minimizing uncertainties in the interpretation of the interferograms arising from plasma gradients along the probe path. The plasma was imaged with a magnification of ~22 by a 25 cm focal length spherical multilayer mirror and relayed to a thinned back-illuminated 1024 × 1024 CCD detector with 12.7 × 12.7 μm^2 pixels. Using the x-ray laser beam with no plasma present, high quality fringes, with visibility $V = (I_{max} - I_{min})/(I_{max} + I_{min})$ of 0.72 ± 0.12, were observed for a 700 × 500 μm^2 (H × V) region.

3. Results

Figure 1 shows a series of interferograms at different probing times relative to the peak of the plasma forming pulse. The electron density, n_e in cm^{-3}, is related to the measured fringe shifts as $N_{fringe} = 6.68 \times 10^{-20}\, n_e\, L$, where L is the length (cm) of the plasma being probed by the 14.7 nm x-ray laser [6]. For the 1mm plasma lengths described here, one fringe shift is equivalent to an electron density of 1.5×10^{-20} cm^3. An estimated uncertainty of 10 - 20 % in the determination of the reference fringes results in an error of $1.5 - 3 \times 10^{19} cm^{-3}$ in the electron density. An additional phenomenon that can introduce errors in plasma probing experiments is the extent to which the probe beam is deflected by refraction in the plasma medium. Raytracing [7] has shown that this effect is negligible for the density gradients and plasma lengths considered here. The contribution of bound electrons to the refractive index within the plasma is calculated to be negligible for the plasma conditions studied here.

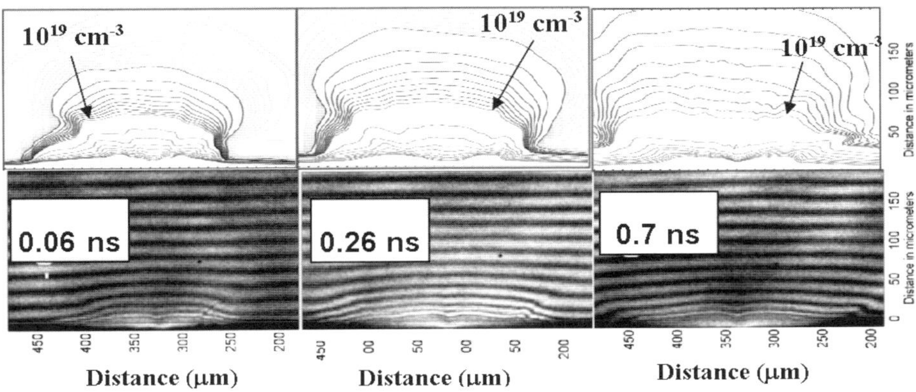

Fig 1. Experimentally obtained interferograms taken through longitudinal probing of an Al plasma at 0.06, 0.26 and 0.7ns after the peak of the 600ps plasma forming pulse. Strong lateral expansion is observed from the original 14μm focal spot. Less than 20μm from the target surface there is a clear formation of a depression in the free-electron density. Shown in the three contour plots above are the calculated two-dimensional electron density profiles from the LASNEX plasma physics code. The extracted electron density contour map is shown in a logarithmic scale with each line connecting regions of equal density. Experimentally obtained spatial and temporal profiles for the irradiation beam were used as inputs into the simulations. The simulations give good agreement for the 2D plasma expansion and show the existence of an on-axis density dip.

Significant lateral expansion from the initial 14 μm focal width is observed at all times. For all the interferograms the sideways expansion is symmetrical about the center of the focusing region. Within 10 μm from the target surface there is a pronounced dip in the measured on-axis electron density profile. The low irradiation intensities ($\sim 10^{12}$ W/cm^2) would suggest pondermotive forces could not explain such an effect [2]. Such a feature would be undetectable using traditional interferometric techniques which employed UV probe wavelengths due to severe refraction effects [7].

Also shown in Fig. 1 are contour plots of the electron density calculated from 2-D LASNEX [8] simulations. The experimentally determined temporal and spatial irradiation conditions were used as input into the code. Such modeling covers all phases of expansion leading to a plasma size much larger than the focal spot. With the 2-D simulations we have looked at several potential different driving mechanisms behind the plasma lateral expansion. Our preliminary findings suggest the dominant mechanism driving the evolution of the plasma is the strong pressure gradients set up by the tight line focus. By increasing the line focus to 50μm FWHM the on axis density dip is not observed.

Picosecond x-ray laser interferometry is a valuable technique in diagnosing plasma evolution within laser produced plasmas. The short sampling time of the probe beam reduces blurring effects and allows for probing to within 10μm of the target surface. In addition, the short wavelength probe minimizes effects associated with refraction and free-free absorption making this diagnostic well suited for studying large, fast evolving, dense plasmas. It has been shown that for the experimental conditions reported within this paper lateral expansion is a significant effect. Two-dimensional plasma physics codes are therefore necessary to model such experiments.

The support of Al Osterheld and Andy Hazi is greatly appreciated. The authors are pleased to acknowledge the technical contributions from Carl Bruns and Al Ellis. This work was performed under the auspices of the U.S. Dept. of Energy by the University of California Lawrence Livermore National Laboratory, through the Institute for Laser Science and Applications, under Contract No. W-7405-Eng-48 and by US Department of Energy Grant No. DE-FG03-98DP00208. This research was sponsored by the National Nuclear Security Administration under the Stewardship Science Academic Alliances program through DOE Research Grant # DE-FG03-02NA00062. The CSU researchers also gratefully acknowledge the partial support of the NSF ERC Center for Extreme Ultraviolet Science and Technology, award number EEC-0310717 and the W.M. Keck Foundation.

References

[1] Smith R F, Dunn J, et al., Phys. Rev. Lett., **89** (6), 065004-1 (2002).

[2] Filevich J, Rocca J J, et al., J. Appl. Phys., vol. 43, No. 19 (2003); Filevich J, Kanizay K, et al., Opt. Lett. 25, 356=96358 (2000).

[3] Wilks S et al., Phys. Rev. Lett. **73**, 2994 (1994).

[4] Filevich J, Rocca J J, et al., Phys Rev E, vol.67, no.5, pp.56409-1-6. (2003).

[5] Dunn J et al., submitted to Optics Letters (2004).

[6] Da Silva L B, Barbee T W, et al., Phys. Rev. Lett. **74**, 3991-3994 (1995).

[7] Smith R F, Shlyaptsev V N, et al., JOSAB, vol.20, no.1, Jan. 2003, pp.254-9. (2002).

[8] Zimmerman G B and Kruer W L, Comments Plasma Phys. Controlled Fusion **2**, 51-61 (1975).

Inst. Phys. Conf. Ser. No 186
Paper presented at 9th Int. Conf. X-Ray Lasers, Beijing, China, 24–28 May 2004

In Situ Observation System for the Simultaneous Irradiation of the X-ray and the Visible Lasers

K. Namikawa[1,2], M. Tanaka[2], R. Z. Tai[2,3], M, Kishimoto[2], M. Ando[4], H. Takenaka[5]

1) Department of Physics, Tokyo Gakugei University, Tokyo 184-8501, Japan.
2) Advanced Photon Research Center, Kansai Research Establishment, Japan ,Atomic Energy Research Institute, Kyoto 619-0215, Japan.
3) Shanghai Institute of Applied Physics, Chinese Academy of Science, Shanghai 201800, P. R. China.
4) Institute of Materials Structure Science, High Energy Research Organization, Ibaraki 305-0801, Japan.
5) NTT Advanced Technology Corporation, Ibaraki 319-1106, Japan.

Abstract. For the purpose of nonlinear experiment by use of plasma based X-ray laser, we constructed a simultaneous irradiation system with a single shot X-ray laser and a single shot visible laser. Performance of the system has been investigated by in situ observation of the evolution of the damages which has been produced on the SnO_2/SiO_2 multi-layers.

1. Introduction

Non-linear optics is an undeveloped big scientific program for the application of X-ray laser[1)]. We have been engaged in the stimulated parametric down conversion of the X-ray laser. By virtue of the stimulation by intense visible laser field, the visible transition rates are accelerated and the phase machining conditions are guaranteed. Here we report on the experimental equipment developed for this novel experiment and on the performance of this simultaneous irradiation system.

2. Experimental system

2.1. Constitution of the simultaneous irradiation system for the X-ray and optical lasers

This system are constituted by a plasma based X-ray laser system, visible laser system, scattering chamber equipped with sample, and in situ monitor system, which are illustrated in Fig. 1. Damage produced by the visible laser on the sample has been monitored by observing the X-ray reflection pattern in situ.

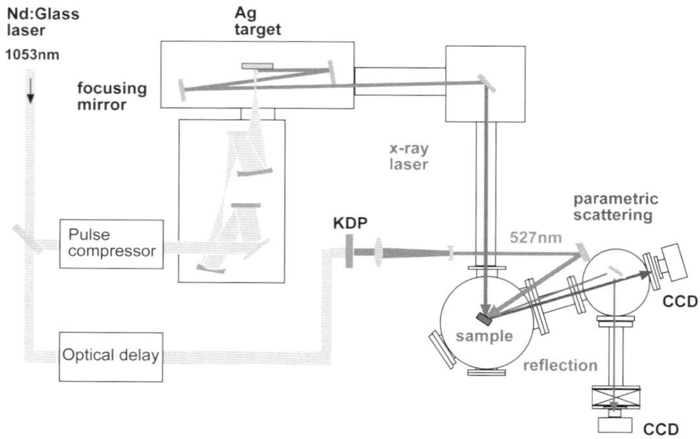

Fig.1. The Experimental set up for X-ray parametric scattering.

2.2. X-ray laser system

Our X-ray laser is Ni-like Ag plasma based single shot pulse laser, 13.9 nm in wave length, 10^{-4} in band width, 10^{12} photons per pulse, 10mrad in divergence, 8psec in pulse duration[2)]. The X-ray image of the source, 50μm in size, is transferred onto the sample, 500μm in size, by a spherical mirror.

2.3. Visible laser system

The visible laser is prepared by use of glass laser system as illustrated in Fig. 1. A part of the IR laser beam transmitted through the beam splitter is amplified to the required intensity. This IR beam, 600psec in pulse duration and 2.5J in intensity at the maximum, converted into second higher harmonics by use of KDP crystal. Conversion efficiency is as small as 1%. The maximum intensity focused on the sample into 1mmϕ is about 15mJ, $10^9 W/cm^2$ in power, which is sufficient to stimulate the process provided that the beam is a single mode.

2.4 Sample

The sample is a $[SnO_2(2nm)/SiO_2\ (8nm)]_{20}$ multilayer. The sample is designed to fulfill the Bragg condition 001 reflection at the scattering angle 96 degree for the 13.9nm X-rays. Reflectivity of the sample is a few % of the incident X-rays at the Bragg condition. Phase matching condition for the stimulated parametric scattering is satisfied at 3.8 degree off the Bragg angle. At this condition, the reflectivity of the specular reflection is one tenth of the Bragg reflection.

2.5 X-ray monitor system

The monitor system is designed to evaluate the intensity of the incident X-rays and to observe focusing image of the X-ray at the sample. The X-ray beam focused on the sample is divergent and reflected from the concaved mirror as an almost parallel beam.

Enlarged focal image is transferred to CCD camera equipped at the down stream of the monitor chamber as in Fig. 1.

3. Performance

3.1 Timing adjustment

We utilized the plasma based continuous emission as indicating the X-ray lasing timing. The X-rays take place at the time corresponding to the rising shoulder of the

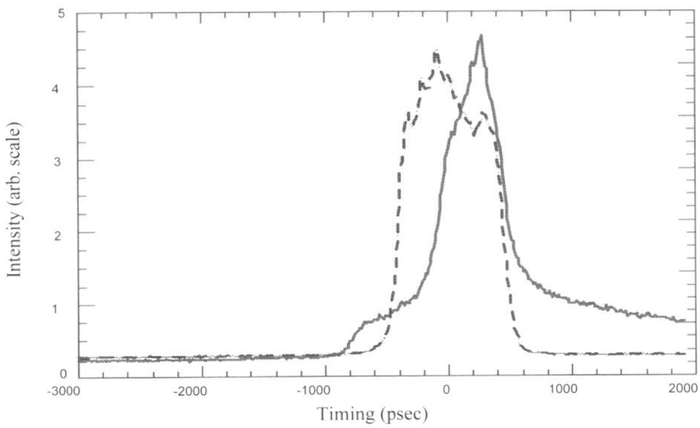

Fig. 2 Timing of the plasma based continuous emission (Solid line) and visible laser (Broken line).

continuous emission. Fig.2 indicates that the X-rays are irradiating simultaneously with the visible laser irradiation. Relative timing is variable with optical delay.

3.2 Confirmation of the spatial pointing

The spatial pointing is confirmed by an in situ observation of the damages produced by the irradiation of an intense visible laser field. Fig. 3 shows the images of the source reflected from the sample. Fig. 3 (a) to (d) shows the evolution of the damages caused by successive irradiation of increasing power of the visible laser pulse. X-ray images exhibiting the evolution of the damages induced by the visible laser guarantee the spatial pointing of both laser beams is coincide within the beam diameter.

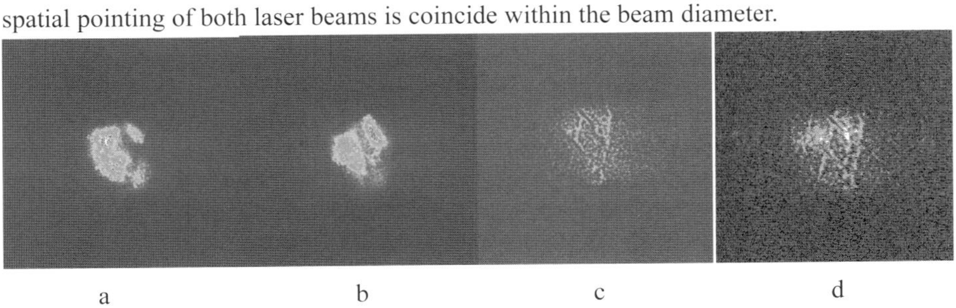

Fig.3. X-ray Reflected images, (a) to (d) show the in situ observation of the evolution of the damage caused by successive irradiation of increasing power of optical laser pulse.

4. Discussion

The intensity of the visible laser in Fig. 3a and Fig.3b is $1x10^8$ W/cm^2 and $3x10^8$ W/cm^2, respectively. While the intensity of the visible laser in both case in Fig. 3c and Fig. 3d is $3x10^9$ W/cm^2. We can estimate the damage threshold is at near the low level of 10^9 W/cm^2. Unfortunately, by the damage threshold we could not get yet a positive signal indicating the stimulated parametric down conversion. The reason that we could speculate is the single mode intensity of the visible laser is not so match strong as we expected because of the multimode nature of the visible laser which we utilized.

References

[1] TESLA Technical Design Report Part V, the X-ray Free Electron Laser, ed. G. Materlik and Th. Tschentscher, DESY 2001.

[2] T. Kawachi, M. Kado, et al., 2002 Phys. Rev. **A66**, 033815

Inst. Phys. Conf. Ser. No 186
Paper presented at 9th Int. Conf. X-Ray Lasers, Beijing, China, 24–28 May 2004

Damage threshold and damage mechanism of Sc/Si multilayer mirrors exposed to intense nanosecond 46.9 nm laser pulses

G. Vaschenko, M. Grisham, C. S. Menoni, and J. J. Rocca
National Science Foundation Engineering Research Center for Extreme Ultraviolet Science and Technology, and Electrical and Computer Engineering Department, Colorado State University, Fort Collins, CO 80523

Yu. P. Pershyn, E. N. Zubarev, D. L. Voronov, V. A. Sevryukova, and V. V. Kondratenko
Metal and Semiconductor Physics Department, National Technical University "KhPI", Kharkov, Ukraine

A.V. Vinogradov and I. A. Artioukov
P. N. Lebedev Physical Institute, Moscow 117942, Russia

Abstract. We investigated damage threshold and damage mechanism of extreme ultraviolet Sc/Si multilayer mirrors by focusing nanosecond pulses of 46.9 nm radiation from a compact capillary-discharge laser. Damage threshold fluences of ~ 0.08 J/cm^2 were measured for coatings deposited on both borosilicate glass and Si substrates. The use of scanning and transmission electron microscopy, and small-angle X-ray diffraction techniques revealed the thermal nature of the damage mechanism. The results are relevant to the use of the Sc/Si mirrors in extreme ultraviolet laser research.

1. Introduction

High reflectivity mirrors for the extreme ultraviolet (EUV)/soft x-ray range with high damage threshold are key elements for enabling numerous applications of the rapidly advancing high power coherent sources at these wavelengths, which include new table-top sources and free-electron lasers. Significant progress has been made in the developing of high reflectivity Sc/Si mirrors for the 35-50 nm range, with measured reflectance values as high as 54 % [1]. However, the damage of these mirrors when exposed to high peak powers of EUV light has not been studied. This problem is now of particular relevance as the peak power and fluence of EUV sources have recently increased significantly (for example, the radiation fluence at the exit of a capillary discharge Ne-like Ar lasers operating at 46.9 nm can exceed 1 J/cm^2).

In this contribution we report results of the study of the optical damage mechanisms and damage threshold for Sc/Si EUV mirrors exposed to high power EUV laser radiation. The study was conducted by focusing the output of a tabletop capillary-discharge Ne-like

Ar laser emitting ns duration pulses at a wavelength of 46.9 nm. The resulting damage of the multilayer coatings exposed to EUV beam was analyzed with scanning electron microscopy (SEM), transmission electron microscopy (TEM), and with small-angle X-ray diffraction (λ=0.154 nm) techniques. Our results show that multilayer coatings on Si and borosilicate glass have similar damage threshold values of ~ 0.08 J/cm^2, compared to the 0.7 J/cm^2 found necessary to damage a bare Si substrate. These values are similar to the thresholds found in Mo/Si, W/C, and W/Si coatings measured at much shorter wavelength [2,3].

2. Sc/Si multilayer coatings and irradiation set-up

The Sc/Si multilayers were deposited by dc-magnetron sputtering at 3 mTorr of Argon pressure on superpolished borosilicate glass (surface roughness σ ~ 0.4 nm) and on silicon wafers (σ ~ 0.6 nm). The multilayers on borosilicate glass consisted of 10 periods of Sc/Si layers, each with a thickness of ~ 26.7 nm, and a ratio of layer thickness H(Sc)/H(Si) ~ 0.7. A top 5 nm thick Si protection layer capped the multilayers. The multilayer coatings deposited on Si consisted of 33 periods of Sc/Si pairs with the same parameters as those deposited on borosilicate glass. In these structures the crystalline Sc layers are always separated from the amorphous Si layers by ~ 3 nm of amorphous ScSi interface layers formed by interdiffusion.

The experimental setup used to irradiate the samples was described in Ref. 4. The laser emission was focused onto the target surface with a spherical R = 10 cm, 30 % reflectivity, Sc/Si multilayer-coated mirror 2.5 cm in diameter positioned at normal incidence. The capillary-discharge Ne-like Ar laser used as a source of EUV radiation was configured to produce pulses with an energy of ~0.13 mJ and a duration of ~1.2 ns FWHM [5]. The far-field laser beam profile had an annular shape with a peak-to-peak divergence of about 4.6 mrad. The Sc/Si multilayer samples were cut to a 3 × 5 mm size and mounted on a 1.6 mm thick brass strip holder positioned at 1.45 m from the exit of the laser. The sample and the holder intercept a small fraction of the laser beam [4], which in that plane has a diameter of ~13 mm. Motorized XYZ translation stage was used for the displacement of the samples that allowed us to vary the distance between the sample and the focal spot (~5 μm in diameter) to select EUV radiation fluences from ~0.01 to >10 J/cm^2, while simultaneously changing the target area irradiated by the beam after each shot.

3. Sc/Si multilayer damage by EUV laser pulses

Figure 1 shows typical SEM images of damaged areas in coatings deposited on a Si wafer resulting from average EUV fluences of 0.13 J/cm^2 (a), 1.5 J/cm^2 (b) and 2.8 J/cm^2 (c). At 0.13 J/cm^2 (Fig. 1(a)), we observe large areas with apparent discoloration that are most likely produced by heat-triggered interdiffusion in the upper layers of the coatings. This surface modification, which already appears at fluences of ~ 0.08 J/cm^2, establishes the damage threshold for the Sc/Si multilayers defined in this work. In comparison, the onset of damage in bare Si substrates measured in this work occurs at a significantly larger irradiation fluence of 0.7 J/cm^2. The areas with larger local fluences (Fig. 1(a)) are molten and covered with cracks resulting from significant mechanical tensile stress generated by thermal expansion and the following cooling down process. At larger fluences (1.5 J/cm^2) some of the molten material is displaced towards the

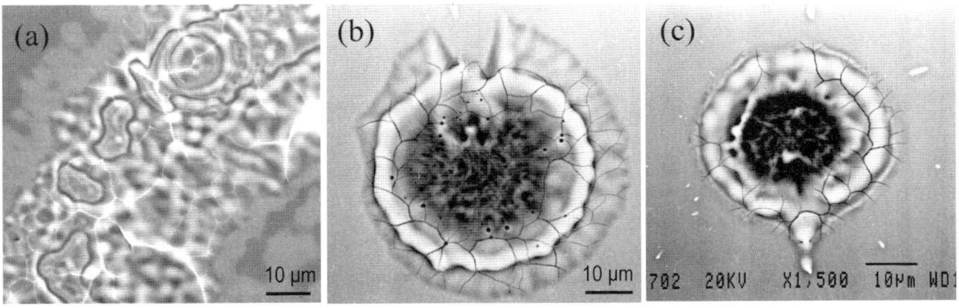

Fig. 1. SEM micrographs of the damaged areas of the Sc/Si coatings exposed to EUV laser beam fluences of 0.13 J/cm^2 (a), 1.5 J/cm^2 (b), and 2.8 J/cm^2 (c). Only a segment of the entire damaged area is shown in (a).

periphery, forming a 'crown' (Fig. 1(b)). Also, a substantial concentration of cracks and micrometer-sized pits resulting from boiling are observed. At even larger fluences of ~ 3 J/cm^2 the coating is fully evaporated from the center of the irradiated spot and the Si substrate is also damaged (Fig. 1(c)).

Small-angle X-ray diffraction analysis of the samples with 2×2 mm^2 area uniformly irradiated with ≥ 0.1 J/cm^2 emission fluence shows a noticeable drop in the intensity of the diffraction peaks with respect to the unexposed areas. However, the diffraction peak's position remains approximately the same, indicating that the coating is only partially destroyed. This evidence suggests that while at these fluences the top layers of the coating are molten, the layers adjacent to the substrate remained unchanged. This interpretation of the X-ray diffraction data was confirmed by cross-section TEM imaging of the sample exposed at 0.21 J/cm^2. The TEM image of Fig. 3(a) shows that the top 700 nm of the coating are molten, while ~180 nm (7 periods) adjacent to the substrate are not destroyed. The top molten layer constitutes an alloy of Sc_3Si_5 and crystal Si as determined from electron diffraction data. Analysis of the surviving multilayer coating beneath the molten layer indicates that changes in layer thickness have occurred within a distance of less than 2 periods from the molten region (Fig. 3), thus the heat affected zone (HAZ) did not exceed ~50 nm.

Comparison of the layer structure in the HAZ with that of isothermally annealed samples indicates that the various stages of structural and phase transformations observed under laser irradiation are the same as in samples annealed at different

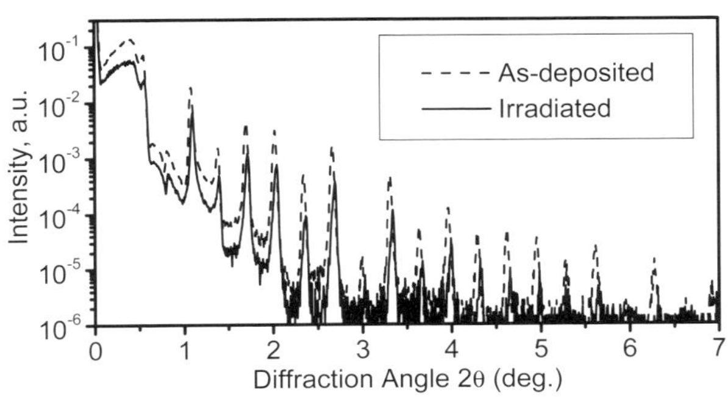

Fig. 2. Small-angle X-ray diffraction patterns in the as-deposited and EUV-irradiated coatings. Laser fluence: 0.21 J/cm^2.

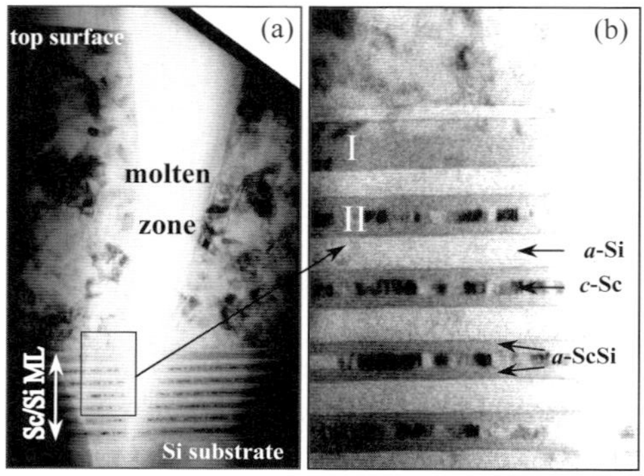

Fig. 3. Cross-sectional TEM image of the molten zone (a) and survived layers (magnified) (b) of a sample irradiated with 0.21 J/cm^2 pulses of 46.9 nm radiation.

temperatures. The changes taking place in the Sc-containing layer nearest to the molten region (indicated by I in Fig. 3(b)) correspond to a stage of formation and crystallization of Sc_3Si_5 silicide that have been previously observed in isothermally annealed coatings at 430 °C after 1 hour [6]. In the next Sc-containing layer (indicated by II in Fig. 3(b)) only minor expansion of the ScSi silicide interface layers is observed, which is a result of solid state amorphization. Similar effects have been observed at annealing temperatures of less than 200°C [6]. The insertion of antidiffusion barriers at Sc-Si interfaces might help to increase the damage threshold of these multilayers, as suggested by the increased heat resistance observed in the isothermally heated structures with such barriers [6].

In summary, we have investigated the damage threshold and damage mechanisms in Sc/Si multilayer EUV mirrors exposed to intense 46.9 nm nanosecond pulses. The study reveals the thermal character of the changes in the mirror coatings underlying the irradiation damage. A damage threshold of ~ 0.08 J/cm^2 was measured in the multilayer mirror coatings deposited on Si or borosilicate glass substrates, compared to a measured value of 0.7 J/cm^2 for bare Si substrates. These results are relevant to the use of these mirrors in combination with newly developed high power EUV sources, and provide a benchmark for their further improvement.

This work was supported by the U.S. Department of Energy, Chemical Sciences, Geosciences and Biosciences Division of the Office of Basic Energy Sciences, and in part by the Engineering Research Centers Program of the National Science Foundation under NSF Award Number EEC-0310717. The CSU researchers also gratefully acknowledge the support of the W. M. Keck Foundation. The authors are indebted to Dr. Yu. S. Kasjanov for helpful discussion.

References

1. Yu. A. Uspenskii, V. E. Levashov, A.V. Vinogradov, *et al.*, Opt. Lett. **23**, 771 (1998).
2. B. J. MacGowan, S. Mrowka, T. W. Barbee, *et al.,* J. X-Ray Sci. Tech. **3**, 231 (1992).
3. F. Leguern, J-M. André, J-P. Lebreton, *et al.*, J. X-Ray Sci. Tech. **7**, 271 (1997).
4. B. R. Benware, A. Ozols, J. J. Rocca, *et al.*, Opt. Lett. **24**, 1714 (1999).
5. B. R. Benware, *et al.*, Phys. Rev. Lett. **81**, 5804 (1998).
6. D. L. Voronov, *et al.*, AIP Conf. Proc. **641**, 575 (2002).

Inst. Phys. Conf. Ser. No 186
Paper presented at 9th Int. Conf. X-Ray Lasers, Beijing, China, 24–28 May 2004

Electron Density Measurements Using Soft X-ray Laser Interferometry

Ch. Wang[1], W. Wang[1], J.R. Sun[1], Z.H. Fang[1], W.X. Ma[1], S.Z. Fu[1], G.P. Zhang[2], W.D. Zheng[2], T.X. Zhang[2], H.M. Peng[2], P. Shao[3], Y. Gu[1], Z.Q. Lin[3], S.J. Wang[1]

[1] Shanghai Institute of Laser Plasma, P.O.Box 800-229, Shanghai, 201800, P.R. China

[2] Beijing Institute of Applied Physics and Computational Mathematics, Beijing, 100088, P.R. China

[3] Shanghai Institute of Optics and Fine Mechanics, P.O.Box 800-211, Shanghai, 201800, P.R. China

Abstract. A soft x-ray laser Mach-Zehnder interferometry is a precise and important way to diagnose the laser plasma electron density profiles. This paper gives an application demonstration of the diagnostics of laser plasma electron density profiles by using a soft x-ray laser Mach-Zehnder interferometry. The maximum density of 95% critical density for drive laser was measured.

1. Introduction

One of major goals of x-ray laser research is the application in diagnostics of laser plasma electron density profiles. When an x-ray laser beam passes through a plasma, the states of laser beam can be changed with the plasma perturbation, and the changes of state indicate abundant information of plasma. A series of ways may be selected by measuring different changes of state, such as deflectometry[1-3] and interferometry[4-8]. By comparison, there are more advantages of interferometry method, because it maybe gives the distribution information of plasma electron density directly.

The first soft x-ray laser interferometry experiments were performed on the Nova laser by DaSilva and co-workers with 15.5nm Ne-like Y x-ray laser using an Mach-Zehnder (M-Z) interferometry equipped with multi-layer beam-splitters at 1994[4]. The maximum electron density of $2\times10^{21}cm^{-3}$ with 2ω green laser irradiation and the proximal position from initial target surface of 25μm were measured. Then Rocca and co-workers developed a novel Mach-Zehnder interferometry using diffraction gratings as beam-splitters for Ne-like Ar x-ray laser at 46.9nm[5,6]. They has obtained well results, however the high electron density may not be diagnosed because the long wavelength of probe beam. Recently, a new scheme of diffraction grating interferometer was developed and optimized by Colorado State University for the ps laser-driven Ni-like Pd x-ray laser at 14.7nm[7,8]. The maximum electron density of $3\times10^{20}cm^{-3}$ were probed.

In this paper, we also report a demonstration of diagnostics of laser plasma electron density profiles using a soft x-ray laser interferometer. Though the scheme was similar to the DaSilva's experiments, more new results and characteristics were obtained. The results give a maximum density of $9.5\times10^{20}cm^{-3}$ and it was very close to the critical density of $1\times10^{21}cm^{-3}$ of ω drive laser.

2. A soft x-ray laser Mach-Zehnder interferometer

Interferometry is a common method to diagnose the refractive index of medium. Wherewith the special advantage of x-ray laser, a soft x-ray laser M-Z interferometry is a precise and important way to diagnose laser plasma electron density profiles.
The soft x-ray laser M-Z interferometer used in our experiments was constructed by two multi-layer beam-splitters and two multi-layer mirrors. The incidence angles of both mirrors and beam-splitters were about 7.3° and each arm of this M-Z interferometer was about 200mm.
Beam-splitters were the key elements of M-Z interferometer. High quality multilayer beam-splitters with size of $10\times10mm^2$ were manufactured for these experiments. We also developed a method based on laser-driven plasma x-ray sources to measuring both reflectivity and transmissivity of the beam-splitter by one shot. Figure 1 was a typical result. The reflectivity was about 19% and the transmissivity was about 24% at the work wavelength of Ni-like Ag x-ray laser at 13.9nm. This result ensured the success of the experiments.

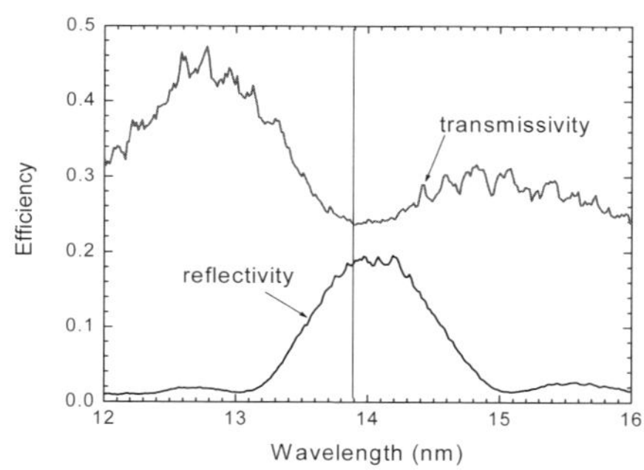

Figure 1. The reflectivity (R) and transmissivity (T) of a beam-splitter measured by using an x-ray source of laser plasma. R~19%, T~24%, R·T~4.5% at 13.9nm.

3. Experimental setup

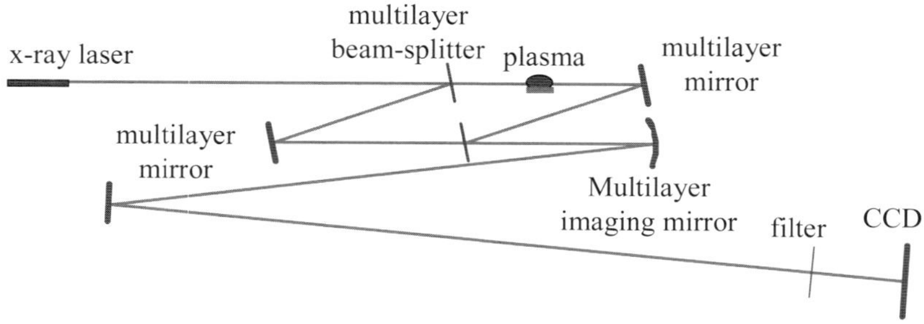

Figure 2. Experimental setup of diagnostics applications of laser plasmas by using soft x-ray laser M-Z interferometry.

The experiment was performed on SG-II laser and the setup was shown in Figure 2. The probe beam was a saturated Ni-like Ag x-ray laser at 13.9nm with about 30ps duration and 3×7mrad^2 angle distributions, driven by SG-II IR lasers (λ=1.053μm). The sampling plasma was created by another SG-II IR laser line-focus irradiation of a C_8H_8 coated-glass flat target. This laser beam was about 100ps duration and divided to pre- and main-pulse with 5% intensity ratio and 3ns temporal separation (Same to the lasers driving x-ray lasers). The line-focus with size 450μm×3.5mm of was produced by a vertical uniform cylinder lens array (CLA) system. So, the traversing length of x-ray laser probe beam was about 450μm. The magnification times of 3.5× gave a spatial resolution of about 5.7μm.

Probe beam passes through the sampling plasma after 1ns of the end of drive laser by adjusting the time correlation of drive lasers.

4. Experimental results

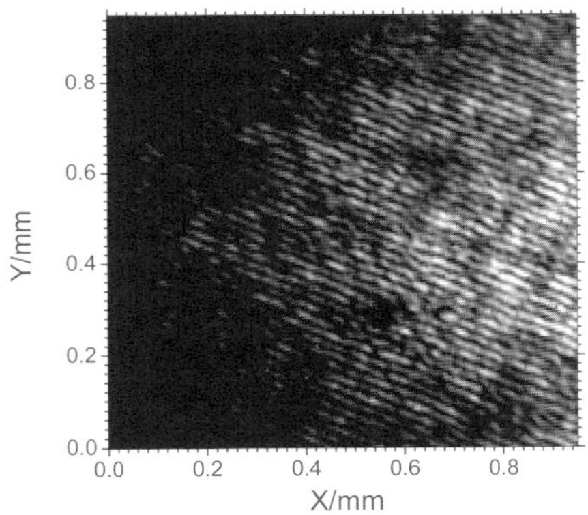

Figure 3. A typical static M-Z interferogram of Ni-like Ag x-ray laser as probe beam recorded in experiments.

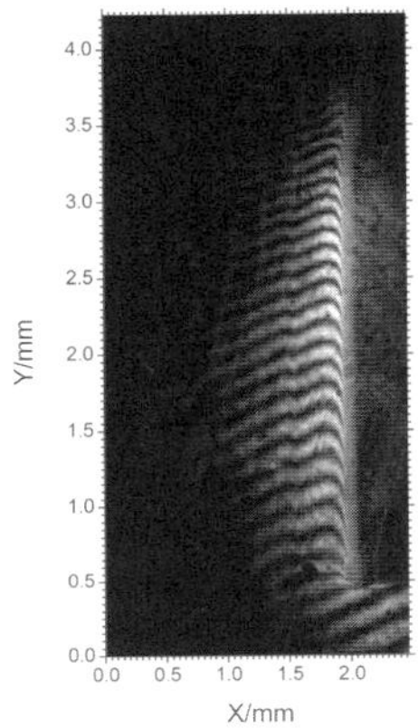

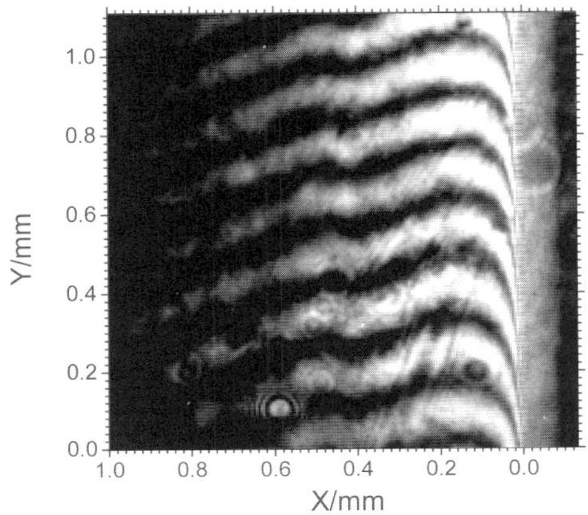

Figure 4. A dynamic M-Z interferogram with plasma perturbation recorded in experiments and its part zoom. The flexure of fringes included abundant information of laser plasma electron density.

Figure 3 showed a typical experimental recorded image of static M-Z interference fringes without plasma perturbation. Figure 4 showed a dynamic M-Z fringes with plasma perturbation. The flexure of stripes included abundant distribution information of laser plasma electron density.
An elementary result of electron density profiles was calculated from the dynamic M-Z interference fringes and shown in Figure 5. The data was average results of several similar fringes. The maximum electron density was about $9.5\times10^{20}cm^{-3}$, and it was very close to the critical density of drive laser ($1.0\times10^{21}cm^{-3}$).
A notable phenomenon was the obvious change of density at about 450μm. It may be relational to the irradiation of pre- and main-pulse laser. However the more detailed explanations have still in hand.

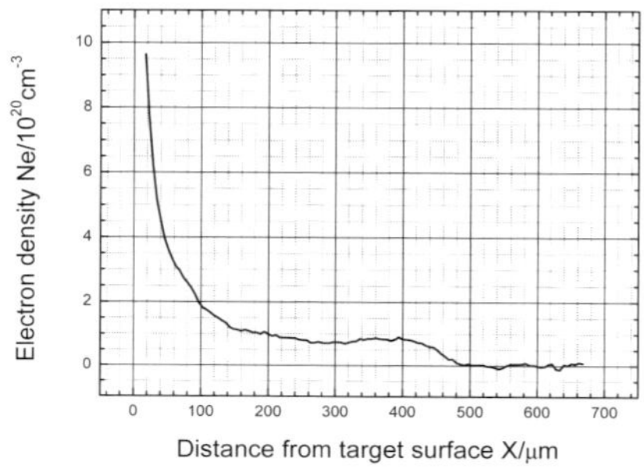

Figure 5. The electron density profiles calculated from the dynamic M-Z interferogram. The resolution of electron density was about $2\times10^{19}cm^{-3}$.

5. Conclusion

An application of the diagnostics of laser plasma electron density profiles was demonstrated by using a soft x-ray laser Mach-Zehnder interferometer. The maximum density was about $9.5\times10^{20}cm^{-3}$, and it was more close to the critical density of drive laser ($1.0\times10^{21}cm^{-3}$) than the other experiments. The steep change data near critical density surface may be very important for the theory studies of laser plasma.

6. Acknowledgments

This work is supported by National Nature Science Foundation of China (Grant No. 69989801) and National High Technology Development Program of China (Grant No. 863-804-7.3). We acknowledge all the members of SG-II operation group, whose excellent works made the experiment going on wheels. We also thank Z.S. Wang's group for providing the beam-splitters, targets and filters, C.S Jin's and K. Yi's group for providing the multilayer mirrors.

References

[1] Ress D et al 1994 Science 265 514.
[2] Wang C et al 2002 Acta Phys.Sin. 51 847.
[3] Kodama R et al. 1999 Rev.Sci.Instrum. 70 543.
[4] DaSilva L B et al 1995 Phys.Rev.Lett. 74 3991.
[5] Filevich J et al 2000 Opt.Lett. 25 356.
[6] Filevich J et al 2003 Phys.Rev.E 67 056409.
[7] Smith R F et al 2002 Phys.Rev.Lett. 89 065004.
[8] Dunn J et al 2002 X-Ray Lasers 2002 AIP 641 481.

Inst. Phys. Conf. Ser. No 186
Paper presented at 9th Int. Conf. X-Ray Lasers, Beijing, China, 24–28 May 2004

Comparison between multiple-energy and two-energy twin image removal algorithms in atomic-resolution x-ray holography

Honglan Xie, Jianwen Chen, Hongyi Gao, Huafeng Zhu, Ruxin Li and Zhizhan Xu

Shanghai Institute of Optics and Fine Mechanics, Chinese Academy of Sciences, Shanghai 201800, China

Abstract. In this work, we compare multiple-energy and two-energy twin image removal algorithms in atomic-resolution x-ray holography by theoretical simulations with a 27-Fe-atom cubic lattice used as the calculated model. The results show that the former is effective only in inside detector x-ray holography and the latter is effective in both inside detector and source x-ray holography. The spatial resolution of atomic images increases with the increase of incident energies in both algorithms. In the former, the wider the range of incident energies is, the better the effect of removing twin images is. Whereas in the latter, the smaller the difference of two incident energies is, the more effectively twin images are removed. However, the latter is not as effective as the former in eliminating background noises.

1. Introduction

In 1986, Abraham Szöke[1] first proposed atomic-resolution x-ray holography called as "x-ray fluorescence holography"(XFH) and suggested that a photoexcited atom within a sample is an ideal source emitting highly coherent outgoing fluorescent x-rays. A hologram is formed by the interference between fluorescent x-rays scattered by neighboring atoms and fluorescent x-rays leaving the sample without scattering. The 3D atomic arrangements in the sample can be reconstructed from the hologram. Because that the source is in the sample, XFH is also called "inside source holography". The resolution of XFH can reach the size of the atoms and is limited by the size and the wavelength of the fluorescent radiation. Utilizing the optical reciprocity theorem, the optical beam paths can be reversed and the positions of the source and detector can be exchanged. This method can be regarded as the inverse of XFH and called "inside detector holography". Since Tegze and Faigel[2] first demonstrated experimentally the XFH method and subsequently many experimental results have been reported for single crystals with atomic structures.

Like conventional holographic imaging methods, atomic-resolution x-ray holography also suffers from the classical problem of twin images, which dates from the original work of Gabor. Barton[3] proposed that properly phased combinations of images reconstructed from holograms recorded at several incident energies can suppress twin image effects. This method can be called "multiple-energy x-ray holography"(MEXH). In 2002, Y. Nishino *et al.*[4] proposed that two holograms recorded at different energies are sufficient to remove twin images in atomic-resolution x-ray holography. In this paper, we discuss and compare in detail the two holographic reconstruction algorithms for removing twin images.

2. Two reconstruction algorithms for twin image removal

The atomic images $U(\boldsymbol{r})$ can be reconstructed from a single hologram $\chi(\boldsymbol{k})$ via[5]

$$U(\boldsymbol{r}) = \iint \chi(\boldsymbol{k}) e^{-i\boldsymbol{k}\cdot\boldsymbol{r}} d\sigma_k \tag{1}$$

Where σ_k is surface element on the hologram. The twin image of each atom is at the same position as the real image of its centrosymmetric counterpart. Even when the condition $kr = \pi(2m+1)/2$ is satisfied, where m is a nonzero integer, a pair of atomic images at $\pm\boldsymbol{r}$ is canceled.

In Barton's multiple-energy twin image removal algorithm, the following phase combination of image functions is taken:

$$U(\boldsymbol{r}) = \sum_k U_k(\boldsymbol{r}) e^{-ikr} \tag{2}$$

If enough holograms recorded at widely distributed wave numbers are added, a random phase sum of twin images approaches to be zero and twin images can be cancelled.

In 2002, Y. Nishino et al proposed the two-energy twin image removal reconstruction algorithm. They take a phase combination of image functions at two wave numbers, k_1 and k_2:

$$V_{k_1,k_2}(\boldsymbol{r}) = U_{k_1}(\boldsymbol{r}) e^{-ik_1 r} - U_{k_2}(\boldsymbol{r}) e^{-ik_2 r} \tag{3}$$

If one records two holograms at two near energies, the reconstructed atomic images don't appear twin images. For example, in inside source holography, two holograms can be recorded at two characteristic fluorescence energies. In inside detector holography, one can choose freely any two near energies above the absorption edge of the emitting atoms.

3. Simulations and discussions

We choose a 27-Fe-atom cubic lattice with a lattice constant $a = 2.90$Å as the calculated model. We choose the centred atom as the fluorescing atom at the origin and other twenty-six atoms around it as the scatterers. Fig. 1(a) is the calculated hologram recorded at the incident energy $k = 8.0$ Å^{-1} above the absorption edge of Fe element.

Fig. 1(b) shows the atomic images reconstructed from the hologram in the plane parallel to the (001) crystallographic plane at $z = 0$. The bright spots represent the atomic images. Only four atoms are shown. It is because that the centred atom is the fluorescing atom and then is not imaged and other four atomic images at (±2.90Å, ± 2.90Å, 0.00Å) for the incident energy 8.0Å^{-1} satisfy the cancellation condition $kr = \pi(2m+1)/2$ when m is 10.

In order to remove the twin images, we calculated eleven holograms recorded at different energies spanning $k = 8.0\,Å^{-1}$ to $10.0\,Å^{-1}$ with $0.2Å^{-1}$ steps. The reconstructed atomic images are shown in Fig. 2(a). Those four atomic images invisible in Fig. 1(b) appear in Fig. 2(a). Moreover, we reconstruct the atomic images shown in Fig. 2(b)

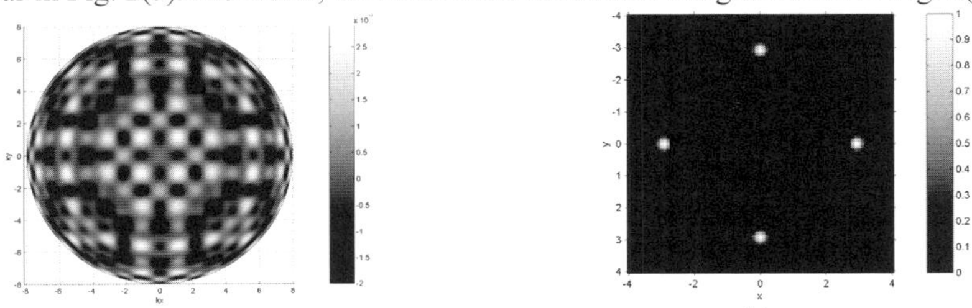

(a) (b)

Figure 1. (a) Thehologram of a 27-Fe-atom cubic lattice recorded at the energy $8.0Å^{-1}$. (b) The reconstructed atomic images in the plane parallel to the (001) crystallographic plane at $z=0$.

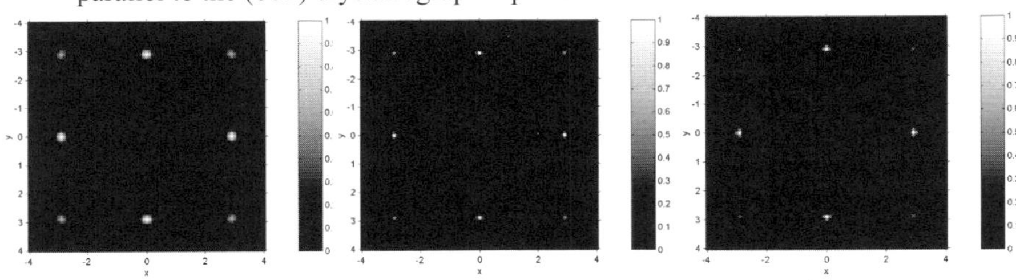

(a) (b) (c)

Figure 2. The reconstructed atomic images of a 27-Fe-atom cubic lattice in the parallel to the (001) crystallographic plane at $z=0$ for (a) $k = 8.0-10.0\,Å^{-1}$, (b) $k = 16.0-18.0\,Å^{-1}$, (c) $k = 8.0-18.0\,Å^{-1}$.

from eleven holograms at $k = 16.0-18.0\,Å^{-1}$ with 0.2 $Å^{-1}$ steps. It shows that the resolution of the reconstructed atomic images increases with the increase of incident energies. Widened the range of incident energies to be $k = 8.0-18.0\,Å^{-1}$ with 1.0 $Å^{-1}$ steps, the atomic images are reconstructed in Fig. 2(c). It can be seen that those faint noises visible in Fig. 2(b) are strongly compressed in Fig. 2(c).

We have discussed multiple-energy twin image removal by numerical simulations and we will discuss two-energy twin image removal here. Fig.3(a) shows the reconstructed atomic images when two incident energies are $8.0Å^{-1}$ and $8.1Å^{-1}$. We can see that four atomic images at (±2.90Å, ±2.90Å, 0.00Å) invisible at single-energy reconstruction appear in Fig. 3(a). However at the same time when the twin images are canceled, many small intensity spots appear.

Now, we consider the influence of different difference between two incident energies to atomic images. Fig. 3(b) and 3(c) show the reconstructed atomic images with the difference of 0.5 $Å^{-1}$ and 0.8 $Å^{-1}$ respectively. It can be seen that the intensities of those four real images at (±2.90Å, ±2.90Å, 0.00Å) diminish with the increase of the difference between two incident energies.

Like multiple-energy x-ray holography, we also consider the influence of the magnitude of incident energies to atomic images in two-energy x-ray holography. Fig. 4 shows the reconstructed atomic images for three cases of two incident energies, i.e., (a) $8.0Å^{-1}$ and $8.2Å^{-1}$, (b) $12.0Å^{-1}$ and $12.2Å^{-1}$, (c) $15.0Å^{-1}$ and $15.2Å^{-1}$. Compared Fig. 4(a),

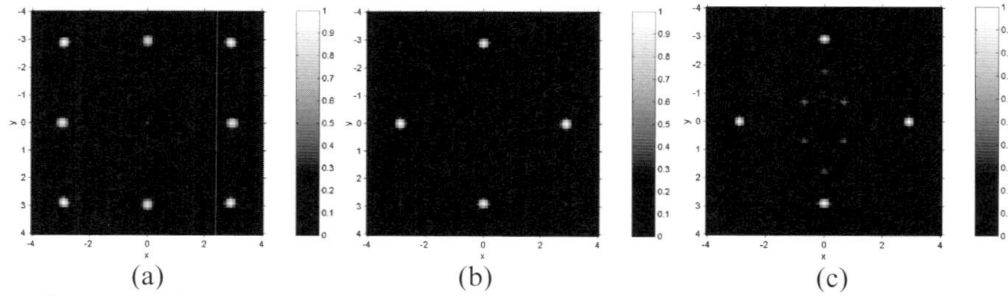

Figure 3. Reconstructed atomic images of a 27-Fe-atom cubic lattice in the parallel to the (001) crystallographic plane at $z=0$ for (a) $k=8.0-8.1\,\text{Å}^{-1}$, (b) $k=8.0-8.5\,\text{Å}^{-1}$, (c) $k=8.0-8.8\,\text{Å}^{-1}$.

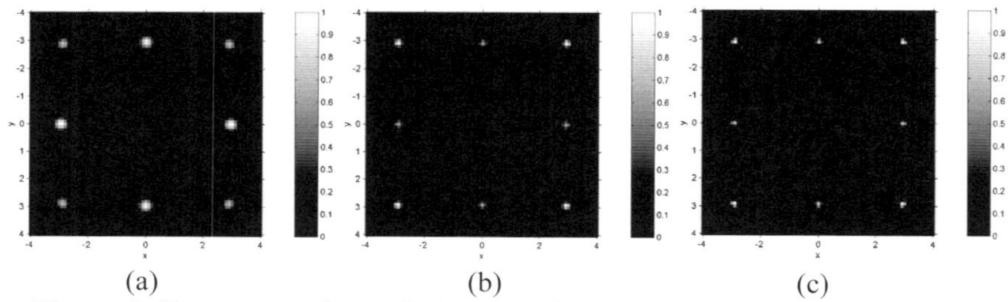

Figure 4. Reconstructed atomic images of a 27-Fe-atom cubic lattice in the parallel to the (001) crystallographic plane at $z=0$ for (a) $k=8.0-8.2\,\text{Å}^{-1}$, (b) $k=12.0-12.2\,\text{Å}^{-1}$, (c) $k=15.0-15.2\,\text{Å}^{-1}$.

4(b) and 4(c), it can be seen that the atomic images are resolved with a higher resolution when incident energies are higher.

4. Conclusion

The conclusion can be obtained that both algorithms are effective to remove twin images in atomic-resolution x-ray holography. The resolution of atomic images increases with the increase of incident energies in both algorithms. The wider the range of incident energies is in the former, the better the effect of removing twin images is. Whereas in the latter, the smaller the difference of two incident energies is, the more effectively twin images are removed. The former is useful only in inside detector x-ray holography, while the latter is useful in both inside detector and source x-ray holography. And the latter removes twin images faster than the former in experimental measurements. However, the latter is not more effective to eliminate background noises than the former.

References:

[1] Szöke A 1986 AIP Conf. Proc. on Short Wavelength Coherent Radiation: generation and Applications, Vol 147 (New York: AIP) p 361

[2] Tegze M and Faigel G 1996 Nature 380 49-51

[3] Barton J J 1991 Phys. Rev. Lett. 67 3106-3109

[4] Nishino Y et al 2002 Phys. Rev. B 66 92105-92108

[5] Barton J J 1988 Phys. Rev. Lett. 61 1356-1359

Inst. Phys. Conf. Ser. No 186
Paper presented at 9th Int. Conf. X-Ray Lasers, Beijing, China, 24–28 May 2004

A numerical study of in-line phase-contrast x-ray imaging and phase retrieval

Honglan Xie Huafeng Zhu Hongyi Gao
Jianwen Chen Ruxin Li Zhizhan Xu

Shanghai Institute of Optics and Fine Mechanics, The Chinese Academy of Sciences, Shanghai 201800, China

Abstract: A theoretical analysis of the in-line method of x-ray phase-contrast imaging with the incident monochromatic plane wave is presented. Based on the Fresnel integral, the phase-contrast images at different object-to-image distances are numerically simulated. According to the transport-intensity-equation, numeric examples of phase retrieval using simulated data are also given and the results indicate that the object-to-image distance is a very important parameter.

1. Introduction

In the x-ray region, the index of refraction, n, deviates only slightly from unity; it can be written as $n = 1 - \delta - i\beta$, where β describes the absorption of x rays and the phase-shift term δ incorporates refractive effects. At typical mammography x-ray energies of 15-25 keV, the phase-shift term can be up to 1000 times greater than the absorption term[1]. Thus it may be possible to observe phase contrast when absorption contrast is undetectable. X rays passing through regions of differing δ pick up different relative phases, which corresponds to being refracted and produces a distorted wave front. These phase differences are detected by the various phase-contrast techniques. Several research groups are exploring ways of exploiting phase information as a source of image contrast. These approaches fall into three broad categories: interferometry[2], diffractometry[3], and in-line holography[4].

2. In-line phase-contrast x-ray imaging

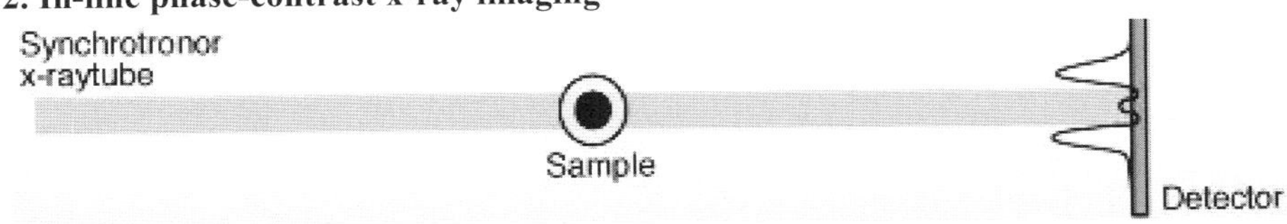

Figure 1. The experimental sketch of in-line phase-contrast x-ray imaging.

If just a detector is in the beam path (figure 1), the x rays emerging from the sample at their various angles will propagate through free space until they reach the detector. With the detector immediately behind the sample, one will get a conventional absorption image. If the source is very highly coherent and the detector is placed at an intermediate distance, one gets Fresnel or near-field diffraction. Here, a combination of imaging and diffraction effects is found, typically involving interference fringes at the edges of features. These fringes improve edge visibility. The optimum positioning of the detector for best enhancement effects varies from sample to sample, depending on the x-ray wavelength and the size of the features of interest.

The monochromatic wave field downstream of the sample can be described as

$$U_0(x,y) = T(x,y)U_i \tag{1}$$

where U_i is the incident monochromatic plane wave and $T(x,y)$ is the complex transmission function of the sample.

$$T(x,y) = \exp\left[-\frac{1}{2}\mu(x,y) + i\phi(x,y)\right] \tag{2}$$

where $\mu(x,y)$ and $\phi(x,y)$ are absorption and phase distribution of the sample respectively.

The intensity distribution at distance z from the object can be represented by

$$I(x,y) = \left|U_0(x,y) ** h_z\right|^2 \tag{3}$$

where

$$h_z(x,y) = \frac{\exp(ikz)}{i\lambda z}\exp\left[i\frac{\pi}{\lambda z}(x^2+y^2)\right] \tag{4}$$

is the Fresnel propagator, the double asterisk denotes two-dimensional convolution, and $k = 2\pi/\lambda$. Let us first consider the Fourier transform of the Fresnel propagator:

$$H_z(\xi,\eta) = \exp(ikz)\exp\left[-i\pi\lambda z(\xi^2+\eta^2)\right] \tag{5}$$

where ξ and η are the spatial frequencies. For small enough λz, the frequency representation of the propagator can be reduced to

$$H_z(\xi,\eta) = \exp(ikz)\left[1 - i\pi\lambda z(\xi^2+\eta^2)\right] \tag{6}$$

Thus the intensity at a sufficiently small distance z becomes

$$I(x,y) \approx \left|\left(1 + \frac{i\lambda z}{4\pi}\nabla^2\right)U_0\right|^2 \tag{7}$$

where ∇^2 is the Laplacian.

With uniform illumination intensity $I_0 = |U_i|^2 = |U_0|^2$, the intensity $I(x,y)$ of a detected image can be expressed in the form:

$$I(x,y) = I_0\left[1 - \frac{\lambda z}{2\pi}\nabla^2\phi(x,y)\right] \tag{8}$$

According to the equation, it is known that the image contrast is proportional to the Laplacian of the phase distortion $\phi(x,y)$ at a proper sample-detector distance z.

The image formula (8) can be rewritten as:

$$k\frac{I(x,y) - I_0}{z} = -I_0\nabla^2\phi(x,y) \tag{9}$$

It is similar to the transport-intensity-equation (TIE) with uniform illumination intensity

$$k\frac{\partial I}{\partial z} = -\nabla\cdot(I\nabla\phi) \tag{10}$$

Therefore, the phase retrieval from the image intensity of in-line phase-contrast x-ray imaging can use the TIE algorithm with Fourier expansion[5-6].

3. Imaging simulation and phase retrieval

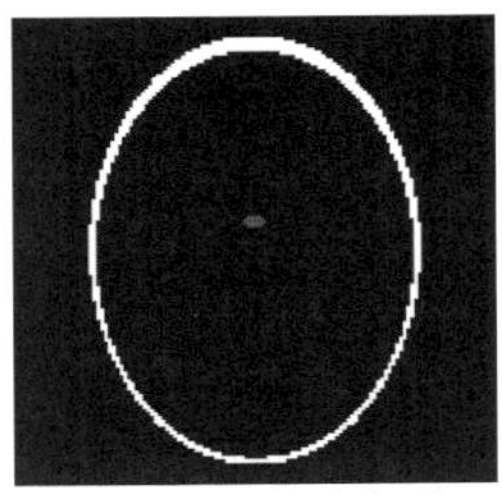

Figure 2. Original pure phase object.

We now present numerical examples with simulated data to help illustrate the above theoretical results. We assumed a perfectly monochromatic x-ray plane wave source with wavelength $\lambda = 0.1$ nm. A pure phase object of size $1280\times1280\mu m^2$ is shown in figure 2.The range of the phase values is approximately equal to [0, 1] rad. When the object-to-image distance is 0.5cm, the image is shown in figure 3(b). For comparison we also plotted in figure 3(a) the calculated distribution of the phase Laplacian. Obviously, the intensity distribution on the image plane and the distribution of the Laplacian of the phase on the object plane appear almost identical, which supports formula (8) in the case of negligible. With the object-to-image distance of 20cm, the image is shown in figure 3(c). Many diffraction fringes appear on the image plane with the increase of object-to-image distance disturb the image.

We use the computed image intensity distribution and TIE phase retrieval algorithm to reconstruct the phase. The results of this reconstruction are shown in figure 3(d-f). Figure 3(d) is reconstruction from Laplacian of the original phase object to verify the effectiveness of the TIE phase retrieval algorithm. The reconstruction result is similar to the original phase object, therefore the algorithm is effective. Figure 3(e) and 3(f) are the reconstructions of figure 3(b-c). Figure 3(e) is in very good agreement with the original phase object. Figure 3(f) is disturbed because of too large object-to-image distance. Therefore, the object-to-image distance is very important to the image in in-line phase-contrast x-ray imaging.

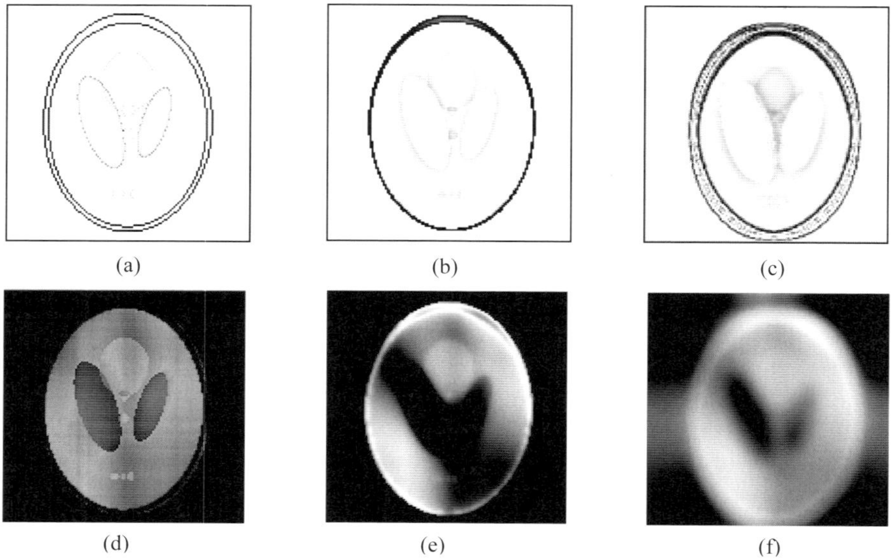

Figure 3. (a) Laplacian of the original phase shown in figure 2.
(b) Simulated image at $z = 0.5$ cm and (c) simulated image at $z = 20$ cm.
(d-f) are phase retrieval from (a-c) respectively.

4. Conclusion

We have presented a rigorous derivation of an equation for the intensity distribution of an image in the Fresnel region of a phase object illuminated by a monochromatic plane wave source. This equation can be regarded as the transport-of-intensity equation (TIE) with uniform illumination intensity. Based on the Fresnel integral, the phase-contrast images at different object-to-image distances are numerically simulated. According to the transport-intensity-equation, numeric examples of phase retrieval using simulated data are also given and the results indicate that the object-to-image distance is a very important parameter.

References

[1] Fitzgerald R 2000 Physics Today 7 23-26.
[2] Momose A 2002 J. Synchrotron Rad. 9 136-142.
[3] Ingal V N and Beliaevskaya E A 1995 J. Phys. D Appl. Phys. 28 231-2317.
[4] Wilkins S W et al 1996 Nature 384 335-338.
[5] Gureyev T E and Nugent K A 1996 J. Opt. Soc. Am. A 13 1670-1682.
[6] Gureyev T E and Wilkins S W 1998 J. Opt. Soc. Am. A 15 579-584.

Inst. Phys. Conf. Ser. No 186
Paper presented at 9th Int. Conf. X-Ray Lasers, Beijing, China, 24–28 May 2004

Study of Pre-main-pulse plasma

Wudi Zheng[1], Guoping Zhang[1], Chen Wang[2], Wei Wang[2], Jinren Sun[2], Zhiheng Fang[2], Jiang Wu[2], Sizu Fu[2], Yuan Gu[2], Shiji Wang [2]
[1]Institute of Applied Physics and Computational Mathematics, Beijing 100088,China
[2]Shanghai Institute of Laser Plasma, shanghai, 201800, China

Abstract: Pre-main-pulse plasma was applied very popularly in X-Ray-Laser (XRL) research, and the developing interference-measure-technique of XRL makes it possible to diagnose electron density (N_e) of dense plasma. In this paper, reconstruction of interference pattern is presented, pattern was gotten experimentally, using Ni-like Ag 13.9nm soft XRL Mach-Zehnder interferometer to diagnose N_e of CH plasma driven by pre-main-pulse. The comparisons with simulation results by 1D code JB19 is included. At last, influence of refraction effect is also considered briefly. The reconstruction shows that, there is a density platform with a width of ~200μm and a height of $7\times10^{19}\text{cm}^{-3}$ exists in distribution of N_e, and a shock wave front appears at 420μm above target surface. Measure result approaches simulation result while flux limit factor f_e=0.1. Although, theory is coincident with experiment only qualitatively, large difference about the height of density platform and the position of shock wave front is apparent.

In the study of collision-excitation-scheme X-ray laser (XRL) driven by laser, pre-main-pulse technique has been employed very popularly. This technique is one of best and creative way to obtain saturated output of XRL[1]. Pre-main-pulse plasma differs much from plasma without pre-pulse drive. According to our numerical simulation, with the arrival of main pulse, a shock wave is excited near critical face, which propagates away from target, then a wide platform of electronic density N_e is formed at the rear of shock wave, and a rapid decline of N_e, appears at the front of shock wave. Moreover, position of wave front is very sensitive to drive-laser intensity and measuring time. To measure N_e of this typical plasma precisely is very helpful for us to understand physical process and gain scheme of XRL, and helpful to verify code. Furthermore, in numerical simulation, flux-limit diffusion approximation is usually adopted, but the value of flux-limit factor is chosen quite arbitrary, experimental measurement will be a good approach to determine the value of it.

The developing technique of interference diagnosis with soft XRL makes it possible to diagnose electronic density of dense plasma. Using soft XRL as probe beam, many types of plasma have been measured [2,3,4,5,6], but to our knowledge, no one else has tried to diagnose pre-main-pulse plasma. Thanks to saturated output of

13.9nm Ni-like Ag XRL on facility of ShengGuang Ⅱ, we measured electronic density of line focused CH plasma driven by pre-main-pulse with M-Z interferometer.

1. Suitable plasma

In order to ensure that electronic density interests us can be measured, suitable plasma should be pre-designed. For diagnosis by soft XRL M-Z interferometer, there are two factors to be considered: number of fringe shift and absorption.

Number of fringe shift is given by $N_f = \frac{1}{\lambda}\int_0^L (1-\eta_{ref})dl \approx \frac{n_e}{2n_c}\frac{L}{\lambda}$, where η_{ref} is index of refraction, λ is wavelength of XRL, N_c is critical density, and the integral is along ray trace, then L is thickness of plasma. Constrained by resolution of CCD, maximum N_f is usually ~50[2]. And, too small N_f leads to large error, we artificially confine: N_f>0.3.

CH plasma is chose in experiment because of its low irradiation. Free-Free absorptions of XRL in CH plasma dominates, because C and H can be easily ionized completely in hotter plasma (>30eV), there is no photoelectric absorption at all. Free-free absorption coefficient is given by: $\alpha \approx 2.44\times10^{-37}\frac{\langle Z^2\rangle n_e n_i}{\sqrt{kT_e}(h\nu)^3}\left[1-\exp(\frac{-h\nu}{kT_e})\right]cm^{-1}$,

$[kT_e]=[h\nu]=eV$; $[n_e]=[n_i]=1/cm^3$.

After considering two factors above, we give out the parameter space suitable for diagnosis in fig.1 (white area), assuming electron temperature of 50eV, ionization degree of 6, and probing beam with wavelength of 13.9nm. It can be seen that to measure N_e of $10^{21}cm^{-3}$, optimum plasma thickness (L) locates at 300~500μm, and to measure density at front of shock wave (~$10^{20}cm^{-3}$), optimum L is larger than 1mm.

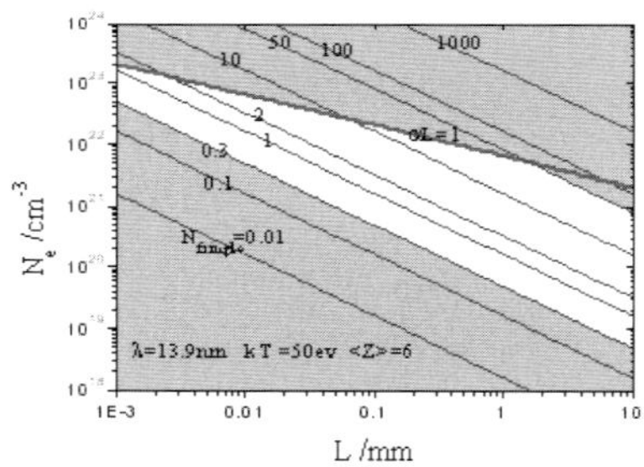

Fig. 1, White area is parameter space accessible for plasma probing by XRL (13.9nm), blue and red line is corresponding to number of fringe shift and absorption αL=1, respectively. L is thickness of plasma

2. Experimental conditions and fringe pattern.

Ni-like Ag XRL(13.9nm) was adopted, its width of 20ps was so short that fringe blurring by movement of plasma could be ignored. Plasma to be probed was produced by irradiating C_8H_8 target with 3.5mm×450μm line focused at a intensity of 6.72×$10^{13}W/cm^2$, focused line partially printed on target(~2.7mm). Parameters of pre-main-pulse are: 1ω, 100ps, 5.2% pre-pulse, 97.4J main-pulse delayed 3.07ns. XRL traveled through plasma, parallel to target surface and vertical to focused line. Plasma was imaged on CCD with magnification of 3.5, pixel size of CCD was 20μm which limited a resolution of 5.7μm. The details of setup were shown in reference [7].

Experimental image of fringe was gotten 1ns after the peak value of main pulse, which is shown in fig.2 with a visibility of ~0.6, there is an angle of ~21° between fringes and normal of target surface. In this experiment, the uncertainty of position of target surface is about 20μm, and measure precision is about 1.5×$10^{19}cm^{-3}$.

3. Reconstruction

It is necessary to get distribution of static fringes while reading the number of fringe shift (phase shift) from pattern, here we extrapolate parallel fringe far away from target surface to whole plane as static fringe. Reconstruction result (2D) of Electronic density shows in fig.3. Target surface locates at x=0. It is seen that density distribution along y direction is not uniform, especially at area of lower electronic density. To make it clear, fig.4 gives out distribution along x direction at different y, large distinction exists especially at edge of the target (y>2500μm). The main reason is that flatness of beam splitters of M-Z interferometer becomes worse. In fact, bad uniformity of laser intensity of focus-line may be another reason. The parallelism and interval of fringes far away from target keep not very well, to take them as reference, large errors produces. The worst thing is that its magnitude can't be given out exactly. Even though, fringes at y=1000~1500μm are rather good, distribution of electronic density of this part is reconstructed credibly, the comparison below between experiment and theory aims at this area. It's obviously that area with higher density have lower relative error. We can see from fig.4 that the positions, at which electronic density declines rapidly, swing a little, about ±15μm. The maximum electron density been measured is about 9.5×10^{20}cm^{-3}.

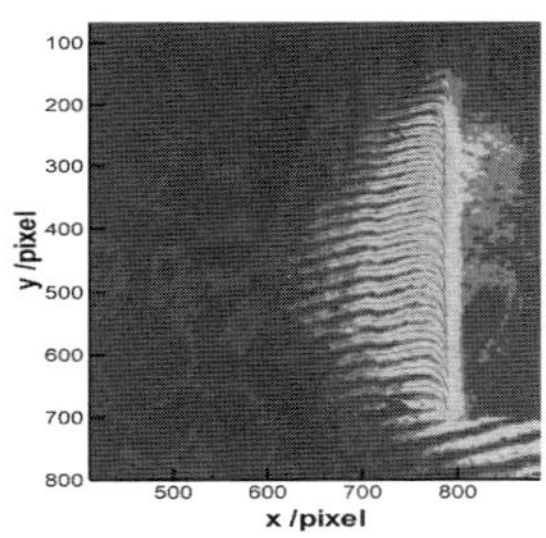

Fig.2 Experimental fringe pattern.

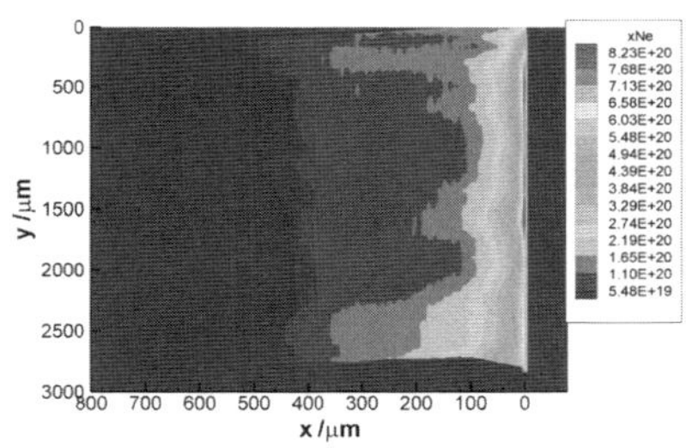

Fig.3 Electronic density reconstruction.

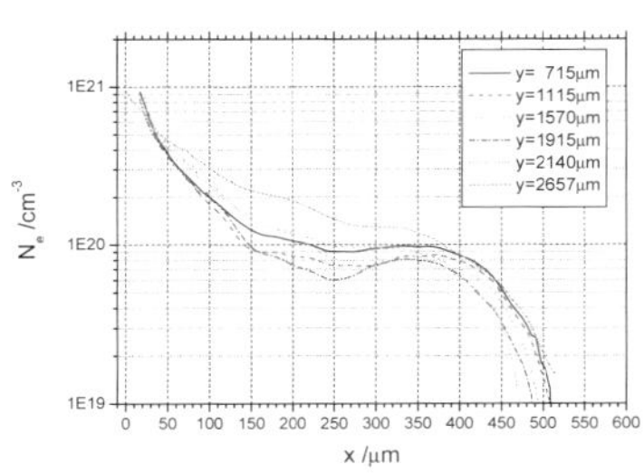

Fig.4, Electronic density distributions at different y.

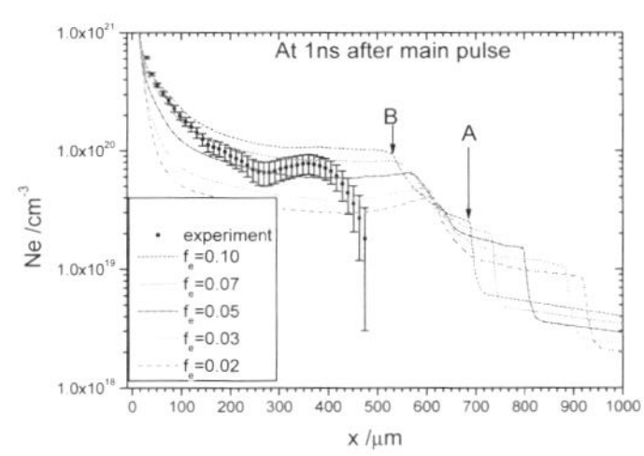

Fig.5, Comparison between theory and experiment.

4. Comparison between experiment and simulation

Numerical simulations are completed by 1D code JB19, fig.5 shows the results. In calculation we have considered side leak of irradiation and pseudo-2D-effect.

Flux-limit diffusion approximation is adopted in code. With the increase of f_e, the gradient N_e declines at dense area, the height of platform heightens, and the velocity of shock wave decreases. Usually, Lower intensity of driven laser matches higher f_e, if intensity increases up to 5×10^{14}W/cm^2, affected by self-generated magnetic field and anomalous transport, the resistance of electronic heat conduction becomes notable, indeed, at area of higher density near critical surface, theoretical result while f_e =0.1 closes to experimental result, but having a higher platform, see fig.5, but the platform looks more like that of theory while f_e =0.07. The platform of experiment has a width of 200μm and a height of 7×10^{19}cm^{-3}. 2D effect of plasma may contribute to this higher platform, plasma may dilute quicker than anticipation. Agreement between theory and experiment is only qualitative.

Though N_e given by experiment become too small to resolve at x>450μm, trend shows that front of shock wave is located at x~420μm. If the thickness of plasma were not 450μm but great than 1000μm, the details of shock wave front would be illustrated clearly, just as described above. Taking f_e =0.1 for example, the shock wave front of theory lies at position signed 'A' in fig.5, which is a relative precise quantity given out by experiment, but excluding target surface uncertainty of 20μm, there is still great difference of at least 200μm between theory and experiment, improperly choosing of equation of state (EOS) may be one of the reasons. In addition, theoretic curve of f_e =0.1 already begins to decline dramatically at the position of 'B' in fig.5, this has relationship with introduction of aggressive artificial viscosity in code.

Because critical density of XRL (13.9nm) is far more than density been probed, refraction effect should be small. Rays which entered at 23μm above target surface (N_e~7×10^{20}/cm^3) traveled through 450μm plasma, produced a reflection angle of 1.26mrad and deflection of 0.6μm, which was equal to 1/10 pixel of CCD. Hence, measure errors induced by reflection were negligible.

5. Conclusion

More improvements would make this kind of diagnosis perfect: (1) Not knowing the distribution of density along incidence direction of XRL, uncertainty was brought, if a snapshot of this direction were taken, more precise result would be gotten. (2) If imaging amplification were increased, influence of bad flatness of beam splitter would be less, and more information of density distribution near critical face could be obtained. (3) Too much uncertainty of position of target surface, weakened the significant of result. (4) Plasma was not thick enough to show the details out of shock wave front.

This work was supported by Chinese National 863 High Technology Plan ICF Field under Contract No. AA841101.

Reference

[1] J.Zhang, A.G..MacPhee et al., 1997 Science, 276, 1097-1100
[2] L.B.Da Silva, T.W.Barbee et al., 1995 Phy. Rev. Lett. 74, 3991-3994
[3] A.S.Wan, T.W.Barbee, et al., 1997 Phy. Rev. E. 55, 6293-6296
[4] J.Filevich, K.Kanizay, et al., 1999 Opt. Lett. 25, 356-358
[5] R. F. Smith, J. Dunn, J. Nilsen, et al., 2002 Phy.Rev.Lett. 89, 065004-1~4
[6] Raymond F.Smith, James Dunn, et al., 2003 J.Opt.Soc.Am.B, 20, 254~259
[7] Wang Chen， Wang Wei et al., have sent to Acta Physica Sinica.

X-ray optics and instrumentation

Inst. Phys. Conf. Ser. No 186
Paper presented at 9th Int. Conf. X-Ray Lasers, Beijing, China, 24–28 May 2004

Fabrication of the beam splitters for 13.9nm x-ray laser applications

Zh Sh Wang, Zh Zhang, F L Wang, W J Wu, H Ch Wang, Sh J Qin, L Y Chen

Institute of Precision Optical Engineering, Department of Physics, Tongji University, Shanghai 200092

Abstract: The x-ray laser Mach-Zehnder interferometer is an important tool to measure the electron densities of a laser-produced plasma near the critical surface. The design of a multilayer beam splitter at 13.9nm for soft x-ray laser Mach-Zehnder interferometer is completed based on the standard of designs of maximizing product of reflectivity and transmission of the beam splitter. The performance of semitransparent beam splitter is simulated using interface roughness and interdiffusion. The beam splitter which is Mo/Si multilayers on 10mm×10mm area Si_3N_4 membrane is fabricated by using the magnetron sputtering. The figure error of the beam splitter has reached the deep nanometer magnitude by using optical profiler and the product of reflectivity and transmission of the beam splitter measured by synchrotron radiation is about 4%.

1. Introduction

Optical interferometry has played in measurements of electron densities in laser-produced plasma. But maximum electron densities are strongly dependent on the wavelength of the probe beam. For these reasons it is needed to developed interferometers working in soft x-ray range. Now it is possible to extend the applications field to x-ray interferometry with the objective to characterize optics and dense plasma because of the success in x-ray optics technology and soft x-ray laser. In 1995, Da Silva and his collaborators developed an amplitude division soft x-ray Mach-Zehnder interferometer with two beam splitters and two high reflectivity multilayers working at 15.5nm[1]. In 1999, Rocca and his collaborators developed an amplitude division soft x-ray interferometer based on diffraction gratings[2]. In the research of nickle-like silver X ray laser at 13.9nm, the depth gain saturation X ray laser output and measured the plasma electron density distribution with the Moire defection method have been obtained in China[3]. In order to make the depth gain x ray laser apply to the dense plasma diagnostics, we were asked to develop the fabrication of beam splitters working at 13.9nm.

2. The design of soft x-ray beam splitters

The key to performing dense plasma diagnostics is a Mach-Zehnder interferometer in soft x-ray region. In visible light, the reflectors and beam splitters in the Mach-Zehnder interferometer are posed with an inclination of 45°. But in soft x-ray region, the situation changes. The maximum reflectivity of the s- and p- polarization components of Mo/Si multilayers versus incidence angles at 13.9nm are shown in Fig.1.

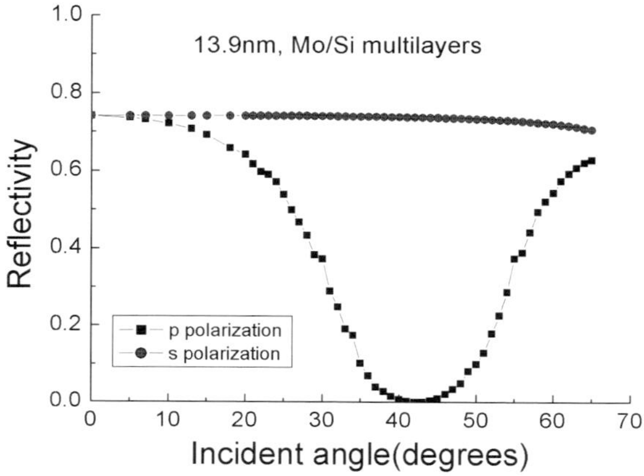

Figure 1. Theoretical maximal reflectivity of s- and p-polarization of periodic Mo/Si multilayers with the ideal interface versus incident angles at 13.9nm

As the incidence angle increases, the maximum reflectivity of s-polarization almost keeps unchanged, but that of the p-polarization reduces gradually to the Brewster angel (42°), and then it increases gradually. When the incidence angle turns larger than 70°, the area of the total reflectance of Mo comes (it is not shown in the figure). The incidence angles of high reflectivity multilayers and beam splitters in the Mach-Zehnder interferometer should be posed at near normal incidence because the soft x-ray laser is non-polarized source. The incidence angles for all multilayers and beam splitters in the Mach-zehnder interferometer is set to 7.2° according to the geometrical structure and the modification requirement of vacuum target. In the Mach-Zehnder interferometer, the reflectivity and transmission of the first beam splitter to the s-,p-polarization are assumed to R_{Fs1},R_{Fp2},T_{Fs1},T_{Fp2} individually, which of the second beam splitter is R_{Fs2},R_{Fp2},T_{Fs2},T_{Fp2} individually. Assumed that the two high reflectivity multilayers and the two beam splitters are fabricated with the same technological parameters, then it is assured $R_{Rs1}=R_{Rs2}=R_{Rs}$, $R_{Rp1}=R_{Rp2}=R_{Rp}$, $R_{Fs1}=R_{F2}=R_{Fs}$, $R_{Fp1}=R_{Fp2}=R_{Fp}$, $T_{Fs1}=T_{Fs2}=T_{Fs}$, $T_{Fp1}=T_{Fp2}=T_{Fp}$. When a beam of soft x-ray enters the Mach-Zehnder interferometer, it is splitted at the first beam splitter, one is reflected, and one is transmitted. After the reflected beam goes through the high reflectivity multilayer and the second beam splitter, it becomes one of the two beam interferometer, the throughput of the s-, p-polarization is $R_{Fs} \cdot R_{Rs} \cdot T_{Fs}$ and $R_{Fp} \cdot R_{Rp} \cdot T_{Fp}$ individually. At the same time the other transmitted beam of soft x-ray, which goes through the first beam splitter, is reflected by the high reflectivity multilayer, then goes through the second beam splitter and becomes the other beam of the interferometer, the throughput of the s-, p-polarization is $T_{Fs} \cdot R_{Rs} \cdot R_{Fs}$ and $T_{Fp} \cdot R_{Rp} \cdot R_{Fp}$ individually. To maximize the throughput of the interferometer, $T_{Fp} \cdot R_{Rp} \cdot R_{Fp}$ should be maximum, which shows that while the design of the high reflectivity multilayer, R_{Rs} and R_{Rp} should be made maximum, and it is concordant with the ordinary periodical high reflectivity multilayer. As to the beam splitter, $T_{Fs} \cdot R_{Fs}$ and $T_{Fp} \cdot R_{Fp}$ should keep maximum namely, the reflectivity-transmission product of beam splitter is maximum.

Mo/Si multilayer is the best material combination operating at 13.9nm of nickel-like silver laser. The beam splitter is deposited on the 100nm silicon nitride membrane. Not only the number of layers but also the thickness of Mo layer and Si layer in each period need to determine in order to design the beam splitter. These parameters should be optimized to make the reflectivity-transmission product of the beam splitter be maximum in design. The real multilayer couldn't have the ideal interface, and the interface roughness and interdiffusion. Using the Sterns model[5], the reflectivity and transmission of the beam splitter can be simulated under the different root-mean-square of interface roughness and interdiffusion. The reflectivity-transmission product of the beam splitter versus the periodic number under different root-mean-square of the roughness shows in Figure 2.

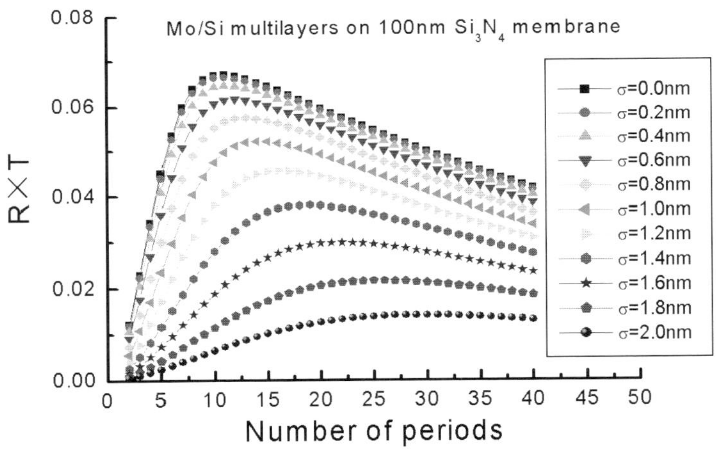

Figure 2. Product of reflectivity and transmission of a Mo/Si beam splitter versus the number of periods in Mo/Si multilayers with different root-mean-square interfacial roughness at wavelength of 13.9nm and incidence angle of 7.2°

The product of the reflectivity and transmission of the beam splitter decreases and the number of period corresponding to the maximal reflectivity-transmission product gradually increase when the root-mean-square of the multilayer interface roughness increases. Therefore, the periodic number of the beam splitter multilayer must be determined by the root-mean-square roughness of the multilayer interface when the soft x-ray beam splitter is fabricated. According to the real condition and processing parameter, the real root-mean-square of the multilayer surface roughness is about 1.0nm. The design result of the beam splitter operating at 13.9nm is following, the sum of the Mo layer thickness and the Si layer thickness is 7.3nm, the ratio of thickness of Mo layer and the periodic thickness is 0.5, the periodic number is 10~12, and the incidence angle is 7.2°.

3. Fabrication of beam splitters

Fabrication of beam splitters consists of two parts, one is the 100nm silicon nitride membrane, which is the substrate of beam splitters, the other is fabricating multilayer on the silicon nitride membrane. The 100nm silicon nitride membrane of clear area of 10mm×10mm is used as the substrate of a beam splitter in the fabrication. First of all, the deposition thickness of Mo and Si per second is required when Mo/Si multilayer is deposited on the 100nm silicon nitride membrane. The thickness of the Mo layer in the beam splitter is 3.65nm, and the one of Si layer is 3.65nm. Mo/Si multilayer is deposited by using magnetron sputtering, and the layer thickness is controlled by time. In order to determine layer thickness per second, we use the stepper and x-ray small angle grazing diffraction method. The base pressure is less than 8×10^{-5}Pa in the process of deposition. The working gas is Ar gas, the purity of which is larger than 99.99%, and the working pressure is 0.27Pa. The distance between targets and substrates is 80mm. The power of Mo and Si target are 20W and 20W respectively. The deposition rate of Mo is 0.47nm/s, and the one of Si is 0.32nm/s at above condition. According to the deposition rate, the time needed to stay in different target is controlled by a computer.

4. Characterization of beam splitters

The silicon nitride membranes size is 10mm×10mm, and the silicon frame size is 14mm×14mm. The thickness of silicon nitride is about 100nm.

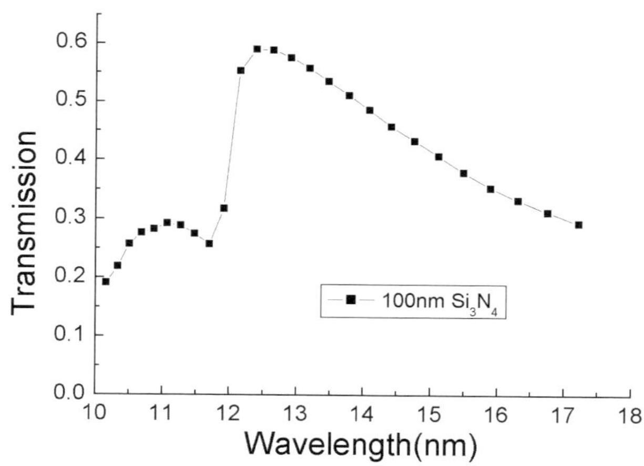

Figure 3. Transmission of the silicon nitride membrane at 7.2° in the 10-17nm wavelength region

Figure 3 shows the transmission of the silicon nitride membrane measured at incidence angle of 7.2° in the 10~17nm wavelength range. The high transmission at 13.9nm (about 50%) confirms that these membranes can be used as support for beam splitters. The specular x-ray reflectivity was measured as a function of the grazing angle with a reflectometer working with the Cu Kα emission (λ=0.154nm). The angular resolution of the reflectometer is better than 0.001°. The reflectivity curve is obtained by varying the grazing angle while tracking the reflected beam. This kind of curve allows the determination of the parameter of the period of multilayers.

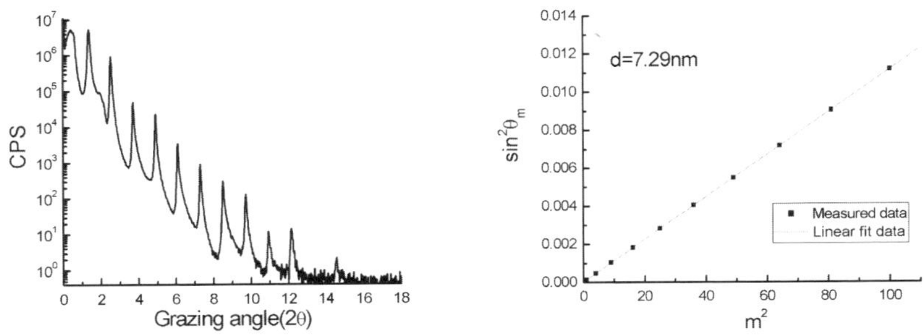

Figure 4. X-ray grazing angle reflectometry curve measured at 0.154nm on multilayer-coated silicon wafer sample and fitted curve for determining the period of multilayer

Figures 4(a) and (b) show the experimental curve and fitted curve for determining the period of multilayer. For the Mo/Si multilayer, the period is 7.29nm, which is in good agreement with the design one. The beam splitters are the key elements in the soft x-ray laser interferometer, and it guarantee not only its definite reflectivity transmission product, but also its surface precision.

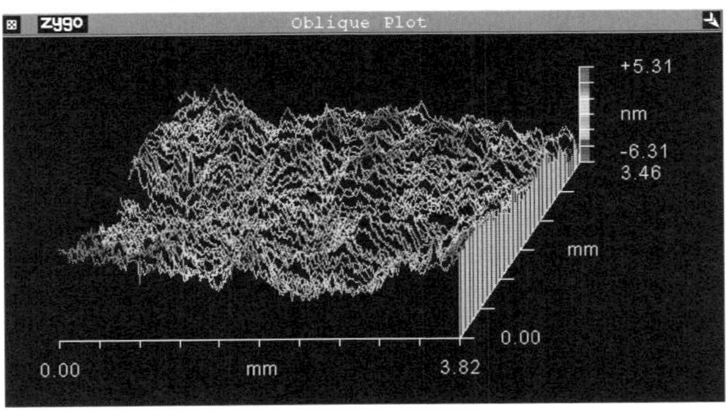

PV	11.622	nm
rms	1.757	nm
Ra	1.426	nm
Size X	3.82	mm
Size Y	3.46	mm

Figure 5. Figure errors of a beam splitter measured by ZYGO profiler

The Figure 5 shown the beam splitter surface precision is measured by the ZYGO profiler, as shown with the tested data, the beam splitter surface shape precision reached the nanometer magnitude in the center region of a beam splitter, which meeting the request of experiment.

Finally, synchrotron radiation measurements at 13.9nm provide a reflectivity of 18% and a transmission of 22% shown in Figure 6.

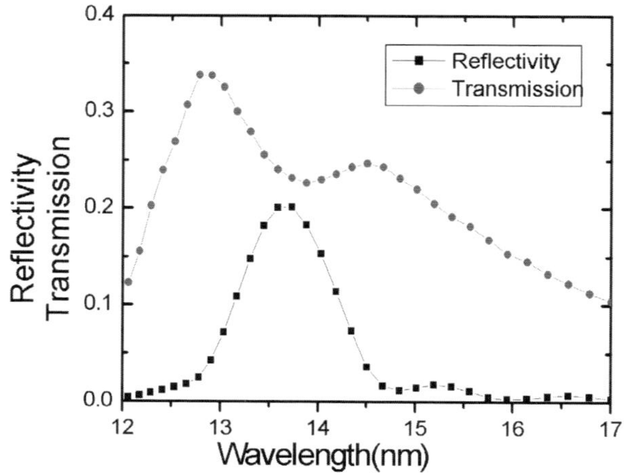

Figure 6. Experimental transmission and reflectivity measured versus soft x-ray wavelength. The angle of incidence was 7.2°

From these measurements the reflectivity and transmission product of the beam splitter is close to 4%. The beam splitters were successfully used in a Mach-Zehnder interferometer presented in this meeting and interference fringes were obtained.

5. Conclusions

The soft x-ray laser for the dense plasma diagnostics is an active research branch, and the beam splitter is one of the key components to realize the soft x-ray interferometer. Because it need not only certain reflectivity and transmission, but also higher surface precision, the high precision substrate material and the accurate layer thickness control device are needed in the process of fabricating the beam splitter. Taking the 100nm silicon nitride membranes as the substrate, beam splitters for an amplitude division interferometer working at a nickle-like silver13.9nm laser have been fabricated by using magnetron sputtering. The measured surface precision of the beam splitters has reached the nanometer platitude, and the reflectivity-transmission product has reached about 4%. The beam splitters were successfully used in a Mach-Zehnder interferometer and interference fringes were obtained.

Acknowledge: This work was supported by the National Natural Science Foundation of China under Contract No. 60378021 and National 863-804-7 sustentation fund.

References

[1] Silva L B Da, Barbee T W Jr, et al, 1995, Phys. Rev. Lett. **74**(20), 3991-3994
[2] Filevich J, Marconi M C, et al., 1999, Proc.SPIE 3776, 232-239
[3] Wang C, Gu Y, et.al.,2002, High power and particle beams, **14**(3):381 (in Chinese)

Inst. Phys. Conf. Ser. No 186
Paper presented at 9th Int. Conf. X-Ray Lasers, Beijing, China, 24–28 May 2004

Design, fabrication and characterization of multilayer mirrors for x-ray laser

H Ch Wang, Zh Sh Wang, Zh Zhang, W J Wu, F L Wang, B Wang, Sh J Qin, L Y Chen
Institute of Precision Optical Engineering, Tongji University, Shanghai 200092

Abstract: Design, fabrication, and characterization of multilayer mirrors for x-ray laser are described. The combinations of different materials are selected by using the available complex refractive indices at different x-ray wavelengths. The optimization of design and fabrication of multilayers aimed at making its reflectivity as high as possible. The performances of multilayers deposited by magnetron sputtering were characterized by x-ray diffraction, cross-section TEM, and the reflectivity at working wavelength was measured by using the synchrotron radiation facility at Beijing.

1. Introduction

The development and application of X-ray laser is one of the significant forward projects in the laser science and technology today. X-ray lasers have widespread application foregrounds with their special features in many fields such as Physics, Biology, Chemistry, Materials Science and Initial Confinement Fusion (ICF) [1-3]. The multilayer optical components, such as high reflectivity mirrors, beam splitters and phase retarders, are of great interest for x-ray astronomy, x-ray lithography and plasma diagnosis [4]. The multilayer for x-ray consists of some thin layers stacked in sequence, and a layer of absorbed material alternates with a somewhat thicker layer of spacing material.

2. Design and Optimization[5]

The reflectivity and transmission of the multilayer mirrors for x-ray laser are usually based on the Fresnel equation. We consider the reflection of a plane monochromatic light with wavelength λ on a multilayer consisting of m layers. The complex refractive index of j-th layer of material is

$$\widetilde{n}_j = n_j - ik_j \tag{1}$$

and the complex refractive index of the substrate is

$$\widetilde{n}_o = n_o - ik_o \tag{2}$$

In x-ray wavelength range, n and k are satisfied with $0<1-n<<1$, $k<<1$, and they usually decreases when the wavelength decreases. The thickness of j-th layer parallel film surface is assumed to be d_j. For the reflective layer, we use the following recurrent equation with amplitude reflectivity,

$$R_j^{n} = \frac{r_j^{n} + R_{j-1}^{n} \exp(-i\delta_j)}{1 + r_j^{n} R_{j-1}^{n} \exp(-i\delta_j)} \tag{3}$$

Where $\delta_j = 4\pi \tilde{n}_j d_j \cos\theta_j / \lambda$ is the phase shift by a return travel of the beam within the j-th layer, and r_j is the Fresnel amplitude reflection coeffiecient of the surface of j-th material in vacuum. The Fresnel coefficients are defined as
For s-component

$$r'_j = \frac{\tilde{n}_{j+1} \cos\theta_{j+1} - \tilde{n}_j \cos\theta_j}{\tilde{n}_{j+1} \cos\theta_{j+1} + \tilde{n}_j \cos\theta_j} \tag{4.2}$$

for p-component

$$r'_j = \frac{\tilde{n}_j \cos\theta_{j+1} - \tilde{n}_{j+1} \cos\theta_j}{\tilde{n}_j \cos\theta_{j+1} + \tilde{n}_{j+1} \cos\theta_j} \tag{4.2}$$

where Φ is the angle of incidence, and Φ_j is the angle of refraction in the j-th material satisfying the Snell's law

$$\sin\phi = n_j \sin\phi_j \tag{5}$$

If there are m layers in the multilayer system, the final amplitude reflectivity is $R_m^{\prime(s,p)}$,

which can be used to calculate the intensity reflectivity $R^{(s,p)} = R_m^{\prime(s,p)} \cdot R_m^{\prime(s,p)*}$, where s represents the s-component, and the p represents the p-component. If the incident angle and wavelength are determined, and the maximized intensity reflectivity R_{max} is required, then the thickness d_j should be optimized. In the model of optimization, we take into account the effect of the rms roughness of the interface between media i and j, and assume a Debye-Waller factor[6]

$$D_{ij} = \exp\left\{-2\left(2\pi\sigma_{ij}\tilde{n}_j \cos\theta_j\right)^2\right\} \tag{6}$$

In this paper the simplex algorithm is used to optimize the thickness for the R_{max}, and the merit function F is defined as

$$F = 1 - (R_s + R_p)/2 \tag{7}$$

When a multilayer mirror uses a period structure with two kinds of material, the thickness of the absorber d_a and the thickness of the spacer d_s should be optimized in the algorithm. The thickness of period d and the thickness ratio Γ are defined as

$$d = d_s + d_a \tag{8.1}$$

$$\Gamma = d_a/d \tag{8.2}$$

The maximized reflectivity for different wavelength of x-ray laser is calculated by using the above method, and the result is summarized in the table I. In the calculations the interfaces between the two materials were assumed to be smooth, abrupt, and without inter-diffusion across the interface. The optical constants for various materials were derived from the atomic scattering factors, and were obtained from the Center of X-ray Optics worldwide web server at Lawrence Berkeley National laboratory [7]. The variation of the reflectivity for Sc/Si multilayer used for 46.9nm x-ray laser with wavelength is shown in Fig.1, and the one for CoCr/C multilayer used for 4.48nm x-ray laser in Fig.2. For the Sc/Si multiplayer, the rms roughness influence the reflectivity slightly, while for the CoCr/C multiplayer, the reflectivity decrease greatly with the increase of the rms roughness. This mainly resulted from the fact that the rms roughness is much less than the wavelength and the thickness of period at 46.9nm, but for the shorter wavelength, the situation is much different. Therefore, the fabrication of the multilayer mirrors for the x-ray laser with a shorter wavelength should make the roughness as small as possible. Selecting suitable material and controlling the parameter of film deposition can resolve the problem.

Produce of the x-ray laser	Wavelength (nm)	Material	Number of Bi-layer Period	R	Thickness of Period (nm)	Γ
Ne-like Ar	46.9	Sc/Si	16	56.1%	25.2	0.42
Ne-like Cr	28.5	Mo/Si	30	29.4%	15.3	0.18
Ne-like Ni	23.1	Mo/Si	40	40.1%	12.1	0.20
Ne-like Ge	19.6	Mo/Si	40	48.4%	10.2	0.25
Ne-like Ni	14.6	Mo/Si	60	73.0%	7.5	0.36
Ni-like Ag	13.9	Mo/Si	70	74.2%	7.1	0.37
Ni-like Sm	7.3	Ru/C	180	35.1%	3.7	0.37
Ni-like Ta	4.48	CoCr/C	150	39.3%	2.3	0.64

Table I. The maximized reflectivity for different wavelength x-ray laser

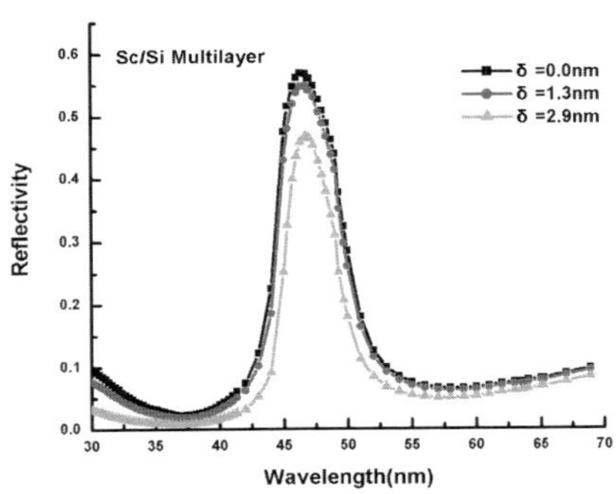

Fig. 1 The variation of the reflectivity of Sc/Si multilayer with the wavelength at different rms roughness for 46.9nm Ni-like Ar x-ray laser (N=16, d=25.2nm and Γ=0.58)

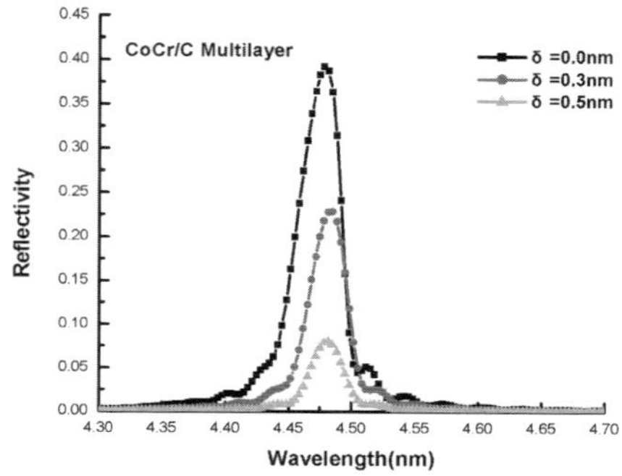

Fig. 2 The reflectivity as a function of the wavelength for CoCr/C multilayer at different rms roughness for 4.48nm Ni-like Ta x-ray laser (N=150, d=2.27 nm and Γ=0.537)

3. Fabrication of the multilayer mirrors

All ML's were deposited with a JGP560C6 DC magnetron sputtering coater. Circular sputtering sources with a 100-mm diameter were used. During deposition the sputter power was hold constant, and adjustable range was 30W~130W. Before deposition the sputtering chamber was evacuated to a pressure of 5×10^{-5}Pa. During deposition, a flow of typically 15-20 standard cm3/min of Ar (99.998% purity) was introduced to the vacuum chamber to maintain a presusure of 200-250 Pa, depending on the sputtering target. Before the ML was fabricated, monolayers of each material were deposited and measured by X-ray diffraction (XRD), which is used to fit the thickness of the film, and to determine the rate of deposition for different materials. Therefore, the time when a sample stays over a target can be used for controlling the thickness of each layer, and a computer was used for controlling a stepping motor determining the time.

1. Characterization of the multilayer mirrors for x-ray laser

The multiplayer mirrors fabricated were characterized by the XRD to fit the thickness of period using the corrected Bragg equation. The reflectivity of some multiplayer mirrors was measured by using the Beijing Synchrotron Radiation Facility. The reflectivity as a function of the energy for W/C multiplayer mirror at different incident angles is shown in Fig. 3. The highest reflectivity can reach more than 6% at about 250-260eV. The structure of W/C multiplayer mirror was also characterized by the TEM, and shown in Fig.4. The interface between the tungsten and carbon is very clear, but there is a little fluctuation on each interface, which is resulted by the roughness of the substrate Si. The reflectivity as a function of the energy for Mo/Si/C multilayer mirror at different incident angles from 42 degree to 48 degree is shown in Fig. 5. The highest reflectivity can reach more than 45% at about 97eV at incident angle of 48 degree. A carbon layer with a thickness about 0.2nm was used to smooth the interface and decrease the roughness of interface. The reflectivity as a function of the energy for Ni/B4C multilayer mirror at different incident angles from 4 degree to 60 degree is shown in Fig.6. The highest reflectivity is only 4.5% at about 163eV at incident angle of 45 degree.

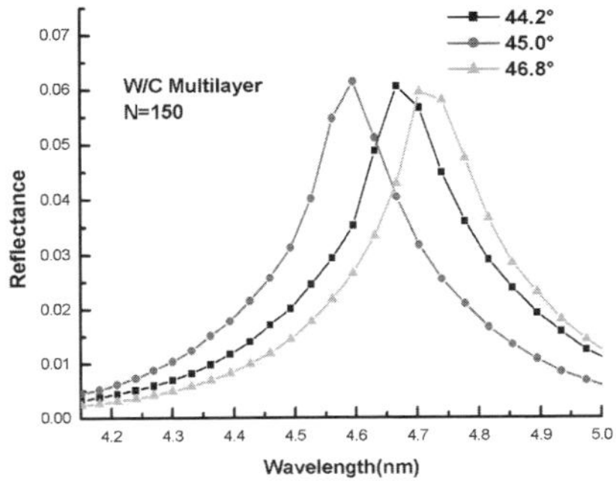

Fig. 3 the reflectivity as a function of the energy for W/C multiplayer mirror, which are measured at incident angle 44.2 degree, 45 degree and 46.8 degree. The thickness of the period d=4.19nm and thickness ratio Γ=0.755

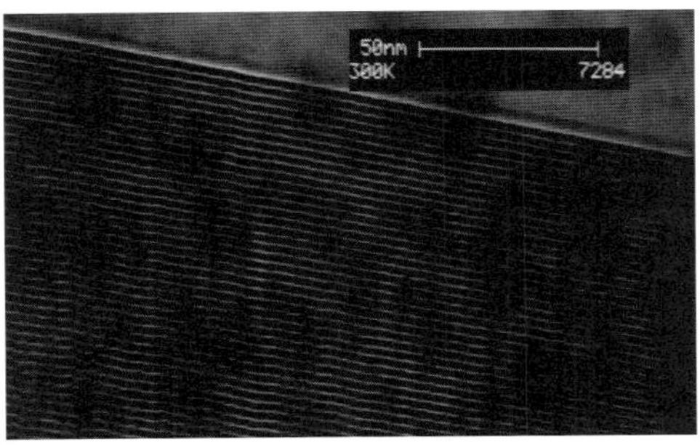

Fig. 4 The image of TEM for W/C Multilayer

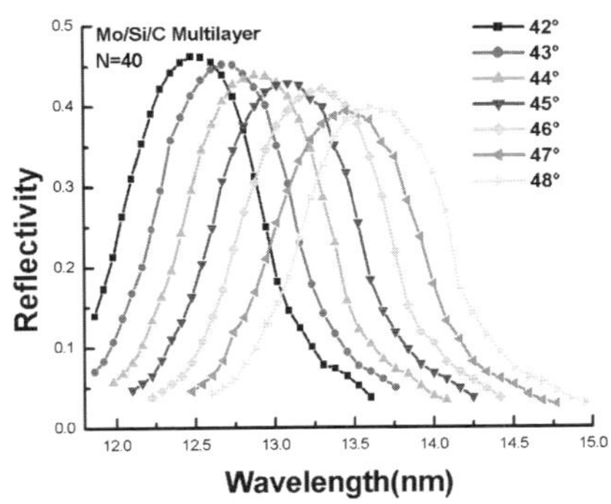

Fig.5 The reflectivity as a function of the energy for Mo/Si/C multilayer mirror at different incident angles from the 42 degree to 48 degree. The thickness of period d=10.02nm and Γ=0.755 for Mo/Si multiplayer, the thickness of the carbon is about 0.2nm

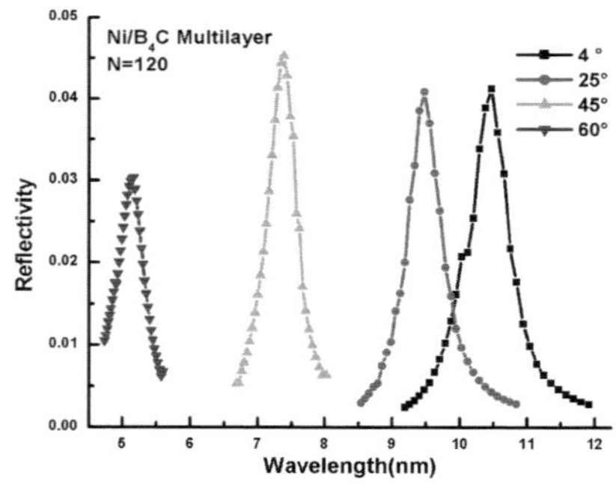

Fig.6 The reflectivity as a function of the energy for Ni/B4C multilayer mirror at different incident angles from the 4 degree to 60 degree. The thickness of the period d=5.47nm and Γ =0.6

4. Summary

In this paper, the principle of designing the multilayer mirrors with a higher reflectivity for x-ray laser is introduced, the reflectivity is calculated at some different wavelength, and the roughness of interface is considered and analyzed at 46.9nm for Sc/Si multilayer, and at 4.48nm for CoCr/C multilayer. The image of the TEM for W/C multiplayer mirror is given, and the reflectivity of W/C, Mo/Si/C and Ni/C multilayer mirrors is measured by using the synchrotron radiation facility at Beijing. All the work of fabrication and characterization has shown the multilayer mirrors can be used for x-ray laser.

Acknowledgments

Financial support from the National Natural Science Foundation of China (No. 60178021) and the Nanometer Technology Special Foundation of Shanghai Science and Technology Committee (No. 0352nm090) are gratefully acknowledged. We also wish to thank the National Laboratory of Beijing Synchrotron Radiation for providing the facility for this work.

References

[1] Trebes J E , Brown S B, Campbell E M. et al., 1987, Science, 238 : 517~519.
[2] DaSilva L B, Trebes J E, Balhorn R. et al., 1992, Science, 258 : 269~271.
[3] Ress D, DaSilva LB, London R A. et al., 1994, Science, 265 : 514~517.
[4] Back C. A., Kauffman R. L, Bell P. M. et al.. 1995, [J]. Rev Sci. Instrum., 66(1):764~766
[5] Bennett H E and Porteus J O, 1961, J. Opt. Soc. Am. , 51:123-129
[6] Yamamoto M. and Namioka T., 1992, Appl. Opt. , 31:1622-1630
[7] http://www-cxro.lbl.gov/optical_constants

Inst. Phys. Conf. Ser. No 186
Paper presented at 9th Int. Conf. X-Ray Lasers, Beijing, China, 24–28 May 2004

Determination of EUV optical constants in reactive materials near the absorption edges of elements

Yu. Uspenskii[a], J. Seely[b], N. Popov[a], I. Artioukov[a], A. Vinogradov[a], D. Windt[c], B. Kjornrattanawanich[d], Yu. Pershin[e] and V. Kondratenko[e].

[a]Lebedev Physical Institute, Leninski pr. 53, 119991 Moscow, Russia
[b]Naval Research Laboratory, Washingthon DC 20375, USA
[c]Columbia University, 550 West 120 St., New York NY 10027, USA
[d]Universities Space Research Association, NSLS, Upton NY 11973, USA
[e]Kharkov State Polytechnic University, Frunze str., Kharkiv 310002,Ukraine

Abstract Data on the complex refractive index n(ω) are necessary for the design of EUV multilayer optics. Atomic scattering factors inadequately describe n(ω) near the absorption edges of elements. At long EUV wavelengths, n(ω) is sensitive to air contamination. We show that our recent method precisely determines n(ω) near the absorption edges even in reactive materials. Transmissivity and reflectivity measurements are performed for a series of identically protected films deposited on photodiodes. To determine n(ω) above and below absorption edges, both thin (~10 nm) and thick (~100 nm) films are used. Data treatment separates the contributions of a capping layer and samples and provides data for a pure material. As an application, the n(ω) of Sc, Ti and La is determined at 17-70 eV, 17-99 eV and 14-438 eV, respectively.

1. Introduction.

Precise data on the complex refractive index $n(\omega)=1-\delta(\omega)+i\beta(\omega)$ are necessary for the design of multilayer coatings used in optics for EUV lasers, EUV lithography, X-ray astrophysics, and many other applications. Atomic scattering factors, which were published for all elements in the interval 10-30,000 eV [1], usually assume a discontinuous description of $\beta(\omega)$ near the absorption edges of elements. Such an approximate description of optical constants introduces significant errors in the design of multilayers. This is highly disappointing, as the best working characteristics of coatings commonly are just below the absorption edges. The problem is that the study of absorption edges requires the films of 5-10 nm to transmit EUV radiation, while the fabrication of thin free-standing films is difficult. At long EUV wavelengths, one more difficulty arises from high absorption by C, N, and O atoms that results in the sensitivity of n(ω) to air contamination. This effect grows many fold and becomes of principal importance when reactive or getter materials are investigated. In view of these difficulties, we invoke for the study of absorption edges our recent method developed for reactive materials [2],.which makes possible to measure thin and thick films with equal ease. Below it is applied to the determination of optical constants of Sc, Ti and La in wide energy intervals, which include the $M_{2,3}$ edges of Sc and Ti and the $O_{2,3}$ and $N_{4,5}$ edges of La.

2. Description of the method.

The theoretical foundations of our method are the exact formulas for the reflectivity *R* and transmissivity *T* of the structure which consists of a capping structure C, a layer of material being studied M with a thickness *d*, and a buffer structure B deposited on a substrate S [2]:

$$R = R_c \left|1 + \xi f \exp(2iD)\right|^2, \tag{1}$$

$$T = T_c \cdot T_b \left| f \exp(iD)\right|^2, \tag{2}$$

were T_c and T_b are the transmissivities of the C and B, R_c is the reflectivity of the C, and $D=2\pi dp/\lambda$ with $p \Leftrightarrow p'+ip''=\sqrt{n^2-\sin^2\theta}$, ($\lambda$ and θ are the wavelength and angle of incidence from vacuum). The complex quantity ξ is expressed through the parameters of the structures C and B, but not the M. The factor f arises from the multiple reflection of radiation inside the M layer. In the EUV region, this reflection is small, so generally $f \approx 1$. For two samples, which differ only in the thickness of the M layer (d_i or d_j). the ratios of their *R* and *T* at normal incidence are:

$$R_i / R_j \approx \left|(1+\xi \exp(4i\pi\, d_i n/\lambda)/(1+\xi \exp(4i\pi\, d_i n/\lambda))\right|^2, \tag{3}$$

$$T_i / T_j \approx \exp(-4\pi\beta(d_i - d_j)/\lambda), \tag{4}$$

where multiple reflection inside M was neglected. The capping and buffer structures are identical in both samples and therefore drop out of Eqs. (3) and (4), remaining only in $\xi = \xi' + i\xi''$. Eq. (4) gives a sufficiently accurate $\beta(\omega)$, which allows the determination of $\delta(\omega)$ from the Kramers-Kronig relation. As an improvement, it is possible to correct $\delta(\omega)$, which was obtained from the Kramers-Kronig relation, utilizing Eq. (3) and doing the fitting of calculated ratios $R_i(\omega)/R_j(\omega)$ to experimental quantities. The use of this procedure decreases the standard difficulties of the Kramers-Kronig method arising from a limited measurement interval $\omega_{min}<\omega<\omega_{max}$. The precision of transmissivity measurements strongly depends on the thickness of a film and an interval of ω. In high-absorption intervals precise data are obtained only with thin films (d~10nm), while in low-absorption interval more accurate measurements are performed with thick films (d~100 nm). These reasons defined the need in both thin and thick samples when a region around the absorption edge is investigated.

3. Samples preparation and measurements.

The six films of each metal were deposited by magnetron sputtering on silicon photodiodes from International Radiation Detectors Inc of the type AXUV-SP2 (Sc), AXUV-100 (Ti) and AXUV-100G (La). The samples were protected with a silicon film of 5 nm (Sc), 8 nm (Ti) and a B_4C film of 7 nm (La). The transmissivity of the films was measured with the National Synchrotron Light Source beamline X24C at Brookhaven National Laboratory in the energy intervals 17-70 eV (Sc), 17-99 eV (Ti) and 14-438 eV (La). The reflectivity measurements of some Sc and La films were also done at these energies.

4. Discussion of results.

The measured transmissivity of Sc, Ti and La films is shown in Fig. 1. The first two metals have low transmission in the interval of intense 3p→3d electron transitions (the region of the $M_{2,3}$ edge), which is placed at 28-52 eV in Sc and at 32-61 eV in Ti. The measured T(ω) spectra of La have two deep minima located at 16-30 eV and 110-140 eV. The minima originate, respectively, from 5p→5d and 4d→4f electron transitions in La (the $O_{2,3}$ and $N_{4,5}$ absorption edges). One more minimum which is seen in the thin films of La at 188-210 eV arises from EUV absorption in a B_4C capping layer and corresponds to the K-edge of boron. In the thick films of La, this minimum is hardly distinguished.

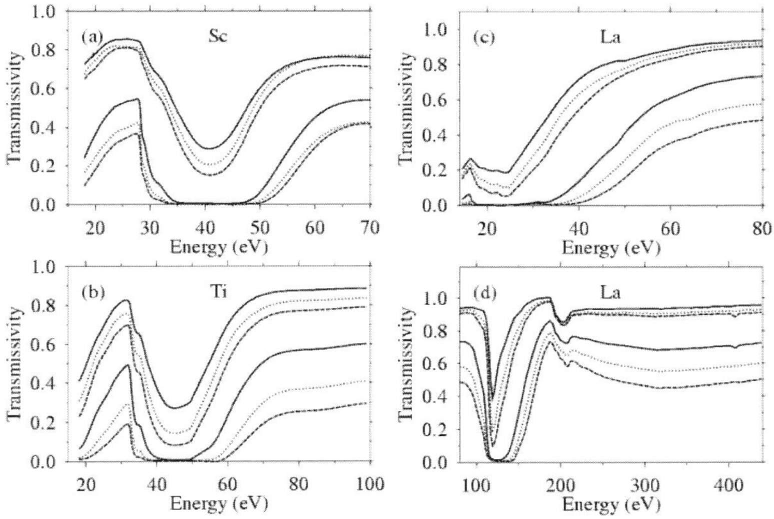

Figure 1. The transmissivity spectra of Sc, Ti and La: (a) for Sc films with the thickness d=7.5 nm, 10 nm, 12.5 nm, 70 nm, 100 nm, and 130 nm (going from the top down), (b) for Ti films with d=10 nm, 14 nm, 18 nm, 40 nm, 70 nm, and 100 nm, (c) and (d) for La films with the thickness d=7 nm, 12 nm, 17 nm, 71 nm, 120 nm and 170 nm.

The obtained δ(ω) and β(ω) of the metals are presented in Fig. 2 together with the data of Ref. 1 given for comparison. It is seen that away from the absorption edges our optical constants closely coincide with those of Ref. 1. But near the edges the difference between the constants is very significant, especially at low energies. In particular, the atomic tables give a different shape of the $M_{2,3}$ edge in Sc and Ti, while our measurements indicate that the shape is identical. The latter is reasonable, as Sc and Ti are neighbors in the Periodic Table. A similar difference in data is observed for the La $O_{2,3}$ edge. Only for the La $N_{4,5}$ edge the agreement with the atomic tables is satisfactory.

At low EUV energies, the observed disagreement with the data of Ref. 1 is likely connected with the effects of air contamination. The maximum of β(ω) in Ti at 23 eV cited in the atomic tables is caused by the reaction of samples with atmospheric N and O atoms, which have the maximal EUV absorption at 20 eV. Our method excludes such side contributions. It is characteristic that the minimum of T(ω) at 188-210 eV, which arises from the capping B_4C layer, perfectly drops out from our δ(ω) and β(ω) of La. An added point in our support is a good agreement obtained between the calculated and

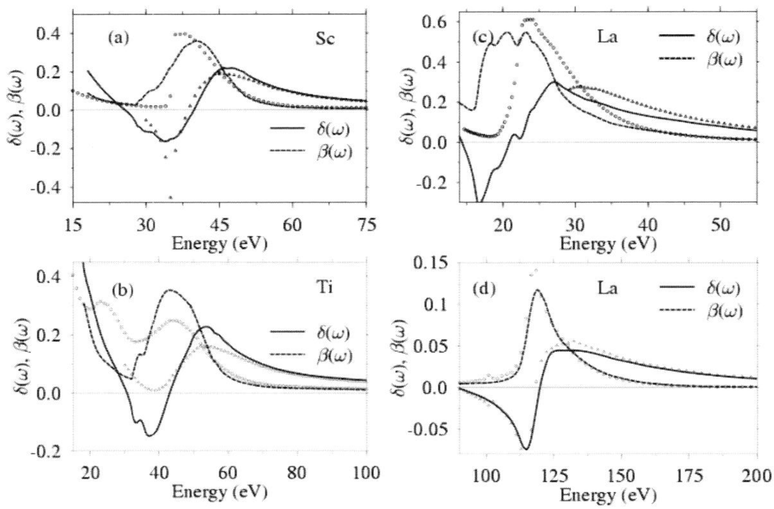

Figure 2. Obtained optical constants δ(ω) and β(ω) (shown by lines) in comparison with the atomic data tables (shown by symbols) for metals: (a) Sc, (b) Ti, (c) and (d) La.

added point in our support is a good agreement obtained between the calculated and experimental ratios of reflectivity of Sc and La. The wide use of Sc/Si multilayers in optics (EUV lasers operating at 46.9 nm [3], astrophysics [4] etc) gives good opportunity for verification of the optical constants of Sc. For these multilayers it was found an excellent agreement between calculated (with our optical constants) and experimental reflectivity [2]. All these facts evidently show the high precision of our optical constants especially close to the absorption edges and their potential for the calculation and design of high-reflective EUV multiplayer mirrors.

Acknowledgements.
This work was supported in part by the US Civilian Research & Development Foundation grant RP-2-2326-MO-02, as well as the grants 1794 and 2297 of the International Science and Technology Center. Three of us (YU, NP and AV) were also supported by the grants N 04-02-17349, N 02-02-16599, and N 03-02-16438 of the Russian Fund for Basic Research as well as by the grants I0859 and B0056 of the Integration Program. Two authors (JS and DW) were supported respectively by NASA grants S-92543-F and NAG 5-12797.

References

[1] B.L. Henke, E.M. Gullikson, et.al., At. Data Nucl. Data Tables, **54,**181-342(1993).
[2] Yu.A. Uspenskii, J.E. Seely, et.al., JOSAA 21, 298-305 (2004).
[3] J.J. Rocca, M. Frati, et.al., Acad. Sci. Paris, 1, Serie IV, 1063-1072 (2000).
[4] J.F. Seely, Yu.A. Uspenskii, et.al.,Appl. Opt. 41, 1846-1851 (2002).

Inst. Phys. Conf. Ser. No 186
Paper presented at 9th Int. Conf. X-Ray Lasers, Beijing, China, 24–28 May 2004

A High Dynamic Range Streak Tube with Plane Focusing Electron-Optical System

Min Chen, Kezheng Zhu, Baosheng Zhao, Hongbo Liu
Xi'an Institute of Optics and Precision Mechanics of CAS,
Xi'an China, Xi'an 710068

Abstract.A high dynamic range streak tube with plane focusing electron-optical system is designed. Its strong ability of focusing can reduce the photoelectron transit time in the tube. The focusing system also can decrease the density of space photoelectrons. Therefore, the space electric charge effect can be restrained greatly. A couple of compensating plates are applied in the system to improve temporal resolution. By computer simulation of photoelectron trajectories, some data such as optimum image position, the deflecting linearity of 1%, the physical temporal resolution of 400 fs, the spatial resolution of 50lp/mm and the MTF in deflection direction are obtained.

1. Introduction

The streak camera as a powerful instrument is applied in the study of dynamic behavior of ultra high-speed optical phenomena, such as multi-channel measurement and characterizing pulse shapes of laser. In the last decade streak cameras have found an increasing number of applications in particle accelerators permitting both precise measurements and instructive visualizations of beam characteristics and behavior that can not be obtained from other beam instrumentation. The developments of streak cameras have always been driven by the need for new studies. In recent years the experiment of the laser nuclear fusion is being carried out actively. Therefore, the streak camera is required to be able to measure with high dynamic range the pulse shapes and timings of many laser beams accurately.

The streak tube is the main part of a streak camera. In order to satisfy such requirements, a streak tube with high dynamic range and good temporal resolution is designed. To improve dynamic range by restraining the space electric charge effect is one of the most effective methods (e.g. [3]).

2. System configuration

According above, plane focusing electron-optical system is applied in the tube to obtain high dynamic range. Figure1 shows the system configuration of the designed tube.

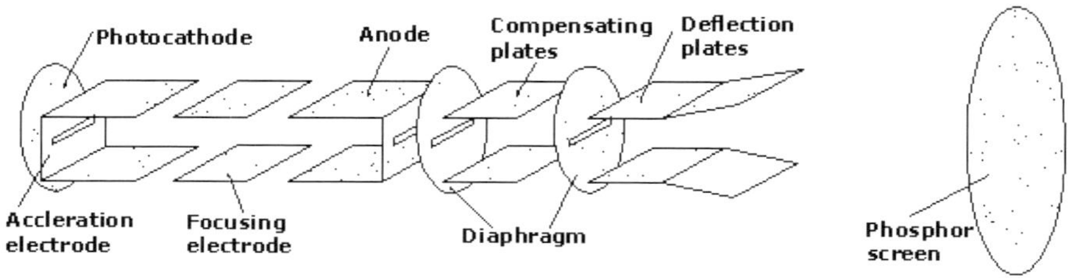

Figure1. Configuration of the tube

As shown in Figure1, light plus is focused onto the photocathode and converted to photoelectrons. The photoelectrons are accelerated by the strong electric field applied between the photocathode and the acceleration. After passing the compensating plates, they are conducted towards phosphor screen. A linear voltage ramp is applied to the deflection plates to produce a linear sweep with respect to time. The photoelectrons are swept at high speed from bottom to top on the screen. Then information in time domain is dispersed in deflection direction as a streak image (e.g. [1] [2]).

In this electron-optical system, a couple of compensating plates are applied to improve temporal resolution. When passing the compensating plates, which are applied the same higher voltage than axis, those photoelectrons far from axis get a greater acceleration than those close to axis. That will help most photoelectrons arriving at phosphor screen simultaneously and decrease temporal dispersion.

The design of deflection plates is also very important. It is better to make them have maximum deflecting sensitivity and minimum capacitance (e.g. [4]).

3. Characteristics of the tube

3.1 Deflecting linearity

The average deflecting distance of 2500 photoelectron trajectories under certain deflecting voltage is calculated. By changing the deflecting voltage, different deflecting distances are obtained. As showing in Figure2, the deflecting linearity is less than 1%.

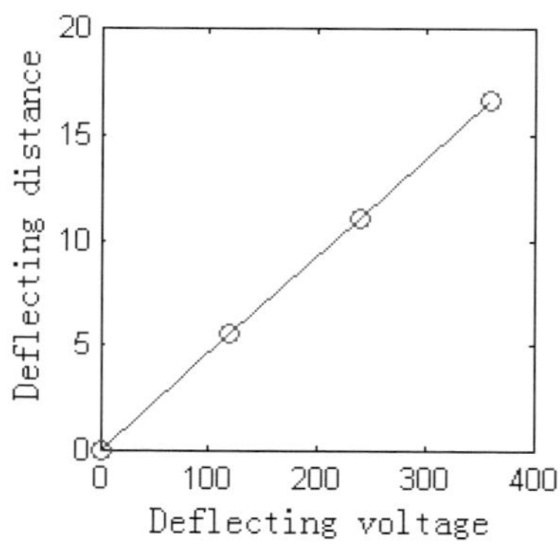

Figure2. Deflecting linearity

3.2 Response of spatial δ function and MTF in deflection direction

Photoelectrons emitted from photocathode at the same point always have different velocities. They always arrive at different positions of phosphor screen. The statistics of position is the spatial spread function. As shown in Figure3, the positions out of axis have a greater spatial dispersion than those on axis. Figure4 show that with the distance from axis increasing, the MTF became worse. Even the position of 0.5 mm from axis, the spatial resolution still can achieve 50lp/mm.

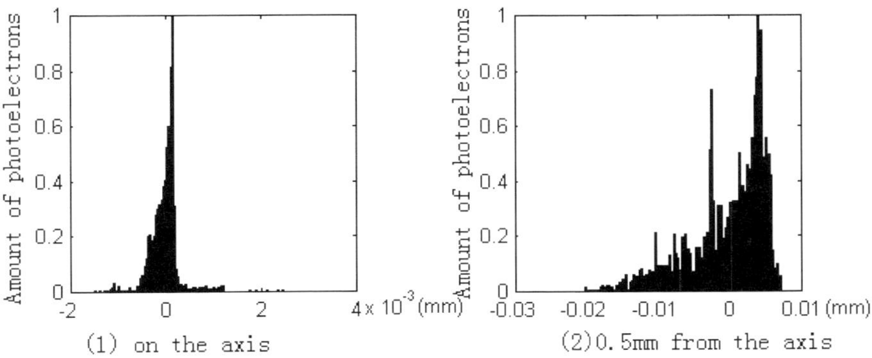

Figure3. Spatial spread function at optimum image plane

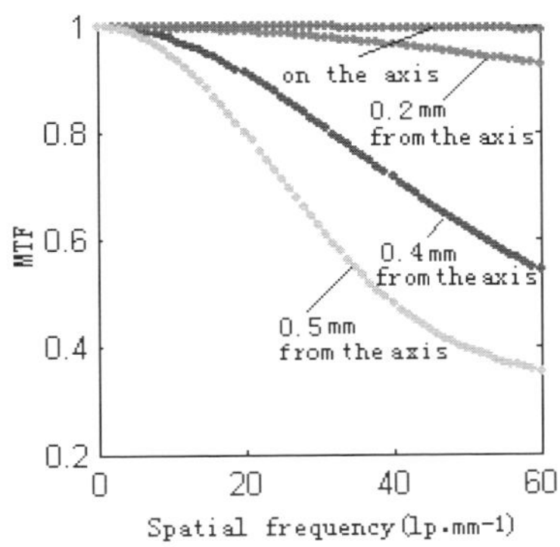

Figure4.MTF in deflection direction

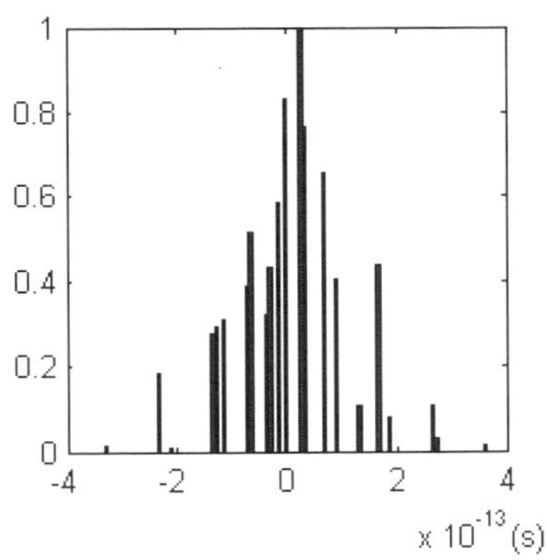

Figure5. Response of time-dependent δ function

3.3 Response of time-dependent δ function

Response of time-dependent δ function is one of the most important parameters. Given photoelectrons emitted from photocathode at the same time, the statistics of the time they achieve phosphor screen is the response of time-dependent δ function. By simulating 14000 photoelectron trajectories, we get the function as shown in Figure5. The physical temporal resolution is about 400fs (FWHM=400fs).

4. Conclusion

We have designed the high dynamic range tube which has a better spatial resolution and temporal resolution. The plane focusing electron-optical system is proposed to enlarge the dynamic range of the tube. This instrument allows us to perform a precise measurement and study high-speed optical phenomena. But, we just get the parameters above by theoretical calculation. The development of the tube is being carried out. Figure6 shows the frame construction.

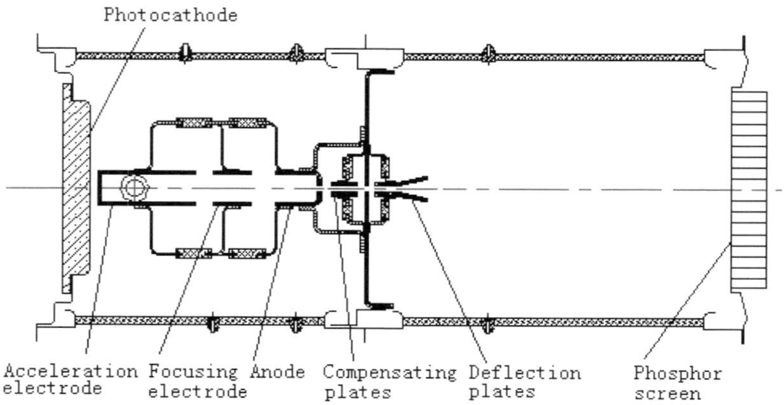

Figure6. Frame construction of the tube

References

[1] T Takeshita, K Suzuki, *et al.* Development of a high dynamic range streak camera [M].2000.139.
[2] Y.Inagaki, T.Takeshita, K.Kinoshita. Preceedings of the SPIE. 2001.Vol.4183
[3] Li Ji, Niu Han-ben Journal of optoelectronics laser, 2002,Vol.13(8)
[4] Hua Zhongyi, *et al.* Electron Optics, FUDAN University publisher(1990).

Inst. Phys. Conf. Ser. No 186
Paper presented at 9th Int. Conf. X-Ray Lasers, Beijing, China, 24–28 May 2004

Passively mode-locked all-solid-state Nd:YVO$_4$ laser with piosecond pulse duration

Y L Jia, J A Zheng, Z Y Wei*
Laboratory of Optical Physics, Institute of Physics, Chinese Academy of Sciences, Beijing, 100080, China

Y G Wang, X Y Ma
Institute of Semiconductors, Chinese Academy of Sciences, Beijing, 100080, China

Z G Zhang
Ultrafast Laser Laboratory, Tianjin University, Tianjin, 300072, China

Abstract. We present an all-solid-state picosecond pulse laser as a potential driver for X-ray lasers. The passively mode-locked laser is built by using a semiconductor absorber mirror, which also acts as an output coupler at the same time. The reached shortest pulse width is about 2.3 picosecond, corresponding to a single pulse energy of 4.3 nJ and a peak power of 1.8 kW. Average mode-locked power of 1.7 W is generated at 9 W pump power.

1. Introduction

All-solid-state lasers with high energy output and picosecond pulse duration has been proven to be a potential driver for X-ray lasers [1]. With the development of passive mode-locked all-solid-state lasers with SESAM (semiconductor saturation mirror) in recent years, picosecond lasers with an average power about 100 W has been obtained at 1 μm wavelength[2]. Boosting this laser to the energy of J level, it will enable us to develop a compact X ray laser source within top-table size. Self-starting passive mode-locking of all-solid-state ultrashort lasers have been demonstrated, which have a high average power[3-4] in several picoseconds(ps)[5-7], but none of them less than 10 ps. Especially with the semiconductor saturable absorber mirrors output coupler (SESAMOC), the new kind of semiconductor saturable absorber, only two were reported with our knowledge. U.Keller *et al* presented a passive Q-switched Nd:YVO$_4$ laser in 2001[8], the pump power on the microchip was 160 mW and the average output power was 7.7 mW with 160 kHz through the SESAMOC, and the pulse width is 143 ps. Recently, Wang *et al* [9] reported a 10 ps mode locked Nd:YAG laser by SESAMOC, with the recovery time of 0.5 ps. In this paper, we show a mode-locked Nd:YVO$_4$ laser with a SESAM as absorber and output coupler. The laser cavity is designed with a three mirrors' configuration, the pump laser is a fiber-coupled diode laser at the wavelength of 808 nm.

* Author to whom correspondence: zywei@aphy.iphy.ac.cn

For easy coupling the pump laser into the Nd:YVO_4 crystal, we placed the gain medium nearby the end mirror, the gain crystal is 12 mm in length and doping in 0.5 % Nd. With a pump power of 9 W, a stable mode-locked power of 1.7 W was obtained; the typical pulse duration is about 2.3 ps. This work supply a possible seeding source to develop an all-solid-state picosecond laser system to drive X-ray laser.

2. Schematic of the all-solid-state laser mode-locked by SESAMOC

Different from the most previous reports, we chose a compact three-mirrors cavity in this experiment and use the SESAM as the output coupler. Figure 1 is the schematic layout. The gain medium is a 12 mm Nd:YVO_4 crystal with 0.5 % Nd^{3+}doping. The diode laser was coupled into the laser crystal through M_1, the flat input mirror coated with 99.8% reflectivity at the laser wavelength (1064 nm) and high transmission (>98%) at the pump wavelength (808 nm). The radius of curvature of the folding mirror M_2 is 100 mm, which focus the laser beam on the SESAMOC with a spot of 50 μm diameter. The transmission of the SESAMOC is about 10% at 1064 nm. For cooling the output coupler, we partly fixed the SESAM on a copper heat sink.

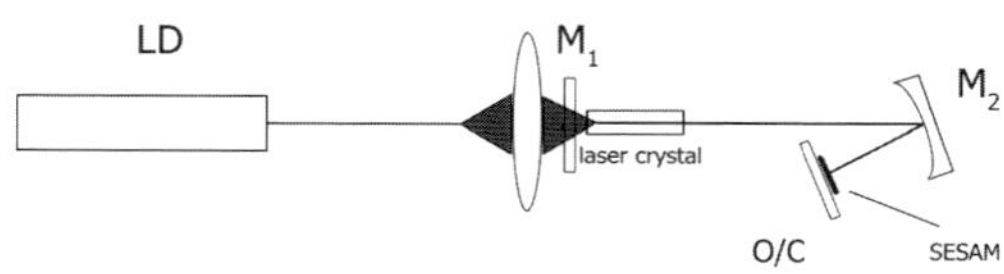

Figure 1. Layout of the all-solid-state laser mode-locked by SESAMOC

3. Result of the experiment

The output power in mode-locked regime is shown as Fig. 2. From the figure we can find the highest slop efficiency is 44.4%, and the highest optic-optic efficiency is 18.9% while the incident power is 9.25 W and the average output power is 1.75 W. The highest stable average output power is 1.7 W, limited by the damage threshold of SESAM. We expect to further increase the threshold to get higher output power in next step. We hope we can improve the result to more than 5 W. The threshold pump power for mode-locking mode is 655 mW, in which the stability of the mode-locking become worse. The damage threshold of the SESAMOC restricted the output power for further increasing, we hope to raise the result by improving the cavity configuration and the structure of the modulation material.

The duration of laser pulse was measured using a non-colinear autocorrelator. The correlation trace is shown in Fig.3, which shows a typical pulse duration of 2.3 ps under the output power of 1.05 W. The spectrum of mode-locked laser pulses were also measured with a spectrometer (Ocean Optics Inc, resolution is about 0.1nm), the corresponding bandwidth is about 0.7 nm which is shown in Fig. 4. Using Fourier transformation, we obtained the time-bandwidth product is 0.56. The typical repetition rate is 400 MHz, correspond to a single pulse energy of 4.25 nJ and a peak power of 1.8 kW.

4. Experiment of second-harmonic-generation

Based on above experimental result, we added a nonlinear crystal KTP in the cavity to observe the second harmonic phenomenon. Because of the absorption of the SESAMOC, the main green laser output from the folding mirror side, output power of about 30 mW as measured when the fundament laser power is 1.05 W. An oscilloscope (520D, Tektronix

Inc.,) was used to monitor the mode-locked signal. Fig. 5 shows the pulse train observed by the oscilloscope. It clearly shows there are 6 pulses per 5ns scale. This means the repetition rate is 1.2 GHz. On the assumption that the output mode is single transverse, the repetition rate should be 400 MHz because the cavity length is about 38 cm, it is only the one-third of the experimental value. We believe the reason is the cross-phase modulation of the fundamental wave and the second-harmonic wave in the cavity, the detailed analysis will be introduced in another paper. We also observed Q-switching phenomenon, but less regular than the mode-locking curve.

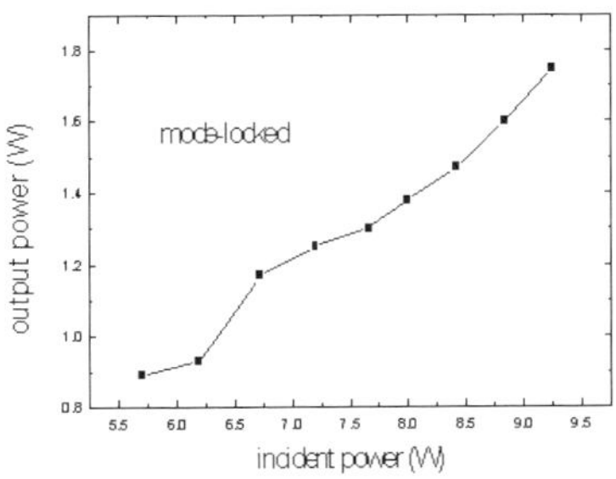

Figure 2. Output power vs incident power

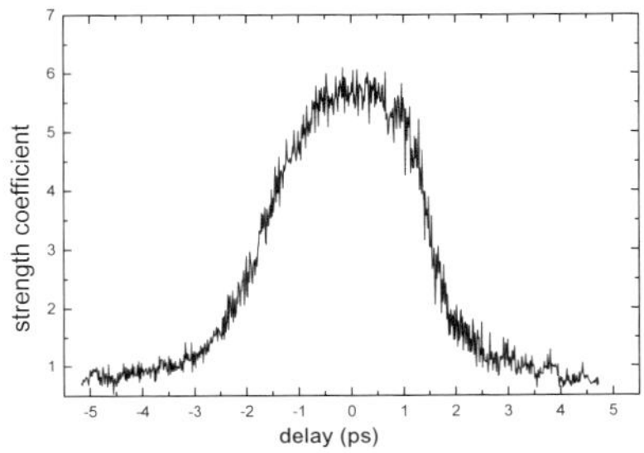

Figure 3. Trace of autocorrelation

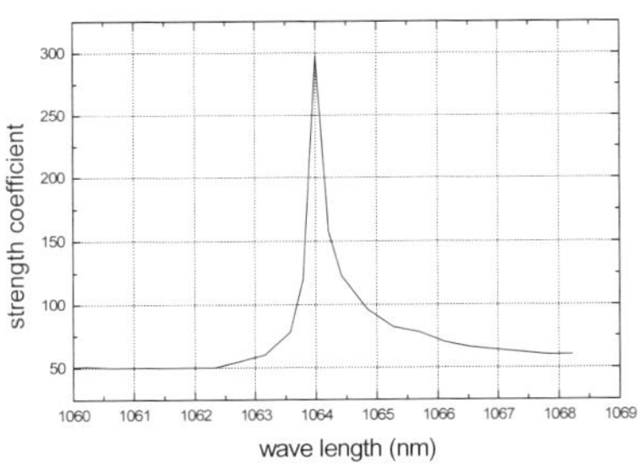

Figure 4. spectrum curve

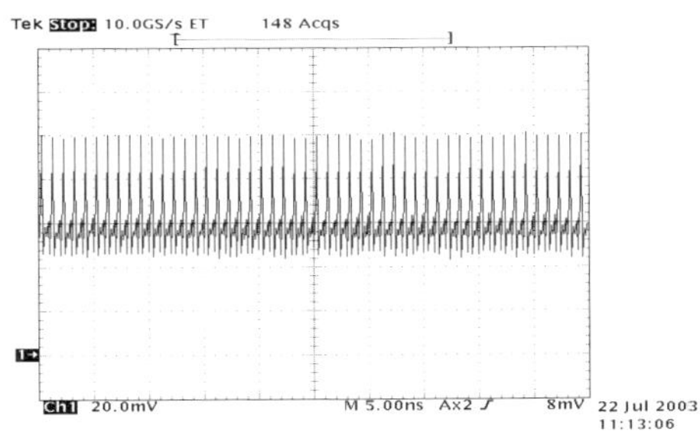

Figure 5. Mode-locking laser pulse trace

5. Conclusions

In conclusion, we have presented an ultrashort Nd:YVO_4 laser, which is mode-locked by a novel SESAMOC. The pulse width is 2.3 ps at an output power of 1.05 W, to our best knowledge, it's the shortest pulse width reported in the world for this kind lasers. The corresponding bandwidth of spectrum is measured to be about 0.7 nm. Using the Fourier transformation, the time-bandwidth product is estimated to be 0.56, less than the theoretical limit 0.35. The high stability was observed under the output average power is 1.7 W. The corresponding peak power is 1.8 kW. The single pulse energy is 4.25 nJ. These results present good characteristics for a possible seeding source to develop an all-solid-state picosecond laser system to drive X-ray laser.

6.Acknowledgements

This work was partially supported by the National Nature Science Foundation of China under Grant No. 10374114, 10176034, 60321003, 10474134, 60128505, 60225005 and the NKBRSF under Grant No. G1999075206, G1999075202.

References

[1] Ozaki T and Kuroda H 2002 Phys. Rev. Lett. 89 253902-1-253902-4.
[2] Krainer L and Keller U 2000 Appl. Phys. Lett. 77 2104-2105.
[3] Burns D and Dawson M D 2000 J. Opt. Soc. Am. B 17 919-926
[4] Chen Y F and Huang K F 2001 Opt. Lett. 26 199 - 201
[5] Krainer L and Paschotta R 2000 Appl. Phys. Lett. 77 2104-2105
[6] Braun B and Zhang G 1997 Opt.Lett. 22 381-383
[7] Graf Th and Bente E 1999 Opt.Comm. 159 84-87
[8] Spühler G J and Keller U 2001 Appl. Phys. Lett. 78 2733-2735
[9] Gang W Y and Ma X Y 2003 Chin.Phys.Lett. 20 1960-1962

Inst. Phys. Conf. Ser. No 186
Paper presented at 9th Int. Conf. X-Ray Lasers, Beijing, China, 24–28 May 2004

A reflective compound lens array scheme in longitudinally pump X-ray lasers

H.M.Li[1,2], Y.J. Li [1,2], X.Lu[1], J. Zhang [1]
[1] Laboratory of Optical Physics, Institute of Physics, Chinese Academy of Sciences, Beijing 100080, China
[2] Department of Physics, China University of Mining & Technology, Beijing 100083, China

Abstract. A scheme of compound lens array to attain an approximately uniform irradiation is proposed and studied to realize the control on the distribution of luminous intensity. The irradiation uniformity along the line focus in longitudinal pump is improved with the compound lens array. The energy attenuation of the laser along the propagation in the plasma is compensated with the modulation of structure of the compound lens array. The proposal of lens array is capable of increasing the focal length as well as the irradiation uniformity along it.

1. Introduction

A very important objective in the development of X-ray laser is to attain a highly coherent and saturation output at short wavelength from a small-scaled facility, the so-called "table-top" system. Gain saturation have been demonstrated in Ne-like Ge[1], Se[2],Y[3], and Ni-like Ag[4] and Sm[5] ion x-ray lasers in laser-produced plasmas, with a driver energy ranging from several kJ to 0.1kJ in pulse lasting for ~0.1 to ~1ns.

In the transient collisional excitation scheme, the first optical laser pulse is usually a long laser pulse with ~ns duration that generates, heats, and ionizes the plasma, while the second pulse is a short pulse with several ps duration that overheats the pre-plasma and excites the ions to produce a very large transient population inversion. To our knowledge, the experimental results and several theoretical analysis of the transient gain scheme indicates considerable efforts should be contributed to overcome several disadvantages of this scheme, mainly the instability of the gain region.

In 1994, to improve the laser irradiation uniformity, Stevenson et al. [6] carried out a research with a binary-phase zone plate. Wang et al. [7] invented the cylindrical lens array line focus system to produce an uniform line focus and realized a operation of the Ni-like Ag x-ray laser at 13.9nm. Experimental results show that the advantage of the application of the lens array in improving the irradiation uniformity is superior to the disadvantage due to the diffraction effect.

In this paper, a novel line focus system for x-ray laser experiments is proposed. In the second section, the model for the longitudinal pumping x-ray laser scheme is described. In the third section, the equation of the compound lens is established. In the fourth section, the result is analyzed. The last section summarizes the results.

2. Model

The experimental scheme for longitudinal pumping is shown below. Firstly, transversely pumping laser pulses irridate on the slab target as a pre-pulse, which produces a Ni-like (Ne-like) pre-plasma at an appropriate temperature, density and large abundance. Then the main pulse longitudinally pumps the plasma and produces the population inversion. In the traditional scheme, the gain length of the x-ray lasers is limited to less than 2mm with the influence of the laser intensity rapidly decreasing from plasma absorption and the refraction of the plasma density gradient and also due to the limited lifetime of the transient plasma. Therefore, it is critical to realize a uniform line focus system with traveling pumping. Considering the influence of the absorption of the longitudinal pumping laser along the focal line, we concentrates our efforts on the characteristics of the axial line-focus of the compound lens array (shown in **Fig**1). If the intensity at each focus can be modulated, the irradiation uniformity of the laser along the axial focal line is capable to be amended to a visible extent.

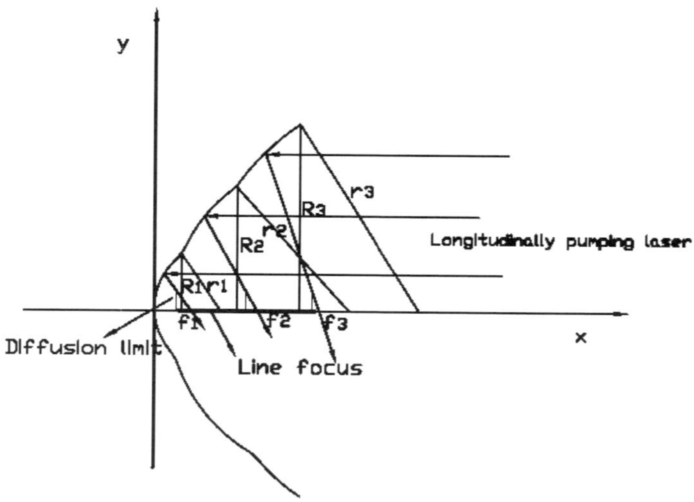

Fig1. The structure of the reflective compound lens array system.r is the radius of curvature, R is the radius of each sub-sphere, l is the line-focus length.

If we assume the optical parameter of the sub-sphere to be $r_1, R_1; r_2, R_2; r_3, R_3$,the diameter of the dispersed spot is D, in conjunction with the corresponding relationship of the triangle, we can obtain the following formula,

$$l_1 = 2l_0 + \frac{r_2 D}{R_2},\ r_3 = \frac{R_3(R_2 + D)}{R_2(R_3 - D)} r_2,\ l_2 = l_1 + \frac{r_3 D}{R_3},$$

where l_0 is the line-focus length for single-lens, l_1 is the line-focus length for the double-lens, l_2 is the line-focus length for triple-lens.

3. Calculation and analysis

To test the theoretical result, some applicative programs are designed and the result is described in the corresponding graph.

Under our practical conditions, the power percentage is set to be 100:10:1, and then we investigate the relation between the line-focus and optical parameter of the reflective compound lens array.

In order to investigate the influence of the radius of the first sub-sphere (r_1) on the line-focus length for a certain energy output percentage, we set R_3=20mm. Relation between r_1 and line-focus length with three combined sub-spheres is demonstrated in Figure2. As shown in the figure, the line-focus length increases with the increase of radius of curvature(r_1).

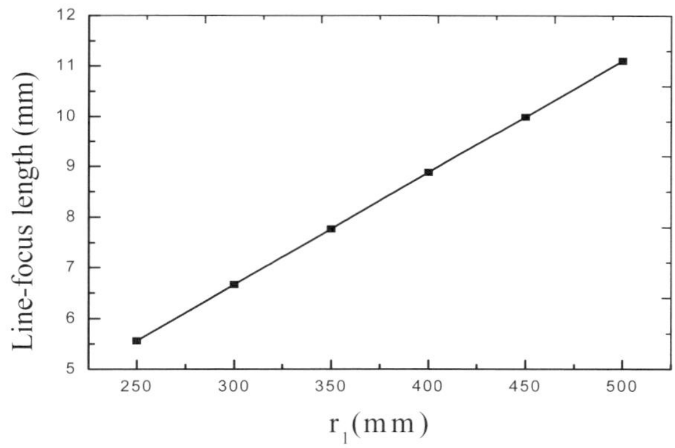

Figure2. Relation between line focus length and the radius of curvature of the first sub-sphere.

Secondly, the influence of the radius R_1 on the line-focus length with certain output percentage is studied. r_1=100mm. As shown in the Figure3, the line-focus length of the lens array decreases as R_1 increases.

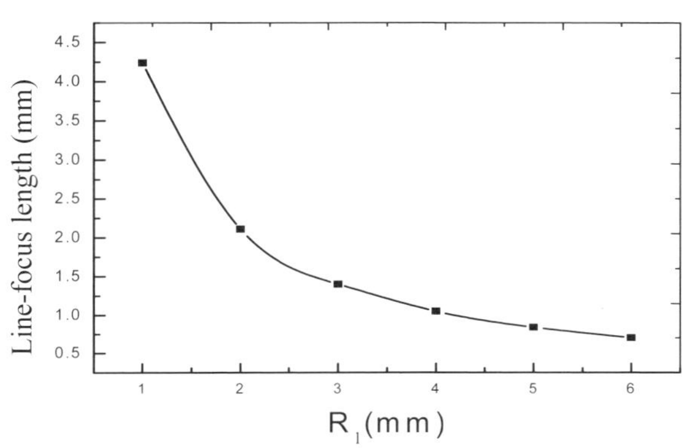

Figure3. Relation between line-focus length and the radius of the first sub-sphere.

4. Conclusion

From the above analysis, we can conclude that the advantage of this approach is: (1) The focal length in a longitudinal pumping laser is improved by using the reflective compound lens array. (2) The energy attenuation of the laser along the propagation in the plasma is compensated by the moderation of the structure of the reflective compound lens array.

As the reflective compound lens array has a flexible design and much bigger modulating atmosphere, the prospective of its research is very bright.

5. Acknowledgement

Project Supported by the National Natural Science Foundation (Grant Nos. 10374114, 10176034, 60321003, 10474134, 60128505) and the NKBRSF under Grant No. G1999075206, and the National Hi-tech ICF Program of China.

References:

[1] A. Carillon, H. Z. Chen, P. Dhez et al., Phys. Rev. Lett. **68**, 2917 (1992) J. Zhang, P. J. Warwick, E. Wolfrum et al., Phys. Rev. A **54**, R4653 (1996).
[2] J. A. Koch, B. J. MacGowan, L. B. DaSilva et al., Phys. Rev. Lett. **68**, 3291 (1992).
[3] L. B. DaSilva, B. J. MacGowan, S. Mrowka et al., Opt. Lett. **18**, 1174 (1993)
[4] J. Zhang, A. G. MacPhee, J. Nilsen et al., Phys. Rev. Lett. **78**, 3856 (1997).
[5] J. Zhang, A. G. MacPhee, J. Lin et al., Science **276**, 1097 (1997)
[6] Stevenson R M, Norman M J, Bett T H, Pepler D A, Danson C N, Ross I N, 1994, Optics Letters **19** 363.
[7] Wang C, Dong G L, Wang W et al., Chinese laser 2002 A **29**(1)

Inst. Phys. Conf. Ser. No 186
Paper presented at 9th Int. Conf. X-Ray Lasers, Beijing, China, 24–28 May 2004

Production of a line-focus in the axial direction using a Fresnel zone plate

Y. T. Li[1], J. Zhang[1], X. Lu[1], Z. Jin[1], D. A. Pepler[2], C. N. Danson[2]
[1]Laboratory of Optical Physics, Institute of Physics, Chinese Academy of Sciences, Beijing 100080, P. R. China
[2]CLRC Rutherford Appleton Laboratory, Chilton, Didcot, OXON, OX11 0QX. UK

Abstract. Generation of a uniform axial line-focus has been demonstrated by a binary-phase Fresnel zone plate coupled with a conventional lens. The optical system is tested using an ultrashort-pulse laser. The results show that such a line focus system can be used in the x-ray laser experiments.

1. Introduction

In x-ray laser experiments a line focus is applied to produce a long plasma column that is necessary to amplify the spontaneous emission. The line focus is generally generated by an off-axis reflected mirror configuration or a cylindrical lens (single or array) configuration traditionally. The two configurations produce a line focus perpendicular to the laser axis.

It is possible to produce table-top size x-ray lasers due to rapid development of picosecond and femtosecond laser technology. However, traveling wave pump scheme has to be used to obtain a high gain when pumped by ultrashort laser pulses [1]. Recently a longitudinal pump scheme has been proposed [2]. Both the scheme and the optical filed ionization scheme [3] need a line focus in the laser axis direction.

An axial line focus is usually generated by axicons [4]. In this paper we will demonstrate a new method using a binary phase Fresnel zone plate.

2. Principles

A Fresnel zone plate can be considered as a lens with a focal length $\pm f_z$

$$f_z = \pm\frac{1}{n}\left(\frac{R_0^2}{\lambda} - \frac{\lambda}{4}\right) \tag{1}$$

where n is the refractive index, R_0 is the radius of the central zone, λ is the wavelength. The focal length of the plate can be adjusted by changing the radius of the zone. If R_0 is changed continually, an axial line focus can be generated.

Based on this idea, an axial line focus system consisting of a zone plate and a conventional lens has been designed. To improve the uniformity of the local laser intensity along the line focus, a random transition instead of smooth transition of the focal length of the zone plate is applied.

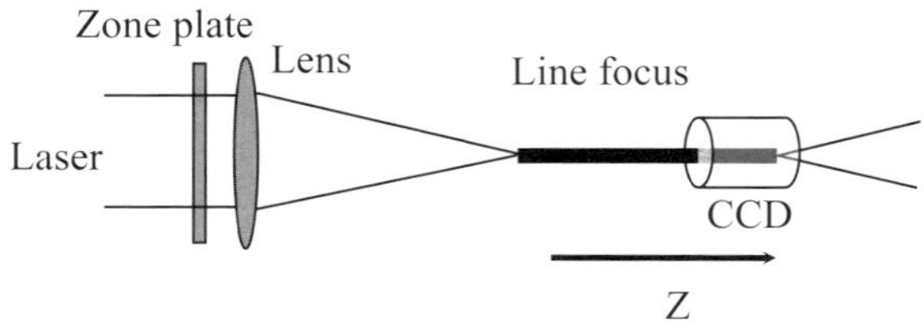

Figure 1. Schematic of experimental setup.

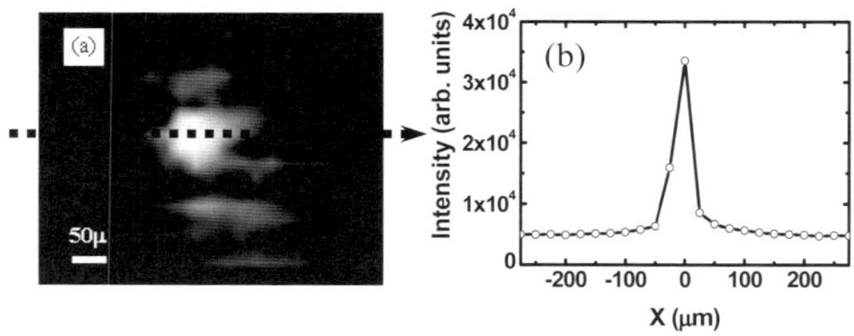

Figure 2. Typical cross-section image of the line focus at the geometrical focus of a *f*=125 mm lens (a) and the line-out along the dash line (b).

3. Optical test

The line focus system was tested using a femtosecond laser with an average power of 1.1 W at 800 nm. The repetition rate of the laser is 82 MHz. The duration of a single pulse is 100 fs. The experimental setup is illustrated in Fig. 1. The laser beam was focused on a 16 bit, 512×512 CCD camera. The size of the CCD pixel is 24μm×24μm. By moving the position of the CCD in the laser axis direction (Z directin) one can obtain a sequence of 2D intensity distribution along the line focus. Two conventional lenses with 90mm and 125mm focal length were coupled with the zone plate in experiments respectively.

Figure 2(a) shows a typical cross-section image of the line focus produced by the zone plate coupled with an *f*=125 mm lens, which located 30 mm in front of the plate. The image was taken at the geometrical focus of the lens. Figure 2(b) shows the line-out

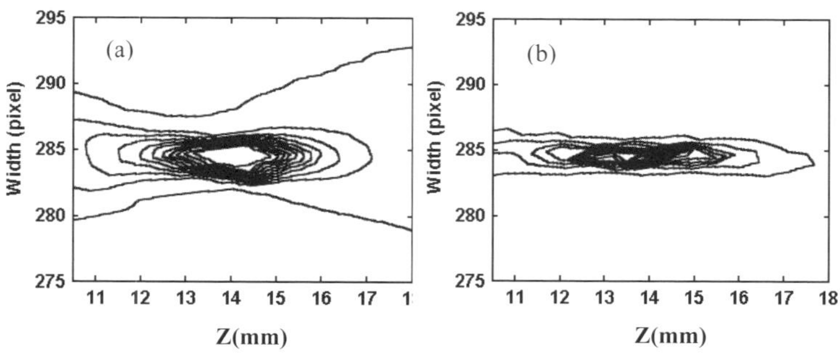

Figure 3. 2D contour plot of the axial intensity distribution of the line focus without the zone plate (a) and with the plate (b).

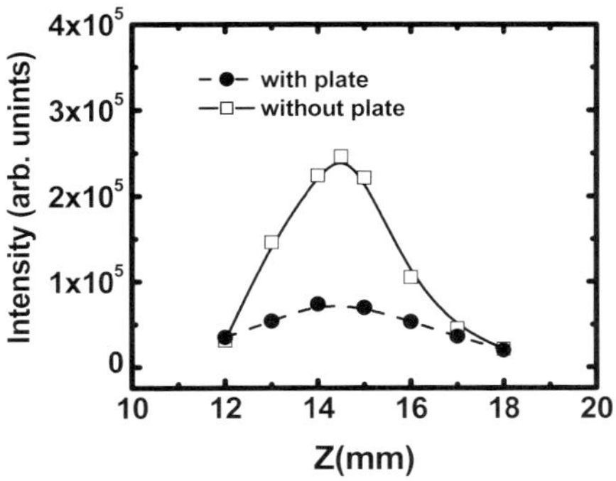

Figure 4. The axial integrated intensity distribution along the line focus.

intensity distribution cross the focus. The bottom width of the line focus is about 100μm. We also note that the intensity around the central spot is increased from ~100 without the zone plate to ~5000 with the plate. The ratio of the peak intensity at the centre of the line focus with the intensity around the line focus is 7:1.This indicates that relatively large part of the incident laser energy is dispersed to the outer region around the focus.

Figure 3 shows the 2D intensity contour without the zone plate (a) and with the plate (b) along the laser axis. The laser is incident from the left. It is obvious that a ~5 mm long line focus is formed when the zone is used.

Figure 4 gives the axial integrated intensity distribution along the line focus. The data are obtained by integrating over an area of the cross section of 75x75 μm^2. After the zone plate is added in the system the intensity on the line becomes relatively uniform.

In x-ray laser experiments a changeable length of the line focus is very desirable. This is easy to achieve using the zone plate method. The length of the line focus can be adjusted using conventional lens with different focal lengths. Figure 5 shows the line focuse produced by the zone plate coupled with two different lenses with focal lengths of 125 mm and 90 mm respectively. The length of the line focus decreases from 5 mm for a f=125 mm lens to 3.5 mm for a f=90 mm lens. One can see the length of the line focus becomes short when a lens with a short focal length is used.

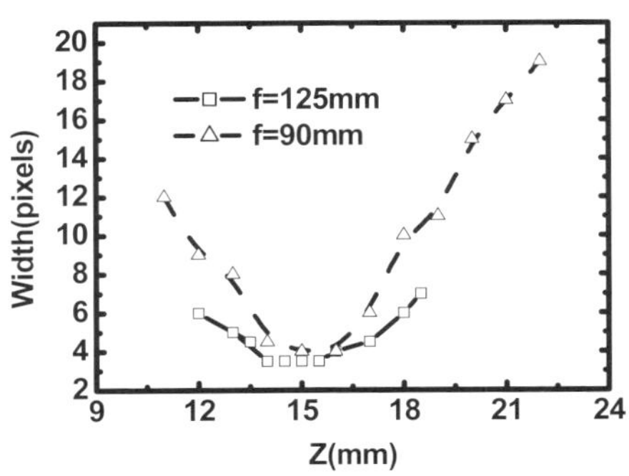

Figure 5. Line focuses produced by lenses with different focal lengths.

4. Conclusions

A uniform axial line-focus, which is important for the x-ray laser amplification pumped by picosecond and femtosecond laser pulses, is generated by a binary-phase Fresnel zone plate. The performances of the plate are tested using ultrashort-pulse laser. The results show that the length and the width of the line-focus are ~5mm and ~100 μm respectively, when coupled with a 125mm focal length lens, and the length of the line focus is changeable.

Acknowledgements

This work was supported by the NSFC under Grant Nos. 10374114, 10374115, 10374116, 60321003, 10474134 and 60128505, and the NKBRSF under Grant No. G1999075206, and the National Hi-tech ICF Program.

References

[1] Moreno J C, Nilsen J, and Da Silva L B 1995 AIP Conf. Proc. **332** 21
[2] Ruxin Li, Tsuneyuki Ozaki, Teruto Kanai, and Hiroto Kuroda 1998 Phys. Rev. E **57** 7093
[3] Sebban S, Haroutunian R, Balcou Ph, et al 2001 Phys. Rev. Lett. **86** 3004
[4] McLeod J H 1960 Journal of the Optical Society of America **50** 166

Inst. Phys. Conf. Ser. No 186
Paper presented at 9th Int. Conf. X-Ray Lasers, Beijing, China, 24–28 May 2004

Ray-tracing Design of a Flat Field Soft X-ray Spectrometer

Y. Q. Liu[1], J. Zhang[1], Z. L. Chen[1], and X. Y. Peng[1], Ch. H. Nam[2]
[1]Optical Laboratory of Institute of Physics, Chinese Academy of Sciences, Beijing, 100080, China
[2]Department of Physics and Coherent X-ray Research Center, Korea Advanced Institute of Science and Technology, Taejon 305-701, Korea

Abstract. Optimization of a grazing incidence soft x-ray flat field spectrometer with toroidal mirror as a relay optics system is performed using a ray tracing code and path function theory. The optimised spectrometer provides a good spectral resolution ($\lambda/\Delta\lambda$) better than 1000 and a spatial resolution better than 40 μm in the spectral range of 4-45nm.

1. Introduction

In the experiments of high field physics [1], x-ray lasers [2], high harmonic generation [3] and laboratory astrophysics etc., plasmas emit abundant soft x rays containing rich information. This information provides a crucial diagnosis of plasma. Traditional soft x-ray spectrometer requires that all of the slits, grating and detectors are arranged along the Roland circle. Therefore some advanced plane detectors, such as charge coupled detectors (CCD), microchannel chip plates (MCP) and streak cameras, can not be well coupled with the Roland circle spectrometers. In 1980s, the HITACHI company has designed a kind of spherical varied-line-space (SVLS) flat field grating that can focus soft x ray on a flat field plane [4]. With this kind of gratings, soft x-ray flat field spectrometers can be built for laser-plasma experiments [5,6,7]. The flat field spectrometers have potential capability for measuring those spectra and plasma diagnosis. Optimum design of a flat field soft x-ray spectrometer is very important.

In order to improve the efficiency of reflection, most of soft x-ray spectrometer adopt the grazing incidence way. The grazing incidence has a disadvantage of low acceptance solid angle [8]. Slitless mode operation with supplementary optical system can improve acceptance solid angle [9]. But for the way of grazing incidence, a nonideal point source can not be imaged by a flat field grating on the same focus in the meridional plane and sagittal plane respectively. Another way of improving acceptance solid angle is to add a relay optical system before the slit and grating. In our flat field spectrometer, we add a toroidal mirror as a relay optical system. Because a toroidal mirror has two-dimensional radius of curvature on the meridional plane and sagittal plane respectively, it can compensate the astigmatism of grazing incidence. The relay optical system not only provides a larger acceptance solid angle, but also enhances spatial resolved ability. In this paper, based on path function theory and a ray tracing code, optimization is carried on the design of the flat field spectrometer.

2. Structure of the spectrometer

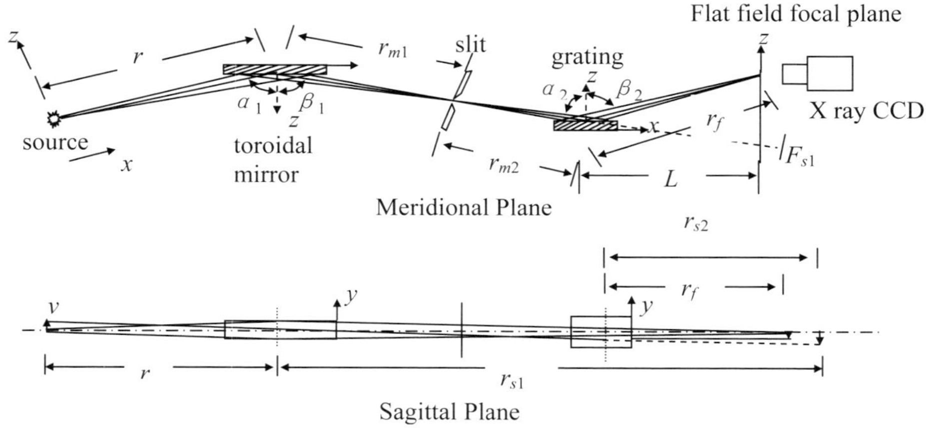

Figure 1 Schematic structure of the spectrometer

As shown in Figure 1, the flat field spectrograph system consists of several basic optical elements, a toroidal mirror, a slit, a SVLS grating, and a soft x ray CCD detector on the spectral plane. In Figure 1, the angle of incidence at toroidal mirror is α_1 and the distance from the source to the center of toroidal mirror is r. The distance from the center of toroidal mirror to the meridional focal position and to the sagittal focal position is r_{m1} and r_{s1} respectively. They are given by

$$\frac{1}{r}+\frac{1}{r_{m1}}=\frac{2}{R_1\cos\alpha_1} \tag{1}$$

$$\frac{1}{r}+\frac{1}{r_{s1}}=\frac{2\cos\alpha_1}{\rho_1}. \tag{2}$$

In the saggittal plane, the grating images a virtual source at r_{s2} after the grating. The spectrum on the spectral plane will suffer large astigmatism for the reason of grazing incidence. In order to compensate astigmatism in the saggittal plane, the virtual source which is formed by the toroidal mirror, will enter the grating with appropriate consideration for the incidence angle α_2 and R_2, given by

$$r_{s2}=\frac{R_2L}{L(\cos\alpha_2+\cos\beta_2)-R_2\left|\sin\beta_2\right|}, \tag{3}$$

where r_{s2} is the sagittal source position from the grating pole. R_2 is the radius of curvature of grating. L is the normal distance from the center of grating to the spectral plane. The diffraction angle β_2 could be calculated from the grating diffraction equation,

$$\sin\beta_2=-\sin\alpha_2+\frac{m\lambda}{\sigma_0}, \tag{4}$$

here σ_0 is the nominal groove spacing and m is the diffraction order. In our spectrometer, we set to correct the spectrographic astigmatism at 100 Å. To meet this requirement, the relation of r_{s1}, r_{m1}, r_{m2} and r_{s2} should satisfy the relation Eq. 5,

$$r_{s1}=r_{m1}+r_{m2}+\left|r_{s2}\right|. \tag{5}$$

3. Ray tracing results

Based on the descriptions of the path function of the toroidal mirror and the flat field grating, we design a ray tracing code to optimally design the spectrometer system. The schematic of the spectrograph as showed in Figure1. The system is a plane symmetry rather than a rotational one. From the source, we generate 5000 rays in the acceptance solid angle of the spectrometer with the Gaussian distribution. All the rays are reflected by the toroidal mirror. The image before the slit is shown in Figure 2. We can find that the image size is about 100μm in the meridional plane. The slit width is 50μm, so more than 70% rays pass by the slit. During the alignment of the spectrometer, we find a curving-moon-like image before the slit, just like the simulated image. At this time, we should make the image be tangent with the slit in order to let most rays passing through the slit. The structure parameters of the spectrograph system are showed in the Table 1.

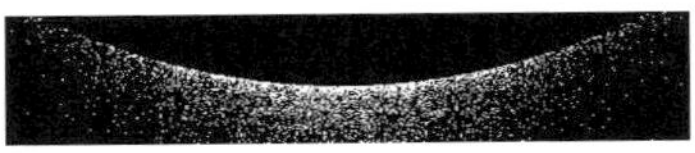

Figure 2 Ray tracing image in the sagittal plane before the slit

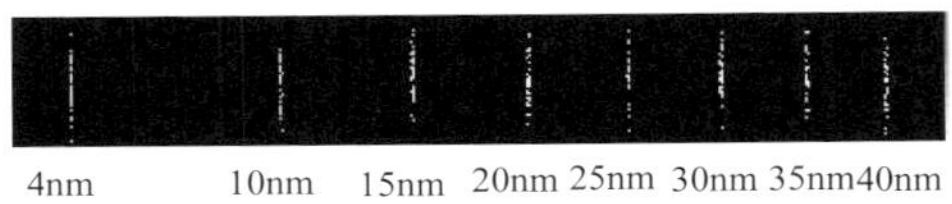

Figure 3 Some typical spectral lines on the spectral plane simulated by the ray tracing code

Table 1 Structure parameters of the spectrometer

Toroidal mirror	Flat field grating
Meridional radius of curvature ρ_1=47.6mm	Incidence angle (degree) α_2=87^0
Sagittal radius of curvature R_1=5194mm	Nominal groove spacing σ_o=1200 line/mm
Incidence angle（degree） α_1=86.3°	Blaze angle (degree) δ=3.2^0
Distance of the source from mirror r=816.9mm	Grating ruling parameters b_2=20,b_3=4.558 $\times 10^2$,b_4=1.184$\times 10^4$
Distance of the meridional focus from mirror r_{m1}=208mm	Radius of curvature R_2=5469mm
Distance of the sagittal focus from mirror r_{s1}=685mm	Distance of the meridional source from grating r_{m2}=237mm
Area of toroidal mirror 150mm×25mm	Distance of the sagittal source from grating r_{s2}=240mm
Width of slit S=50μm	Normal distance of the spectral plane from grating L=235mm

Those rays passing by the slit are collected by the SVLS grating, then diffracted on the spectral plane. The simulated spectrums by the ray tracing code on the spectral plane are showed in the Figure 3. The spatial resolution of the whole spectral range is the almost same, which gives about 36μm at the wavelength of 5nm and gives about 38μm spatial resolution at the wavelength of 40nm for a point plasma source. As to the grating, we just consider the first order diffraction and assume that all the rays incident on the grating are diffracted by without any lost. Our spectrometer system has 15 mrad of the acceptance angle in the meridional plane and 30mrad in the sagittal plane. Using this code, the effect of source displacement along the x, y and z is simulated. Figure 4 shows that if the source displacement along y direction is less than 1.5mm, the spatial resolution of the system will be less than 80μm. If the plasma source is displaced along x direction, the system has a tolerance of 20cm. But the displacement in the sagittal plane along z direction should be not more than 70μm in order to let enough soft x rays pass through the slit and minimize the aberration. The spectral resolution of the spectrometer is shown in Figure 5. Because we correct the spectrograph astigmatism at the wavelength of 10nm, the spectral resolution at 10 nm is the best of the whole spectrum range. The wave-front theory [10, 11] developed by Chrisp has been adopted to analyze a flat field soft x-ray spectrometer successfully [6]. Comparing with the wave-front theory, the ray tracing code can give us intuitive spectrum on the spectral plane. The spectral and spatial resolutions calculated by the wave-front theory agree with the ray tracing results very well.

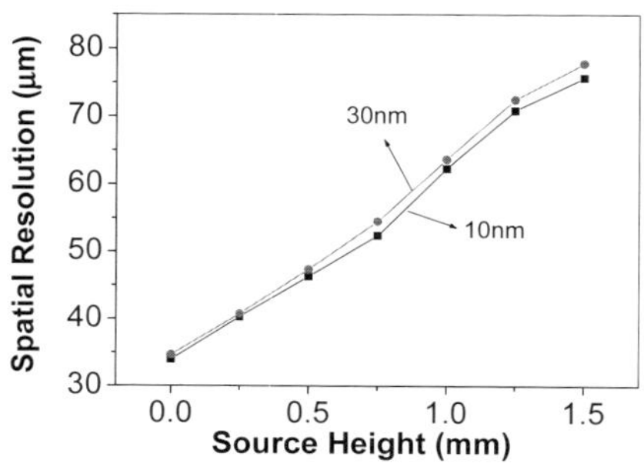

Figure 4 Spatial resolution at wavelength of 10nm and 30nm for different source height

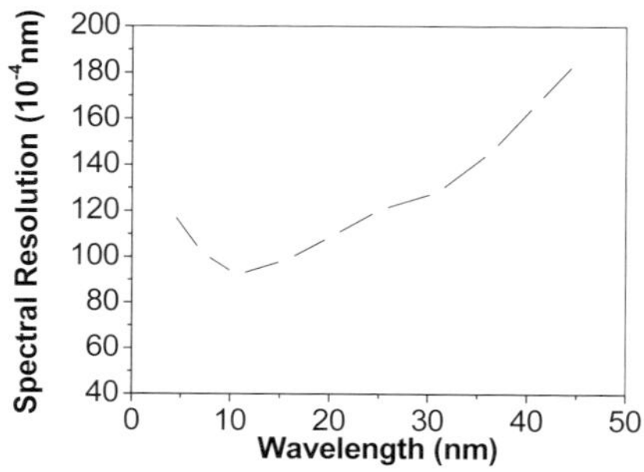

Figure 5 Spectral resolution of the spectrometer at spectral plane

4. Conclusion

We design a flat field soft x-ray flat field spectrometer using a ray tracing code and path function. Our spectrometer has a relay optical system with a toroidal mirror that has two dimensions of meridional radius and sagittal radius. In our spectrometer system, the magnification in the meridional plane is r_{m1}/r=0.255 and the magnification in the sagittal plane is $(r_{s1}/r)\times(r_{s2}/r_f)\approx0.85$ respectively. For this kind of structure, it is very easy to calibrate the source position of the spectrometer with plasma source. The code calculation and the wave-front theory show that the spectrum resolution of the spectrometer is better than $\lambda/\Delta\lambda>1000$ and spatial resolution is better than 40μm.

Acknowledgements

This work was supported by the NSFC under Grant Nos. 60321003, 10474134 and 60128505, and the NKBRSF under Grant No. G1999075206, and the National Hi-tech ICF Program.

References

[1] Zhang J 1997 Physics 26 643 (in Chinese)
[2] Zhang J et al., 1997 Science 276 1097
[3] Chang Z et al., Phys. Rev. Lett., 1997 79 2967
[4] Harada T et al., 1980 Appl. Opt., 19 3987
[5] Luca Poletto and Glamplero Naletto 2001 Opt. Eng., 40 (2) 178
[6] Choi W et al., 1997 Appl. Opt., 36 1457
[7] Jaehoon Kim and Dong-Eon Kim 2000 Jpn. J. Appl. Phys., 39 6062
[8] Kita T et al., 1983 Appl. Opt., 22 512
[9] Li Y J et al., 1998 HPLPB 10, 409
[10] Chrisp M P 1983 Appl. Opt., 22 1508
[11] Chrisp M P1983 Appl. Opt., 22 1519

Inst. Phys. Conf. Ser. No 186
Paper presented at 9th Int. Conf. X-Ray Lasers, Beijing, China, 24–28 May 2004

Co/C sliced multilayer grating for the 4.4-5 nm wavelength interval

R M Feshchenko, A V Vinogradov
P.N. Lebedev Physical Institute, 119991 Moscow, 53 Leninskii Pr., Russia

E A Bugaev, D L Voronov
Kharkov Polytechnical Institute, 310002 Kharkov, 21 Frunze St., Ukraine

V P Petukhov
Nuclear Physics Institute of Moscow State University, 119899 Moscow, Russia

Abstract. An example of the new type of X-ray spectral elements – sliced multilayer grating (SMG) for 4.4-5 nm interval has been produced and tested. A novel experimental approach based on a flow proportional counter was introduced. The properties of the SMG are discussed.

1. Introduction

The development of X-ray multilayer optics for the last twenty years has enabled creation of normal incidence multilayer mirrors working in the broad range 3-50 nm [1]. The interval of 4.4-5 nm is of a particular interest because carbon has the K absorption edge at 4.4 nm. In this region dry biological objects and other organic materials are relatively transparent so one can expect that such a radiation would be useful in the microscopy and spectromucroscopy, material science, biology and astronomy. In this wavelength range multilayer mirrors with reflectivities up to 15% have been produced [2] but the problem of creation of the effective spectral elements remains. Although different types of ruled, holographic and etched multilayer gratings have been proposed they work mainly in the soft X-ray region with wavelengths longer than 10 nm [3-4]. Ones that work in the shorter wavelength interval usually utilize grazing incidence design and have relatively low dispersion restricted by fabrication techniques.

Nevertheless it has been know for at least a decade that gratings can be produced by cutting a multilayer coating at some angle. The resulting structure is a functional analog of asymmetrical crystal and is called sliced multilayer grating – SMG. Since the period of the initial multilayer coating can be very short, the dispersion of the SMG can be as high as one of the ruled gratings with number of lines per mm of 20000 or more. The first SMG was produced more than ten years ago [5] for the wavelengths 13-25 nm and was based on a *Mo/Si* multilayer coating. Several plasma spectra were obtained with its aid [6]. A few papers were published recently proposing utilization of such grating in the hard X-ray region [7, 8]. An extensive theoretical work was published two three years ago, which provides comprehensive calculation methods for the SMG, describes their fabrication and proposes to use a gradient of the period to focus radiation [9].

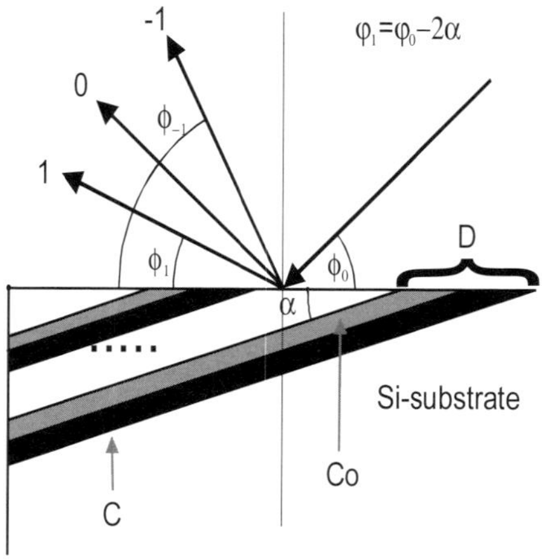

Fig.1 The scheme of sliced multilayer grating.

In the present paper the creation of an SMG working in the wavelength interval 4.4-5 nm is reported. The results of measurements of its diffraction efficiency using a novel experimental technique are presented and discussed. The further prospects of such gratings for the spectroscopy near carbon K absorption edge are considered.

2. Fabrication and properties of the SMG

The fabrication technique was described in details in [9]. Eventually the structure (SMG) with N periods shown in Fig. 1 is made. It is evident from this figure that the main grating equation can be written as:

$$D(\cos\varphi_n - \cos\varphi_0) = n\lambda\,,\ D = d/\sin\alpha \tag{1}$$

where d is the multilayer period.

The parameters of *Co/C* SMG that is discussed here are in Table 1. The cobalt comprises about 40% of the multilayer period. In the real coating the cobalt is presented in the form of Co_2C (as electron microscopy showed) carbide with the density of 8.0 g/cm^3, which is lower than one of the metallic cobalt – 8.9 g/cm^3. The size of the grating is about 12*0.06 mm^2.

Table 1. Parameters of the *Co/C* SMG grating discussed in the paper, where d – the multilayer period, D – the SMG period, N – the number of periods, L=D*N – the width of SMG, φ_0 – the resonant angle, R – the calculated reflectivity.

d, nm	D, nm	N	*α, degree*	*L, μm*	φ_0, degree	R, %
10.7	61	1000	10	61	23	20

3. Experimental setup and measurements

For diffraction efficiency measurements a tabletop setup with the novel design was assembled (see Fig.2). The setup is placed in a vacuum chamber (V) with basic pressure 10^{-5} torr. The distinct feature of this setup is utilization of a proportional counter that separates a desired line or spectral band [10].

The electron gun is powered with 3 kV voltage and provides an electron beam, which is focused onto the carbon target exciting K_α characteristics line with wavelength 4.47 nm. Optics, soft X-ray and hard bremsstrahlung radiation are also emitted but the K_α line dominates in range 3-10 nm. The collimator comprises one or two slits with 0.1 mm width which deliver a beam to the sample with angular divergence better then 10 mrad.

The sample SMG and flow proportional counter are mounted in the compact goniometer (G) with ability to perform θ-2θ scans. The angular step of the goniometer is 0.01 degree. The goniometer allows independent rotating of the grating and counter by two step electrical motors. The sample SMG diffracts X-ray radiation in the direction of the flow proportional counter. The counter has an input window with the area 8*15 mm^2. Its internal volume is filled with $Ar\text{-}CH_4$ gas mixture with pressure 10 torr. The detector has energy resolution about 70% at the wavelength 4.4 nm, which is sufficient to separate the carbon K_α line from soft and hard X-ray radiation. Every measurement usually takes several seconds and the number counted photons is about 10^3. Since the incident beam is bigger than the grating it was impossible to determine the absolute diffraction efficiency. The precession of diffraction efficiency measurements is about 3%, which is enough for purposes stated.

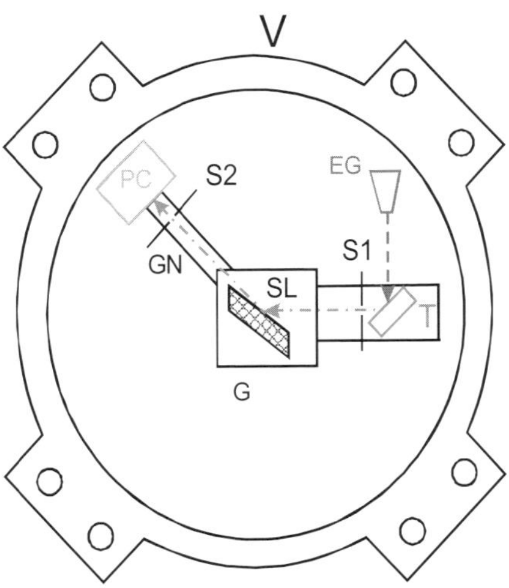

Fig.2 The scheme of the experimental setup (V – vacuum chamber, EG – electron gun, T – carbon target, S1 – entrance slit, G – goniometer, SL – SMG, S2 – exit slit, PC – flow proportional counter)

The measured rocking curve – dependence of the grating efficiency on the incident angle φ_0 for the SMG (see Table 1) is shown in Fig.3 by filled circles. The solid line represents theoretically calculated one [9]. The measurements were performed for the first order only since the minus first order is severally obscured by the external total reflection. To match the theoretical and experimental results the theoretical curve was shifted by $\Delta\theta$=1.25 degree. Such a discrepancy is probably caused by misalignment between the SMG and silica plate due to nonuniformity of the glue layer. Since absolute measurements weren't possible all curves depicted in Fig.3 were normalized to unity and are shown in arbitrary units. Such normalization enables comparison of the curve shapes.

From Fig.3 one can see that experimental curve has relatively strong wings (background) that can be explained by noise or weak reflection from silicon wafers. To remove these wings a constant value that is equaled to the average background was subtracted from the experimental curve. The results of subtraction of the background are shown in Fig.4 by filled triangles. The subtracted curve is in a reasonably good agreement with the theoretical one.

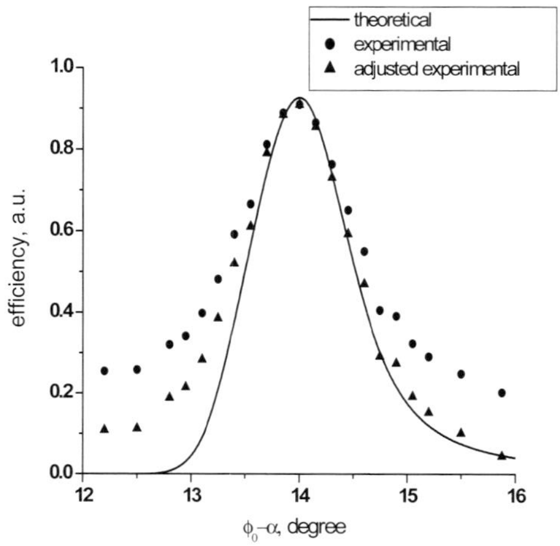

Fig.3 The experimentally measured rocking curve of the SMG described in this paper (filled circles) and the same results with the background subtracted (filled triangles). The solid line is the calculated diffraction efficiency. All three curves were normalized to the unity. The theoretical curve was shifted by 1.25 degree to match experimental ones.

4. Conclusion

A sliced multilayer grating working in the 4.4-5 nm wavelength range was designed, produced and tested. The experimental setup with a novel design based on a flow proportional counter was assembled and a new experimental technique for measurements of diffraction efficiencies of SMG was proposed. Such a setup enables measurements without a complicated monochromator thus improving the throughput and sensitivity of the system. The setup was used for measurements of diffraction efficiency (rocking curve) of the SMG mentioned above. The experimental results are in a reasonably good agreement with theoretical calculations.

The SMG considered in this paper was the first sample of new series of gratings specially designed for the spectral range 4.4-5.0 nm, which holds a great potential for material science, astrophysics, biology etc. The next generation of gratings will have shorter multilayer period (around 5 nm), more layers (4000-8000) and smaller slicing angle (approximately 5°). It'll lead to greater angular dispersion, bigger aperture size and better spectral resolution.

The research described in this publication was made possible in part by Award No. RP2-2326 of the U.S. Civilian Research & Development Foundation for the Independent States of the Former Soviet Union (CRDF). The present work was also supported under the ISTC project 1794 and RFBR (Russian Foundation of Basic Research) project 03-02-16438a.

References

[1] Schaefers F 2000, Physica B 283 119–124
[2] PXRMS Multilayer Survey Results, http://www-cxro.lbl.gov/cgi-bin/mldata.pl
[3] Seely J F, Uspenskii Yu A, et.al. 2002 Applied Optics 41 1846-1851
[4] Patchet B E, Harrison R A, et.al. 1989 ESA SP-1104 39-49
[5] Levashov V E and Vinogradov A V 1993 Applied Optics 32 1130-1135
[6] Levashov V E, Zubarev E N, et.al. 1994 Optics Communications 109 1-4
[7] Wilson M M and Zukic M 1995 Proc. SPIE 2517 133-143
[8] Wilson M M, Zukic M, et.al. 1994 Proc. SPIE 2283 189-199
[9] Fechtchenko R M, et.al. 2002 Opt. Commun. 210 179-186
[10] Kondrashov P E, et.al.1998 Diamond and Related Materials **7** 1647-1650

Inst. Phys. Conf. Ser. No 186
Paper presented at 9th Int. Conf. X-Ray Lasers, Beijing, China, 24–28 May 2004

Narrow bandwidth C/Si multi-layer mirror in soft X-ray and extreme ultraviolet range

Shuji Qin, Zhanshan Wang, Wenjuan Wu, Hongchang Wang, Zhong Zhang, Fengli Wang, Lingyan Chen
Institute of Precision optical engineering, Department of Physics, Tongji University, Shanghai, 200092

Abstract: The optical elements with high reflectivity and spectral selectivity in the waveband from 12.4nm to 40 nm are desired. Theoretical calculation indicates that the bandwidth of C/Si multi-layer is narrower than the one of Mo/Si multi-layer used frequently, while its reflectivity is lower than that of Mo/Si multi-layer. 0.154 nm X-ray glancing incident diffraction was used for measuring the structure of C/Si and Mo/Si multi-layers, and synchronous radiation for measuring their reflectivity and bandwidth in the operating waveband. The measured results indicate that the bandwidth of C/Si multi-layer is almost a half of Mo/Si multi-layer, but its reflectivity is lower than that of Mo/Si multi-layer. The experimental results are consistent with the theoretical calculation ones.

1. Introduction

As the performance of optical sources used in the soft X-ray waveband is improved gradually, their applications need to develop the correlated optical elements. The multi-layer with narrow bandwidth in some applications is desired. One way to decrease the bandwidth is that the special designed multi-layer replaces the routine multi-layer consisting of two materials with high and low atomic number respectively [1], and an alternative way is to design a multi-layer consisting of two materials with low atomic number [2].

The recent development of ultraviolet photolithograph technique at 13nm promotes the investigation in the multi-layer. Now the reflectivity of Mo/Si multi-layer at 13nm has reached to 70% [3,4], but its bandwidth is wider than desired, and increases quickly as the wavelength extends to the longer direction. However, C/Si assemble is a good one in the waveband from 12.4nm to 40nm according to their optical constant, and the bandwidth of C/Si multi-layer is about a half of routine Mo/Si multi-layer.

In present paper, we calculated and measured the performance of C/Si multi-layer. The results verify the advantage of narrower bandwidth of C/Si multi-layer.

2. Theoritical reflectivity of C/Si multi-layer

There are two ways to calculate the reflectivity of a multi-layer. One is the matrix method, and the other is the iteration method. However, the iteration method is more convenient when the interface roughness and inter-diffusion of a multi-layer are considered. Under considering the interface roughness, inter-diffusion and Nevot coefficient, the reflectivity calculation of a multi-layer is as following:

$$R = R_0 \exp[-\frac{8\pi^2 \cos\theta_{j-1} \cos\theta_j \sigma^2}{\lambda^2}] \qquad (1)$$

In the formula σ is root mean square value of roughness, θ_{j-1} and θ_j incident angle at j-1 and j layer respectively. The reflectivity of multi-layer will vary with wavelength, and the optical constants at different wavelength are given by the literature [5].

Fig.1 shows that the calculated reflectivity of Mo/Si and C/Si multi-layer with 30 periods for the soft X-ray of s polarization incident at 45° varies with wavelength. The reflectivity and bandwidth of Mo/Si multi-layer reaches 74% and 1.1nm, and the ones of C/Si multi-layer 54% and 0.64nm respectively. The reflectivity of Mo/Si multi-layer is larger than that of C/Si multi-layer, but the bandwidth of C/Si multi-layer is narrower than that of Mo/Si multi-layer. Therefore, according to different applications we need to make a compromise between the reflectivity and the bandwidth, and choose a suitable multi-layer.

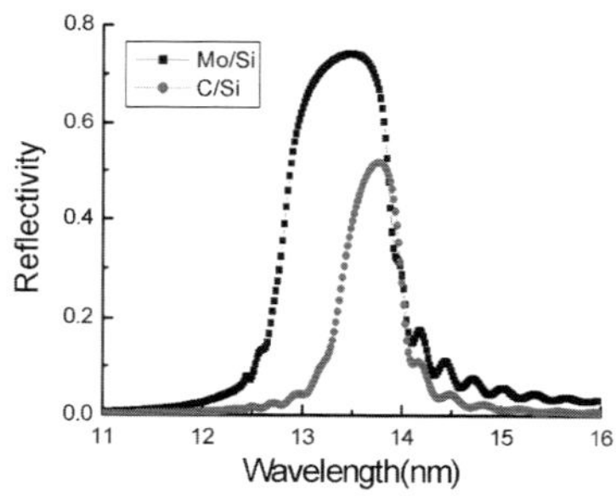

Fig.1 the calculated reflectivity of Mo/Si and C/Si multi-layer

3.Experimental process

The multi-layer of a reflective mirror was deposited by using a high vacuum DC magnetron sputtering coater model JGP560C6. The background pressure of vacuum chamber was smaller than 5.0x 10^{-5} Pa, and operation pressure 0.36 Pa while depositing C/Si multi-layer. The DC constant power mode was used for operation. The accurate calibration and control of the thickness for each layer are key factors. The former was determined by using an atomic force microscope and a glance incident X-ray diffraction instrument, and the latter was controlled by the time a substrate stayed above the target. It is noted that the targets should be pre-sputtered for 10~20 minutes before depositing a multi-layer for removing the surface contaminants and keeping the targets in a good thermal equilibrium state.

The structures of Mo/Si and C/Si multi-layers were measured by a X-ray diffraction instrument with high resolution (<0.001°) using Kα line of Cu (λ =0.154 nm) . The X-ray emitted by the optical source passed through a multi-layer paraboloidal mirror for collimation, and then was incident to the measured multi-layer sample. The width of both incident and emergent slits was 100 μm

The reflectivity of the multi-layers in soft X-ray waveband was measured by a soft X-ray reflectivity-meter on the 3WB1 beam line of synchrotron at Beijing. The beam line uses a variable space grating for spectra, and its energy resolution is about 2.8 eV round 100 eV. The accuracy of the goniometer used for reflectivity-meter is smaller than 0.1° , and the soft X-ray detector photoelectric diode.

4. Experimental results and discussion

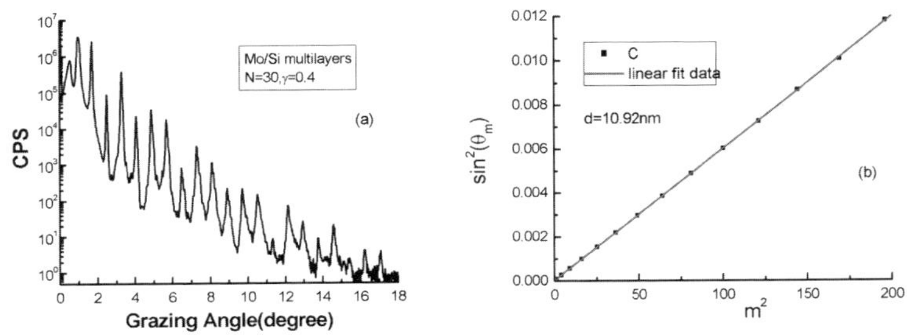

Fig.2 X-ray diffraction spectrum of Mo/Si multi-layer at grazing incidence (a) and the results of data fitting (b)

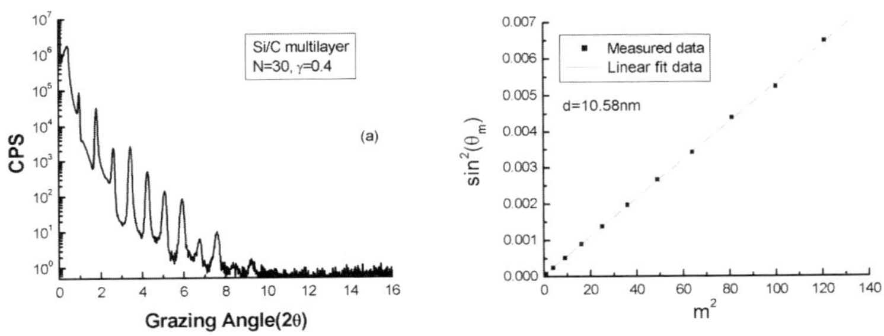

Fig.3 X-ray diffraction spectrum of C/Si multiplayer at grazing incidence (a) and the results of data fitting (b)

Fig.2 (a) and fig.3 (a) are the X-ray diffraction spectra of Mo/Si and C/Si multi-layers at grazing incidence. The multi-layer used for X-ray can be understood as an artificial crystal with a lattice constant, which is equal to the period thickness of a multi-layer. Therefore, the diffraction of soft X-ray on a multi-layer also obeys the Bragg's law of reflection. However, a modified Bragg's formula is needed for determining the period of a multi-layer due to the diffraction effect, and as following:

$$\sin^2\theta_m = (\lambda/2d)^2 m^2 + 2\delta \qquad (2)$$

In the formula m is the diffraction peak order, θ_m grazing angle, λ wavelength of X-ray. The positions θ_m of diffraction peak in X-ray diffraction spectrum can be used for determining the period d of multi-layer and the attenuation δ of average refraction index of multi-layer. The least square fit for the positions of multi-order diffraction peak can decrease the error, and obtain more accurate values of d and δ. If formula (2) is rewritten as the following form:

$$Y = A + BX \qquad (3)$$

then formulas for calculating d and δ are as following:

$$d = \lambda/(2\times\sqrt{B}) \qquad (4)$$

$$\delta = A/2 \qquad (5)$$

Fig.2 (b) and Fig.3 (b) are the fitting curves obtained according to the positions of multi-order diffraction peak. The period d and the attenuation δ of average refraction index of multi-layer are the slope and the intercept of fitting strait-line respectively. The fitting values of period d for Mo/Si and C/Si multi-layers are shown in Fig.2 (b) and Fig.3 (b).

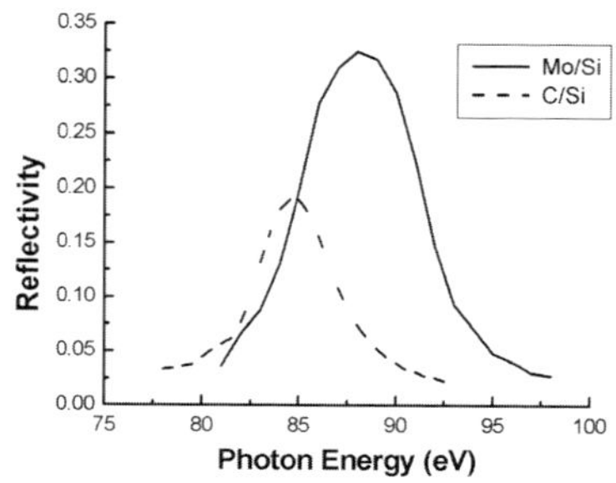

Fig.4 the reflectivity of Mo/Si and C/Si multi-layer measured with synchrotron

Fig.4 shows the reflectivity of Mo/Si and C/Si multi-layer measured by using synchrotron as optical source varies with photo energy. The maximum reflectivity of Mo/Si multi-layer at 89 eV is 33%, while the one of C/Si multi-layer at 85 eV 19%. However, the bandwidth of C/Si multi-layer is 4.5 eV, and the one of Mo/Si multi-layer 8 eV. The results indicate that the maximum reflectivity of Mo/Si multi-layer is larger than that of C/Si multi-layer, but the bandwidth of C/Si multi-layer is much smaller than that of Mo/Si multi-layer.

5. Conclusion

C/Si multi-layer is a quite useful one in the ultraviolet and soft X-ray waveband. Its reflectivity isn't as high as Mo/Si multi-layer, but it has a narrower bandwidth than Mo/Si multi-layer. Therefore, it is more suitable for the applications requiring a narrow bandwidth. Our work demonstrated the distinctive advantage of C/Si multi-layer in both theoretical and experimental respects.

Acknowledgements

Financial supports from the National Natural Science Foundation of China (approval №60178021), the Science and Technology development Foundation of Shanghai (№ 022261049) and the Nanometer Technology Special Foundation of Shanghai Science and Technology Committee (№ 0352nm090) are gratefully acknowledged.

Reference:

1. Delaboudiniere J P, Hochedez J F, et al, Proc SPIE, 1992, 1742:296
2. Hotta Y, Furudate M, Yamamoto M et al, Surface Review and Letters, 2002, 9(1):571-576
3. Lohmann M, Klabunde F, et al., Thin Solid Films, 1999, 342: 127-135.
4. Bajt S, Stearns D G, Kearney P A, J Appl Phys., 2001, 90: 1017-1025
5. Henke B L, Gullikson E M, et al., At Data Nucl Data Tables, 1993, 54:181-182

Inst. Phys. Conf. Ser. No 186
Paper presented at 9th Int. Conf. X-Ray Lasers, Beijing, China, 24–28 May 2004

Experimental measurement of soft X-ray multilayer beam splitter's optical properties in one laser shot

J.R. Sun, Z.H. Fang, Ch. Wang, W. Wang, S.Z. Fu, Y. Gu, S.J. Wang
Shanghai Institute of Laser Plasma, P.O.Box 800-229, Shanghai, 201800, P. R. China

Abstract: A new experimental method to measure the soft X-ray multilayer beam splitter's reflected and transmitted spectrum has been proposed and carried out with laser-produced plasma x-ray source, flat field grating spectrometer and soft x-ray CCD. Two grazing incident mirrors were employed to eliminate the high-order diffraction disturbance. Reflectivity and transmissivity of the multilayer beam splitter at the designed central wavelength 13.9 nm was measured in one laser shot.

1. INTRODUCTION

Soft X-ray multilayer mirrors have been rapidly developed in recent years, which are fabricated in some laboratories and widely used in many fields, such as surface science, astronomy, lithography and plasma diagnostics. In 1982, Ping Lee[1] proposed the theoretical concept of soft X-ray multilayer beam splitters. The first experiment using a soft X-ray multilayer beam splitter as the X-ray laser-cavity mirror was carried out at Lawrence Livermore National Laboratory (LLNL) in 1986[2]. In recent years, we have employed the soft X-ray multilayer mirror as the spectroscope in plasma diagnosis experiments. It is desirable to develop a convenient method to measure the specifications of soft X-ray multilayer beam splitters, because they are key optical components in the Mach-Zehnder interferometry experiments and very difficult to be made.

A new method to measure soft X-ray multilayer beam splitter's optical properties, especially the reflected and transmitted spectrum at the given incidence angle in one laser shot has been proposed in this paper. We used the laser-produced plasma source as the compact source of soft X-ray radiation, and measured the reflectivity and transmissivity of the multilayer beam splitter at the designed central wavelength 13.9 nm in one laser shot without contaminating the splitter surface.

2. EXPERIMENT

In order to generate the soft X-ray, a flat silver target was irradiated on the near normal incidence with a pulsed Nd: YAG laser beam of 1053nm in wavelength, 300ps in pulse width and 10J energy on the target. The laser beam was focused with an aspherical lens of 280 mm focal length to a spot of 50 micrometer in diameter. The experimental arrangement is shown in Fig.1 and the actual setup in the target chamber is shown in Fig.2. The reflected and transmitted spectrum of the multilayer beam splitter was measured with a flat field grating spectrometer which had a concave varied-spacing diffraction grating of 1200 grooves/mm. The spectrum was divided by the vertical filter and the vertical split and recorded with a soft X-ray CCD. The filter was placed in the back of the slit. The grazing incident mirrors were employed to eliminate the high-order disturbance.

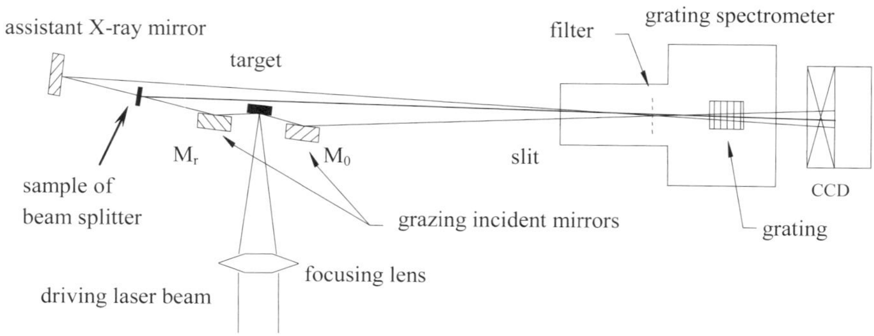

Figure 1. Experimental arrangement to calibrate soft X-ray multilayer beam splitters

The intensity distribution of the soft X-ray is symmetrical about the target normal[3], and in inverse proportion to the square of optical path length from the target to the CCD surface. The reflectivity of multilayer beam splitter can be expressed as

$$R(\lambda)=\frac{r_r^{\ 2}\cdot I_r(\lambda)\cdot R_0(\lambda)}{r_0^{\ 2}\cdot I_0(\lambda)\cdot R_r(\lambda)}$$

where r_0 is the above mentioned optical distance of the referenced beam, r_r is the optical distance of the reflected beam. I_r and I_0 are the intensity readings of above two beams on CCD, R_r and R_0 are the reflectivity of the grazing incident mirrors M_r and M_0, respectively.

The data treatment related to the reflectivity of multilayer beam splitters was proposed in the paper written by Wei Wang et al[4-7]. The transmissivity is determined by dividing the transmissive spectrum with the reflected spectrum of the assistant mirror.

Fig.2 Picture of the actual setup in the target chamber

3. RESULTS AND CONCLUSION

The measured spectrum of the tested soft X-ray multilayer beam splitter is shown in Fig.3, which was recorded by the soft X-ray CCD. The tested splitter was of 12 layer-pairs made from Mo/Si coatings, where the period thickness was 7.3nm. The incidence angle to the splitter in test was 7.24°.

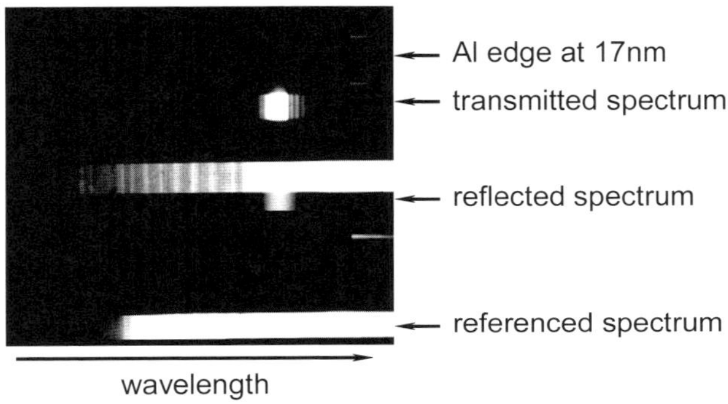

Fig. 3 The CCD image of soft X-ray beam splitter spectrums for incident angle 7.24°

The reflectivity and transmissivity of the multilayer beam splitter at incident angle 7.24° have been quantitatively determined by treating the CCD data and shown in Fig.4. It is concluded that the reflectivity at wavelength 13.9nm is about 18% and meanwhile the transmissivity is about 23%. Therefore, the product of reflectivity and transmissivity at wavelength 13.9nm is about 4.1 %. This result is satisfactory to the requirement of plasma diagnosis experiments by means of Mach-Zehnder interferometry which can favour us to get clear interferograms of laser plasmas.

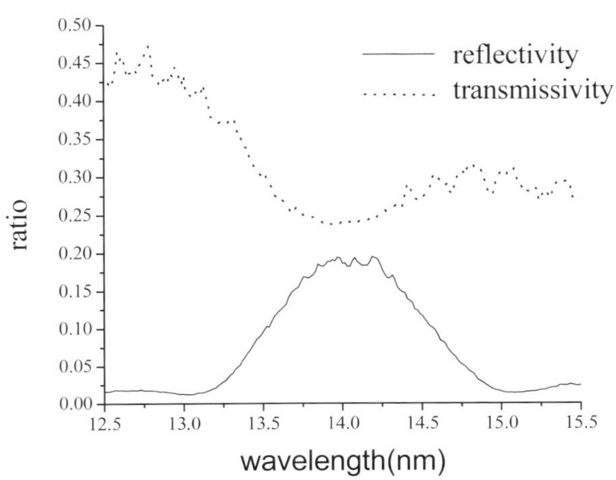

Fig.4 The reflectivity and transmissivity of the tested soft X-ray multilayer beam splitter

ACKNOWLEDGEMENTS

This work was supported by National Science Foundation of China (Grant No. 69989801) and National High Technology Development Program of China (Grant No. 863-804-7-3).

The authors are obliged to Tongji University and Changchun Institute of Optics and Fine Mechanics, Chinese Academy of Sciences, for their contributions to supply samples of the soft X-ray multilayer beam splitter. The stuff of SG-II group exerted excellent efforts to fulfil the experiment well.

REFERENCES

[1] P. Lee, (1982) Opt. Commun.: **43**(4), 237-241

[2] Hawryluk A., Ceglio N., Stearns D., Danzmann K., Khune M., Muller P., Wende B., (1986) Proc. SPIE, **688**, 81-89

[3] Gerritsen H. C. etc (1986) J. Appl. Phys.: **59**(7), 2337-2344

[4] W. Wang, Y.L. Ni, B.G. Wan, J.R. Sun, J. Wu, Ch. Wang, Y.Q. Sun, G.L. Zhou, Y. Gu, S.J. Wang, (2001) High Power Laser and Particle Beams:**13**(5),525-528

[5] Y. Gu, S. Chunyu, Y. Li, Q. Zhang, Y. You, (2000) High Power Laser and Particle Beams: **12**(1), 58-60

[6] H. Azuma, Y. Watanabe, Y. Kato, T. Motohiro, S. Noda, K. Murai, (1992) Jpn. J. Appl. Phys.:**31**(2B), L203-L205

[7] T. Haga,. Marcia C.K. Tinone, M. Shimada, T. Ohkubo, A. Ozawa, (1998) J. Synchrotron Rad.: **5**,690-692

Inst. Phys. Conf. Ser. No 186
Paper presented at 9th Int. Conf. X-Ray Lasers, Beijing, China, 24–28 May 2004

The Uniform Line-Focus System Used for Special Laser Beam Intensity Distribution of SG-II laser facility

Ch. Wang[1], W. Wang[1], Y. Gu[1], G.L. Huang[2], Z.Q. Lin[2], S.J. Wang[1]
[1] Shanghai Institute of Laser Plasma, P.O.Box 800-229, Shanghai, 201800, P.R. China
[2] Shanghai Institute of Optics and Fine Mechanics, P.O.Box 800-211, Shanghai, 201800, P.R. China

Abstract. A satisfying line-focus was not obtained with the classical CLA system under some special laser conditions, such as SG-II laser system. So, a novel unequal-width-element cylindrical lens array (UCLA) line-focus system was designed and manufactured. The experiments of x-ray laser had shown the obvious improvement of uniform characteristics of the line-focus.

1. Introduction

A uniform line-focus is an important condition for the experimental studies of x-ray laser. Several schemes have been advanced along with the development of x-ray lasers[1-5], such as Single cylindrical system (SC), Pair of crossed cylindrical system (PCC), Off-axis spherical mirror system (OSM), Segmented wedge arrays system (SWA), Pair of crossed cylindrical lens and wedge system (PCCW) as well as Cylindrical lens arrays system (CLA). Among them, the CLA line-focus system was a good choice because of its uniformity and high efficiency. There were many successful experiments proved the validity of CLA systems.

However, a satisfying line-focus was not obtained with the classical CLA system under some special conditions, such as the special beam intensity distribution of SG-II laser facility. So the system would be redesigned. Aiming at the conditions of SG-II laser beam, a novel unequal-width-element cylindrical lens array (UCLA) line-focus system was designed and manufactured. Simulation and subsequent experiments of x-ray laser had shown the obvious improvement of uniform characteristics of the line-focus.

2. A classical CLA line-focus system

A CLA line-focus system is made up of a cylindrical lens array (CLA) and an aspherical lens. Dividing the wave front of drive laser beam by arrays of cylindrical lens, a series of child line-focus would be created on focus surface by each of cylindrical lens elements (CLE), like the behaviour of a single cylindrical lens system. Each child line-focus was asymmetry, but the integrative results of all the child line-focus were very uniform along the direction of line-focus. The scheme was shown in Figure 1.

The classical CLA was so-called that was made of several cylindrical lens elements with the same width and focus. Generally, the number of elements was chosen as four or five.

The very good line-focus would be obtained by using it under usual beam of drive laser, and the uniformity along the whole line-focus was less then ±6% (see Figure 2). On the other hand, the efficiency of drive energy was almost 100%.

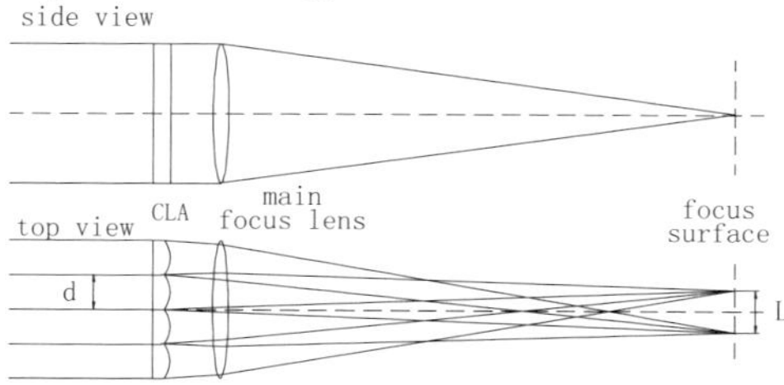

Figure 1. Scheme of a four elements CLA line-focus system.

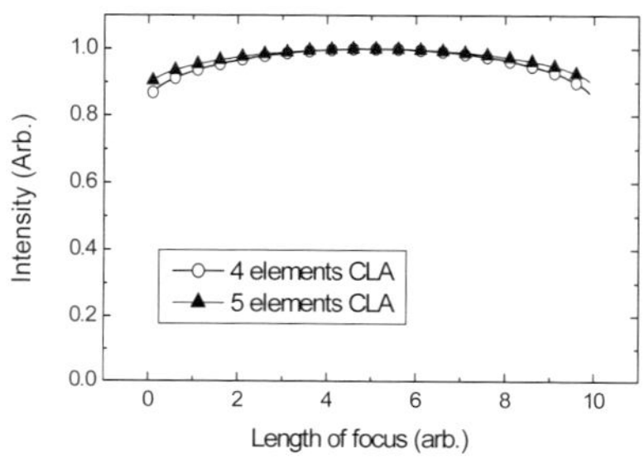

Figure 2. The simulated intensity distribution of line-focus using classical CLA systems.

3. Probable unsuitability of CLA system

However, the classical CLA system could not work very well under some conditions, especially with special intensity distributions of wave fronts of drive laser, such as SG-II laser system (see Figure 3).

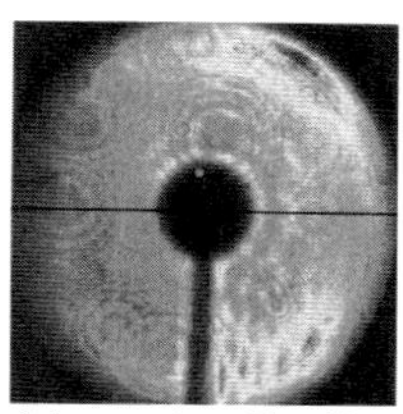

Figure 3. The special intensity distributions of wave fronts of SG-II laser.

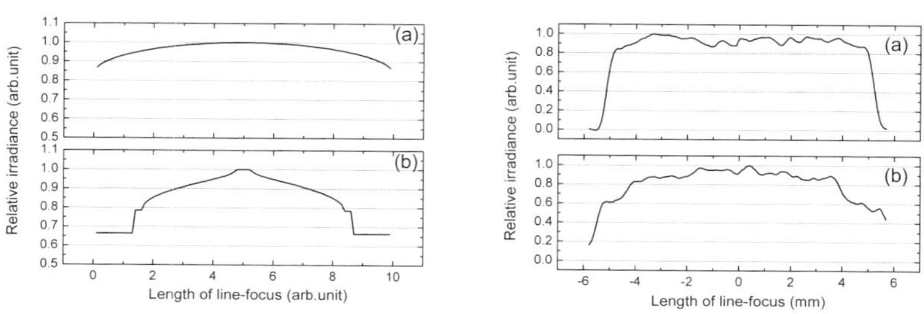

Figure 4. The intensity distribution of line-focus with four elements CLA system. The left was simulated and the right was measured with He-Ne laser as input. Input laser: (a)circle beam, (b)misshapen beam like SG-II.

Figure 4 showed the intensity distribution of line-focus with four elements CLA system. The left was simulated and the right was measured with He-Ne laser as input. The curves of (a) and (b) were respectively correspond the conditions of circle input laser beam and misshapen beam like SG-II. It was obvious that the misshapen beam of SG-II caused a bad line-focus, and it was not suit for experiments of x-ray lasers. In early experiments, the output of x-ray laser was always very small. The line-focus was an important reason.

4. A novel UCLA line-focus system

So, a novel unequal-width-element cylindrical lens array (UCLA) line-focus system was designed and manufactured based on the special conditions of SG-II laser. The main improvements were as below (see Figure 5):

- Choosing five elements instead of four elements CLA, in order to minish the effects of mid hole at two ends of line-focus.
- Choosing a novel configuration of unequal-width-element CLA. The width of middle elements was the same as the diameter of the mid hole, and the other four was the same. It can farther avoid the effects of mid hole.
- Circumvolving the pole of wave fronts of drive laser to horizontal directions by adjusting reflectors of target area (this was easy to do), in order to eliminate the effects of the pole.

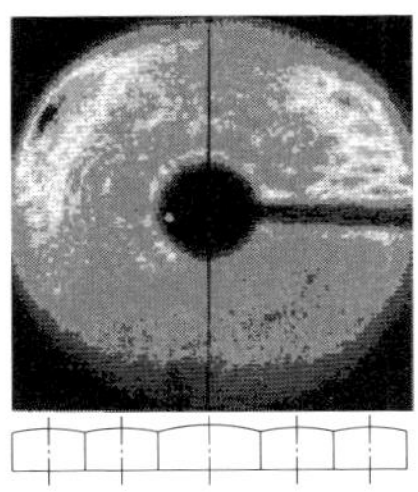

Figure 5. The wave fronts with horizontal pole and configuration of five elements UCLA.

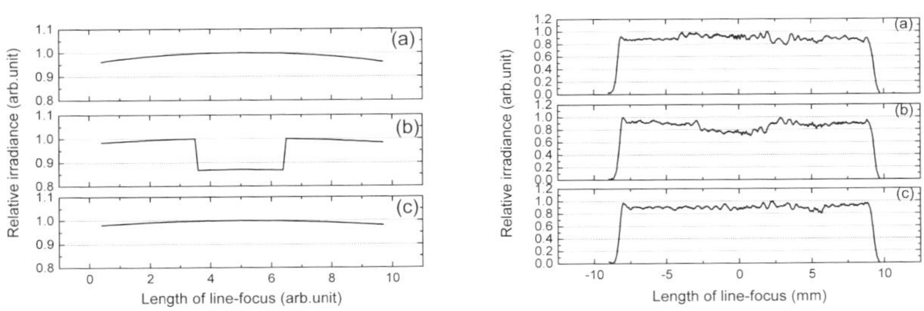

Figure 6. The intensity distribution of line-focus with five elements UCLA system. The left was simulated and the right was measured with He-Ne laser as input. Input laser: (a) circle beam, (b) misshapen beam with vertical pole, (c) misshapen beam with horizontal pole.

Figure 6 showed the intensity distribution of line-focus with five elements UCLA system. The left was simulated and the right was measured with He-Ne laser as input. The curves of (a) (b) and (c) were respectively correspond the conditions of circle input laser beam, misshapen beam like SG-II with vertical pole and misshapen beam with horizontal pole. It was obvious that the improvement of line-focus of (c) would provide a very uniform line-focus for experimental studies of x-ray lasers.

Double slit camera was a good instrument to inspect the uniformity of line-plasma[6]. With x-ray laser experiments, we measured the characteristic of UCLA line-focus system (Figure 7). The results were shown in figure 8. Uniform x-ray emissions indicated a uniform line-plasma and a good performance of UCLA line-focus system. The following experiments of saturation output of Ni-like Ag x-ray laser also proved the efficiency of UCLA systems.

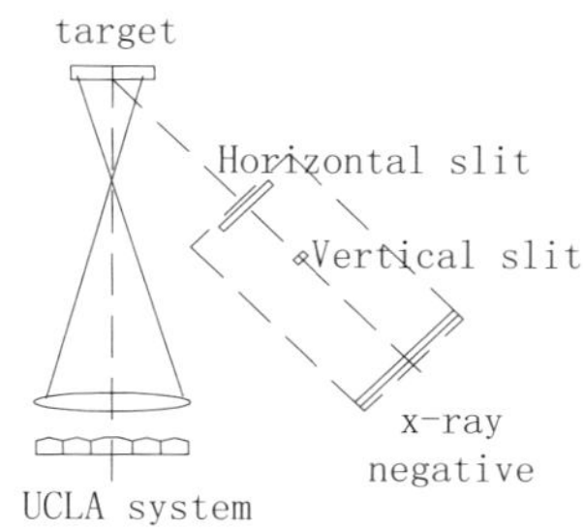

Figure 7. Scheme of measurement of line-focus with a double slit camera.

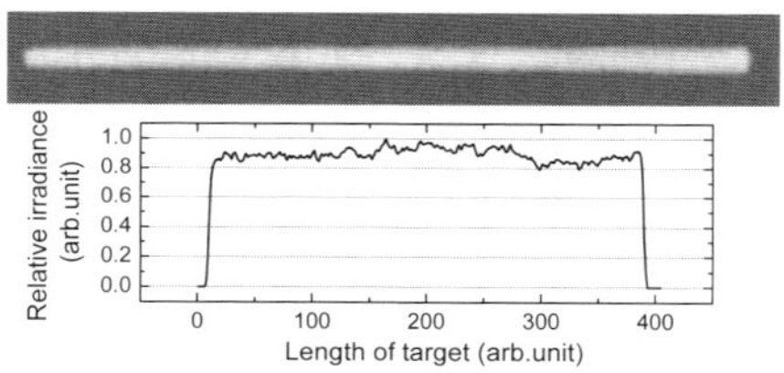

Figure 8. Results of line-focus measured by a double slit camera.

5. Conclusion

The special intensity distributions of SG-II laser system caused invalidation of a classical CLA line-focus system. So, a novel unequal-width-element cylindrical lens array (UCLA) line-focus system was designed and manufactured. Simulation, measurement and subsequent experiments of x-ray lasers also proved the obvious improvement of line-focus.

6. Acknowledgments

This work is supported by National Nature Science Foundation of China (Grant No. 69989801) and National High Technology Development Program of China (Grant No. 863-804-7.3). We acknowledge all the members of SG-II operation group, whose excellent works made the experiment going on wheels.

References

[1] Ross I N et al 1985 J.Phys.(E): Sci.Instrun. 18 169~172
[2] Ross I N et al 1987 Appl.Opt. 26 1584~1588
[3] Ileneuve D M V et al 1991 Opt.Commun. 81 54~58
[4] Chen W N et al 1991 Acta Optica Sinica 11: 829~833(in Chinese)
[5] Huang G L et al 1999 Acta Optica Sinica 19: 713~717(in Chinese)
[6] Fu S Z et al 1991 High Power Laser and Particle Beams 3 79~84(in Chinese)

Inst. Phys. Conf. Ser. No 186
Paper presented at 9th Int. Conf. X-Ray Lasers, Beijing, China, 24–28 May 2004

Wave-front correction of high-intensity femtosecond laser beams by using a closed-loop adaptive optics system

Zhaohua Wang, Zhan Jin, Jiaan Zheng, Peng Wang, Zhiyi Wei +, Jie Zhang

Laboratory of Optical Physics, Institute of Physics, Chinese Academy of Sciences, Beijing 100080

Abstract: The wave-front distortion of an intense femtosecond laser beam from our multi-TW laser facility was studied. By using an adaptive optics system to correct the wave-front distortion, beam quality with 1.5 diffraction limit was demonstrated. This result not only provides an intensity of 10^{19}W/cm^2, the spatially cleaned beam also prevent undesired effects from the energy deposited outside the central focal spot.

1. Introduction

The chirped pulse amplification (CPA) technique has greatly accelerated the development of ultra-short pulse, ultra-intense laser technology, the peak power as high as 0.85-PW has been obtained with table-top laser systems [1]. The rapid progress enables us to carry out experiments on ultrafast x-ray sources, electron acceleration, and relativistic plasma physics etc. Up to now, most efforts have been concentrated on increasing the energy, reducing the pulse duration, and temporally cleaning the pulses from the CPA systems to boost the peak power. However, for those experiments of interaction of high intensity laser with matter, the power density on targets plays a critical role rather than the peak power. Therefore, it is very important to pursue a good beam quality for high intensity lasers. In general, the wave-front becomes worse after multi-amplification in the high power laser system. This leads to an undesirable beam profile and focus ability. By using an adaptive optics system, O. Albert *et al* corrected the distortion of wave-front in the kHz CPA Ti:sapphire lasers. A focal spot as small as 1μm was achieved[2]. This provides a way to generate the relativistic laser intensity with only mJ energy. In this paper, we report the wave-front correction for our 20TW Ti:sapphire laser system (Jiguang II)[3], beam quality with 1.5 diffraction limit was obtained with an adaptive optics system. This result not only provides an intensity of 10^{19}W/cm^2 for many experiments[4-6], the spatially cleaned beam also prevents undesired effects from the energy deposited outside of the central focal spot[7].

2. Adaptive optical system

The configuration of the adaptive optical system consists of a wavefront sensor, which is

+ Author to whom correspondence: zywei@aphy.iphy.ac.cn

based on Shack-Hartmann principle of wavefront measurement, a deformable bimorph mirror as a wave-front corrector, a control unit for bimorph mirror and control and analysis softwares.

2.1 Shack-Hartmann wavefront sensor

Shack-Hartmann wavefront sensor is composed of a mini lens array and a CCD camera[8]. It splits a laser beam into many mini parts, and then focused them onto a focal plane, shown as Fig.1 (a). If the laser beam has an ideal plane wavefront, the focal spots on the CCD camera is a group of regular spot distribution, shown as Fig.1 (b). While the real laser beam is not a plane wavefront, this causes some aberration. Thus the focal spots is a group of irregular spot distribution, shown as Fig.1 (c). We can evaluate the beam quality by calculating the difference between the ideal and real spot distribution.

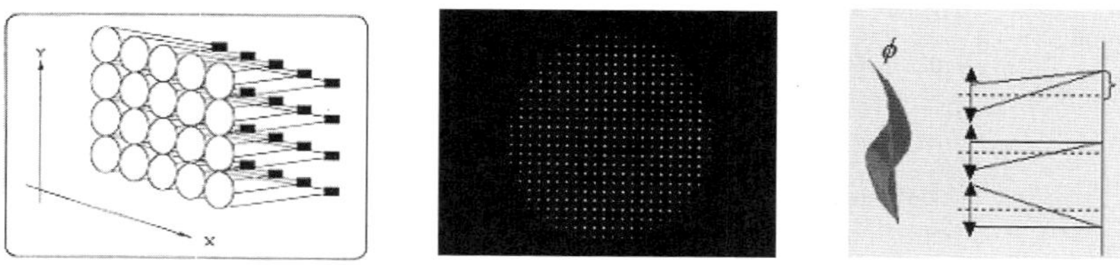

(a) (b) (c)

Fig.1 The principle of Shack-Hartmann wavefront sensor, the array of lenticule (a), the spot array of plane beams (b), and the spot of aberration beams (c).

2.2 Deformable bimorph mirror

The bimorph mirror consists of a glass, copper or quartz substrate firmly glued on a plate actuator disk made from piezoelectric ceramic. Applying the electrical signal to the piezoceramic plate causes tension of the piezodisc. Glued substrates prevent this tension, and this results in deformation of the reflective surface. To reproduce different types of aberrations with the help of such corrector the outer electrode is divided into several controlling electrodes, which have the shape of a part of a sector. The size as well as the number of such electrodes depends on the number and the type of the aberrations to be corrected[9]. The scheme of deformable bimorph mirror is shown in Fig.2.

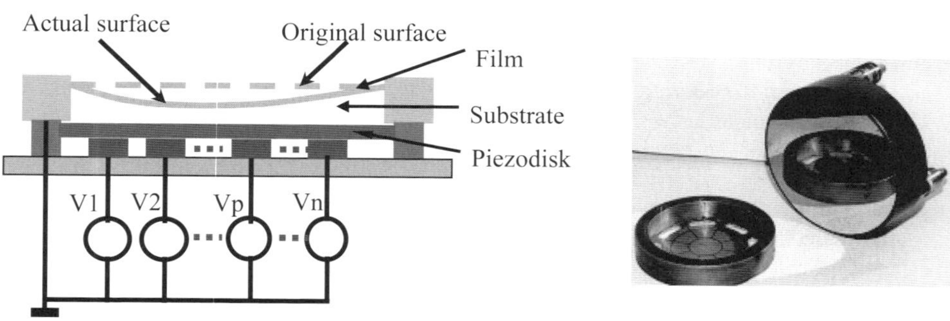

Fig.2 Scheme of the deformable bimorph mirror.

3. System design

We install the system between the compressor box and the target chamber, as shown in Fig.3. The laser beam is injected onto the deformable mirror, and is reflected by a mirror and a splitter. The reflection distribution is about 99:1. The transmission beam is accepted by a wavefront sensor, which gets the aberration of the beam. According to the aberration, the deformable mirror changs the reflected surface to correct the wavefront of the beam. Thus, the wavefront of the laser beam after the compressor is corrected before entering the target chamber.

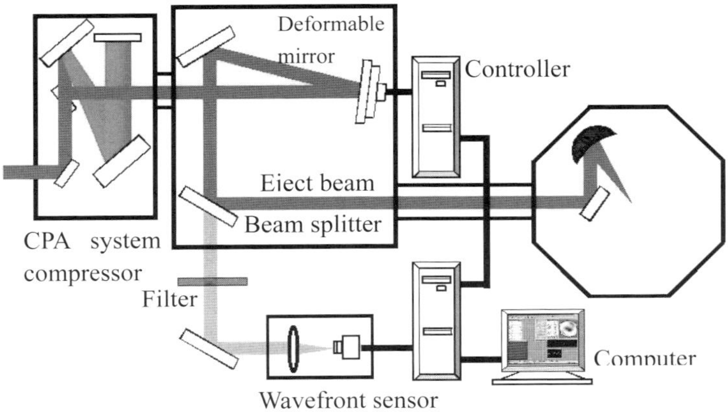

Fig.3 Scheme of the system setup

We apply some voltage to the deformable mirror and use the SHW to measure the wavefront of the laser beam. Then the voltages to be applied to all electrodes of the bimorph mirror are calculated. And these voltages multiplied by a coefficient are applied to electrodes of the deformable mirror. In this case, the wavefront is not absolutely compensated and the residual distortions are measured by the SHW. Again voltages to be applied to corrector electrodes are calculated and applied with the coefficient until reaching a good wavefront[10].

4. Experimental results

As a preparation work, we must know the aberration of the laser beam, which is the difference between the ideal wavefront and the actual wavefront. The wavefront is measured by the Shack-Hartmann wavefront sensor. Fig.4 (a) is an ideal wavefront of the laser beam, fig.4 (b) was an actual laser beam wavefront.

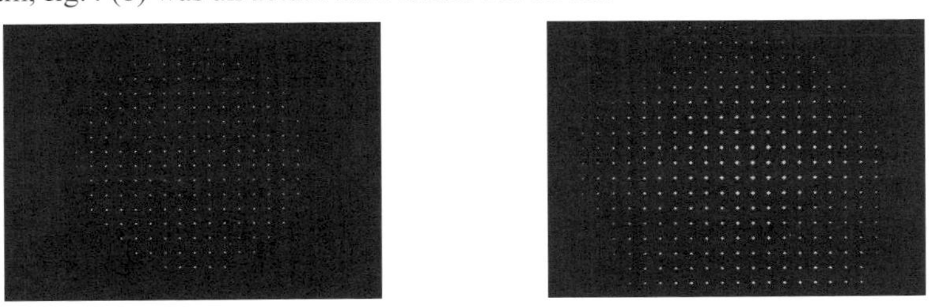

(a) (b)

Fig.4 Wavefront of laser beams using Shack-Hartmann wavefront sensor.

According to the aberration measured by the SHW, we apply a correct voltage to the deformable mirror. The distortion of the beam is then corrected. We measure the focusing spot size without and with the wave-front correction system. The beam diameter of the

20TW laser is about 30mm. The typical distorted laser wave-front is about 20 wavelengths of P-V values, and the focusing spot size is about 432μm using f/133.3 focus lens. This corresponds to a beam quality of about 3.5 diffraction limits (shown as Fig.5 (a)). By comparison with the adaptive optics system, the wave-front is about 0.15 wavelengths of P-V values after the distortion is corrected. Using the same focus lens, the focusing spot size is about 200μm. This corresponds to a beam quality of 1.5 diffraction limits (shown as Fig.5 (b)).

(a) (b)

Fig. 5 Focal spot of the laser beam
(a) without the wavefront correction (b) with the wavefront correction

5. Conclusions

In summary, the wave-front correction with an adaptive optics system has resulted in significant improvement in the focusability of the 20TW laser beam. Beam quality with 1.5 diffraction limit has been obtained. An even better beam with near diffraction limit should be possible by improving the optical quality, such as beam splitter and reflector, in the adaptive optics system.

AcknowledgementsThis work was supported by the NSFC under Grant Nos. 10374114, 10374115, 10374116, 60321003, 60225005, 60308001, 10474134 and 60128505, and the NKBRSF under Grant No. G1999075206, and the National Hi-tech ICF Program.

References

[1] M. Aoyama, K. Yamakawa, Y. Akahane et al.. 28 (17): 1594~1596
[2] O. Albert, H. Wang, D. Liu, Z. Chang, G. Mourou, Opt Lett, 2000, 25(15):1125-1127
[3] Z.Y. Wei, J.Zhang, J.F. Xia et al.. Science In China 2000, 43 (10): 1083~1087
[4] Y.T. Li, J. Zhang, Z.M. Sheng et al, Phys Rev Lett, 2003 **90(16)** 165002(1-4).
[5] G. Malka. Phys Rev E , 2002 **66**, 066402(1-8).
[6] T. Hosokai. Phys Rev E , 2003 **67**, 036407(1-8).
[7] S. Matsuoka, K. Yamakawa al.. J.Opt.Soc.Am.B 2000 17 (4): 663~667
[8] J. Primot al.. Appl Opt 1993 32 (31): 6242~6249
[9] E. Zeek, K. Maginnis, S. Backus et al.. Opt Lett, 1999, 24 (7): 493~495
[10] F. Druon, G. Cheriaus, J. Faure et al.. Opt Lett, 1998, 23 (13): 1043~1045

Investigations of X-ray laser media

Inst. Phys. Conf. Ser. No 186
Paper presented at 9th Int. Conf. X-Ray Lasers, Beijing, China, 24–28 May 2004

Measuring the Ultrafast Evolution of Electron Density in Plasma Channel Generated by Ultrashort Laser in Air via Diffractometer

Bin Shuai[1], Jiansheng Liu[1], Zuoliang Duan[1], Yi Zhu[1], Zhinan Zeng[1], Shi'an Zhang[2], Ji Zhang[2], Jingxin Ding[2], Liang'en Ding[2], Zengrong Sun[2], Ruxin Li[1], and Zhizhan Xu[1]

[1]Laboratory for High Intensity Optics, Shanghai Institute of Optics and Fine Mechanics, Chinese Academy of Sciences, P.O. Box 800-211, Shanghai 201800, China

[2]Department of Physics, East China Normal University, Shanghai 200032, China

Abstract. The evolution of the electron density of the plasma column generated by the self-channelling of femto-second laser pulses in the air was investigated. Based on the typical pump-probe technique, the electron density in the plasma channel can be extracted from the diffraction pattern of the probing laser beam. By employing a pump-probe back longitudinal diffractometer, the measurable electron density scale was extended to 10^{15} cm^{-3}. The ionization and decaying process on the hundreds of pico-second time scale was measured. A novel technique termed as linear-chirped longitudinal spectral diffractometer is developed. Via this method the evolution of the density on a time scale of tens of femto-seconds can be re-constructed from the spectra of the diffraction pattern of a linear-chirped laser pulse. By this method we realized the single-shot measurement of the electron density evolution of femtosecond time scale.

1. Introduction

Plasma channels in air formed by self-guided ultra-fast laser pulses has been drawing a lot of interest due to its potential applications [1-5]. The investigation of the electron density evolution in the channel will be of much help to the research of laser-plasma interaction, which is of tightly related to X-ray lasing and ICF. The electron density can be measured with Michelson [6] or Wollaston [7] interferometer, or by directly measuring the conductivity [8]. However, via these methods, the electron density at the level of 10^{15} cm^{-3} cannot easily be measured. In this paper a novel approach of diffractometry measurement was proposed. Via this method, the measurable density level was extended to $\sim 10^{15}$ cm^{-3}, and the evolution of the electron density was measured with backward longitudinal diffractometer. Furthermore, a technique termed as linear-chirped longitudinal spectral diffractometer was developed and was adopted to measure the ultra-fast evolution of the electron density on a time scale of tens of femto-second.

2. The principle of longitudinal diffractometer and theoretical analysis

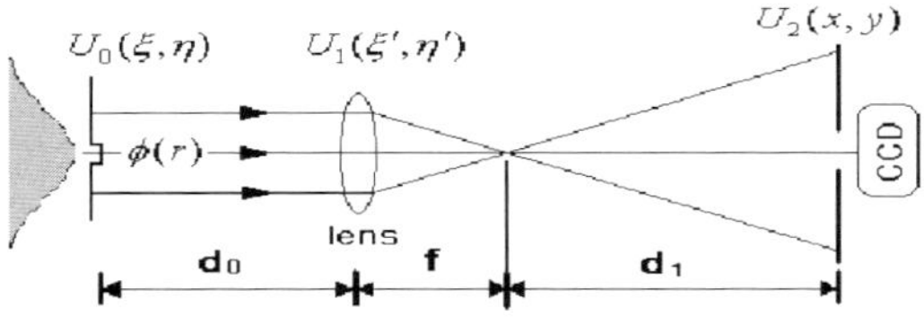

Figure 1. Description of the diffractometry method.

As is shown in Fig. 1, the probe beam which has the field distribution of $U_0(\xi,\eta)$ propagates trough the plasma channel thus suffers phase shift $\phi(r)$) and will be diffracted. The diffracted field at CCD plane can be expressed by Fresnel diffraction integral [9]:

$$U_2(x,y) \propto U(x,y) = \iint\iint U_0(\xi,\eta)\exp(-i\beta(\xi^2+\eta^2))\exp(ia_0[(\xi'-\xi)^2+(\eta'-\eta)^2])\times \exp(ia_1[(x-\xi')^2+(y-\eta')^2])d\xi d\eta d\xi' d\eta' \tag{1}$$

where $\beta=\pi/\lambda f$, $a_0=\pi/\lambda d_0$, $a_1=\pi/\lambda d_1$. After simplification,

$$U(x,y) = \exp(iA_0(x^2+y^2))\bullet \iint U_0(\xi,\eta)\exp(iA_1(\xi^2+\eta^2))\exp(-iA_2(x\xi+y\eta))d\xi d\eta \tag{2}$$

where $A_0=(\pi/\lambda d_1)(1-1/(1+d_1/d_0-d_1/f))$, $A_1=(\pi/\lambda d_0)(1-1/(1+d_0/d_1-d_0/f))$, $A_2=2\pi/(\lambda(d_0+d_1-d_0d_1/f))$. Assuming the original beam has the Gaussian profile $U_0\propto\exp(-ar^2)$, the phase shift $\phi(r)=\phi$ when $r<r_c$, and $\phi(r)=0$ when $r>r_c$ (i.e., r_c is the radius of the channel), the intensity distribution at the CCD plane can be deduced:

$$I(r) = -sign(A_1)C\pi r_c^2 \frac{1}{\sqrt{a^2+A_1^2}}\exp(-b_0r^2)\frac{J_1(A_2r_cr)}{A_2r_cr}\cos(b_1r^2-\phi/2)\sin(\phi/2) \tag{3}$$

where $b_0=A_2^2a/4(a^2+A_1^2)$, $b_1=A_2^2A_1/4(a^2+A_1^2)$, $sign(A_1)$ is the sign of A_1, J_1 is 1st order Bessel function and C is a constant.

From (3) we can see immediately that the diffraction pattern is sensitive to both the plasma density and the radius of the plasma channel. After the phase shift ϕ is extracted the electron density can be calculated using the relationship

$$\phi = (2\pi/\lambda)(n_e/2n_c)l \tag{4}$$

where n_e is the electron density, $n_c=m\omega^2/4\pi e^2$ is the critical density, l is the length of the channel.

3. Backward longitudinal diffractometer experiments

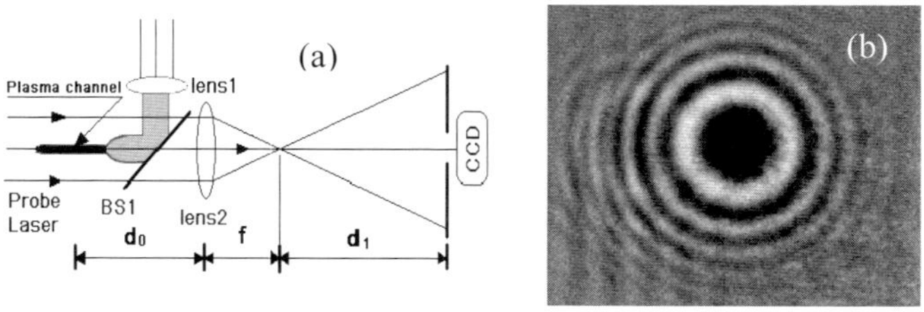

Figure 2 Experimental setup (a) and typical diffraction pattern recorded by CCD (b)

Fig. 2 shows the experimental setup for this section and a typical diffraction pattern recorded by the CCD. The laser source used in this experiment is a compact Ti:S CPA laser system at SIOM. The output of the laser is 0.6mJ/pulse, with the duration of 45fs and the repetition rate of 1 kHz. The output beam was split to two parts: 90% as the pump beam and 10% as the probe beam. The image should be analyzed with the help of (3) to extract the phase shift and the radius of the channel, as is described in Fig. 3.

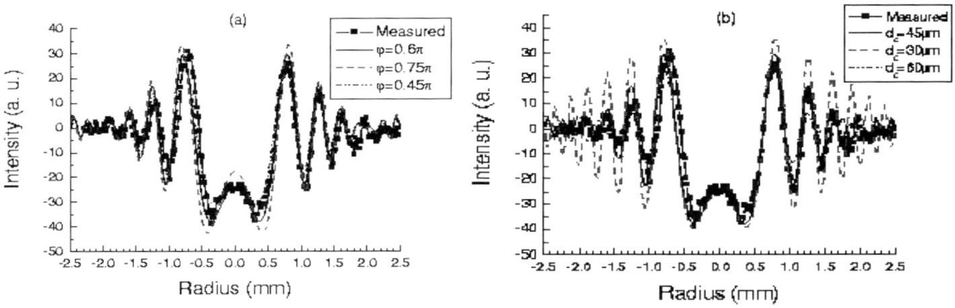

Figure 3. Extracting the electron density (a) and the radius of the channel (b) by fitting the experimental data with the calculated curve with Eq. (3)

From Fig. 3 we can see that the phase shift ϕ=0.6π and the radius r_c=45μm can be extracted simultaneously from a single image (Fig. 2 (b)). By varying the time delay between the probe and pump pulse, the density evolution on a time scale of hundreds of picosecond was measured, which is shown in Fig. 4.

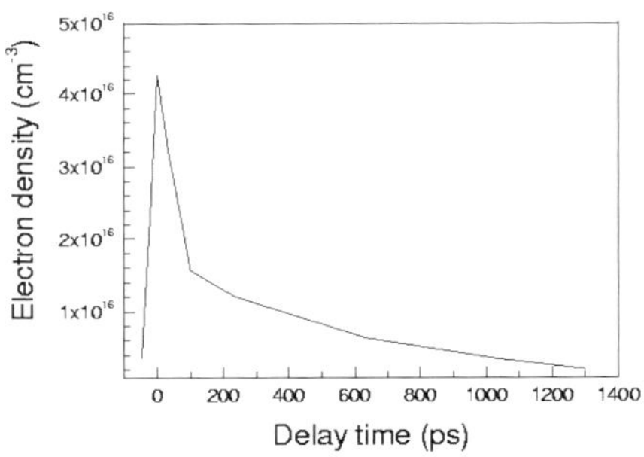

Figure 4. Evolution of the electron density in the channel versus time measured by backward longitudinal diffractometer

4. Linear-chirped longitudinal spectral diffractometer

Adopting a linear-chirped pulse beam as the probe beam, different spectral component will meet the plasma channel at different time and thus see different electron density. By recording the diffracted light with a spectrometer of which the entrance slit is placed to the centre of the diffracting rings, the electron density versus the light frequency (thus versus time) can be extracted.

Based on (2), adopting Fourier analysis, after deducing and simplification, the spectra of the diffraction pattern recorded by the CCD can be expressed as:

$$I(y,\omega) \propto \exp(-b_0 y^2)\frac{J_1(A^2 r_c y)}{A^2 r_c y}\cos(b_1 y^2 - \phi(\omega)/2)\sin(\phi(\omega)/2)F^2(\omega) \tag{5}$$

where y is the coordinate in the direction perpendicular to the spectral dispersion direction and $F(\omega)$ is the Fourier transformation of the temporal envelope of the pulse. The time dependent $\phi(t)$ can be obtained by extracting $\phi(\omega)$ with (5) and utilizing the linear chirp relation $\omega(t)=\omega_0+\alpha t$, where ω_0 is the central frequency and α is the chirp rate.

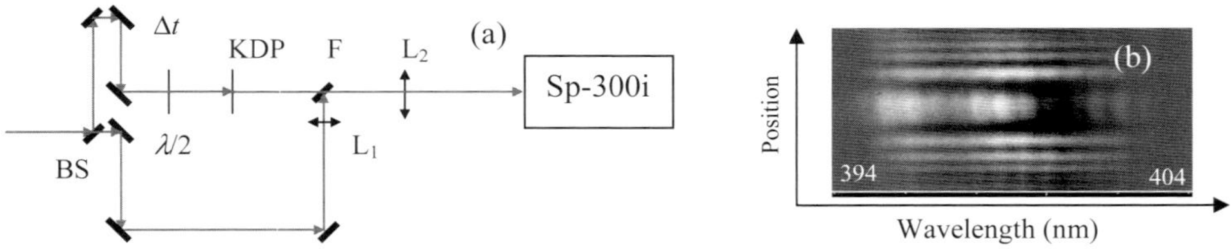

Figure 5. Experimental setup of linear-chirped longitudinal spectral diffractometer (a) and recorded spectra at the wavelength of 400nm (b).

The experimental setup for this section is shown in Fig. 5 (a) and the spectrum of the diffracted pump beam is shown in Fig. 5 (b). The 24mJ, 40fs laser pulse output from a 0.5 TW Ti:S laser was stretched to a linear chirped pulse with the duration of 300fs by detuning the compressor, and split into two beams. The pump beam was focused by lens L_1 and formed a plasma channel to the right of the filter F. The probe beam was frequency-doubled in order to be separated easily from the pump beam. The radius of the channel and the phase shift at various wavelengths was extracted by fitting the vertical density profile of Fig 5 (b) with the theoretically calculated curve using Eq. (5). From the extracted values at various wavelengths the ultra-fast electron density evolution on a time scale of tens of femtosecond was figured out as is shown in Fig. 6.

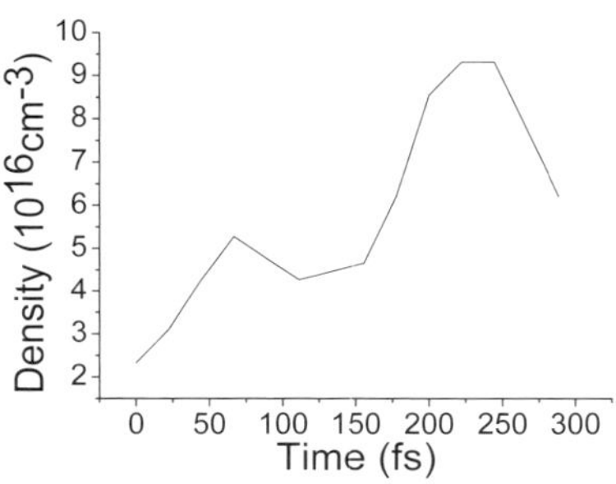

Figure 6. Electron density evolution obtained by analyzing Fig. 5 (b)

5. Conclusion

We have introduced the principle and the theoretical analysis of the diffractometry method to measure the electron density in a plasma channel formed by self-guiding of femtosecond laser pulses. Via this method, the measurable electron density was extended to as low as 10^{15} cm^{-3}. The electron density evolution in hundreds of picosecond was measured by backward longitudinal diffractometer. A novel technique termed as linear-chirped longitudinal spectral diffractometer was developed and the measurement of the tens-of-femtosecond scaled density evolution was enabled via this method.

References

[1] Kasparian J, Rodiguez M, Mejean G et al. 2003, Science 301, 61-64
[2] Pepin H, Comtois D, Vidal F et al. 2001 Phys. Plasmas 8, 2532
[3] Rodriguez M, Sauerbrey R, Wille H et al. 2002 Opt. Lett. 27, 772
[4] Kosareva O G, Kandiddov V P, Brodeur A et al. 1997 Opt. Lett. 22
[5] Braun A, Kom G, Liu X, Du D et al. 1995 Opt. Lett. 20, 73-75
[6] La Fontaine B, Vidal F, Jiang Z et al. 1999 Phys. Plasmas 6, 1615-1621
[7] Yang H, Zhang J, Li Y et al. 2002 Phys. Rev. E 66, 016406
[8] Tzortzakis S, Prade B, Franco M, Mysyrowicz A 2000 Opt. Comm. 181, 123
[9] Goodman J W, Introduction to Fourier Optics, 1996 the McGraw-Hill Company

Inst. Phys. Conf. Ser. No 186
Paper presented at 9th Int. Conf. X-Ray Lasers, Beijing, China, 24–28 May 2004

Nickel-like Silver Spectrum (AgXX) from a Fast Capillary Discharge Plasma

A. Rahman [1], J.J. Rocca [1,2] and J.-F. Wyart[3]

[1]National Science Foundation Engineering Research Center for Extreme Ultraviolet Science and Technology and Department of Electrical and Computer Engineering, Colorado State University
[2]Department of Physics,
Colorado State University. Fort Collins, CO 80523, USA
[3]Laboratoire Aimé Cotton, CNRS (UPR 3321),
Centre Universitaire, 91405-Orsay, France

Abstract. In connection with the development of efficient Ni-like Ag lasers operating at 13.9nm the spectra of Ni-like AgXX have been observed and analyzed in the 13.7-20.5nm region. A high-current capillary discharge system was employed to create a highly ionized silver plasma. Forty-three AgXX transitions have been identified with the assistance of calculations performed using the Slater-Condon method with generalized least-squares fits of the energy parameters. The average difference between the measured transition wavelengths and the theoretical values is 0.0026 nm.

1. Introduction

The development and optimization of practical soft x-ray lasers based on transitions of Nickel-like ions generates significant interest in the spectra of these ions. Recently a detailed spectroscopic study of Ni-like Cd has been reported utilizing a capillary discharge plasma[1].

A similar discharge was employed to study Ni-like silver spectra in the spectral region between 12.7 and 20.5 nm. Highly ionized Ag plasmas were obtained with fast capillary discharge excitation of a silver vapor filled capillary channel with discharge current pulses up to 200 kA peak amplitude. The rate of increment of the current was $\sim 1.5 \times 10^{13}$ A/sec. Previous work in laser-created plasmas [2] has identified five AgXX lines in the 17.0-19.0 nm spectral region as 4p-4d transitions. In the present spectra, all previously reported 4d-4p and 4d-4f transitions of AgXIX [3] and AgXX were observed as strong emission lines. This was an incentive to extend the classification to weaker lines.

2. Spectral Measurements and Interpretation

Silver vapor was generated by discharge ablation and was injected into a 4cm polyacethal capillary channel of 5mm diameter. A very high current pulse of 160-200kA peak amplitude and 12~15 ns rise time generated the highly ionized silver plasma of interest . It followed a low current pre-ionizing current pulse. A detail description of experimental setup is provided else where[1]. Figure 1 shows a typical spectra of the capillary discharge silver plasma for the 18.1~19.5 nm wavelength region. Calibration of the different spectral region were performed using known transitions of OVI, OVII, ArVIII, ArIX through ArXIV and FVI [4] and were verified by comparing with previously known lines of AgXIX [3] and AgXX [2]. From these comparisons, from the difference between the wavelength of other known lines and those obtained applying the calibration curve, and from shot to shot variations of the measured wavelengths, the error in the measurement of the transition wavelengths reported herein can be conservatively estimated at 0.01 nm.

The same theoretical methods previously used to classify lines from CdXXI [1] was used for AgXX. The advantages of the Generalized least-squares (GLS) techniques for improving the reliability of multicharged ions predicted energies in the framework of the Racah-Slater method has been discussed in ref [1]. Isoelectronic constraints were applied to the scaling factors of radial parameters of the four excited configurations $3d^9 4l$ (l=s,p,d,f). The scaling factor is the ratio of an 'adjustable' radial parameter to its *ab initio* (HFR) value, which is derived from the RCN, RCN2 codes by R.D. Cowan [5]. The experimental 4s and 4p levels used in the GLS fits were those of [2] and earlier references. For $3d^9 4d$, the set of E_{exp} values was extended from 228 values in [2] to 239 after inclusion of the new AgXX levels in the final fit. The predicted levels of AgXX are reported in Table I. As for the case of CdXXI, the (J_1,j_2) coupling is a better approximation than is LS and the changes in leading components are small in those close elements.

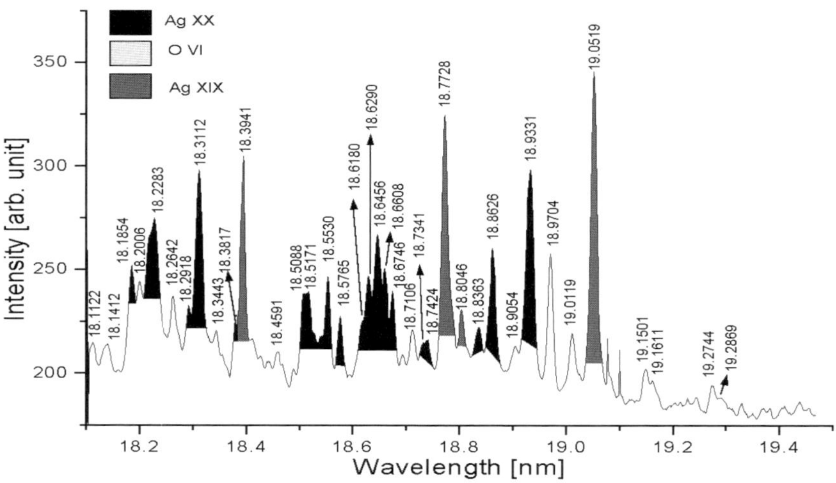

Figure1. Spectra of capillary discharge silver plasma for the 18.1-19.5 nm wavelength region. Spectra corresponds to a single discharge shot and was acquired with a time resolution of about 5 ns. The capillary diameter was 5 mm and the discharge current pulse had a peak amplitude of 164 kA. Each spectral line is identified with the experimental wavelength corresponding to that particular discharge shot.

Table I.

Classification of lines of Nickel-like AgXX. The first column shows the calculated wavelengths (λ_{cal} in nm) as they are derived from the "best" experimental values. The second and third column show the experimental wavelengths (λ_{exp}) and wavenumber. Int is the measured relative intensity. The level designations imply the J-value and index N_{th} which numbers the levels from the lowest energy in the same J-value and configuration as used in [2]. The emission transition probability gA (in 10^9 s^{-1}) in length form is derived by means of Cowan codes for E_{GLS} level values with no C.I. effects included.

λ_{cal}(nm)	λ_{exp}(nm)	λ (cm^{-1})	Int	gA	$\lambda_{exp}-\lambda_{cal}$	J_o N_{th}	J_e N_{th}	E_{odd}	E_{even}	Comment
13.4269	13.4250	744879	1	16	-0.0019	4p 1 1	4d 0 2	3181814	3926590	
13.8811	13.8830	720305	7	153	0.0011	4p 1 2	4d 0 2	3206188	3926590	
16.5377	16.5390	604617	2	198	0.0013	4p 2 2	4d 2 3	3175562	3780240	
16.5924	16.5919	602704	2	46	-0.0005	4p 3 1	4d 3 2	3129393	3732080	S, AgXXI
16.6898	16.6898	599168	3	249	0.0000	4p 2 1	4d 3 1	3122417	3721590	
16.7105	16.7122	598365	3	100	0.0017	4p 1 1	4d 2 3	3181814	3780240	S, AgXXI
16.8863	16.8870	592171	4	286	-0.0007	4p 3 1	4d 3 1	3129393	3721590	
16.9038	16.8995	591733	4	347	-0.0043	4p 2 1	4d 2 1	3122417	3714000	
17.0684	17.0704	585809	5	517	0.0020	4p 2 2	4d 3 3	3175562	3761440	
17.1055	17.1088	584495	2	73	0.0033	4p 3 1	4d 2 1	3129393	3714000	S, AgXXI
17.2005	17.2039	581264	6	797	-0.0034	4p 3 1	4d 4 1	3129393	3710770	
17.3709	17.3752	575533	1	125	0.0043	4p 1 1	4d 1 3	3181814	3757490	
17.4200	17.4152	574211	3	94	-0.0048	4p 1 2	4d 2 3	3206188	3780240	
17.6174	17.6213	567495	3	97	0.0039	4d 3 3	4f 3 4	3761440	4329060	S
17.7075	17.7074	564736	4	210	-0.0001	4p 2 1	4d 1 1	3122417	3687150	
17.7029	17.7074	564736	4	99	0.0045	4d 1 2	4f 1 2	3719060	4283940	S, AgXXI
17.8070	17.8070	561577	1	85	0.0000	4p 1 1	4d 0 1	3181814	3743390	Tentative, Blend
17.8079	17.8070	561577	1	28	-0.0009	4p 2 2	4d 2 2	3175562	3737110	Tentative, Blend
17.8336	17.8377	560610	2	140	0.0041	4d 4 1	4f 4 2	3710770	4271600	
17.8709	17.8688	559635	2	95	-0.0021	4p 2 3	4d 3 3	3201870	3761440	
17.9689	17.9670	556576	1	78	-0.0019	4p 2 2	4d 3 2	3175562	3732080	
18.0190	18.0216	554890	1b	161	0.0026	4d 5 1	4f 5 2	3719430	4274400	
18.1182	18.1182	551931	2b	248	0.0000	4d 2 1	4f 3 1	3714000	4265930	
18.1815	18.1773	550137	3	194	-0.0042	4d 3 1	4f 4 2	3721590	4271600	
18.2209	18.2205	548832	4b	747	-0.0004	4d 3 3	4f 4 3	3761440	4310260	Tentative, Blend
18.2209	18.2205	548832	4b	442	-0.0004	4d 2 3	4f 3 4	3780240	4329060	Tentative, Blend
18.2299	18.2205	548832	4b	315	-0.0094	4d 3 1	4f 4 1	3721590	4270140	Tentative, Blend
18.2862	18.2868	546849	2	219	0.0006	4p 4 1	4d 4 2	3191211	3738070	
18.3092	18.3094	546174	5	902	0.0002	4d 4 1	4f 5 1	3710770	4257030	
18.3665	18.3803	544040	1b	122	0.0138	4d 3 2	4f 3 2	3732080	4276550	Tentative, Blend
18.3709	18.3803	544040	1b	227	0.0094	4d 3 1	4f 3 1	3721590	4265930	Tentative, Blend
18.3942	18.3941	543653	5	521	-0.0001	4d 3/2	4f 5/2	892277	1435920	AgXIX 18.3942
18.5048	18.5088	540284	2b	60	0.0040	4d 3 4	4f 3 4	3788660	4329060	Tentative, Blend
18.5060	18.5088	540284	2b	142	0.0028	4p 3 3	4d 3 4	3248295	3788660	Tentative, Blend
18.5110	18.5152	540097	2b	126	0.0042	4p 0 1	4d 1 4	3229410	3769630	
18.5350	18.5309	539639	1b	436	0.0041	4d 3 2	4f 4 2	3732080	4271600	q

18.5378	18.5378	539438	1b	373	0.0000	4d 2 2	4f 3 2	3737110	4276550	q
18.5530	18.5530	538996	2	450	0.0000	4d 2 4	4f 3 3	3783160	4322160	
18.5770	18.5770	538300	2	640	0.0000	4d 3 4	4f 4 4	3788660	4326960	
18.6134	18.6166	537155	1b	43	0.0032	4p 1 2	4d 1 2	3181814	3719060	
18.6290	18.6290	536797	3b	894	0.0000	4d 4 3	4f 5 3	3777140	4313940	
18.6451	18.6451	536334	4	747	0.0000	4d 4 2	4f 5 2	3738070	4274400	
18.6599	18.6599	535909	4	1060	0.0000	4d 5 1	4f 6 1	3719430	4255340	
18.7319	18.7341	533786	1b	64	0.0022	4d 3 2	4f 3 1	3732080	4265930	q
18.7441	18.7415	533575	2b	83	-0.0026	4d 3 4	4f 3 3	3788660	4322160	
18.7753	18.7719	532711	7	705	-0.0034	4d 5/2	4f 7/2	904224	1436839	AgXIX 18.7753
18.7995	18.8046	531786	2	205	-0.0051	4p 1 3	4d 2 4	3251230	3783160	blend AgXIX 18.8073
18.8352	18.8368	530876	2	82	0.0016	4p 1 2	4d 2 2	3206188	3737110	
18.8605	18.8626	530150	4	271	0.0021	4p 2 3	4d 3 2	3201870	3732080	Tentative, Blend
18.8654	18.8626	530150	4	405	-0.0028	4p 2 4	4d 3 4	3258589	3788660	Tentative, Blend
18.9092	18.9092a	528843	7c	693	0.0000	4p 3 3	4d 4 3	3248295	3777140	
18.9315	18.9315a	528220	7c	851	0.0000	4p 4 1	4d 5 1	3191211	3719430	
18.9330	18.9230a	528457	7c	496	-0.0100	4p 3 2	4d 4 2	3209891	3738070	
18.9674	18.9704	527137	4	15	0.0030	4p 3 2	4d 2 2	3209891	3737110	S
19.0632	19.0519	524882	8	147	-0.0112	4p 2 4	4d 2 4	3258589	3783160	blend AgXIX 19.0543
19.1502	19.1501	522190	2	159	0.0000	4p 3 2	4d 3 2	3209891	3732080	
19.4980	19.4956	512936	5	103	-0.0024	4p 1 2	4d 1 2	3206188	3719060	blend AgXIX 19.4981
19.5427	19.5420	511718	1	98	-0.0007	4p 3 2	4d 3 1	3209891	3721590	

Notes:
- a — Wavelength from ref [1]. The present spectra lead to an unresolved peak with global intensity 7
- S — line is too strong with regard to the gA value
- S, Ag XXI — This Ag XX transition is probably masked by a AgXXI transition.
- blend Ag XIX — blend with a classified line of Cu-like Ag XIX with wavelength from[3].
- Ag XIX — AgXIX lines with wavelengths from [3].
- q — this line did not appear in all spectra
- b — broad line
- c — only combined intensity can be assigned for all three transitions

3. Conclusions.

A highly ionized silver plasma generated by a fast high current capillary discharge was used to identify forty-three transitions of Ni-like Ag. The average deviation between the measured and theoretical GLS transition wavelength values is 0.0026 nm. The newly identified lines can be of use in the plasma diagnostic necessary to optimize performance of Ni-like Ag lasers operating at 13.9 nm.

Acknowledgement. This work has been supported by DARPA grant G-DAAD 19-99-1-0279 and by the NSF ERC for Extreme Ultraviolet Science and Technology

References

1. A. Rahman et al. Phys. Scripta, **67**, 414 (2003).
2. S.S. Churilov, A.N. Ryabstev, et.al. Phys. Scripta, **38**, 326, (1988).
3. J. Reader, N. Acquista, et.al., J. Opt. Soc. Am. **73**, 1765 (1983).
4. R.L. Kelly, L.J. Palumbo, *Atomic and Ionic Emission Lines Below 2000 Angstroms*, NRL Report 7599 (1973).
5. R.D. Cowan, *The Theory of Atomic Structure and Spectra*, University of Calif. Press, Berkeley (1981) and computer codes.

Inst. Phys. Conf. Ser. No 186
Paper presented at 9th Int. Conf. X-Ray Lasers, Beijing, China, 24–28 May 2004

Ablation mechanism for silica and silicon with double femtosecond laser pulses

Y-P Deng, X-H Xie, H Xiong, R-X Li, and Z-Z Xu
Shanghai Institute of Optics and Fine Mechanics, Chinese Academy of Sciences

Abstract. The optical breakdown thresholds (OBTs) of dielectric and semiconductor were measuring using both single and double 40-fs laser pulses with various energies. By measuring the OBTs with different energy and the delay time between two pulses, we found opposite transformation rules for fused silica and silicon. We suggest that the difference is caused by two factors that affect the ablation of the material. One is the energy gap being shrunk due to the electron-hole interaction if electrons are excited from the valence band and another factor is the diffusion of the electron-hole plasma.

In recent years, the interaction between ultrashort laser pulses and solid has been paid extensive attention [1-8]. It has been recognized as potential applications in micromaching by femtosecond laser pulses [3-4, 7-8]. When femtosecond laser pulses irradiate the solid target, the laser energy is quickly absorbed by electrons, and then transferred to the lattices, which restrains the thermal conduction effect and induces cool ablation. So the precision of laser machining can be well improved.

In this paper, we use pump-probe method to investigate the effects of CBE on the band gap and the diffusion of CBE by measuring OBT of fused silica and silicon irradiated by laser pulses at the wavelength of 800 nm and with the pulse duration of 40fs.The output parameters of the Ti: Sapphire laser used in our experiments is 40fs, 0.7mJ/pulse, 800nm and 1 Hz. Horizontally linearly polarized intense laser pulses are directed to a beam-splitter (BS), where each single pulse is split into two temporally identical pulses with variable delay from 0 to 1ps. The time delay is precisely set from 0 to 1ps monitored by a single shot autocorrelator (SSA). The collinearity of the beams is achieved by adjusting their interference pattern at zero time delay. The energies of the two pulses can be individually adjusted by the half wave plates and the polarizers. The collinear beams are focused onto the front surface of the samples in normal incidence. The diameter of the focus spot is measured to be about 45 μm . With a long-working-distance microscope and a charge-coupled device camera (CCD), a magnified image of the target surface is obtained. Experiments are performed for samples of fused silica (FS, 0.5 mm thickness) and silicon (Si, 0.5 mm thickness),which have been optically polished.

With the experimental setup described above, we first measure the single pulse OBT. For a 40fs 800nm pulse, the OBT was measured to be 2.0J/cm2 for fused silica and 0.2J/cm2 for silicon with the error of approximately 8%. In the double pulse OBT measurements, the pulse energies of both laser pulses were set lower than the single pulse breakdown threshold.

Fig. 1(a) shows the breakdown thresholds of the second pulse versus the time delay between the two pulses, with different energies of the first pulses for fused silica. As the silica sample is irradiated by the laser pulses, the breakdown threshold of the second pulse rises quickly in the time delay range of 100-125fs when the first pulse energy is relatively low, and rises in the time delay range of 100-180fs with higher first pulse energy, but more slowly. In the case of silicon sample Fig. 1(b), the phenomenon is different. The breakdown threshold of the second pulse keeps rising in the time delay range of 100-400fs, but the slopes are less steep than in the case of silica sample.

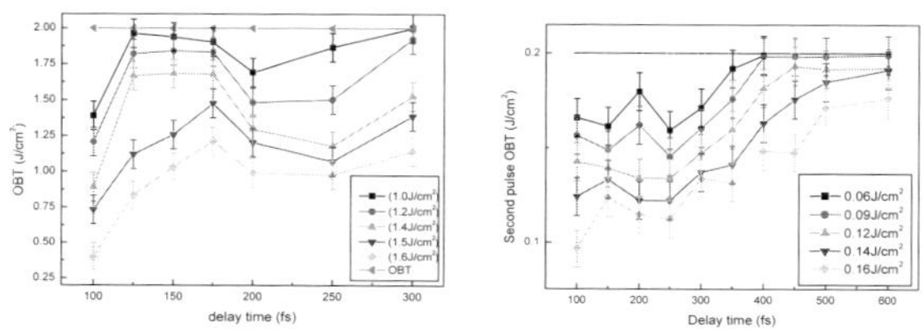

Fig.1a. The second pulse OBT vs. the time delay between two pulses for silica, the energy of the first pulse is 1.0J/cm2, 1.2J/cm2, 1.4J/cm2, 1.5J/cm2, 1.6J/cm2, respectively

Fig.1b. The second pulse OBT vs. the time delay between two pulses for silicon, the energy of the first pulse is 0.06J/cm2, 0.09J/cm2, 0.12J/cm2, 0.14J/cm2, 0.16J/cm2, respectively

The OBT is associated with a threshold density of CBEs $n_{th} = \omega m / 4\pi e^2$ [6].

The typical equation to calculate the density of CBEs in dielectrics is given by [1, 6, 9]:

$$\frac{dn(t)}{dt} = \alpha I(t) n(t) + \sigma_k I(t)^k \quad , \tag{1}$$

When the femtosecond laser pulses interact with silicon, the diffusion of CBEs in the laser incident direction is considered in the equation of the CBEs density [10], which is different from Eq.(1), and generally the diffusion depth of the electron is much less than $2\,\mu m$. Because the focus spot is much larger than the spatial size of electron diffusion, the electron diffusion in the vertical direction of focus spot can be described as follows [10]

$$\frac{dI(z,t)}{dz} = -[\alpha_0 + \alpha_{fca}(z,t) + \beta I(z,t)] I(z,t)$$

$$\frac{dn(z,t)}{dt} = [\alpha_0 + \frac{1}{2}\beta I(z,t)] \frac{I(z,t)}{\hbar\omega} \tag{2}$$

In double-pulses experiment, CBEs excited by the first pulse will relax during the time delay between two pulses. The relaxation times of CBE for silica and for silicon are about 200fs [2] and 400fs, respectively. CBEs relax to the bottom of conduction band 100fs after the laser pulse passes, and the process keeps the thermal equilibrium generally. Therefore, when the time delay exceeds 200fs for the silica sample and 400fs for the silicon sample, the energy is transferred to the crystal lattice by the electron. Then the conduction band electrons almost relax to the defect state and may recombine with the holes with phonon emission. Therefore, when the time delay is less than 200fs for silica, some electrons excited by the first pulse are still in the conduction band, which leads to the enhancement of the absorption rate of laser energy by the material as well as the decreasing of ablation threshold for the second pulse.

Considering the loss of conduction band electrons generated by the first pulse, as shown in Eq. (3) Ming Li et al. introduced another term to Eq. (1) to explain the experiment result [2],

$$\frac{dn(t)}{dt} = \alpha I(t)n(t) + \sigma_k I(t)^k - \frac{n(t)}{\tau} \tag{3}$$

Though the addition of this term greatly improved the fit of the result in the range of 200fs-1ps, it cannot fit the result very well in the relaxation time.

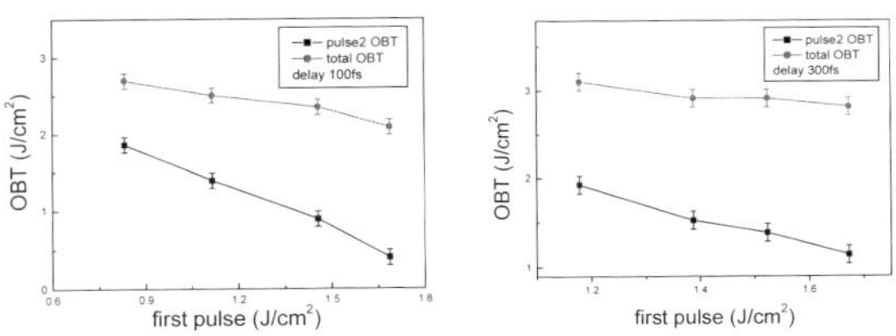

(a) (b)

Fig. 2. Total ODT and the second pulse OBT vs. the fluence of the first pulses for fused silica. (a) time delay is 100fs. (b) time delay is 300 fs

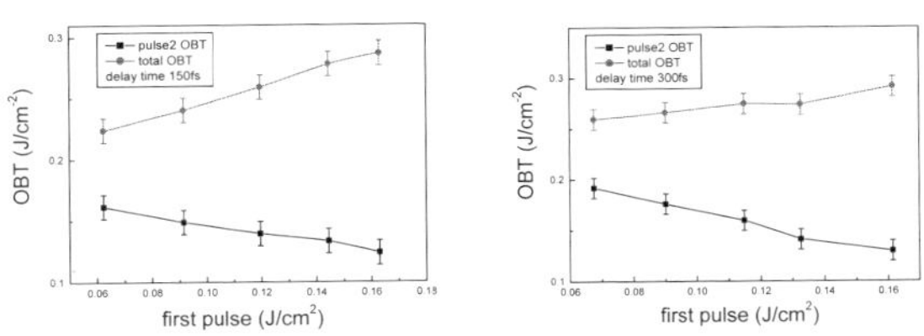

(a) (b)

Fig. 3. Total ODT and the second pulse OBT vs. fluence of the first pulses for silicon. (a) time delay is 150fs. (b) time delay is 300 fs.

In order to further investigation of CBE dynamics mechanism in the range of relaxation time, the first time we measure the breakdown threshold of the second pulse with different first pulse energies and time delay specified, and he results are shown in Fig. 2 and Fig. 3. From the results, we may see the obvious difference in the variation of the total thresholds for silica and silicon samples. The phenomena are interesting that the total threshold decreases for silica, but increases for silicon with the increase of the first pulse energy, and such phenomena have not reported in the literature to our knowledge.

They might be induced by the relaxation characteristics of the CBEs, which cause the modification on the band-structure and the diffusions of the electrons.

We need to note that the results of Eq. (3) disagree with our results show in Fig. (2). According to Eq. (3), as the energy of the first laser pulse increases, the density of electrons generated by the first laser pulse increases as well as the loss of the electrons increases, consequently the total ablation energy increases. This is in contradiction to our experimental result. In the equation (3), it has not been taken into account the influence of the CBEs generated by the first pulse on the energy gap. But as mentioned in Ref. [10, 13], when the electron-hole density becomes large, the band gap will decrease because of the interaction among electron-hole pairs. Therefore, the action of the second pulse will be influenced by the electron-hole pairs generated by the first pulse and the threshold will be affected by the change of the energy gap. It is our opinion that CBEs affect band-structure and hence the ablation threshold of the second pulse.

In conclusion, we reported the results of double femtosecond laser pulses ablation in silica and silicon. In the experiment, we found that the change of the total threshold is different in silica and silicon when the energy of the first pulse changes. Generally, the ablation will be affected by two factors: the energy gap being shrunk due to the electron-hole interaction and the diffusion of the electron-hole plasma in solid. Our work is helpful to understand the damage and micromachining by temporally shaped femtosecond laser pulses, and optimize the parameters of laser for machining of different material.

References:

1. B. C. Stuart, M. D. Feit, et.al., Phys. Rev. Lett. 74, 2248 (1995)
2. Ming Li, Saipriya Menon, et.al., Phys. Rev. Lett. 82, 2394 (1999)
3. R. Stoian, M. Boyle, et.al., App. Phys. Lett. 80, 353(2002)
4. E. N. Glezer and E. Mazur App. Phys. Lett., Vol. 71, 882 (1997)
5. S. I. Kudryashov and V. I. Emel'yanov, Journal of Experimental and Theoretical Physics, 94, 94 (2002)
6. B. C. Stuart, M. D. Feit,et.al, Phys. Rev. B 53 1749(1996)
7. Chris B. Schaffer, André Brodeur, et.al., Opt. Lett. 26,93(2001)
8. J. Bonse, S. Baudach, et.al., Appl. Phys. A, 74, 19 (2002)
9. M. Lenzner, J. Krüger, et.al., 80, Phys. Rev. Lett. 4076(1998)
10. K. Sokolowski-Tinten and D. von der Linde Phys. Rev. B 61, 2634(2000)*
11. B. A. Nechay, U. Siegner, et.al., Rev. Sci. Instrum. 70, 2758 (1999)
12. P. P. Pronko, P. A. VanRompay, et.al., Phys. Rev. B, 58, 2387(1998)
13. C.F. Klingshirn, 'Semiconductor Optics,' Springer-Verlag (1995, ISBN 3-540-58312-2)

Inst. Phys. Conf. Ser. No 186
Paper presented at 9th Int. Conf. X-Ray Lasers, Beijing, China, 24–28 May 2004

Collapse of area theorem for attosecond pulses propagating through a two-level system

Shangqing Gong, Xiaohong Song, Ruxin Li, Zhizhan Xu
Shanghai Institute of Optics and Fine Mechanics, Shanghai 201800, China

Abstract: We investigate the propagation of small area attosecond laser pulse in a dense media of two-level systems by using an iterative predictor-corrector finite-difference time-domain. We show that different from femtosecond pulse case, when subfemtosecond pulse is considered, the area theorem will collapse even for small area pulse.

1. Introduction

The advent of subfemtosecond or attosecond light pulses will open new fields of extreme nonlinear optics. Recently, experimental results on the production and measurement of a single 250-asec pulse were reported by Krausz et al [1]. This opens up new possibilities for studying and controlling electron dynamics in atoms and molecules in asec time scale. Recently, Drescher et.al [2] has demonstrated that a laser-based sampling system, consisting of a few-femtosecond visible light pulse and a asynchronized sub-femtosecond soft X-ray pulse, allows us to trace the relaxation dynamics directly in the time domain with attosecond resolution. Moreover, photoionization and high-order harmonic generation (HHG) in molecular systems under the action of combined infrared femtosecond and ultraviolet attosecond laser pulses were investigated by Bandrauk and Shon [3]. It was shown that punctual "turn-on" of an asec pulse can trigger the ionization process, which in turn leads to significant enhancement of HHG. Despite the extensive work in this area, propagation of subfemtosecond laser pulse has rarely been investigated.

The interaction of intense laser light with a collection of two-level atoms has been an area of active research for many years [4]. For example, the famous area theorem [5, 6] is one of the most interesting features, which can predict and explain many fascinating effects, such as self-induced transparency and pulse compression. In femtosecond time scale, it has been demonstrated that for small area pulse, self-induced transparency results are still recovered with minor modification [7, 8]. While for large pulse areas, the area under individual carriers might themselves cause Rabi flopping, which can lead to carrier-wave reshaping and significantly higher spectral components [9] and even soft-x-ray generation [10]. Very recently, we investigated the spectra effects for two-color ultrashort pulses propagating in a two-level medium [11]. It showed that higher spectral components can be produced even for small area pulses due to the interference effect of the two laser pulses. Moreover, these higher spectral components depend crucially on the relative phase ϕ of the two pulses. Continuum and distinct peaks spectra can be achieved by adjusting the relative phase to equal to 0 and π.

Here we investigate the propagation of small area attosecond pulse in a dense media of two-level systems. By using an iterative predictor-corrector finite-difference time-domain method to solve the full Maxwell-Bloch system numerically, we found different from femtosecond pulse case, when subfemtosecond pulse is considered, self-induce transparency cannot be recovered even for 2π pulse, carrier reshaping can be clearly found during the course of propagation, obvious higher spectra components can occur during the course of propagation.

2. Theoretical model

Consider the propagation of an attosecond pulse along the z-axis in vacuum to an input interface of a two-level medium at z=0, the propagation property can be modeled using the full-wave Maxwell-Bloch equations. Further details are given in Ref [10]. Basically, we solve for the nonlinear polarization $P_x=2Nd\mathrm{Re}[\rho_{12}]$, where N is the density of two-level medium, d is the dipole moment and $\rho_{12}=(u+iv)/2$ is the off-diagonal density matrix element. We consider a hyperbolic secant functional form for the initial electric field, which can be written as

$$\Omega(t=0,z)=\Omega_m \sec h[1.76(z-z_0)/c\tau_p]\cos[\omega_p(z-z_0)/c], \tag{6}$$

where Ω_m is the maximum Rabi frequency, τ_p is the full width at half maximum of the pulse intensity envelope. We take the carrier frequency ω_p to be equal to the chosen level transition frequency ω_0. The corresponding pulse areas is $A=\Omega_m\tau_p\pi/1.76$, and $\Omega_m=1\mathrm{fs}^{-1}$ corresponds to the electric field of $E_x=2.5\times10^7$ V/cm or an intensity of $I=1.6\times10^{12}$ W/cm^2. Several gaseous atoms have well isolated resonances, and electric dipole transitions frequency between low-lying atomic levels can be very large, such as, in hydrogen atom, the 1s-2p transition is 10.2eV, corresponding to 15.48fs^{-1} [12]. For pulse intensity, recent study have shown that the simple two-level system can still serve as reference point in the regime of extreme nonlinear optics [13], and a few important publications can be found using two-level model in these regime (see Refs [9, 10, 12]). In the numerical analysis, the medium is initialized with $u=v=0$, and the population difference between the upper and lower states takes $w_0=-1$ at $t=0$. The choice of z_0 ensures that the pulse penetrates negligibly into the medium at $t=0$. We employ a standard finite-difference time-domain approach [14] for solving the full-wave Maxwell equations, and predictor-corrector method to solve the Bloch equations, and adopt the following pulse and material parameters: $\tau_p=0.6$ fs, $\omega_0=\omega_p=10$ fs^{-1}, $z_0=1.8$ μm, $d=2.65$ eÅ, $N=8\times10^{19}$cm^{-3}. We set the population and polarization relaxation constants $\gamma_1^{-1}=\gamma_2^{-1}=1$ns. The time and space increments Δt and Δz are chosen to ensure $c\Delta t\leq\Delta z$ [15].

3. Results

Ideal SIT of femosecond pulses have been simulated in two-level mediums [7-9]. It was shown that the time-derivative behavior of the input field may cause local carrier modification but the pulse envelope is essentially unchanged. However, it not true for attosecond pulse propagating in a dense two-level medium. Figure 1(a) shows the Rabi frequency of the propagating the 2π attosecond pulse at the respective distances of 0, 18, and 30μm. It is evident that the carrier wave is modulated, and this modulation becomes more and more evident with further propagations. Moreover, the pulse envelope also changes and the ideal SIT disappears. The corresponding spectrum at z=54μm is presented in Fig.1 (b), higher spectral components can obviously be seen. These features are absent in the standard area theorem. Hence, when attosecond pulses are considered, the area theorem will collapse even for small area pulses.

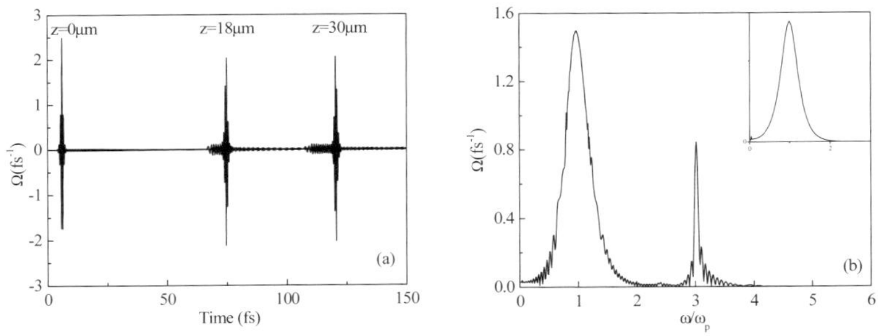

Figure1. The 2π attosecond pulse propagation through a dense two-level medium.
(a) The Rabi frequency at the respective distances 0, 18, and 30μm.
(b) The corresponding spectrum at z=54μm. The initial spectra are shown in the inset.

These phenomena can be qualitatively interpreted as following. For attosecond pulse, the individual carriers will have a more profound effect due to the much shorter pulse duration, and this effect is related sensitively to the pulse area, the larger the area is, the more evident this effect is. The carrier-wave Rabi flopping [9] can occur even for small area attosecond laser pulse. Moreover, it has been demonstrated for ultrashort pulse, that the refractive index follows the electric field instantaneously [9]; hence, the self-phase modulation will be much stronger for attosecond pulse. All these will result in strong local carrier reshaping and subsequently to the production of significantly higher spectral components on the propagating pulse.

4. Conclusions

In conclusion, by using an iterative predictor-corrector finite-difference time-domain method, we investigate the propagation of small area attosecond pulse in a dense two-level media. We find when attosecond pulse is considered, that the standard area theorem will break down even for small area pulses: ideal self-induced transparency can not occur even for 2π pulse, Significantly higher spectra components can occur on the propagating pulse due to strong carrier reshaping.

Acknowledgements

The work is supported by the National Key Basic Research Special Foundation of China (NKBRSFC) (Grant No. G1999075200), the National Natural Science Key Foundation of China (Grant No. 10234030), and the Natural Science Foundation of Shanghai (Grant No. 03ZR14102).

References

[1] Kienberger R, Goulielmakis E, et.al. 2004 Nature 427 817.
[2] Drescher M, Hentschel M, et.al. 2002 Nature 419 803.
[3] Bandrauk Z D and Shon N H 2002 Phys. Rev. A 66 031401(R).
[4] Allen L and Eberly J H 1987 *Optical Resonance and Two-level Atoms* (Dover, New York).

[5] McCall S and Hahn E L 1969 Phys. Rev. 183 457.
[6] Lamb G L, Jr. 1971 Rev. Mod. Phys. 43 99.
[7] Xiao J, Wang Z Y, and Xu Z Z 2002 Phys. Rev. A 65 031402(R).
[8] Ziolkowski R W, Arnold J M, and Gogny D M 1995 Phys. Rev. A 52 3082.
[9] Hughes S 1998 Phys. Rev. Lett. 88 3363.
[10] Hughes S 2000 Phys. Rev. A 62 055401.
[11] Song X H, Gong S Q, Jin S Q and Xu Z Z 2004 Phys. Rev. A 69 015801.
[12] Ishikawa K and Midorikawa K 2002 Phys. Rev. A 65 031402 (R).
[13] Tritschler T, Müche O D, and Wegener M 2003 Phys. Rev. A 68 033404.
[14] Ziolkowski R W, Arnold J M, and Gogny D M 1995 Phys. Rev. A 52 3082.
[15] Taflove A and Brodwin M E 1975 IEEE Trans.Microwave Theory Tech. 23 623.

Inst. Phys. Conf. Ser. No 186
Paper presented at 9th Int. Conf. X-Ray Lasers, Beijing, China, 24–28 May 2004

Soft X-ray emission from ethanol droplets irradiated by moderate femtosecond laser pulses

Z. Jin, X.Y. Peng, Z.L. Chen, T.J. Liang, Q.Z. Yu, Z.Y. Zheng, J.Y. Zhong, Y.T. Li, Z.M. Sheng, J. Zhang[@]
Laboratory of Optical Physics, Institute of Physics, Chinese Academy of Sciences, Beijing 100080, China

Abstract: We investigate the soft X-ray emission and angular distribution of hot electrons from ethanol droplets irradiated by linearly polarized 150 fs laser pulses at an intensity of 10^{16}W/cm^2. The Strong soft x-ray emission is detected from the droplet plasma. Two hot electron jets symmetrically with respect to the laser propagation direction are observed within the polarization plane. This agrees with 2D particle-in-cell simulations. An atomic kinetics code suite FLY is used to simulate the soft X-ray emission spectrum. The strong X-ray emission indicates the interaction between liquid droplets and intensity laser would be a possible way to generate X-ray lasers.

1. Introduction

Over the past several years, the interaction between high-intensity femtosecond (fs) lasers and different materials was widely studied. Energetic particles and radiations such as fast electrons, high-energy ions, neutrons, and x-rays were generated in laser-solid and laser-cluster interactions.[1-2] Recently, the interaction between high-intensity fs lasers and droplets was proved to have some unique characteristics. In contrast to continuous solid targets, there is no possibility of electron or radiation cooling to the bulk of target from the overdense plasma region while the high laser absorption is comparable to solid targets. As a result of the geometric properties and the size effects of microplasmas produced by laser-droplets interaction, a number of new phoemena are expected.[3-4]

In our experiments, we measured the angular distribution of hot electrons and the soft X-ray spectrum from ethanol droplets irradiated by linearly polarized 150 fs laser pulses at intensity around 10^{16}W/cm^2. The absorption mechanism was discussed and the plasma properties were investigated by analysis of the soft X-ray spectrum.

2. Experimental setup

The experiment was undertaken using a Ti:saphire laser operating at 800nm at a repetition rate of 10Hz. The laser delivered 5-mJ energy in 150 fs pulses into a focal spot with a diameter of 20 μ m and produced a peak intensity of 10^{16}W/cm^2.

A pulsed valve was used to generate ethanol droplets. High-pressure nitrogen (at 2 MPa) pushed the ethanol in a liquid reservoir to a solenoid-driven pulsed valve with a 100 μm nuzzle. The valve was pulsed for 460 μs at a repetition rate of 10 Hz. Thus we

[@] Corresponding author: jzhang@aphy.iphy.ac.cn

obtained high-density, polydisperse sprays containing several-micron-size ethanol droplets in vacuum. The temperature of the valve was about 300K.

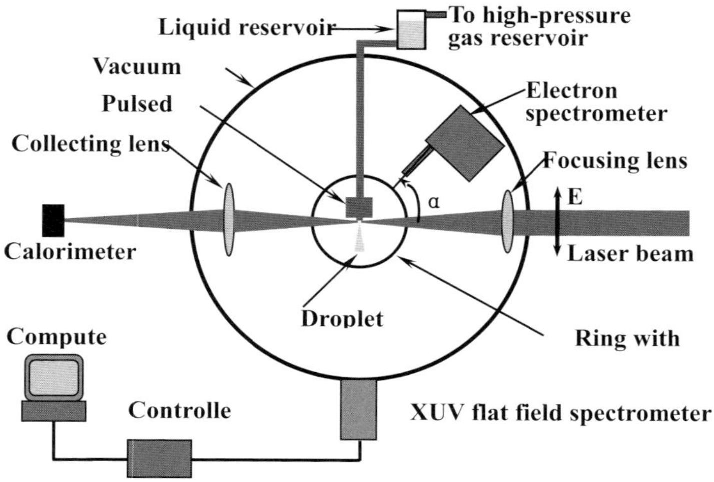

Fig.1. The experimental setup

The angular distribution of hot electron emission was measured with an array of LiF thermoluminescent dosimeter (TLD) pieces (type GR-200F) attached inside a ring around the focus spot in the polarization plane of the laser beam. The TLD detectors were wrapped with a 20 μ m thick aluminum foil overcoat to block ions, scattered laser lights, soft X rays, and electrons with energies below 50 keV.

The droplet spray was back-illuminated with the laser beam and the shadows were magnified and imaged to a charge coupled device (CCD) camera to measure the droplet size. The mean droplet size is measured to be about 4 μm. The number density and the atomic number density of the spray were $\sim 2.1\times10^8$ droplets/cm^3 at 3 mm under the nozzle where we focused the laser.

An XUV flat field spectrometer was deployed to acquire the XUV emission from the laser-droplet interaction. A grazing incidence cylindrical mirror was positioned in front of the grating to collect soft X-rays. This cylindrical mirror and a 1200-lines/mm aperiodically ruled grating were oriented so that their grazing incidence reflections gave perpendicular astigmatic line images of the focus of the laser beam at the detector plane.

3. Results and discussion

The angular distribution of hot electrons with energies over 50 keV recorded by the TLDs is shown in Fig. 2 (a). Two jets of hot electrons were found symmetric with respect to the laser axis. 2D particle-in-cell (PIC) simulations are performed to explain this double-jet structure (Fig. 2 (b)). At about 50th laser cycle, longitudinal and the transverse electric fields near the critical surface reaches the maximum (Fig. 2 (c), (d)). These results suggest that the resonance absorption may be the main mechanism for the laser energy deposit and generation of the double jet of the hot electrons.[5]

A typical spatially integrated spectrum is shown in fig 3. Comparing with the Ar cluster under the same laser conditions, the ethanol droplets shows much stronger soft X-ray emission. The spectral components are mainly from O V, O VI, C V and C VI ion stages.

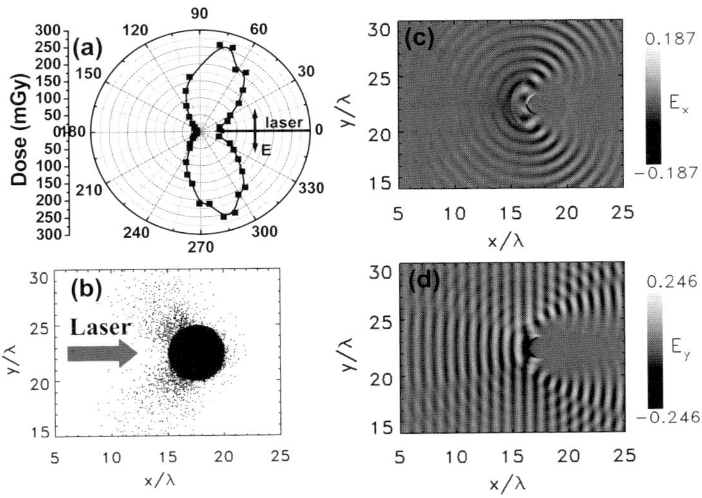

FIG. 2. (a) Angular distribution of hot electrons with energies over 50 keV. (b) (c) (d) 2D PIC simulation results for a micro droplet irradiated by linearly polarized laser pulses at the 50th laser cycle. (b) Plots (x, y) of the electron positions. (c) (d) Plots of the longitudinal and transverse component of the resonantly excited electric field.

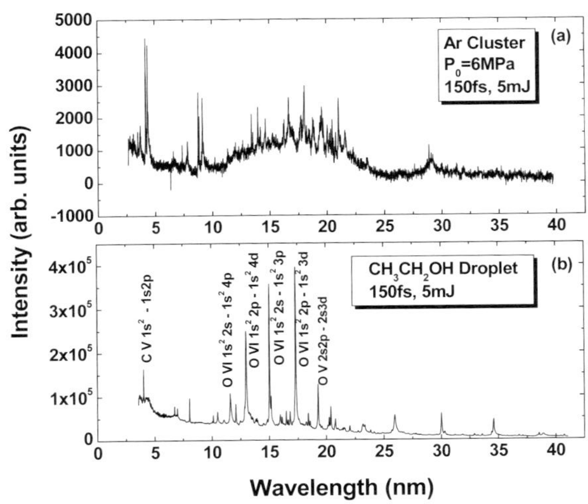

Fig. 3. A typical spatially integrated spectrum. The spectrum of Ar cluster illuminated by 150fs 5mJ laser is shown in (a). And (b) is the ethanol droplets spectrum under same conditions.

An atomic kinetics code suite FLY was used to simulate the emission spectrum. With assumption of non-LTE steady-state, we chose O VI lines to calculate as they are the strongest lines in our experimental spectrum. The simulation result is shown in Fig. 4 which fits well with the experimental data. Thus a mean electron density of 8×10^{20} was achieved. And the average electron temperature was estimated to be about 90 eV.

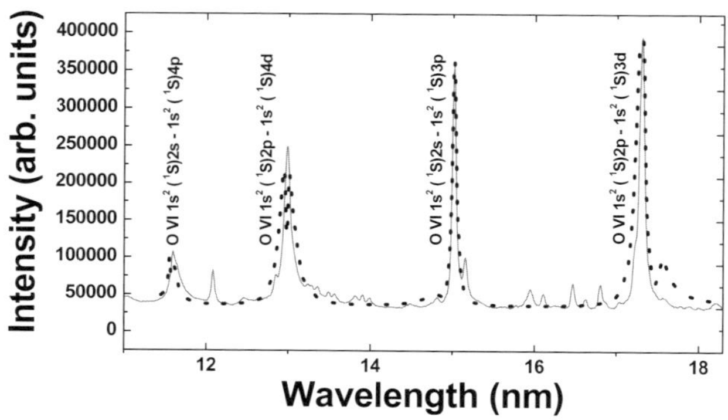

Fig. 4. O VI emission spectrum from the laser-droplets interaction. The solid line is the experimental data while the dots are results from the code FLY.

4. Conclusions

We investigated the angular distribution of hot electrons and soft X-ray emission from an ethanol micron droplet spray irradiated by 150 fs laser pulses at an intensity of 10^{16}W/cm^2. Particle-in-cell simulation suggests that the resonance absorption at the critical surface of the droplets is mainly responsible for the generation of the double jet of the hot electrons. The atomic kinetics code suite FLY was used to calculate the emission spectrum. An electron density of 8×10^{20} and an average electron temperature of 90 eV were estimated by calculating the intensity ratio of different ionization states. Comparing with clusters, the much stronger X-ray emission indicates the interaction between liquid droplets and intensity laser would be a possible way to generate X-ray lasers.

Acknowledgements

This work is jointly supported by the NSFC (under Grant Nos. 10374114, 10176034, 60321003, 10474134, 60128505) and the NKBRSF (under Grant No. G1999075206).

References

[1] T. Ditmir, T. Donnelly, R. W. Falcone, and M. D. Perry, Phys. Rev. Lett. **75**, 3122 (1995).

[2] T. Ditmir, J. W. G. Tisch, E. Springate, M. B. Masion, N. Hay, R. A. Smith, and M.H. R. Hutchinson, Nature **386**, 54(1997).

[3] E. T. Gumbrell, A. J. Comley, M. H. R. Hutchinson, and R. A. Smith, Phys. Plasmas **8**, 1329 (2001).

[4] S. Karsch, S. Düsterer, H. Schwoerer, F. Ewald, D. Habs, M. Hegelich, G. Pretzler, A. Pukhov, K. Witte, and R. Sauerbrey, Phys. Rev. Lett. **91**, 015001 (2003)

[5] X. Y. Peng, J. Zhang, Z. Jin, T. J. Liang, Z. M. Sheng, Y. T. Li, Q. Z. Yu, Z. Y. Zheng, Z. H. Wang, Z. L. Chen, J. Y. Zhong, X. W. Tang, J. Yang , C. J. Sun, Phys. Rev. E 69, 026414 (2004)

Inst. Phys. Conf. Ser. No 186
Paper presented at 9th Int. Conf. X-Ray Lasers, Beijing, China, 24–28 May 2004

Transparency induced by chirped femtosecond laser pulse in a two-level medium

Xiaohong Song, Shangqing Gong, Ruxin Li, Zhizhan Xu
Laboratory for High Intensity Optics
Shanghai Institute of Optics and Fine Mechanics
Shanghai 201800, China

Abstract: We investigate transparency induced by chirped femtosecond laser pulse in a two-level medium. It is demonstrated that the initial chirp can substantially modify the behavior of the pulse and spectra. By controlling the chirp rate, the medium can become transparent even for pulses whose area are not integer multiples of 2π.

1. Introduction

Self-induced transparency (SIT) [1] is a process in which a short optical pulse propagates with anomalously small energy loss through a dielectric medium composed of two-level resonantly interacting quantum systems (atoms, molecules, etc.), which have important implications for optical logic and for other photonics applications. However, the area theorem sets tight restrictions on the circumstances under which SIT can occur, limiting the usefulness of SIT for practical applications. In many ways the modification of SIT by the presence of a control field is reminiscent of electromagnetically induced transparency (EIT), in which the properties of nonlinear optical interaction become strongly modified by the presence of coupling fields. Recently, Park et al. [2] have shown that it is possible to control the properties of SIT in a novel and useful fashion by applying an additional control field to the material system.

Recently, the advances in generation and characterization techniques of ultrashort laser pulses made it practical to prepare precisely defined laser pulses, these open up new possibilities for studying and controlling dynamical processes in femtosecond time scale. Hitherto, frequency-synthesized or chirped laser pulses have been widely applied to high-order harmonic generation (HHG) [3], population transfer [4], multiphoton transitions between Rydberg states [5], ect. Both the interference of two-color pulses and the nontrivial time dependence of chirped pulse laser field may lead to very interesting results. Moreover, recent studies have manifested that chirping can also have much impacts on the laser pulses as it propagates through different medium. For example, the propagation properties of chirped soliton pulses in optical nonlinear Kerr media were investigated by Desaix et al. [6]. The result indicates that the properties of asymptotically emerging solitons depend not only on the magnitude, but also on the form of the initial chirp. In particular, the splitting of an initial pulse into separating soliton pulse pairs only occurs for certain classes of initial chirp functions. The influence of pulse shape and frequency chirp on stability of optical solitons was studied by Klaus and Shaw [7]. They found that destabilization occurs when eigenvalues of an associated Zakharov-Shabat system, which move along the positive imaginary axis with increasing chirp parameter, either are absorbed into the lower half plane or collide with a nother eigenvalue. Furthermore, the signal and group velocities for chirped pulses propagating through a GaAs cavity were measured by Centini et al. [8]. They showed that the chirp substantially modifies the group velocity of the pulse, but leaves the signal velocity unaltered.

In this paper, we investigate transparency induced by chirped femtosecond laser pulse in a two-level medium. Full Maxwell-Bloch (M-B) equations are solved without invoking any of the standard approximations such as the slowly-varying-envelope approximation [9] and the rotating-wave approximation [10]. It is shown when the chirp rate is adjusted to appropriate values, that the medium can become transparent even for pulses whose areas are not integer multiples of 2π, such as 3.1π pulse.

2. Theoretical model

Consider the propagation of a few-cycle pulse laser along the z-axis in vacuum to an input interface of a two-level-atom medium at z=0, the propagation property can be modeled using the Maxwell-Bloch equations:

$$\partial_t H_y = -\frac{1}{\mu_0}\partial_z E_x, \tag{1}$$

$$\partial_t E_x = -\frac{1}{\varepsilon_0}\partial_z H_y - \frac{1}{\varepsilon_0}\partial_t P_x, \tag{2}$$

$$\partial_t u = -\gamma_2 u - \omega_0 v, \tag{3}$$

$$\partial_t v = -\gamma_2 v + \omega_0 u + 2\Omega w, \tag{4}$$

$$\partial_t w = -\gamma_1 (w - w_0) - 2\Omega v, \tag{5}$$

where E_x and H_y are the electric and magnetic fields, γ_1 and γ_2 are the population and polarization relaxation constants, respectively. ω_0 is the transition frequency of the two-level-atom medium, and $\Omega=dE_x/\hbar$ is the Rabi frequency. The macroscopic nonlinear polarization $P_x=Ndu$ is related to the off-diagonal density matrix element $\rho_{12}=(u+\mathrm{i}v)/2$ and the population difference $w=\rho_{22}-\rho_{11}$ between the upper and lower states, N is the density of the medium, d is the dipole moment. The refractive index is determined by the real part of ρ_{12} and the gain coefficient is proportional to the imaginary part of ρ_{12}. We consider a hyperbolic secant functional form for the initial chirped electric field, which can be written as

$$\Omega(z=0,t) = \Omega_m \,\mathrm{sec}\,h[1.76(t - z_0/c)/\tau_p]\cos(\omega_p t + \frac{1}{2}\chi t^2), \tag{6}$$

where ω_p is the carrier frequency, Ω_m is the maximum Rabi frequency, τ_p is the full width at half maximum of the pulse intensity envelope and χ is the chirp rate. In the numerical analysis, the medium is initialized with $u=v=0$, $w_0=-1$ at $t=0$. The choice of z_0 ensures that the pulse penetrates negligibly into the medium at $t=0$. We employ a standard finite-difference time-domain approach [11] for solving the full-wave Maxwell equations, and predictor-corrector method to solve the Bloch equations. The material and laser parameters are consistent with those in Ref. [12]: $\tau_p=5$fs, $\omega_0=\omega_p=2.3\text{fs}^{-1}$, $z_0=15\mu$m, $d=2\times10^{-29}$ A s m, $N=4\times10^{18}\text{cm}^{-3}$, $\gamma_1^{-1}=\gamma_2^{-1}=1$ns. The corresponding pulse areas is $A=\Omega_m\tau_p\pi/1.76$, and $\Omega_m=1\text{fs}^{-1}$ corresponds to the electric field of $E_x=5\times10^9$V/m or an intensity of $I=6.6\times10^{12}\text{W/cm}^2$. The time and space increments Δt and Δz are chosen to ensure $c\Delta t\leq\Delta z$ [13]. The results to follow can of course be scaled to various laser and material parameters.

3. Numerical results and analysis

By virtue of the area theorem [1], the area of 3.1π pulse will increase to 4π through pulse broadening. Moreover, for ultrashort pulses, as investigated by Xiao et al. [12], the area theorem can still predict the profile of the area evolution of a few-cycle optical pulse. However, the variation of the few-cycle pulse area is caused by the pulse splitting but not by the pulse broadening or the pulse compression. Here we show that when chirped femtosecond pulses are considered, the medium can become transparent even for 3.1π pulse. Figure 1 (a) depicts the electric field profile of 3.1π pulse with chirp rate equal to $0.025 fs^{-2}$ at different propagation distances, and the corresponding spectra are shown in Fig1 (b). They show that the medium becomes approximately transparent to the pulse, no frequency variation can be found.

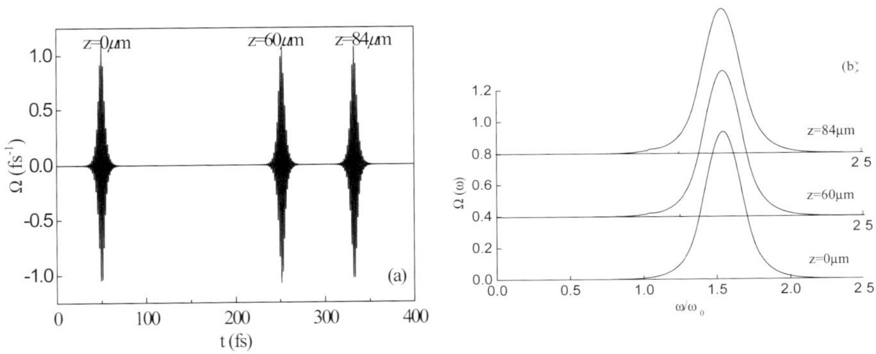

Figure 1. (a) the electric field profile of 3.1π chirped pulse at the respective distances of $0\mu m$, $60\mu m$, and $84\mu m$ for $\chi=0.025 fs^{-2}$, (b) the corresponding spectra at the same propagation distances.

These phenomena can be qualitatively interpreted by the intrinsic chirp which is produced in the course of propagation. Similar to that near-resonant case [14, 15], an initial chirped pulse can also bring an intrinsic chirp during the propagation in the medium, which pulls the carrier frequency toward the resonant frequency. When adjusting the initial chirp rate to appropriate value, the intrinsic chirp will become approximately equal to the initial chirp, spectral changes can hardly be observed, and the medium seem to be approximately transparent to the pulse.

4. Conclusion

In conclusion, we investigate propagation and spectral effect of an initial chirped few-cycle pulse in the two-level-atom medium by solving the full Maxwell-Bloch equations. It is demonstrated that the initial chirp can substantially modify the behavior of the pulse and spectra. By controlling the chirp rate, the medium can become transparent even for pulses whose area are not integer multiples of 2π.

The work is supported by the National Key Basic Research Special Foundation of China (NKBRSFC) (Grant No. G1999075200), the National Natural Science Key Foundation of China (Grant No. 10234030), and the Natural Science Foundation of Shanghai (Grant No. 03ZR14102).

Reference

[1] McCall S L and Hahn E L 1967 Phys. Rev. Lett. 18 908; 1969 Phys. Rev. 183 457.
[2] Park Q -H and Boyd R W, 2001 Phys. Rev. Lett. 86 2774.
[3] Kim H T, Lee D G, et.al. 2003 Phys. Rev. A 67 051801 (R).
[4] Cheng J and Zhou J Y 2001 Phys. Rev. A 64 065402.
[5] Conover W S, Doogue M C,et.al. 2002 Phys. Rev. A 65 033414.
[6] Desaix M, Helczynski L, et.al. M 2002 Phys. Rev. E 65 056602.
[7] Klaus M, and Shaw J K 2001 Opt. Commun. 197 491.
[8] Centini M, Bloemer M,et.al. 2003 Phys. Rev. E 68 016602.
[9] Born M and Wolf E, 1975 Principles of Optics 5th ed. (Pergamon, Oxfod), Sec. 10.4.
[10] Lamb W E, Jr, 1964 Phys. Rev. A 134 1429.
[11] Ziolkowski R W, Arnold J M, et.al.1995 Phys. Rev. A 52 3082.
[12] Xiao J, Wang Z Y, and Xu Z Z 2002 Phys. Rev. A 65 031402 (R).
[13] Taflove A and Brodwin M E 1975 IEEE Trans. Microwave Theory Tech. 23 623.
[14] A. M. Alhasan, J. Fiutak, and W. Miklaszewski, Z. Phys. B **88,** 349 (1992).
[15] Diels J C 1970 Phys. Lett. A 31 111.

Inst. Phys. Conf. Ser. No 186
Paper presented at 9th Int. Conf. X-Ray Lasers, Beijing, China, 24–28 May 2004

Simulations of keV x-ray spectra in laser-produced plasma of Samarium

J. Y. Zhong[1,2], J. Zhang[1], and G. Zhao[2]
1 Laboratory of Optical Physics, Institute of Physics, Chinese Academy of Science, Beijing 100080, People's Republic of China
2 National Astronomical Observatories, Chinese Academy of Sciences, Beijing 100012,People's Republic of China

Abstract. We have studied the characteristics of x-ray emissions in the 6.5-10 Å range from plasmas produced by the irradiation of slab targets of Sm with a laser pulse of 600-ps duration at intensities near 4×10^{13} Wcm^{-2}. The prominent lines of Ni-, Co-, Fe-, and Mn-like Sm ions arising from 4f-3d radiative transition are identified. A simple collisional radiative model has been used to simulate these spectra. The intensity ratios of the resolved Ni-like Sm lines are shown to be useful tools for electron density and temperature diagnosis.

1. Introduction

The study on keV x-ray spectra of laser-produced plasmas has been progressed for many years. Recent improvement in laser power densities together with the development of very high resolution x-ray crystal spectrometers have provided new insights into the nature of such plasmas [1]. A two-step collisional-radiative model was developed to interpret the high-resolution x-ray spectra of ions include cerium, xenon, and barium [1,6-8]. In that model the autoioniztion and dielectronic capture processes were included for the main radiative transitions observed in the experimental spectrum arisen from the doubly excited configurations. In this paper, keV spectra in the 6.5-10 Å range from plasmas produced by the irradiation of slab targets of Sm with a single 600-ps laser pulse at intensities near 4×10^{13} Wcm^{-2}. These spectra are diagnosed by two space-resolving, time-integrating potassium acid phthalate (KAP) curved-crystal spectrometers and recorded on InstaSpec IV x-ray charge-coupled device (CCD) detectors. The prominent lines of Sm transitions visible are due to 4f-3d transitions from Ni-, Co-, Fe-, and Mn-like ions respectively.

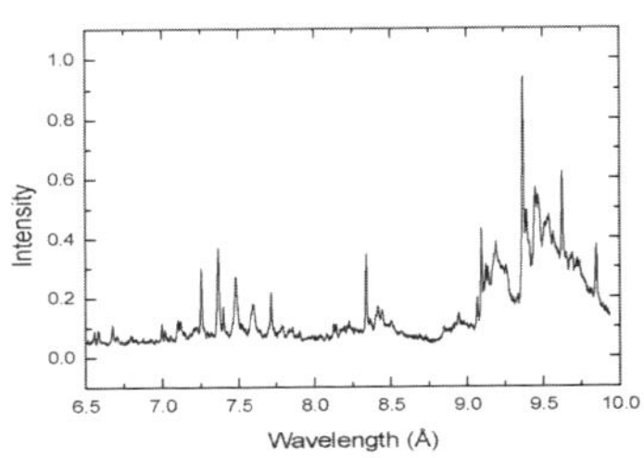

FIG.1. KeV x-ray spectrum of Sm plasmas emission by the irradiation with a single 600-ps laser pulse at intensities near 4×10^{13} Wcm^{-2}.

We identify the distinct lines of experimental spectra based on the oscillator strength. Then a simple collisional radiative model has been used to simulate these spectra. The Co- and Ni-like ions are included in this model and their ionization abundances inferred from the relative intensities of the 4f-3d emission in the experimental spectra.

2. Experiment

The experimental set-up is similar to that described in Ref. [10, 11]. Three beams of the VULCAN Nd:glass laser at 1.05 μm at the Rutherford Appleton Laboratory were used in a standard off-axis focus geometry, which provided a line focus with 25 mm length and 100 μm width, giving an irradiance of ~ $4.0\text{x}10^{13}$ Wcm^{-2} on targets. The temporally resolved spectra were recorded on HP5 films (Figure 1).

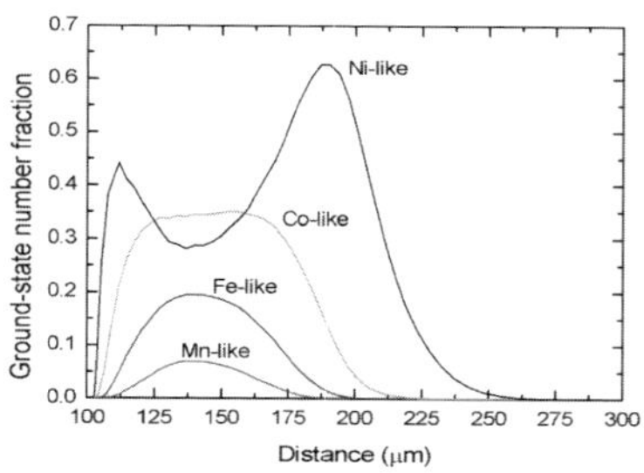

FIG. 2. The number fraction of ground state for Mn-, Fe-, Co-, and Ni-like ions at the peak of the main laser pulse driven by a single 600 ps pulse at $4.0 \times 10^{13}\text{Wcm}^{-3}$.(by MED103)

3. Identifications and simulations of spectra

3.1. Identification of spectra lines

The oscillator strength of Ni-like, Co-like, Fe-like and Mn-like lines are also shown in figure 3, based on the calculations of the Flexible Atomic Code (FAC) [13]. The calculations are carried out including all coupling of states with a given number of n=3, 4 and n=5 electrons, including 249, 508, 1281 and 2138 levels for Ni-like, Co-like, Fe-like and Mn-like ions, respectively. With the help of calculations, the spectra can be identified the line at 0.911 nm as combined lines of 4f-3d Co-like transitions. The oscillator strength of this transition is about two orders of magnitude larger than that of the Co-like 4d-3p transition. Our identification also agrees with the identification for a similar element in Ref. [14,15].

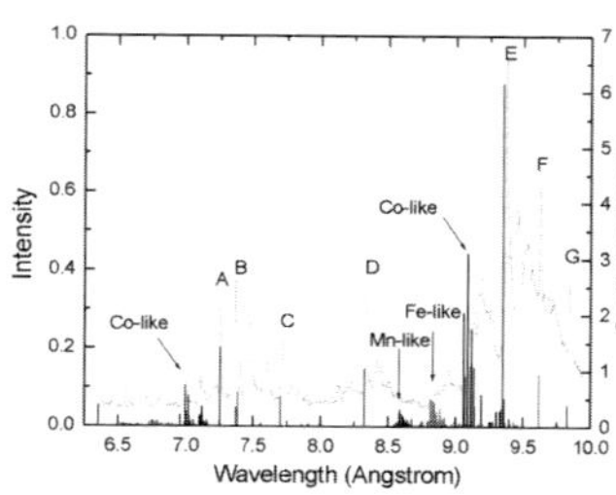

FIG.3. fit the experimental ratio of the 4f-3d transitions from different ionisation stages. Labels by alphabet correspond to the transition identification in Table 1.

6 prominent lines are identified to be the transitions of Ni-like ions listed in Table I (ftp://159.226.36.237/keV-sm). As a zero-order approximation, we can compare the ionisation abundances inferred from the relative intensities of the 4f-3d emission with the ionisation distribution predicted by the one-dimensional hydrodynamic code MED103 [9] simulation. Figure 2 shows the number fraction of ground state for Mn-, Fe-, Co-, and Ni-like ions at the peak of the main laser pulse. The lines are results by a single 600 ps pulse at 4.0×10^{13} Wcm^{-3}. In order to obtain a similar ratio of different ionisation stages

as in the experimental data, we have multiplied the Co-like, Fe-like and Mn-like lines by factors of 0.4, 0.14 and 0.055 respectively to produced figure 3. It fits the experimental results fairly well except the intensity ratio of A and B line in Ni-like Sm ion. This indicates some other model should be constructed in order to interpret it.

3.2. Theoretical simulation of spectra by CR Model

Ignoring the autoionization and dielectronic capture processes, the simplest collisional-radiative model is constructed where all collisional excitations and deexcitations and all radiative decays among the singly excited levels are taken into account. The corresponding set of rate equations for the ground and singly excited levels of the emitting ion can be written as

$$\frac{dn_i}{dt} = n_e \sum_{j \neq i} n_j C_{ji}(T_e) + \sum_{j>i} n_j A_{ji} - n_i \left(n_e \sum_{j \neq i} C_{ij}(T_e) + \sum_{j<i} A_{ij} \right) \tag{1}$$

where n_i represent the density of ions in singly excited levels i, n_e represents the electron density. A_{ji} denotes the Einstein coefficient for radiative decay from j to i, and $C_{ij}(T_e)$ denotes the rate coefficient for collisional excitation or deexcitations.

Since the spectrum is time integrated and the time evolution of the plasma is not well known, a steady-state model is employed.

$$n_e \sum_{j \neq i} n_j C_{ji}(T_e) + \sum_{j>i} n_j A_{ji} - n_i \left(n_e \sum_{j \neq i} C_{ij}(T_e) + \sum_{j<i} A_{ij} \right) = 0 \tag{2}$$

At first singly ions rate equations of Ni-like Sm are solved and the obtained populations of 249 levels are normalized. Then we let the ratio of these two ions Ni-like and Co-like ground state $n_{co\text{-}like}/n_{ni\text{-}like}$ equal 0.4, which inferred from the relative intensities of the 4f-3d emission with the ionization distribution predicted by the one-dimensional hydrodynamic code MED103 simulation, and solve the Co-like Sm rate equations. The populations of 508 levels Co-like Sm are obtained. Finally, the populations of the 149+508 levels are used for calculation the line intensities I_{ji} arising from decays between singly excited levels

$$I_{ji} = n_j A_{ji} \tag{3}$$

All lines are assumed here to have a uniform Gaussian profile to match the narrowest experimental line widths.

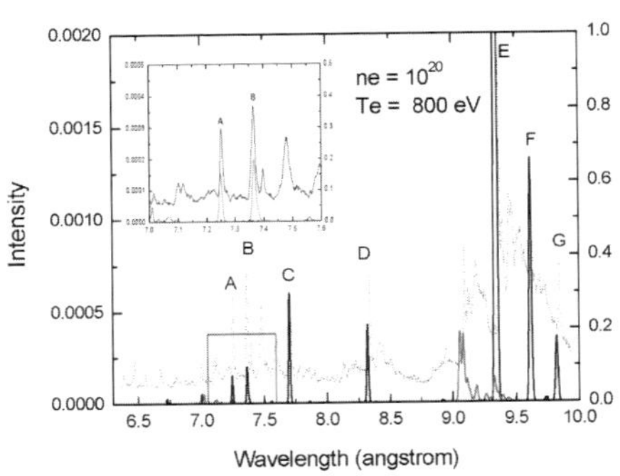

FIG. 4 KeV x-ray experimental spectrum of Sm plasma compared to the theoretical spectrum calculated for $n_e=1\times10^{20}$ cm^{-3} and kT_e=800 eV. Small figure is the amplifier of 7.0-7.6 Å wavelength range in order to shows the difference of the zero-order approximation and the CR model.

Figure 4 shows that the theoretical spectrum fits the experimental results best when the electron temperature and electron density are set to be 800 eV and 1×10^{20} cm^{-3}.

4. Conclusions

The keV spectra in the 6.5-10 Å range from plasmas has been produced by the irradiation of slab targets of Sm with a single 600 Pico-second laser pulse at intensities near 4×10^{13} Wcm^{-2}. A simple collisional radiative model has been used to simulate these spectra. The intensity ratios of the resolved Ni-like Sm lines are shown to be useful tools for electron density and temperature diagnosis.

Acknowledgments

The authors are indebted to the support and assistance of the VULCAN laser facility. This work is jointly supported by the NSFC under Grant Nos. 10176034, 10173014, 10374114, 60321003 and the NKBRSF under Grant No.G1999075206, No.G19999075406 and the National Hi-tech ICF program.

References

[1] R. Doron, E. Behar et al., Phys. Rev. A 58, 1859(1998)
[2] A. Zigler, P. G. Burkhalter et al. Appl. Phys. Lett. **59**, 777 (1991)
[3] W. H. Goldstein, A. Zigler et al., Phys. Rev. E **47**, 4349 (1993).
[4] R. Doron, M. Fraenkel et al., Phys. Scr. **58**, 19 (1998)
[5] A. Bar-Shalom, J. Oreg et al., Phys. Rev. A **40**, 3183 (1989)
[6] R. Doron, E. Behar et al., Phys. Rev. A **59**, 188 (1999)
[7] R. Doron, E. Behar et al., Phys. Rev. A 62, 052508(2000)
[8] R. Doron, E. Behar et al., JQSRT 65, 161(2000)
[9] J .P. Christiansen, D. E. T. F. Ashby et al., Computer Phys. Commun., 7, 271 (1974).
[10] J. Zhang, A. G. MacPhee et al., Phys. Rev. Lett., 78, 3856 (1997).
[11] J. Zhang, A. G. MacPhee et al., Science, 276, 1097 (1997).
[12] C.L.S. Lewis et al., X-Ray Lasers 1992, E.E. Fill, Editor, IOP Publishing, Conf. Series No. 125, 23 (1992).
[13] M. F. Gu, Astrophys. J. 582, 1241 (2003)
[14] P. G. Burkhalter, D. J. Nagel et al., Phys. Rev. A **9**, 2331 (1974).
[15] J. Bailey, J. D. Kilkenny et al., Phys. Rev. A **35**, 2578 (1987).

Developments of X-ray laser drivers

Inst. Phys. Conf. Ser. No 186
Paper presented at 9th Int. Conf. X-Ray Lasers, Beijing, China, 24–28 May 2004

High Power Laser Development at CCLRC Rutherford Appleton Laboratory for X-ray Laser Research

Colin Danson
CCLRC Rutherford Appleton Laboratory, United Kingdom

Abstract. The Vulcan Nd:glass laser at the Central Laser Facility (CLF) has been a leading facility in the study of X-ray laser development for over 20 years. During this time several important technological improvements to the facility have been key to the research programme, including: focusing geometries using off-axis spherical optics; pre-pulse generation and travelling-wave geometries. In addition, the technique of Chirped Pulse Amplification (CPA) has been implemented and used for the study of transient schemes culminating in the commissioning of the Vulcan Petawatt (10^{15} Watts) interaction facility.

1. Introduction

The development of the Vulcan facility for X-ray laser research has principally been concerned with generating the right plasma conditions. This includes generating the long line focus and optimising the pulse fidelity in terms of its duration, energy and shape. As the requirements became more sophisticated multi-pulse geometries were required – particularly true of the applications driven experiments. Developments to the facility, which have not necessarily been driven by the X-ray laser programme, have also been of significant benefit. These include the generation of multi-terawatt pulses and the development of Phase Plate focusing techniques.

2. Line-foci generation

The first major development of the facility specifically for X-ray lasers was to provide a technique that could produce uniform line-foci. The traditional technique, which is still used in many laboratories today, uses a cylindrical lens in tandem with a spherical lens. This technique suffers by usually producing a non-uniform focal profile and can be expensive to implement. The alternative technique was to use an off-axis spherical mirror in conjunction with a standard focusing optic [1]. The aberration generated when the spherical mirror is used off-axis produces a line focus. A layout of such a scheme is shown in figure 1.

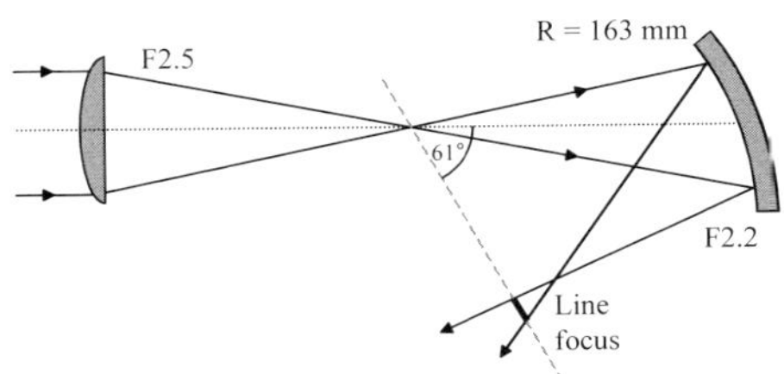

Figure 1. Off-axis spherical mirror used in the generation of line foci.

Figure 2. A view of the vacuum target interaction chamber used for X-ray laser research, in Vulcan's Target Area East.

The geometry can be designed such that adjusting the focal lengths of the two focusing optics and the off-axis angle of the spherical optic can generate the required length of line-focus. The uniformity of the plasma generated was reported in 1987 [2] where the technique was tested on both fibre and foil targets. To date this has been the standard method of generating line-foci for X-ray laser research on Vulcan.

In 1985 a target area was commissioned on Vulcan which specialised in long pulse interactions including line focus geometries for X-ray lasers, cluster geometries principally for shock wave generation and multi-beam hohlraum irradiation. Figure 2 shows the vacuum target chamber set-up for a multi-beam X-ray laser irradiation experiment. In this geometry twin targets are each irradiated by three beams. The extension tubes holding the principle focusing lenses for three of the beams that irradiate one of the targets can be seen on the extreme left of figure 2.

3. Pre-pulse generation

Pre-pulses have been specified for optimising the plasma conditions in X-ray lasers since the mid 1980's. The first implementation consisted of a Fabry-Perot, dual mirror type arrangement at a beam aperture of 100 mm. These proved inflexible and very difficult to optimise during the course of an experiment. The requirements for multi-pulses for time resolved plasma diagnostics led to the development of an more effective pulse stacker [3], which used two coupled dog-legs, at a few millimetre aperture, to produce four equal intensity pulses with an efficiency of 50%.

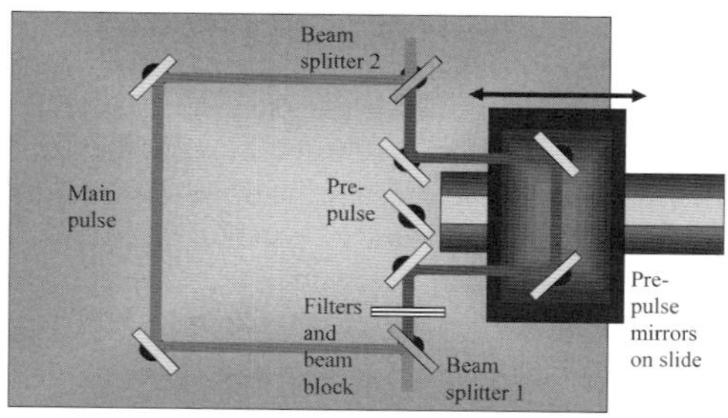

Figure 3. A custom pre-pulse generator developed to optimise plasma conditions.

Variations on the multi-pulse stacker device were then developed which could be configured to produce a range of optical pulses, a layout of such a device is shown in figure 3. These devices are generally used at beam apertures up to 25 mm. They have the advantage that they can have <50% insertion loss, be able to generate pre-pulses with zero picoseconds delay, can be easily adjusted to provide different delays and are easy to align. These developments proved invaluable in the development of the Germanium X-ray laser at 19.6 nm where optimised multi-pulse irradiation to target improved the X-ray laser output by four orders of magnitude [4].

4. Ultra-short pulse generation

The wish to develop ultra-short (picosecond) X-ray lasers requires that the drive pulse be of the order of a picosecond itself. This could be realised with the development of Chirped Pulse Amplification (CPA) [5] implemented on high energy laser systems such as Vulcan [6]. The technique uses a pair of diffraction gratings to stretch the sub-picosecond pulse to the nanosecond regime, prior to amplification. The amplified chirped pulse, whose energy is no longer limited by Self Phase Modulation, can then be recompressed in a similar manner just prior to target focusing. To avoid the beam breaking up due to B-integral effects, the compressed pulse is required to propagate to target in vacuum and is also focussed with a reflective off-axis parabola rather than the more traditional lens focusing.

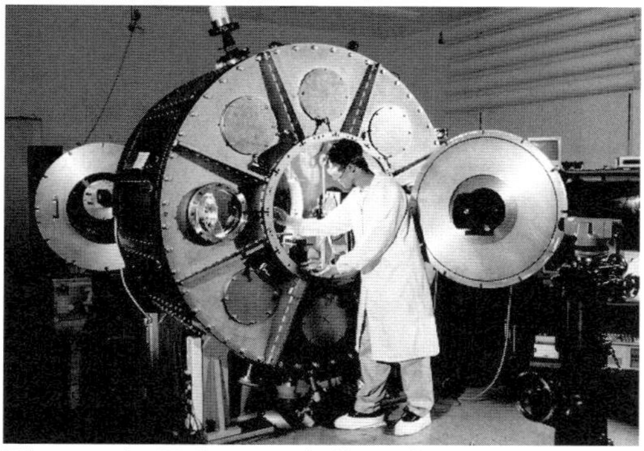

Figure 4. Vulcan's dedicated 100 TW CPA interaction area.

This was implemented in a dedicated target area on Vulcan whose interaction chamber is shown in figure 4. The laser could deliver 100 TW in a single short pulse beam-line and had the significant advantage of being able to deliver synchronised high energy nanosecond pulses as well as an additional CPA probe. The chamber was designed such that a range of complicated irradiation geometries could be used.

The transient X-ray laser schemes required the implementation of a travelling wave along the target length. Since the upper-state lifetime of these lasers is only picoseconds in duration, maintaining the optimum gain conditions along the target length was essential for efficient operation. A diffraction grating was inserted prior to the CPA compressor and its angle tuned to provide a tilted wave front that would produce a travelling wave propagating along the target at the speed of light. A schematic of this set-up is shown in figure 5. The travelling wave grating was installed in the Vulcan rod amplifier chain (at a beam aperture of 45 mm) allowing a small, relatively inexpensive grating, to be used.

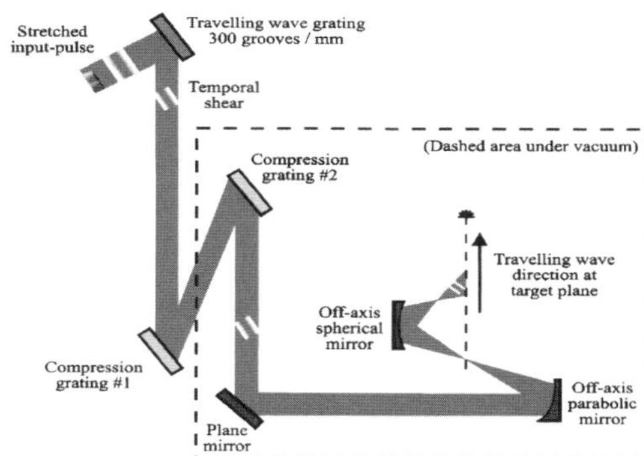

Figure 5. Travelling-wave implementation for transient X-ray laser schemes.

5. Phase Plate Development

A random phase plate (RPP) [7] is a binary diffractive optic which has a glass substrate with a surface relief either etched directly into the surface or into a thin film covering the surface of the glass. The surface is tessellated with a basic shape, initially a square, with an etching depth of $\lambda/2(n-1)$; where n is the refractive index of the etched medium and λ

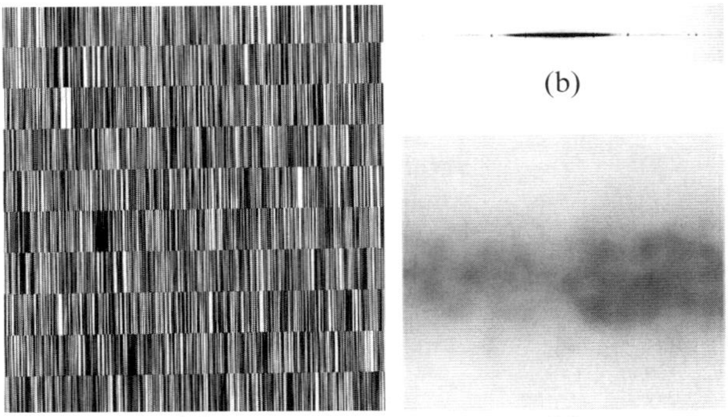

(a) (b) (c)

Figure 6. (a) The RPP mask used to generate a 100:1 aspect ratio line focus, (b) A far-field image of the line focus, (c) A magnified portion of the far-field image.

the operational wavelength. Each square acts a diffraction aperture which when used with a principle focusing optic, of focal length f, produces a focal profile which is the superposition of diffracted light from all the apertures. The focal profile consists of a speckle pattern with a spot size of fλ/d FWHM where d is the square element size. The pattern is randomly formed with half the squares producing a 0 phase shift and the other half, a π radian phase shift. The equal areas of 0 and π destructively interfere on axis eliminating any coherent spike. The small-scale speckle contains mainly high spatial frequencies, which in many laser plasma interaction experiments will be smoothed by energy transport in the plasma.

The manufacturing process used at RAL is relatively simple [8], using standard photolithographic techniques to produce a phase plate whose binary structure is imprinted into an e-beam photo-resist. The RPP technique can be used to generate a broad range of focal profiles. An example of an asymmetric pattern to produce a 100:1 aspect ratio line focus is shown in figure 6(a). When this phase p

late is used in conjunction with a focussing optic the resultant far-field can be seen in figure 6(b). A magnified image of a section of the line focus is shown in figure 6(c) where the speckle can clearly be seen. This is an issue when trying to generate a uniform plasma column for X-ray lasers. A potential solution to this is the use of one-dimensional Smoothing by Spectral Dispersion (SSD) [9]. Its implementation on Vulcan for non-line foci is described in [8] although this is equally applicable to line foci.

6. Longitudinal Line focus

Traditional methods to generate a line-focus along the axis of an optical system have centred on the use of axicons [10]. However, these devices are relatively expensive in large aperture beamlines (> 100 mm diameter), they do not produce uniform intensity along the line focus (with top-hat intensity beams) and they have a fixed focal length, limiting their use to a particular configuration. These limitations can be overcome by the use of binary phase diffractive optics. A modified design of a binary phase Fresnel zone plate [11] was used that has an azimuthal variation in focal length to achieve the desired focal intensity distribution.

Figure 7. The mask used for a variable focal length Phase Zone Plate in order to generate a longitudinal line focus.

A diffractive optical device, a Fresnel zone plate can be thought of geometrically as a lens with a focal length $\pm F_Z$. The focal length is proportional to the square of the radius R_O of the central zone. It is clear then, that modifying the radius of the zone can change the focal length and that an axial line focus can be generated by a continual change in the radius of the zone structure as one sweeps around the plate, shown schematically in figure 7.

Although the plate manufactured using this structure did produce a line focus, as one traversed along the line, a cusp of focused light was seen swirling around the optic axis and was of a sufficient intensity that it would have made any useful experimentation

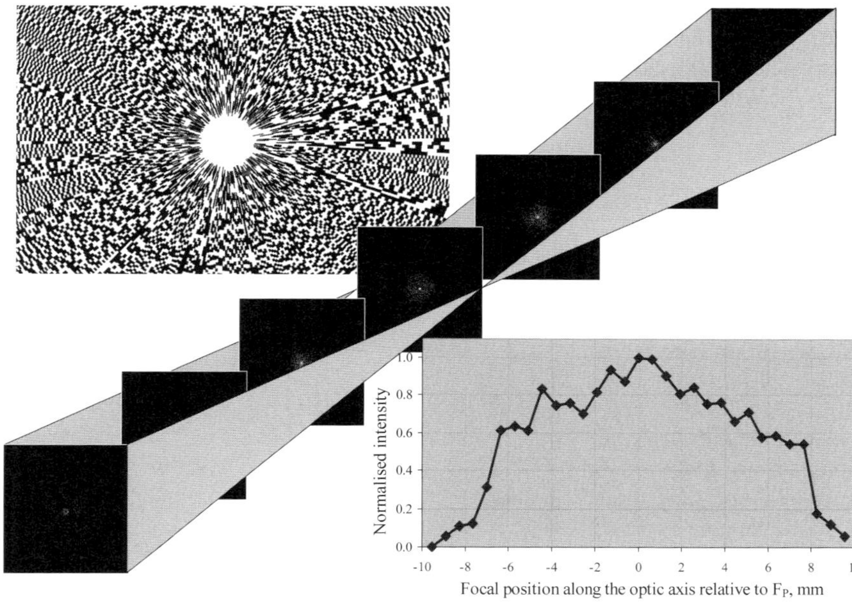

Figure 8. Images of the focal spot taken along the optic axis, produced by the modified Fresnel zone plate (top left). The normalised axial intensity profile obtained is shown in the graph (bottom right).

impossible. This cusp is effectively a result of the whole beam collapsing from a 2D plane into the 3D optic axis and so the design was modified in order to break up this structure. This was achieved by using a random rather than a smooth transition in focal length as one sweeps around the plate. The same extremes of F_z were used but each 1.5° segment has a randomised focal length. This is shown in figure 8, along with images of the focal spots at various points along the optic axis and a graph indicating the relative intensity profile of the longitudinal line focus.

7. Vulcan Petawatt Facility

In 1998, the science programme based on the Vulcan facility was achieving an unprecedented level of success with high profile scientific publications across a wide range of topics. In particular, the CLF was recognized as a world-leading centre for high irradiance interaction experiments using its 100 TW CPA beam-line. In response to user demand, a CLF team devised an upgrade strategy for Vulcan with the aim of delivering Petawatt power from an enhanced CPA beam. The specification was for a beam delivery of 670 J in the stretched pulse with 500 J on target in the compressed pulse of 500 fs, giving a peak irradiance of 10^{21} W.cm^{-2} in a <10μm focal spot. The Petawatt beamline was funded through an Engineering and Physical Sciences Research Council (EPSRC) Standard Research Grant (GR/M48383) and was made possible through the transfer of equipment from Lawrence Livermore National Laboratory (LLNL) following the closure of the NOVA laser facility.

The Petawatt performance is not achievable by amplification solely by Nd:glass amplifiers, since gain narrowing would reduce the pulse bandwidth to a value below the ~4 nm limit which is required to support a 500 fs pulse duration. The approach taken for the Petawatt was to use an Optical Parametric Chirped Pulse Amplification (OPCPA) [12] pre-amplification system [13], which generates pulses of ~10 mJ at large bandwidth (1mJ/nm) for injection into the Vulcan rod and disc amplifier chain. The system uses a customized 2 J, 1053 nm, 10 Hz YAG laser to generate a 4.5 ns frequency doubled 200-300 mJ pump pulse which is approximately "top-hat" in both space and time. This pumps a 3-stage optical parametric amplifier that is seeded by the stretched oscillator pulse. The non-linear medium used for the three amplification stages is *ß - Barium Borate* (BBO). The amplifier chain is optimized for bandwidth using a combination of Nd:phosphate and Nd:silicate amplifying media.

Figure 9. The Vulcan Petawatt Interaction area; prior to the installation of radiation shielding.

The additional laser energy is obtained from an additional disc amplifier stage on Vulcan comprising three ex-NOVA 208 mm aperture disc amplifiers with gain isolation provided by a 208 mm aperture Faraday rotator and a single polariser at the stage input. The amplifiers were measured to have a small signal gain of ~2.2, narrow bandwidth 1053 nm operation, falling to approximately 2.0 under saturation conditions for 670 J total output. The extended amplifier chain was fully commissioned and the maximum beam energy delivered was 792 J, well in excess of the energy requirements for the beam-line. The Faraday rotator has been commissioned and found to give an extinction of 0.5% operating at 20 kV.

The final pulse compression is accomplished used gratings of 940 mm in diameter with 1740 lines per mm separated by 13 m. The compression optics are completely enclosed in a vacuum vessel which is shown in figure 9.

8. Focusing and characterisation

The use of a large aperture, 120 mm, adaptive optics (AO) module is essential to the Petawatt facility. It is required both to improve the wavefront quality to produce the optimum focal spot on target and also to achieve the best possible recompression of the stretched pulse. A mirror was designed with 61 addressable elements in a 2D array suitable for correcting slowly varying aberrations. The specification of the adaptive optics system for use with Vulcan is for a cycle time > 1Hz, correction to 1.5 times the diffraction limit and a high level of immunity to electrical noise. This is of significant benefit to the science programme at CLF, the physics of which is very sensitive to irradiance.

A 620 mm diameter, 1.8 m focal length, F#3.1 off-axis parabola was installed in the Petawatt interaction chamber. The parabola was silver coated with a magnesium fluoride protective overcoat shown in figure 10.

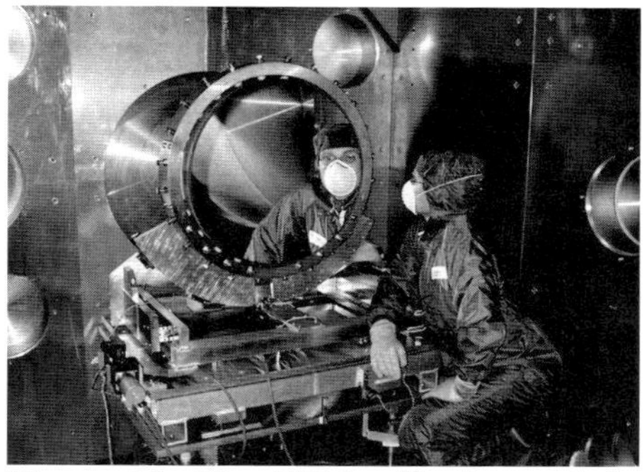

Figure 10. The F#3.1 off-axis parabola installed in the Petawatt target interaction chamber.

X-ray imaging was used to measure the emission from the focal region. Figure 11 shows a 20 KeV X-ray image which has a FWHM of 11 μm indicating a smaller optical spot. The shot energy was measured at 240 J with a 630 fs pulse giving an irradiance of $3x10^{20}$ Wcm^{-2}. This was consistent with our most optimistic predictions. Full shot laser far-field measurements were not possible due to optical imperfections in the laser diagnostic chain. An optical compensator has now been designed and ordered and will be installed in the system in 2004 to allow far-field measurements to take place.

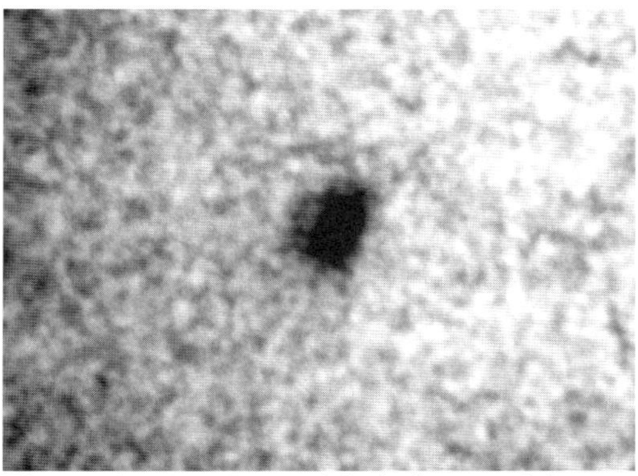

Figure 11. An X-ray pinhole image giving an image size of 11 microns FWHM.

A number of options are being investigated to use the Petawatt facility for X-ray laser experiments. The method of an off-axis spherical mirror to generate a line focus is shown in figure 12. In this case a spherical mirror with 400 mm focal length is used, with an off-axis angle of 15.4° degrees, which generates a line focus length of 10 mm. The possibility of superimposing a long pulse pre-former can be realised by synchronising a narrowband long pulse with the stretched CPA pulse and amplifying them together. In the compressor the CPA pulse will be compressed and the pre-forming pulse will maintain its pulse-length. The important consideration is possible damage to the gratings. If the combined energy of the short and long pulses does not exceed the design Petawatt energy of 670 J in to the compressor then this should not be an issue.

An alternative scheme is to use a phase plate placed in the focusing beam of the normal off-axis parabola. The geometry is such that to prevent the incoming 600 mm beam being clipped the phase plate has to be placed only 360 mm from focus. This presents an issue as to whether the phase plate will survive. For the ultra short pulse (500 fs) the damage threshold is estimated at 2.3 Jcm^{-2} [14] and at maximum energy (500 J on target) the average flux in the beam at the phase plate position is 4.4 Jcm^{-2}. It is therefore unlikely that the phase plate would survive without significant damage and it will need to be replaced on a shot by shot basis. Ideally, the phase plate pattern would be coated on the thin glass substrate of a target debris shield, which itself is regarded as a "disposable" optic.

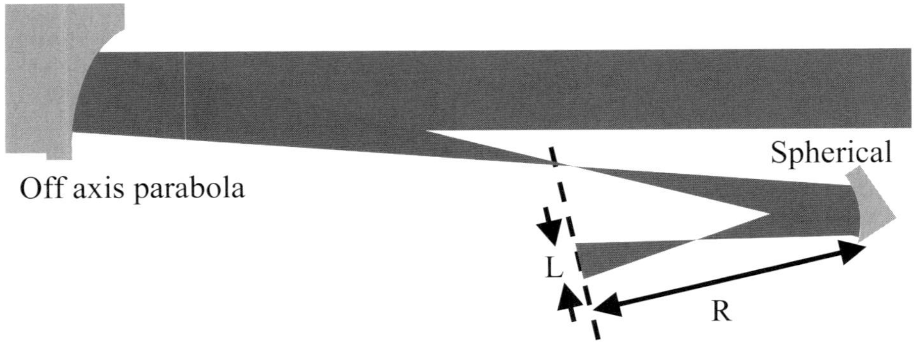

Figure 12. A possible X-ray laser geometry in the Vulcan Petawatt interaction facility where R is the radius of the spherical mirror and L the length of the line-focus.

9. Conclusion

X-ray lasers have been an important aspect of the research activity of the Central Laser Facility (CLF) since its opening in the late 1970's. As part of this programme several areas of laser development were necessary to provide facilities to enable the research to be conducted. These developments include the design of focusing geometries using off-axis spherical optics; multi-pulse and pre-pulse generators; and the use of diffraction gratings to generate travelling-wave geometries. The technique of Chirped Pulse Amplification (CPA) has been implemented and used for the study of transient schemes culminating in the commissioning of the Vulcan Petawatt (10^{15} Watts) interaction facility which can generate a peak focal intensity of up to 10^{21} Wcm^{-2}.

References

[1] Ross I N and Hodgson E M, 1985 J. Phys. Sci. Instrum., Vol 18
[2] Ross I N *et al*, 1987 Applied Optics, Vol. 26, No. 9
[3] Danson C N, Edwards C B and Ross, I N 1985 Optics and Laser Technology
[4] Warwick P J *et al*, 1997 Opts Comm 144, 192-197
[5] Strickland D and Mourou G 1985, Opts Comm., Vol.56, No 3, 219
[6] Danson C N *et al*, 1998 Journal of Modern Optics., Vol 45, No 8, 1653
[7] Kato Y *et al*, 1984 Phys Rev Lett., Vol. 53, No. 11, 1057-1060
[8] Pepler D A *et al*, 1993 SPIE Vol 1870, 76-87
[9] Skupsky *et al*, 1989 J of Appl Phys, Vol 66, No 8, 3456-3462
[10] McLeod J H, 1960 Journal of the Optical Society of America 50, 166-169
[11] Stevenson R M *et al*, 1994 Optics Letters 19, 363-365
[12] Ross I N *et al*, 1997 Opts Comm 144, 125
[13] Collier J *et al*, 1999 Applied Optics, Vol 38, No. 36, 7486
[14] Stuart BC *et al*. 1995 Phys Rev Lett, 74, 12, 2248-2251

Inst. Phys. Conf. Ser. No 186
Paper presented at 9th Int. Conf. X-Ray Lasers, Beijing, China, 24–28 May 2004

LASERIX : the French X-Ray Laser Facility

G. Jamelot, D. Ros, A. Klisnick, S. Kazamias, M. Pittman, F. Plé, S. Chapuisat, J-C. Lagron, J. Dubau, D. Benredjem, K. Cassou, and O. Guilbaud
Laboratoire d'Interaction du rayonnement X Avec la Matière, UMR n° 8624 Université Paris-Sud 11 / CNRS, Bât 350, 91405 Orsay cedex, France

J-P. Chambaret, S. Sebban, and P. Zeitoun
Laboratoire d'Optique Appliquée, UMR n° 7639 ENSTA / Ecole Polytechnique / CNRS, Chemin de la Hunière, 91761 Palaiseau cedex, France

B. Rus
Institute of physics of the ASCR, Department of X-ray Lasers, Na Slovance 2, CZ 18221 Praha 8, Czech Republic

Abstract. This paper gives the present state of the x-ray laser facility under construction at the LOA by the University Paris-Sud 11. This facility will deliver transient collisionally excited x-ray lasers for scientific and technological applications, but it is also intended to study new x-ray laser schemes for shorter wavelengths and better optical qualities.

1. Introduction

The first actual x-ray lasers were driven by high power lasers intended to inertial fusion research. The low access and low repetition rate of these facilities could not contribute to a fast improvement of short wavelength lasers and to the development of their applications. After the reduction of the driving energy from kilojoules to few hundred joules, transient pumping, optical field ionised (OFI) pumping and capillary-discharge pumping have contributed to progress towards "tabletop" x-ray lasers. The LIXAM team is now building LASERIX, a laser facility able to drive transient x-ray lasers of wavelength comprised between 10 and 40 nm, which can be improved and used for applications. LASERIX will be also able to develop new x-ray lasers based on alternative pumping schemes such as innershell pumping.

In this paper we will shortly describe x-ray laser drivers, then we will precise the specifications of the new LASERIX facility and describe the present state of its construction.

2. X-ray laser drivers

Many inversions mechanisms have been considered to produce lasing action in the XUV range. Mainly the so-called recombination scheme [1] and collision scheme [2,3] have been extensively studied, but only the latter has been able to produce large gain coefficients in long plasma filaments, resulting in x-ray laser beams. All actual x-ray lasers are 3-level lasers pumped in dense and hot plasmas by electron collisions. Emitters must be stable in a large range of plasma temperature, so the best candidates are close-shell or subshell ions such as Ne-like (transition $1s^22s^22p^53s$ – $1s^22s^22p^53p$), Ni-like (transition …$3d^94p$ - …$3d^94d$), and Pd-like ions (transition ….$4d^95d$ - $4d^95p$)

The first claim for an actual x-ray laser amplifier was made 20 years ago by using the huge energy of NOVETTE ($5x10^{13}$ Wcm^{-2} of frequency doubled Nd:glass laser with 450 ps pulse), the 2-beam prototype of the NOVA laser at the Lawrence Livermore National Laboratory (LLNL) [4]. Gain-length products up to 12 have been obtained.

Then, the main challenge was to reducing the driver energy in order to develop x-ray lasers in smaller installations. An important milestone has been the realisation of a saturated x-ray laser with the VULCAN laser at the Rutherford Appleton Laboratory (RAL) due to the double-pass in the active medium using a Mo:Si half-cavity [5]; then, low-energy prepulses were employed to preheat the target and make the plasma filament wide enough to give rise to large amplification in high-gain lines $J = 0 \rightarrow 1$ [6,7]. The delay between the prepulse and the main pulse is a major parameter, as shown by recent results with 50 ns compared to previous ones with 10 ns [8]. Finally, reducing the driving pulse duration to $\cong$ 100 ps permit to reduce the driving energy, and also to use efficiently the Ni-like ions, which give smaller lasing wavelength for a given driving energy [9,10].

All these x-ray lasers can be described by quasisteady state (QSS) models. The development of high-energy, ultrashort-pulse lasers using chirped pulse amplification (CPA) gave rise to the new class of "transient collisional excitation" (TCE) x-ray lasers [11] : the plasma production and the population inversion pumping are decoupled in two separate operations, a first "long" pulse ($\cong$ 500 ps) ionising the target, while a "short" one ($\cong$ 1 ps) heats the plasma free electrons. TCE reduces by more than one order of magnitude the required driving energy. The duration of the population inversion is shorter than in the QSS case, so generally the energy front of the short pulse must be tilted by 45 ° in order to produce a travelling pumping wave along the plasma filament [12]. A single "shaped" (i.e., with a low-level pedestal) laser pulse of 6 ps may also pump TCE x-ray laser [13].

Other types of drivers have been developed since a few years : capillary discharges have shown their capability to drive x-ray lasers and give laser beams of high optical quality in Ne-like argon at 46.9 nm [14] ; ultrashort ($\cong$ 30 fs) CPA pulses, circularly polarised, produce plasmas with high-energy free electrons by optical field ionisation (OFI), in which TCE x-ray lasers take place [15,16]. These different pumping schemes may be combined in hybrid x-ray laser schemes [17].

3. Objective of the LASERIX project

Among this abundant variety of x-ray laser drivers may be operated, some of them are very large-scale facilities intended for inertial fusion and high-field physics studies. The short time devoted to x-ray lasers in these installations makes very slow the progress of x-ray laser control and optimisation. This question is much more difficult for application experiments, which require great numbers of laser shots; then, mainly demonstrations of feasibility have been performed up to now. As an example, we had to wait 5 years between the construction of an experimental device for XUV interferometry of dense plasmas and its utilisation in proper conditions. This experiment was planned to be used at the Laboratoire pour l'Utilisation des Lasers Intenses (LULI) in Palaiseau in 1998 [18], but the first records were obtained in 2003 at the Advance Photon research Center (APR) in Kizu (Kyoto) [19].

Installations specially devoted to x-ray lasers are absolutely necessary to make these sources as useful tools for scientific and technological purpose. Capillary discharge systems are very compact and could be easily reproduced, but they operate presently at rather long wavelengths and give long x-ray laser pulses. The OFI x-ray lasers, which are presently produced at the LOA, are also of long wavelength and low energy per pulse, and they could not be substituted to TCE x-ray lasers for any application.

In France, only the 100-TW, CPA laser of the LULI [20] is capable to produce TCE x-ray lasers, but it is used for many other purposes and the particular geometry of x-ray laser experiments implies long setup periods prior to experiments. Another installation aimed to produce and develop new applications of intense and short x-ray beams might be operating permanently. With this aim, we are developing LASERIX, a new facility capable to produce high quality and high intensity TCE lasers in the wavelength range comprised between 10 and 40 nm. This beamline will be equipped with a two-beam driving laser, each of them with 10 J, one in a long pulse of 500 ps, the second in a "short" pulse, of duration comprised between 500 fs and 100 ps.

4. Driver technology

When we began the study of a driver, the question of the laser technology has cropped up differently than for the two x-ray laser facilities built several years ago with similar specifications, COMET at the LLNL [21] and the x-ray laser laboratory of the APR 22]. Both use titanium-doped sapphire (Ti:Sa) front end (oscillator and regenerative amplifier) and neodymium-doped glass (Nd:glass) amplifying chain. Besides this class of lasers, the pure Ti:Sa technology was emerging for laser energies in the range of few tens of Joules, owing to the recent development of a new large-crystal-growing technology (Fig. 1). A first experiment at LLNL permitted to extract an energy of 15 J from a large Ti:Sa crystal

Figure 1. The large titanium-doped sapphire crystal ($\phi = 10$ cm) of the LASERIX-driver amplifier (Crystal Systems).

pumped by JANUS [23]. Petawatt scale Ti:Sa lasers are presently under development or in project at the APR [24] and at the University of Michigan [25]. With very short pulse and fine focusing, these systems are aimed to high-field interactions with matter, but also they could be very efficient drivers for X-ray lasers. On the other hand, Ti:Sa technology has recently emerged and appears to be more risky and hazardous, and also more expensive than the well-experienced Nd:glass one. We have made a comparative study of both laser technologies before deciding the LASERIX driver.

4.1. Repetition rate

Owing to the large thermal conductivity of sapphire, Ti:Sa lasers may be operated at much higher repetition rate than Nd:glass lasers. The nominal laser shot rate at the APR, where both long and short pulses are amplified in the same glass rods, is about one shot per 15 minutes, whereas it is one shot per 4 minutes in COMET. It is unlikely to get a larger shot rate with Nd:glass without large degradation of the laser wavefront. On the contrary, a water-cooled sapphire crystal may suffer much larger shot rate. The pump laser of the crystal is the main factor limiting the shot rate. For Ti:Sa-driven LASERIX, we have considered Nd:glass laser designs able to emit 100 J in the second harmonics (0.53 μm) at the rate equal to or larger than 0.1 Hz.

4.2. Pulse duration

Due to their large line width, Ti:Sa lasers may emit much shorter pulses (≅ 30 fs) than Nd-glass ones (≅ 500 fs). That is not important for TCE x-ray lasers, which requires pulse duration comprised between one and a few picoseconds. But short driver wavelengths should be of advantage to some alternative x-ray laser schemes, which are one of the 3 objectives of LASERIX, besides TCE x-ray lasers and applications. In particular, simulations of innershell pumping for lasing at short wavelength show the need of femtosecond pulses.

4.3. Wavelength

The most of the x-ray lasers until now has been pumped using radiation at 1.06 μm, the wavelength of Nd:glass laser. Ti:Sa lasers deliver emissions at ≅ 0.8 μm, so we have studied the effect of the driver wavelength on the gain coefficient that one can expect in the Ne-like and the Ni-like cases. Figure 2 shows an example of results obtained for the Ne-like iron laser, which emits at 25.5 nm. The local-gain maximum is slightly higher for the Nd:YAG, but some parameters, such as the delay between the long and the short pulses may have a contrary effect : a longer delay results in wider amplifying zone, in which the x-ray amplification may be more efficient than in thin inverted layers. From these results, we don't expect large difference between x-ray lasers pumped by wavelengths of 1.06 μm and 0.8 μm. Simulations of Ni-like silver x-ray lasers emitting at 13.9 nm leads to similar conclusions

An interesting feature of the Ti:Sa lasers comes from the fact that they work at the same wavelength of 0.8 μm from the oscillator to the last amplifier. Consequently the front-end beam, which is naturally collinear with the amplified beam and which works at 10 Hz, is intense enough to be used for alignment of various beams up to the target. In the case of Nd:glass lasers, alignments require either a part of amplifiers with low repetition rate or an auxiliary continuous laser that must be previously aligned on the main laser beam.

5. The LASERIX driver

Taking into account the technological risks of a Ti:Sa driver, but also its large advantages in terms of shot rate and pulse duration compared to the technical limitations of Nd:glass,

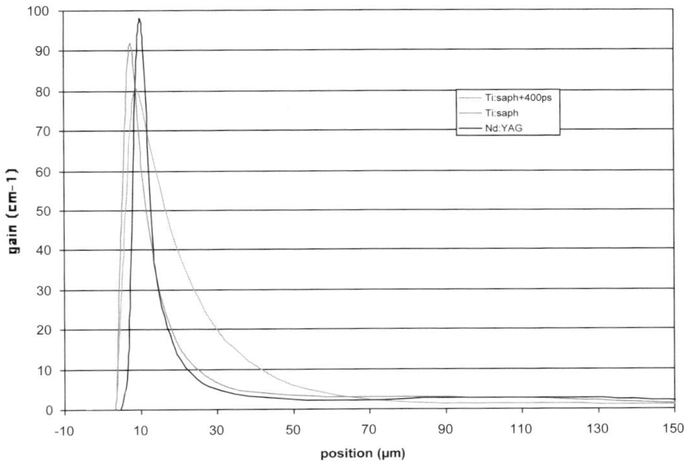

Figure 2. Neon-like iron x-ray laser emitting at 25.5 nm. Calculated distribution of the gain coefficient for driver wavelengths of 1.06 µm (Nd:YAG) et 0.8 µm (Ti:Sa) for a 200 ps delay between the long and the short pulse, and for 400 ps in the case of Ti:Sa.

we have opted for this new technology to equip LASERIX. In addition, as LASERIX will be located at the LOA, it will also contribute to progress toward the petawatt laser in collaboration with the ELF team. Indeed, the planned location of LASERIX is a new building of the LOA in Palaiseau (bâtiment “P”), which is to be constructed. Meanwhile the termination of this building, the driver assembling is presently in progress in an available room near the future location (salle turquoise).

The general architecture of the Ti:Sa laser is schematically represented in figure 3. The front end is designed as a customised laser based on standard modules. It is set on three optical tables, one for the shaping and pre-amplification of the oscillator pulse, the two

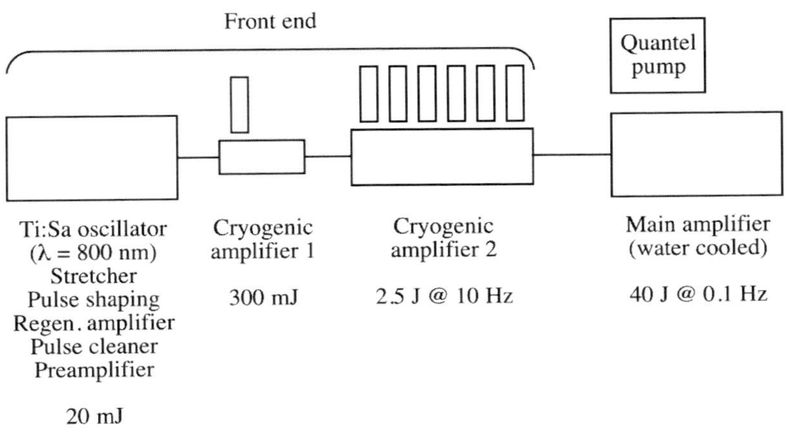

Figure 3. Schematic of the LASERIX driver architecture.

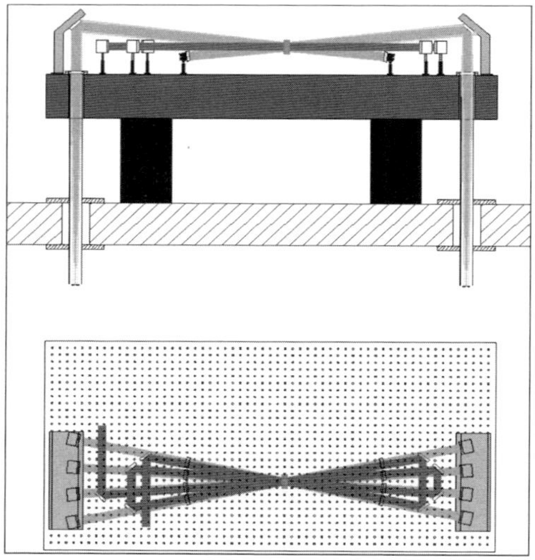

Figure 4. Configuration of the final amplifier of the LASERIX driver.
Upper part : vertical view; the pumping green laser beams come from the lower floor to illuminate uniformly the Ti:Sa crystal.
Lower part : horizontal view showing the 4-pass pumping of the crystal.

other for two cryogenic amplifiers. The output energy at the front end is 2.5 J, and the repetition rate of 10 Hz. The front-end beam can pass trough the final crystal and determines the direction of the high-energy beam. Then, when the final amplifier is not pumped, one can use the front end at 10 Hz to perform precise alignment and adjustment of the focalisation optics and the target

The front-end beam is then injected in the main amplifier, which is based on the large Ti:Sa crystal (diameter 100 mm) shown in Fig. 1. The crystal is pumped by a 4-module Nd:glass laser (Quantel) delivering 100 J of 2ω green light. The energy deposition on each side of the crystal is homogenised using lens arrays. After four passes through the crystal, the expected output before compression is ≅ 40 J. The organisation of the amplifier in the final arrangement is presented in figure 4. The high-energy pumping laser will be isolated in a lower room, while the front end, the final amplifier and the rest of the x-ray laser experiments will be in the upper room (15 m x 9.4 m). Before the construction of the "bâtiment P", all these systems will be on the same level in the provisory installation.

6. New prospects opened by Ti:Sa technology

The initial LASERIX project consisted only of two 10 J beams, respectively of 500 and 1 ps. The Ti:Sa option opens large experimental possibilities related to the ultrashort pulse durations that can be realised. In particular the zero-order rejected by the compressor may be itself compressed to give a beam of ≅ 1 J in 30 fs pulse perfectly synchronised with the main beam. Such beam is sufficient to produce either high-order harmonics (HOH) or TCE x-ray lasers in OFI plasmas. LASERIX will then be an ideal tool to amplify the HOH emission in TCE plasmas produced by the high-energy beam, as that has been done recently at the LOA with OFI plasmas [26]. One will be able also to realise density diagnostics of plasmas produced by the low-energy beam [19].

The opportunity to use femtosecond pulses, in addition to classical driving 500- and 1-ps pulses for TCE X-ray lasers, lets consider other types of X-ray laser schemes that need high pumping power. In particular, pumping in innershell may give rise to x-ray laser of short wavelengths and open the way to lasing in the water window. LASERIX will be a perfect tool to investigate such prospects.

7. Acknowledgments

LASERIX will be an x-ray laser facility of the Université Paris-Sud 11. The support of the Conseil Général de l'Essonne and the Ministère de la Recherche under the CPER 2000-2006 is gratefully acknowledged. It will be located at the LOA UMR n° 7639 (ENSTA, Ecole Polytechnique, CNRS), Palaiseau.
We thank Patrick Georges (LCFIO), Pascal D'Oliveira (CEA-DRECAM) and Claude Rouyer (CEA-CESTA) for valuable discussions.

References

[1] Gudzenko L I *et al* 1965 Sov. Phys. Doklady **10** 147
[2] Elton R C 1975 Appl. Optics **14** 97
[3] Vinogradov A V *et al* 1973 Sov. Phys. JETP **36** 1115
[4] Matthews D L *et al* 1985 Phys. Rev. Lett. **54** 110
[5] Carillon A *et al* 1992 Phys. Rev. Lett. **68** 2917
[6] Cairns G F *et al* 1994 in "X-ray Lasers 1994" D.L. Matthews and D. Eder Eds AIP Proceedings **332** 289
[7] Rus B *et al* 1997 Phys. Rev. A **55** 3858
[8] Rus B *et al* these proceedings
[9] Lewis C L S *et al* 1999 in "X-ray laser 1998" Y. Kato H. Takuma and H. Daido Eds. IoP Conf. Series N° 159 p. 1
[10] Daido H *et al* 1999 in ref 9 p. 83
[11] Nickles P V *et al* 1997 Phys. Rev. Lett. **78** 2748
[12] MacPhee A G *et al* 1999 in ref. 9 p. 75
[13] Janulewicz K A *et al* 2003 Phys. Rev. A **68** 051802(R)
[14] Rocca J J *et al* 1994 Phys. Rev. Lett. **73** 2192
[15] Lemoff B E *et al* 1995 Phys. Rev. Lett. **74** 1574
[16] Sebban S *et al* 2001 Phys. Rev. Lett. **86** 3004 and Sebban S *et al* 2002 Phys. Rev. Lett. **89** 253901
[17] Janulewicz K *et al* 2001 Phys Rev A **63** 033803
[18] Hubert S *et al* 1998 Rapport LULI (Ecole Polytechnique) p. 103 (*in French*)
[19] Tang H *et al* 2004 Applied Phys B **78** 975-7
[20] Klisnick A *et al* 2004 *these Proceedings*
[21] Dunn J *et al* 1999 in ref. 9, p. 51
[22] Kato Y *et al* 2001 in "X-ray Lasers 2000", G. Jamelot, C. Möller, A. Klisnick Eds., J. Phys IV France **11** Pr2-3
[23] Bonlie J D et al 2000 Appl. Phys. B **70** S155
[24] Tajima T 2003 REEFS **1** 11/28
[25] Mourou G 2004 Workshop on Technological Bottlenecks for High Intensity Petawatt Class Titanium Sapphire Lasers, Paris (France) 1 - 4 April
[26] Faivre G *et al*, 2004, *these Proceedings*

Inst. Phys. Conf. Ser. No 186
Paper presented at 9th Int. Conf. X-Ray Lasers, Beijing, China, 24–28 May 2004

Optimization of a multi-TW Ti:sapphire laser system for X-ray laser studies

Zhiyi Wei, Zhaohua Wang, Weijun Ling, Peng Wang, Jie Zhang
Laboratory of Optical Physics, Institute of Physics, CAS,
Beijing,100080, China

M. Suzuki, H. Kuroda
Institute for Solid-State Physics, University of Tokyo,
Kashiwa, Chiba, 277-8581, JAPAN

Abstract. We present the design and construction of a 20TW Ti:sapphire laser system in this paper. With the optimized parameters for each stage, higher amplified efficiency and superior performances are demonstrated.

1. Introduction

The rapid progress of ultraintense and ultrashort laser techniques has greatly accelerated the study on laser matter-interaction. Since G.Mourou *et al* invented the chirped pulse amplification (CPA) in 1985[1], the available peak power has been increased to a PW level during the last decade [2,3]. Especially, TW Ti:sapphire laser system based on CPA technique have enabled us to establish the top-table platforms for high field physics in university laboratory scale. Many applications, such as the ultrafast x-ray sources, optical field ionisation, electron acceleration, relativistic plasma physics and generation of hot electrons can be done with such experiment platforms [4]. To pursue those challenging research works, we developed a 1.4TW and 20TW Ti:sapphire lasers in 1997 and 2002 [5], respectively, with pulse durations of 25fs and 31fs,respectively. More recently, a new 20TW laser facility with kHz pre-amplifier was developed, the improved stability makes it not only a compact driver for X-ray lasers, but also a desirable source for the research of ultrafast spectroscopy. In this paper, we report the design and construction of the 20TW laser, main specifications in each stage are discussed.

2. Optimized design of 20 the TW Ti:sapphire laser

A typical CPA laser consists of four parts, a femtosecond oscillator, a stretcher, an amplifier and a compressor. In general, one stage amplifier is enough for an output energy of mJ. However, for those high peak power laser with an output of multi-terawatt level, it is desirable to have two or even more multi-pass amplifiers to boost the energy [2,3]. The additional amplifiers introduce more material dispersion and B-integral, leading a wider pulse duration and a poor beam quality. To overcome these problems, we use only two amplifier stages in our design of 20TW laser facility. The schematic diagram of the laser system is shown as fig.1. The oscillator is a typical Kerr-Lens Mode-Locking Ti:sapphire laser, an Öffern stretcher is used to extend the seeding pulses to about 300ps for safety amplification. Following the stretcher, we designed a regenerative amplifier to boost the laser energy to 2mJ and then further amplify the energy to 1J by a main amplifier. Finally, the amplified laser pulses are injected into a vacuum compressor. With the optimised design, we obtained a laser output with an energy of 640mJ and a pulse width of 31fs, corresponding to a peak power of 20TW.

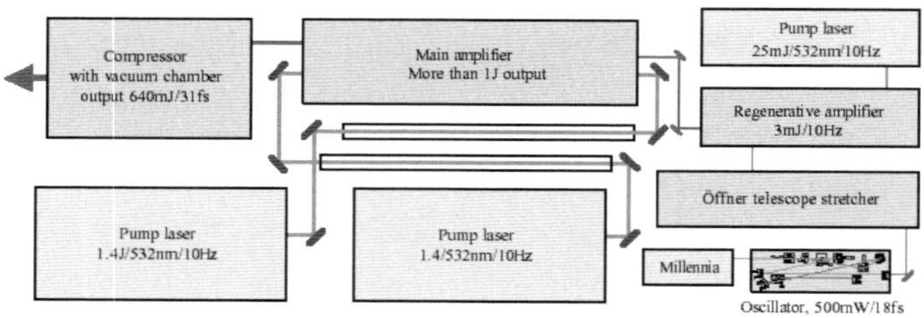

Figure 1. The schematic diagram of 20 TW laser facility. It consist in an oscillator, an Öffern stretcher, a regenerative amplifier, a multi-pass amplifier and a vacuum compressor.

2.1. *Oscillator*

In the design of the Ti:sapphire oscillator, we choose the length of the gain medium as 4mm, the shorter high-doped Ti:sapphire crystal (Shanghai Institute of Optics and Fine Mechanics, SIOFM) can support a high output power and shorter pulse durations. To compensate the self-phase modulation (SPM) in the gain medium, we use a pair of fused silica prisms to balance the dispersion and SPM, the effective light path between the two prisms from tip to tip is about 54cm. This design makes the oscillator in a very compact size of about 500×200mm and shows a superior power stability. With 4.2W 532nm pumping laser (Millennia, Spectra-Physics Inc.), stable mode-locking pulses of shorter than 18fs are produced with an average power of 500mW by optimizing alignment, we measure the fluctuation of mode-locking power of less than 1%. Fig.2 (a) and (b) show the pulse autocorrelation trace and spectrum.

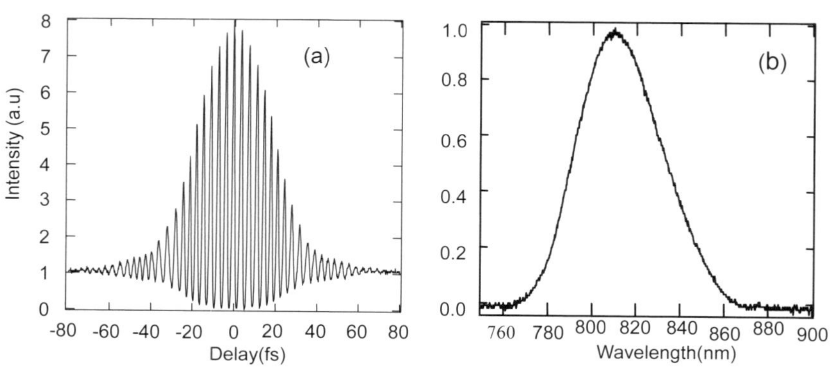

Figure 2. Autocorrelation trace (a) and spectrum (b) of the laser pulse from oscillator.

2.2 *Stretcher*

The stretcher was designed with a configuration for Öffner telescope [6], which consists of a holographic diffraction grating (Spectrogon AB Inc., Sweden) with the groove frequency of 1200/mm, a concave mirror with 1m radius of curvature and a convex mirror with –0.5m radius of curvature. Both the concave and convex mirror are coated with gold film, the distance between them is 0.5m. To realize a large extend for seeding pulse, we simulated the dispersion with ray tracing method, the extended pulse duration was calculated based on the dispersion characteristics. Figure 3 shows the stretched pulse duration vs the distance from the grating to the convex mirror at the Littrow angle. With the optimized parameters, the pulse duration can be stretched to about 300ps.

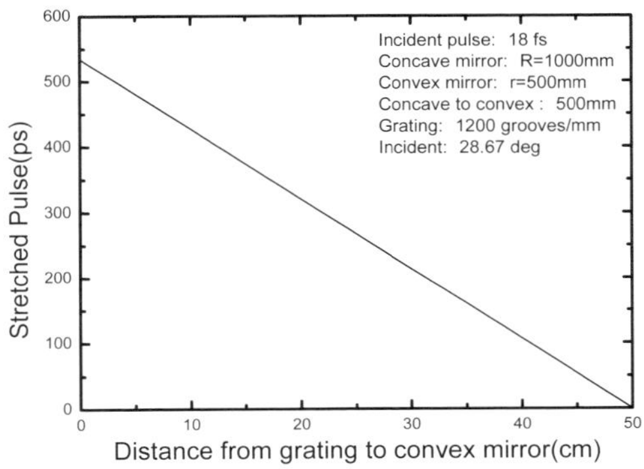

Figure 3. The stretched pulse duration vz the distance between the grating and the convex mirror in the Martinez telescope stretcher. The grating groove is 1200line/mm, laser incident at the Littrow angle of 28.67^0.

2.3 Amplifier

In general, the energy of seeding pulses is only about nJ after stretcher. To boost the chirped pulse energy for the application of high field physics, normally a pre-amplifier is first used to supply a large gain for seeding pulse. The pre-amplifier may be multi-pass or regenerative configure, both can amply the pulse with gain of around 10^6. With a multi-pass as the pre-amplifier, the seeding pulse will suffer low material dispersion and gain narrowing effect, however, the alignment will be a little difficult if pursue an efficient pump, beam quality is also worse. For this reason, we used a typical regenerative resonator as the pre-amplifier in this work, the seeding pulse is rotated at S polarization and then coupled into the cavity by the surface reflection of Ti:sapphire rod. Controlled the timing delay and the voltage on the Pockels cell with the synchronous electronics, we amply the seeding pulse with 30 trips inside the regenerative cavity and dump out by the thin film polarizer (TFP). With 25mJ pump energy (Spectra-Physics, LAB-130-10), the optimized output energy is about 3mJ. A photo diode was set behind the back mirror to detect the leakage light, connecting the output electrical signal to oscilloscope enable us to watch and optimise the amplification process, figure 4 is a typical pulse trace on oscilloscope.

The performance of pre-amplifier is very critical for obtaining an excellent output pulse form the main amplifier. To improve the stability of TW laser, we also constructed a kHz regenerative amplifier as the pre-amplifier and then decrease the repetition rate to 10Hz with a Pockels Cell and Glan prism. Figure 5 is the experimental layout for down repetition rate, a self-design electrical circuit was used to synchronize and delay the 1kHz laser with the 10Hz amplifier, then we enlarge the beam diameter and inject into the main amplifier further amplification.

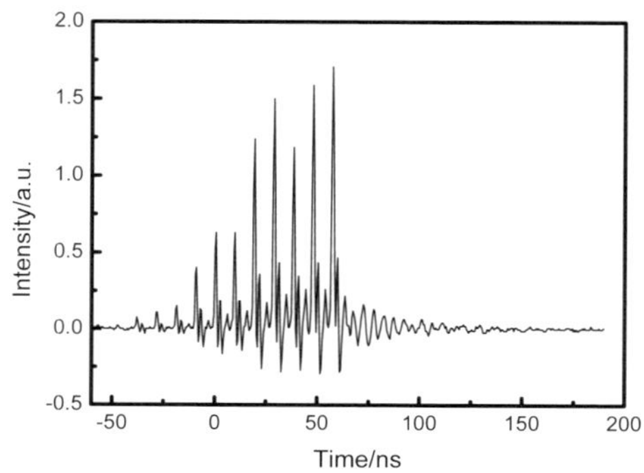

Figure 4. The oscillogram trace of amplification process, monitor with a photodiode behind the back mirror of regenerative amplifier.

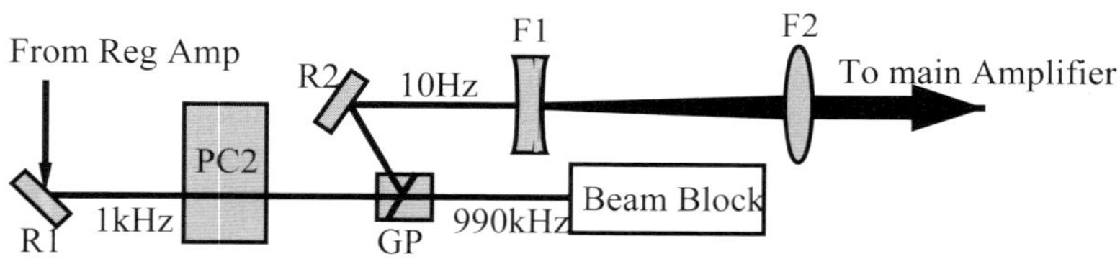

Figure 5. The experimental layout for the repetition rate divider and the beam enlarge system. The 1kHz repetition rate is divided into 10Hz and 990Hz with S and P polarization respectively. Then the 10Hz laser beam is enlarged to 14mm in diameter.

Difference from the conventional configuration with two or three multi-pass stages, we only designed one stage to boost the mJ pulse energy to J level, that greatly decrease the material dispersion and support even shorter compressed pulse. For a typical multi-pass amplifier, the Frantz-Nodvick simulation has shown the optimized conversion efficiency occur around the pump energy fluence of 1J/cm^2 [7], higher fluence will lead the gain saturation and optical damage, in principle, a better mode match between pump and amplifying laser should be also support the efficient output with one stage main-amplifier. Based on above consideration, we enlarge the laser beam to 14mm with a Galilean telescope before the multi-pass amplification, the Pockels cell between two amplifiers not only down the repetition rate to 10Hz, but also eliminate the ASE and rotate the polarization from S to P. Pump with two 1.4J 532nm Nd:YAG lasers (Spectra-Physics Inc., Pro-350-10), we obtained an amplified output at the energy of more than 1J after 6 passes, corresponds to efficiency of 36% .Higher output seems still possible with accuracy alignment for beam overlap and an careful control for synchronous timing. For keep a stable and safety running, we have not further optimize the output energy,

We compared the energy stability after main amplifier under two modes of 10Hz and kHz regenerative as pre-amplifier, the typical result with 1kHz regenerative amplifier is about ±1%, the fluctuation is obvious decreased compare the ±5% with 10Hz regenerative amplifier.

The relay imaging pipes can well improved the beam pattern and energy of amplified laser pulses. We compared the output characteristics with and without relay imaging. Without the pipes, the amplified energy is only about 700mJ, beam pattern shows un-uniform distribution and easy damage the last mirror in the amplifier

2.4. Compressor

We set the compressor with a standard configuration based on Treacy's design[8] and covered inside a vacuum chamber. Consider the damage threshold of holographic grating is only about 300mJ/cm^2, we once again enlarge the beam diameter to 30 mm before input the window of vacuum chamber. Refer to the total dispersion of the stretcher and transparence optical components; we stimulate the optimized parameters for the compressor combined with ray tracing method and Treacy's equation. Figure 6 is the numerical curves of pulse profiles with the optimized incident angle and distance between two gratings, We simulate the result with the grating groove of 1480line/mm, the laser pulse trips 30 passes in the regenerative amplifier with a 15mm Ti:sapphire rod and 20mm KDP, 6 passes in the multi-pass amplifier with a 20mm Ti:sapphire. The total material is 20mm fused silica for telescope and 20mm KDP for the Pockels cell before the main amplifier. The simulated result shows the compressed pulse with shorter duration and higher contrast ratio in leading can be realized by using the mismatch grating pair which is the different groove from the stretcher.

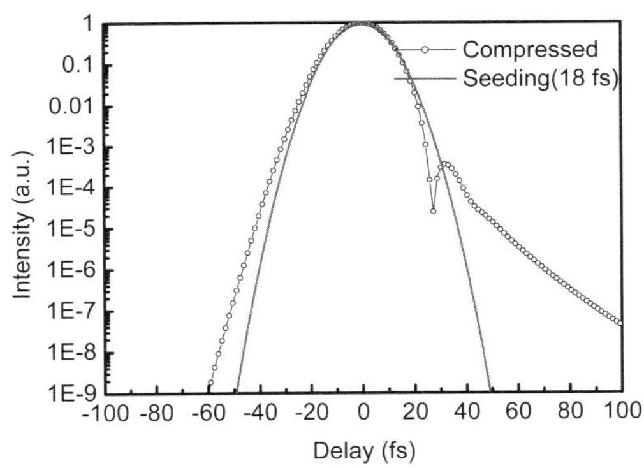

Figure 6. The simulated pulse profile with optimized compressor design. All the material dispersion has been considered.

With the simulated calculation, we used a pair of grating with groove of 1480line/mm (Spectrogon AB Inc.,) in the compression experiment, the sizes of first and second gratings are 50×80mm and 120×80mm respectively. After first compression, a pair of rectangle mirror was used as an elevator to reflect the laser beam to the grating pair for second compression again, the height of 80mm is enough to support twice diffractions on top and bottom of the gratings. By optimizing the simulated distance between two gratings and the incident angle with experiment, we realized 31fs compressed pulse with output energy of 640mJ. Figure.7 is the typical autocorrelation trace (a) and the corresponding spectrum (b), the time bandwidth is about 0.43 with sech^2 assumption, typical energy stability is 5%. Compare to the conventional CPA techniques, the new design with only two stages amplifier well demonstrate a 20TW Ti:sapphire laser facility with a simplified configuration and compact size, which support shorter pulse duration, higher stability, better beam patter and even lower cost.

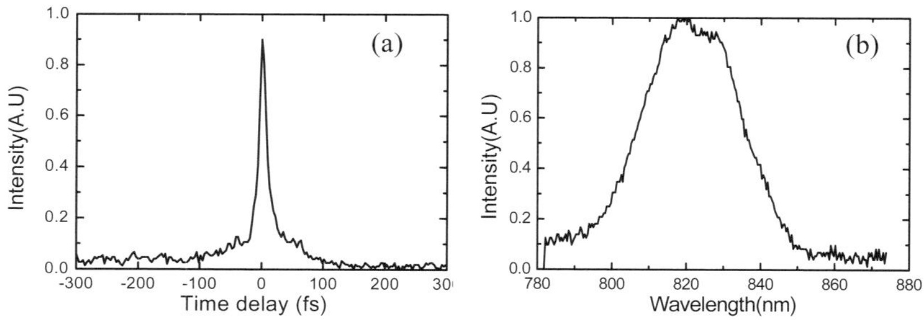

Figure 7. The autocorrelation trace (a) and spectrum (b) of the compressed laser pulse.

3. Conclusion.

We successfully constructed two 20TW femtosecond Ti:sapphire lasers only with two stage amplifiers, the stability was obviously improved by using 1kHz regenerative amplifier. The optimized design of the CPA laser system not only supports high efficiency, less material dispersion and B-integral, but also result in compact size and low cost.

Acknowledgements: We acknowledge the helpful discussion with Dr Jinhua Sun and Jinrong Tian. This work was jointly financed by National Nature Science Foundation of China under Grant Nos. 60225005, 60308001, 60321003, and the NKBRSF under Grant No.G1999075202.

References

[1] Strickland D and Mourou G 1985, Opt. Commun. **56**,219
[2] Perry M D, Pennington D, Stuart B C, Tietbohl G, Britten J A, Brown C, Harman S, Golick B, Kartz M, Miller J, Powell H T, Vergino M, and Yanovsky V, 1999, Opt.Lett. **24**, 160
[3] Aoyama M, Yamakawa K, Akahane Y, Ma J, Inoue N, Ueda H, and Kiriyama H 2003, Opt Lett. **28**, 1594
[4] Mourou G, Barty C P J, Perry. M D 1998, Physics Today, Jan: 22~28
[5] Wei Zhiyi, Zhang Jie, Xia Jiangfan 2000., Science in China, **43,** 1083
[6] Chambaret J P, Blanc C Le, Chériaux G, Curley P, Darpentigny G, Rousseau P, Hamoniaux G, Antonetti A, and Salin F 1996., Opt Lett, **21**, 1921.
[7] Yamakawa K, Aoyama M, Matsuoka S, Takuma H, Fittinghoff D, and Barty C P J 1998. IEEE. J. of Selected Topics in Quantum Electron., **4**, 385.
[8] Treacy E B1969, IEEE Journal of Quantum Electronics, **5**, 454

Inst. Phys. Conf. Ser. No 186
Paper presented at 9th Int. Conf. X-Ray Lasers, Beijing, China, 24–28 May 2004

Extracavity femtosecond pulse compression by the use of bulk media

Yi Zhu, Xiaowei Chen, Yuxin Leng, Jun Liu, Lihuang Lin, Ruxin Li and Zhizhan Xu

Laboratory for High Intensity Optics, Shanghai Institute of Optics and FineMechanics, Chinese Academy of Sciences,
P.O. Box 800-211, Shanghai 201800, P.R.China

Abstract. The compression of relatively high-energy femtosecond laser pulse is experimentally investigated. We compressed 0.26mJ 50fs laser pulse to >0.1mJ 19fs by focusing the laser beam into the bulk media. The time domain compression is realized by a 45^0 apex angle prism pair, and the whole energy transmission is about 40%. This technique can, in principle, be used to compress much higher energy laser pulses to few- cycle level with relatively high transmission.

1.Introduction

Ultrashort and ultraintense pulses have been obtained with the development of mode-locking and Chirped Pulse Amplification (CPA) technology in the last 20 years[1]. Such ultrashort pulses with high energy have wide applications in high-order harmonics[2], plasma channel[3] , attosecond pulses[4] and so on.

However, due to the limitation of gain narrowing, the direct output pulse duration of the present CPA systems is near 30fs, which is too long for many applicants and makes the extracavity pulse compressing technology very important. The present compressing technology is mainly based on spectrum broadening introduced by SPM and the corresponding dispersion compensation [5]. At first, normal optical fiber is used as spectrum broadening medium [6], but it is mainly suitable for amplifying the pulses with the energy of nJ level. M. Nisoli et al [7] used hollow fiber filled with noble gas as the broadening medium and obtained pulses of about 10fs with the energy of mJ level. However, the noble gas in the capillary still can be ionized when the intensity of the laser pulse is strong enough, which will affect the spectrum broadening. In addition, vacuum system is necessary and the capillary should be strictly straightened, which makes it difficult to realize.

Besides fiber, bulk medium can also be used to broaden the spectrum. In 1988 year, P.B.Corkum[8] focused the laser into the bulk medium, and then compressed a 90fs laser pulse to 24fs 100μJ. But the energy efficiency is very low. In 2003, E.Constant[9] compressed the pulses of 42fs 480μJ to 14fs 220μJ and obtained higher energy efficiency.

During our experiment, we put the bulk media at a special place after the focus and the laser pulse of 50fs with the energy of 0.26mJ was compressed to 19fs with the energy more than 0.1mJ. The temporal pulse shape is quite good.

2. Experimental setup

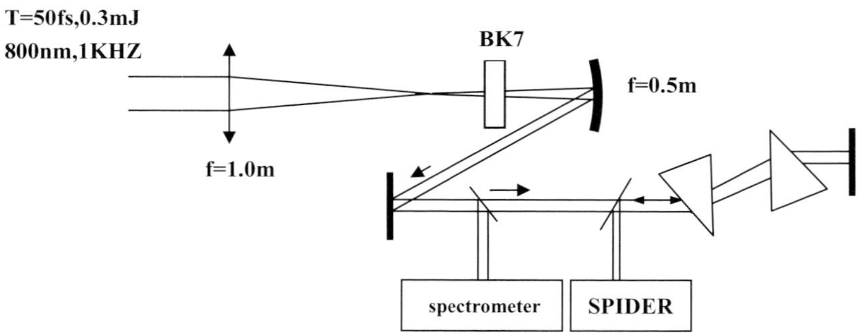

Fig1. Schematic of the experimental setup

The experimental setup is shown in fig1.The femtosecond laser system used in the experiment is Spitfire from the Spectra Physics Company. The laser pulse is centered at 800nm with the duration of 50fs at the repetition rate of 1KHz. The maximum pulse energy is about 0.5mJ .We let the laser pulses pass through a half-wave plate and a polaroid so we can adjust the laser energy continuously. The laser pulse was focused by a lens with a focal length of 1m. We placed a 3-mm-thick BK7 glass as the spectrum broadening medium at the point of 17.5cm after the focus. The output laser from the glass was collimated by a silver-coated concave mirror with a focal length of 50cm. After passing the silver-coated mirror and beam splitter, the pulse is measured by spectrometer and by SPIDER[10] to get the pulse duration after being compressed by a prism pair. The prisms pair used is quartz prism with 45^0 apex angles and space interval 1.18m. The laser pulse can pass the prism pair twice by putting a reflecting mirror after the back prism.

3. Experimental result and analysis

When the energy of the injected pulse is 0.26mJ, a plasma channel was formed near the focus. At the input plate, the laser pulse power is about 5×10^9W, which is much more than the self-focusing threshold $P_{cr}=1.8\times10^6$W. If the glass is put forward, there will be supercontinum induced by SPM.

There are mainly two reasons why we put the glass after the focus. Firstly, we found the pulse is narrowed a little after being focused in the air. The pulse duration is compressed from 52fs to 44fs while the temporal shape still remains good. This method is expected to generate better compressed pulses. Secondly, we found the beam quality much better after the pulse passes the glass, as shown in Fig2. The non-uniform light intensity distribution on the cross section changed to a quite uniform Gaussian distribution.

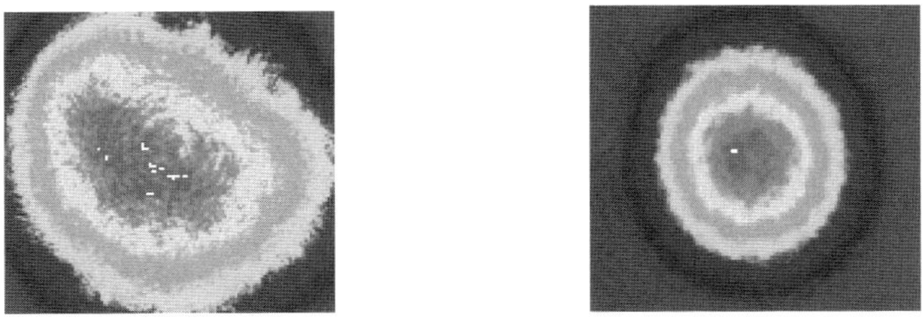

Fig 2 . The pulse intensity distribution before and after passing the glass

When propagating in the glass, the pulse spectrum was broadened and chirped by temporal SPM effect combined with dispersion and self-steeping effect, which is shown in Fig.3. The fluctuations on the spectrum are the characteristics of SPM effect.

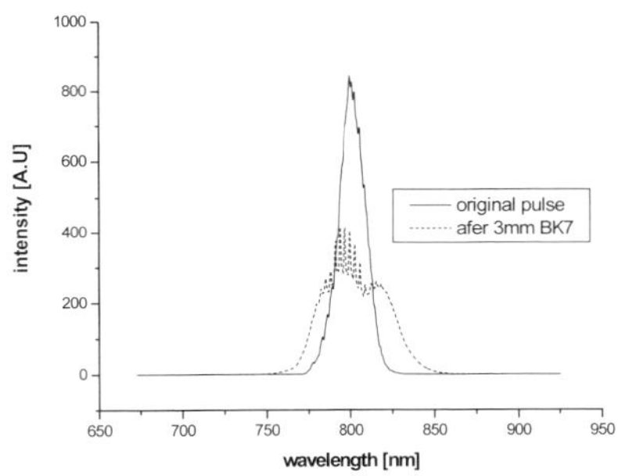

Fig 3 .The laser spectrum after passing the K9 glass

Fig4 shows the temporal shape of the initial pulse and the pulse after compression by the prism pair, which is measured by SPIDER. We can see the FWHM of the laser pulse is changed from the initial 52fs to 19fs and the pulses shape is quite good.

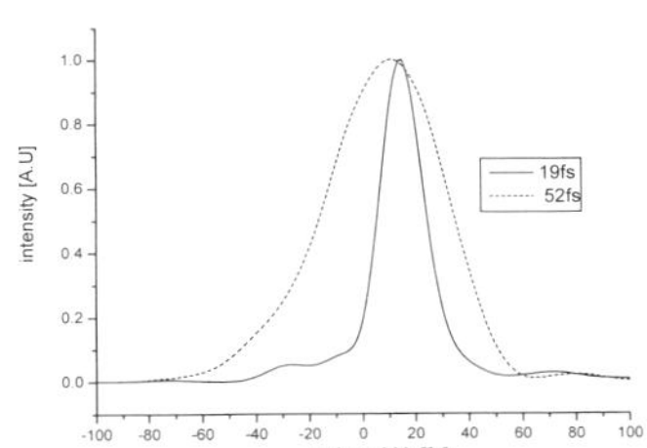

Fig 4. The pulse waveform before and after compressing

We used a quartz prism pair as the dispersion compensating devices. The phase before and after passing the prism pair are shown Fig5. We can see the pulse carries obvious chirp after being broadened by the bulk medium, and is unified after being compensated by the prisms which have negative GVD. But the prism pair can only compensate second-order positive chirp, so the pulse still carried high-order chirp.

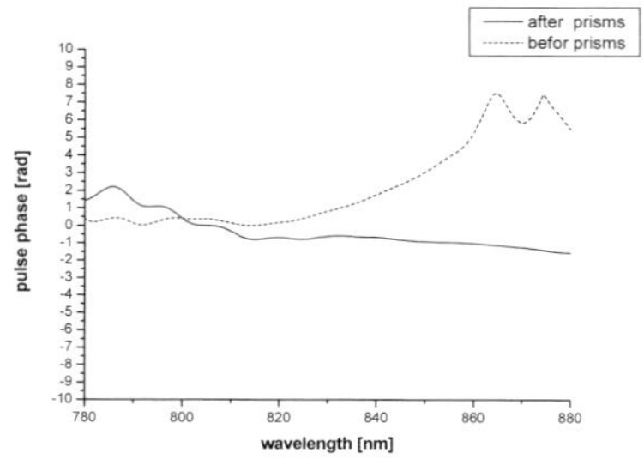

Fig 5. The phase diagram of the pulse before and after the prism pair

The injected pulse on the glass has the energy of 0.204mJ. The shortest pulse has the energy of 0.104mJ after being compensated by the prism pair.

4. Further improvements

Theoretically, the spectrum can be broadened greatly as long as the bulk medium is put on the appropriate position. We believe that this method can be easily scaled to the compression of tens of mJ or even higher energy femtosecond pulse

The dispersion compensation device used in the experiment can not compensate high order dispersion. We believe shorter pulses can be obtained if better compensation devices were used. We can also greatly increase the energy efficiency by using 800nm high transmission film coated prism pair to decrease the reflecting energy loss on the prism surface.

5. Conclusion

In this paper, we realized a novel extracavity pulse compression method for femtosecond laser. We compressed the laser pulse of 50fs at the repetition rate of 1KHz level with the energy of 0.3mJ to 19fs with the energy of 0.1mJ by using third-order nonlinear effect of bulk medium. The compressed pulses have quite good pulse quality .We believe that using this method and improving the experimental condition, pulses shorter than 10fs with a few mJ, which have wide applications in high-order harmonics and attosecond physics, can be obtained.

References

[1] Brabec T and Krausz F 2000 Reviews of Modern Physics vol.72 547-567

[2] Huiller et al 1993 Phys. Rev. Lett. 70 774-777

[3] Braun A and Korn G 1995 Opti. Lett. 20 73-75

[4] Hentschel M and Krausz F 2001 Nature 414 509-513

[5] G. P. Agrawal, Nonlinear Fiber Optics, 2nd ed.(Academic, San Diego, Calif., 1995).

[6] Nikolaus B and Grischkowsky D 1983 Appl. Phys. Lett. 43 228-230

[7] Nisoli M and Stagira S 1997 Appl. phys. B vol 65 189-196

[8] Rolland C and Corkum P B 1998 J.Opt. Soc. Am. B 5 641-647

[9] Mevel E and Constant E 2003 J. Opt. Soc. Am. B 20 105-108

[10] Iaconis C et al 1998 Opt. Lett. 23 792-794

Author Index